**HANDBOOK
OF INDUSTRIAL
MEMBRANES**

1st Edition

HANDBOOK OF INDUSTRIAL MEMBRANES

1st Edition

K. Scott

ISBN 1 85617 233 3

Copyright © 1995 ELSEVIER SCIENCE PUBLISHERS LTD

All rights reserved

This book is sold subject to the condition that it shall not by way of trade or otherwise be resold, lent, hired out, stored in a retrieval system, reproduced or translated into a machine language, or otherwise circulated in any form of binding or cover other than that in which it is published, without the Publisher's prior consent and without a similar condition including this condition being imposed on the subsequent purchaser.

Other books in this series include:
Hydraulic Handbook
Seals and Sealing Handbook
Handbook of Hose, Pipes, Couplings and Fittings
Handbook of Power Cylinders, Valves and Controls
Pneumatic Handbook
Pumping Manual
Pump User's Handbook
Submersible Pumps and their Applications
Centrifugal Pumps
Handbook of Valves, Piping and Pipelines
Handbook of Fluid Flowmetering
Handbook of Noise and Vibration Control
Handbook of Mechanical Power Drives
Industrial Fasteners Handbook
Handbook of Cosmetic Science and Technology
Geotextiles and Geomembranes Manual
Reinforced Plastics Handbook

Published by

Elsevier Advanced Technology
The Boulevard, Langford Lane, Kidlington, Oxford OX5 1GB, UK
Tel 010 44 (0) 865-843000
Fax 010 44 (0) 865-843010

Preface

This remarkable manual contains necessary and useful information and data in a easily accessible format relating to the use of membranes. It is a vital contribution to modern industry and is indispensable for engineers, designers, managers, sales and marketing professionals and indeed anyone using membranes in the course of their work.

Membranes are among the most important engineering components in use today, and each year more and more effective uses for membrane technologies are found – for example: water purification, industrial effluent treatment, solvent dehydration by pervaporation, recovery of volatile organic compounds, protein recovery, bioseparations and many others.

The pace of change in the membrane industry has been accelerating rapidly in recent years, occasioned in part by the demand of end-users, but also as a result of the investment in R & D by manufacturers.

To reflect these changes the author, Keith Scott, has obtained the latest information from some the leading suppliers in the business, and both he and the Publishers are very grateful for their assistance.

In one complete volume this unique handbook gives practical guidance to using selected membrane processes in individual industries while also providing a useful guide to equipment selection and usage.

The Handbook of Industrial Membranes is a welcome addition to the Elsevier Science industrial engineering handbook programme and will prove as valuable as established titles such as the Filters and Filtration Handbook (3rd Edition), the Pumping Manual (9th Edition) and the Seals and Sealing Technology Handbook (4th Edition).

The Publishers

Membrane Technology
an international newsletter

Every month, in just 12 succinct pages, Membrane Technology newsletter brings you an up-to-date international digest to follow all the news and developments affecting Industrial membranes and membrane technology.

Your monthly snapshot of world wide news

Each issue is packed with essential information... from the latest news & views to case studies, and covering the entire range of membrane technologies – from micro filtration to reverse osmosis.

In every issue

- Latest news and views on the development and application of Industrial membranes
- Case studies
- New product launches
- The latest patents – designs and inventions
- Research
- Events

For more information contact:

Elsevier Advanced Technology
The Boulevard
Langford Lane
Kidlington
Oxford OX5 1GB

Tel: (+44) (0) 1865 843842
Fax: (+44) (0) 1865 843971

Contents

SECTION 1 – Introduction to Membrane Separations
    1.1   Introduction 3
    1.2   Contamination, Particle Size and Separation 16
    1.3   Membrane Separation Processes 23
    1.4   Polarisation and Fouling 71
    1.5   Module Designs 94
    1.6   Membrane Process Equipment 110
    1.7   Electrodialysis Cell Stacks and Design 163
    1.8   Laboratory Equipment 175

SECTION 2 – Membrane Materials, Preparation and Characterisation
    2.1   Introduction 187
    2.2   Characterisation of Membranes 227
    2.3   Electrodialysis and Ion Exchange Membranes 257

SECTION 3 – Gas Separations 271

SECTION 4 – Air and Gas Filtration and Cleaning 309

SECTION 5 – Separation of Liquid Mixtures/Pervaporation 331

SECTION 6 – Separation of Organic Vapour/Air Mixtures 355

SECTION 7 – Microfiltration 373

SECTION 8 – Analytical Application of Membranes 433

SECTION 9 – Water Desalination 489

SECTION 10 – Water Purification 521

SECTION 11 – Industrial Waste Water and Effluent Treatment 575

SECTION 12 – Absorption, Desorption and Extraction of Membranes 633

SECTION 13 – Waste Water Treatment and Liquid Membranes 643

SECTION 14 – Biotechnology and Medical Applications 655

SECTION 15 – Medical Applications 683

SECTION 16 – Recovery of Salts, Acids and Bases 691

SECTION 17 – Food Industry 725

SECTION 18 – Membranes for Electrochemical Cells 773

SECTION 19 – Electrokinetic Separations 793

SECTION 20 – Appendix 803

SECTION 21 – Indices 879

Acknowledgements

The Publisher and the Author would like to thank the following companies for their help in compiling this book;

A/G Technology	Inceltech
Allied Signal	Ionics
Altenburger	Koch Membrane Systems
Amicon	Kuboto
Anderman & Co	Le Carbonne Lorraine
APV Pasilac	Lucas
Aqualytics	Lurgi
Aquilo Gas Separation	Membrane Products Kiryat Weizmann
atech Innovations	Membrane Technology & Research (MTR)
Asahi	Microdyn Modulbau
Asea Brown Boveri	Millipore
Berghof	New Logic
Bioseparators	NWW Acumen
Cartridge-Seitz	Osada
COPS	Osmonics
Costar	Osmota Membratechnik
Culligan	Pall Corporation
Cyanara	Parker
DDS	PCI Membrane Systems
Delair	Refractron
domnick hunter	Rennovex
Dorr Oliver	Rhone Poulenc
Dow	Schenk
Du Pont	SCT
Electrosynthesis Coompany	Sartorius
Eurodia	Schleicher & Schuell
Fairey Industrial Ceramics	Seitz-Filter-Werke
FILMTEC	Serck Baker
Filterite	Siemens
G.Hertz	Stantech
Gelman	Tech Sep
Gore	Tokuyama Soda
Grace Membrane Systems	Toray
Graver	UBE Industries
GFT	UF Membrane Systems
Hach & Co	Union Filtration
Hoechst	VSEP
Holland Industrial Ceramics	X-Flow
Hydrocarbon	Zander
IBM	Zenon Environmental

Recommended Reading

In the compilation of the Handbook of Industrial Membranes, the author has made use of figures and photographs from other previously published titles. The author and Elsevier Science Ltd would like to thank the following publishers for their kind permission in allowing reproduction of this material;

>BOC Priestly Conference Proceedings
>Published by The Royal Society of Chemistry. 1986.
>
>Crossflow Filtration by R.G.Gutman
>Published by IOP Publishing Ltd (under the Adam Hilger imprint). 1987.
>
>Industrial Electrochemistry by Pletcher & Walsh
>Published by Blackie Academic. 1990.
>
>Membrane Processes by Rautenbach & Albrecht
>Published by John Wiley & Sons Ltd. 1989.
>
>Membrane Processes in Separation and Purification by Crespo & Boddeker
>Published by Kluwer Academic Inc. 1994.
>
>Polymeric Gas Separation by Paul & Yampolski
>Published by CRC Press Inc. 1994.
>
>Polymeric Gas Separation Membranes by Kesting & Fritzsche
>Published by John Wiley & Sons Ltd. 1993.
>
>Reverse Osmosis by Z. Amjad
>Published by Van Nostrand Reinhold Inc. 1993.
>
>Ultrafiltration Handbook by M. Charyan
>Published by Technomic Publishing Inc. 1986.

Related Titles Published by Elsevier Science

The following publications were all referred to by the author while compiling the Handbook of Industrial Membranes;

Books
Effective Industrial Membrane Processes by M.K.Turner. 1991.
Filters & Filtration Handbook (3rd Edition) by Dickenson. 1991.
Inorganic Membranes by Burggraff and Cot. 1996.
Inorganic Membranes for Separation and Reaction by Hsieh. 1996.
Membrane Separation Technology by Noble & Stern. 1995.
Pervaporation Membrane Separation Processes by Huang.
Structure and Dynamics of Membranes by Lipowsky and Sackmann. 1995.

Reports
A Profile of the International Filtration & Separation Industry (2nd Edition)

Magazines
Filtration & Separation

International Newsletters
Membrane Technology

Journals
Journal of Membrane Science
Desalination
Gas Separation and Purification
Microporous Materials

Conference Proceedings
Gas Separation Technology. 1989.
Industrial Wastewater Treatment and Disposal. 1990.
Industrial Water Technology: Treatment, Reuse and Recycling.
Membrane Technology in Wastewater Management. 1992.
Separation Technology. 1993.
Wastewater Sludge Dewatering. 1992.

For details of these titles and Elsevier Sciences' other membrane and filtration publications, please contact:-

Elsevier Science Ltd, PO Box 150, Kidlington, Oxon. OX5 1AS, UK
Telephone: +44 (0) 1865 843 699
Facsimile : +44 (0) 1865 843 911

SECTION 1
Introduction to Membrane Separations

INTRODUCTION

CONTAMINATION, PARTICLE SIZE AND SEPARATION

MEMBRANE SEPARATION PROCESSES

POLARISATION AND FOULING

MODULE DESIGNS

MEMBRANE PROCESS EQUIPMENT

ELECTRODIALYSIS CELL STACKS AND DESIGN

LABORATORY EQUIPMENT

FOR UPDATES ON A REGULAR BASIS READ:

The international magazine for the Filtration & Separation Industry

Filtration & Separation is read by thousands of engineers, specifiers, designers and consultants across a range of industries:
• process engineering • chemical engineering • food and beverage • petrochemical • biotechnology • water supply & treatment
and they rely on it to keep them up to date with developments in their industry

Coverage includes:
• Industrial Pollution Control • Pressure Filters: Belt, Vacuum and Drum • Centrifuges and Cyclones • Municipal Water & Waste Treatment • Utilities • Filter Presses • Metal Wire Cloth • Screens & Plates • Air Filters • Contamination • Dust Control • Clean Rooms • Filtration Media (& Fabrics) • Overview of the World Filtration Industry • Environmental Protection • Toxic Waste • Settling, Flotation and Gravity Filter Systems • Food & Beverage Filter Processes • Membranes • Separation processes in chemical & allied Industries • Cartridge & Bag Filters

Plus a World Directory of Filtration & Separation Equipment

Filtration & Separation is published 10 times a year, and is available from Elsevier Advanced Technology — see address on order form.

Because we are so confident of the value of **Filtration & Separation**, we offer a no-nonsense guarantee — if you are not satisfied, we will refund the cost of all unmailed issues without question!.

☐ Please bill me for _____ subscription(s) to **Filtration & Separation** @ £75/US$115

Name: .... Position: ....
Organization: ....
Address: ....
....
.... Post/Zip: ....
Country: ....
Tel: .... Fax: ....

Return to:
Elsevier Advanced Technology • PO Box 150 • Kidlington • Oxford OX5 1AS • UK
Tel: +44 (0)1865 843841; Fax: +44 (0)1865 843971

INTRODUCTION TO MEMBRANE SEPARATIONS

SECTION1.1 – INTRODUCTION

Membranes can be used to satisfy many of the separation requirements in the process industries. These separations can be put into two general areas; where materials are present as a number of phases and those where species are dissolved in a single phase.

A membrane is a permeable or semi-permeable phase, polymer, inorganic or metal, which restricts the motion of certain species.. This membrane, or barrier, controls the relative rates of transport of various species through itself and thus, as with all separations, gives one product depleted in certain components and a second product concentrated in these components. The performance of a membrane is defined in terms of two simple factors, flux and retention or selectivity. Flux or permeation rate is the volumetric (mass or molar) flowrate of fluid passing through the membrane per unit area of membrane per unit time. Selectivity is a measure of the relative permeation rates of different components through the membrane. Retention is the fraction of solute in the feed retained by the membrane. Ideally a membrane with a high selectivity or retention and with a high flux or permeability is required, although typically attempts to maximise one factor are compromised by a reduction in the other.

Membranes are used for various separations; the separation of mixtures of gases and vapours, miscible liquids (organic mixtures and aqueous/organic mixtures) and solid/liquid and liquid/liquid dispersions and dissolved solids and solutes from liquids.. The main uses of membrane separations in industry are in the:

i   The filtration of micron and submicron size particulates from liquid and gases (MF).
ii The removal of macromolecules and colloids from liquids containing ionic species (UF).
iii The separation of mixtures of miscible liquids (PV).
iv The selective separation of mixtures of gases and vapour and gas mixtures (GP and VP).
v   The selective transport of only ionic species (ED).

vi The virtual complete removal of all material, suspended and dissolved, from water or other solvents (RO).

The main feature which distinguishes membrane separations form other separation techniques is the use of another phase, the membrane. This phase, either solid, liquid or gaseous, introduces an interface(s) between the two bulk phases involved in the separation and can give advantages of efficiency and selectivity. The membrane can be neutral or charged and porous or non-porous and acts as a permselective barrier.

Transport of selected species through the membrane is achieved by applying a driving force across the membrane. This gives a broad classification of membrane separations in the way, or mechanism, material is transported across a membrane. The flow of material across a membrane is kinetically driven, by the application of either mechanical, chemical, electrical or thermal work. The important membrane processes, together with the general classification of membranes used are listed in Table 1.

TABLE 1 – Membrane separations and materials.

Membrane Separation	Membrane Type	Driving Force	Applications
Microfiltration	Symmetric and Asymmetric microporous	Hydrostatic pressure	Clarification, sterile filtration
Ultrafiltration	Asymmetric microporous	Hydrostatic pressure	Separation of macro-molecular solutions
Nanofiltration	Asymmetric	Hydrostatic pressure	Separation of small organic compounds and selected salts from solutions
Reverse Osmosis or Hyperfiltration	Asymmetric, composite with homogeneous skin	Hydrostatic pressure	Separation of micro-solutes and salts from solutions
Gas permeation	Asymmetric or composite, homogeneous or porous polymer	Hydrostatic pressure concentration gradient	Separation of gas mixtures
Dialysis	Symmetric microporous	Concentration gradient	Separation of micro-solutes and salts from macromolecular solutions
Pervaporation	Asymmetric, composite non-porous	Concentration gradient, vapour pressure	Separation of mixtures of volatile liquids
Vapour permeation	Composite non-porous	Concentration gradient	Separation of volatile vapours from gases and vapours
Membrane Distillation	Microporous	Temperature	Separation of water from non volatile solutes
Electrodialysis	Ion exchange, homogeneous or microporous polymer	Electrical potential	Separation of ions from water and non-ionic solutes
Electrofiltration	Microporous charged membrane	Electrical potential	De-watering of solutions of suspended solids
Liquid Membranes	Microporous, liquid carrier	Concentration, reaction	Separation of ions and solutes from aqueous solutions

TECH-SEP,
the clean processes specialist in Cross-Flow Filtration
Microfiltration
Ultrafiltration
Nanofiltration
Reverse Osmosis

TECH-SEP manufactures **mineral** and **organic membranes** and **modules** and proposes expertise in processes for all industrial applications.

Membranes : a wide range of membranes on high performance supports : **PLEIADE®** organic plate membrane of different polymers ; **CARBOSEP®** tubular mineral membrane ; **KERASEP™** monolithic ceramic membrane ; **FLOSEP™** hollow fiber modules range ; and spiral modules.

Modules : from laboratory to production our modules are adaptable to whatever field of application : range of micro modules, range of pilot unit and production modules.

Applications :
Our wealth of knowledge of membranes and modules in every established application is at your disposal along with the capability of test piloting new processes.
- Purification of potable and industrial water : clarification of water, process water.
- Environmental protection : industrial and municipal waste water, decontamination radioactive effluents, bioreactors, heavy metals...
- Process optimization : chemical and mechanical industries, paper, textile, sugar, petroleum industries and biotechnology
- Enhancement of finished products : food and pharmaceutical industries.

References : 50 000 m² installed in 44 countries concerning industrial applications

TECH-SEP
GROUPE RHÔNE-POULENC

5, chemin du Pilon - St Maurice de Beynost - F-01703 Miribel Cdx -
tel (33) 72 01 27 27 - fax : (33) 72 25 88 99 -

ISO 9001 CERTIFIED
AFAQ 1994 / 2176

SCEPTER™ stainless steel membranes thrive where other membranes fail.

- 0.001 to 1.0 micron removal
- Sustained temperatures up to 177°C (350°F)
- Pressures up to 1,000 psig
- Superior chemical stability, pH 0–14
- High solids, high viscosities
- Unlimited steam and chemical sanitization
- Custom systems and modular components

TYPICAL APPLICATIONS:

- Food and beverage processing
- Waste treatment
- Resource recovery
- Caustic recycling and recovery

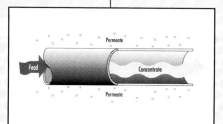

New SCEPTER filtration systems use a patented stainless steel membrane and coating technology to provide years of trouble-free, reliable filtration under extreme process conditions.

SCEPTER's tubular membranes allow processing of "difficult" streams, including dirty or hostile fluids, over a broad range of conditions that could destroy conventional polymeric materials. The systems are impervious to harsh chemicals, and designed for high pressures and temperatures. The durable all-stainless construction offers virtually unlimited steam and chemical sanitization. Plus, innovative form-in-place membrane coatings allow a single system to be adapted to variable feed streams, or even to be used for multiple separations.

SCEPTER stainless membrane systems are manufactured by Graver Separations, specialists in high efficiency purification and separation products.

For more information, call **302-731-1700**. In the U.S., call toll-free, **1-800-249-1990**.

GRAVER SEPARATIONS

Graver Separations
200 Lake Drive
Glasgow, DE 19702-3319
Phone: (302) 731-1700
1-(800) 249-1990 (U.S.)
Fax: (302) 731-1707
Email: graverchem@aol.com

©1995 Graver Chemical Company. Scepter is a trademark of Graver Chemical Company.

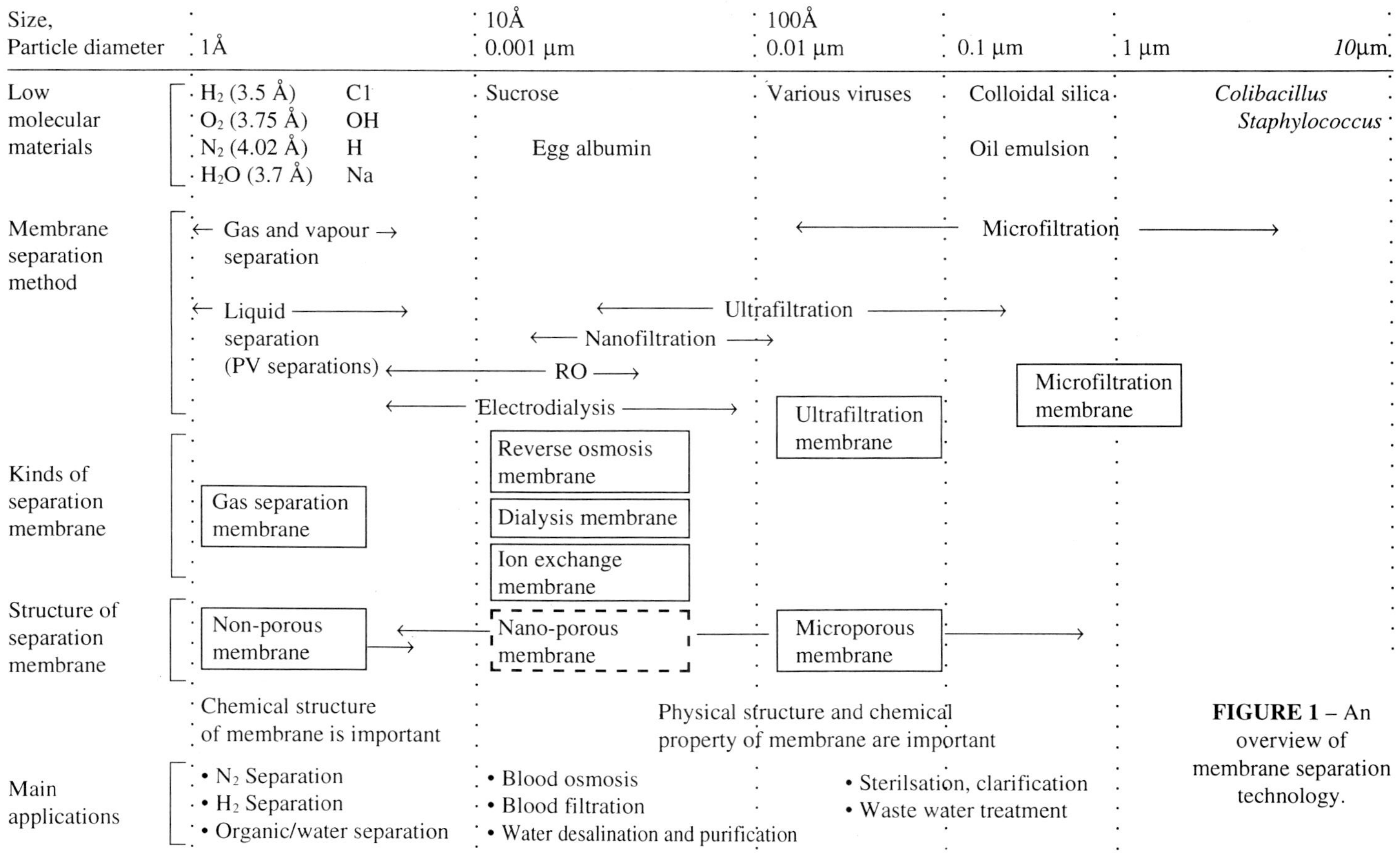

Size, Particle diameter	1 Å	10 Å 0.001 μm	100 Å 0.01 μm	0.1 μm	1 μm	10 μm
Low molecular materials	H_2 (3.5 Å) Cl O_2 (3.75 Å) OH N_2 (4.02 Å) H H_2O (3.7 Å) Na	Sucrose Egg albumin	Various viruses	Colloidal silica Oil emulsion		Colibacillus Staphylococcus
Membrane separation method	← Gas and vapour separation ← Liquid separation (PV separations) ← Electrodialysis →	← RO → ← Nanofiltration →	← Ultrafiltration →	← Microfiltration →		
Kinds of separation membrane	Gas separation membrane	Reverse osmosis membrane Dialysis membrane Ion exchange membrane	Ultrafiltration membrane		Microfiltration membrane	
Structure of separation membrane	Non-porous membrane	↑ Nano-porous membrane Chemical structure of membrane is important	Microporous membrane Physical structure and chemical property of membrane are important			
Main applications	• N_2 Separation • H_2 Separation • Organic/water separation	• Blood osmosis • Blood filtration • Water desalination and purification		• Sterilisation, clarification • Waste water treatment		

FIGURE 1 – An overview of membrane separation technology.

The driving force is either pressure, concentration, temperature or electrical potential. The use of driving force is not a satisfactory means of classification because apparently different membrane processes can be applied for the same separation, for example electrodialysis, reverse osmosis and pervaporation in the desalination of water. From the view of applications, classification in terms of suspended solids, colloids or dissolved solutes, etc is preferred (see Fig 1). Thus the techniques of microfiltration, ultrafiltration, employed in the category of suspended solid separation. All these processes use membranes which are microporous in nature. These are the most simplest form of membrane regarding mode of separation and consist of a solid matrix with defined pores ranging from 100 nm to 50 micron in size.

Microfiltration (MF), in combination with ultrafiltration (UF), can solve almost any separation problem involving particulate material and macromolecules. Major technical advantages of these filtrations are that they are well suited to temperature sensitive materials and are not chemically altered as in competitive procedures such as precipitation and distillation. Membrane filtrations offer relative simplicity of operation and low costs in comparison to competition such as centrifugal separation, vacuum filtration and spray drying. The market areas for ultrafiltration are in the food and dairy industries, biotechnology, water purification and effluent treatment. The latter of these is a developing market for membrane separations as a whole. The largest market share of membrane separations is held by microfiltration and is used for clarification and sterile filtration in a wide range of industries including food and biochemical. Typical systems consist of cartridges where membranes offer absolute filtration capabilities.

A second classification of membranes under a heading homogeneous films encompass the separations; gas permeation, pervaporation, vapour permeation, reverse osmosis and nanofiltration. Separation in these cases is related directly to the transport rate of species in the membrane, determined by their diffusivity and concentration in the membrane phase. These membranes are often in the form of composites of a homogeneous film on a microporous support as used in hyperfiltration and pervaporation. The last two processes are used for similar separations, the removal of water and the concentration of solutions of ionic or organic solutes.

The membrane separations of reverse osmosis (or hyperfiltration) is not restricted to aqueous based solutions, but can in principle be applied to organic based solutions. Hyperfiltration is used in the same industries as microfiltration and ultrafiltration although a major application is in desalination to product potable water. The operating pressures of reverse osmosis are an order of magnitude grater than those of ultrafiltration and microfiltration ie 10 - 100 bar. Competition is with separations such as evaporation and distillation, where membranes score heavily because they do not involve a change in phase and do not expend energy in the latent heat of evaporation. The operating costs of membrane separations are therefore often much lower than competitive separations.

Gas permeation uses homogeneous membranes which separate species in terms of diffusivity and concentration in the membrane. This membrane technology has only recently been applied commercially to separate individual components from mixtures of gases. The membranes are non-porous thin layers on porous substrates. The technical breakthrough, in terms of selectivity and rate of separation, in the membrane separation

GelmanSciences

Over 6000 microfiltration products

For industrial process, healthcare and OEM, and laboratory filtration applications

Fluids Microfiltration

Chemically resistant, high flow rate cartridge and capsule filters and 316L stainless steel housings are designed to meet production needs in high purity water processing, fine chemicals, and beverage production. Choose from high flow rate filter cartridges in 0.1 to 0.65 µm, polypropylene depth filters, and microfiber glass prefilters.

Healthcare and OEM Applications

The wide variety of membranes produced by Gelman Sciences are available in roll-stock form, or may be custom cut and configured into devices for specific medical and OEM applications.

Analytical Labs and Environmental Sampling

A complete line of products featuring membranes, Acrodisc® syringe filters, and accessories for applications including microbiology, analytical sample preparation, and environmental analysis.

Biotechnology/Pharmaceutical Applications

A full line of filtration products for research, scale-up, and final production. Choose from binding membranes, sterile Acrodisc syringe filters, centrifugal MWCO devices, UF membranes, validated pharmaceutical-grade capsules and cartridges, and sanitary-grade stainless steel housings.

For more information, call: **01604.765141**

more products, better ideas, the best solutions

for your filtration and membrane separation needs.

Gelman Sciences Inc.
World Headquarters
600 South Wagner Road
Ann Arbor, MI 48103-9019
USA
Tel: 313.665.0651
800-521-1520
FAX: 313-913-6119

Australia - Sydney, N.W.S.
Gelman Sciences Pty. Ltd.
Tel: 02-428 2333
FAX: 02-428 5610

Canada - Montréal, Quebec
Gelman Sciences Inc.
Tel: 514-337-2744
1-800-435-6268
FAX: 514-337-7114

China - Beijing, P.R.
Gelman Sciences Inc.
Tel: 010-4911960
FAX: 010-4993034

France - Marle la Vallée
Gelman Sciences S.A.
Tel: 1-64 68 30 81
FAX: 1-64 68 29 56

Germany - Roßdorf
Gelman Sciences (Deutschland) GmbH
Tel: 06154-9075
FAX: 06154-83519

Ireland - Dublin
Gelman Ltd.
Tel: 01-284-6177
FAX: 01-280-7739

Italy - Milan
Gelman Italy s.r.l.
Tel: 02-69006109
FAX: 02-69006110

Japan - Tokyo
Gelman Sciences Japan, Ltd.
Tel: 03-3844-5411
FAX: 03-3844-5433

Puerto Rico - Humacao
Gelman Sciences Inc.
Tel: 809-850-1790
FAX: 809-852-6882

Russia - Moscow
Gelman Sciences
Tel: 095-265-48-45
FAX: 095-261-67-81

Sweden - Gothenburg
Gelman Sciences
Tel: 031-129214
FAX: 031-7750190

United Kingdom - Northampton
Gelman Sciences Ltd.
Tel: 01604-765141
FAX: 01604-761383

Acrodisc is a registered trademark of Gelman Sciences Inc.

©Gelman Sciences, 1995 GN 95.0522

Wide experience in engineering as well as application knowledge combined with an innovative approach to the latest developments within the cross-flow filtration technology make our membrane plants attractive possibilities to traditional processes.

Interesting cost savings in production processes can be obtained owing to:
- Less water intake resulting from recycling of water
- Product recovery from process streams/effluents
- Reduction of and less polluted waste streams
- Energy efficient processing and reduced consumption of chemicals

We manufacture plants for process water and treatment of industrial waste streams from a.o. the textile, galvanic, metal-working, beverage, dairy, and pharmaceutical industries.

To meet the ever-growing demand for customized equipment for uncommon applications we also offer:
- Technical consultations and cost/benefit analyses
- Laboratory and pilot-scale trials
- Commissioning, training, and service support

Union Filtration a/s will be delighted to be also your supplier of reverse osmosis, nanofiltration, ultrafiltration, and microfiltration plants.

Union Filtration a/s, Sandvikenvej 7, DK-4900 Nakskov, Denmark.
Tel: +45 54 95 1300   Fax: +45 54 95 1301.

Separating - Filtrating - Concentrating
Check out our potential!

- Softening plants
- Partial demineralization plants
- Full demineralization plants
- Seawater demineralization
- Brackish water demineralization
- Ultrapure water production
- Boiler feed water treatment

- Coffee concentration
- Yeast concentration
- Juice concentration
- Alcohol absolution
- Clear filtration

- Electrodip paint recycling
- Degreasing bath recycling
- Aquaeous paint recycling
- Pigment concentration
- Solvent filtration

- Acis and caustic recovery
- Acid filtration
- Hardening salt recovery
- Recovery of pickling acids

- Oil/water emulsion splitting
- COD/BOD reduction
- Leachate treatment
- Environmental protection
- PCB separation
- Separation of radioactive substances

OSMOTA is specialized in most membrane separating processes. We offer you the economical and ecological system solution for your problem.

OSMOTA Membrantechnik GmbH, Jahnstraße 4/1
D-70809 Korntal-Münchingen, Telefon 0711/83 10 91, Fax 0711/83 47 55

INTRODUCTION TO MEMBRANE SEPARATIONS

of gases made gas permeation competitive with cryogenic separation, adsorption etc.. Other market areas are now opening in organic vapour separations and hydrocarbon separations. Pervaporation is a membrane process which can essentially replace fractionation by distillation, although applications are restricted to the more difficult separations which typically involve azeotrope formation, e.g. ethanol/water. This process is somewhat unique within membrane separations involving a change in phase from feed (liquid) to permeate (vapour).

The third classification of membranes are those that are electrically charged. These membranes carry either fixed positive or negative charges and separate by exclusion of ions of the same charge as carried in the membrane phase. The separation which selectively removes ions is called electrodialysis. Market areas are again similar to the membrane filtrations with a principle application in the desalination of brackish water. It is used in a number of applications in desalting foods, in effluent treatment and notably in Japan for the manufacture of salt. Competition for electrodialysis comes from separations such as evaporation and also from hyperfiltration. The economic advantage gained with electrodialysis is through its specificity and efficiency of separation achieved at low temperatures. A major application of ion-exchange membranes is as electrolytic cell separators in for example the production of chlorine and caustic soda.

The advances made in membrane technology over the last few decades have seen

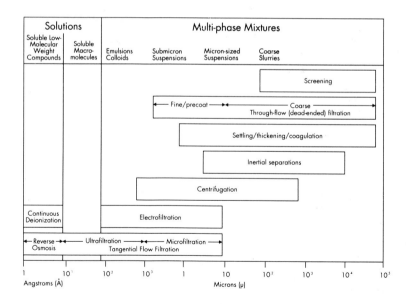

FIGURE 2 – Separation processes based on size of suspensions and solutes.

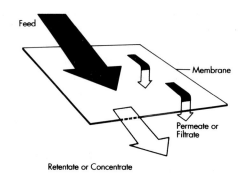

FIGURE 3 – The tangential or crossflow process.

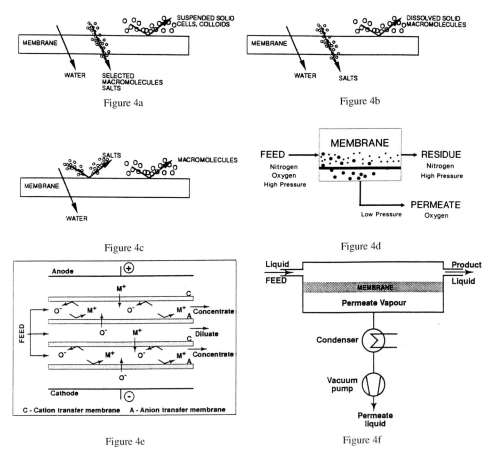

FIGURE 4 – Schematic representation of crossflow membrane separations.
a) microfiltration, b) ultrafiltration, c) reverse osmosis, d) gas permeation, e) electrodialysis, f) pervaporation, g) liquid membrane.

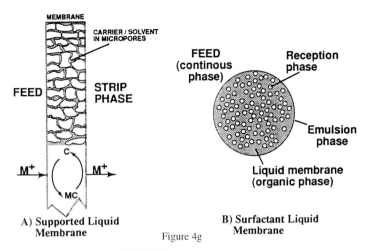

A) Supported Liquid Membrane

B) Surfactant Liquid Membrane

Figure 4g

FIGURE 4 (continued).

applications expand in many industrial sectors; chemical, petrochemical, mineral and metallurgical, food, biotechnology, pharmaceutical, electronics, paper and pulp, and water etc. Membrane separations are in competition with physical methods of separation such as selective adsorption, absorption, solvent extraction, distillation, crystallisation and cryogenic gas separation. The process of membrane filtration separate components and suspensions on the basis of size. Fig 2 shows size-based separations of liquid solutions and suspensions and the typical particle size and molecular size over which they are more effective. Generally membrane filters can separate solutions and suspension species which have diameters below 10 μm when operated in a tangential, or cross, flow mode as shown in Fig 3. The particles or solutes retained at the membrane surface are continuously removed in the retentate which flows tangentially across the membrane surface. The clarified solution flows through the membrane as the permeate. Clearly from Fig 2, the membrane separations operate at the low range of the particle size spectrum, approximately 10 μm and lower. The first of the membrane processes, referred to as microfiltration, is used to remove small suspended particulate from solution (see Fig 4) by the application of a differential pressure. The process itself can be assisted by the application of an electrical field as in the process of electrfiltration.

Ultrafiltration (UF) is the second of the membrane separation processes which are grouped together under pressure driven processes. UF covers the region between MF and RO and is used to remove particles in the size range 0.001 to 0.02 micron. Solvents and salts of low molecular weight will pass through the membranes whilst larger molecules are retained (Fig 4). Thus the principal application of UF is in the separation of macromolecules with a size retention in the molar mass range of 300 to 300,000. UF membranes are permeable to molecules of molar masses of a 1000 and thus exhibit low rejection of salts.

Ultrafiltration is typically applied in the separation of macromolecular solutes and colloidal material from macromolecular solutes and solvents. There are many analytical applications on the laboratory scale. These include the concentration of proteins, enzymes, hormones etc and in biochemical and clinical analysis. The wide variety of industries utilise UF (see Table 1a) including: Chemical and nuclear; in the treatment of waste water and effluents, Automobile; for recovery and recycling in electropaint baths, Pulp and paper, Food and diary; for the clarification of juices and wines, milk concentration sterile filtration etc., Biological and pharmaceutical; for the manufacture of antibiotics, removal of pyrogens and the treatment of blood and plasma.

Hyperfiltration or Reverse osmosis (RO) is another pressure driven process for the separation of ionic solutes and macromolecules from solutions. The method of dissolved salt removal is not just a physical process based on size difference of solute and solvent. The RO process is the reversal of the natural process of osmosis. The application a pressure to a salt solution forces the flow of water through the membrane when the pressure exceeds the osmotic pressure.

Nanofiltration is essentially a form of RO, i.e. a pressure driven process, which is applied in the area between the separation capabilities of RO membranes and UF membranes, that is in the separation of ions from solutes such as small molecules of sugars. This a achieved with suitable "loose" membranes which are typically thin film non-cellulose materials. Nanofiltration is used when high sodium rejection, typical of RO, is not needed but where other salts such as Mg and Ca (ie divalent ions) are to be removed

The process of gas permeation (GP) is used to separate gas mixtures by virtue of differences in molecular size and gas solubility in the membrane. Gases of smaller molecular size have larger diffusion coefficients and in a convection free environment in the pores of a membrane can be suitably separated by virtue of the different mobilities.

Pervaporation is a membrane process for the separation of miscible liquid mixtures into more concentrated products of the constituents. Interest in this process is growing due to the practical limitations of reverse osmosis in many potential separations where otherwise extreme pressures would be required. Separation is achieved by applying a lower pressure (vacuum) to the permeate side of the membrane whilst the other side is exposed to the liquid to be separated. The partial pressure of the permeate is thus kept lower than the saturation pressure and provides the necessary force for separation. The commercial application of PV is being investigated in the chemical and biochemical industries. The recovery of low concentrations of organics, eg alcohols, from fermentation broths and the removal of small quantities of solvents from water is typical.

Electrodialysis (ED) is a method which uses a direct electrical current to transport ions through sheets of ion selective membranes. The process of electrodialysis is used for the concentration of electrolyte solutions, and for the dilution, or de-ionising, of solutions. The process of electrodialysis is used to perform several operations:

- The separation of salts, acids, and bases from aqueous solutions
- The separation of ionic compounds from neutral molecules
- The separation of monovalent ions from ions with multiple charges
- The introduction of ionic moieties to generate new species

Electrodialysis competes with other separation processes, such as reverse osmosis, ion-

TABLE 1a – The Application of Ultrafiltration.

Process	Separation/Application
Electrophoretic paint	Process rinse water, recycle paint to dip tank, allow reuse of rinse water
Cheese whey	Concentrate/fractionate proteins from lactose and inorganics
Juice clarification	Remove haze components from apple juice
Textile sizing agents	Recover polyvinyl alcohol after scouring of woven goods
Wine clarification	Remove of haze components from red and white wines
Oil/water emulsions	Metal cutting oils (lubricants) concentrated from waste water for incineration
Polymer latex	Latex emulsions concentrated from wastewater
Dewatering	Separation of wax components from lower paraffins
Desphalting	Solvent recovery/recycle for deasphalting of heavy crudes
Egg-white preconcentration	Partial dewatering before spray drying
Fermentation broth	Separate low molecular weight organics/therapeutic agents from cells or cell debris
Kaolin concentration	Partial dewatering or clay slurry before centrifugation
Affinity membranes	Retain ligand complex from noncomplexed proteins
Reverse osmosis pretreatment	Retain colloidal silica, bateria Concentration of leachate from landfill sites
Water treatment	Concentration before sludge dewatering Treatment of wool scour effluent for water recovery and re-use
Biological waste treatment processes	Pig slurry concentration Biomass separation in biological pre-treatment of landfill leachate Biomass separation in municipal sewage treatment
Textile industry	Treatment of dye effluents
Cutting tool lubricants/coolants	Separation and recovery of oils from waste oil emulsion and from oily wash waters
Pulp, paper and rayon	Recovery of lignosulfonate fractions from waste waters
Food	Treatment of food processing waste water
General rinsing and washing	Treatment of parts washer effluent Treatment of compressor condensates Cosmetic manufacturing rinsewater Dishwasher effluent Car wash effluent Laundry waste water

exchange, etc., in many applications, such as water purification and can offer in many cases several significant advantages.

Overall in membrane separations the membrane is clearly the most important part of the separation process. Membrane material science has produced wide range of materials of different structure and with different ways of functioning (see Table 2). Generally these materials can be classified into three types:-

- synthetic products; a vast number of polymers and elastomoners
- modified natural products; cellulose based
- Inorganic, ceramic and metals

A membrane material should ideally posses many of the following properties to be effective for separation;

- appropriate chemical resistance,
- mechanical stability,
- thermal stability,
- high permeability,

TABLE 2 – Type and structure of synthetic membranes.

Membrane type	Membrane structure	Applications
Asymmetric CA, PA, PS, PAN	Homogeneous or microporous, "skin" on a microporous substructure	UF and RO (MF) GP, PV, VP
Composite CA, PA, PS, PI	Homogeneous polymer film on a microporous substructure	RO, GP, PV, VP
Homogeneous S	Homogeneous polymer film	GP
Ion exchange DVB, PTFE	Homogeneous or microporous co-polymer film with positively or negatively charged fixed ions	ED
Microporous: Ceramic, metal Glass	0.05 to 20 mm pore diameter 10 to 100 mm pore diameter	GP F (molecular mixtures)
Microporous: Sintered polymer PTFE, PE, PP	0.1 to 20 mm pore diameter	F (suspensions, air filtration)
Microporous: Stretched polymer PTFE, PE	0.1 to 5 mm diameter	F (air, organic solvents)
Microporous: Track-etched PC, PEsT	0.02 to 20 mm pore diameter	F (suspensions, sterile filtration)
Symmetric microporous phase inversion CA	0.1 to 10 mm pore diameter	Sterile filtration, water purification, dialysis

PTFE - polytrafluoroethylene. CA - cellulosic esters. PVC - polyvinylchloride. PA - polyamide. PE - polyethylene. PS - polysulfone. PP - polypropylene. S - silicon rubber. PC - polycarbonate. PEst - polyester. PAN - polyacrylonitrile. PI - polyimide. DVB - divinylbenzene. UF - ultrafiltration. RO - reverse osmosis. GP - gas permeation. MF - microfiltration. ED - electrodialysis. F - filtration. PV - pervaporation

- high selectivity or retention,
- stable operation and low cost.

All these properties are obviously relative in terms of individual processes and the respective capital and operating costs. Chemical resistance relates more to the operating lifetime of the membrane. A gradual deterioration of the membrane can occur over months and years with perhaps only a relatively small loss of selectivity.

TABLE 3 – Types and Structures of Membranes.

Modified natural products
Cellulose acetate (cellulose-2-acetate, cellulose-2,5-diacetate, cellulose-3-acetate), ceuulose acetobutyrate, cellulose regenerate, cellulose nitrate
Synthetic products
Polyamide (aromatic polyamide, copolyamide, polyamide hydrazide), polybenzimidazole, polysulphone, vinyl polymers, polyfuran, polycarbonate, polyethylene, polypropylene, PVA, PAN, polyether sulphone, polyolefins, polyhydantoin, (cyclic polyurea), polyphenylene sulfide), silicone rubber, PTFE, PVDF, Nylon.
Miscellaneous
Polyelectrolyte complex, porous glass, graphite oxide, ZrO_2-polyacrylic acid, ZrO_2-carbon, Al_2O_3, SiC, Metals (Pd, Ag, Stainless Steel, Al).

The membrane function will depend on its structure as this essentially determines the mechanism of separation and thus the application. Two types of structures are generally found in membranes (solid material) symmetric or asymmetric. Symmetric membranes are of three general types; with approximate cylindrical pores, porous and non-porous (homogeneous). Asymmetric membranes are characterised by a non uniform structure comprising an active top layer or skin supported by a porous support or sub-layer. There are three types porous, porous with top layer and composites. There are several methods for producing membranes, discussed in detail in Section 2.

Electrochemical Processes

The electrochemical industry is a major user of membranes principally as a separator between an anode and cathode in electrochemical cells. Electrosynthesis of many inorganic and organic chemicals relies on membranes to achieve high selectivities and efficiencies in specific reactions. The membrane serves to restrict the transport of certain species from anode and cathode compartments. The most significant process is for the production of chlorine and caustic soda, where ion selective (cation) membranes ensure transport of sodium ions from anolyte to catholyte, which combine with the hydroxide ions, which do not pass through the membrane, to produce sodium hydroxide. The membrane in this case is a fluorocarbon based material.

Electrochemical cells utilising membranes are also used in several effluent treatment and recycling applications and in electrochemical power generation devices, ie batteries and fuel cells. Ion specific membrane electrodes are used in chemical analysis and chemical diagnostics, where the membrane gives selective ion transmittance to suitably chosen detector electrodes.

Electrofiltration

Electrofiltration is membrane separation driven by the application of a differential pressure and an electrical potential gradient across the membrane. This potential facilitates the transport of mobile ions and liquid through the pores of the membrane. The mechanism involves the formation of an electrical double layer at the surface due to the ability of the membrane to acquire a charge when immersed in an aqueous solution. This double layer also has an associated mobile diffuse layer of opposite charge which on experiencing an electrical potential gradient moves through the pores. Applications of this process are mainly in dewatering, in the treatment of colloidal suspensions and sludges in effluent and waste streams.

Liquid Membranes

Separation using liquid membranes can be likened to that of conventional solvent extraction and stripping in which a thin liquid films is used to transport the solute from the feed to the product side. The use of a liquid film offers the possibility of much higher separation rates than in polymer films because of the higher diffusion rates in the former.

There are two basic forms of liquid membrane processes. The first is where a liquid is supported in the pores of a suitable microporous membrane, referred to as carrier impregnated, immobilised liquid membrane (ILM) or supported liquid membranes. The other form of membrane, surfactant liquid membrane (SLM) combines surfactant with the membrane liquid which coats the discontinuous droplets of an emulsion. Solute transport can then typically proceed from say a continuous aqueous phase across the membrane and into the droplet phase.

Section 1.2 – CONTAMINATION, PARTICLE SIZE AND SEPARATION

The removal of contaminants from fluid is the major application of membranes. This is because of how we define a contaminant and the fact that they are present to some degree in all fluids. A contaminant is normally defined as any matter present in the fluid which must be removed or controlled to prevent chemical, physical or biological interference with the product quality. Contaminants are generally considered as solid, semi-solid or liquids particles that can be separated from the carrier fluid by some physical process, typically some form of barrier. However contaminants can also be species which are completely miscible or mixed with the carrier phase in for example the contamination of fuels by water, the presence of acid gases ie SO_2, CO_2, H_2S, etc is waste gases and natural gases as well as dissolved in aqueous fluid and the contamination of waters by organics (fertilisers, pesticides, solvents etc) and inorganic species (nitrates, phosphates, heavy metals etc). This rather open ended definition of contaminants means that all membrane separation processes may be considered as contaminant removers. Their uses however go

INTRODUCTION TO MEMBRANE SEPARATIONS

TABLE 1 – Relative size of small particles.

Separation	Mol.Wt.	Å	μm	Particles/Items
General Filtration		10^7	1000	
			800	• Sewing Needle Diameter
			600	
			400	• Razor Blade Thickness — Beach Sand — Drizzle
			200	
			105	• 150 Mesh
		10^6	100	
			80	• Human Hair Diameter
			60	• Smallest Visible Particle — Mist — Pollens
			40	
			37	• 400 Mesh
			20	• Ragweed Pollen — Carbon Black
		10^5	10	
Microfiltration			8	Bacteria — White Light Microscopy
			6	Red Blood Cell — Emulsions (Latex)
			4	
			2	Syrups
		10^4	1	Yeasts & Fungi
		8,000	0.8	
			0.6	• *Serratia marcescens*
		4,000	0.4	• *Pseudomonas diminuta*; DOP — Mycoplasm
			0.2	
	1,000,000	1,000	0.1	Colloids
		800	0.08	Tobacco Smoke
Ultrafiltration	500,000		0.06	Electron Microscopy
		400	0.04	Endotoxins (Pyrogen) — Proteins
	300,000		0.02	
	100,000	100	0.01	
		80	0.008	Virus
	50,000		0.006	• Albumin (60,000 MW)
	30,000	40	0.004	
			0.002	
	10,000	10	0.001	
Reverse Osmosis	500			
				Soluble Salts (Ions) — Metal Ions
	50			

Å, Angstrom = 10^{-8} cm
μm, Micrometer (micron) = 10^4 Å
1 mil = 0.001 in = 25.4 μm

Differential pressure *increases* with smaller μm pore sizes; dirt holding capacity and relative flow rates *decrease* with smaller μm pore sizes.

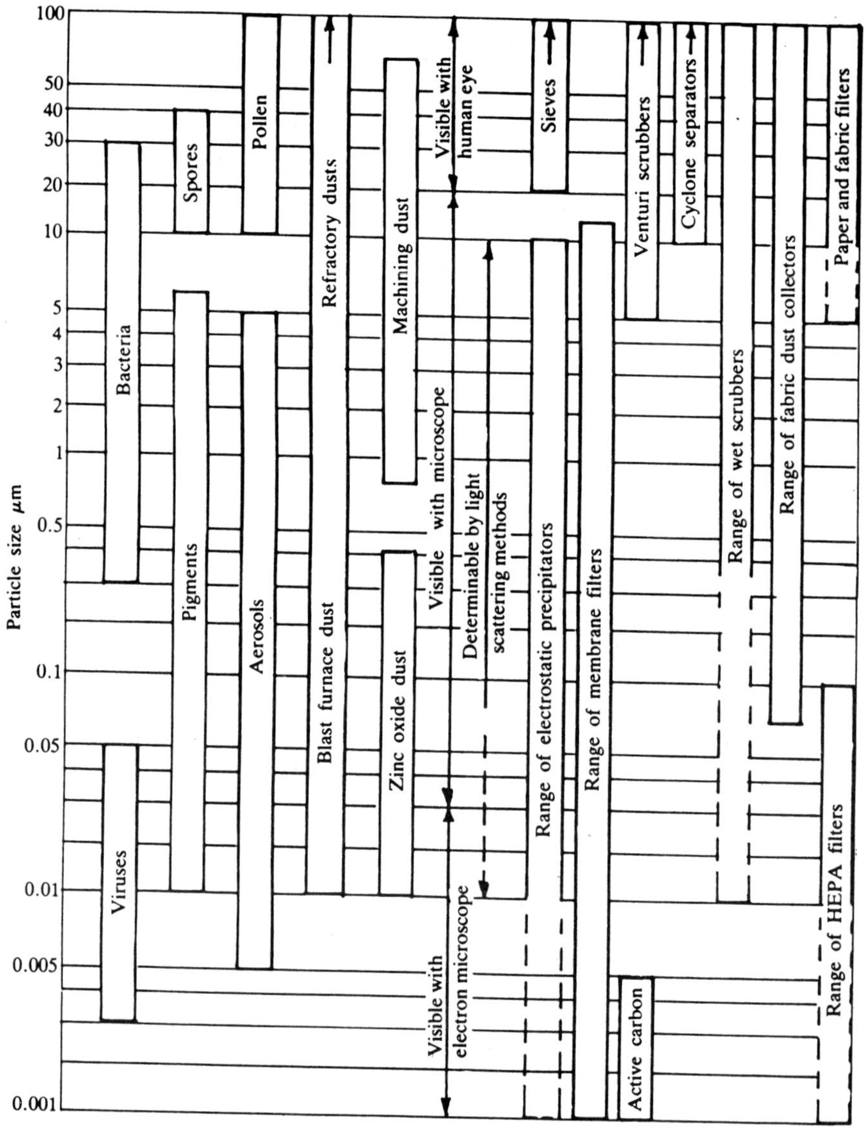

TABLE 2 – Contaminant sizes, visibility and separation.

far beyond this as is demonstrated in this manual. The removal of solids, liquid and gases may typically form part of a processing strategy where the residual fluid is of only secondary interest eg a diluent such as water.

Contaminants in the form of particulates occur in a wide range of sizes in gases and liquids (see Table 1). The largest individual particle visible to the naked eye has a size of approximately 40 μm. Groups of smaller particles are "visible"/detectable as dusts, suspensions, turbidity. Individual particles with a size of around 0.2 μm are visible by conventional white light microscopy and below this size methods such as light scattering and electron microscopy are needed. Membranes separations largely operate in the size range of 10 mm and lower ie below the visible detectable limit of single particles. Such particles have a settling velocity in still air of approximately 0.2 m/min and 0.004 m/min in water. This sized particle is therefore generally removed from a fluid by means other than gravity settling. Naturally not all particulate separations are carried out by membranes, several other physical or component separations may be used as indicated in Table 2.

The choice of the separation or the degree of contaminant removal can only be made from an appreciation of the contamination and its environment, the filtration or separation method and of the effect of conditions on the separation. The properties of the fluid are important, notably the fluid viscosity. This parameter is particularly important for liquids as it is a measure of the resistance to flow, higher values of viscosity means a higher pressure differential is required to achieve the same flowrate through a given membrane for example. In certain cases for more viscosous fluids than water, which has a viscosity of 1 centipoise (cP) at 25°C, viscosity can be decreased by increasing the temperature. For certain "fluids" such as syrups and resins, a modest increase in temperature (of a few degrees) will cause an order of magnitude reduction in viscosity, with corresponding benefits on pressure requirements. In certain cases the rheological behaviour of the fluid should be identified for example thixotropic fluids such as colloidal gels (paints and inks) are free flowing under an applied pressure (shear force), but on coming to rest return to a gel state with a high viscosity many thousands of centipoise. Other factors which must be considered are pH, chemical compatibility and surface tension.

In the case of gases the compressibility of this phase must be designed for. Whether the gas is inert (or essentially non-reactive) or reactive must be allowed for in the selection of the separation medium. The presence of moisture in the gas stream can lead to problems in separation. Operation below the dew point can cause condensation which may wet and block the pores of hydrophilic filters. In such instances hydrophilic filters/membranes are preferred which are resistant to blocking by moisture.

Typical contaminants of interest are bacteria and bacterial fragments, crystals, colloids and manufacturing debris. The characteristics of the particulate contaminant (shape, size, type phase) are a major factor in separation procedure.

Shape - Outline or contour of the contaminant's external surface. Whether a contaminant maintains its contour (spherical or rod shaped) depends on its type (hard, soft, or liquidMmost particles are not spherical. In filtration, the controlling dimension of a rod-shaped particle depends on how it challenges the filter media. If it collides end-on, its

TABLE 3 – Relative Sizes of Bacterial Contaminants.

Organism	Rod Length (μm)	Rod or Coccus Diameter (μm)	Significance
Acholeplasma laidlawii	2.0-5.0	0.3	Test organism for retention of 0.1 μm membranes
Alcaligenes viscolactis	0.8-2.6	0.6-1.0	Causes ropiness in milk
Bacillus anthracis	3.0-10.0	1.0-1.3	Causes anthrax in mammals
Bacillus stearothermophilus	2.0-5.0	0.6-1.0	Biological indicator for steam sterilization
Bacillus subtilis	2.0-3.0	0.7-0.8	Biological indicator for ethylene oxide sterilization
Clostridium perfringens	4.0-8.0	1.0-1.5	Produces toxin causing food poisoning
Clostridium tetani	4.0-8.0	0.4-0.6	Produces exotoxin causing tetanus
Erwinia aroideae	2.0-3.0	0.5	Causes soft rot in vegetables
Escherichia coli	1.0-3.0	0.5	Indicator of fecal contamination in water
Gluconobecter oxydans subsp. melanogenes	1.0-2.0	0.4-0.8	Strong beer/vinegar bacterium
Haemophilus influenzae	0.5-2.0	0.2-0.3	Causes influenza and acute respiratory infections
Klebsiella pneumoniae	5.0	0.3-0.5	Causes pneumonia and other respiratory inflammations
Lactobacillus delbrueckii	2.0-9.0	0.5-0.8	Causes souring of grain mashes
Leuconostoc mesenteroides		0.9-1.2	Causes slime in sugar solutions
Mycoplasma pneumoniae (PPLO)		0.3-0.5	Smallest known free-living organism
Pediococcus acidilactici		0.6-1.0	Causes mash spoilage in brewing
Pediococcus cerevisiae		1.0-1.3	Causes deterioration in beer
Pseudomonas aeruginosa	2.0-3.0	0.5-1.0	Cause of burn would infection
Pseudomonas diminuta	1.0	0.3	Test organism for retention of 0.2μm membranes
Salmonella enteritidis	2.0-3.0	0.6-0.7	Causes food poisoning
Salmonella hirschfeldii	1.0-2.5	0.3-0.5	Causes enteric fever
Salmonella typhimurium	1.0-1.5	0.5	Causes food poisoning in man
Salmonella typhosa	2.0-3.0	0.6-0.7	Causes typhoid fever
Sarcina maxima		4.0-4.5	Insolated from fermenting malt mash
Serratia marcescens	0.5-1.0	0.5	Test organism for retention of 0.45μm membranes
Shigella dysenteriae	1.0-3.0	0.4-0.6	Causes dysentery in man
Staphylococcus aureus		0.8-1.0	Causes pus forming infections
Streptococcus lactis		0.5-1.0	Contaminant in milk
Streptococcus pneumoniae		0.5-1.25	Causes lobar pneumonia
Streptococcus pyogenes		0.6-1.0	Causes pus forming infections
Vibrio percolans	1.5-1.8	0.3-0.4	Test organsim for retention of 0.2μm membranes

INTRODUCTION TO MEMBRANE SEPARATIONS

diameter will be the significant dimension; if sideways, its length will be significant. Spherical contaminants are the most difficult to remove.

Size - Particles of concern in filtration can range from viruses of less than 0.1 mm in diameter to cells of more than 10 mm all the way up to biological materials, such as pollen, at 100 mm. Microfiltration is concerned with the removal of particles in the 0.01 mm to 60 mm range.

Type

Hard: Contaminants that are rigid and will not deform under pressure or driving force. These include crystals and most manufacturing debris.

Soft: Contaminants that are deformable and can change shape under pressure or driving force. Deformation can alter the shape of the particle enough to allow it to pass through a filter. Filtering at low pressure can help minimize distortion and prevent passage of the contaminant through the filter. A number of factors influence the retention of soft contaminants; these include the number of membrane layers, the amount of applied pressure, the pore size rating of the membrane, and the concentration of the contaminants in the solution to be filtered.

Liquid: Contaminants that are miscible or immiscible with another liquid. Immiscible fluids, such as water and oil, can often be separated by microfiltration.

Table 3 gives a list of typical bacterial contaminants, their shape and significance for removal. These bacteria can be removed quite effectively from a range of fluids using filtration. Areas of application are many as is typically illustrated in Table 4.

TABLE 4 – Application guide to pore size rating of filtration

$0.1\ \mu m$	$0.2\ \mu m$	$0.45\ \mu m$	$0.65\ \mu m$	$0.8\ \mu m$	
Semiconductor chemicals	Sterilization of solutions and gases	Bacterial reduction	Bacterial reduction	Yeast removal	
Optical films	Acid filtration	Large volume parenterals	Cosmetics	Serum prefiltration	
Acid filtration	High purity water	Wine sterilization	Wine filtration	Beverage filtration	
Photoresists High purity water	Mineral water				
$1.2\ \mu m$	$3\ \mu m$	$5\ \mu m$	$10\ \mu m$	$30\ \mu m$	$60\ \mu m$
Mold removal	Plating solutions	Reverse osmosis prefiltration	Magnetic media	Visible particle removal	Agglomerate removal
General clarification	Waste water	Deionizing resin trap	Solvent prefiltration	Syrup filtration	Gel particle removal
Plating solutions/	Laboratory	Final polishing	Mineral water	Viscous food	Lubricating

TABLE 5 – Impurities present in air.

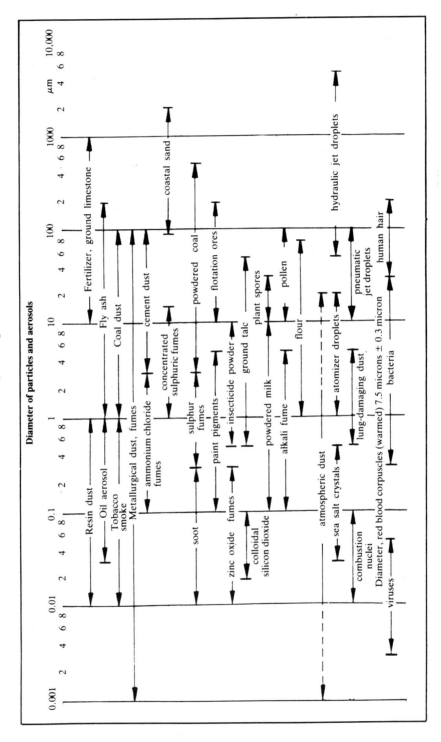

INTRODUCTION TO MEMBRANE SEPARATIONS

These filtrations may be carried out using non-membrane separations, for example depth filtration using a range of filter materials. The use of alternative filtration techniques is described at length in the Filters and Filtration Handbook - Elsevier Publications. For convenience summarised information on the range of contaminant removal and alternative separation technologies is given in the Appendix. A great deal of these applications deal with impurities present in air. Particles present in air are generated from many sources, typically dusts from a range of manufacturing operations, machining, chemical manufacture, combustions, aerosols, bacteria, viruses etc (see Table 5). In air outside specific process operations 90% of the impurities are in the size range of 0.1 to 10 mm, local environment will however greatly affect the concentration of particulates. The particles in general will be freely suspended in the air due to convective currents, thermal buoyancy forces which are inevitably present. These particles contaminants can have detrimental effects on human health, the operation of machines, the quality of water and all manufacturing and processing operations. Particulates which are emitted to air can eventually find their way into water and a range of other liquids. Conversely operations with liquids can release contaminants into the air, for example oil mists generated from compressor lubricants, water vapour and droplets from cooling towers.

Contaminants present in liquids will generally be of the same type and size as these present in air (see Table 5). These particulates will tend to accumulate if the contact between air borne material and liquids is not controlled. Water is an excellent media for the absorption of many gases generated from combustion, for organic solvents and other organic species, for solid material and colloidal species which will remain suspended or emulsified in the liquid. Water is also a suitable environment for the growth of bacteria, especially where suitable nutrients are present. Particulate material will be generated by the natural wear of materials in contact with liquids. This will be enhanced by mechanical operations, pumping, the presence of contaminants, the effect of corrosion and erosion, leaching of extractables etc. Particulates can also be generated by the separation process itself, designed to control the contaminant problem in for example the shedding of materials of construction of filters and from the grow through of bacteria. In these cases the final solution may require absolute filtration /(contaminant removal) at the point of use of the fluid, in for example high purity water rinsing operation in the electronics industry, with preceding filtrations used to restrict and control the growth of bacteria.

SECTION 1.3 – MEMBRANE SEPARATION PROCESSES

Microfiltration

Filtration is a process for the separation of solid from a liquid, or a gas, stream by the mechanical means of sieving. A pressure differential is applied across the filter to maintain fluid flow through the filtration media. The resultant filtrate or permeate flowing though the filter should ideally be devoid of suspended solid. With conventional filters particle fragments and filter medium can escape during filtration.

In filtration the flow of fluid through the filter medium is proportional to the applied pressure difference (ΔP) across the membrane. The rate of flow is dictated by the resistance of the filter to the flow of fluid any resistance associated with the trapped particulate. The filter medium characteristic is defined in terms of its resistance or permeability, K. The flow, or the flux, J, through the filter is thus given by

$$J = K \Delta P / \mu \quad \ldots \ldots (1)$$

The flowrate or flux is thus also dependent on the viscosity, m of the fluid being filtered.

Filters for membrane microfiltration (MF) are typically made from thin polymer films with 'uniform' pore size and a high pore density of approximately 80%. The principle method of particle retention (Fig 1) is 'sieving', although the separation is influenced by interactions between the membrane surface and the solution. The high pore densities of the filters generally mean hydrodynamic resistance is relatively low and hence high flowrates, or membrane flux rates, (cubic metres of permeate per square metre of membrane area per hour, m/h), result at modest operating differential pressures up to 2 bar.

FIGURE 1 – Separation using microfiltration membranes.

The irregularity of the pores of most membranes and the often irregular shape of the particles being filtered mean there is not a sharp cut off size during filtration. With symmetric membranes some degree of in-depth separation could occur as particles move through the tortuous flow path. To counter-act this effect, asymmetric membranes, which have surface pore sizes much less than those in the bulk of the membrane, can be used These entrap the particles almost exclusively at the surface (the membrane skin) whilst still offering low hydrodynamic resistance. This technique has also enabled inorganic membranes to be used in sveral applications.

Microfiltration is widely applied in a dead-end mode of operation. In this the feed flow is perpendicular to the membrane surface and the retained (filtered) particles accumulate on the surface forming a layer of retained solid or a filter cake. The thickness of this cake increases with time and the permeation rate correspondingly decreases as the resistance of the cake increases. Eventually the membrane filter reaches an impractical (uneconomic), low filtration rate and is either cleaned or replaced. Typical filters come in the form of readily replaceable screw-in cartridges.

The effect of a build-up of solid particle cake on the membrane surface can be reduced by the use of cross flow, or tangential flow. Cross flow separation can be seen (fig 2) as a process in which a feed stream flows along, tangential to a membrane surface. As a result of the application of an appropriate driving force, i.e. differential pressure, a permeating species passes through the membrane. This permeate is then collected as a second

We're mostly water.

But we also design membranes for any fluid on earth.

As pioneers in the field of fluid filtration, Osmonics has the technology to make all kinds of processes more profitable and efficient. Examples include a new paint filtration system for the Ford Motor Company and membrane elements that increase the quality of high fructose corn syrup. Chances are we already stock the membrane you need. If not, we can test and custom-design one that fits your application.

Call us in the USA at 612/933-2277 or FAX 612/933-0141

- POTABLE WATER • ULTRA PURE WATER • WASTE WATER TREATMENT •
- EFFLUENT TREATMENT • SULPHATE REMOVAL • GAS TREATMENT •

SERCK BAKER

MEMBRANE SEPARATION SYSTEMS

INTEGRATING SERCK BAKER'S EXPERTISE AND PRODUCT RELIABILITY WITH THE LATEST MEMBRANE TECHNOLOGY

CAMBERLEY Tel: (44) 1276 64411
Fax: (44) 1276 63806

HOUSTON Tel: (1) 713 586 8400
Fax: (1) 713 586 9604

CARACAS Tel: (58) 2 976 6250
Fax: (58) 2 976 6130

GLOUCESTER
380 Bristol Road
Gloucester GL2 6XY
England
Tel: (44) 1452 421561
Fax: (44) 1452 423414

SINGAPORE Tel:(65) 256 3616
Fax: (65) 256 4410

MOSCOW Tel: (7) 095 207 7715/
5604/4855
Fax: (7) 095 207 6243

WITH MAXCELL™ HOLLOW FIBERS, SCALING UP YOUR SEPARATIONS PROCESS WAS NEVER SO EASY.

MaxCell™ hollow fiber and tubule cartridges provide high volume capability in a compact, lightweight design. Cartridges containing up to 140 ft² (13 m²) can be manifolded as close as 7 inches (18 cm) on center. Systems as large as 7000 ft² (650 m²) have been configured to meet the most demanding industrial and bio/pharm applications.

Features
- Rugged cartridge design
- High pressure capability (UF) to 50 psig (3.4 bar)
- Retrofit competitive cartridges
- 30% more area than competitive cartridges
- Easily changed by a single operator
- Wide range of UF/MF pore sizes

Applications
- Bio/pharmaceutical processing (e.g., cell harvesting)
- Oil/water separation
- Machine coolant recycle
- Alkaline bath cleaner recycle
- Wine clarification

MaxCell™ LARGE PROCESS SCALE CARTRIDGES

A/G Technology Corporation
101 Hampton Avenue, Needham, MA 02194 USA
(617) 449-5774 • FAX (617) 449-5786

"product" phase. As a result of this permeation, the feed is gradually reduced in concentration of the permeating species along the membrane. The retentate, or retained solution above the membrane, may also be a product stream which can go to further processing. Cross flow velocities of several meters per second can be used in practice to minimise the impact of the accumulation of particulate material. Even so there is general a decline in fluxrate because the microfiltration membranes suffer from the problem of fouling, ie the accumulation of material at the surface of the membrane. Procedures are usually required to clean and sterilise the membrane, which must be able to withstand the associated mechanical, chemical and thermal stresses. Thus as well as organic polymers, inorganic materials such as ceramics, carbon, metals are used as membranes although these are generally thicker than organic, although usually asymmetric in structure.

The selection of an appropriate membrane is a crucial factor in microfiltration. Adsorption phenomena can play an important role in fouling for example hydrophobic membranes (eg PTFE) generally show a greater tendency to foul, especially by proteins. Another disadvantage of hydrophobic membranes is that water will not generally flow through the structure at low pressure unless they are pretreated prior to use with for example alcohol. Generally microfiltration membranes will require cleaning at some stage in use. This may involve using appropriate chemicals and thus the membrane material must exhibit appropriate chemical resistance to reagents. In the operation of microfiltration several methods are used to reduce the influence of polarisation and fouling; e.g. vortex and pulsatile flows, ultrasonics, vibrational shear and electric fields.

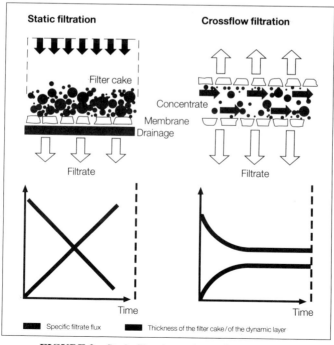

FIGURE 2 – Static filtration and crossflow filtration.

Overall microfiltration is applied in both production and analytical applications which can be summarised as:
- filtration of particles from liquid and gas streams for chemical, biological, pharmaceutical and food industries
- clarification and sterile filtration of heat sensitive solutions and beverages
- production of pure water
- product purification
- waste water treatment

Ultrafiltration

Ultrafiltration is a membrane process which is similar to microfiltration in operation, but which uses asymmetric membranes to carry out 'tighter' filtrations. The membrane top layer pore size is in the range of 0.05 μm to 1 nm. Hence, ultrafiltration is a process of separating extremely small particles and dissolved molecules from fluids. The primary basis for separation is molecular size although secondary factors such as molecule shape

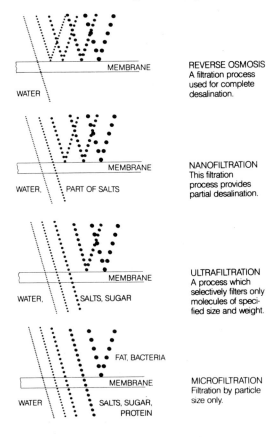

FIGURE 2a– Pressure driven separation process.

INTRODUCTION TO MEMBRANE SEPARATIONS

and charge can play a role. Materials ranging in size from 1,000 to 1,000,000 molecular weight are retained by ultrafilter membranes while salts and water will pass through. Colloidal and particulate matter can also be retained.

Ultrafiltration membranes can be used to purify, and collect, both the fluid material passing through the filter and material retained by the filter. Materials smaller than the pore size rating pass through the filter and thus can be depyrogenated, clarified and separated from high molecular weight contaminants. Materials larger than the pore size rating are retained by the filter and can be concentrated or separated from low molecular weight contaminants.

Ultrafiltration membranes are often operated in a tangential flow mode – feed material sweeps tangentially across the upstream surface of the membrane as filtration occurs – thereby maximizing flux rates and filter life. These systems offer the advantage of long life because ultrafilter membranes can be repeatedly regenerated with strong cleaning agents.

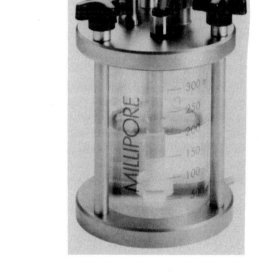

Performance

Osmotic effects in UF membranes are small and the applied pressure, of the order of 1 to 7 bar, is primarily to overcome viscous resistance of liquid permeation through the porous network of the membrane. Commercial UF membranes are asymmetric, with a thin skin some 0.1 to 1 micron thick, of fine porous texture exposed to the feed side. This skin is supported on a highly porous layer some 50 to 250 micron thick, which combines to give the unique requirement of high permeability and permselectivity. Although most UF membranes are polymeric, inorganic, ceramic membranes are now breaking into the market place. Typical membrane materials are polysulfone, polyethersulfone, polyacrylonitrile, polyimide, cellulose acetate, aliphatic polyamides and ceramics, eg zirconium and aluminium oxides.

The principle of operation of ultrafiltration is analogous to microfiltration and is based on "fine sieving". The main hydrodynamic resistance of the membrane is in the membrane top layer, the supporting porous sublayer (or sometimesmacrovoid layer) offers minimal hydraulic resistance. The value of permeability constant K for UF membranes is much smaller than that for microfiltration membranes, in the general range of 0.1-10 m/day. The value of K depends upon many structural properties. For pure water (or other liquids) there is a linear correspondence between flux and transmembrane pressure. With real solution there is a tendency for flux to reach an asymptotic value with increasing pressure. This is a result of several factors, including concentration polarisation, gelation, fouling and osmotic effects.

Table 1 gives a range of water flux measurements for hollow fibre cartridge units with polysulphone membranes. These membranes are macrovoid free membranes. Values of flux are also dependent upon temperature, values generally increase with increasing temperature as a result of the change in viscocity (see Fig 3). For water the recirculation rate (cross flow velocity) will have little, if any, effect on flux rate. This however is not the case with real process fluids where build up of solutes at the membrane surface induce concentration factors which reduce membrane flux. Although the flux can be increased by increasing cross flow velocity, the pressure losses in the feed channel (see Fig 5) may become too high and result in large changes (reduction) in transmembrane pressure (driving force) which is counterproductive. Optimisation between cross flow velocity and membrane length may then be necessary.

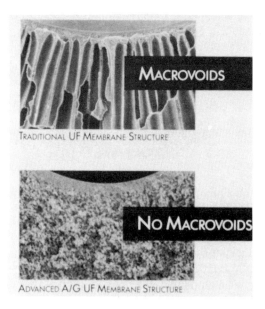

Polysulphone membranes with and without macrovoids.

INTRODUCTION TO MEMBRANE SEPARATIONS

TABLE 1a – Nominal water flux values for UF polysulphone membranes.

Membrane Area as a Function of Housing & Fiber/Tubule Internal Diameter				
Cartridge Housing Identifier	Fiber/Tubule Internal Diameter Code	Fiber/Tubule Internal Diameter (mm)	Cartridge Membrane Area	
			(sq. ft)	(sq. m)
3	C	0.5	0.16	0.015
	D	0.75	0.10	0.009
	E	1	0.08	0.007
4	C	0.5	0.70	0.065
	D	0.75	0.50	0.046
	E	1	0.35	0.032
4X2TC	H	2	0.8	0.073
	K	3	0.5	0.046
5	C	0.5	3	0.28
	D	0.75	2	0.19
	E	1	1.5	0.14
6	C	0.5	6	0.56
	D	0.75	4	0.37
	E	1	3	0.28
	H	2	2.4	0.22
	K	3	1.6	0.15
8	C	0.5	6.7	0.62
	D	0.75	5	0.46
	E	1	3.75	0.35
9	C	0.5	15	1.4
	D	0.75	10	0.93
	E	1	7.5	0.7
	H	2	6	0.55
	K	3	4.9	0.45
10	H	2	12.1	1.1
	K	3	8.8	0.82
35,35A 35STM	C	0.5	14	1.3
	D	0.75	11	1
	E	1	8.5	0.8
55, 55A 55R, 55STM	C	0.5	36	3.4
	D	0.75	27	2.5
	E	1	23	2.1
	H	2	14	1.3
	K	3	10.6	1
75, 75R	C	0.5	65	6
	E	1	40	3.7
	H	2	27	2.5
	K	3	20	2

TABLE 1b – Nominal water flux values for UF polysulphone membranes.

This table of nominal water flux is only valid with "clean" water. Particulates, bacteria and dissolved metals in the feed water will all lower membrane cartridge productivity values.

			Nominal UF Permeate Flow Rate with Clean Water Feed Stream (Liters/min)								
Pore Size	Fiber/ Tubule ID (mm)	Average TMP (psig) @25°C	A/G Technology Housing Size Identifier								
			3	4	4X2	5	6	9	10	35	55	75
5K	0.5	10	0.010	0.05	—	0.20	0.40	1.0	—	0.9	2.4	—
	1	10	0.004	0.02	—	0.08	0.16	0.4	—	0.5	1.2	2.1
10K	0.5	10	0.015	0.06	—	0.28	0.55	1.4	—	1.3	3.3	5.6
	1	10	0.006	0.03	—	0.12	0.25	0.6	—	0.7	1.8	3.2
30K	0.5	10	0.025	0.11	—	0.47	0.95	2.4	—	2.2	5.7	10.0
	1	10	0.017	0.04	—	0.16	0.32	0.8	—	0.9	2.4	4.2
	2	10	—	—	0.04	—	0.13	0.3	0.6	—	0.7	1.4
	3	10	—	—	0.03	—	0.09	0.2	0.3	—	0.4	0.8
100K	0.5	5	0.017	0.07	—	0.32	0.63	1.6	—	1.5	3.8	6.8
	1	5	0.007	0.03	—	0.14	0.28	0.7	—	0.8	2.1	3.7
	2	5	—	—	0.04	—	0.13	0.3	0.6	—	0.7	1.4
	3	5	—	—	0.02	—	0.06	0.2	0.3	—	0.4	0.8
500K	0.5	5	0.025	0.10	—	0.47	0.95	2.4	—	2.2	5.7	10.0
	1	5	0.010	0.04	—	0.20	0.40	1.0	—	1.1	3.0	5.2

Liters/min ÷ 3.785 = US gal/min   psig x 0.06895 = barg

IMPORTANT NOTES:
1. Values at noted average TMP, 25°C.
2. Values are nominal.
3. Values to be adjusted for actual pressure.
4. Values to be adjusted for actual temperature.
5. Permeate flow rates only valid at low pressure.

Permeate flow rates do not provide accurate membrane permeability when normalized on a unit area basis.

Transmembrane pressure
$= (P_{inlet} + P_{outlet})/2 - P_{permeate}$

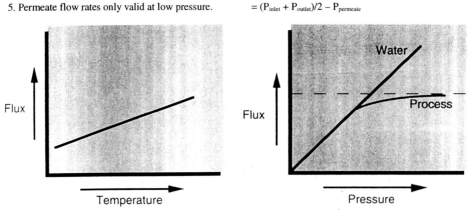

FIGURE 3 – Variation flux with temperature and with pressure for UF membrane.

INTRODUCTION TO MEMBRANE SEPARATIONS

In ultrafiltration membrane performance is defined in terms of its ability to retain certain molecules of a specific size. This is measured by the rejection R defined as

$$R = 1 - \frac{C_p}{C_f}$$

where Cp and Cf are the solute concentrations in the permeate and feed respectively.

Values of rejection are not absolute values for any single membrane or solution, but depend upon conditions of operation, the concentration of the feed solution and flowrates (hydrodynamics) ie concentration polarisation at the membrane surface. Rejection coefficients vary with molar mass of solute. This typically will see values of R vary from 0 to 1.0 over one (tight) membrane or two or greater (loose) membrane orders of magnitude in molar mass, for any one membrane.

The selection of a membrane for ultrafiltration will require determining the molar mass of the species to be separated and selecting a membrane with a limiting rejection ($\rightarrow 1.0$) under anticipated conditions of operation. Small scale application tests will generally need to be performed. As with microfiltration factors of chemical compatibility of materials with solution will need to be addressed.

The separation mechanism is broadly considered to be one of sieving where an increase in applied pressure increases the flux. However the effect of concentration polarisation limits the practical fluxrates due to a build up of solute on the feed side of the membrane in the concentration boundary layer. At sufficiently high pressures gelation of the macromolecules can occur resulting in the formation of a thin gel layer on the surface which can also act as a secondary membrane. In addition the performance of UF can be significantly affected by fouling.

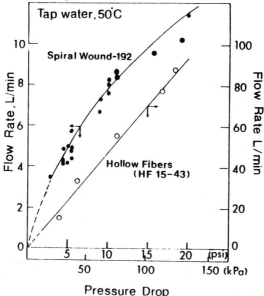

FIGURE 5 – Typical variation of flux with pressure drop during UF.

Ultrafiltration is typically applied in the separation of macromolecules, colloids from solutions. There are a wide range of applications in the laboratory, in processing and in analysis.

Pore Size Selection

Ultrafiltration membranes are rated in terms of their Nominal Molecular Weight Cut-off (NMWC). There are no industry-wide standards for this rating, hence manufacturers use different criteria for assigning UF pore sizes. Therefore if changing to another manufacturer, it is advisable to test more than one membrane cut-off to determine the preferred membrane type from a new supplier. For example for protein concentration, the protein should be larger than the molecular weight cut-off of the membrane by a factor of 2 to 5. The greater the difference (ie tighter the membrane pore size), the higher the protein yield. If protein passage is desired, then an order of magnitude difference between the membrane's NMWC and the protein's size is suggested. In practice, a 500,000 NMWC UF membrane or a microporous membrane will be required to affect significant protein yields.

The protein shape, in addition to its molecular weight, plays a role in determining its retention by the membrane. The more globular the protein, the greater its retention. while linear proteins may require a tighter membrane for high recoveries. Moreover, protein shape may be effected by solution pH or salinity.

Flux declines with time, even with "clean" water. The influence of time on the rate of flux decline may, however, be insignificant compared to the effect of concentration. A rapid flux decline, while processing a stream in total recycle (ie no concentration) indicates either the circulation rate is too low or foulants are present. Flux decline as a function of time may also occur with a process stream due to gel layer compaction.

Concentration factors in membrane systems are expressed either in terms of volumetric concentration factor C_v

$$C_v = \frac{\text{Original Feed volume}}{\text{Original Feed volume} - \text{Volume of Permeate Collected)}}$$

or the system conversion y (or recovery factor) given by

$$y = 1 - \frac{1}{C_v}$$

Fig 6 shows the typical effect of time on the concentration of a fermentation broth with two types of UF membrane hydrophilic polysulphone and polyaramide. Typically the initial loss of flux is relatively rapid whilst at greater times the decline in flux is less severe. The difference in the particular membrane flux behaviour is due to in this case the greater tolerance of the very hydrophilic polyaramide membrane to fouling. Fig 7 shows a typical variation of flux with concentration factor for a polyaramide UF membrane and a polyethersulphone UF membrane during the concentration of an oily wastewater.

Flowrate and Fouling

Recirculation Flow Rates or the feed stream flow rate has a major effect on permeate flux.

INTRODUCTION TO MEMBRANE SEPARATIONS

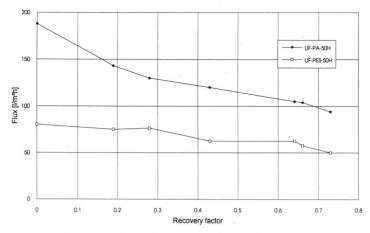

FIGURE 6 – Variation of flux with conversion for the ultrafiltration of an oily flowrate.

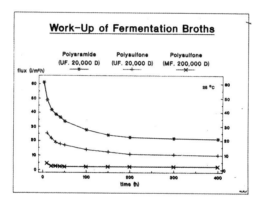

FIGURE 7 – Variation of flux with time during concentration of fermentation broth.

Fouling streams do not respond well to feed dilution and tend to reach low, steady-state flux levels which are less dependent on feed concentration. Higher feed flow rates, exhibiting a shear rate of at least 8,000 sec-1, should be utilized. Low-fouling streams exhibit stable flux rates over time with low circulation rates. The flux of low-fouling streams is basically concentration dependent. Thus upon feed stream dilution, the permeate rate will increase and approach the starting performance level. A flow rate which provides an intermediate shear rate, on the order of 4,000 sec–1, is a good starting point for processing low fouling streams. Shear sensitive streams contain fragile components (eg cells, proteins) which may be damaged by high circulation rates or high temperature. Flow rates which provide shear rates on the order of 2,000 sec–1 are recommended for shear sensitive streams.

The resistance to permeation of a membrane is a function of the membrane pore size, feed stream components, and the degree to which gel layer formation and fouling layer formation occur. Increasing the feed stream circulation rate will, as a general rule, reduce gel layer thickness and increase flux. Operation within the turbulent flow regime, may

TABLE 0 – Typical relationship between shear rate and feed flowrate.

Housing Size	Lumen ID (mm)	Sheer Rate ~2000 sec-1 (liters/min)	Shear Rate ~4000 sec-1 (liters/min)	Shear Rate ~8000 sec-1 (liters/min)
3	0.5	0.05	0.1	0.2
	0.75	0.1	0.2	0.4
	1	0.15	0.3	0.6
4, 4X2TC	0.5	0.15	0.3	0.65
	0.75	0.4	0.8	1.5
	1	0.6	1.2	2.5
	2	1.8	3.5	6.5
	3	3	6	12
5, 6	0.5	0.75	1.5	3
	0.75	1.8	3.5	7
	1	2.2	4.3	8.5
	2	6.3	12.5	25
	3	11	22	45
8, 9, 10	0.5	2.2	4.3	8.5
	0.75	5	10	20
	1	7.5	15	30
	2	18	35	70
	3	30	60	120
35, 35SMO, 55, 55R, 55SMO, 75, 75R	0.5	5	10	20
	0.75	10	20	40
	1	15	30	60
	2	38	75	150
	3	65	125	250
45, 65, 85	0.5	10	20	40
	1	35	65	125
	2	75	145	285
	3	130	260	520

Shear rates and flow rates are directly proportional. Highly fouling streams may require flow rates equivalent to a shear rate in the range of 10,000 to 12,000 sec^{-1}. Shear rates based on viscosity of 1 cp.

significantly enhance permeation by reducing both gel layer and fouling layer thickness through improved transfer of solids from the membrane surface back into the bulk stream.

Reverse Osmosis and Nanofiltration

Reverse osmosis is a process to remove low molecular weight solutes, e.g., microrganic salts, small organic molecules (glucose for example) from a solvent, typically water. The process is sometimes referred to as hyperfiltration. As the name reverse osmosis implies, it is the reversal of the natural process of osmosis, by the application of a pressure on the more concentrated solution in contact with the membrane.

INTRODUCTION TO MEMBRANE SEPARATIONS

Reverse osmosis (RO) is a pressure driven process aimed at the separation of ionic solutes and macromolecules from aqueous streams. The method of dissolved salt removal is different to that of microfiltration and is not just a physical process based on size difference of solute and solvent. Such species are of similar molecular size and of a size comparable to the wide range of pore spaces in the polymeric RO membrane. Membrane pore size is of the order of 10nm and lower.

To visualise the RO process first consider the process of osmosis (see Fig 8). Osmosis occurs when a suitable semi-permeable membrane is used to separate two solutions of equal volume, one water and the other a dilute salt solution. Water is transferred from the water side of the membrane to the dilute solution side until an osmotic equilibrium is reached, at which point a hydrostatic pressure, the osmotic pressure, has built up in the solute solution side. By applying a pressure to the salt solution side, the flow of water through the membrane can be arrested and if the pressure exceeds the osmotic pressure the flow is reversed. This is termed reverse osmosis or hyperfiltration in which the concentration of salt is increased by the flow of water (or solvent) from a more concentrated solution to a dilute solution.

Reverse osmosis membranes can essentially separate all solute species, both inorganic and organic from solution (Fig 8). The mechanism of separation of species are based on processes relating to their size and shape, their ionic charge and their interactions with the membrane itself. This mechanism can be visualised as a thermodynamically controlled partitioning, analogous to solvent extraction. The operating principle, referred to as the solution-diffusion model, is that a surface layer of the membrane is a relaxed region of amorphous polymer in which solvent and solute dissolve and diffuse. To overcome the molecular friction between permeates and membrane polymer, during diffusion, large operating pressures are required in the range of 30 to 100 bar.

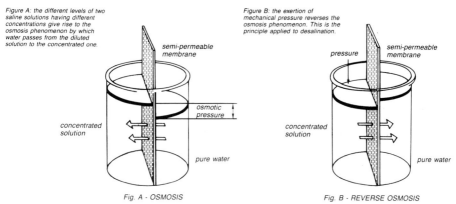

FIGURE 8 – Separation by reverse osmosis.

With ionic species the membrane exerts an electrostatic free energy barrier against ionic movement into the membrane. Thus the mobility of ionic species in the membrane is much less than that of the water molecules and the degree of separation depends on ionic charge, feed solution, ionic composition and the size of hydrated ions. For organic species of non-ionic solutes the separation is determined by their affinity with the membrane and also their molecular weight. The non-ionic species are usually taken up by the membrane and separation is only achieved because of their relative low mobility in the membrane compared to that of water. Thus typically we find the exclusion of non-ionic species by RO membranes is not as good as that of ionic species. The experiences of membrane technologists in applying RO to organic species has led to other explanations and mechanisms of solute transport in RO membranes which can at least qualitatively explain reverse osmosis behaviour of organic species.

The membranes used for hyperfiltration are either asymmetric or composite which typically have a < 1 mm thick, dense top layer supported by a 50 to 150 μm thick porous sublayer. The top layer imparts the intrinsic separation characteristics and the thinness of this layer ensure high fluxrates. Typical membranes are made by phase inversion of cellulose esters eg cellulose triacete, aromatic polyamide and from poly (either urea) using interfacial polymerisation.

The major application of RO membranes is in the processing of aqueous solutions containing inorganic solutes. A common membrane is of asymmetric cellulose acetate some 100 micron thick, with an active dense layer 20 to 50 μm thick, to achieve adequate fluxes at what are relatively high operating pressures in comparison to microfiltration.

The particle size range for applications of RO is approximately 0.0001 to 0.001 micron (1 to 10 A) and with solutes of molar masses greater than 300 Dalton complete separation is achieved. RO has principally has seen a wide range of applications in the processing of aqueous solutions in the following areas:
- Desalination of brackish water and sea water.
- Production of pure water for a variety of industries.
- Concentration of solutions of food products, pharmaceutical solutions and chemical streams.
- Waste water treatment.
- The use of RO is generally increasing as more resilient membranes emerge.

Performance Criteria

The major objective of reverse osmosis as a separation process is to primarily remove water or perhaps another solvent from a solution to produce a more concentrated product and/or a 'pure' water stream. Thus the permeate water flow and any passage of salts or solutes are the key performance indicators.

The water flow or flux is defined in the same way as for microfiltration and ultrafiltration. The flux depends upon the membrane permeability, the pressure driving force and the osmotic pressure difference across the membrane Δp, according to

$$J = K \cdot (\Delta P - \Delta \pi,)$$

In practice the membrane will be permeable to some low molecular mass solutes and the real osmotic pressure difference across the membrane must be corrected by using a reflection coefficient, σ i.e the osmotic pressure difference $= \sigma \Delta \pi$ ($\sigma \leq 1$). Typical values of permeability coefficient are in the range 5.10^{-3} to 5.10^{-5} mh^{-1}bar^{-1}.

INTRODUCTION TO MEMBRANE SEPARATIONS

The flow of water through the membrane will invariably carry solute(s) with it to a lesser or greater degree. This solute flux, Js is proportional to the solute concentration difference across the membrane ΔCs i.e.,

$$Js = Ks\, \Delta Cs \quad (mol\ m^{-2}\ h^{-1})$$

where Ks is the solute permeability coefficient. Values of Ks lie in the range of 5.10^{-3} to 10^{-4} mh^{-1}, the lowest value is for high retention (or rejection) membranes. Rejection is given by equation ( ). The ratio Cp/Cf is sometimes referred to as the salt passage (expressed as a percentage). The values of permeabilities of water and solute are both inversely proportional to the membrane (or membrane skin) thickness and proportion to the appropriate diffusivity of water or of a solute.

Reverse osmosis typically operates at high pressure from 20 to 100 bar, much higher than ultrafiltration. The values of both permeabilities effectively determines the retention of the membrane, which can be expressed as

$$R = \frac{K_s/K\,[\Delta P - \Delta \pi]}{(1 + K_s/K\,(\Delta P - \Delta \pi))}$$

The choice of membrane material is thus very important in defining performance characteristics. In reverse osmosis an increase in the operating pressure raises the water flux, and also causes an increase in quality of the permeate as the solute passage through the membrane falls (see Fig 9). Thus the attraction of high pressure is both a higher water productivity and improved water quality, which is of course at the expense of increased cost.

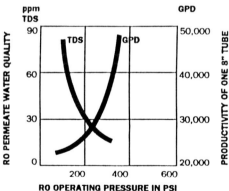

FIGURE 9 – Effect of applied pressure on flux and rejection.

Conversion

The conversion in reverse osmosis is the ratio of permeate water flow to feed flow. Conversion has a crucial effect on solute passage and product flow through the membrane. As the conversion increases, the degree of increase in concentration of solutes, in the feed (retentate) increases. This is referred to as the concentration factor, the ratio of solute in the retentate to that in the feed. An increase in conversion therefore results in an increase

TABLE 3 – Performance characteristics of reverse osmosis, nonofiltration and ultrafiltration spiral wound permeates.

Designation	Nominal Molecular Weight Cutoff (MWCO) 100 / 1K / 10K / 100K / 1000K	% NaCl Rejection	Operating Pressure in Psig (bar) Recommended	Operating Pressure in Psig (bar) Maximum	Recommended pH Range	Max. Temp. °F(°C)	Recommended Max. Free Chlorine
Polymer Type: SEPA-S							
SS		≥98	400 (27.6)	1000 (69.0)	2-8	122 (50)	2 ppm
ST		≥95	400 (27.6)	1000 (69.0)	2-8	122 (50)	2 ppm
SR		≥92	400 (27.6)	1000 (69.0)	2-8	122 (50)	2 ppm
SF		≥85	300 (20.7)	1000 (69.0)	2-8	122 (50)	2 ppm
SX		50-70	200 (13.8)	500 (34.4)	2-8	122 (50)	3 ppm
SV		30-50	200 (13.8)	400 (27.6)	2-8	122 (50)	3 ppm
SP		20-40	200 (13.8)	300 (20.7)	2-8	122 (50)	3 ppm
SG		—	100 (6.9)	200 (13.8)	2-9	140 (60)	3 ppm
SN		—	50 (3.5)	150 (10.0)	2-9	140 (60)	3 ppm
SZ		—	25 (1.7)	100 (6.9)	2-9	140 (60)	3 ppm
Polymer Type: SEPA-J							
JY		—	100 (6.9)	20 (13.8)	2-9	140 (60)	3 ppm
Polymer Type: SEPA-M							
MS		>98	200 (13.8)	1000 (69.0)	2-12	176 (80)	<0.1 ppm
MX		50-70	100 (6.9)	1000 (69.0)	2-12	176 (80)	<0.1 ppm
MQ		30-50	100 (6.9)	1000 (69.0)	2-12	176 (80)	<0.1 ppm
MG		—	150 (10.4)	400 (27.6)	2-12	176 (80)	<0.1 ppm
MW		—	100 (6.9)	400 (27.6)	2-12	176 (80)	<0.1 ppm
MY		—	100 (6.9)	200 (13.8)	2-12	176 (80)	<0.1 ppm
MC		—	25 (1.7)	100 (6.9)	2-12	176 (80)	<0.1 ppm
MJ		—	25 (1.7)	100 (6.9)	2-12	176 (80)	<0.1 ppm

1. CELLULOSIC (SS–JY)
2. POLYAMIDE (MS–MJ)

INTRODUCTION TO MEMBRANE SEPARATIONS

Category	Polymer Type	Code	Pore size (μm)	Pressure psi (bar)	Max Pressure psi (bar)	pH range	Temp °F (°C)	Test conc.
	SEPA-B	BQ		30-50	100 (6.9)	1000 (69.0)	0.5-13	212 (100)	25 ppm
		BP		20-40	100 (6.9)	1000 (69.0)	0.5-13	212 (100)	25 ppm
3. POLYSULPHONE	SEPA-H	HP		—	100 (6.9)	400 (27.6)	0.5-13	212 (100)	25 ppm
		HG		—	100 (6.9)	400 (27.6)	0.5-13	212 (100)	25 ppm
		HW		—	100 (6.9)	400 (27.6)	0.5-13	212 (100)	25 ppm
		HN		—	50 (3.5)	300 (20.7)	0.5-13	212 (100)	25 ppm
		HZ		—	25 (1.7)	200 (13.8)	0.5-13	212 (100)	25 ppm
4. FLUOROPOLYMER	SEPA-A	AG		—	100 (6.9)	200 (13.8)	1-11	176 (80)	25 ppm
		AN		—	50 (3.5)	200 (13.8)	1-11	176 (80)	25 ppm
5. ACRYLIC	SEPA-R	RG		—	100 (6.9)	400 (27.6)	0.5-10	212 (100)	25 ppm
		RZ		—	25 (1.7)	200 (13.8)	0.5-10	212 (100)	25 ppm
6. POLYPROPYLENE	SEPA-Y	YL		—	25 (1.7)	200 (13.8)	0.5-13	212 (100)	25 ppm
		YB		—	25 (1.7)	200 (13.8)	0.5-13	212 (100)	25 ppm
		YK		—	25 (1.7)	200 (13.8)	0.5-13	212 (100)	25 ppm
		YC		—	25 (1.7)	200 (13.8)	0.5-13	212 (100)	25 ppm
	SEPA-W	WK		—	50 (3.5)	200 (13.8)	1.0-10	212 (100)	50 ppm
		WC		—	25 (1.7)	200 (13.8)	1.0-10	212 (100)	50 ppm
		WD		—	25 (1.7)	200 (13.8)	1.0-10	212 (100)	50 ppm
		WH		—	25 (1.7)	200 (13.8)	1.0-10	212 (100)	50 ppm

Pore size scale (μm): 0.001 — RO | 0.01 — NF | 0.1 — UF | 1.0 — MF

TABLE 2 – Summary of some standard test conditions for RO.

Test	Device	Solution	Pressure (psig)	Recovery (%)	Temp (°C)
High Pressure Brackish water	Spiral	2 gdm^{-3} NaCl	400-420	10-15	25
Low Pressure Brackish	Spiral	2 gdm^{-3} NaCl	225	10-15	25
Seawater	Spiral	30-35 gdm^{-3} NaCl	800-1000	30-35	25
High Pressure Brackish	Fibre	1.5 gdm^{-3} NaCl	400	75	25
Seawater	Fibre	30-35 g dm^{-3} NaCl	800-1000	30-35%	25

in salt concentration, which through the effect of concentration polarisation, ultimately causes an increase in salt concentration at the membane surface and hence an increase in salt flux. This increased concentration also increases the osmotic pressure difference which correspondingly results in a lower permeate flow. Overall this scenario leads to a greater impact of the effect of concentration polarisation and may also increase the extent of membrane fouling.

Membrane Performance

There are many factors which affect the performance of a reverse osmosis membrane in addition to pressure. The effect of temperature, for example, affects both water flow and osmotic pressure. The permeability coefficient increases with temperature. The osmotic pressure also increases with temperature therefore tending to decrease flux. Overall a rule of thumb effect of temperature is that membrane capacity increases by approximately 3% per °C increase in temperature (above a typical standard of 25°C).

The product flow and salt rejection are dependent on membrane material, membrane thickness, feed water quality as well as operating conditions. Membrane manufacturers will specify particular device capacity and salt rejection based on performance measurements with solutions of sodium chloride under standard conditions. The standard conditions depend on the device configuration (e.g., fibre, spiral) and the intended use e.g., seawater, brackish (see Table 2). Typical salt rejections are greater than 90% for brackish water and greater than 99% for seawater, Table 3 gives a range of different characteristics for typical brackish water spiral wound permeates.

In addition to NaCl the membrane rejection of other salts (and solutes) is important. It is typical to see the passage (100 -% rejection) of divalent ions ($Ca^2,+ SO_4^2, - Mg^{2+}$) to be 1.5 to 2 times that of monovalent ions (Na^+, Cl^-). However these values vary widely from material to material as seen in Table 4, and between manufacturers. Particular features which consequently influence the choice of membrane are tolerance to free chlorine (present in tap water) biological resistance and resistance to fouling.

An additional factor to be considered is membrane compaction. Compaction is caused

TABLE 4 – Membrane Characteristics and Recommended Operating Ranges.

	CA	HFF	TFC
Salt rejection	90%-97%	90%-96%	96%-98%
Silica rejection	85%	85%	98%
Nitrate rejection	85%	85%	92%
Chlorine tolerance (ppm)	0.2-0.5	0	0
Maximum SDI	5	3	5[1]
Temperature range	32-95°F	32-95°F	32-113°F
Nominal organics rejection range	300+ mwt	300+ mwt	200+ mwt
Typical operating pressure, psig	400[2]	400[2]	250[2]
pH range	3.0-6.5	4-11	2-11
Flux, rate, gpd/ft^2	12-16	2-4	15-20
Biological resistance	Poor	Good	Excellent
3rd year compaction	20%	20%	0
Hydrolysis	2x SP@3 yr	None	None

[1]Less than 3 strongly preferred   [2]Typical system values

by creep deformation of polymeric membranes with time and depends upon material, pressure, temperature. Increasing pressure and temperature, increases the tendency to creep, which tightens the membrane top (rejection layers) and reduces the water flux. The effect is more pronounced in asymmetric homogeneous membranes. To offset this affect, and other factors which decrease flux, membrane modules are designed with excess capacity in terms of lower initial operating pressures. This enables pressure to be increased with time to retain required flux rates throughout operation.

Membrane Selection

The quality of the water is a major factor in membrane selection. The quality will determine the type and the number of pre-treatment steps required to obtain the best performance from the RO membrane. Membrane selection and pretreatment requirements go hand in hand in system design. An ideal membrane should exhibit high flux and rejection, be tolerant to oxidants notably chlorine, be resistant to biological attack and fouling by colloidal and suspended material, exhibit mechanical strength, chemical stability, thermal stability and be inexpensive.

Water pretreatment methods are required to remove suspended and colloidal material, to prevent membrane scaling by precipitation of sporingly soluble salts and to prevent fouling from biological growth, colloids, organic species metal oxides and hydroxides. Membrane and system manufacturers have recommended cleaning methods for pretreatment of feed water which is dependent on feed water quality and desired final water product quality. Colloidal and suspended solids pretreatment is determined by the silt density index (SDI) of the feed which is calculated from the microfiltration performance of the feed using a 0.45 mm filter. The SDI is calculated from the ratio of the time required for 500cm3 of feed to pass through the filter initially and after 15 minutes of continuous use, at a constant pressure.

The three common membrane materials used in RO are cellulose acetate (spiral wound), polyamide (hollow fibre, spiral wound) and thin film composites. Cellulose acetate membranes are made from cellulose diacetate and cellulose triacetate. The increased acetyl content imparts improved chemical stability and salt rejection but decreases flux. Consequently blends of cellulose acetates are used in membranes to optimise different characteristics. Typical quoted behaviour of blended (CA – CTA) membranes are – 3 year life treating water with a pH of 5.9 at 50°F. Rejection and flux fall continuously with operation time. C.A. membranes are more tolerant to foulants than other membranes. Salt rejection is not as good and consequently they are not used for high purity applications.

They are not recommended for treatment of warm feeds due to the increased chance of biological attack. CA membranes are however chlorine tolerant and thus biological degradation is unlikely in this environment. Aromatic polyamide (polyaramid) mem-

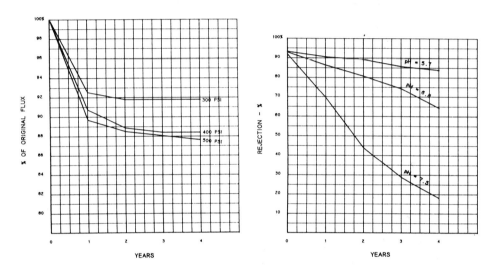

FIGURE 10 – Salt rejection and flux characteristics of RO membranes as a function of operating time.

branes have no chlorine tolerance and feed must be dechlorinated in use. They have good chemical stability (composed to CA membranes) and can be used over a wider pH range (4-11) and temperature range (0°C – 35°C). Polyamide fibres are characterised by a dense skin 0.1 to 1.0 mm thick. They are quite susceptible to fouling and biological attack: Polyamide membranes are used for treatment of brackish water (deep well) supplies and seawater. Salt rejection of these membranes is enhanced by posttreatment of fibres with, for example, poly vinyl methyl ether.

Thin Film Membranes

Thin film composite (T.F.C) membranes consist of a dense ultrathin barrier layer typically 0.2 mm thick on top of a microporous polsulfone support. The advantages of these membrane are that they operate at higher flux and lower pressure, have greater chemical stability, have higher salt rejection, they are not biodegradable, they have higher rejection of other materials (silica, nitrate, organics). Operating ranges of these membranes are pH of 2 to 12 and temperatures of 0°C to 40°C.

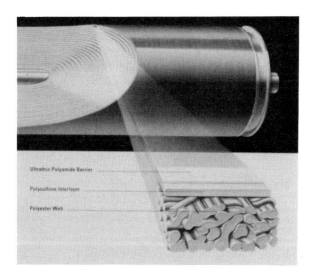

Thin film composite membrane.

There are two widely used types of thin film composites: polyamide and polyetherurea. Polyamide membrane is widely used as a thin film coated over polysulfone. It is a synthetic polymer and thus is not subject to biological attack. It is, however, highly susceptible to biological fouling. This is due to two characteristics:

- Surface Charge: The polyamides exhibit a moderate to strong anionic surface charge. For this reason, the surface of polyamides will exhibit a strong affinity for organics and bacteria.
- Surface Morphology: Viewed with a scanning electron microscope, the polyamide membranes have an extremely irregular surface with fissures, hooks and depressions compared with, cellulose acetate and polyetherurea which are both very smooth. Bacteria and organics become caught in the irregular surface of the polyamide film and are difficult to remove or require more frequent cleaning.

Polyetherurea membrane is used as a thin film coated over polysulfone. Polyetherurea membrane has a surface charge that is considered neutral to slightly cationic and offers minimal problems with high fouling, organics or bacteria laden waters as its surface is very smooth. It has been used in municipal sewage reclamation projects where higher rejection and resistance to fouling is required. Overall, biological degradation does not occur with synthetic thin films. Biological fouling can occur, and with polyamide thin films, it is often severe and difficult to remedy.

The tolerance of TFC membranes to fouling is lower than cellulose actate. They are preferred for use in high purity systems and often for large RO installations where they offer lower energy consumptions. Due to their superior pH stability, higher pHs can be employed when cleaning thin film membranes. This is especially useful for organics or oil fouling where high pH solutions is required for effective cleaning . Although thin films can be cleaned aggressively, it is often preferable to use membranes that do not require frequent cleaning. The surface charge and texture of the polyamide membranes often require the most frequent and vigorous cleaning.

Surface waters may contain high levels of organics and/or suspended solids. These may be clarified with heavy cationic polymers, often at high dosage rates. Rapid membrane fouling, loss of flux, and high operating pressures can result if excess polymer comes in

TABLE 5 – Comparison of Characteristics of think film membranes

Characteristic	Polyamide Thin Film TFCL®	Polyether Urea TFC®	Cellulose (Blend) ROGA®
pH Stability Range (operating)	4-11	5-10	3-7
Surface Charge	Moderate to high anionic	Neutral to slight cationic	Neutral
Surface Morphology	Very irregular	Smooth	Smooth
Flux (GFD): LP (200 psi net) HP (400 psi net)	23 15	22 12.5	20 23
Normal Operating Pressure: LP MP HP or SS	225 psi 420 psi 1,000 psi	225 psi 420 psi 800-1,000 psi	225 psi HR or SD - 420 psi —
Maximum Temp	113°F	113°F	104°F
Chlorine Tolerance	1,000 ppm hours	–0–	Continuous @ up to 1 ppm
Biofouling	High	Low	Very low
Particulate Fouling	High	Low	Very low
Chloride Rejection: LP MP HP or SS	98.5% 98.5% 99.4%	97.5% 99.0% 99.4%	75.0% SD - 95.5$ HR - 98.0%
Nitrate Rejection: MP	98.0%	94.0%	HR - 85.0%
Silica Rejection: MP	98.0%	95.0%	HR - 90.0 - 93.0%

NOTE 1: Polyetherurea TFC® membrane is available for a variety of pressure conditions:
    LP: Low Pressure Brackish: Lower pressures
    MP: Medium Pressure Brackish: Operates at higher pressure for higher salinity feedwaters.
    SS: High Pressure Operation for Seawater Applications
NOTE 2: Cellulose diacetate/cellulose triacetate blend ROGA® membrane is available in three distinct types:
    LP: Low Pressure: 75.0% chloride rejection, 95.0% hardness rejection
    SD: Standard Rejection: 95.5% chloride rejection at higher flow
    HR: High Rejection: 98.0% chloride rejection, treated to achieve higher rejection at lower flows
NOTE 3: Polyamide TFCL® thin film membrane is generally provided in a variety of types:
    LP: Lowe Pressure Brackish Water
    MP: Medium Pressure Brackish Water
    HP: High Pressure Seawater
NOTE 4: Rejection of most ionic species, including silica and nitrate, is dependent on total water chemistry and membrane properties. Generally, multivalent ions are very highly rejected while weakly ionized species, such as silica, are less well rejected.

INTRODUCTION TO MEMBRANE SEPARATIONS

contact with the membrane. If the membrane is a polyamide thin film, fouling will be rapid and not easily removed.

A features of polyamide membrane is its higher flux which allows operation at lower pressures and low energy requirements, although many factors influence or even over-ride these energy savings. Many waters may not be appropriate for polyamide membrane. Rapid fouling and subsequent ever-decreasing response to cleanings is often observed when treating municipal wastewater. Clarified surface water is also very difficult to treat with polyamide membrane and the use of cationic coagulant polymers compounds the fouling problems . As a result of their propensity for fouling, polyamides need stricter attention paid to pretreatment and then fouling rates can be unacceptable.

As the RO system becomes fouled, higher operating pressures are required. In several Gulf Coast trials with polyamide thin film, the frequency of cleaning is every ten days. Prior to cleaning, pressures climb, as does the power requirement. The high cleaning expense and lost capacity will often negate any energy saving. Table 5 shows a typical comparison of characteristics of thin film membranes with cellulose acetate membranes.

RO System Design

Reverse osmosis system design encompasses all aspects of the process for purifying the

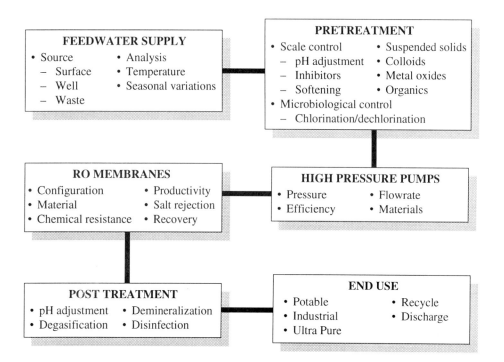

FIGURE 11– Components in the design of a RO system.

water; from the feedwater supply to the end use. The components of the system have been summarised in Fig 11. Factors to be considered include feedwater supply (source and analysis), pretreatment requirements, pumping requirements, membrane type and configuration, cleaning, pressure (energy) recovery, post treatment (eg degasification, ion-exchange) and overall operation and maintenance.

Nanofiltration

Nanofiltration is similar to RO and is a pressure driven process applied in the area between the separation capabilities of RO membranes and UF membranes, that is in the separation of ions from solutes such as small molecules of sugars. It has only recently achieved success due to developments in thin film non-cellulose membranes. Membranes can be formed by interfacial polymerisation on a porous substrate of polysulphone or polyethersulphone. Generally this opens up the possibilities for process efficiency improvements and the production of new products particularly in the food and biotechnology industries. Nanofiltration systems typically operate at lower pressures than RO (eg 5 bar) but yield higher flow rates of water albeit of a different quality to RO.

Nanofiltration is used when high sodium rejection, typical of RO, is not needed but where other salts such as Mg and Ca (ie divalent ions) are to be removed. The molecular weight cut off the NF membrane is around 200. Typical rejections are (5 bar, 2000ppm solute) 60% for NaCl, 80% for calcium bicarbonate and 98% for magnesium sulphate, glucose and sucrose.

Several companies , have developed a range of polyamide membranes with low salt rejection capabilities. These have applications in the processing of salty cheese wheys (diafiltration) and pharmaceutical preparations. Some other specific applications are removal of colour, removal of TOC and trihalomethane precursors form surface water, removal of hardness, radium and TDS from well waters. In electroless copper plating the separation of a Cu-EDTA complex from bproduct salts formed as part of processing is another application. One membrane in particular can be used to give an almost 100% rejection of sulphate (sodium or calcium) from seawater. Table 6 shows a comparison of one company's range of nanofiltration membrane compared to a RO membrane.

TABLE 6 – Rejections of Electrolyte Solutes by Nanofiltration Membranes.

	Rejection, %			
Solutes	FT-30	NF-70	XP-45	XP-20
NaCl	99.5	75	50	20
MgCl2	>99.5	70	83	—
MgSO4	>99.5	97.5	97.5	85
NaNO3	90	50	<20	0
Ethylene glycol	70	13	24	11
Glycerol	96	25	44	15
Glucose	99	93	95	60
Sucrose	100	100	100	89

Pressure for inorganic solutes, applied to give a flux of 10 μm/s (=20 gfd); pressure for organic solutes, unknown.

Gas Separation

Separation of mixtures of gases is possible using either porous or non porous membranes although quite different mechanisms of transport are involved. Separation in porous membranes is through the difference in Knudsen flows (diffusion) of the components in the pores which are of a size less than the mean free paths of the molecules. The flux of individual gases is inversely proportional to the square root of the relative molar mass for a given membrane temperature and pressure driving force. Hence generally low separation factors are obtained with most gas mixtures except those containing hydrogen and helium. For example for a H_2/CH_4 mixture a separation factor of 2.83 has been obtained. Such separation factors are much less than achievable with non-porous membranes and so no commercial applications of non-porous membranes exist for common gases. Only one commercial application of gas membrane separation is known, the enrichment of uranium hexa fluoride using porous ceramic membranes. This involves the separation of 235 UF_6 from 238 UF_6 with an ideal separation factor of 1.0043, using a cascade arrangement of membranes.

Non-porous membranes are used in many commercial applications for the separation of gas mixtures. Separation of gases through non porous membranes depends on the differences in permeabilities of the constituent gases. Gas permeation (GP) is the only means by which membranes can be used to separate gas mixtures without a change in phase. Separation of different gases is achieved by virtue of differences in molecular size and gas solubility in the membrane. Gases of smaller molecular size have larger diffusion coefficients and in a convection free environment in the pores of a membrane can be suitably separated by virtue of the different mobilities.

The solubility of gaseous components in the membrane will combine with diffusion to determine the permeability and selectivity of separation. This is particularly true of asymmetric membranes which have a thin dense skin layer which controls performance. The flow rate of a gas through a membrane of thickness, l, is based on Fick's Law of diffusion and can be simply expressed by

$$J = D\Delta C/l$$

where D is the gas diffusion coefficient in the membrane and ΔC is the concentration difference of the gas in the membrane.

On the assumption that the membrane gas concentration and external gas partial pressure, p, are linearly related (according to Henry's Law), ie $C = SP$, where S is the solubility coefficient, the gas flowrate is given by

$$J = \frac{P(\Delta P)}{l}$$

where P is the permeability coefficient (= DS) for a gas.

The selectivity of these membranes is represented by the ratio of the permeabilities of any two components in the membrane. This factor is a specific characteristic of a membrane and generally varies inversely with gas permeability. This therefore means that

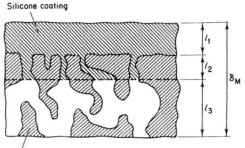

FIGURE 12 – Composite gas permeation membrane.

to achieve a high permselectivity requires the membrane to operate with a low permeability.

In general real separation factors and operating separation factors are not equal due to several factors, including non-ideal gas solubilities, plastisation, where a permeating gas has a high chemical affinity to the membrane, and the influence of the pressure ratio. Plasticisation and high pressure ratios (feed: permeate) generally decrease selectivity.

Membranes for gas separations are usually either asymmetric or composites to minimise resistance to gas flow in comparison to symmetric membranes.

A major problem with asymmetric membranes was found to be their susceptibility to faults and pin holding which caused a drastic fall off in selectivity. This problem was solved by coating the membrane skin with a thin layer of silicone polymer for example, which exhibits high permeability and low selectivity, which effectively sealed the faults.

Membranes for gas permeation are either one of two types: elastomer eg polydimethylsiloxane and polymethylpentene, or glassy polymer eg polyimide or polysulfone. Elastomers generally show rather low selectivities for some separations whereas glassy polymers exhibit higher selectivities but lower permeabilities. Microporous ceramic and metal membranes are also used, particularly to separate isotopes of uranium in the nuclear industry. The latter membranes exhibit high permeabilities compared to homogeneous membranes, but are accompanied by lower selectivities.

The production of high purity hydrogen (< 1 ppm impurity) is required for example in the semiconductor industry. Gas permeation is achieved using non-porous membranes of Palladium, usually alloyed with 23% silver. Alloying improves the tolerance to fluctuations in thermal stress and also improves the permeability in comparison to Pd alone. These membranes are very sensitive to damage by reaction with trace quantities of species such as chlorine, mercury, arsenic and sulphur compounds and are poisoned by unsaturated hydrocarbons or oxygen in organic compounds.

Membrane Performance

The requirements of an ideal gas separation membrane can be summarised as
1. The selective layer is very thin to minimise gas flow.
2. The selective layer is defect free to ensure gas transport is purely by solution-diffusion and not partly with, poorer selectivity, by flow through pores.

INTRODUCTION TO MEMBRANE SEPARATIONS

3. The support structure should contribute negligible resistance to gas flow.
4. The support structure should be mechanically strong to support the thin active layer during high pressure operation.

There are generally five principle types of membrane structures used to achieve these requirements for gas separations; integrally skinned, multicomponent, single-layer and multi-layer composites and asymmetric composites.

The main requirement for selective relative rates of diffusion of gases (and vapours) through the membranes is met by selection of the polymer active layer. In general the types of polymers are divided into two areas for gas separation

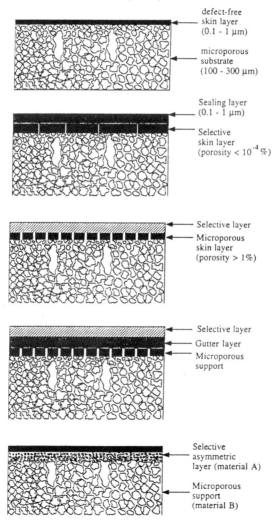

FIGURE 13 – Structure of gas separation membranes.

1. Glassy polymers (amorphous materials below the glass transition temperature) for separation of permanent gases eg N_2 from air.
2. Rubbery polymers (amorphous materials above the glass transition temperature often crosslinked) for separation of organic vapours from air).

The permeability of permanent gases such as CO_2 is generally much higher in rubbery materials (elastomers) than in Glassy polymers (eg polyimide see Table 7). The reason glassy polymers are frequently used in permanent gas separations is that they generally exhibit much better selectivities. The relative poor permeabilities can be overcome by good engineering design of thin layers.

TABLE 7 – Permeabilities of Carbon Dioxide and Methane.

	PCO_2 (Barrer*)	P_{CO_2}/P_{CH_4}
Polytrimethylsilylpropyne	33100	2.0
Silicone rubber	3200	3.4
Natural rubber	130	4.6
Polyamide	0.16	11.2
Polysulfone	4.4	38
Polyether sulfone	7.4	32
Cellulose acetate	6.0	31
Poly etherimide	1.5	45
Polyimide	0.2	64

* 1 Barrer = 10-10 cm3 (STP) cm cm-2 s-1 (cm Hg-1)

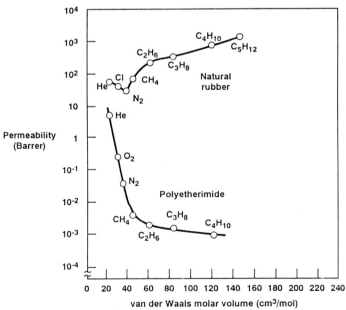

FIGURE 14 – Permeability of gases in a rubbery and a glassy polymer.

INTRODUCTION TO MEMBRANE SEPARATIONS

The separation of H_2 from hydrocarbons is one of the major applications of gas permeation and a successful membrane in this separation is silicone (coated onto polysulphone). The reason for this is the much higher permeabilities achieved with this composite membrane even though other (Glassy) membranes exhibit higher selectivity.

Overall the permeability of a gas is very dependent upon the choice of the polymer and different gases exhibit vast differences in permeabilities in the same polymer, which can be by several orders of magnitude. The choice of separating membrane layer depends upon achieving the correct balance between permeability and selectivity – a highly selective membrane is of little use if permeability is very low as module costs are uneconomic whereas high permeability is a little use if selectivity cannot be achieved without resorting to multiple module operation.

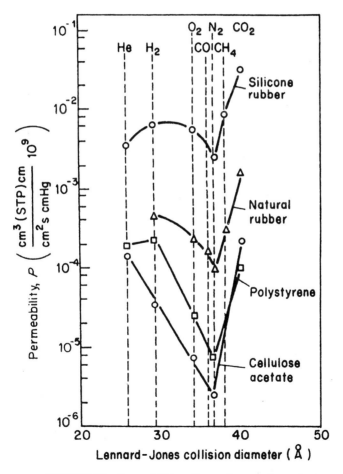

FIGURE 15 – Permeabilities of gases in several membranes.

The differences in permeabilities of vapours and permanent gases in given membranes can vary greatly, 4-5 orders of magnitude are not untypical. Higher permeabilities of organic vapours is attributed to a much higher solubility of organics in the polymer, which tends to make the polymer chains more flexible increasing the free volume for diffusion of the molecules. In the separation of organic vapours from gases (eg air) the selection of a membrane material depends on the relative contributions of diffusion and sorption to the overall process. The ratio of diffusion coefficients of organic to air will always be less than 1.0, due to the differences in molecular size predominantly. The relative sorption coefficients or organic to air will normally be grater than 1.0, especially for rubbery polymers. The different diffusion and sorption characteristics of rubbery and Glassy polymers has meant that rubbery polymers are used to selectively remove organics from air. In glassy polymers the diffusion coefficients of permeant gases are much higher than organics, making these materials generally air selective.

In the separation of minor amounts of organic vapours from air rubbery membranes are

TABLE 8 – Permeabilities of vapour and gases in polydimethylsiloxane.

Component	Permeability (Barrer)
nitrogen	280
oxygen	600
methane	940
carbon dioxide	3200
ethanol	45,000
methylene chloride	168,000
chloroform	284,000
carbon tetrachloride	200,000
1,2-dichloroethane	248,000
1,1,1-trichloroethane	247,000
trichloroethylene	614,000
toluene	1,460,000

TABLE 9 – Selectivities of the MTR-100 Membrane to Common Industrial Solvent Vapors, Measured at Ambient Temperature.

Vapor	Membrane selectivity
Octane	90-100
1,1,2-Trichloroethane	60
Isopentane	30-60
Methylene chloride	50
CFC-11 (CCl_3F)	23-45
1,1,1-Trichloroethane	30-40
Isobutane	20-40
Tetrahydrofuran	20-30
CFC-113 ($C_2Cl_3F_3$)	25
Acetone	15-25
CFC-114 ($C_2Cl_2F_4$)	10
Propane	10
Halon-1301 (CF_3Br)	3

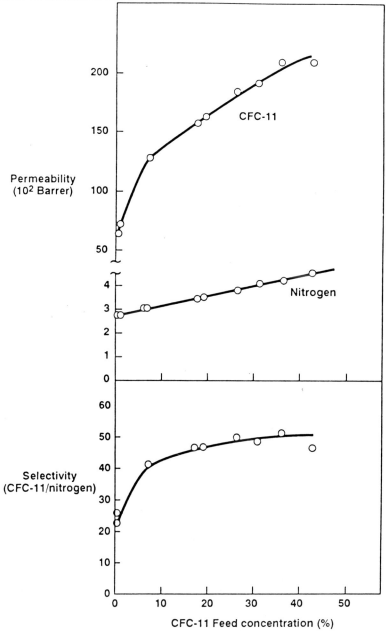

FIGURE 16

commonly used. A typical example of this separation as discussed in section 6 is the recovery of halocarbon from air. Selectivities of greater than 20 are achieved even at very low halocarbon concentrations. The selectivity of the membrane generally increases with increasing boiling point of the vapour associated with the increasing sorption of the vapour in the membrane material.

TABLE 10 – Performance of State-of-the-Art Asymmetric Membranes Made by Various Phase Inversion Processes for Oxygen/Nitrogen Separation ($T = 25°C$)

Membrane material	Geometry	Oxygen permeance ($10^{-6}cm^3/[cm^2\ s\ cmHg]$)	O_2/N_2 Selectivity	Ref.
Polyetherimide[a]	Flat sheet	2	9.1	125
Polyetherimide[a]	Hollow fiber	0.5[b]	7.9	98
Tetrabromo-bisphenol A polycarbonate	Hollow fiber	19	7.0	108
Polyethersulfone[a]	Flat sheet	5	7.0	80
Polyimide (Matrimid 5218)	Hollow fiber	10	6.8	126
Polysulfone	Flat sheet	28	6.0	83
Polyimide (6FDA-IPDA)	Flat sheet	82	5.4	83
Polysulfone[a]	Hollow fiber	43[c]	5.2	98
Polycarbonate	Flat sheet	52	5.1	83
Poly(phenylene oxide)	Flat sheet	121	4.2	128
Poly(phenylene oxide)	Hollow fiber	50	4.0	127
Poly(vinyltrimethyl silane)	Flat sheet	21	3.8	129
Ethylcellulose	Hollow fiber	138	3.5	124

[a] Multicomponent ("caulked") membrane   [b] $T = 30°C$ [c] $T = 50°C$

TABLE 11 – Gas Separation Performance of State-of-the-Art Thin-Film Composite Membranes for Oxygen/Nitrogen Separation ($T = 25°C$).

Microporous support	Sealing layer	Selective layer	Oxygen permeance ($10^{-6}cm^3/[cm^2\ s\ cmHg]$)	O_2/N_2 selectivity	Ref.
Polypropylene	—	Poly(trimethyl silylpropyne)	1360	1.4	181
Polyimide	—	Poly(kitoxime organosiloxane)	1500	2.2	182
Polyetherimide	—	Polydimethylsiloxane	330	2.1	183
Polysulfone	—	Polyaminosiloxane	103	2.4	138
Polysulfone	—	Polyaminosiloxane	110	3.9	184
Polysulfone	—	Ethylcellulose	26	3.8	150
Polysulfone	—	Poly(4-methyl-1-pentene)	24	4.3	186
Polysulfone	Ethyl cellulose	Poly(4-methyl-1-pentene)	90	4.0	157
Polysulfone	Polyamino-siloxane	Poly(phenylene oxide)	34	4.3	138
Ceramic	—	Polyimide [6FDA-IPDA]	38	4.8	149
Cellulose acetate	—	Poly(siloxane amide copolymer)	39	5.9	185
Polysulfone	—	Tetramethylbisphenol-A polyester	14	5.9	152

The selection of a particular membrane for a separation depends upon the gases in questions and on the current state of the art of membranes which are available. A variety of composite asymmetric membranes are for example suitable for oxygen/nitrogen separation which would not necessarily be suitable for other separations.

Composite Membrane of Siloxane Containing Polyurethanes

For use in an artificial lung a multilayered hollow fibre membrane has been developed

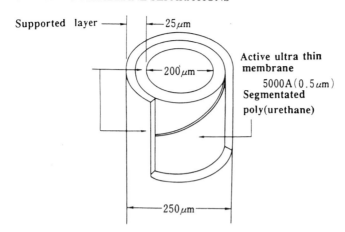

FIGURE 17 – Three layered hollow fibre membrane.

consisting of an ultra thin dense layer sandwiched between two support layers as shown in Fig. The dense active permeation layer (5000 A) is siloxane containing polyurethane prepared by melt spinning and stretching or by solvent casting. The porous layer (25 μm thick) is of poly(4-methyl pentane 1). Gas permeation rates for O_2 are between 10^{-3}-10^{-4} cm^3 (STP) cm^{-2} s^{-1} (cm Hg)-1 depending upon temperature. These values are over two orders of magnitude better than those for a thick single siloxane – containing polyurethane dense film.

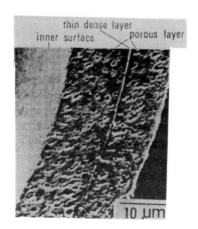

Three layered hollow fibre membrane.

Gas permeation membranes find their major applications in the chemical and petrochemical industries. The separation and recovery of hydrogen, from refinery gas and ammonia, is the major application. Other applications are in the purification of natural gases and methane recovery from biogas, O_2/N_2 separations from air and in dehydration of gases.

Pervaporation

Pervaporation is one of only a few membrane processes which can separate a mixture of two or more miscible liquids into more concentrated products of the constituents. The analogous non membrane based process is distillation although the mechanism of separation is quite different. In the pervaporation process a liquid feed of say two constituents is contacted with one side of an appropriate membrane. If the opposite (permeate) side of the membrane is kept under a reduced pressure, which must be lower than the saturation pressure, then transport of species will occur from the feed across the membrane into the permeate. This transport involves a change of phase with the permeate appearing as a vapour (see Fig 18). The vapour is then condensed to produce a liquid permeate product.

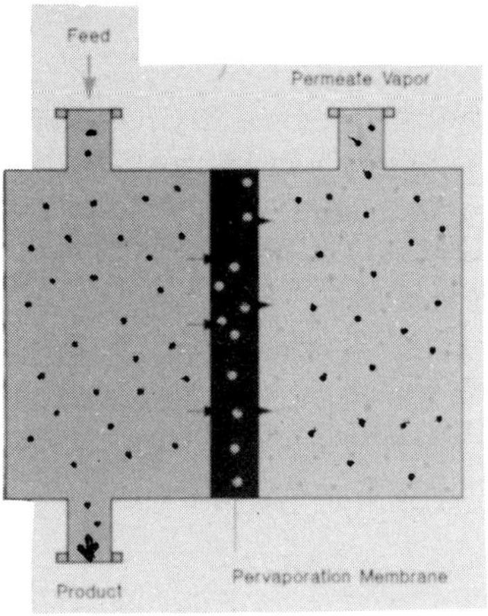

FIGURE 18 – The pervaporation membrane separation process.

There are two methods of producing a reduced pressure in the permeate, one by continuous pumping under reduced pressure (vacuum pervaporation) or by a stream of carrier gas.

The process of pervaporation is complex involving simultaneous mass and heat transfer – latent heat is required to "pervaporate" the liquid. The principle of separation is not similar to that of distillation, where separation is largely dependent upon vapour/liquid equilibria, rather it is dependent on the solubility and the diffusivities of the constituents in the membrane. Transport across the membrane, which is non-porous, is in a sequence of three steps

i) partitioning (selective sorption) of the feed constituents into the retentate side of the "swollen" membrane
ii) selective diffusion of penetrants through the membrane
iii) desorption of the permeate at the membrane surface into the vapour phase

The result of this process is a permeate which has a composition significantly different to that which would be established if the vapours was evolved at a membrane free liquid-vapour interface at equilibrium. This difference with the process of distillation is the prime driving force for the use of pervaporation. Pervaporation can effectively shift the vapour/liquid equilibrium to positions at which a higher fraction of product vapour is produced in a single step. The process is not limited by the formation of azeotropes and the latter is a major application area in for example dehydration of azeotropes of ethanol and water. The relatively high costs of pervaporation equipment in achieving this phase change in a single step generally means that target areas of pervaporation are separations which themselves involve high costs. Other applications may involve the use of hybrids of pervaporation and (azeotropic) distillation.

In pervaporation three basic steps are required in the separation, selective sorption in to the membrane on the feed side, diffusion through the membrane and desorption into the vapour phase permeate. The last step of desorption is not generally a limiting factor in determining the transport rate. Transport is described by a solution and diffusion mechanism in the same way as gas permeation. The process of pervaporation is much more complicated than gas permeation, involving a phase transition where heat must be supplied to vaporise the permeate. In addition liquids show a greater affinity to polymers than gases so that solubility is much higher. This affects the solubilities and diffusivities of species which depend upon concentration. Thus permeability is very dependent on the composition of the liquid mixture.

Liquid will generally swell the membrane to some extent during pervaporation. The swelling varies across the membrane because the concentration of the penetrant varies across the membrane. Thus across a membrane the diffusivity also varies. The swelling of the membrane by the desired penetrant also enhances the diffusivity of the other penetrants, which has the effect of reducing the separation factor. The separation factor aAB is defined as

$$aAB = \frac{Y_A/Y_B}{X_A/X_B}$$

where X_A, X_B, Y_A and Y_B denote the weight fraction of components A and B in the feed and permeate respectively.

Membranes for pervaporation are non-porous as in the case of gas permeation. The ideal structure is asymmetric with a thin dense top layer and open porous substructure. The open porous substructure should prevent capillary condensation of vapour and offer minimal resistance to flow, whilst enabling the application of a thin dense skin. Preparation methods for asymmetric pervaporation membranes are similar to those for gas permeation; dip-coating, plasma polymerisation and interfacial polymerisation. The selection of a polymer is highly application dependent and generally elastonomers are no more permeable than glassy polymers due to the effect of swelling by the high penetrant concentrations.

The degree of swelling should be in the range of 5 to 25% (by wt.) to give an optimum between high flux (large swelling) and high selectivity (low swelling).

A basic principle in membrane separation of organic liquid mixtures is to use a polar membrane to permeate a polar liquid and vice-versa. Consequently in the permeation of water in dehydration applications, hydrophilic membranes are employed. Typical materials which have been considered are polyacrylonitrile (PAN), polyvinylalcohol (PVA), poly (vinyl pyrrolidone), polyacrylic acid, and polyethersulphone (PES). Typical values of flux and selectivity in the separation of 90% by weight ethanol in water with 50mm thick membranes are:

	Flux (kg m-2h-1)	α
PAC	0.03	12,500
PVA	0.38	140
PES	0.72	52

The effect of penetrant concentration on flux is very significant as shown in Fig 19 for the separation of water-ethanol mixtures with a PVA membrane. The figure demonstrates the much greater selectivity of this membrane for water.

A range of composite membranes have been developed for a number of separations. Notable among these are the composites of cellulose acetate (support layer matrix) incorporating polyphosphates and PVA. The degree of cross linking of the active layer affects the performance, in that improvements in selectivity are generally at the expense of flux. High flux membranes are often restricted in terms of the maximum water feed content at which separation can be attained. In practice these membranes are supported on

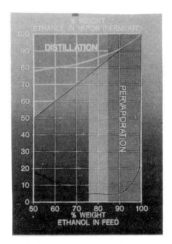

FIGURE 19 – Typical separation curve, compared to vapor-liquid equilibrium curve (ethanol/water mixture).

polyester felt for improved mechanical strength. The good termperature stability of these membranes, up to 120 degrees Centigrade, is a significant advantage.

The separation of water from for example alcohol is relatively easy because of the large differences in size and chemical properties. Separation of chemically similar components is much more difficult and separation factors are typically much lower.

INTRODUCTION TO MEMBRANE SEPARATIONS

Separation of organic species from water selectively requires the use of organophilic membranes such as polydimethlysilicone (PDMS) (so called silicone rubber (SR)). Such material permeates alcohol freely.

TABLE 12 – Typical positive azeotropic mistures (characterized by a minimum boiling temperature) which can be separated by pervaporation through a permselective membrane obtained by grafting polyvinylpyrrolidone onto a thin polytetrafluoroethylene film[1]

A-B Azeotropes (A=Fast component)	T_b (°C)	Azeotrope characteristics		Selectivity		Permeate flux (kg/h m2)
		T_b(°C)	c(%)	α	β	
A: Chloroform B: n-Hexane	61.2 69.0	60.0	72.0 28.0	3.9	1.25	2.65
A: Ethanol B: Cyclohexane	78.5 81.4	64.9	30.5 69.5	16.8	2.89	1.10
A: Butanol-1 B: Cyclohexane	117.4 81.4	78.0	10.0 90.0	23.5	7.23	0.30
A: Water B: Ethanol	100.0 82.8	78.2	4.4 95.6	2.9	2.68	2.20
A: Water B: t-Butanol	100.0 82.8	79.9	11.8 88.2	41.0	7.17	0.35
A: Water B: Tetrahydrofurane	100 65.5	63.8	5.7 94.3	19.1	9.24	0.94
A: Water B: Dioxane	100.0 101.3	87.8	18.4 81.6	18.1	4.36	1.33
A: Ethanol B: Ethylacetate	78.5 77.2	71.8	31.0 69.0	2.4	1.67	0.95
A: Methanol B: Acetone	64.7 56.2	55.7	12.0 88.0	2.9	2.36	0.65
A: Ethanol B: Benzene	78.5 80.1	67.8	32.4 67.6	1.3	1.18	2.90

The selectivities a and b are defined by the following ratios, respectively:

$$\alpha = \frac{(c'_A/c'_B)}{c_A/C_B} = \frac{c'(1-c)}{c(1-c)'}, \quad \beta = \frac{c'}{c}$$

where c and c' are the weight concentrations of the faster permeant (A) in the feed (c) and the permeate (c'), respectively. T_b denotes the normal boiling point of an organic compound or of an azeotropic mixture (at 1atm). Pervaporation temperature: T - 25°C.

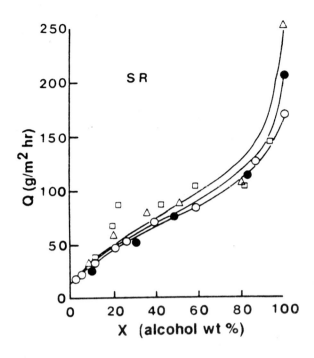

(●)MeOH; (○) EtOH; (△) 1. ProH; (□) 2. PrOH.

FIGURE 20 – Permeation of alcohols through S.R. membranes.

Vapour Permeation

Vapour permeation is a membrane process for the separation of saturated mixed vapours, with no change of phase involved in its operation. Thus, compared to pervaporation, the addition of heat equivalent to the enthalpy of vaporisation is not required in the membrane unit itself. Operation in the vapour phase virtually eliminates the effect of concentration polarisation prevalent in liquid phase separation such as pervaporation.

Vapour permeation can be used to separate a vapour from either non-condensable gases or from a mixture of vapour compounds. Preferred membranes are non-porous and materials are similar to those for pervaporation. Consequently organophilic membranes such as PDMS are used to separate organic species from air and other gases. For the separation of mixed vapours a range of membrane materials are possible depending upon the desired penetrant. Thus in the latter case PVA (or PVA/PAN composite) membranes are used to separate water from for example alcohols.

A potential attraction of vapour permeation is its possible use in a hybrid separation with distillation where, the overhead vapour in the latter could be further separated. It is desirable in vapour permeation to avoid condensation of vapour, caused by variations in feed compositions, pressure or temperature. Thus one potential application is the removal of contaminating organic solvents from air streams.

TABLE 13 – Separation of organic solvents by vapour permeation (air or N_2 if stated).

Solvent	Membrane	Selectivity	Reference
Methanol	Polyimide	221	25
	Silicone	38	25
Ethanol	Polyimide	297	25
Ethanol/N2	Vycor glass	2-400	4
Acetone	PDMS	11-25	18
		47.7	25
		158	19
Acetone/N2	Vycor glass	2-300	4
Hexane	Polyimide	32	25
Benzene	Polyimide	51	25
Toluene	Polyimide	180	25
	PDMS	83	19
p-Xylene	Polyimide	460	25
	PDMS	68	19
m-Xylene	Polyimide	513	25
1,2-Dichloromethane	PDMS	142	19
Chloroform	Polyimide	24	25
Carbon tetrachloride	Polyimide	32	25
1,2-Dichloroethane	Polyimide	52	25
	PDMS	103	19
1,2-Dichloropropane	Polyimide	57	25

Thermopervaporation

An alternative method to achieving a pervaporation type of separation is to replace the vacuum (or carrier gas) induced driving force by a thermal driving force. This can be done by isolating the evacuating pump, after a steady state for the pervaporation is to be achieved, and inducing a high rate of condensation sufficient to maintain a low pressure of vapour at the permeate side of the membrane. Permeation will thus continue until a new steady operating regime is achieved.

Membrane Distillation

The process of pervaporation makes use of non-porous membranes for separation whereas a similar process of membrane-distillation makes use of porous membrane materials. The material in this case is not wetted by the liquid feed and thus liquid penetration and transport across the membrane is prevented, provided that the feed side pressure does not exceed the minimum entry pressure for the pore size distribution of the particular material. Separation is by virtue of vaporisation at the mouths of the pores and vapour transport through the pore network of the membrane. The membrane exerts little influence on the separation (fractionation) of the liquid, as the vapour-liquid equilibrium is not disturbed.

At the permeate side condensation of the vapour is induced at the surface of the membrane in a "Direct Contact Membrane-Distillation". Thus vapour is locked inside the membrane and mass transport is driven by the application of, usually, a temperature driving force. The feed is at a higher temperature than the liquid condensate. Efficient separation relies on convective mass transport of vapour in the pores and the presence of inert gases in the feed can be detrimental to operation as they will accumulate in the pores. Vapour transport in this case would eventually shift to one of diffusion through a stagnant gas, a much slower process. De-gasification of feed is therefore generally advised in operation.

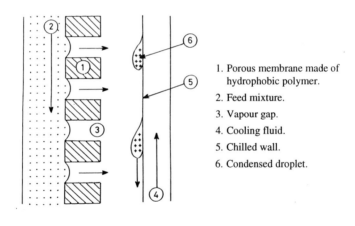

1. Porous membrane made of hydrophobic polymer.
2. Feed mixture.
3. Vapour gap.
4. Cooling fluid.
5. Chilled wall.
6. Condensed droplet.

FIGURE 21 – Fractionation by membrane distillation.

An alternative mode of operation in membrane-distillation is to allow vapour free access to the permeate side outer surface of membrane and condense the vapour at an additional cooled surface relatively close to the membrane. The techniques is sometimes referred to as "low pressure membrane distillation", as the driving force for vapour transport is the pressure differential. It is also feasible to use an inert sweep gas to remove the vapour permeate and maintain the differential pressure.

As already mentioned the extent of fractionation of the feed is theoretically equivalent to that of a single-stage vaporisation and thus a high separation factor is not achieved. This has been experimentally verified. This puts practical limitations on the types of separations to which membrane distillation is used. The major advantage of membrane

INTRODUCTION TO MEMBRANE SEPARATIONS

distillation is that, with compact modules equipped with hollow fibres, a high surface area per unit liquid volume for mass transport is accessible and thus high overall permeation rates are attainable.

The major practical limitation in membrane distillation applications is the requirement pressure upstream of the membrane which must be less than the penetration pressure of membrane, P, which can be calculated form the WASHBURN equation.

$$P = \frac{4 \sigma \cos Q}{d}$$

where s is the surface tension of the contacting liquid Q is the contact single between the liquid and the surface (of the membrane) and d is the maximum pore diameter of the membrane.

The feasibility of a membrane for a particular separation will be affected by two factors associated with the feed

i) the proportion of water in the aqueous phase
ii) the presence of surface active agents.

In the case of ethanol-water mixtures the critical ethanol concentration is between 42% to 75% by weight and thus membrane distillation is not used for ethanol rich feeds.

Applications of membrane distillation have predominately focused on aqueous based feeds using hydrophobic membranes, with characteristics shown in Table 14.

TABLE 14 – Hydrophobic porous membranes used for membrane distillation.

Membrane	Porosity	Mean pore diameter (nm.)	Thickness ($\mu m.$)	Penetration pressure (bars)	Critical ethanol concentration* (for water-ethanol mixtures) % by weight
Polypropylene CELGARD 2400	0.38	20	25	> 10	42
Polytetrafluoroethylene PTFE TF 200	0.60	200	60	3	75
Polypropylene hollow fibers MITSUBISHI rayon engineering	0.45	36	22	12.5	
Polyvinylidene fluoride PVDF ACCUREL capillaries	0.75	550		3	

Tab. 2: Characteristics of several hydrophobic porous membranes used in membrane distillation.
 * Critical concentration = concentration at which penetration pressure vanishes.

The materials are typically polypropylene, PTFE and PVDF with submicron pore size, which have penetration pressures of several bar. These membranes are applied in other processes, such as microfiltration and as supports for liquid membranes etc. The establishment of the separation process in the market place may eventually lead to the production of specific membranes for membrane distillation with optimised characteristics.

Applications of membrane distillation are in the following areas:
- Water purification and demineralisation of sea-water, brackish water and waste water
- Extraction of ethanol from fermentation media
- Concentration of aqueous salt solution and acids

Liquid Membranes

The term "liquid membrane" is used to describe a process of separation which does not rely on inherent chemical characteristics of a thin, solid (semi permeable) barrier. The closest equivalent non-membrane separation process is liquid extraction and consequently liquid membrane processes are referred to as "liquid pertraction" although other names are used eg carrier-mediated extraction and facilitated transport. The equivalence between liquid membranes and liquid extraction arises from the use of a multiple (2 or 3) phase system.

The principle of liquid membrane operation, shown in Fig X, is relatively simple; two homogeneous, miscible liquid, one the feed (donor) the other strip (acceptor) are spatially separated by a third phase, the membrane. The membrane phase consisting of solvent and carrier is immiscible and practically insoluble in the donor and acceptor solutions. Separation (transport) of solute from the donor to the acceptor solutions is due to favourable conditions created at the two interfaces between the three phases. The thermodynamics at the donor/membrane interface favour extraction of solute into the membrane while simultaneously the thermodynamics at the membrane/acceptor interface favour the reverse transport, i.e. stripping.

The membrane phase itself in most applications is an organic liquid although aqueous based membranes can be used for separation of organic solutions. The attraction of liquid membrane for extraction processes lies in several features
- molecular diffusion in liquids is generally several orders of magnitude faster than in solids eg polymer membranes
- pertraction can be intensified by eddy diffusion in the mobile membrane phase
- liquid membranes can be designed to be highly selective to specific solutes
- a maximum driving force can be created, thus multistage processes commonly used in liquid extraction are not needed
- the organic phase, which contains the liquid membrane carrier can be selected from a wide range of inert, insoluble and harmless organic liquids
- relatively small quantities of carrier are needed and therefore highly selective, relatively expensive agents can be used

A major disadvantage of liquid membranes, in comparison to solid membranes is the

INTRODUCTION TO MEMBRANE SEPARATIONS

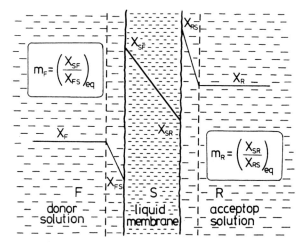

FIGURE 22– Principle of liquid membrane operation.

greater degree of complexity in setting up the membrane phase and maintaining its stability.

Pertraction Systems

Liquid membranes can be put into two classifications: methods with phase dispersion and methods without phase dispersion. Excluding bulk liquid membranes, which are used for fundamental studies methods without phase dispersion, these include supported liquid membranes (SLM) and liquid film pertraction (LEP).

Supported liquid membranes are based on the use of a porous solid membrane (polymer or ceramic) which supports or hold the liquid membrane phase (see Fig). The pore of the thin solid membrane are completely filled with the membranes and this impregnation produces, relatively stable heterogeneous solid/liquid membranes. In some applications the liquid membrane phase may be sandwiched between two porous solid supports

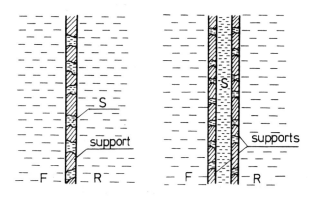

FIGURE 23 – Supported liquid membrane process.

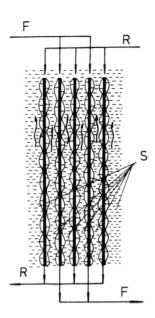

FIGURE 24– Liquid film pertraction process.

(contained liquid membrane). The typical supports used are thin flat sheets or hollow fibres, manufactured from oleophilic polymers, wettable by the membrane liquid. In application the membranes are put in spiral wound or hollow fibre bundle modules. The attraction of liquid membranes lies in the small amount of liquid required to produce a large interfacial area eg 1 m² from 10 cm³ in a 20 μm thick, 50% porous polymer.

The liquid film pertraction process has the three solutions in continuous motion: the donor and acceptor solution flow along vertical supports alternatively arranged at small distances in the membrane solution. Hydrophilic membrane supports enable stable and uniform film flow of donor and acceptor solutions down the solid support.

Emulsion Liquid Membranes

A second form of liquid membrane process is based on the formation of a double emulsion and are referred to as emulsion (surfactant) liquid membranes (ELM). This is a three phase system which is stabilised by an emulsifier, which can be up to 5% or more of the membrane liquid. The membrane liquid in the form of relatively large droplets (0.1 – 1 mm diameter) encapsulates the acceptor solution which in effect is a fine dispersion of smaller drops (1 – 10 μm). This emulsion is very stable due to the size of the drops and the high content of emulsifier or surfactant. In operation the membrane emulsion phase is then dispersed into the donor feed solution and the process of pertraction or mass transfer can take place. This process is quite rapid due to the high interfacial areas (1000 – 3000 m⁻¹) produced by the emulsion droplets and the microdrops in the acceptor phase. This feature makes ELM attractive for the treatment of low concentrations of permeating species.

INTRODUCTION TO MEMBRANE SEPARATIONS

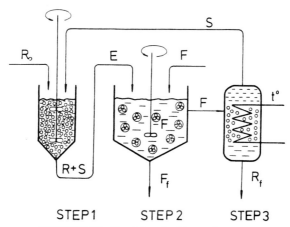

FIGURE 25 – Emulsion liquid membrane process.

There are four essential steps in the process:
1. *Emulsion preparation.* Here an aqueous reception phase is emulsified into the organic membrane phase. The aqueous phase will be either acid or base depending upon the extraction system. The organic phase will contain a surfactant to stabilise the emulsion and may contain an appropriate carrier (ion exchanger) for certain separations.
2. *Solute pertraction.* Here the membrane emulsion is contacted (dispersed) with the feed (donor) solution, as a continuous phase, for extraction to take place.
3. *Membrane and feed separation.* The emulsion membrane phase (and internal strip phase) is then separated from the spent feed.
4. *Emulsion break-up.* Here the water in organic emulsion is separated to enable recycling of the organic membrane phase and recovery of solute from the aqueous reception phase.

In the ELM process the main problems relate to emulsion stability – poor stability incurs partial rupture of the membranes which reduces overall efficiency – high stability brings problem is the breaking of the emulsion in the final stages of the process.

In view of the above problems associated with membrane stability alternative forms of phase dispersion without surfactants have been proposed which utilise porous polymer membranes. One example uses a dispersion and hollow fibre technique.

In this the aqueous reception phase is dispersed in the organic phase contained within the lumen of a porous oleophilic capillary membrane. The aqueous donor phase flows over the outer surface of the capillary membrane and mass transport takes place through the organic phase, which is also contained within the pore walls, into the dispersed aqueous reception phase.

Membrane Phase Contactors

Processes such as absorption and extraction involving phase contact usually involve dispersion of one fluid phase, as droplets, bubbles etc, into other phase. After mass transport the dispersed phases are then separated by a method utilising the difference in their phase density. Equipment designs can have limited interfacial areas and limited versatility for process changes. Membrane technology which uses modules, with for example microporous hollow fibre, can establish a consistent, stable interface between the phases. The interface is immobilised at the pores by the hydrophobicity of the membrane and by controlling the differential pressure between the phases. The separation principle differs from other membrane separations such as filtration and gas separation as there is no convective flow through the pores, instead, the membrane acts as an inert support to facilitate diffusive transfer. The two immiscible phases are in direct contact without dispersion and the mass transfer between the two phases is governed entirely by the equilibrium chemistry of each phase. Because of the hollow fibre pore structure and porosity, very high surface area per unit volume can be achieved.

Membrane high efficiency phase contact technology offers additional advantages over conventional, dispersion technology based on its ability to provide a constant, fixed interfacial area and allow for predictable scale up from single module to multi-module systems. There are many applications of hollow fibre membrane technology in phase contacting, from waste recovery, food and pharmaceutical industries to analytical and medical applications.

Electrodialysis

Electrodialysis (ED) is a membrane process which gives selective separation of ions of one charge from ions of an opposite charge. The electrodialysis units use membranes which are ion selective and are hence called ion exchange membranes. Such membranes have fixed charged groups bound into the polymer matrix to which mobile ions with opposite charge (counter ions) are attached. Two general types of ion exchange membranes are used either heterogeneous or homogeneous. Heterogeneous membranes are prepared from ion exchange resins and generally exhibit high electrical resistance and relatively poor mechanical strength especially at high swelling with water. Homogeneous membranes are made by bonding the ionic group into a polymer film which is a crosslinked copolymers, based on divinylbenzene with polystyrene (or polyvinylpyridine) or polytetrafluorethylene and poly(sulfonyl) fluoride-vinyl ether. Ion exchange groups are primarily sulfonic acid or carboxylic acid groups for cation exchange membranes and quaternary ammonium salts for anion exchange membranes. When placed in an electrolyte solution and an electrical current is passed through them by the motion of the mobile ions, the ions with the same charge as the mobile ions are free to flow from one face of the membrane to the other under the influence of a potential gradient. Thus cationic polymeric membranes are permeable to cations and almost impermeable to anions and vice versa for anionic membranes.

In an electrodialysis unit (Fig 2.6) both types of membranes are employed. Each unit consists of many flat membrane sheets, typically 150 to 400, arranged alternatively as cation and anion exchange membranes. This membrane stack is sandwiched between two electrodes, a cathode and anode. When a direct electrical current is passed through the stack, cations will try to move towards the cathode and anions will try to move towards the anode. The presence of ion selective membranes restricts the movement of charged species, such that in one compartment both cations and anions will be removed by charge transport in opposite directions to adjacent chambers, thus diluting the solution in this chamber. Meanwhile the concentration of ions in the adjacent chamber increases. Overall the electrolyte concentration in alternate cells of the stack decreases. In operation, electrolyte flows in narrow gaps formed between the membranes by spacers which intensity mass transfer and act as membrane supports. Two electrolyte streams are produced continuously; one a concentrate in which the salt concentration increases and the other a diluent in which the salt concentration decreases.

The transport rate of species through an ion exchange membrane can be written as

$$J_i = -K\, U_i\, d\,\mu_i/dx$$

where mi is the chemical or electrochemical potential or related property, U_i is the mobility of the penetrant in the membrane, and K is a coefficient of transport.

This equation assumes that in the simultaneous transport of species there is no interaction, i.e. the transport of one species does not influence the transport of any other species. In practice the equation only holds for low concentrations of penetrants and in other cases deviations occur when the displacement of the different species become mutually dependent. The ion-exchange membranes can be distinguished by a supporting skeleton of usually organic macromolecules with functional groups firmly anchored to the pore walls. These functional groups are capable of dissociation in aqueous media, for example sulphonic or carboxylic acids which can split off hydrogen ions and are characteristic of cation exchange membranes. The mobilities of these bound groups is one and they are assumed uniformly distributed through the membrane.

When a membrane bearing a group such as $-SO_3H$, or $-CH_2\, N^+R_3Cl^-$ in the case of anion exchange membranes, is put in water it swells sufficiently such that the ionisable group will release the small counter ion, e.g. H^+ or Cl^-, and the oppositely charged group, called the fixed ion, remains covalently bonded to the skeleton. When the membrane is placed in an aqueous electrolyte some salt will enter the membrane. The sorbed ions which have a charge similar to the fixed ions are called co-ions. The concentration of these co-ions increases with the concentration of the electrolyte. When an electrical current passes through the membrane the counter ions can enter into it from one side and can leave it from the other side, which results in the formation of a concentration gradient. On increasing the current density a point is reached at which the concentration of counterions at the interface approaches zero and the system is then polarised. The swelling of membranes when placed in electrolyte solutions occurs throughout their structure due to the binding of the water of hydration. The insertion of the water among the ions enables some counter ions to diffuse away from the fixed ions thereby creating an osmotic pressure which draws more water into the membrane and increases the swelling. An equilibrium is eventually reached when the polymer chain internal elastic forces are balanced by osmotic pressure forces.

The performance of ion exchange membranes is frequently discussed in terms of the transference number and permselectivity. As an approximation ion transport in can be expressed in terms of a simple phenomenological equation based on electrical potential as the major driving force

$$J_i = C_i^m \cdot U_i^m \cdot \frac{d\Phi}{dx}$$

where m represents the membrane phase.

In electrodialysis the transfer of electric charges is due to the transport of ions and the mass flux is directly proportional to the electric current. The relative fluxes of the different ions are denoted by transport numbe, t which is the ratio of the electric current conveyed by that ion to the total current. The transport number (or the transference number) are a measure of the permselectivity of an ion-exchange membrane.

The membrane permselectivity, j defines the degree to which it passes an ion of one charge and prevents the passage of an ion of the opposite charge and is defined by

$$\phi^c = \frac{t_+^{mc} - t_+}{t_-}$$

and

$$\phi^{ma} = \frac{t_-^{mc} - t_-}{t_+}$$

Where j is the permselectivity of a membrane, t is the transference number, the superscripts mc and ma refer to cation- and anion-exchange membranes and the subscripts + and − to cation and anion respectively.

The permselectivity of an ion-exchange membrane relates transport of electric charges by specific counter ions to total transport of electric charges through the membrane. For example an ideal permselective cation exchange membrane would transmit only positively charged ions, i.e. for $t_+^{mc} = 1$ is $f^{mc} = 1$. The transference number of an ion in the membrane is proportional to its concentration in the membrane. This is a function of its concentration in the solutions in equilibrium with the membrane phase, due to the Donnan exclusion as discussed earlier. Typical values of membrane transport numbers and membrane resistances are given in section 2.

The major applications for ED are in concentration of electrolyte solutions or in the diluting or de-ionising of solutions. The latter application has over the years been the dominant application in the desalination of brackish water. Electrodialysis is also used extensively for desalting and concentrating sea water in salt production. In principle the technique has many potential applications in the removal or recovery or ionic species. Other applications have been realised in the food, pharmaceutical and metal plating industries for effluent treatment and chemical regeneration from salt solutions.

A relatively new and important application of the use of ion exchange membrane is as so called bipolar membranes. These are composites of anion exchange and cation exchange membranes which under the application of an applied potential split water into hydrogen ions and hydroxide ions. The bipolar membrane is thus an alternative method to electrolysis for generation of H+ and OH- ions which can be used to regenerate acid and base from salts, without the production of oxygen and hydrogen gases.

INTRODUCTION TO MEMBRANE SEPARATIONS

SECTION 1.4 – POLARISATION AND FOULING

Polarisation and fouling are phenomena which occur in membrane system which result in a reduction in performance in comparison to certain ideal situations. The reduction in performance is seen typically as a fall in flux with time, the extent of the effect varies with the type of separation. For example membrane processes which are primarily filtrations are much more susceptible to polarisation and fouling as particulate, macromolecules etc build-up at the membrane surface. However with separation of gases and liquid mixtures eg gas permeation prevaporation, polarisation and fouling are much less of a problem, assuming feeds are free of contaminants.

The general effect of polarisation, fouling and other factors which cause a reduction in flux, is to increase the resistance to fluid flow through the membrane, through the formation of an additional barrier. These factors include, pore blocking, adsorption and gel layer formation.

Concentration Polarisation

Consider the case of a cross flow membrane process with a feed containing a solvent and solute or suspended solid flowing over the surface of the membrane. The fluid flow in a channel, say formed in a fibre or tubular membrane (or spiral wound) has the characteristics of having a higher velocity in the centre line than near the surface. This behaviour is due to the shear stress effects, due to fluid viscosity, which are initiated by the presence of the stationary membrane surface. At the surface the fluid velocity is zero and close to the surface there is a small region (boundary layer) of variable velocity (see Fig). This boundary layer formation has great significance in membrane separation performance.

In a membrane separation such as reverse osmosis water is flowing in two directions. One flow direction is across the membrane surface, this is the feed/concentrate stream. The other flow direction is through the membrane, this is the permeate stream.

As permeate passes through the membrane, most of the dissolved salts in the water are spopped at the membrane surface. These salts must be carried away in the feed/concentrate stream and hence they must move out of the boundary layer into the faster moving feed/concentrate stream. Salts move through water by diffusion out of the boundary layer and for this to occur, the concentration of salts at the membrane surface must increase above the concentration in the faster moving bulk stream above the surface. This phenomenon has the technical name of concentration polarisation which means there is a higher concentration of salts at the membrane surface.

Different salts diffuse away from the membrane surface at different rates as the rate of diffusion is affected by the size of the ion or particle as well as its charge and concentration. For example, multivalent ions, such as calcium and sulphate, move much more slowly than monovalent ions, such as sodium or chloride.

The concentration of solute at the membrane surface depends upon the flux through the membrane, the retention of the membranes, the diffusion coefficient of the solute D and the approximate thickness of the concentration boundary layer d ie the region near the membrane in which the concentration of solute varies.

$$C_m \Big/ C_b = \frac{\exp(J/K)}{R + (L-R)\exp(J/K)}$$

where Cm and Cb are the solute concentrations at the membrane surface and in the bulk feed respectively.

It is usual to refer to the ration of D/d as the mass transfer coefficient, k, for the solute. This coefficient is a function of the physical properties of the feed and the geometry of the system and the fluid flow. High values of k indicate relatively low extent of concentration polarisation.

Table 1 – Effect of Concentration Polarisation on Membrane Separations.

Membrane Separation	Influence	Factors
Microfiltration and Ultrafiltration	Strong	Small k and large J mean Cm is high.
Reverse Osmosis	Moderate	High k but low J give only moderate Cm values.
Gas Separation	Very Low	High k (high diffusivity) and low J.
Pervaporation	Low (Moderate)	Low J, relatively small k. Low conc. of permeating species is more susceptible.
Electrodialysis	Strong	Small k and large ionic flux.
Dialysis	Low	Low J and modest k.

The effect of concentration polarisation varies for different membrane separations. With, for example, salts accumulation at the membrane surface, the observed retention will be lower than the real (intrinsic) retention (R). In the case of macromolecules, a second

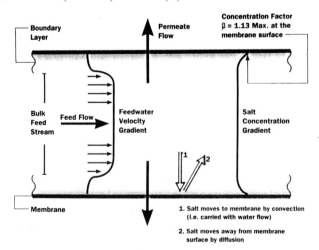

FIGURE 1– Boundary layer formation at a membrane surface.

(dynamic) membrane may be formed which can actually increase retention as experienced in ultrafiltration. The flux will be lower due to the increased resistance to fluid flow caused by accumulate solute or particulates. The effect is particularly sever in microfiltration and ultrafiltration. Table 1 summarises the impact of concentration polarisation on the common membrane separations.

The effect of concentration polarisation has an influence in membrane module design. This is in particular the case in MF and UF where a range of methods have been studied and applied as a means of reducing the effect of concentration polarisation and increasing mass transport.

TABLE 2 – Methods to reduce polarisation at membrane surfaces.

Increase cross flow velocity	limited due to high pressure drop
Turbulence promotors	inert inserts induced eddies, pressure drop can be high
Corrugated membranes	eddy formation induced by irregular membrane surface
Flow instabilities	pulsation of flow with baffles induce vortices
Vortices	rotational flows. Taylor and Dean vortices
Rotating Membranes	rotating cylinders or discs
Module Design	rotating modules, cross flows in hollow fibre modules
Vibrating Membranes	vibratory shear enhanced units
Ultrasonics	ultrasound induced cavitation at surface
	Complex module design
Gas Sparging	inert gas induces turbulence
Electrical Charge	electric fields influences charge on macromolecules
Chemical Methods	charge modification and hydrophilising membrane surface

The simplest procedures either increase membrane cross flow velocity or use inserts or turbulence promotors to increase mass transport. The limitation is that high pressure losses can result, in the case of turbulent flow pressure drop is proportional to the square of the flow velocity.

Chemical modification of membrane surface is used industrially although this often has the effect of limiting membrane fouling. Examples of chemical modifying are:-
– polymerisation of hydroxypropyl acrylate onto PVDF – base membrane in situ
– deposition of hydroxypropyl cellulose onto a polysulphone membrane
– conversion of polyacrylonitrile to a very hydrophilic surface

A variety of methods of depolarising incorporate new module design concepts or modified flows. These include rotating cylinder units, rotating disc units (the membrane is stationary) and annular fluid flow with vortices. Because of additional costs of many of the techniques they are generally limited to small scale production. Overall the techniques have demonstrated much improved flux rates in comparison to straight forward cross flow velocity increase.

The use of cross flow in membrane filtration was introduced to reduce concentration polarisation by utilising hydrodynamics forces to hinder deposition of solute onto mem-

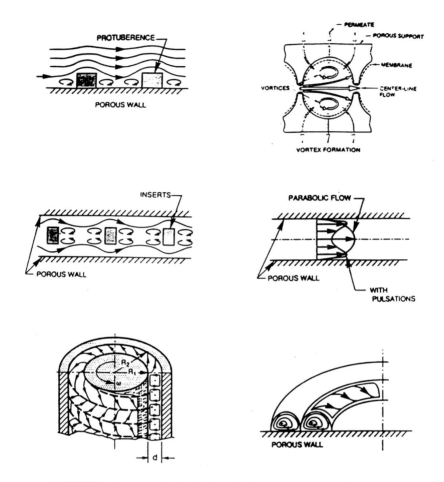

FIGURE 2– A range of methods for reducing concentration polarisation in membrane separations.

branes. The velocities which can be used are however limited and this together with solute/membrane interactions means concentration polarisation still occurs.

Enhancement of flux can be obtained in certain applications by carrying out cross flow filtration of aqueous suspensions in a dc electrical field. This applies to low-shear cross flow filtration and to high-shear filtration. In plate-and-frame filters this is arranged by placing the microporous membrane on a porous metallic support connected to the negative pole of the dc source. The anode is placed on the opposite side of the filter channel, facing the filter medium. there are two fundamental phenomena that cause the increase in flux;

1. *Electrophoresis*. The particles in suspension are in most cases repelled from the cathode and collected on the anode (if the particles are positively charged, the polarity of the electrode reversed). With the cathode behind the microporous

medium, all deposit on the medium is avoided. The particles are collected on the anode surface and swept away by the flow.
2. *Electro-osmosis*. Water flow is intensified toward the cathode, which means that the filtrate flux is intensified again.

Overall the two reasons for flux enhancement in electric field, the increased water flux due to electro-osmosis and the decreased hydraulic resistance of the media due to electrophoresis, generally act together.

Electrochemically Enhanced Cross Flow Filtration

The application of a superimposed electric field can prevent, or reduce, deposition of solid on the membrane, and subsequently reduce polarisation and enhance membrane fluxrates. The technique can be applied to microfiltration and ultrafiltration with order of magnitude increases in fluxrate achieved. The membrane materials which can be used in this can be polymer, metallic or ceramic. The application of a electrical potential field can be used to control the concentration polarisation in pressure driven membrane separations. An enhancement in the membrane flux can often thus be achieved due to the influence of the electrophoretic migration of the different species. With the use of suitable electronically conducting "mineral" membranes this method of enhancing the rate of ultrafiltration has been demonstrated. These membranes, are typically alumina, coated with a metallic layer of nickel, nickel/tungsten alloy or with a RuO_2-TiO_2 mixture. The application of a potential, field particularly with the RuO_2-TiO_2 membranes, can produce a significant enhancement in the fluxrate. For example, an approximate doubling in the membrane fluxrate is achieved on the application of a potential field of 15 V. The magnitude of the voltage applied depends on the design of the electro-ultrafilter and the conductivity of the solution feed. Thus with a compact design practical voltages may be much lower.

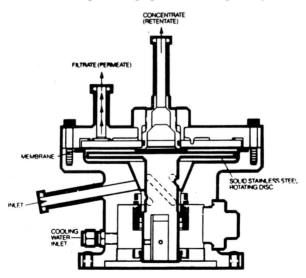

FIGURE 3– Rotating disc laboratory scale membrane filter.

The use of electrofiltration is attractive in for example the ultrafiltration of shear sensitive fluids where hydrodynamics are limited to laminar flow conditions. In the use of electrofiltration the relative improvement in flux must be assessed against several criteria including
1. the relative total energy consumed by the filter and pump per unit volume of filtrate produced compared to that without the electric field
2. the relative increases in cost of the electrofilter and associated equipment
3. the reduced requirement for membrane cleaning and back flushing.

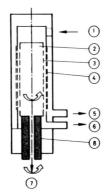

1. Suspension inlet.
2. Rotating cylinder.
3. Rotating filter medium.
4. Stationary filter medium.
5. Filtrate outlet.
6. Concentrate outlet.
7. Filtrate outlet from hollow shaft.
8. Rotor.

FIGURE 4 – Rotating cylindrical filter.

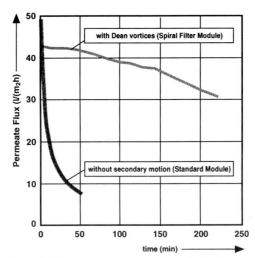

FIGURE 5 – Enhanced of flux due to concentration polarisation in spiral flow element with Dean vortices.

INTRODUCTION TO MEMBRANE SEPARATIONS 77

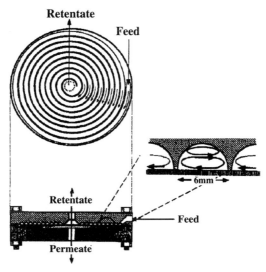

FIGURE 6– Spiral flow membrane unit.

A major consequence of concentration polarisation in membrane processes arises through problems of membrane fouling.

Membrane separations which are severely affected by concentration polarisation are ultrafiltration and electrodialysis. The influence of concentration polarisation in electrodialysis is discussed in section . In ultrafiltration concentration polarisation is a significant factor due to both the high flux rates which can be experienced and the relatively low diffusivities of the macromolecules. This causes very high concentrations of species at the membrane surface which can reach maximum values where gelation of the macromolecules may occur. The gel concentration depends on the chemical and physical structure and degree of solvation of the macromolecule. The occurrence of gelation introduced a "gel-layer" onto the surface which is a cause of an additional resistance to fluid flow. Occurrence of gelation at the membrane surface causes the ultrafiltration performance to take on limiting flux characteristics. That is as the transmembrane pressure differences is increases, the flux does not increase linearly, according to the membrane permeability model, but rather tends to approach an asymptotic value as the impact of the gel layer resistance sets in. On further increasing pressure difference after the solute concentration reaches the maximum (gel) concentration (Cg), the thickness of the gel layer increases or the layer becomes more compact. Thus the flux becomes completely controlled by the gel layer flow resistance. This behaviour is simply modelled in terms of a series of flow resistances at the membrane surface, one for the membrane, Rm, effectively the inverse of permeability and one for the gel layer, Rg. The flux rate in ultrafiltration is therefore given by;

$$J = \frac{\Delta P}{\mu (R_m + R_g)} = K . \ln (C_g / C_b)$$

Note in this expression that as the gel layer forms and the thickness Rg will dominate and that Rg will be approximately proportional to DP and thus limiting flux will occur. The above model is attractive and useful in its relative simplicity, but is limited by several factors which will not be addressed further here. One factor already mentioned is that due to increased osmotic pressure affects at the membrane surface.

Membrane Fouling

Fouling is a major cause of reduction in flux during several membrane separations and in particular microfiltration and ultrafiltration, where feed pretreatment is not used (cf reverse osmosis and elctrodialysis). Fouling is the essentially the deposition of retained particles (colloids, salts, macrolecules etc) onto the membrane surface or in the pores. The build up of this deposit causes a continuous decline in flux with time, which in microfiltration particularly can fall to values an order of magnitude less than the initial values. The fouling behaviour is specific to the system of interest and consequently varies according to contaminant of foulant, whether organic macromolecules, biological, inorganic salts, particulates. Characterisation of the fouling phenomena is often based on

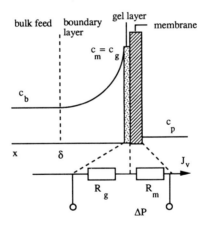

FIGURE 7– Formation of gel layer and gel layer resistance model at a membrane surface.

simple single parameter models using; silting index (SI), plugging index (PI), silt density index (SDI), fouling index (FI, MFI). A simple view of fouling is taken from the analogy with filtration theory where a cake of solid forms at the surface which adds an additional hydraulic resistance to flow. This cake layer builds up continuously during operation and resistance therefore correspondingly increases. The model can be successfully applied in many microfiltration applications but is limited in many respects. The crucial factor with fouling is to identify the source of the problem (see Table 3) or the occurrence and to take steps to either minimise the effect and/or correct the problems.

INTRODUCTION TO MEMBRANE SEPARATIONS

TABLE 3 – Example of foulants in membrane separations.

Large suspended particles	Particles can block the flow channels of hollow fibre and spiral wound modules. The particles may be present in the original feed or formed as a result of scaling during processing.
Smaller colloidal particles	Colloidal particles of ferric hydroxide are frequently encountered in the reverse osmosis of brackish waters, and give rise to slimy, brown fouling layers. In the recovery of cells from fermentation broths the cells, whose surfaces are lined with sticky polymer products, are often observed to foul the membranes.
Macromolecules.	A gel formation on ultrafiltration membranes. Macromolecular fouling within the porous structure of microfiltration membranes.
Smaller molecules	Certain small organic molecules have very strong interactions with plastic membranes. Polypropylene glycols, which are used as anti-foaming agents during fermentation, cause rapid fouling of certain plastic ultrafiltration membranes.
Proteins	Interaction with UF and MF membranes is influenced by shape, size, zeta potential, hydrophibicity of the surface.
Chemical Reactions	Precipitation of salts, hydroxides due to concentration increase or pH change in RO and ED.
Biological	Bacterial growth on membrane surface and excretion of extracellular polymers.

Fouling Control and Correction

Although fouling cannot generally be avoided in membrane separations there are several method which can be used to control the extent of fouling or to correct the fouling, by cleaning. The methods can be summarised as, feed pretreatment, adjustment of membrane properties, membrane cleaning and modification of operating conditions. The latter are generally the methods used to reduce concentration polarisation.

Pretreatment of feed is extensively used in reverse osmosis desalination of brackish water and sea water. There are several methods used depending upon the quality (chemical analysis) of the feed. The formation of scale due to the increase in concentration of calcium and magnesium bicarbonate during reverse osmosis is frequently overcome on desalination plants by a coagulation and sedimentation process. Lime and sodium carbonate are added to the feed water to precipitate calcium ions and bicarbonate ions as calcium carbonate and precipitate magnesium ions as magnesium hydroxide. The precipitated material is then removed by sedimentation and the clarified water is usually further treated with acid to replace residual bicarbonate ions by more soluble chloride or sulphate ions. A sequestrant, (e.g. sodium hexametaphosphate) is often added to the pretreated water to retard the crystallisation of scaling compounds within the reverse osmosis modules. On small desalination plants the lime-soda process is often replaced by ion exchange softening process, in which the calcium and magnesium ions are replaced by sodium ions.

The above lime softening and base exchange pretreatment processes also remove colloidal particles from the feed water. Silica scale may partly be removed as silicate precipitate in the lime softening process. Base exchange softening aids in prevention of scaling due to calcium sulphate. After softening the effluent is usually further clarified, for example by sand filtration.

Micro-organisms are frequently found in surface water feedstream. They are generally destroyed by chlorine dosing or in some cases ultraviolet sterilisation or a combination of both. Polyamide membranes are generally damaged by residual free chlorine and hence if such membranes are used the feed water is usually dechlorinated. This dechlorination is achieved through an activated carbon column or by sodium bisulphite treatment, upstream of the cartridge filters (5-10 µm) rated installed to protect the reverse osmosis modules and high-pressure pumps.

Pretreatment stages may also be required if the feed water contains ferrous ions or hydrogen sulphide. The ferrous ions are usually oxidised with permanganted ions or by exposure to air to precipitate ferric hydroxide. Hydrogen sulphide is usually removed by degassing and chlorine oxidation.

Membrane Cleaning

The methods for cleaning membranes after fouling has occurred are based on hydraulic, mechanical, chemical and electrical methods. The method(s) used depend upon the separation process and the configuration of the module. The hydraulic cleaning of the membrane is typically achieved with back flushing of the permeate through the membrane. The process is carried out by reversing the direction of flow of the permeate, usually for a few minutes, at a pressure which can be as large as the filtration pressure. This effectively dislodges the foulant from the membrane and restores the flux to a value close to the initial (or previous high) value. Back flushing is carried out repeatedly at regular intervals and leads to a saw-tooth type of flux behaviour, as shown in Fig 8. The average flux rate generally still shows a gradual decline with time and membrane cleaning will still be necessary. The procedure is routinely carried out with MF membranes and hollow fibre and tube UF membranes, both types are able to withstand the reversal in pressure difference. The method is effective with many kinds of foulants, particularly larger particles, but less so with material which forms adherent films. Ceramic membranes are particularly suitable for backflushing being mechanically very strong. An alternative method of hydraulic cleaning is by back-pulsing, ie short bursts of back pressure, alternative pressurising and depressurising and reversing the feed flow direction with the permeate exit closed (this induces back pressure).

Mechanical membrane cleaning is generally limited due to the sensitivity to damage of the membrane surface. One method for the cleaning of tubular modules is by forcing foam balls (polyurethane foam) down the tubes. The foam balls are slightly smaller than the bore of the tube, and are forced down the tubes at high velocity by the application of fluid pressure. which creates turbulence at the membrane surface as the ball passes, and helps to dislodge fouling matter. The foam balls are either injected into one end of a membrane tube and collected from the other end, or are pushed back and forth along the module from storage channels at either end of the module.

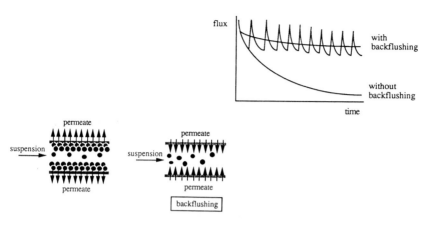

FIGURE 8 – Effect of back flushing on membrane fluxrate.

Electrical methods of cleaning can utilise electrical pulsing which results in the movement of charged species (particles, molecules) away from the surface. The process is carried out while the membrane separation is in process, although special module designs are required to introduce the charge to the membrane surface, which is generally metal.

Another electrical method is referred as direct membrane cleaning. Direct membrane cleaning (DMC) is a technique which enhances the performance of cross flow filtration processes. In practice, for cross flow filtration, solid and colloidal materials, etc, foul the membrane surface and result in a decline in flux with time. By application of periodic in-situ generation of microscopic bubbles (H_2 or O_2), from the conducting membrane surface, by electrolysis, this fouling deposit is dislodged and the flux rate is restored. This results in an overall increase in flux rate and allows lower cross flow velocities to be used. Major economies in pumping energy are therefore made, together with reduced membrane area requirements. The technology has been resulted in units which use stainless steel filters. With a direct membrane cleaned unit there are additional costs associated with the electrical connection to the "electrode membrane", the introduction of the counter electrode, and the supply of equipment for the current pulsing. Furthermore the material must be stable under the operating conditions employed.

A method which is currently under experimental research is the use of ultrasonic cleaning.

Chemical cleaning is the most important and extensively used method for controlling fouling in membrane separations. The choice of agent(s) depends upon the type of membrane, the type of foulant and the severity of fouling. Membrane manufacturers generally recommend appropriate agents for their specific products. The types of agents are:
- acids (strong or weak)
- alkalis (NaOH)

- detergents
- enzymes
- complexing agents (EDTA)
- disinfectants

Chemical cleaning is usually performed as "clean in place" (CIP) techniques by filling the retentate channels of the membrane module with cleaning solution from a separate cleaning tank. The module is generally exposed to the cleaning solution for a period of several hours. In most cases the plant feed and circulation pumps are used to recirculate the cleaning liquid though the module at high velocity and low operating pressure. The permeate side of the module is either completely full or empty of liquid. Modules that can be back washed may have the cleaning solutions introduced from the permeate side of the membrane. The function of the cleaning solution is to loosen and in some cases dissolve fouling matter that strongly adheres to the surface of the membrane or is lodged within the porous substructure of the membrane. The composition of the cleaning is strongly dependent on the nature of the fouling and the properties of the particular membrane. Individual plant manufacturers have developed several cleaning solutions for the most frequent encountered foulants and applications. For new process application the preffered cleaning solution(s) can only be established by experiment.

Many desalination plants use ammoniated citric acid solutions to remove fouling layers containing ferric hydroxide and also calcium solutions. Microbial and organic foulants cleaning solutions compromise a mixture of surface active agents, e.g. carboxymethylcellulose, sequestering agents, e.g. polyphosphates and ethyl-enediaminetertra-acetic acid (EDTA) and sterilising agents, e.g. formaldehyde. In the food industry, cellulose acetate reverse osmosis membranes are usually cleaned with enzymatic detergent solutions, which are able to break down protein layers, followed by sanitisation with an oxidising solution such as sodium hypochlorite or hydrogen peroxide.

The majority of ultrafiltration and microfiltration membranes, and also some of the newer (noncellulose acetate) types of reverse osmosis membranes can be cleaned with more aggressive and cheaper chemicals. Membranes in the food and dairy are frequently cleaned sequentially with hot strong alkali (NaOH at 50°C) and hot acid (HNO_3 at 45°C).The membranes are water washed and finally sanitised with an oxidant e.g. sodium hypochlorite, ethylene oxide or hydrogen peroxide. The sanitisation step will generally destroy most, but not all, micro-organisms in the system. Certain modules can as an alternative be steam sterilised at 120°C which is likely to destroy all micro-organisms in the system. The cleaning process takes several hours. In hygienic applications cleaning it is likely to be carried out every day and results in a considerable downtimeof the plant. In non-hygienic applications cleaning is less frequent, and membrane lifetime can be improved. Cleaned membranes often foul more rapidly than new membranes which suggests that the surface properties are altered during cleaning.

An example of a basic flux recovery procedure is given in Table , for polysulphone fibre and tube membrane modules. Optimisation of the procedures in term of chemical concentration, recirculation time, temperature and pH will typically be performed on a case-by-case basis. Even with stringent cleaning procedures permeate flux will not

INTRODUCTION TO MEMBRANE SEPARATIONS

necessarilly recover to values for new cartridge. It is important that the permeate water flux after cleaning be higher than the average initial process permeate flux.

Cleaning formulations and the frequency and duration of cleaning cycle depend on the stream being processed, the degree of fouling, the extent of the concentration, etc. In general, cleaning should be performed at low pressure and high velocity, at temperatures of 40 to 50°C. Typical cleaning formulations for processing of biologicals and for general process applications are listed in Appendix (see A/G Tech Brochure). Alternative cleaning regimes for each process are provided in the tables.

Table 4, gives a selection of cleaning procedures for UF and MG hollow fibre/tubular systems.

Membrane Properties

The properties of particular membranes can be modified to reduce the extent of adsorption and fouling. For example the modification of membranes which are normally hydrophobic, and tend to strongly adsorb proteins, by introducing hydrophilic characteristics is used in several cases. This can be achieved by blending hydrophilic and hydrophobic polymers eg polyvinylpyrolidone with PVDF, or by pretreatment with hydrophilic surfactant or

TABLE 4 – Typically cleaning procedure for polysulphone membrane.

1	*Flushing.* before cleaning, flush residual feed from cartridge with clean warm (50°C) water, saline or buffer solution. Use buffer solution to prevent precipitation of solutes (eg proteins) during flushing. After cleaning, flush residual cleaning agent from the cartridge. Flushing should be performed in a non-recirculating mode, such that the flush water does not re-enter the system.
2	*Water Quality.* Cleaning and flush water should contain <0.05 ppm iron, <25 ppm calcium and magnesium and no colloidal silica. Water should be free of particulate matter, oil and grease. Ideally, ultrafiltration permeate or reverse osmosis permeate should be used.
3	*Temperature.* Cleaning at room temperature (ie 20°C) is NOT recommended. Preferably, cleaning should be performed at 50°C, to decrease the strength of foulant/membrane surface bonds and to improve solubility of residual feed constituents. Higher temperature cleaning (>60°C) is not recommended due to potential cleaning chemical/membrane interactions.
4	*Time.* Nominal cleaning time is defined for each cleaning step in the procedure tables. These times should be used as guidelines. Shorter or longer times may be required depending on the extent of the membrane fouling. In many cases, soaking the membranes overnight improves the effectiveness of the cleaning cycle. Flushing time depends on the cleaning chemical, the membrane pore size and the total hold-up volume of the system.
5	*Chlorine Wash Cycles.* Chlorine dissipates with time and is rapidly depleted in very dirty operations. A chlorine test kit should be used to check chlorine levels and additional chlorine should be introduced as needed.
6	*Safety.* Caustic, acid, bleach and other cleaning chemicals should be handled with care. Operators should take appropriate precautions to prevent contact with eyes and skin.

enzymes or by surface modification of the membrane. Surface modifications include, plasma treatment, polymerisation or grafting of the surface initiated by heat, chemicals or UV light, interfacial polymerisation and introduction of ionic groups ($-SO_3H$) or polar groups (-OR, -F) by chemical reaction.

Santisation

For santisation, thoroughly clean and rinse the membrane cartridges, then use any of the following agents:
1. Up to 100 ppm sodium hypochlorite. If properly cleaned, 10 ppm should be sufficient. Circulate 30 to 60 minutes.
2. Up to 3% formalin. Circulate 30 to 60 minutes.
3. Up to 0.5 N hydroxide. Circulate 30 to 60 minutes.
4. 100 to 200 ppm peracetic acid. Circulate 30 to 60 minutes.
5. Up to 70% ethanol in water.
6. Autoclave.

Storage

Ultrafiltration cartridges must be stored wet or reglycerised. Before storage the cartridges should be thoroughly cleaned and rinsed with water. For short-term storage, up to two weeks, cartridges need only be water-wet. For storage up to 1 month, cartridges may be filled with a storage solution and sealed at all endfittings and permeate ports or submerged in a storage bath. Acceptable storage solutions are:
1. Water with 5 to 10 ppm active chlorine (10 to 20 ppm sodium hypochlorite). Monitor levels weekly.
2. 0.1 N sodium hydroxide.
3. Up to 3% formalin.
4. 30% ethanol in water.
5. Up to 1% sodium azide.

For storage of longer than 1 month, check periodically to be certain that the membranes remain wetted. Prior to reuse it is recommended that the cartridge be flushed with a 100 ppm sodium hypochlorite solution and thoroughly flush all storage solution prior to use.

Microfiltration cartridges may be stored dry, after cleaning. It is advisable to clean and sanitise the microfiltration cartridges prior to reuse. If necessary, to fully wet the membranes after extended storage, expose the membranes (inside and outside) to 70% ethanol for one hour. Drain, wet with water and flush.

Depyrogenation

For depyrogenation, thoroughly clean, sanitise and rinse the membrane cartridges, then circulate either of the following for 30 minutes at 30°C to 50°C.
1. 100 ppm sodium hypochlorite, pH 10 to 11.
2. 0.1 N sodium hydroxide, pH 13.

After 30 minutes, thoroughly flush with non-pyrogenic water.

TABLE 5 – Methods to Control or Correct Membrane Fouling

Feed Pretreatment	
Pre-filtration with 1 m filter to remove particles	GP, VP, RO, ED
Activated carbon filtration to remove higher HC's	GP, VP
Adsorption on molecular sieve to remove H_2O	GP
Sand or candle filter to remove trace organics	PV
pH adjustment to reduce protein fouling	UF
chemical comlexation	RO
pH adjustment	RO, ED
Active carbon adsorption	RO
Chlorination	RO
Membrane Cleaning	
Hydraulic cleaning	
Permeate back flushing	MF, UF (hollow fibre)
Mechanical cleaning	
Foam ball	Tubular membranes
Chemical cleaning	Most membranes
(acids, alkali, detergents, disinfectant, sequesterants)	
Electrical cleaning	MF (UF) with
(electrical pulsing electrochemical gas evolution)	conducting membranes
Membrane Modification	
Surface charge	UF
Hydrophilicity or hydrophobicity	UF
(by pretreatment, polymerisation chemical reaction)	

Reverse Osmosis Membrane Cleaning

There are significant costs in RO operation associated with feed pretreatment and with the replacement of the membranes. The economic feasibility of RO depends on maintaining relatively high permeation rate and maximising membrane life. The efficiency of RO depends on the maintenance of the membrane in an unfouled condition. Fouled membranes generally cause reductions in flux rate and operating efficiency. Fouling may eventually lead to unscheduled shutdown, lost production time and replacement of membranes. The prevention and control of membrane fouling is one of the most important aspects of RO. The control of membrane fouling in RO systems has focused on feed-water pretreatment although feed-water pretreatment alone is inadequate and therefore, regular cleaning is an integral part of system operation.

In practical operation fouling problems are often not initially apparent because system performance appears to be adequate. However if minute quantities of foulants are allowed to accumulate on the membrane, a gradual, but serious, deterioration of performance will

Fouled RO membrane.

occur and if left uncorrected the membrane will become irreversibly fouled, necessitating replacement.

If there is an indication of membrane failure, deposit analysis is a key factor in identifying the problem and introducing appropriate remedies.

Techniques for removing (cleaning) deposits from fouled membranes include chemical (use of chemical agents), mechanical (ie, direct osmosis, flushing with, spongeball or brush cleaning, air sparging, etc), and a combination of mechanical, ultrasonic, and chemical cleanings. However the most routinely applied method is chemical cleaning using specially formulated membrane cleaners. Membrane cleaning involves the use of chemicals to react with deposits, scales, corrosion products, and other foulants. Chemical agents can be classified as acids, alkalies, chelants and formulated products.

A variety of inorganic, organic and mixed organic acids have been used for the removal of deposits. The most commonly used include hydrochloric, phosphoric, and sulphuric acids, organic acids (eg oxalic, citric) and acid salts such as sodium and ammonium acid citrates.

Although acids are effective in removing calcium-based scales (eg calcium carbonate and calcium phosphate) iron oxides, and metal sulfides they have limited effect on scales formed by silica and metal silicates and in removing biological suspended solids and other organic foulants.

Alkaline cleaning solutions (eg phosphates, carbonates, and hydroxides) are used to loosen, emulsify, and disperse deposits. The detergent effect of alkaline cleaners is usually enhanced by the addition of surfactants. Alkaline cleaning is often alternated with acid cleanings to remove deposits such as silicates.

In addition to acids and alkalies, chelants are also used to remove deposits from fouled membranes. Common chelants include ethylene-diamine tetra-acetic acid (EDTA), phosophonocarboxylic acid, gluconic acid, citric acid, and polymer-based chelants. Generally gluconic acid is effective in chelating ferric ion in strong alkaline solution.

The commodity chemicals are often limited in the ability to clean membranes and thus membrane manufactures have produced nonproprietarry formulations (see Table) for use as membrane cleaning agents.

TABLE 6.1 – Generic Membrane Cleaning Formulations

Cleaner	Scale/ Metal Oxides	Colloidal/ Particulate	Biological	Organics
Hydrochloric acid 0.5% (wt)	X			
Citric acid 2% (wt) and ammonium hydroxide (pH 4.0)	X			
Phosphoric acid 0.5% (wt)	X			
Sodium hydroxide pH 11-11.9		X	X	
Trisodium phosphate or sodium tripoly phosphate 1% (wt), sodium salt of ethylenediaminetetraacetic acid 1% (wt), and sodium hydroxide-pH 11.5-11.9		X	X	
Sodium hydrosulfite 1% (wt)	X			
BIZ* 0.5% (wt)			X	
Citric acid 2.5% (wt) and ammonium bifluoride 2.5% (wt)	X	X		

* Trademark of Proctor and Gamble, USA, for a detergent sold in the United States.

In addition several companies have developed a number of propiertary formulated cleaners for removing deposits (Table 6.2). The non proprietary and formulated membrane cleaning agents are superior to the single-function cleaners such as citric acid (used for removing calcium carbonate scaling) and laundry detergents with enzyme additives (used for removing biofoulants and certain organics).

Overall the effect of foulants on the performance of RO units can be quite varied (see Table) and require different actions to remedy or control the situation. When system performance changes at unanticipated levels troubleshooting will be required. Troubleshooting will generally require one or all of the following procedures: verification of plant operation, reviewing operating data, evaluating potential mechanical and chemical problems and identifying foulants. The identification of foulants can be achieved by nondestructive testing of suspect elements or destructive testing. Following foulant identification the appropriate steps can be taken to return the plant to its desired operation by;

 – cleaning the unit
 – modification of feed pretreatment system
 – replace the elements
 – change the operating procedures
 – rejuvenate the elements (and sanitisation).

TABLE 6.2 – Customised Membrane Cleaning Compounds

Cleaner	Scale/ Metal Oxides	Colloidal/ Particulate	Biological	Organics	Source[1]	Membrane Type(s)[2]
Arro-Clean 2100	X	X		BFG	B	
Arro-Clean 2200		X	X		BFG	B
Arro-Clean 2300	X	X	X		BFG	B
Arro-Clean 2400				X	BFG	B
Bioclean 103A	X				AS	CA
Bioclean 107A		X	X		AS	CA
HPC 303	X				AS	CA
HPC 307		X			AS	CA
Bioclean 511			X		AS	PA
IPA 403	X				AS	PA
IPA 411		X				
Poly Blue 202				X	AS	B
Filtra Pure Iron Remover	X				HBF	B
Filtra Pure CA		X	X		HBF	CA
Filtra Pure		X	X		HBF	PA
Filtra Pure HF		X	X		HBF	PA
Monarch Membrane Acid 23	X				HBF	B
Monarch Enzyme 95, DDS		X	X		HBF	B
Monarch Enzyme 96		X	X		HBF	PA
Monarch RO Cleaner 115		X				
Monarch HPH-RO		X				
Monarch Enzyme		X	X		HBF	CA
Monarch Chelate	X				HBF	B

[1] Information source and provider:
  BFG = Arrowhead Industrial Water, Inc. A subsidiary of the BFGoodrich Co, Lincolnshire, Illinois
  AS = Argo Scientific, San Marcos, California
  HBF = HB Filter Co, Monarch Division, Minneapolis, Minnesota

[2] Membrane types:
  CA = Cellulose acetate
  PA = Polyamide/thin fim composite
  B = Both CA and PA

INTRODUCTION TO MEMBRANE SEPARATIONS

TABLE 6.3 – Summary of Foulants and Likely Performance Effects (RO book)

Foulants	Description	Effects on RO Performance	Method of Control/Cleaning
(General order of appearance)			
Scale	Precipitate of sparingly soluble salts (minerals) caused by the concen-tration of salts in the feed/brine solution during passage across the membrane surface. Example: $CaCO_3$, $CaSO_4$, $2H_2O$, $BaSO_4$, $SrSO_4$, SiO_2	• Major loss of salt rejection • Moderate increase in differential pressure • Slight loss of production • The effects usually occur in the final stage	• Lower recovery • Adjust pH • Use scale inhibitor • Clean with citric acid or EDTA-based solution • Clean silicate-based foulants with ammonium bifluoride-based solutions
Colloidal Clays/Silt	Agglomeration of suspended matter on the membrane surface. Example: SiO_2, $Fe(OH)_3$, $Al(OH)_3$, $FeSiO_3$	• Rapid increase in differential pressure • Moderate loss of production • Moderate loss of rejection • The effects usually occur in the first stage	• Filtration • Charge stabilisation • Higher feed-brine flows • Clean w/EDTA, STP, or BIZ-type detergents at high pH • Clean silicate-based foulants with ammonium bifluoride-base solutions • Lower recovery
Biological	Formation of bio-growth upon membrane surface. Example: Iron reducing bacteria, sulfur reducing bacteria, mycobacterium, *Pseudo-monas*	• Major loss of production • Moderate loss of salt rejection • Possible moderate increase in differential pressure • The above effects occur slowly, steadily	• Sodium bisulfite addition Chlorination with or without activated carbon filtration • Clean with BIZ-type detergents or EDTA-based solutions at high pH • Shock disinfection program with formaldehyde, hydrogen peroxide, peracetic acid, etc
Organic	Attachment of organic species to the membrane surface. Example: poly-electrolytes, oil, grease	• Rapid and major loss of production • Stable or moderate increase in salt rejection • Stable or moderate increase in differenital pressure	• Filtration with GAC • Cleaning is rarely successful but isopropanol or proprietary solutions have been effective

Rejuvenation procedures are recommended for RO membranes when cleaning treatments have little impact in restoring membrane salt rejection characteristics. Salt rejection may be lost as a result of surface defects, abrasion damage, chemicals attacks and/or hydrolysis. There are three membrane rejuvenation agents in use as illustrated in Table 6.4.

TABLE 6.4 – Rejuvenation Agents.

Trade Name	Chemical Name	Comments
Colloid 189*/zinc chloride	Polyvinyl acetate copolymer in an ammoniacal solution	• Hole plugging • Proven effective on most CA-type spiral wound membranes that have been well cleaned • Applied in situ • Temporary improvement (1 month to 6 months) • Erractic improvements on poly(ether) urea and poly amide membranes • Flux loss generally <15 percent • Differential pressure increase may be unacceptable • Cleaning treatments will remove the coating • Usually very effective when salt rejection is above 75 percent
PTA/PTB (Lutonol M40 form BASF)	Polyvinyl methyl ether/tannic acid	• Proven effective on DuPont B-9 and B-10 fibre permeators • Applied in situ • Either has been applied individually with some success • Surface coating • PTB wears away with time • PTA is generally durable • Cleaning treatments may remove PTA and will readily remove PTB • Usually very effective when salt rejection is above 80 percent • Flux loss below 5 percent • Little, if any, increase in differential pressure
Vinac/Gelva	Polyvinvly acetates	• Surface treatment • Proven effective on CTA hollow-fibre membranes • Applied in situ • Temporary (1 – 12 month improvement) • Flux loss generally above 5 percent • Cleaning treatments will remove the coating • An increase in differential pressure may be encountered • Usually very effective when salt rejection is above 75 percent

Two mechanisms have been proposed to explain the rejuvenation process: "surface treatment", which is, in effect, a surface coating on the membrane, and "hole plugging". Membrane rejuvenation can typically increase salt rejection to at least 94%. Successful application of these agents requires thorough agents cleaning of the RO membranes. Rejuvenation will not work when physical damage of the membrane is large. If salt rejection is below 75%, successful rejuvenation is unlikely and if below 45%, rejuvenation is completely ineffective.

Biofouling

Biofouling is the result of complex interactions between the membrane material, dissolved substances, fluid flow parameters and micro-organisms and is basically a problem of biofilm growth. The main source of microbial contamination is the feed water. Surface waters with high micro-organisms content are bound to result in microbial problems. The operation of RO with high surface areas of the membrane coupled with the tendency for cells to be transported towards the membranes increases the probability of contact of the micro-organism and the membrane. Pretreatment steps are also sources of bio fouling.

Flocculants, used to assist suspended solid removal, provide a suitable habitat for microbial growth. Conditioning agents eg sodium hexametaphosphate are potential sources of micro-organisms and of nutrients for the biofilm (eg breakdown to orthophosphates). Sodium thiosulphate, used to neutralise chlorine, is also a source of nutrient for bacteria.

Chlorination of seawater can also instigate biofouling by the degradation of humic acid into biologically assimilated material, which support biofilm growth. Oil and other hydrocarbons which may enter the RO system through leaks or spills serve as a nutrient.

Another source of microbial contamination is the piping holding tanks and treatment systems prior to RO, eg ion exchangers, granulated active carbon filters, degasifiers, which offer surfaces for biofilms and which can release micro-organisms. Biofilm mode of growth enables micro-organisms to survive and multiply in very low nutrient habitats and provides protection against biocides. Problems associated with biofilms can occur in high purity industrial waters with total organic carbon content as low as 5 to 100 $\mu g/dm^3$.

Biofouling is a comparatively slow process and its effects are seen in a gradual decline in the water fluxrate, a gradual increase in the transmembrane and differential pressure and a gradual decrease in mineral rejection. Thus the overall effects are reduced membrane lifetime, increased cleaning and maintenance costs and a deterioration in quality of product water. Biofouling may also cause secondary mechanical deformation of spiral wound modules. Mineral deposition (inorganic scaling) frequently accompanies biofouling.

The flux decline associated with biofilm formation often occurs in two phases: a rapid decline over a period of a few weeks and then an a symptotic approach to an equilibrium value, which may be 60 to 80% of the initial permeation rate.

Pressure difference increases in RO operation can be a result of biofilm attachment to the membrane surface and to the Vexar spacer. The flow is affected in at least three ways, a reduced cross section area for flow, increased roughness of the surface and increased drag by virtue of viscoelastic properties of the biofilm. The reduced mineral rejection properties of a biofould RO membrane are attributed to:-

1. The biofilm acts as a more or less dense gel layer. The dissolved minerals accumulate in this layer and increase concentration polarisation.
2. Direct or indirect promotion of the decomposition of the RO membrane polymer.

There are a number of factors that can increase the risk of biofouling and include the plant design, feed water characteristics and operational characteristics (see Table 7).

Biofouled membrane.

TABLE 7 – Factors in Promoting Biofouling

Plant Design	Extended piping systems Access of light Dead ends Fissures Armatures with dead zones Non disinfected holding tanks
Feed-water characteristics	High temperature (> 25C) High amounts of organic and inorganic nutrients Large amounts of cells (>10^4 colony – forming units per ml) High SDI
Operational characteristics	Infrequent monitoring of performance Use of microbially contaminated pretreatment chemicals Relatively slow cross-flow Long storage periods

The prediction of the potential for biofouling is desirable in order to prepare for the inevitable problems, to implement effective control or to implement preventative measures. If biofouling is expected, the microbial content of the water is checked, typically with a cultivation method. Effective pretreatment is then an obvious method to try to reduce the extent of membrane cleaning. The usual pretreatment scheme can include

1. Addition of biocide, preferably chlorine or its compounds. Chlorine sensitive membrane will however require dechlorination. Chlorination will not eradicate all bacteria and thus surviving bacteria will occur. Chloramine is an alternative biocide which has the advantage of not degrading humid acid.
2. Flocculation, precipitation and sedimentation.
3. Adsorption with granulated activated carbon.
4. Cartridge or micron filters.
5. Addition of scale inhibitor.

However all these steps have their limitations. Pretreatment chemicals should be checked for their microbiogical quality. Activated carbon is a potential site for colonising micro-organisms. Prefiltration with standard RO filters, rated from 5 to 20 μm will not efficiently retain bacteria. Thus adequate monitoring of biofouling is required in practical RO operation. Systems capable of directly detecting biofouling are not used in practice, rather indirect detection is employed based on symptoms of biofouling on plant performance parameters. Generally biofilm formation occurs quite quickly, within the 2 weeks of initial operation, before plant performance parameters respond and thus monitoring should ideally be carried out as directly as possible, that is as close to the biofilm as possible.

A suggested and reasonable approach to detect biofouling is the use of "sacrificial elements" as monitors. Mini RO probes which are integrated on-line in the system may be used. Such probes can be removed from time to time for a direct check on biofouling. Other indirect biofilm monitoring systems can be adapted from different applications, such as biofouling control in heat exchangers. It is also reported that accumulations of biomass in prefilters generally precedes any significant biofouling of the membrane. Thus inspection of the prefilters may enable biofouling to be detected at an early stage. Overall an effective biofouling monitoring system should include

- Monitoring of plant performance parameters
- monitoring the microbial quality of feed water and all added pretreatment chemicals
- and monitor biofilm formation on representative surfaces.

In systems suffering from biofouling the killings of the micro-organisms present is not always the solution to the problem. In many cases it is the actual presence of the biofilm, that causes the problem and a biologically inactive biofilm may continue to cause biofouling. The removal of the biofilm can then be crucial for effective sanitation. Biofilms also serve to protect their members against biocides. Regions which are inaccessible to biocides may bear biofilms which can act as inoculum for the cleaned system, which can induce rapid bacterial after growth. The effectiveness of biocide depends upon several factors:

- the types of biocide and concentration
- biocide demand
- interference with other dissolved substances
- contact time

- pH and temperature
- the type of organisms and their physiological state
- presence of biofilms
- speed of detection of biofilms

Even after the use of a biocide the presence of dead suspended bacteria and dead biofilm should be considered. Dead bacteria may attach to surfaces, dead biofilms provide substratum and substrate for subsequent cells, oxidising biocide may react with dead suspended bacteria and bacteria may be injured and survive, resulting in an inoculum for the cleaned system.

SECTION 1.5 – Module Designs

Module design is crucial to the operation of a membrane separation process as this is the unit which must operate at a technical scale with large membrane surface areas. The module design is based around two general types of membranes, flat sheet or tubular. The method of packing the membrane and the relative membrane size lead to a range of module designs.

There is a wide range of membrane equipment available at the present time covering all aspects of membrane separations from small laboratory scale test units to full scale modules and associated plant. The key part of the plant from the view of the membrane separation is the membrane module and the ancillary equipment. There are many speciality type of modules for specific applications which are described throughout this handbook. Module design here is described with regard to the cross flow separations of RO, UF, MF, ED, GP and PV.

The two important aspects of a membrane module are the actual material of the membrane and its eventual manufactured form and shape and thus the configuration of the module. This section therefore discusses membrane module design while membrane materials and classification are discussed in Section 2. The two factors of material and configuration, for any particular application, are closely linked and cannot really be considered as separate entities. The design requirements of a particular separation and the mechanism of transport through the membrane will dictate, and frequently limit, material choice, thickness, pore size etc. These limitations must be accommodated into an appropriate configuration and membrane module design to withstand the imposed conditions of operation.

Membrane Module Designs

Membrane modules are available in five basic designs, hollow fibre, spiral wound, tubular, plate and frame and capillary. Due to the particular operating requirements of individual membrane separations and the applications to which they are put, the choice of module design is limited. Generally the suitability of a module design for a particular type of separation can be classified as follows:

Forget about old fashioned filtration systems.

Conventional filters are subject to fouling, high cost, and overall ineffectivness. The next time you consider filtration remember V✧SEP® (Vibratory Shear Enhanced Process) the filtration alternative that works.

V✧SEP® filtration technology is simple. By vibrating the filter module the filters remain clean. **The age old problem of fouling has been solved thereby eliminating the need for frequent replacement or servicing of the filter!** Unlike cross flow filtration, the shear needed to keep the filter clean is created at the filter surface. Because of this unique design even high solids applications are no problem.

✔ Create crystal clear permeate and a concentrated sludge in a single pass.

✔ Replace expensive, traditional processes such as flocculation, sedimentation, vacuum filtration, centrifugation and evaporation.

✔ Perform solid/liquid separations ranging from low molecular weights (RO) through 30 microns.

✔ Join the many V✧SEP® users who are, for the first time, using filtration as an inexpensive option for their tough separation applications.

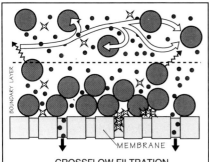

CROSSFLOW FILTRATION

Traditional crossflow filters plug and foul because the majority of shear created by the turbulent flow is away from the membrane boundary layer and cannot efficiently remove retained particles. This inefficient use of shear accounts for the eventual loss of flux experienced in traditional systems over time.

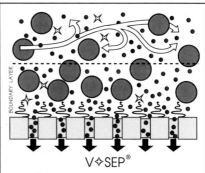

V✧SEP®

V✧SEP's® vibrational energy focuses shear waves at the membrane surface repelling solids and foulants within the boundary layer. This patented method allows for high solids concentrations while maintaining long term sustained rates up to ten times higher than conventional filtration systems.

✔Call a New Logic application engineer today to discuss a <u>free</u> trail run on your material.

NEW LOGIC INTERNATIONAL
1295 67th Street, Emeryville, California, U.S.A. 94608
(510) 655-7305   FAX-(510) 655-7307

CROSS-FLOW FILTRATION
MEMBRANE TECHNOLOGIES
Microfiltration - Ultrafiltration

DESIGN AND MANUFACTURE OF :

* Pilot systems and industrial plants fitted with :
  - tubular ceramic membranes
  - hollow fibre or spiral cartridges
* Steam-in-place and sanitary units
* Automatic systems with backflush option
* Fermentation-filtration integrated systems

FIELDS OF ACTIVITY :

* Biotechnology
* Food industry
* Pharmacy
* Fine Chemicals
* Environment
* University

For further information, please don't hesitate to contact us...  ...and ask us for our brochure.

INCELTECH France
15, Allées de Bellefontaine
31100 Toulouse

Tel. (33) 61 40 85 85
Fax (33) 61 41 51 78

INCELTECH UK
22A Horseshoe Park, Pangbourne
Berkshire RG8 7JW

Tel (44) 01734844888
Fax (44) 01734841677

INCELTECH USA
4377 Adeline Street, Emeryville,
CA 94608-3311

Tel. (510) 652 2295
Fax (510) 652 2011

INTRODUCTION TO MEMBRANE SEPARATIONS

Spiral wound	RO	PV	GP	(UF)		
Hollow fibre	RO	PV	GP			
Plate and frame	(RO)	PV		UF	ED	MF
Tubular		(RO)		UF		MF
Capillary				(UF)		(MF)

This situation has arisen because of a variety of design requirements such as the need for easy cleaning, high packing densities, cost effective operation and membrane replacement. Also in design, liquid based separations factors such as frictional losses, hydrodynamic properties and material properties impose different design requirements compared to gas or vapour based separations. Thus needless to say a plate and frame module designed for say PV will be quite different to one designed for ED. The basic features of each of the modules is now considered and differences in designs for particular types of separations discussed.

Hollow Fibre Modules

Hollow fibre modules occupy one of the largest scale applications of membranes, that is in RO desalination. A unit shown schematically in Figure 1 consists of a bundle of very fine membrane fibres packed into a cylindrical housing or shell.

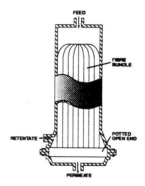

FIGURE 1 – Hollow fibre module design.

The external diameter of fibres is in the range of 80 to 200 micron and wall thickness can be as little as 20 micron, although for the larger diameters, thickness is greater to withstand large membrane pressure differences, which can be as high as 80 bar. The individual fibres are placed in the module with both open ends fixed to the permeate head plate to facilitate easier manufacture. Feed is external to the fibres, as is the active membrane surface, and permeate flows internally along the fibres.

Packing densities for fibre modules can be as high as 10,000 m^2/m^3. A variety of manufacturers are active in this field, particularly for RO modules, and polyamide fibre modules have taken a predominant position. Other modules are manufactured with membranes of cellulose triacetate and sulphonated polysulphone. Although adequate for desalination and similar applications these materials suffer from a limited stability at extremes of pH and temperature etc.

A major requirement of hollow fibre module operation is that the feed stream should be clean. The flow of feed in the narrow channels formed in the mass of fibres is laminar and as such is particularly susceptible to fouling. The cost of restoring a plugged module is high and feed pretreatment is therefore essential as is an awareness of problems which may result from precipitation of salts etc during concentration of the feed.

Modules sizes for RO range from 0.1 to 0.2 m in diameter and 0.6 to 1.2 m in length. Productivity is of the order of 800 $m^3/d/m^3$. In contrast the module size for GP has a greater length to diameter ratio and uses fibres of a greater diameter. Also fibres are not potted to one common end plate, but are straight and open at only one end or at both ends. Transmembrane pressure differences can typically be much greater than for RO.

The arrangement of flows of permeate and retentate can be an important parameter in determining module performance, particularly regarding selectivity. The variations of local pressure and concentration are factors which in particular affect performance. Generally module performance is better when permeate and retentate are in countercurrent flow or when feed flow is in radial, crossflow, brought about by a porous distributor plate along the axis of the module.

The design of hollow fibre units for PV is still in its infancy in comparison to RO and GP. Impetus in this area regarding applications may accelerate such designs, but a lot depends on the availability of suitable membrane materials. A major factor to be considered in PV is maintaining the established low partial pressure along the length of the fibre.

Capillary Modules

The capillary module is essentially of hollow fibre design utilising "fibres" which have diameters greater than those for RO. They are mainly applied in UF where much lower transmembrane pressures of operation permit larger diameter and enable fibres to be internally pressurised. Thus feed flow is arranged inside the fibre, as is the active skin layer, giving better control of concentration polarisation.

Modules come housed in shells some 0.8 to 1.0 metre in diameter and approximately 1.0 m in length. Packing densities are an order of magnitude lower than RO units (due to the larger diameter) and transmembrane pressures are limited to 2 to 3 bar, although 10 bar can be achieved in certain situations. Fibres are open at both ends with the tube-fibre bank sealed in epoxy blocks. Ultrafiltration capillary membranes are available in a wide range of polymers including polysulphone, polyacrylonitrile and chlorinated polyolefins.

Capillary UF designs, like RO modules, require feed pretreatment to remove particles greater than 50 micron in diameter because of a susceptibility to fouling. However fouling can be alleviated by a physical process called back flushing. In this the permeate flow is reversed and is thus used to dislodge the fouling material from the surface of the membrane. This procedure is applied, for only a few minutes, at regular prescribed intervals during operation to maintain flowrates within a required band. Occasionally other fluids such as air can be used. The procedure is also used for microfiltration. If physical methods like this or the foam ball method fail then recourse to chemical cleaning may be necessary. The chemicals adopted vary from application to application and may either dislodge or loosen the fouling material or dissolve it. Clearly the cleaning solution must be compatible with the membrane and not affect stability and selectivity. Ultrafiltration (and MF) membranes are generally more chemical resistant than RO membranes and often hot dilute acid or alkali can be used. Frequently mixtures of surfactants and sequestering agents are used to remove organic and microbial foulants. Sterilising solutions such as formaldehyde can also be used.

Chemical cleaning processes are often recommended by manufactures as many applications are well researched. Newer process applications will require a series of tests to determine the most effective and cheapest method.

Capillary modules are also used in certain MF applications and UF module manufacturers now produce such equipment. Fibre sizes tend to be somewhat greater than those for UF, starting at approximately 600 micron.

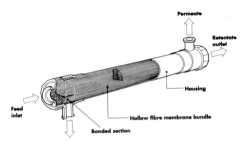

FIGURE 2 – Capillary membrane module.

Spiral Wound Modules

Spiral wound modules (see Figure 3) consist of a sandwich of flat sheet membranes, spacers and porous permeate flow material wrapped around a central permeate collecting tube. Feed solution passes axially along the sandwich in the channels formed by the spacers. This channel is of the order of 1.0 mm in thickness and the cross flow velocities used generally give laminar flow, although the spacer material may act as a turbulence promoter and thus reduce concentration polarisation. The permeate flows through the membrane in cross flow to the feed solution, that is radially inwards towards the central collecting tube. The diameter of this tube is small to maximise membrane specific areas, typically 600 m^2/m^3.

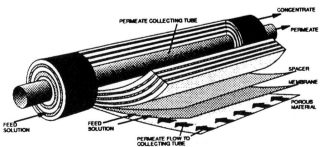

FIGURE 3 – Spiral wound module.

In practice two or more modules are fitted in series and suitably sealed into a pressure housing into which feed solution is introduced at one end and retentate collected at the other.

For RO applications modules are typically 1 m long and 0.2 m in diameter giving flowrates of up to 28 m^3 per day, equivalent to productivity of 900 $m^3/d/m^3$. Operating pressures are up to 40 bar and pressure losses are approximately 1 bar. Typical membrane materials are cellulose acetate and newer polyamides and as such impose the usual limitations associated with membrane stability regarding pH. In common with hollow fibre modules spiral wound modules cannot be mechanically cleaned and thus have a low tolerance to particulate material. This has therefore limited the application in UF although a few manufacturers have produced modules for UF in the food industry. Such modules are less compact than the RO counterparts due to the increased width of the feed channel (between membranes) to give greater tolerance to particulates.

Spiral wound modules have potential in PV and GP, although due to the currently limited number of commercial applications, module designs aimed specifically at these separations is at an early stage. Commercially available modules for GP are also available.

Tubular Modules

The typical tubular module is shown schematically in Figure 4.

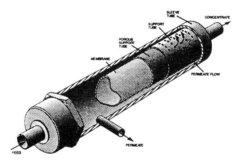

FIGURE 4 – Tubular membrane module.

The basic concept of a tubular module is a straight membrane tube surrounded by a porous support layer and support tube. Feed flows internally along the tube and permeate passes though the membrane into the porous support layer and through suitable holes in the support tube. Tube diameters are in the range of 1.2 to 2.4 cm and a number of tubes are placed in one pressure housing to increase module productivity. Systems resemble shell and tube heat exchangers.

Principle applications of tubular membranes are with feeds which cannot be pretreated to remove potential foulants and when very hygienic conditions are required. The relatively large tube diameters permit in situ mechanical cleaning methods. Cross flow velocities in tubular modules are usually large, generally giving turbulent flow, and thus operation with recycle is generally required to give reasonable water (permeate) recoveries. The high velocities also give rise to large pressure losses, up to 4 bar. Owing to the tube size specific membrane area is very much lower (< 100 m^2/m^3) than in spiral wound and fibre modules, and thus productivity is an order of magnitude lower, typically 100 $m^3/d/m^3$. Principle applications are in UF and MF. Developments in inorganic membrane materials in tubular form are of particular significance in UF applications. They enable aggressive conditions of cleaning to be used eg steam sterilisation, and generally extend the range of applications. The greater cost of inorganic materials will demand that operating life of the membranes must be much greater than for organic materials.

Plate Modules

Plate and frame modules using membranes in the form of flat sheets feature in two principle industrial applications, UF and ED. Because of the different driving forces used in the separations, individual designs are somewhat different and thus are considered separately.

Ultrafiltration (or RO)

In modules used for UF, the membrane sheets are stacked either one on top of the other or side by side. The sheets are either in the form of circular discs, elliptical sheets or rectangular plates and equipment design resemble plate heat exchangers or plate and frame filters. There is a variety of industrial equipment designs, but they all have the same basic components ie membranes separated by alternate porous permeate support plates and feed flow spacers. Applications of plate modules are in the same areas as tubular modules. The feed channels are often less than 1 mm and although more sensitive to fouling are easier to clean as no mesh support is used.

Typical modules with vertical elliptical membrane and vertical rectangular membrane have areas of 35 and 50 square meters respectively and are characterised by relatively easy membrane replacement. Productivity of these plate and frame modules is of the order of 80 $m^3/d/m^3$. An alternative design uses membrane cartridges (0.14 m square, 0.5 m long); eight in all installed in one pressure vessel.

For laboratory and pilot scale applications a wide range of equipment is currently available in plate form.

Electrodialysis

Electrodialysis is carried out in modules with vertically oriented membranes separated from one another by flow spacers. The module, or cell stack, consists of cell pairs (up to 300 to 500) comprising a cation selective membrane, a diluent flow spacer, an anion selective membrane and a concentrate flow spacer. The application of a potential field drives ions of positive charge through the cation selective membranes and ions of a negative charge through the anion selective membranes. Thus in the processing of electrolyte solutions two product streams result: one concentrated in ions and the other depleted in ions.

In addition to the cell pairs, stacks contain two electrodes and electrode compartments. Electrode rinse solutions pass through these compartments and serve as electrode depolarisers for the Faradaic reactions which inevitably occur as a consequence of the applied potential field. There are a wide range of ion exchange membrane manufactured with different properties, although a low electrical resistances is an important requirement

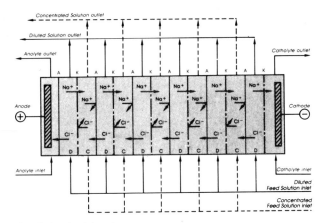

A : Anionic Membrane
K : Cationic Membrane
D : Diluted Compartment
C : Concentrated Compartment

FIGURE 5– Schematic diagram of electrodialysis module.

of a membrane, to minimise energy consumption. Equally important are requirements such as good chemical, mechanical and thermal stability and high permselectivity. Ion exchange membranes are manufactured as two general types:

- *Heterogeneous* – from a mixture of powdered ion exchange resin and polymer (polypropylene or polystyrene)
- *Homogeneous* – by block condensation/block polymerisation of polyelectrolyte.

The flow channel or spacer design in ED stacks is an important factor and must realise the essential design features below:

- small inter membrane distance to give low electrical resistance of electrolyte
- maintaining the cell dimensions
- uniform fluid flow distribution
- high mass transfer rates to minimise concentration polarisation

Commercial ED cell designs use spacers some 1 to 1.5 mm thick, which give low electrical resistance but only allow relatively small velocities to keep pressure losses low. To try to improve mass transport and to give uniform fluid flow spacers are used, aimed at inducing some degree of turbulence. Two types are generally used:

- *Tortuous flow path design*: flow starts at one corner of the membrane and follows a defined serpentine path across the membrane to the adjacent corner. This design gives good flow distribution and high mass transfer rates at the expense of high pressure drop.
- *Single flow path design*: flow starts at a number of points from one edge of the membrane and fluid traverses to exit ports at the opposite edge. Pressure losses in

ns design are much lower than in the tortuous path design at the same fluid flowrate although mass transport and flow distribution are not as good.

There are physical limitations to the stack size imposed by the requirements of flow distribution and low pressure drop etc. Individual membrane sizes tend to be in the range of 0.5 to 2.0 m^2 in area. Flow in alternate diluent and concentrate compartments is in opposite directions through appropriate internal flow ports. Operating current densities are typically in the range of 10 to 100 A/m^2.

Membrane function and permselectivity can be particularly affected by two factors, concentration polarisation and fouling. Concentration polarisation is a particularly crucial factor in the diluent compartment when concentrations of ions are reduced near to cell exit ports. Poor mass transport can cause mineral ion starvation at the membrane surface and result in water dissociation. This is identified with a change in local pH at the surface (and in the exit stream), an increase in electrical resistance and loss of current efficiency.

Fouling can be caused by any material which settles on the membrane surface and deactivates the ion diffusion ability. This may be from organic macromolecules present in the water or scaling due to salt concentration. Polarity reversal, in which current flow is reversed through the stack, can be used to alleviate such problems.

Pervaporation/Vapour Permeation

For PV, current industrial equipment is also available based on the plate and frame concept. This configuration is adopted primarily due to a requirement for a relatively large vapour flow channel to give the necessary control of the vapour partial pressure for separation. In comparison the feed side channels are narrow.

System Design

The selection of the membrane module is mainly determined by economic considerations. These economic considerations must account for all the cost factors and not just the cost of the module. The cheapest configuration is not always the best choice because the type of application, and thus the functionality of a module is also an important factor. The characteristics of all the modules which must be considered in a system design, include packing density, investment cost, fouling tendency, cleaning, operating costs and membrane replacement are compared qualitatively in Table 1.

The cost of the various modules varies appreciably, the tubular module is the most expensive per installed membrane area. Although the most expensive configuration, the tubular module is suited for applications with "a high fouling tendency" because of its ease of operation and of membrane cleaning. In contrast, the much smaller diameter hollow fibre modules are very susceptible to fouling and are difficult to clean. Pretreatment of the feed stream is therefore an important factor in hollow fibre systems.

Generally specific modules will tend to find their own field of application, although it is often possible to choose between two (or more) different types in specific applications. For example hollow fibre and spiral-wound modules in seawater desalination and gas separation.

TABLE 1 – Qualitative comparison of various membrane configurations.

	tubular	plate-and-frame	spiral-wound	capillary	hollow fiber
packing density	low	—————————————→			very high
investment cost/installed area	high	—————————————→			low
fouling tendency	low	—————————————→			very high
ease of cleaning	good	—————————————→			poor
operating cost	high	—————————————→			low
membrane replacement	yes/no	yes	no	no	no

The design of membrane systems vary significantly because of the different and varied scale of applications. The module is the focus of a membrane installation as it is the prime separation unit. When several modules are connected together in series or parallel, to give a specified performance, this is called a stage. The design of this staged arrangement of modules is carried out to give an optimum performance.

The simplest system design is the dead-end operation, as applied in MF, where all the feed is forced through the membrane. In this system the concentration of rejected components in the feed or region above the membrane increases with time and consequently there is a risk to the quality of the permeate with time. This concept is used very frequently where a complete filtration of all the feed is required, e.g. sterilisation, and so there is minimal loss of the product liquid permeate.

For many industrial applications, a cross-flow operation is often preferred because of the lower fouling tendency compared to the dead-end mode. In the cross-flow operation, the feed flows across the membrane surface and the feed composition inside the module varies with distance in the module. The feed stream is separated into two streams: a permeate stream and a retentate stream. Flux decline is relatively small with cross-flow,

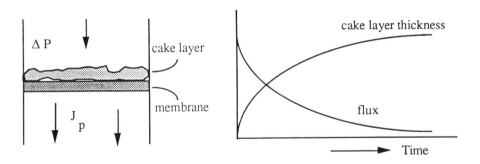

FIGURE 6 – Flux decline in dead-end filtration.

INTRODUCTION TO MEMBRANE SEPARATIONS

in comparison the dead end mode of operation, and can be controlled and adjusted by good module choice and suitable pressure differentials and cross-flow velocities. In dead-end filtration, the cake grows with time and consequently the flux decreases with time.

Cross-flow operations

Membrane modules operate with cross flow of feed as either a necessity for continuous operation or to reduce concentration polarisation and fouling as far as possible. For a given module design and feed solution, the cross-flow velocity is the main parameter that determines mass transfer in the module. Various cross-flow operations are possible:
- co-current -feed/retentate flow in parallel with the permeate
- counter-current- feed/retentate flow in opposite direction to permeate.
- cross-flow of feed with perfect permeate mixing
- perfect mixing of both feed and permeate

In the cross-flow mode with perfect permeate mixing , it is assumed that plug flow occurs on the feed side whereas mixing occurs so rapidly on the permeate side that the composition remains the same. With the cross-flow operations, counter-current flow generally gives the best performance followed by cross-flow and co-current flow, respectively. The poorest performance is obtained with perfect mixing as the composition is constant in the module and equal to that of the exit stream. In practice, systems often operate in the cross-flow mode with perfect mixing of permeate.

The arrangement of flow in and around the module is one of the principal variables determining the separation achieved. Three basic methods can be used in a single-stage or a multi-stage process: i) the "single-pass system" ii) the "recirculation system" and iii) a batch system. Batch systems are used for certain small-scale applications. A typical batch system will operate with a recycle of feed through the module, as shown in Fig 7. The concentration in the feed thus gradually increases with time, while the permeate is drawn off. The single-pass and recirculation systems are shown schematically in Fig 8.

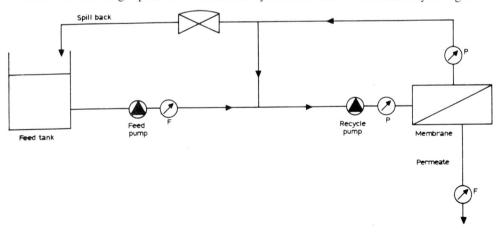

FIGURE 7 – Schematic diagram of a batch system.

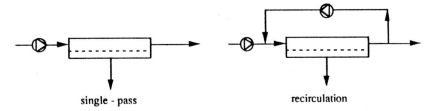

single - pass   recirculation

FIGURE 8 – Schematic representation of the single-pass and recirculation systems.

Cascade Operation

In single-pass operation the feed passes only once through an individual module, ie there is no recirculation and hence the volume of the feed decreases with path length. To achieve practical operation to accomodate the loss in volume and to achieve high separation yields a multi-stage single-pass is used with the modules in a "tapered design" as shown in figure 9. In this arrangement the cross-flow velocity is almost constant in each and the total path length and the pressure drop are large. The volume reduction factor, ie the ratio between the initial feed volume and the volume of the retentate, is determined mainly be the configuration of the modules.

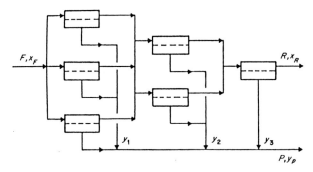

FIGURE 9 – Single-pass system (tapered cascade or "Christmas tree".

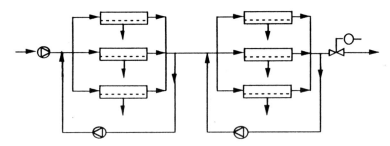

FIGURE 10 – Recirculation system.

In the recirculation system or "feed recycle system" (see figure 10), the feed passes several times though one stage, consisting of one or more modules. Recycle in each stage is via a recirculation pump which maximises the hydrodynamic conditions whilst the pressure drop over each single stage is relatively low. The feed recycle system offers greater flexibility than the single-pass system and is often preferred when severe fouling and concentration polarisation are likely to occur in for example microfiltration and ultrafiltration. A disadvantage is that the feed becomes more concentrated with an increase in the amount of recycling. However, with relatively simple applications such as the desalination of seawater the single-pass system is more than suitable as the retentate water is of little value.

As noted above a single-stage does not always give the desired product quality and consequently the retentate or permeate stream must be treated in a second stage. This is achieved in a cascade operation, employing more than one unit, where the permeate (or retentate) of the first stage is the feed of the second stage and so on. An example of a two-stage operation process is given in Figure 11 where permeate purity is not sufficient after one separation stage. The system design depends on whether the permeate or the retentate is the desired product and the degree of purification or separation desired. It is possible to achieve a high product separation with a large number of units although the control and optimisation of the system is more complex due to the large number of variables involved.

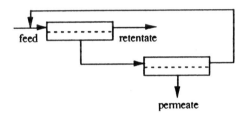

FIGURE 11 – Two-stage membrane module arrangement.

Examples of three-stage processes in which the permeate is recycled are common in the separation of natural gas (CO_2/CH_4 separation).

Characteristics of Membrane Modules

Module design characteristics clearly vary according to application, membrane material, manufacturer, scale of operation, etc., and comparisons are therefore often difficult. Nevertheless it is instructive to appreciate pertinent characteristics in specific cases as this will lead to an understanding of why a particular module design is used in a particular application. For example fine fibre membrane modules are used for RO and GP because they withstand high pressures and particulates, etc., which would block the fibres, are excluded. Consequently fine fibres are never considered for MF although other module designs are routinely used.

To illustrate the different characteristics Table 2, compares tubular, hollow fibre, plate and spiral wound modules for ultrafiltration. In certain membrane separations there is a move towards a particular type of membrane configuration. In the case of microfiltration in bio-processing this move appears to be towards hollow fibre and tube modules.

Cross flow Microfiltration in Bio-processing

Tangential flow microfiltration is now an established bio-process technology for cell harvesting, cell washing and lysate clarification. Historically, flat sheet microporous membranes configured in a plate and frame geometry were utilised as a result of the availability of flat sheet dead-end disc filtration. The varied technical success of these units, combined with the relatively high capital an operating costs, have increased interest in alternative microfiltration membrane geometries in hollow fibre, tubule/tubular and spiral wound configurations. Although module designs vary from manufacturer to manufacturer, the points to note relative to each configuration are listed in table 2.

As advancements in fermentation result in sustantially higher cell densities the membrane module must have open feed channels. Thus, screened flat sheet devices and spiral wound modules, both of which are vulnerable to plugging and difficult to clean and sanitise, are not suitable for cell processing. Tubular devices (> 6 mm ID) are typically viewed as an unfavourable option due to their high pumping requirements, high capital costs and inability to be bubble point tested. Overall, the options for bioprocessing are, open channel plate and frame, hollow fibres and tubules.

With the introduction of microporous hollow fibres and tubules, the choice of configuration has become a significant issue in terms of membrane performance, ease of operation, cleaning and sanitisation and economics.

TABLE 2 – Comparison of membrane modules for ultrafiltration of bioproducts.

Parameter	Spiral Wound	Screened Plate & Frame	Open Plate & Frame	Hollow Fiber	Tubule	Tubular
Fully Assembled, Self-Contained Modules	No	No	No	Yes	Yes	Yes
Factory QC of Entire Membrane Assembly	No	No	No	Yes	Yes	Yes
Simple and Reliable Integrity Confirmation	No	No	No	Yes	Yes	Some
Open Feed Channels with no Screens to Plug	No	No	Yes	Yes	Yes	Yes
Self-Supporting, not Prone to Particle Shedding	No	No	No	Yes	Yes	Some
No "Dead" or Stagnation Spots	No	No	No	Yes	Yes	Yes
Compact, Lightweight Design	Yes	No	No	Yes	Yes	No
Translucent Module-Visible Membranes	No	No	No	Yes	Yes	Some
No Bulky Housings or Holders Required	No	No	No	Yes	Yes	Some
Backflushable Design	No	No	No	Yes	Yes	Some
Cost Effective	Yes	No	No	Yes	Yes	No
Integral Housing Minimizes Contact with Process/Cleaning Solutions	No	No	No	Yes	Yes	Some
Simple, Quick Membrane Replacement/Addition	Yes	No	No	Yes	Yes	Some
Excellent Scaleability from Lab to Production	No	No	No	Yes	Yes	No

TABLE 3 – General Comparison of Characteristics of Membranes.

Tubular Membranes

1. Tubular modules have relatively large channel diameters, and are capable of handling feed streams and slurries containing fairly large particles. The general rule of thumb is that the largest particle that can be processed in a membrane module should be less than one-tenth the channel height. Thus feed streams containing particles as large as 1250 m can be processed in 1.25 mm tubular units.

2. Tubular units of 1.25 to 2.5 cm diameters are operated under turbulent flow conditions with recommended velocities of 2-6 meters per second. Flow rates are 15-60 litres per minute per tube, depending on the tube diameter. Reynolds Numbers are usually greater than 10,000.

3. Pressure drop averages 2-3 psi per 8-12 foot tube. Thus, typical pressure drops for 0.5-1.0" tubes will be approximately 30-40 psig (2-2.5 atm) for UF units operating in parallel flow under these flow conditions. This combination of pressure drop and high flow rates gives high energy consumption

3. The open tube design and the high Reynolds Numbers make it easy to clean by standard clean-in-place techniques. It is also possible to insert scouring balls or rods to help clean the membrane.

4. Tubular units have the lowest surface area to volume ratio of all module configurations..

In certain modules the individual membranes can be replaced fairly easily in plant resulting in considerable savings in transportaion costs and membrane costs

Tubular module costs vary widely from about $90-700 per m2 for replacement membranes of cellulose acetate, polysulphone or composites.

Hollow Fibre Modules

1. The recommended operating velocity in the UF hollow fibre system is around 0.5-2.5 m/sec. This results in Reynolds Numbers of 500-3000. Hollow fibres thus operate in the laminar flow region.

2. Shear rates are relatively high in hollow fibres due to the combination of thin channels and high velocity. Shear rates at the wall mare 4000-14,000 sec^{-1}.

Hollow fibres have the highest surface area-to-volume ration. Hold-up volume is low, typically 0.5 litres in a typical "short" 3" cartridge of 1.4-1.7 m^2 membrane area.

4. Pressure drops are tpically 0.3 to 1.3 bar depending on the flow rate. The combination of modest pressure drop and flow rates make hollow fibre modules very economic in energy consumption.

5. Hollow fibres have only a modest maximum pressure rating of about 1.8 bar. The short-short (30 cm) cartridge can withstand pressures up to about 2.4 bar at low temperatures (less than 30°C). Sseveral process streams that are dilute enough to permit UF operation at pressures much higher than the present 1.7 bar limiting transmembrane pressure. In addition, since the flow rate is proportional to pressure drop, flow rates are limited since the inlet pressure cannot exceed 1.7 bar. This can be problematic with highly viscous solutions, especially withe long cartridges

The small fibre diameters make them susceptible to plugging at the cartridge inlet. To prevent this the feed should be prefiltered to at least 100 m.

Hollow fibres are suitable for "back-flushing" because the fibres are self-supporting. This vastly improves performance due to cleaning in situe potential its cleanability.

9. Replacement membrane costs are relatively high. Deamage to one single fibre out of the 50-3000 in a bundle generally means the entire cartridge has to be replaced However it is possible to repiar membrane fibres in situe in certain cases.

The cost is about $600 per 7.5 cm industrial cartridge, regardless of surface area. Replacement cost is about $200-300 per square metre.

TABLE 3 (continued).

Plate

1. The typical plate channel height is between 0.5 and 1.0 mm. UF systems operate under laminar-flow, high shear conditions. The channel length (the distance between the inlet and outlet posts) is between 6-60 cm. The Grober equation agrees reasonably well with experimental in the Reynolds Number range of 100-3000 for slits of channel height 0.4-1.0 mm

2. The permeate from each pair of membranes can be visually observed in the plastic tubing coming from each support plate. This is convenient for several reasons, eg detection of leaks in a particular membrane pair, if samples need to be take for analysis, or if flux measurement as a function of capacity needs to be made.

Replacement of membranes on site is relatively easy provided that care is taken when closing the stack of plates together The previously embedded grooves of the unreplaced plates match exactly as they were previously, or else leakage of feed can occur.

In horizontal modules, the flow is parallel through all channels at velocities of about 2 m/sec. For a stack of 30 plates, this can result in a pressure drop of about 10 bar. Plate-and-frame systems tend to be intermediate between spiral-wound and tubular systems in energy consumption for recirculation

Membranes are currently sold in USA for about $100 per square metre for cellulose acetate, $200/m^2 for non-cellulosic RO membranes, and $120/m^2 for polysulfone membranes.

Surface area-to-volume ration is fairly high, averaging about 200-300 ft^2/ft^3.

Spiral Wound

1. In spiral-wound modules the feed channel height is controlled essentially by the thickness of the mesh-like spacer in the feed channel. Spacers of 0.76 mm or 1.1 mm are most common. The advantage of a narrow channel height is that much more membrane area can be packed into a given pressure vessel

2. A larger channel height, while reducing the surface area-to-volume ratio slightly, may be more desirable to minimise pressure drops and reduce feed channel plugging. The general rule of prefiltering to one-tenth the channel height is modified for the spiral-wound unit due to the presence of the spacer which reduces the free volume in the channel. Prefiltration of the feed down to 5-25 m is recommended for the o.76mm spacer-module, and 25-50 m for the 1.00 channel.

Lengths of individual membrane assemblies vary from 1 foot to 6 feet. When calculating the surface area of a spiral-wound membrane, it is convenient to consider it as two flat-sheets, although the *effective* membrane area of spiral-wound modules must allow for gluing the membrane sandwich, for fixing the fourth side to the permeate collection tube and the outer periphery

4. The hydrodynamics in the spiral-wound module is not too clear. The velocity in spiral-wound units range from 10 to 60 cm/sec, being higher for the large mesh spacers. These are "superficial" velocities, however, since the volume occupied by the mesh-like spacer in the feed channel is neglected. These velocities correspond to Reynolds Numbers of 100-1300. Technically, this is in the laminar flow region, but the additional turbulence contributed by the spacers means that the flow is in the turbulent region..

Surface area-to-volume ration is fairly high, averaging about 200-300ft^2/ft^3.

TABLE 3 (continued).

Pressure drops in the feed channel are relatively high due to the effect the spacer. At a superficial velocity of 25 cm/sec the pressure drop is around 1-1.4 bar. This high pressure drop can give rise to a "telescoping" effect at high flow rates, ie the spiral pushes itself out in the direction of flow. This can damage the membrane and so anti-telescoping devices are used at the downstream end of the membrane element to prevent this.

The combination of low flow rates, pressure drops, and relatively high turbulence makes this an economic module in terms of power consumption. .A problem with the mesh spacers is the creation of "dead" spots directly behind the mesh in the flow path. This may cause particles to "hang up" in the mesh network, resulting in chelaning problems. This makes it difficult to process feeds containing suspended particles, especially if it is a concentrated slurry and a high recovery of the particles is required. Spiral modules work best on relatively clean feed streams with a minimum of suspended matter.

Capital costs are also quite low. The membrane element can be recovered from the pressure vessel and returned to the factory for reassembling new membranes. Replacement membranes are priced at typically $30-120 per square meter for cellulose acetate, vinylidine fluoride, and polysulfone membranes.

Plate and frame modules are made up of a series of membrane "packets" which are stacked between heavy support plates. Large, sometimes tortuous path o-rings and gaskets must be positioned by the operator to assemble the stack prior to torquing the module to proper compression. Quality control of the entire plate and frame membrane module must be made on-site as incorrect assembly may lead to leakage of biologicals during operation.

Unlike plate and frame membrane stacks, hollow fibre and tubule cartridges are available in fully assembled, compact, easy to connect cartridge housings which are 100% integrity tested at the factory. Furthermore, since the hollow fibre and tubule membranes are provided in translucent housings, they offer rapid, visual confirmation of integrity.

Inevitably membranes will require periodic replacement. With plate and frame devices, operators need to disassemble and then reassemble the stack. Viruses and toxogenic materials must be completely inactivated to prevent operator exposure to live strains. The self-contained hollow fibre and tubule cartridges are easily changed by one operator without exposure to the feed liquid.

Hollow fibres and tubules are self-supporting geometries which have fewer components in their modular form than do plate and frame or cassette modules and do not require support webbing of the membrane or screens for the permeate collection channel. The feed channels of hollow fibre and tubule membranes are straight cylindrical paths readily contacted by the cleaning fluid while, plate and frame modules may have areas of restricted flow where reduced cleaning solution velocity occurs and may even have "dead" or stagnation spots where cleaning and sanitisation is difficult.

All three membrane configurations are available in autoclavable and steam-in-place versions. When hollow fibre and tubule modules are steamed, they can be secured in a stainless steel housing for operator safety.

Microfiltration membranes operate at low transmembrane pressures and therefore, as a first approximation, permeate flux on a membrane area basis are equivalent for any configuration used. On this basis, capital and operation cost comparisons both favour the hollow fibre and tubule geometries over the plate and frame option. This move toward hollow fibre technology becomes greater as the process flux differential increases. Operator labour would also be reduced during *in-situ* integrity testing and at membrane change-out. Open channel, hollow fibre and tubule microfiltration membrane geometries are the state-of the art choice for fermentation cell broth processing. Hollow fibre and tubule membrane cartridges are sanitary, compact, fully assembled and factory quality controlled. These devices are more operator friendly than flat sheet configurations and offer significantly lower installed capital costs as well as economical membrane replacement costs.

SECTION 1.6 – MEMBRANE PROCESSES EQUIPMENT

Microfiltration

Filtration is a process of physically removing suspended matter from a liquid or a gas by forcing the material through a porous mechanical barrier, the filter.

A pressure gradient is maintained across the filter to maintain fluid flow through the filtration media. This pressure may arise from gravitational forces, centrifugal forces, pneumatic or hydraulic mechanical forces. The resultant filtrate or permeate flowing through the filter should ideally be devoid of suspended material, at least of a specified size.

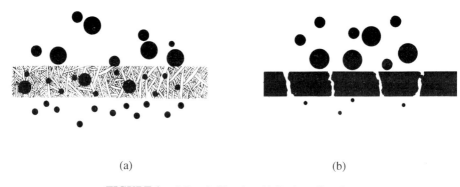

FIGURE 1 – a) Depth filtration. b) Surface filtration.

The action of the filter facilitates the extraction and analysis of the material separated from the fluid or gas. The efficiency of the filtration process is defined as the ability of the media to distinguish between particles of different specific sizes. It is a function of the characteristics of the filter media, properties of the fluid or gas, and the operating conditions.

There are generally two types of filtration media: depth filters and membrane filters.

Depth Filters are fibrous materials which provide a series of channels with no clearly

INTRODUCTION TO MEMBRANE SEPARATIONS

defined or regular pore structure. As the fluid passes through a depth filter, particulate matter is randomly trapped within the body of the media. Depth filters are often used as prefilters or in coarse filtration and clarification techniques where the tight specification of pore sizes is not necessary. They are less efficient than membrane filters, but their thicker construction and higher porosity result in higher flow rates and dirt loading capacity. These characteristics are highly desirable in fluid filtration systems containing expensive fine filters such as those used in the electronics and pharmaceutical industries. A typical fibre used is a microfiber glass depth filter available in three forms: pure glass fibre, and containing either an acrylic or a cellulosic binder. The binder increases the wet strength of the fiber mat. Because of these physical properties and their lower cost, depth filters are also often used as prefilters to protect finer membrane filters used in microfiltration, ultrafiltration and reverse osmosis.

Membrane filters are routinely used in applications for fluid sterilization which demand the ability of the membrane to separate suspended matter of a defined size out of a fluid. They are manufactured from a variety of polymeric materials using different processes including solvent casting, stretching, or nuclear particle track-etching. The resulting membranes are thin, strong microporous materials with well-defined pore size, pore structure, pore density, bubble point and strength characteristics. The inherent advantage of a membrane filter lies in the precision with which such characteristics can be defined and tailored to the application and process conditions.

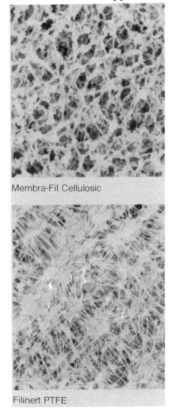

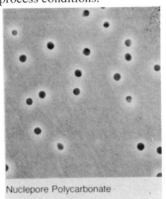

TABLE 1 – Comparison of Depth and Membrane (Sieve) Filtration.

	Depth	Membrane
Separation Method	Process that traps contaminants both within the matrix and on the surface of the filter media.	Process that traps contaminants larger than the pore size on the top surface of the membrane. Contaminants smaller than the pore size may pass through the membrane or may be captured within the membrane by some other mechanism.
Material Type	*Depth filters* are composed of random mats of metallic, polymeric, or inorganic materials. These filters rely on the density and thickness of the mats to trap particles, and generally retain large quantities or contaminants within the matrices. Media migration, which is the shifting of the filter medium under stress, and particulate unloading, however, can be common problems.	*Membrane filters* are generally polymeric films approximately 120 mm thick with a narrow pore size distribution. Membranes normally capture particles on their upstream surface.
Advantages	Lower cost High throughputs High dirt-holding capacity Protects final filters Removes variety of particle sizes	Absolute sub-micro pore size ratings possible Bacterial and particle retentive Low extractables Integrity testable
Disadvantages	Media migration Nominal pore size Particulate unloading	Lower flow rates than depth media More costly than depth media

In membrane microfiltration (MF) the filter is generally thin with a uniform pore size and a high pore density of approximately 80%. The principle method of particle retention is characterised as sieving although the separation is influenced by interactions between the membrane surface and the solution. The high pore densities of the filters coupled with the thin material generally mean high flux rates, expressed as cubic metres of permeate per square meter of membrane area per hour (m/h or m/s), result at modest operating differential pressures up to 2 bar.

The irregular nature of the pores of the membrane and the often irregular shape of the particles being filtered mean there is not a sharp cut off size during filtration. With symmetric membranes some degree of in-depth separation could occur as particles move through the tortuous flow path. To counter-act this effect, asymmetric membranes, which have surface pore sizes much less than those in the bulk of the membrane, have been introduced. These entrap the particles almost exclusively at the surface layers (the membrane skin) whilst still offering low hydrodynamic resistance.

Microfiltration is operated either in a dead-end mode or a cross flow mode. Microfiltration is most widely applied in a dead-end mode of operation where the feed flow is perpendicular to the membrane surface and the retained (filtered) particles accumulate on the surface forming a filter cake. The thickness of this cake therefore increases with time and the permeation rate correspondingly decreases. Eventually the membrane filter reaches an impractical or uneconomic low filtration rate and is either cleaned or replaced. Typical filters come in the form of readily replaceable screw-in cartridges.

To reduce the effect of the build-up of solid particle cake on the membrane surface an alternative cross flow operation of filtration can be used. A cross flow separation can be seen as a process in which a feed stream flows along, almost parallel to, a membrane surface through the membrane. The shearing effect of the feed flowing tangentially along the membrane surface reduce the build-up of potentially fouling layers on the membrane, reduces concentration polarisation, adsorption and cake layer formation. Cross flow velocities of several meters per second can be used in practice to minimise the impact of the accumulation of particulate material. Cross flow filtration is frequently used to process fouling feeds. Even so there is often a decline in fluxrate during continued operation because the microfiltration membranes will suffer from the problem of fouling, ie the accumulation of material at the surface of the membrane. Procedures are usually required to clean and sterilise the membrane, which must be able to withstand the associated mechanical, chemical and thermal stresses. Thus as well as organic polymers, inorganic materials such as ceramics, carbon, metals and glass are used for microfiltration membranes. Inorganic membranes are generally thicker than organic and hence are asymmetric in structure.

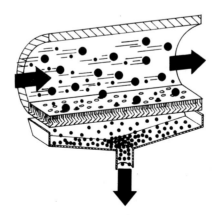

FIGURE 2 – Cross flow operation of microfiltration.

The selection of an appropriate membrane is crucial factor in microfiltration, adsorption phenomena can play an important role in fouling. Hydrophobic membranes (eg PTFE) generally show a greater tendency to foul, especially in the case of proteins. Another disadvantage of hydrophobic membranes is that water will not generally flow through the structure at low pressure unless they are pretreated prior to use with for example ethanol. Generally at some stage microfiltration membranes will require cleaning, typically using appropriate chemicals and thus the material must exhibit appropriate chemical resistance to reagents. In the operation of microfiltration there have been several methods used to reduce the influence of polarisation and fouling which include vortex and pulsatile flows, ultrasonics, vibrational shear and electric fields. One particular interesting filter uses a flexible, collapsible membrane, which after build up on filter cake, is squeezed between rollers to remove the deposited cake.

Microfiltration is employed in both production and analytical applications. The technologically important applications are summarised as:
- removal of particles from liquid and gas streams for chemical, biological pharmaceutical and food industries
- clarification and sterile filtration of heat sensitive solutions and beverages
- production of pure water in the electronics industry
- product purification, gas filtration, process solvent recovery in the chemical industry
- waste water treatment

Factors in Filtration

The performance of microfiltration membrane is affected by several factors associated with the membrane medium, the particulate material and the fluid carrier phase. The selection of a particular material and conditions of operation will be made with an understanding of these factors and their interactions. As with all separations, there will be several possible solutions to the problem, although frequently one solution is better than the others. This selection or optimum system will trade-off the various characteristics of the alternatives. These characteristics relate to the particulate (size, shape, concentration, distribution, zeta potential, viable or non viable) filter medium structure (pore size, rating, asymmetry) filter medium stability (chemical, mechanical, thermal, hydrolytic, extractables, shedding) retention mechanisms, absorption, adsorption, impingement, cake retention (see Table 2).

The selection of the membrane is critical in any separation. The material will determine relative stability criteria and some retention characteristics. The Fabrication method will determine its structure and also retention characteristics. Fig 3 shows the different morphologies possible in microfiltration membranes. Sintered membranes will generally have a low porosity and variable pore size distribution. Stretched membranes will have a relatively high porosity and a wide pore size distribution. Track etched membranes have a low porosity and narrow size distribution. Phase inversion produces membranes with a high porosity and narrow size distribution. In the latter case the characteristics relate to the underlying support structure and surface layer respectively. Symmetric membranes, apart

INTRODUCTION TO MEMBRANE SEPARATIONS

TABLE 2 – Operating Factors in Microfiltration.

Particulate
Shape – The contour of the material whether near spherical or rod shaped will have a bearing on which is the critical size for separation. Particles may be rigid or soft or liquid and thus deform upon filtration to affect separation behaviour.
Concentration – Amount of contaminant in the fluid per unit volume. This can be expressed as: Number of particles/fluid volume Particle weight/fluid volume – mg/L or ppm Particle volume/fluid volume – % v/v Particle weight/fluid weight – % w/w
Distribution – The frequency of occurence of a contaminant in a fluid with a respect to a certain property such as size. Distribution must be considered when selecting filter media type. It is a factor in whether the medium should be nominal or absolute rated, in whether prefiltration is called for, in whether the system design should be parallel or series, and in whether to expect particulate loading.
Zeta potential – Electrical charge potential across the interface of the contaminant and the filter media in liquid filtration. Most particles carry a charge while most filters are naturally negative positively charged (naturally or artifically boosted) to enhance adsorption and improve retention efficiency.
Viable – Contaminants that are living organisms, such as bacteria, viruses, moulds, and fungi.
Non-viable – Contaminants that are not alive, such as manufacturing debris, rust scale, sand, and both dead bacteria and viruses.

Filter Structure
Pore size _ Filter medium is identified by the diameter of the particle that it can be expected to retain to some degree of efficiency. Pore sizes are usually stated in micrometers (µm).
Morphology – Is the pore structure uniform, foam or bubble like, symmetric, asymmetric, with a top layer.
Ratings – Membrane ratings refer to the size of a specific particle or organism retained by the filter to a specific degree of efficiency. *Absolute:* Under strictly defined conditions, the test filter medium retains 100 percent of the challenge organism or particle. Among the conditions that must be specified are: test organisms (or particle), challenge pressure, concentration, and detection method used to identify contaminant. *Nominal:* Describes the ability of the filter medium to retain the majority of particulates at the greater than the rated size. Process conditions such as operating pressure and concentration of contaminant have a significant affect on the retention efficiency of the filter.

Continued

TABLE 2 (continued)

Filter Stability
Thermal Stability – Ability of the filter medium to maintain its integrity and funtionality at an elevated temperature and pressure. Stability is measured by determining the maximum operating temperature and pressure under specified conditions. This indicates whether the media can be *in situ* (in-;ine) steamed, autoclaved or hot-water sanitised – or whether it can be used in a process with elevated temperatures. **Chemical Resistance** – A measure of the filter's ability to endure exposure to a chemical agent without shedding and without adding extractables. Exposure time, temperature, pressure, and concentration are important factors in determining chemical resistance. **Mechanical Strength** – Ability of the filter medium to maintain its integrity under stress during installation, sterilisation, and dynamic-use conditions. Sometimes a filter medium is sandwiched between layers of support material or cast on a substrate during manufacture to increase its mechanical strength. Support materials are often woven or non-woven synthetic fabrics. **Shedding** – Particulate matter can be flushed from the filter media during filtration and contaminate the filtered fluid. The lower the incidence of particle shedding, the better. **Extractables** – Substances present in the composition of the filter media or the filter manufacturing process that may be leached into the fluid as it is filtered, thereby affecting its purity. Extractables may include manufacturing debris, surfactants, and adhesives. Extractables can be minimised with sufficient preflushing. **Hydrophilic** – Having an affinity for water. A hydrophilic filter can be wetted with virtually any liquid. **Hydrophobic** – Repels water. Hydrophobic filters cannot be wetted by aqueous solutions or by liquids with high surface tensions without first prewetting with agents at low surface tensions, such as alcohol, or subjecting them to high pressure. Hydrophobic filters are best suited for gas filtration and venting. **Hydrolytic Stability** – Ability of the filter media to maintain integrity when exposed to aqueous solutions under specified conditions. Hydrolysis is a chemical reaction in which a compound decomposes by splitting a bond and adding H^+ and OH^- ions. Hydrolysis of filter media can cause filter degradation and loss of integrity. **Biosafety Tests** – Series of tests performed on substances that come in contact with solutions which then come in contact with body fluids. Tests such as the United States Pharmacopoeia (USP) XXI, Class VI Plastics Test – as well as other internationally recognised tests – assure that there will be no adverse reactions to fluids exposed to the materials used in the construction of a filter. **Pyrogenicity** – Property of a substance that, when injected, causes a rise in body temperature. Filtration materials that come in contact with injectable fluids must be non-pyrogenic. Pyrogenicity can be determined by such standard tests as the Limulus Ameobocyte Lysate (LAL) test.

TABLE 2 (continued)

Retention
Absorption – Process by which a contaminated fluid penetrates and occupies the inner structure of the filter media. In filtration, the pores or interstices soak up the fluid and entrap the contaminants, similar to a sponge.
Adsorption – Process by which contaminants are collected on the surface of the filter media due to a cariety of molecular interactions. The pore walls of the filter capture particles as they encounter and adhere to its surface. The adherence is caused by electrostatic attractions, van der Waals forces, and other interactions. These forces can be affected by pH, ionic strength, flow rates, pressure, and in gas filtration processes, relative humidity.
Impingement or Inertial Impact – Capture by virtue of collision or impact on or within the filter. This is a filtration method generally used for recapturing large mass particles, since the larger the particle, the greater its resistance to changing direction, and the more likely it is to strike the filter's surface. The momentum and velocity of the particles are extremely important in gaseous streams for effective retention.
Sieve Retention – A mechanical retention method also known as straining or screening. In filtration, a particle is retained when its dimensions are larger than the size of the filter pores.
Cake Retention – Another mechanical capture method which relies on a bridging effect caused by partial blockage of pores on the filter surface. As particles begin to build up on this bridge, they form a cake which restricts flow through the filter. The cake functions as a dynamic filter and improves filter efficiency.
Binding – Property of a filter to interact with particular substances. Binding can be a desirable characteristics, as in the case of nucleic acid or protein binding on certain microporous membranes, or an undesirable characteristics, as in the case of protein binding during filtration of protein solutions which causes a loss of active ingredients during filtration.
Dirt Loading/Unloading – *Loading:* Ability of the filter media to capture particulate matter and retain it. *Unloading:* Release of captured particulate matter. This can be caused by excessive differential pressure or a change in flow across the media. Unloading is highly undesirable.

from tracketched will tend to offer depth filtration characteristics. The membrane surface will tend not to be smooth which is not beneficial to cross flow filtration. Asymetric membranes, with a relatively dense top layer, and an open support structure are well suited to cross flow microfiltration.

A membrane which is hydrophilic (eg PTFE) will have a greater tendency to foul, especially by proteins. Hydrophilic membranes are not wetted by water and will not pass water through them under normal operating conditions. They require wetting with, for

SYMMETRIC MICROFILTRATION MEMBRANES

pore structure	straight pores	soap bubble-like (foam-like)	coral-like (tortuous)	stretched
production technique	track-etching/ anodising processes	casting + leaching/evaporation		film-stretching

ASYMMETRIC MICROFILTRATION MEMBRANES

pore str.	finger-like substructure sieve-like toplayer	foam-like substr. nodular toplayer	double toplayer	sintered ceramic spheres
production technique	phase inversion	phase inversion	phase inversion	sintering/ slip casting

FIGURE 3 – Morphology of microfiltration membranes.

example, alcohol prior to filtration of water based solutions. They are consequently good filtration media for gases. Hydrophilic membranes commonly used are made from polytetrafluoroethylene, poly vinylidenefluoride, polypropylene. Hydrophilic membranes commonly encountered are cellulose ester, polycarbonate, polysulfone and poly-

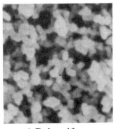
a) Polysulfone.

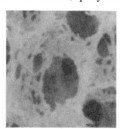
b) Nylon.

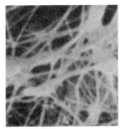
c) PTFE.

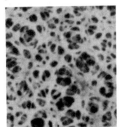
d) Acrylic.

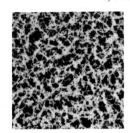
e) Polypropylene.

Membranes structure.

INTRODUCTION TO MEMBRANE SEPARATIONS

TABLE 3 – Typical characteristics and applications of microfiltration membranes.

Material	Characteristics	Typical Applications
Polysulfone	An inherently hydrophilic polysulfone membrane with excellent flow rates, low extractables, broad chemical compatibility, high mechanical strength and temperature resistance.	Food and beverages Pharmaceuticals Semiconductor water Serum
Nylon	Hydrophilic membrane with high tensile strength. Very high flow rates, long life, and low extractables. Offers excellent chemical compatibility.	Semiconductor water Chemicals Beverages.
PTFE	Naturally hydrophobic membrane laminated to a polypropylene support for extra durability and strength. Superior chemical and temperature resistance.	Air and gases Pharmaceuticals Aggressive chemicals
Acrylic Copolymer	Inherently hydrophilic copolymer with strong nonwoven polyester fabric support. Offers high flow rates, low differential pressures, and low extractables.	Semiconductor water Pharmaceuticals Food and Beverages
Polypropylene	Naturally hydrophobic membrane and chemically inert. Broad pH stability, high temperature resistance and high flow rates	Chemicals Microelectronics Pharmaceuticals
Glass	Nominal 1.0 mm fine borosilicate glass fibre. High flow rates at differential pressures. Good wet strength and high dirt-holding capacity.	Chemicals Serum Beverages
Polycarbonate	Hydrophilic membrane unique pore structure and capture, strong, flexible, high flow rate, thermal stability, non extractable.	Pharmoceuticals. Air pollution. Laboratory analysis.
Cellulose	Hydrophilic membrane General purpose, with limited thermal and mechanical stability. Some extractables.	Air pollution Microbiology Foods and Pharmaceutical

ethersulfone, nylon and polyimide. More recently ceramic membranes mainly alumina (Al_2O_3) and zirconia (ZrO_2) have become routinely used in more demanding applications. Membranes from glass, carbon and metals are used for special applications. Typical characteristics and applications of some of these membrane in microfiltration, particularly dead-end are summarised in Table 3. Further details of characteristics and applications can be found in the appendix.

Membrane Performance Characteristics

Membrane Performance characteristics vary considerably from manufacturer to manufacturer and vary due to material type, method of manufacture and physical structure, and application. It is useful, however, to illustrate some typical behaviour.

The important performance indicator in microfiltration is the volume flow through the membrane, which is defined in terms of the membrane Flux J. The flux J for a clean system is directly proportional to the applied pressure difference across the membran ΔP.

$$J = K \Delta P/\mu$$

Where K is the permeability constant, which is a function of membrane material and structure and μ is the fluid viscosity.

For most membrane materials flux values are quoted for particular conditions of temperature, applied pressure and for specific Fluids. Fluids frequently used are water, air and methanol. Typical values of fluxrates or flowrates are given in Table 4 for the membrane materials, polycarborate (track etched), PTFE (stretched) and mixed cellulose esters. Values of Flux clearly decrease with pore size and are typically; for water with a 0.2 filter, 15-20 ml\min\cm² and for air with 1 µm membrane, 7-30 l/min/ cm²

TABLE 4a – Microfiltration membrane performance characteristics.

Rated Pore Size (µm)	Rated Pore Density (pores/cm²)	Nominal Wt. (mg/cm²)	Nominal Thickness (µm)	Bubble Point[3]		Typical Flow Rates	
				(psi)	(bar)	Water[1] (ml/min/cm²)	Air (l/min/cm²)
12.0	1 x 10⁵	1.0	8	< 1	>0.07	3000	85[2]
10.0	1 X 10⁵	1.0	10	> 1	>0.07	2500	65[2]
8.0	1 x 10⁵	1.0	7	3	0.21	2000	40[2]
5.0	4 x 10⁵	1.0	10	3	0.21	2000	55[2]
3.0	2. x10⁶	1.0	9	7	0.48	1500	50[2]
2.0	2 x 10⁶	1.0	10	9	0.62	350	22
1.0	2 x 10⁷	1.0	11	14	0.96	250	25
0.8	3 x 10⁷	1.0	9	18	1.24	215	24
0.6	3 x 10⁷	1.0	10	29	2.00	115	10
0.4	1 x 10⁸	1.0	10	42	2,90	70	11
0.2	3 x 10⁸	1.0	10	82	5.65	20	4.0
0.1	3 x 10⁸	0.6	6	>100	>6.90	4.0	1.5
0.05	6 X 10⁸	.06	6	>100	>6.90	0.7	0.70
0.015	6 X 10⁸	0.6	6	>100	>6,90	<0.1	0.024

1. Typical flow rate using water or air at 10 psi (0.7 bar). 2. 5 psi (0.35 bar). 3. Water bubble point

TABLE 4b – Membra-Fil® Mixed Esters of Cellulose Membrane Specifications.

Pore Diameter (mean pore size) (μm)	Minimum Bubble Point[1] (bar)	(psi)	Thickness[2] (μm)	Mean Flow Rate[3] Water (ml/min/cm²)	Air (l/min/cm²)	Refractive Index	Extractables (Wt/Wt%)	Wetting Time (sec)	Porosity (%)
5.0	0.42	6	175	700	78	1.51	<1	<3	84
3.0	0.7	10	175	490	55	1.51	<1	<3	83
1.2	0.83	12	175	330	45	1.51	<1	<3	82
1.2 grid	0.83	12	175	330	45	—	<1	<3	82
0.8	0.97	14	175	290	33	1.51	<1.5	<3	82
0.8 grid	0.97	14	175	290	33	—	<1.5	<3	82
0.8 black	0.28	4	175	290	33	—	<1.5	<3	82
0.8 black grid	0.28	4	175	290	33	—	<1.5	<3	82
0.8 green	0.28	4	175	290	33	—	<1.5	<3	82
0.7 white grid	1.24	18	150	160	16	—	<1.5	<3	82
0.65	1.31	19	150	150	15	1.51	<1.5	<3	81
0.45	2.07	30	150	65	10	1.51	<1.5	<3	79
0.45 grid	2.07	30	150	65	10	—	<1.5	<3	79
0.45 black	0.55	8	150	65	10	—	<1.5	<3	79
0.45 black grid	0.55	8	150	65	10	—	<1.5	<3	79
0.45 green	0.55	8	150	65	10	—	<1.5	<3	79
0.45 green grid	0.55	8	150	65	10	—	<1.5	<3	79
0.45 hydrophobic edge	2.07	30	150	65	10	—	<1.5	<3	79
0.30	2.54	37	150	34	6	1.51	<1.5	<3	77
0.22	3.45	50	125	15	4	1.51	<2.0	<3	75
0.22 grid	3.54	50	125	15	4	—	<2.0	<3	75
0.22 hydrophobic edge	3.45	50	125	15	4	—	<2.0	<3	75
0.10	2.41	35	100	5	1	1.51	<2.0	<3	74

1. The bubble point is defined as the differential pressure required to push air through a membrane filter previously wet with water (except 0.1μm and all green and black membranes which are wet with silicone oil).
2. Thickness is held within a tolerance of ±25 μm of the listed value.
3. Flow rates represent typical initial flow rates using clean water or air at 10 psi (0.7 bar).

TABLE 4c – Filinert™ PTFE Membrane Specifications.

Pore Size (μm)	Methanol Bubble Point (bar)	(psi)	Flow Rates (10 psi, 0.7 bar) Methanol (ml/min/cm²)	Air (l/min/cm²)	Water Intrusion Pressure bar	(psi)	Porosity (%)
0.20	0.91	13.0	15	2	2.81	40	70
0.45	0.49	7.0	40	3	1.41	20	85
1.0	0.21	3.0	90	7	0.49	7.0	85

Table 6 – Membrane Filtration Selection Criteria.

Criteria	Characteristics
Fluid Properties	What liquid or gas is being filtered? What are of the fluid properties (pH, viscosity, temperature, surface tension, stability etc)? What are the important chemical components and their concentrations? What pretreatment has been given to fluid? What is the desired minimum and maximum flow rate? What is the product batch size?
Pressure Characteristics	What is the maximum inlet pressure? What is the maximum allowable differential pressure? Is there a required initial differential pressure? What is the source of pressure (centrifugal\positive displacement pump, gravity, vacuum, compressed gas etc)?
Sterlisation\Sanitisation	Will the filtration system be steamed or autoclaved? Will the system be sanitized with chemicals or hot water? How many times will the system be sterilized or sanitized? What are the sterilized\sanitization conditions?
Hardware	Is there a restriction on the material for the housing? Is there a recommended housing surface finish? What are the inlet and outlet plumbing connections? Is there a size or weight restriction?
Filter	What is the size of particles to be retained? Will the filter be integrity tested; if so, how? Will this be a sterilizing filtration? Is there a minimum acceptable level of particle removal? Is there a recommended filter change frequency?
Temperature	What is the temperature of the fluid. Temperature affects the viscosity of liquids, the volume of gases and the compatability of the filtration system
Configuration	How will the filtration systems be configured- in series or in parallel. *Parallel flow arrangement:* uses several filters of equal pore size simultaneously to either increase flow rates, extend filter service life or lower differential pressure. It also permits filter changeout without system shutdown. The total flow rate and differential pressure is equally distributed across each filter. For any given flow rate, the differential pressure can be reduced by increasing the number of filters in parallel. Series flow arrangement: uses a group of filters of descending pore sizes to protect the final filter when the contaminant size distribution indicates a wide range or a high level of particulates that are larger than the final pore size. You can also use additional filters of the same pore size in series to improve particle removal efficiency, to protect against the possible failure of a unit within the system, and to add an extra measure of safety in any application.

From the typical data shown in Table 4, it is possible to calculate the permeability constant for the membrane for a particular fluid. In principle, this should be independent of the Fluid if there are no interactions between the membrane and Feed. The important factor in microfiltration is not the clean 'solvent' Fluxrate but the performance during actual Filtration duty. Here the membrane is required to achieve a specified degree of separation of solute in the solvent. Performance is hence affected by several solute related parameters and specifically concentration polarisation and fouling. Generally MF membranes should exhibit a high surface porosity and a narrow pore size distribution to maximise Flux.

Membrane Filtration System Selection

The selection of a microfiltration system requires that specific information or data is known relating to the application. The system selection will specify the type of membrane, the type of filter, how many filters and the size. To do this the typical information required (See Table) is:
- Fluid and contaminant/solute properties
- Pressure characteristics
- Conditions for sterilization/sanitization
- Hardware/housing requirements
- Membrane characteristics
- System configuration
- Temperature range

The fluid components must be compatible with all the components of the membrane filter. This is done by consultation with chemical compatibility charts (see Appendix). In using these charts, other factors must be identified which include, the nature and concentration of all fluid ingredients, filtration temperature, fluid pH, filtration pressure and exposure time. Chemical compatibility is measured against the solution's effect on the most vulnerable system component. For instance, an ingredient of a solution may be compatible with a filter cartridge sleeve and membrane, but not its O-ring. In many cases there may not be a 'perfect' choice, but there is always a best alternative that can be optimized by adjustments in other variables,, such as pressure, temperature, volume of through-put, or exposure time.

In certain cases it may be necessary to know the source of the fluid components and whether they are variable, for example in surface waters or are constant for example in well water. From a knowledge of the contaminant (variable, non-variable, soft, hard, spherical etc) select the pore size. In many cases, clarification and prefiltration stages upstream of the final filter greatly increase filter life by removing most of the particles that would clog the subsequent filter. The effective final differential pressure across the cartridge (change-out pressure drop) is a function of the particle/filter interaction. In cases of fluids containing deformable particles, the final differential pressure should be lower than the maximum differential pressure stated in the literature.

A multistage filtration may be suitable to enhance overall system preformance. A prefilter can extend the life of more expensive final filters.

Filters are usually blocked by particles close in size to their pore size rating. Larger particles can quickly load the membrane surface and curtail the life of the filter. A correctly designed prefiltration system can provide substantial cost savings as well as a more efficient system. However, since the differential pressures of the filters are additive in a series arrangement, care must be taken to select the appropriate filters so that the total differential pressure and resultant flow rate are in agreement with processing requirements and equipment limitations.

In system design, it should be determined whether or not sterilisation (autoclaving, in situ steam, ethylene oxide) or santization (chemical, hot water) is required and if so the duration required. To size the membrane area the required throughput or flowrate is required. During operation these will be degradation in flow, eg at a fixed pressure, flux will fall or at a fixed flowrate pressure wil have to be increased (see Fig 4). Typical guidlines for microfiltration membrane sizing are:

Liquid 20 l/min/m^2 or less – final filters
10 l/min/m^2 or less – pre filters

Air 85NMH (50 SCFM) at initial differential pressure of 0.07 – 0.15 bar for a 25 cm cartridge

As an approximation the operating specification can be based on a calculated clean differential pressure 20% to 25% of the minimum available pressure.

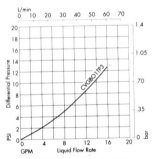

Durapore CVGB 0.22 μm Hydrophobic Cartridge, Liquid Flow Rate at 25°C.

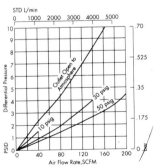

Durapore CVGB 0.22 μm Hydrophobic Cartridge, Air Flow Rate.

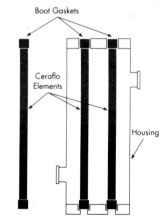

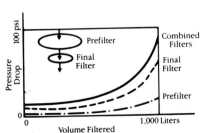

FIGURE 4 – Degradation of flow during membrant filtration.

Membrane Filters

Dead-end microfiltration is primarily carried out with flat sheet membranes. There is an abundance of suppliers of flat sheet membranes either as standard disc diameters or as rectangular sheets (or rolls).

The cross-section area of the membrane for filtration vary enormously form less than 1 cm^2 to around 3 m^2 achieved in pleated cartridge designs. The membrane filter holders or housings vary depending upon the application, membrane size and individual manufacturers design but typically 4 types are used; flat sheet, stacked disc, cartridge and capsule.

1. For small volume filtrations, very small filter devices consisting of no more than a single membrane disc sealed into a plastic support are used. Luer type connections typically enable the filter to fit into the end of a syringe.
2. Larger filtration areas are obtained using stacked disc devices consisting of a parallel arrangement of "doughnut" membrane sheets. The membranes are fitted around a central core and sealed to plastic supports. With suitable sealing of the individual membranes, the required overall filtration area is obtained fluids entering from the outside with filtrate collected in the central core.
3. Larger membranes areas are achieved using pleated cartridge units, which are cylindrical modules. In basic cartridge design the membrane is sandwiched between two non-woven fabric supports, pleated then formed into a cylinder and sealed at the two ends. In operation the fluid to be filtered enters from the outside and passes through the membrane to the inner core as filtrate.

A commonly used membrane filter is either the cartridge or capsule design for dead end operation.

Cartridges

Cartridges are pleated filter devices designed to provide high performance and economy in large volume applications. The pleated medium increases the filter's effective surface area, offering high throughput and high flow rates at low differential pressures.

Capsules

Filter capsules are convenient, self contained disposable filtration devices used for smaller volumes than cartridges. Capsules incorporate the pleated filter medium, the housing and the fittings in one unit. Capsules are both time-saving and economical, since they are quickly and easily changed. No capital investment in filter holders or housings is necessary.

Cartridge and capsules are available in a wide ragne of sizes to meet specific applications. Further details of this type of membrane filter is given in the section on microfiltration.

A typical cartridge is available in 5, 10, 20, 30 and 40 inches (12,7, 25.4, 50.8, 76.2 and 101.6cm). Diameters range from 2.0 to 2.9 inches (5.1 to 7.4 cm). Capsules are available in sizes and shapes that optimize their performance. Membrane to membrane seam seals and membrane and pleat pack-to-endcap seals are thermally bonded or sealed with

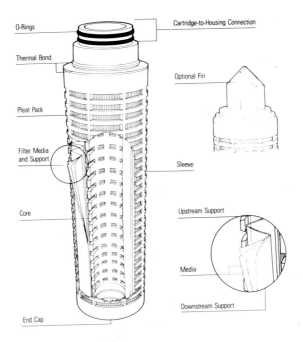

FIGURE 5 – Cut out view of cartridge and capsule filters.

FIGURE 6 – Stacked disc membrane filters.

urethane adhesives. Multiple-length cartridge modules are usually thermally bonded. Cartridges seal into housings with compression-type fittings, such as O-rings or gaskets.

Pressure Filter Holders

Pressure filter holder systems for filtration offer modest filtration areas for single membrane disc applications. Filtration areas typically vary from 16cm^2 to 560cm^2. Housings are typically stainless steel or PTFE. Applications are in many areas: biochemistry, foods, cosmetics, medicine etc (see Table 6).

The typical filter holder consists of an upper and a lower pressure plate. The nozzles and a ventilating valve are mounted on the exterior of the upper pressure plate. A seamless back pressure filter is located in the inside. It protects the membrane filter from back pressure and simultaneously provides for an even distribution of the liquid to be filtered. The outlet nozzles are in the lower pressure plate. In the inside there is a weldless sieve at the very bottom. It serves as a support and a drainage facility for the filter support, which is above it, as well as providing for a simultaneous rapid draining-off of the filtrate. The filter support, which is also seamless, is a screening seive made of spring-hardened stainless steel. The entire inner chamber is absolutely free of any threads and grooves. All the specified interior components are easy to remove. The equipment is particularly easy to clean. The upper and lower pressure plates are tightly connected by means of clamps.

The teflonized filter support prevents the baking-on of any membrane filter during the autoclaving sequence. With the standard models the R^3/$_8$ inch internal thread which is screwed into the inlet and the outlet, accommodates the corresponding connnection – which can be supplied in various designs.

These filters are limited in use because of poor fluid distribution, large heavy housings, inconvenient to scale-up and difficult to clean and sterilize.

Typical performance characteristics are show in Fig 8 for a range of membrane filters for air and water as fluids.

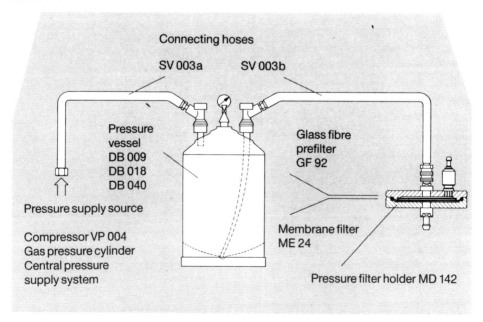

FIGURE 7 – Schematic of a pressure filtration system.

TABLE 6 – Applications of Pressure Filter Holder.

Biochemistry	Electronics	Medicine
Sterile filtration of tissue culture media Clarification filtration of buffer solutions Purification of reagent solutions	Preliminary filtration of heavily contaminated process liquids, ultra-purification of VE water, ultra-purification of rinsing liquids, ultra-purification of solvents, ultra-purification of acids, separation of particles from photographic varnishes and developers, separation of traces of water and particles in pure gas lines	Clarification filtration of serums Clarification filtration of buffer solutions Ultra-purification of reagent solutions Sterile filtration of air Sterile filtration of blood plasma Sterile fisltration of physiological solutions
Biology, microbiology Separation of bacteria from virus suspensions Sterile filtration of nutrition media Clarification of soil suspensions Clarification of plant extracts Clarification of serums	**Precision engineering** Purification of lubricating agents Ultra-purification of measuring and regulating liquids Ultra-purification of solvents for ultrasonic baths Separation of traces of water and particles in pure gas lines	**Food stuffs industry** Sterile filtration of water for industrial use Sterile filtration of worts in breweries
Blood donor services Sterilisation of plasma fractions	**Clinical chemistry** Enrichment, isolation and concentration of proteins, enzymes, hormones, poly-saccharides, viruses, clarification filtration, sterile filtration of liquor, ascites, urine, column eluates	**Pharmacy** In-line filtration of solutions for removing particles, clarification filtration and sterile filtration of alcohols, infusion solutions, injection solutions, nasal drops, ophthalmics, oral agents, medical preparations which contain protein, serums, tinctures, virus injection serums, vitamin solutions
Chemistry In-line filtration of solutions and gasses for the removal of particles Purification of polymer solutions Clarification filtration of acids and alkali Clarification filtration of photographic varnishes Ultra-purification of solvents Separation of traces of water and particles in pure gas lines	**Cosmetic industry** Clarification filtration of alcohols Sterile filtration of ointment bases	Sterile filtration of de-ionized water for ampule-rinsing machines Separation of traces of water and particles in pure gas lines Sterile air evacuation and ventilation of filtrate containers (jars) with the sterile solution
	Environment For the filtration of heavy metals of waste and ground-water	

Cartridge filters are available for pre-filtration use, sterile filtration of gas and liquid, in single membrane and multi media configuration for may applications. The major disadvantages are they require a housing with hold-up volume and create core hold up

INTRODUCTION TO MEMBRANE SEPARATIONS

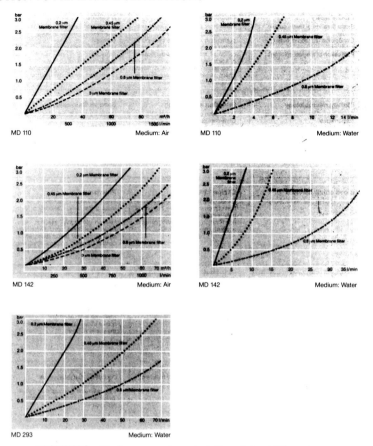

FIGURE 8 – Performance characteristics of sheet filters.

volume and require several materials of contruction for seals, O-rings, etc. The major attractions are:

- Prefabricated
- Large surface area
- Low operating pressures
- Easy scale up
- Reliable redundant O-ring/groove seal
- Good fluid distribution
- Rapid installation and replacement
- Thorough sterilization
- Audit trail
- Pretested
- Economical
- Disposable filter

Filtration Housings

Different applications require different housing configurations. General industrial filtration requirements differ from sanitary filtration requirements. The chemical nature of the fluid to be filtered may dictate a specific housing style or material of construction.

Sanitary Housing

Designed for ultra clean or sterile filtration, these housings meet the rigorous 3A standards that are often followed by the pharmaceutical industry. These include:

300-series all-stainless steel construction

No contact between the fluid and threaded connections

Sanitary flange connections on all inlets and outlets
- Able to be completely disassembled for thorough cleaning
- All welds must be ground smooth and flush
- Completely drainable
- Surface finish for all wetted parts must be equal to or better than 150-grit polished

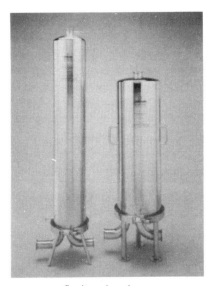

Sanitary housing.

General Industrial Housings

There are no industry standards for these housings; quality, design and materials of construction vary with the application. Industrial housings generally cost less than sanitary housings. Typical elements of an industrial housing design include:
- Materials of construction vary from high-grade plastics to 316L stainless steel, used for critical applications
- Threaded or raised-face flange connections

- O-ring or gaskets seals for filter connection
- Surface finishes which can vary from painted to 180-grit polished; a 10 Ra finish is used for critical applications in the microelectronics industry.

Materials

The material of choice for most housings is stainless steel because of its superior strength, corrosion and thermal resistance, and compatibility with a broad range of chemicals. Low-carbon 316L-grade has become the standard for the phamaceutical, as well as segments of the microelectronics and food and beverage industries.

Alternatives to stainless steel for industrial applications, especially where certain chemicals demand it, include:
- Polypropylene and PVC for low cost
- Fluoropolymer for chemical resistance and low extractables
- Pure polypropylene for low cost and low extractable levels

Combination designs are used in applications where housing strength and high resistance to corrosion are both critical, stainless steel housings with internal polymer-coatings of PEA, PVDF, or ECTFE are available.

Size

Selecting the proper size housing is critical to the performance of the filtration system. Housing size is referred to by the 'number round' (how many 10inch filters can be stacked together). That is a 7 round, 4 high housing would hold seven 40 inch filters.

Surface Finishes – Minimum standards call for use of a 150-grit polishings compound for sanitary housing though the finer 180-grit followed by electropolishing is more typical and is becoming standard for other highly critical applications.

Electropolishing is an electrochemical process that further smooths a metallic surface. Electropolishing reduces surface roughness, minimizing the possibility of contaminants adhering to the surface, improving cleanability, and improves corrosion resistance. Surface finish ratings are based on a measure of the average roughness of the surface (Ra). Welds on sanitary housings must be ground completely flush to reduce the risk of bacteria or contaminant entrapment.

Connections

There are two connections to be considered in the filters, housing-to-Filter Connections and housing inlet/outlet connections. Housing to filter connections in sanitary housings require O-ring connections in the common -222 size, or the -226 locking style for vent or bi-directional flow applications. Other speciality seals include -020 and -216/-218. The pharmaceutrical industry routinely uses a double O-ring on every seal. Industrial housings for non-critical applications can use gasket seals, though these are not suitable for submicron filtration applications or differential pressure greater than 2.7 bar (40psid).

Housing Inlet/Outlet Connections in sterile filtration applications require sanitary flange connections to facilitate cleaning and to reduce the area that attract bacteria. Small industrial housings are normally threaded (NPT) or have the same raised face (RF) flanges

TABLE 7 – Typical filter housing product lines

Housings	Pharmaceuticals					Chemicals				Microelectronics				Foods & Beverages				
	Water	Parenterals	Vent/Gas	General	Ink	Acid/Etchants	Solvents	Plating	General	Water	Vent/Gas	Acid/Etchants	Photoresists	Water	Beer	Wine	General	Vent/Gas
Industrial Stainless Steel	X		X	X	X		X		X	X	X		X	X	X	X	X	X
Sanitary Stainless Steel	X	X	X	X						X	X		X	X	X	X		X
Heavy Duty Polypropylene	X			X	X	X	X	X	X			X	X	X				
Pure Polypropylene	X			X	X	X	X	X	X	X		X	X	X			X	
Polycarbonate	X													X			X	
Stainless Steel ASME-code Stamped	X		X	X	X		X	X	X	X	X		X	X	X	X	X	X

as large industrial housings. Vacuum flanges are necessary for vessels under negative pressure.

Styles

Housings are typically available in three flow patterns; T-style, In-line, or L-shaped (fig 10). Selection depends on installation requirements or the desired differential pressure through the housing.

– T-style – Fig 9(a): Places the filter perpendicular to the proces line. Depending on design, this style may have higher pressure drops than other housings. If offers ease of filter change-out and is normally easier to plumb into existing systems.
– In-Line – Fig 9(b): Places the filter directly into the process line with the flow running in one end and out the other in a straight-through flow pattern. Typically, this style offers the lowest pressure drops. This style is most commonly used in pharmaceutical vent applications.
– L-shaped – Fig 9(c): Found only on multi-round housings, this style directs flow in the side and out the bottom. Pressure drops will vary with the design since a diverter plate is often required internally at the inlet to protect filter elements. Most often used in industrial lines.

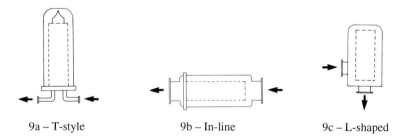

9a – T-style   9b – In-line   9c – L-shaped

FIGURE 9 – Typical industrial filter housings.

Cross Flow Microfiltration and Ultrafiltration Modules

The crossflow principle of operations is applied to the difficult filtration of solutions and suspension where high sheer and good mass transport is beneficial in reducing the build up of particulates or macromolecules at the membrane surface. Membrane filters can separate solutions and components with diameter below 10 mm when operated in tangential flow. Membrane modules are available in a wide choice of configuration (Hollow fibre, Tubule, tubular, plate and frame and spiral wound) from a relative host of manufacturers). It is generally the characteristics of the feed which determines the module configuration for a particular application although other factors are an issue.

In comparing membrane module configuration, one must consider several issues. First it is important to realize that there is little difference between an open rectangular feed channel and an open round feed channel with respect to the separation process. The use of properly scaled channel heights (diameters) and equivalent channel lengths will allow

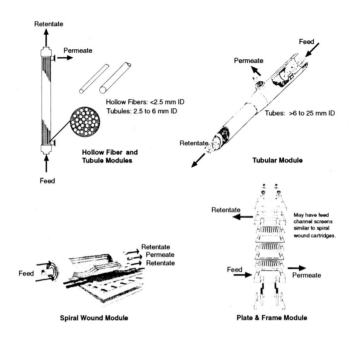

FIGURE 10.

the same module aspect ratio to be maintained. Thus, fluid shear rate at the membrane wall, the most critical scaling parameter, can be maintained, regardless of which channel geometry is utilized.

The preferred channel height will be the smallest opening which will allow concentration to proceed without plugging of the channel inlets by cells for example biological uniform and cell debris. The consistent feed channel diameter of hollow fibre and tubule configurations are advantageous in providing a consistent feed velocity to the entire membrane cartridge. With for example the trend in fermentation toward higher cell densities, larger feed channel openings may be required. This makes the tubule configuration attractive as it becomes more difficult to maintain channel height consistency in large gap plate and frame modules. Furthermore, the larger the plate and frame circumference, the more difficult it is to maintain channel height consistency and heavier, more rigid plates are required and compression consistency around the entire plate is harder to achieve.

Pressure capabilities are often of minor concern in cross flow microfiltration applications. Operation is at low inlet pressures and very low trans-membrane pressures to minimize build-up of particulates at the membrane surface. Thus, the bulky and costly supporting structure of plate and frame modules is superfluous to the separation process.

INTRODUCTION TO MEMBRANE SEPARATIONS 135

The impact of membrane material in microfiltration applications can often be secondary. Unlike ultrafiltration, where protein adsorption to the membrane surface may be an issue, the factors to consider in microfiltration are that the membrane is tough and durable enough to withstand the rigors of chemical cleaning and/or heat sterilization by autoclaving or steam and that the membrane is sufficiently self-wetting so that it is not 'blanked off' by air bubbles trapped within its pore structure.

Plate Units

Although the plate module units offer the lowest membrane area per unit volume of module of any of the configurations, there is a considerable market for their use. The modules come either as plate and frame units in which individual membranes are located between frames or as units with individual membrane cassettes or cartridge packs. The attraction of the cartridge approach is a much easier replacement of the membranes and less problems in sealing. The similarities between the general design principles makes it unnecessary to describe all manufacturers' device; brief details of suppliers are given in the Appendix. This also applies to all other module types.

Disc Plate and Frame UF

A typical geometry is shown in Figure 12. The module is composed of elliptical membranes and polysulphone support plates, which have a pair of matching holes (which form the inlet and outlet ports) near either end of the long axis of the ellipse. Membrane sheets are situated on either side of a polysulphone plate to form a sandwich. The polysulphone support plates are ribbed, and during operation the membranes deform under pressure forming a series of elliptical flow channels between the inlet and outlet

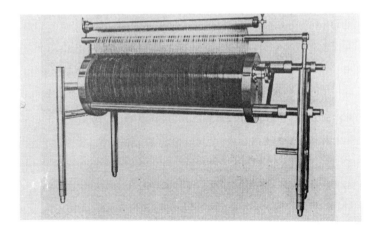

FIGURE 12– Plate and frame module circular vertically orientated membranes.

ports. Sealing rings around the inlet and outlet ports of the sandwich help to separate the membranes of one sandwich from another.

The flow channel depth is less than 1 mm and flow is laminar. For high-viscosity streams the difference in path length of the different elliptical channels can result in non-uniform flow distribution and fouling, although a modified module is available. In this module the shortest flow channels are blanked off and the depth of the remaining flow channels is varied to maintain a constant hydraulic resistance.

The individual membrane pairs of the basic module offer typically $0.15m^2$ of area. The number of pairs installed can be varied up to a maximum module area of $35m^2$. Module size and flow arrangement can be varied within wide margins. Insertion of blank discs into one of the ports of a support plate enables sections of the module to be joined in series rather than parallel. In operation the module is mounted horizontally, with its support plates resting on support lugs so that the elements can be moved easily during assembly and disassembly. Permeate is withdrawn individually from each membrane sandwich, which enables individual faulty membranes to be detected and replaced without completely dismantling the stack.

Membranes for the module can be cast directly on support paper (polyester or polypropylene) or cast separately and supported separately. Typical membranes are cellulose acetate, polysulphone and thin film composites.

Rectangular Plate and Frame

Figure 13 shows one type of rectangular design plate and frame unit. The modules are made up of sub-assemblies fitted together on a frame giving between 9 to 54 square metres

FIGURE 13 – Rectangular plate and frame module.

INTRODUCTION TO MEMBRANE SEPARATIONS 137

in membrane area. The membrane holding plates have transverse grooves on their faces, on to which the membrane is placed and held by two pairs of sealing rings and a sealing gasket on the periphary of the plate. The grooves help to drain the permeate towards the collection channels on their edge of the plate, from which the permeate is removed, and the grooves increase the turbulence in the feed channel. The rectangular module has 1.5mm feed channels and flow through its channels can be turbulent for low viscosity feeds. For viscosities above about 5 centipoise, flow is basically laminar but the ribs on the polysulphone support plates provide some turbulence. The module is mounted horizontally, has separate permeate offtakes for each membrane pair, and membrane replacement can be performed without completely dismantling the stack. Blank spacer plates are used to introduce parallel flow configuration within the module.

Each membrane plate is 0.35 square metres in surface area . The hold up (dead) volume is about 1.3 litres /m2 of the membrane area. Maximum pressure for the module is 6 bar. Pressure drops are about 0.67 bar per sub assembly for a recirculation rate of 50-60 litres/ minute.

Cartridge Units

Cartridge units are different to plate and frame units, typically several membrane units are sandwiched together and placed in a pressure vessel. The individual membrane sandwiches consist of two rectangular membrane sheets cast onto porous supports, mounted in turn on a permeate mesh sheet. The sandwich is formed by heat sealing the three-layer composites to rigid plastic plates, which have a central hole to withdraw permeate. Adjacent membrane sandwiches are separated by the projecting lips of the central hole and the peripheral rim of the plates. Two types of cartridges are available;
i) A wide channel (1,5mm) configuration with 20 plates spaced 2.5mm apart providing 1.3m^2 of active membrane area.
ii) A thin-channel configuration (2,5mm) consists of 32 elements spaced 1.0mm apart for 2.1m^2 of membrane area.

FIGURE 14– Construction of cartridge membrane unit.

The cartridge when fully assembled is 35 cm² cross section. The cartridges are placed in stainless steel rectangular housings containing up to 12 cartridges in series. The permeate is collected in a single common header and the feed is pumped parallel to the membrane usually under turbulent flow.

Cassettes

In a typical a crossflow module a number of pairs of ultrafilter membranes are arranged in parallel and sealed together in a rectangular frame. The liquid moves over the thin channels between the filters and the filtrate collecting layers. Alternating ports located along the top and the bottom of the module serve as either the inlet for the retentate or the outlet for the permeate. The size of modules vary considerably, for example small mini units of 0.1m² filtration area, with up to 5 modules can be used in parrallel and larger modules have filtration areas of 0.7 m² and as many as 7 modules can be used in parallel. The cassette modules are generally used for smaller scale operation. Membrane cassette materials are typically available in polysulfone, polyethersulfone and cellulose triacetate and regenerated cellulose.

Membrane cassette system for cross flow filtration.

Microfiltration Plate Modules

Crossflow filtration units used for ultrafiltration are readily adapted for microfiltration by the insertion of the appropriate polymeric membranes. Many manufacturers supply cross flow modules with this option. Cassette modules are available for small scale applications which utilize single membrane elements consisting of bonding membrane sheets and support screens. Alternative designs are based on stacked plates with open channels. The modules are solvent bonded and contain no adhesive or compression dependent seals. Modules with 2, 4 and 10 membrane packs are available with effective membrane areas up to 0.84m². With PVDF membranes they are steam sterilizable for at least 20 cycles.

INTRODUCTION TO MEMBRANE SEPARATIONS

Tubular Modules

Several companies manufacture tubular modules (see Appendix). The characteristics of typical tubular systems are relatively large open channels with internal diameters ranging from 12.5mm to 25mm and length varying from 0.6 to 6.4 metres. The actual tube size used (which may be as low as 1.0mm diameter) are a compromise between energy consumption and the cost of membranes and support tubes. Figure 16 shows a tubular module design consisting of 12.5mm or 25mm inner diameter epoxy re-inforced fibreglass tubes iwith the membrane cast inside. The individual tubes can be placed inside PVC or stainless steel 'sleeves' for the smaller units and connected in series , the permeate is collected in the housing/shroud and removed under pressure. They tubes can be used in bundles of 3-14 tubes, kept in place with appropriate end plates. In a slightly different approach, the membrane is cast on the inside of a synthetic paper tube while the tube is being continuously manufactured in a helical-spiral design. The paper membrane inserts fit inside perforated stainless steel support tubes, about 14mm diameter. This produces an inner (flow channel) diameter of approximately 12.5mm. In the module 18 of these tubes

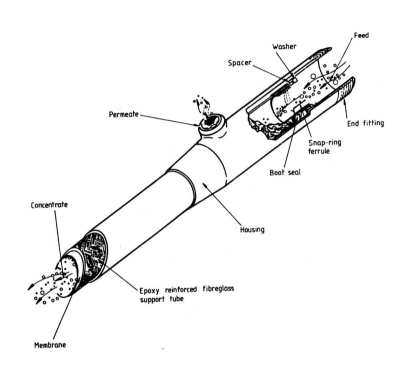

FIGURE 16 – Tubular UF module design.

INTRODUCTION TO MEMBRANE SEPARATIONS

Moslep unit

FIGURE 18 – UF plant showing stacking of modules.

are mounted together in a shell-and-tube arrangement either in series or parallel (Figure 17). Series flow is used when low flow rates can be used, as in RO applications and parallel flow is typically used for UF. A variation in the tubular design of a typical multi-stage UF plant is the membrane inside the tube. Units have the membrane cast on the exterior of a porous tube and this tube is then placed inside another tube so that feed is pumped at high velocity through the annulus and the permeate is removed from the central core

Tubular Inorganic and Ceramic Modules (MF and UF)

Tubular modules are also available with inorganic membranes. Modules incorporating carbon membrane typically use membranes tubes which are 1.2m in length, with an internal diameter of around 6mm. The membrane construction features a thin, porous layer of carbon (10 m m thick) applied to the internal surface of narrow-diameter support tubes, which are built from a composite of carbon and carbon fibre (Figure 18). Great mechanical strength is provided by the carbon fibre matrix, enabling the support tubes to

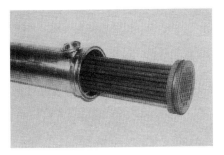

withstand pressures greater than 50 bar, even though wall thickness is only 1.5mm. The tubes have a 6mm internal diameter and are mounted together in bundles with the package size varying according to the membrane required.

Each membrane module consists of tubes mounted within a steel shell, with the complete unit resembling a small shell-and-tube heat exchanger (see Figure 21). A stainless steel shell is used for processes involving high pH values, up to 14, while PTFE-lined steel is utilised for low pH applications, down to 0. The porosity of the tube's highly permeable inner carbon layer governs the filtration cut-off point of the membrane with 0.1 μm currently the lowest achievable limit. The monolithic tube bundle is sealed in place without gaskets and the tube ends are embedded in end plate material without any possible bypass. Sealing between the tube bundle and the shell is by an O-ring. Modules can accommodate up to 782 tubes with a membrane area of $18m^2$. They can operate under extreme conditions, highly turbulent flow, pressures up tp 40 bar and temperature of up to 165°C. They can withstand many thousands of back flushing oprations (8-12bar) for membrane cleaning. Plant scale and laboratory cross flow units are available.

Membranes with cut off thresholds of 0.8 and 0.2 μm are also available. The corrosion performance of the modules, combined with their temperature and pressure handling abilities, also allows regular chemical washing of the membranes. This should be

FIGURE 20 – Photomicrograph of the membrane showing the upper active carbon layer and the carbon fibre/composite support.

sufficient to maintain performance, without replacement over several years. The insolubility of the carbon membrane renders it suitable for cleaning by steam sterilisation.

Inorganic and ceramic ultrafiltration and microfiltration modules are available from a number of companies.

The structure of one module design is produced by machining a series of 4mm diameter holes in porous hexagonal logs. The pores in the basic log are quite large (15 µm), but by successively depositing finer layers (eg of a α- and then γ aluminium oxide) onto the basic support, an ultrafiltration membrane is produced. The ends of the composite log are sealed into a stainless steel housing with elastomeric gaskets. The standard hexagonal log contains 19 membrane tubes, each 4mm in diameter and 0.85m long. The width is about 18mm and membrane area is $0.2m^2$. Larger modules containing up to 19 hexagonal logs are also available.

Microfilters

The microfilter is an symmetric ceramic structure composed of 99.96% pure alumina. It is able to withstand a pH range from 0-14, temperatures from 0°C to 300°C, and pressures up to 8 bar with standard module and gasket designs. The pore size distribution is narrow and well-controlled for selective filtration. High strength sintered ceramic bonding of filtration layers to the substrate support permits high pressure backpulsing as a flux enhancement method. Harsh chemical cleaning and steam sterilisation are also possible.

Microfiltration membranes are available in the following pore sizes: 0.2, 0.5, 0.8, 1.4, 3.0 and 5.0 micron. The multichannel element is available in an 850mm length with 4mm or 6mm diameter channels. Single tube elements are available in 7mm diameter and length of 250mm.

Ultrafilters

Ultrafilter elements have the same geometric multi-channel support design as the Microfilter elements. The Gamma Alumina Ultrafilter has a service range of pH 5 to 8, and temperatures from 0°C to 300°C. Gamma Alumina Ultrafilter Membranes are available in pore sizes of 50, 100, 500 and 1,000 Angstroms. Ultrafilters are also available with Zirconia membrane layers. The Zirconia Ultrafilters have a working pH range of 0.5 to 13.5 and temperatures from 0°C to 300°C. They are available in 200, 500, 700 and 1,000 Angstrom pore sizes. Ultrafilters are also available in 850mm length with 4mm to 6mm diameter and 7mm diameter single tube elements in 250mm length.

The typical fluxrate performance of the membranes is indicated by the value of the clean water flux, i.e., of the upper limits of permeate flow through a membrane at a given tranmembrane pressure. Permeate flux is a function of transmembrane pressure as shown in Figure X. Figure Y shows the effect of temperature and viscosity on flux with water at 20°C (1.002 centipoise), water at 50°C (0.54 centipoise) and a polyethylene glycol solution at 8.7centipoise (20°C).

Ceramic Membrane Systems offer many advantages over conventional organic polymer membranes, including:
- Suitable for steam sterilization at up to 130°C (226 F).

- Highly resistant to solvents
- Compression resistant design permits effective use even in exceptionally high pressure applications.
- Cleanable with powerful chemical solutions at high temperatures.
- Backwashable at high pressures
- Extruded, multi channel design imparts great strength and resistance to thermal and mechanical shock.
- Easy to clean design results in minimal loss of production time and minimal energy consumption.

Alternative configurations utilise cylindrical membrane blocks for both SiC and Al_2O_3 base layers. The standard membrane has a length of 900mm and is available as single channel, 7 channel and 19 channel designs with various pore diameters (0.05, 0.1, 0.2, 0.4, 1, 3, 6, 8 μm). Membrane areas of the modules vary from $0.1m^2$ to $3.8m^2$.

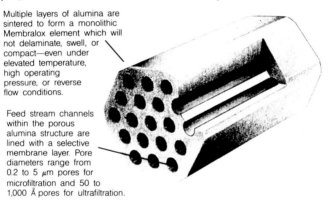

Multiple layers of alumina are sintered to form a monolithic Membralox element which will not delaminate, swell, or compact—even under elevated temperature, high operating pressure, or reverse flow conditions.

Feed stream channels within the porous alumina structure are lined with a selective membrane layer. Pore diameters range from 0.2 to 5 μm pores for microfiltration and 50 to 1,000 Å pores for ultrafiltration.

FIGURE 21– Sectional view of a 7 channel ceramic element.

FIGURE 22 – SEM of a membrane.

INTRODUCTION TO MEMBRANE SEPARATIONS

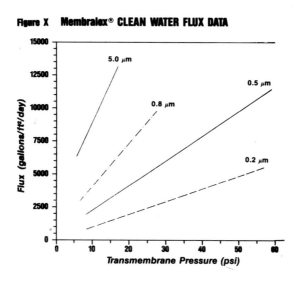

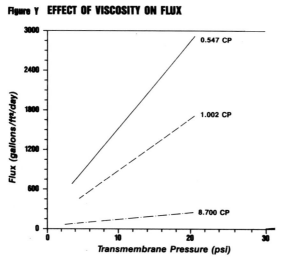

FIGURE 23 – Typical flux performance of the ceramic modules.

TABLE 8 – Typical characteristics of the inorganic membrane modules.

Housing	Sanitary	*Light Industrial	@ASME Coded	Surface Area ft²/m²
1T1-70	X	X		0.18/0.016
1P19-40	X	X		2.1/0.2
1P19-60	X	X		3.2/0.3
7P19-40	X	X		15.0/1.4
3P19-60	X	X		9.7/0.9
19P19-40	X	X		40.8/3.8
12P19-60	X	X		38.7/3.6
36P19-40			X	75.6/7.2

Code = 7 P 19 40

- 7 = Number of Ceramic Elements In Each Module
- 19 = Number of Channels in Each Element
- 40 = Channel Diameter in 0.1 mm

* Light industrial has a 150 pound class primary service pressure rating
@ The ASME-Coded module has a maximum allowable working pressure of 150 psig

Product Characteristics	Microfilter	Ultrafilter	Ultrafilter
Membrane Composition	Alpha Alumina	Gamma Alumina	Zirconia Alumina
Available Pore Size	0.2 to 5μm	50 to 1000Å	200 to 1000Å
Burst Pressure	Limited by the Maximum Housing Operating Pressure		
Maximum Operating Pressure	120 psig (Standard, higher pressures optional)	120 psig (Standard, higher pressures optional)	120 psig (Standard, higher pressures optional)
Water Permeability 20°C			
0.2μm Pore Ø	2000 l/h•m²•bar		
40Å Pore Ø		10 l/h•m²•bar	
500Å Pore Ø			850 l/h•m²•bar
Resistance to Corrosion	Can be Washed With NaOCl 2% NaOH 2% HNO₃	Limited	Same as Microfilter
*Steam Sterilizable	Yes	Yes	Yes

*The Membralox Steam Sterilization Procedure must be followed.

INTRODUCTION TO MEMBRANE SEPARATIONS

Hollow fibre modules

Hollow fibre modules for ultrafiltration are available with a range of non-cellulosic polymeric membranes, typically polysulfone and polyacrylonitrile. These membranes can be internally pressurised and thus during operation the feed can flow down the inside of the tubes. A typical module uses membranes in the form of self supporting tubes with the skin on the inside. Each hollow fibre has a fairly uniform bore; several different diameter fibres are available, ranging from around 0.19 to 1.25mm in diameter. Fibres have a thickness of about 200 μm. For UF applications, the feed is pumped through the inner core of the tube, unlike hollow 'fine' fibers used for RO.

Tube bundles can contain between 50-3000 individual fibres (depending on the diameters of the fibres and the cartridge shell) sealed into hydraulically symmetrical shells ('shell-and-tube') and bonded on each end in an epoxy tube sheet. The cartridge shells are typically constructed of clear polysulfone or translucent PVC. Each cartridge is provided with process (feed) inlet and outlet, and a pair of permeate outlets on either end of the cartridge.

The burst strength of the basic 'single skin' membranes is generally a maximum of 2 bar, which limits operational flows to avoid excessive pressure drops in the flow channels. Operating temperature limits of the modules is around 75-80°C. Modules cannot be steam sterilized but may be sanitized with a range of chemicals or hot water (up to 95°C).

Alternative hollow fiber modules use a tough, smooth double skinned fiber with a dense internal layer. This membrane has a uniformly dense skin on both walls of the fiber and can withstand high trans membrane pressure (up to 3 bar) in either direction. The double skin improves the efficiency of solute removal. Modules are available with membrane areas up to $4.7m^2 - 7.8m^{2)}$ using 89mm diameter housing, usually made from polysulfone or PVC.

The small diameter of hollow fibre ultrafiltration membranes means that pretreatment to remove particles above about 50-100mm in diameter is required. The fibres are also sensitive to fouling during the processing but however, have the advantage that they can

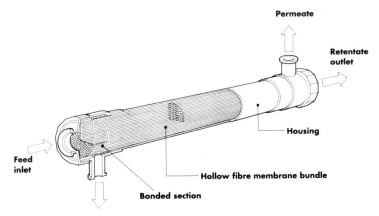

FIGURE 25 - Cross section through hollow fibre UF module.

TABLE 9 – Typical Operating Parameters of UF hollow fibre modules

		Scale	Pilot				Production			
Part number			AIP-2013	ACP-2013	ACP-2053	AHP-2013	AIP-3013	ACP-3013	ACP-3053	AHP-3013
Performance	MWCO Permeate flow[1]	daltons L/h	6,000 50	13,000 160	13,000 90	50,000 460	6,000 160	13,000 750	13,000 360	50,000 2,250
Dimensions	Fibre inner diameter Membrane area Module length Module diameter[2]	mm m^2 mm mm	0.8 1.0 552 60	0.8 1.0 552 60	1.4 0.6 552 60	0.8 1.0 552 60	0.8 4.7 1,129 89	0.8 4.7 1,129 89	1.4 3.1 1,129 89	0.8 4.7 1,129 89
Operating conditions	Maximum inlet pressure Maximum differential transmembrane pressure Maximum permeate side pressure Maximum operating temperature pH range	bar g bar bar g °C 	5 3 3 50 2-10	5 3 3 50 2-10	5 3 3 50 2-10	4 2 2 50 2-10	5 3 3 50 2-10	5 3 3 50 2-10	5 3 3 50 2-10	4 2 2 50 2-10
Materials	Membrane Housing Membrane bundle sleeve Guard Potting material Gasket		Polyacrylonitrile Polysulphone Polyethylene Polypropylene Epoxy resin Silicone							

[1] Initial clean water permeate flow at 25°C and 1 bar average transmembrane pressure.
[2] Nominal shell diameter excluding headers and permeate ports.

be backflushed. The hollow fibre membranes are sufficiently strong to withstand external pressures of up to 1.7 bar.

Spiral Wound Modules

Spiral wound modules are essentially flat sheets arranged in parallel to form a narrow slit for fluid flow. The feed channel dimension is controlled by the thickness of the mesh like spacer in the feel channel. This configuration is compact and relatively inexpensive but is prone to particulate fouling thus pre-filtration may be required. In a typical construction two flat membrane sheets are placed together, with active sides facing. They are separated by a thin mesh spacer and glued together on 3 sides. The 4th open side is fixed around a perforated centre tube. A second spacer is placed on one side of the envelope, to form the feed channel, and the whole assembly is roled around the centre tube in a spiral. The membrane assembly is then fitted inside appropriate housing made from stainless steel or

PVC. Several of this spiral assemblies can be fitted to a single housing. The ultrafiltration modules are produced in standard module sizes, 0.05, 0.1, 0.2m diameter, lengths between 0.15m to 1.2m giving membrane areas up to 15m^2 depending upon channel width. Membrane can be operated at high pressures of around 10 bar.

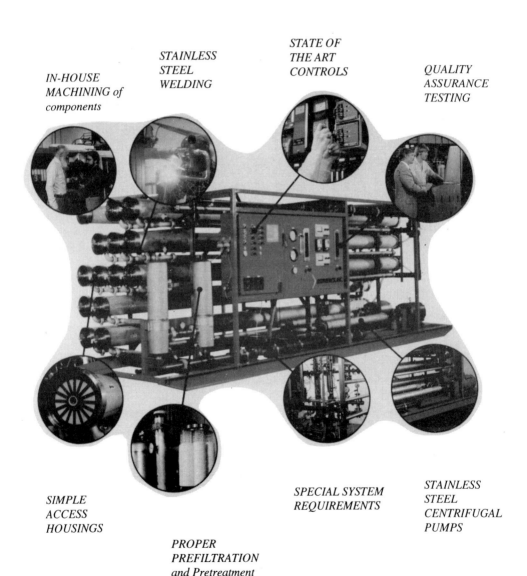

IN-HOUSE MACHINING of components

STAINLESS STEEL WELDING

STATE OF THE ART CONTROLS

QUALITY ASSURANCE TESTING

SIMPLE ACCESS HOUSINGS

PROPER PREFILTRATION and Pretreatment

SPECIAL SYSTEM REQUIREMENTS

STAINLESS STEEL CENTRIFUGAL PUMPS

INTRODUCTION TO MEMBRANE SEPARATIONS 151

Reverse Osmosis Modules

Reverse Osmosis modules are availabe in four configurations, Hollow fibres, spiral wound, flat sheet and tubular. For water purification and other applications the hollow fibre and spiral wound modules are predominantly used. Flat sheet and tubular modules are less frequently used, the typical applications are liquids with a high level of suspended solids or highly viscous liquids in for example, food products, beverages, pharmaceuticals and waste waters.

Hollow Fibre

The construction of hollow fibre modules for reverse osmosis is based on bundles of short loops of fine fibre membranes inside a pressure vessel. The fibres are about the diameter of a human hair, 80mm, and with 20mm wall thickness. Fibres are produced as continuous lengths by melt spinning or solvent casting. The continuous fibre is looped over both sides of a flat porous web, which is then wound around a pipe which is the central feed distributor of the final module. As the web is wrapped around the distributor epoxy resin is applied to one end of the bundle to seal both ends of the loops of fibres. After setting, the epoxy adhesive block is then sectioned, producing an epoxy header. Permeate collects in the fibre bores and is withdrawn from the module via the header. The general construction is illustrated in figure 28. A typical module comprises a bundle of aromatic polyamide hollow fibres installed within a fibre glass reinforced pressure vessel. Feed is introduced at the centre of the pressure vessel, flows radially out through the fibre bundle, and is withdrawn at the periphery.

Modules are available with diameters of 0.1 and 0.2m and lengths of 0.6 or 1.2m and

FIGURE 28 – Hollow fine fibre membrane.

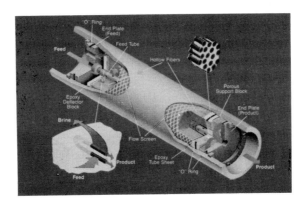

FIGURE 29 – Hollow fibre module for RO.

contain about a million or more fibres and can typically process about for example $6m^3d^{-1}$ of brackish water. Hollow fibre modules have a vast membrane area in a given volume, typically $8000m^2/m^3$.

The flow in the packed hollow fibre bundle is however laminar and thus severe concentration polarisation would occur if high fluxrates are used. The polyamide hollow fibres operate with a flux of about $0.1md^{-1}$. Overall however the hollow fibre module's processing rate for a given equipment volume is still high at $800d^{-1}$. Typically an individual module will process approximately $6m^3/day$ of brackish water.

Spiral Wound Cartridge

Spiral wound cartridges are widely used in brackish water, seawater and water purification applications. The structure is almost identical to that used for ultrafiltration. Membrane materials frequently used are cellulose acetate, polyamide and composite polyamide. Spiral devices are made from flat film membranes wound around a perforated polyvinylchloride or polypropylene centre permeate tube. Pairs of membrane sheets supported and separated by polyester fabric, or thin plastic net are sealed at the edges by epoxy and polyurethane adhesives, applied as the membrane sandwich is wound around the centre tube. The sandwiched mesh or fabric, allows permeate passing through the membranes to be conducted towards the remaining unsealed edge of the sandwich. This unsealed edge is located in a slot in the central mandrel, through which permeate is removed from the final, assembled module. A sheet of mesh is placed against one face of the sandwich and the whole multilayer assembly is rolled up around the central mandrel to produce a cylindrical module that can be installed within a pressure tube. Feed is introduced at one end, and a gasket seal around the outer circumference of the module prevents by-passing of the module. Support discs located at the the module ends prevent telescoping of the module under the applied pressure.

An alternative design is to introduce the feed at the pressure vessel wall, which travels in a spiral pattern to the centre, ie parallel to the permeate, where is is removed from the edge of the membrane envelope. The spiral is wrapped in fibreglass , tape or shrink wrap to give mechanical strength. In typical operation individual modules may operate with only 8-10% recovery, and hence four to seven elements are connected together in series in a single pressure vessel which may be up to 7m in length to achieve a 50% recovery. The common size of module has a membrane area of around 5m^2 in a 0.1m diameter, 1m long cylinder. Each module has a productivity of around 7m^3/d. Both large and small diameter modules are available; 0.2 and 0.3m and 0.025m and 0.005m diameter.

The feed flow in the spiral channel (typical size 1mm) is generally laminar with some turbulence induced by the spacer. Consequently there may be a tendancy for concentration polarisation and thus extensive pretreatment is required to prevent problems with particulates.

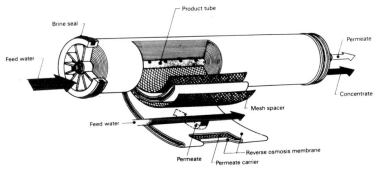

FIGURE 30– Spiral wood membrane parameter.

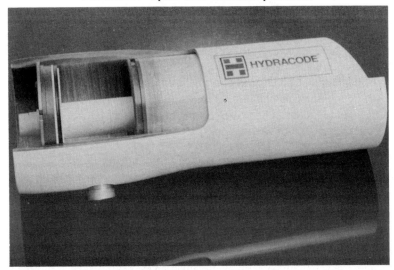

Spiral wound element.

Nanofiltration

The modules for nanofiltration are essentially the same construction as reverse osmosis elements and utilise appropriate elements with required rejection properties, for example for bivalents ions.

Tubular RO Modules

A typical tubular module uses individual membrane tubes 12.6mm in diameter, produced by casting the membrane onto a porous plastic support tube. This tube is installed within a perforated stainless steel tube, capable of withstanding operating pressures up to 70bar. The stainless steel support tubes are welded to header blocks, to which the membranes are sealed with rubber grommets. Modules typically contain 18 tubes, connected in series and are available in lengths up to 3.7m with membrane areas up to 2.6m^2. Flow in the tubes is highly turbulent, Reynolds numbers are over 30,000.

Alternative designs use support tubes of fibreglass, ceramic, etc. which must in operation be capable of withstanding the pressure. The tubes may be connected in series using external U-tube connectors to increase the overall recovery, which is generally low. The modules have a low membrane area 100m^2m/m^3 and relative low production capabilities compared to spiral wound and hollow fibre units. The benefits associated with tubular membranes is the ability to deal with feeds containing suspended solids, a low tendency to foul, easy to clean, and the membrane can be removed and reformed.

Flat Sheet Modules

Flat sheet membrane modules of plate and frame design utilise disc membranes. Designs are similar to those of UF modules using sets of alternating frames, which support the membrane on the permeate side and separate the membrane on the feed side. One design of reverse osmosis module utilises annular membrane discs 0.3m in diameter. Pairs of discs are located on either side of a polysulphone support plate, in which channels are cut for permeate flow. The membrane/support plate sandwiches are separated from one another by polysulphone spacer plates. These plates have central and peripheral holes to direct feed over the surface of the membranes, passing radially inwards then outwards over successive membranes. Consequently the liquid velocity and degree of turbulence varies during passage over each membrane element. The entire stack is mounted vertically, stack height is 1.7m, and can contain up to 19m^2 of membrane area. To replace a faulty membrane the entire stack has first to be disassembled. The module can operate at up to 70bar and 80∞C.

Flat sheet and tubular modules are generally used in similar applications. The flat sheet systems have much narrower feed channels, generally less than 1mm wide, in comparison to tubular modules. They are therefore more compact but are more sensitive to fouling and the presence of large particles. Although a flat sheet module has a large number of seals, the feed channels are not obstructed by a mesh spacer and are much easier to clean

Gas Permeation and Pervaporation Module Design

Membrane modules for gas permeation are either spiral wound or hollow fibre designs. As with other membrane processes the hollow fibre design is more compact offering a

INTRODUCTION TO MEMBRANE SEPARATIONS

higher area per unit module volume. Hollow fibre diameters are typically 150-200 mm. An important factor in module design is the pressure drop inside the permeator. Pressure loss in the direction of flow can be significant due to the narrow channel size. The pressure loss can be in the feed side and/or in the permeate side depending upon the flow arrangement. The feed and permeate flows may be either in skew flow (spiral wound) i.e. at right angles, parallel flow or with cross flow. Feed is usually introduced on the shell side in hollow fibres as this will result in a lower pressure drop. However if a high separation factor (or cut) is required feed may be introduced into the fibre lumen as most of the gas will then permeate into the shell.

Hollow fibre modules can operate with low pressure across the membrane of approximately 8.5 bar, the flow arrangement is crossflow with the feed flowing radially outwards across the membrane to minimise problems of pressure variations. Short fibres are used to give low pressure losses in the permeate side.

Alternative designs operate at high pressure differences up to 115 bar with feed and permeate flow countercurrent. The high pressures enable long fibres 3 metres in length to be used without experiencing difficulties associated with pressure losses. Packing densities are approximately 800 m^2/m^3.

The performance of gas permeation membranes is affected by a number of parameters notably feed composition, temperature and operating pressure. Generally in module design for GP the flow characteristics are not too crucial due to the usual inherent high selectivities. Thus it has been possible to adopt a crossflow arrangement as opposed to the more efficient counter-current flow. However as a rule complete mixing should be avoided.

For the majority of membranes an increase in temperature will usually decrease selectivity whilst increasing fluxrate. Thus selection of operating temperature for isothermal operation becomes one of selecting the economic optimum with the proviso of meeting required purities.

However an assumption of isothermal operation is not always accurate due to temperature changes which can result from expansion of ideal gases when pressure is reduced (across the membrane) i.e. the Joule Thompson effect. In design this factor should not be overlooked. Permeate flux is usually increased by an increase in pressure difference across the membrane. It can be more productive to achieve this flux increase by decreasing permeate side pressure rather than by increasing feed side pressure. However there are practical limitations in decreasing permeate side pressure due to an increased importance of frictional pressure losses.

In normal circumstances the influence of gas phase mass transfer is not significant in gas permeation due to the much higher diffusion coefficients of species in gases compared to solids and liquids. Exceptionally gas phase mass transfer may have to be considered if permeation flux is high when very thin membranes are used.

Pervaporation

In the design of modules for pervaporation the membrane mass transfer characteristics must be considered in conjunction with possible concentration polarisation on the feed side liquid, frictional losses at the permeate side and heat transfer resistance.

In general however friction losses at the feed side have little affect on selectivity and flux, which is in marked contrast to GP or RO. However friction losses of permeate flow are extremely important because of the requirement of low partial pressures of permeating species. Modules must be designed for low pressure losses in the permeate side, otherwise the increased driving force produced by application of a partial vacuum is offset by the pressure loss itself. Consequently any design of PV module based on hollow fibre would require large fibre diameters. Commercially available modules are not surprisingly based on the tubular or the plate and frame concept with relatively large flow conduits for vapour and small flow conduits for feed.

Pervaporation is a process which involves a change of phase i.e. evaporation of the feed and thus a heat source is needed. This heat is drawn from the feed itself and in practice requires some preheating. This is best accomplished in external heat exchangers.

Temperature gradients naturally develop across the membrane in the direction of permeate flow as well as along the membrane. Design calculations must therefore compensate for these temperature gradients and in particular the effect on permeate flux which is markedly influenced by temperature. High temperatures will give increased membrane fluxrates as well as an increased concentration in the feed of the more permeable component.

The influence of polarisation in pervaporation is generally small due to the relatively low permeation rates experienced. However the effect of feed side concentration polarisation does depend on the composition of the mixture to be separated, the imposed flow hydrodynamics and the actual permeation rate through the membrane. The more rapid permeating species is usually the minor constituent in PV separation. Thus as the concentration of this component declines the boundary layer influences become more prominent especially if thin or more highly permeating membranes are used. In these cases greater attention to good feed hydrodynamics is required. This is certainly the case in the use of pervaporation with organophilic membranes for the recovery of small (or trace) quantities of organics.

Commercial module design for PV is typically based on the plate and frame concept, although designs with spiral wound configurations are available. Hollow fibre modules although available for small scale applications would suffer from problems of concentration polarisation inside the fibres.

Pervaporation plant, with modules based on the plate and frame design use membranes 0.5 metres square. Each module is of 5 square metres or greater active area. The modules are staged in one common pressure vessel sharing a common vacuum pump and condensing system. The membrane used for dehydration is a porous support of PAN with a PVA active layer onto a non woven fabric. For other applications, polysulphones, polyetherimides or regenerated cellulose are available.

In a typical plant the vacuum for performing pervaporation can be obtained by selecting the temperature of the condenser for permeate, and the vacuum pump is operated to non-condensable gases from the vacuum system.

A PV module of tubular configuration is also manufactured. Membrane tubes, 4 mm inside diameter and 0.8 m long, of sintered ceramic, are mounted in a tube bundle to give a typical area of 0.2 square metres with 19 tubes. Feed flow is inside the tube. The active material deposited onto the ceramic is based on an acrylic salt. The first installation was at a pharmaceutical company who are using the plant (six modules of 5 m^2 total area) to dehydrate an isopropanol solution of an intermediate used in antibiotic production.

Vapour Permeation

Vapour permeation is a process which has similarities with both pervaporation and gas permeation. Separation is achieved by a solution and diffusion mechanism through the membrane which are of a type used for PV. The feed exposed to the membrane is already in the vapour phase and thus potential applications appear to be in areas where the feed is already in the form of a vapour, say from the top product of a distillation column. Two main disadvantages are that membrane performance is susceptible to feed side pressure losses and that of the limits of temperature of operation.

Vapour permeation outperforms pervaporation in terms of specific membrane flux and hence lower specific membrane requirements. The much higher volumetric flowrates and the greater diffusion coefficients in the vapour phase mean that concentration polarisation is less significant than in PV.

Vapour permeation is typically performed in plate and frame modules. One design consists of a membrane module and a permeate condenser in one module. Vapour feed is introduced into the module via an upper internal distribution channel and passed downwards across the membrane and exits via an internal retentate collector channel. Permeate vapour from the back of the membrane passes through a supporting sieve plate to a profiled permeate collector plate. Permeate flows through a short internal leader to the condensation cells where it is liquefied.

The low pressure in the permeate side of around 10-30 mbar is maintained by efficient permeate condensation utilising a closed loop cooling system.

Complete permeate vapour condensation is maintained with consequently little organic vapour emission via the vacuum pump, which primarily removes any uncondensables and inert gases.

The overall design of permeator is based on commercial plate heat exchangers utilising stainless steel; an ideal material for high temperature operation.

Operating conditions are typically, 95-100°C, 3 bar feed side pressure and 10-30 mbar permeate side pressure. The profile of the heat exchanger plates ensures reliable sealing. Plate sizes vary from 0.2 m by 1.0 m to 0.8 m by 1.6 m (1.7 m^2 area). The wide diameter internal flow channels are ideally suited to the high volumetric flowrates of vapour through the modules. Module sizes vary depending upon the number of membranes used; units with 120 double membrane cells are in operation. A typical application is based on a plate and frame module for the dehydration of ethanol. The duty is 30 m^3 per day of ethanol concentrated from 94% to 99.9%. The plant consists of an initial feed evaporation stage prior to separation in a three stage vapour permeation unit with two vapour recompressors.

Although VP is energy intensive in the feed evaporation the operating advantages of

Vapour permeation membrane plates.

isothermal operation, higher concentration gradients and favourable fluid dynamics compared with PV have added to its attractive economics and seen many applications develop.

Dialyser Modules

In the process of dialysis solute trasport is by diffusion across the membrane due to a concentration gradient. Separation is by virtue of the difference in diffusion rates of solutes asociated with the difference in molecular size. The membranes are non porous and thus need to be thin to achieve respectable diffusion rates. To reduce the diffusion resistance the membranes are highly swollen which effectvely increases the diffusion rate significantly in comparison to unswollen membranes.

The membrane materials which are used in the dialysis of neutral molecules are thin hydrophilic polymers such as regenerated cellulose materials (cellophane and cuprophane), polyvinylalcohol, polymethylmethacrylate and other co-polymers.

The transport in a dialysis membrane is described by a simple diffusion equation which is similar to that applied in reverse osmosis. In addition the actual transport rate is significantly affected by the mass transfer resistances at the membrane boundary layers.

The design of dialysers is based on two configurations, either the plate and frame type and the hollow fibre type. The design of dialysers varies somewhat according to the intended application either in medical use or for industrial use. In the case of medical use the dialyser is packaged in a sterile form and is intended for single use (or single patient use). Industrial dialysers are intended for re-use and are subject to periodic cleaning, probably in the caustic solutions. Clearly the scale of individual operations has significant bearing on design, industrial applications generally requiring much larger modules than hemodialysers where the economics associated with throw away modules means that the size (and thus cost) are very carefully minimised with little overdesign capability.

Industrial Modules

Hollow fibre dialysers developed first for hemodialysis, have been modified for industrial applications based on regenerated cellulose hollow fibre membrane 200 mm inside

INTRODUCTION TO MEMBRANE SEPARATIONS

diameter and 16 mm wall thickness. The fibres are interwoven with thread and grouped into three bundles contained in the annular space between two stainless steel tubes. These are encapsulated with polyurethane at each end, which forms tubesheets – the inner tube acts as a conduit feeding the shell side of the module, the flow leaves at one end of the outer tube. Two overall units of 28 cm effective fibre length are connected in series, to give a module of 22.5 m^2 membrane area (membrane area per unit volume approximately 2200 m^{-1})

Modern hollow fibre dialysers with better chemical resistance for application in aqueous/organic solvent extraction use regenerated cellulose and poly(acrylonitrile), both of 200 mm inside diameter. Fibres encapsulated in epoxy, forming a cartridge that has ends provided with radial O-ring seals. The cartridge is installed in a housing together forming a module which has fibre packing volume from 1.5 cm^3 to 12 cm^3 in size. A 12 litre module packed with typical regenerated cellulose dialysis fibres has a membrane area of 65 m^2.

An example of the plate-and-frame design is the Graver dialyser. This design uses a PVC membrane support frame and separator. A "repeat unit" in such a dialyser consists of two such frames – one for dialysate and one for feed solution – two sheets of membrane, and the necessary gaskets. The Graver dialyser consists of 150 such units. The solution flows are in countercurrent operation. The lower density dialysate flows downward and the higher density feed flows upward in adjacent compartments. Thus, the increasing density of the dialysate as solute is transferred to it and the corresponding decrease in density of the feed both serve to facilitate a balanced flow distribution among the cells. Hydrostatic pressure in the dialysate cells is maintained at a higher level than in the feed cells, which ensures that the membranes are pressed against the support structure of the feed frames.

Overall the hollow fibre module has dominated the industrial market due to higher volumetric capacity and ease of construction.

Medical Use

Hemodialyser designs are now generally of the hollow fibre type, much similar to shell and tub heat exchangers.

A bundle of fibres is contained in a housing and encapsulated at each end forming tubesheets. At each end, a gasket and endcap form headers to direct blood flow in and out of the lumens of the fibres. Adjacent to each tubesheet is a circumferential header, which directs dialysate flow in and out of the shellside space. The materials used typically for housing and endcaps, are transparent polymers such as polycarbonate or polystyrene-co-acrylonitrile.

The encapsulant that forms the tubesheet is polyurethane and the gasket between the endcap and tubesheet elastomer, such as silicone rubber.

Other less frequently used dialysers are based on the plate and frame or coil type of design. The plate and frame design is based on interleaved flat membranes and support plates. The coil design is based on tubular (cellophane) membranes. The tubes are flattened and rolled into a coil with a sheet of plastic mesh. The blood flows inside the flattened tube following the direction of the coil, while the flow of dialysate is axially in the space formed by the plastic mesh.

Alternative Microfiltration Configurations

There are several membrane filtration systems which incorporate special features to either enhance flux by increasing shear rate and/or to reduce the effect of fouling from particulates. These devices include high shear rotating cross flow designs, vibrating shear units and collapsible membrane systems. Details of some of these devices are described in other sections in relation to specific applications. One device, the cross flow rotational membrane filter, is in essence an adaption of the disc membrane stack system and has several applications.

Cross Flow Rotational Membrane Filter

The membrane filter system is built up from independently functioning modules, each of which comprises a number of membrane cells stacked one on top of the other. Each cell has two filter membranes, the material of which does not affect the principle of operation but does influence the degree of separation achieved. The separation process depends on the pressure difference across the filter cell membranes, which must no be allowed to become too high.

The cells are designed in the shape of squares, the corners of which are used for inlets and outlets, just as in plate heat exchangers. The dimensions of the module are typically 0.8 x0.8x1.5m, although larger area modules with a larger diameter are also available. The filtrate fluxes normally are between 100 and 1000 litre/m^2h and a module can produce 2-20m^3/h filtrate at a pressure about 3 bar, but normally not higher than 5. The cells are made of plastic material and if the media are properly selected the filtration process can be carried out at temperatures up to 90° and within a broad pH-range of 1-12.

Between the surfaces of the two membranes is the rotor which generates the tubulent flow (figure 2). The rotor generates a powerful shearing force in the wastewater inflow which is made to flow parallel with the filter surface. This high shearing force minimises clogging of the filter and the pressure rise is insignificant. As the shearing forces increase – translating into a faster flow of the wastewater – the filtering capacity also increases.

A variable-speed AC motor next to each module drives the vertical rotor shaft. The motor rating may be varied between 3 and 18 kW or 40 and 55 kW depending on requirements. Two standard modules are available. The CR500 module has between 3 and 40 membrane cells and a total filter area of 1 to 14m^2; the CR 1000 has up to 60 membrane cells and a maximum filter area of 80m^2.

Filtering capacity for a filter with 80m^2 area is 24m^3/h. The number and size of cells and modules can be varied to obtain a large range of filtering capacity. By arranging the right number of filters in series it is also possible to achieve volume reduction factors of 50:1 to 100:1.

The factors determining the degree of separation are the composition of the wastewater and the pore size of the membrane. Membrane material can be selected to suit the application – microfiltration and ultrafiltration are possible.

A special novelty of the system is the use of dynamic membranes, here called 'microprecoat' because of some similarities with the traditional precoat. This microprecoat has three objectives: to enhance the filtrate flux, to minimize its decline and to protect the

INTRODUCTION TO MEMBRANE SEPARATIONS

filter media. The layer of the precoat also protects the media against plugging and absorbs particles. When, after a long operation time, often several days or weeks, the flux begins to decline too fast, the microprecoat has to be removed and renewed, which can be done without dismantling the filter and with only a short interruption of the filtration process (10-30mins).

The fields of application are very numerous; dewatering of metal hydroxides (metallurgical and nuclear industry) purification of bleaching water and other purification tasks in the paper and cellulose industry, water clarification (in the nuclear power, electronic, offshore drilling platforms, pharmaceutical and food industries), purfication of hot and valuable effluents, dewatering of dyestuff suspension etc.

Applications of Cross Flow Rotational Membrane Filters

The rotational high shear membrane filter has features which make industrial use attractive in many applications. A selection of these examples are listed below:

Pulp Bleaching Effluent:

The first alkaline extraction stage in a pulp mill produces large quantities of organic

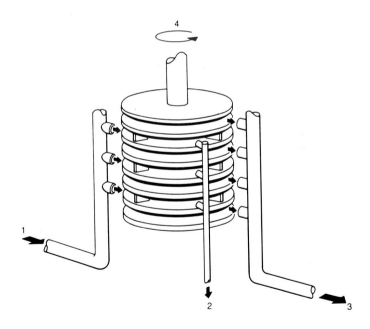

FIGURE 31a– Schematic of a cross-rotational membrane filter. The filter comprises a number of membrane cells, here four, stacked one of top of the other. Each cell has two membranes which can be any conventional material: 1 – process water inlet; 2 – concentrate; 3 – filtrate; 4 – rotor (stirrer).

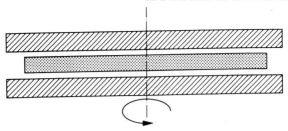

FIGURE 31b– Rotors located between the membrane surfaces generate a turbulent flow in the water which creates shearing forces.

chlorine compounds. The wastewater flow largely comprises particulate matter and high molecular weight solutes which are suitable for separation by the membrane filter. At a Sulphate Mill in northern Sweden, the total content of organic chlorine compounds is not allowed to exceed 1.2kg per tonne of pulp. To achieve this target four 1000 filters with a capacity of 50m^3/h were installed in 1988 and a further installation is now planned.

COD and BOD pollutants: A Swedish board manufacturer had to reduce COD and BOD pollutants in its wastewater to comply with new national legislation. To ensure high product quality the company must removed three tonnes of dry substances from the process each day. using two CR 1000 filters . Part of the white water flow is filtered by the automatically controlled filter plant: the white water – cleaned, still hot, and free from particulate matter can be profitably used for high pressure sprays and as washing water; the concentrate is used, after evaporation, as fuel in an existing bark fired boiler.

Uranium dioxide grinding debris: In a nuclear fuel factory the water from grinding machines used in the production of pellets for nuclear fuel contains small uranium dioxide

TABLE 10 – Examples of high fluxes obtained with filter medium.

Liquid	Filter media	Pressure (bar)	Flux (cm/h after 60-120 min)	Filtrate purity (%DS)
Kaolin 0.9% 90% <2 µm	Acropor 0.45 µm	2.0	39	clear
	Dutch woven	2.0	31	0.7-0.1
Encapsulated dye solution <10 µm 18% DS	Sepa UF 50 KCA	4.5	22	
	Acropor 0.45 µm	4.5	15	
	Dutch woven textile	4.5	1.5	
Black liquor 43% DS <2 µm	Acropor 0.45 µm	5.0	71	29
	Non-woven polyester Du Pont 2470	5	39	35
Al-hydroxide 5-8% DS	Acropor 0.45 µm	4	20	
	Woven textile	4	13	

particles which are separated from the water using a CR 500 filter. Cleaned water is recycled to the grinding machines.

Water contaminated with oil: A southern Swedish waste disposal company, separates oil from contaminated water collected from industry, controlled tips, ports, service stations and sources. After screening and sedimentation in a basin the water passes through two CR 500 filters in a final treatment stage.

SECTION 1.7 – ELECTRODIALYSIS CELL STACKS AND DESIGN

There is some variation in the design of electrodialysis cell stacks depending upon the module manufacturer, the particular application and the scale of operation. Laboratory modules are available with single cell effective membrane area of 200 cm^2, in which 5 cells are incorporated. Full scale plant modules will either utilise square membranes, each 1 m^2 area or rectangular membrane 46 x 102 cm in size. The standard membrane thickness is around 0.14 to 0.5 mm, although special heavier membrane of around 1.0 mm are also available. The materials of construction of the stack will vary for individual industrial applications – stacks for pharmaceutical and food applications in particular must avoid contamination to the products. There is even a variation regarding the orientation of the membranes between supplies, some stacks are mounted with vertical membranes and electrodes, some stacks are mounted with horizontal membranes and electrodes.

Fig 1 shows a schematic of a horizontally operated cell stack in which repeating cation membranes and anion membranes can be seen. The spacers between the membranes

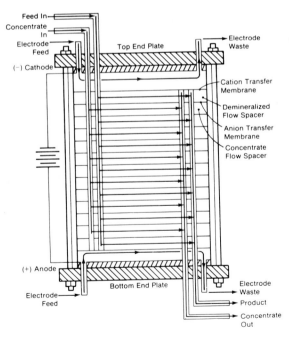

FIGURE 1 – Schematic of electrodialysis cell stack.

represent the flow paths of the demineralised and concentrate streams formed by plastic separators, called demineralised and concentrate water flow spacers respectively. These spacers are arranged in the membrane stack so that all the demineralised streams are manifolded together and all the concentrate streams are manifolded together. A repeating

ED cell stacks.

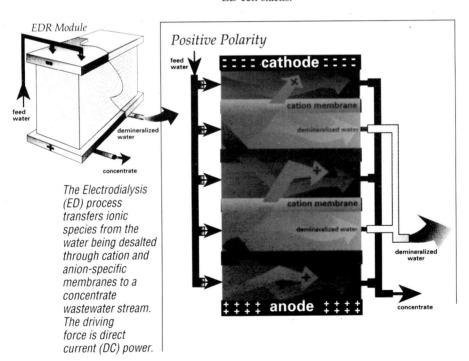

The Electrodialysis (ED) process transfers ionic species from the water being desalted through cation and anion-specific membranes to a concentrate wastewater stream. The driving force is direct current (DC) power.

The electrodialysis process.

section called a *cell pair* is formed consisting of a Cation Transfer Membrane, Demineralised Water Flow Spacer, Anion Transfer Membrane, and Concentrate Water Flow Spacer.

A typical membrane stack may have from 300 to 500 cell pairs.

Figure 1 also illustrates the flow of the two streams through the membrane stack. One stream enters the membrane stack and flows, in parallel, only through the demineralising compartments, while another stream enters the membrane stack and flows in parallel only through the concentrating compartments. As the water flows across the membrane surfaces, ions are electrically transferred through the membranes from the demineralised stream to the concentrate stream under the influence of a DC potential.

Water from the two electrode compartments does not mix with the demineralised and concentrate streams. Upon exiting the membrane stack, the combined electrode stream is sent through a degasifier to vent off reaction gases.

The top and bottom end plates are steel blocks which are used with tie rods to compress the entire stack, thus sealing the membranes and spacers to provide discrete concentrate and demineralised flow paths and prevent leakage from the sides of the membrane stack.

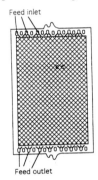

FIGURE 2 – Schematic flow spacer design.

Spacers are used to form the concentrating and demineralising flow paths within the membrane stack and to separate individual membranes. There are two general types of spacers used in electrodialysis stacks either the tortuous path or sheet flow spacer (Fig 2). The distinction refers to the flow path taken by solution through the compartments. The flow path in the tortuous flow cell is much greater than that in the sheet flow system. The velocities in the tortuous flow path design are relatively high between 15-50 cm s^{-1} which still gives significant solution residence times whilst reducing the effect of membrane polarisation.

Tortuous Flow

Spacers are made of low density polyethylene sheets with die-cut flow channels. In the manufacture of a spacer, two sheets of plastic are die-cut and then glued together to form a turbulence producing, "tortuous path (Figure 3). The stream entering the spacer (A) is split into two streams (B, C) each of which is divided into three or four parallel flow paths which wind their way through the spacer until they exit at the outlet manifold (D).

FIGURE 3.

A cross section of the spacer is shown in Fig 4 which shows an array of cross-straps, only one-ply thick, which will force water to flow in an over-and-under manner. This flow causes turbulence which promotes mixing and aids transfer of ions to the surface of the membranes allowing higher electric current flow per unit area, and therefore more efficient use of the membrane area in the stack.

Spacer Models

There are various models and sizes of tortuous path spacers available as listed in Table 1. The spacers are 0.040 inches (1 mm) in thickness.

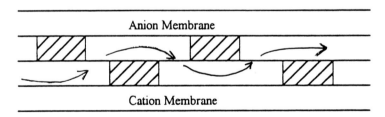

FIGURE 4 – Cross section of spacer design illustrating the flow and spacer design. This diagram depicts one spacer sandwiched between two membranes to form a flow compartment.

The number of flow paths in a spacer determines the length of an individual flow path. A Mk 111-3 spacer will have a longer flow path (488 cm) than a Mk 111-4 spacer (366 cm). Flow path length determines water velocity for a given pressure drop in the membrane stack. The water velocity determines the degree of turbulent mixing within the flow path and the amount of time an increment of water is exposed to the DC field and hence the amount of desalting per trip, or pass, through a cell pair. Water velocity and pressure drop are design parameters which will result in the choice of a spacer model for each application and provide design flexibility of the membrane stack. The pressure drop through a membrane stack is the sum of the pressure drop through each of the stages which is dependent upon the spacer type, the number of cell pairs in each stage and the flowrate per stage.

Table 1 – Tortuous flow path spacer designs.

Spacer Model	Dimensions inches (cm)	No. of Flow Paths	Flow Path Length inches (cm)	Effective Transfer Area Inches ^{2} (cm^2)
Mk 111-4	18 x 40 (46 x 102)	4	144 (366)	459 (2960)
Mk 111-3	18 x 40 (46 x 102)	3	192 (488)	462 (2980)
Mk 11-4	18 x 20 (46 x 51)	4	140 (356)	223 (1440)
Mk 1	9 x 10 (23 x 25.5)	1	137 (348)	35.5 (230)

Electrode Compartments

The electrode compartments are normally located at the top and bottom of the membrane stack. An electrode compartment is formed by an electrode, an electrode water flow spacer, and a heavy cation membrane. The electrode spacer is similar in design to electrolyte spacers but is six-ply rather than the normal 2-ply spacer thickness. This gives an overall thickness of 0.120 inches (3 mm) and enables a larger volume of water to flow by the electrodes thereby reducing electrode scaling and fouling. The heavy cation membrane is used to withstand the slight pressure differential which is maintained between the electrode stream (approximately) 2 psi lower) and the main flow streams in the membrane stack.

Stack Design and Staging

Each ED system is designed for the particular needs of the application. The capacity of the system, that is, the quantity of treated water needed, determines the sizes of the ED unit, pumps, piping and the stack size. The fraction of salt to be removed determines the configuration of the membrane stack array. The manner in which the membrane stack array is arranged is called staging. The purpose of staging is to provide sufficient membrane area and retention time to remove a specified fraction of salt from the demineralised stream. Two types of staging are used, hydraulic staging and electrical staging. The single membrane stack described in Fig earlier is an example of a one hydraulic, one electrical stage stack. This means that each increment of water, upon entering the stack, makes one pass across the membrane surface between one pair of electrodes and then exit.

A typical maximum salt removal for any hydraulic stage is 55-60 % and with normal design values at around 40-50 %. To increase the amount of salt removed in an ED system requires additional hydraulic stages. In systems where high capacities are required, additional hydraulic stages are made by simply adding more stacks in *series* to achieve the desired water purity (Figure 5). In this arrangement each stack has only one electrical stage (one anode and one cathode).

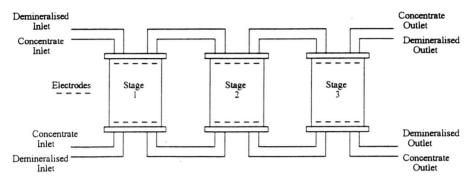

FIGURE 5 – Series staging of electrodialysis stacks.

An alternative is to use systems where additional hydraulic stages are incorporated within a single membrane stack. In this one of more interstage membranes are used. This membrane is a heavy cation membrane with all of the properties of the regular cation membrane but is twice as thick (1.0 mm) as the normal cation (0.5 mm) to withstand a greater hydraulic pressure differential than that of a normal membrane. The heavy cation membrane has only two manifold cutouts in contrast to the four manifold cutouts in a regular membrane. The heavy cation membrane is included as one of the components which makes up a cell pair since it performs the same ion transfer functions as the regular cation membrane. A hydraulic stage is formed by placing the heavy cation, or interstage membrane, at an appropriate place in the membrane stack.

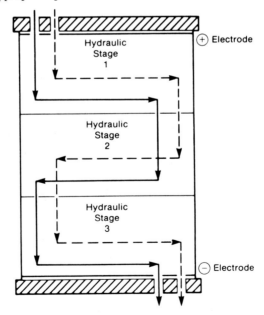

FIGURE 6 – Internal hydraulic staging of membranes.

INTRODUCTION TO MEMBRANE SEPARATIONS

In the staging arrangement of Fig6 there is a problem regarding the requirements of salt removal in each stage. The salt removal from a given volume of water is directly proportional to the current and inversely proportional to the flow rate through each cell pair. Higher currents will transfer greater amounts of salt. Higher flow rates will cause smaller amounts of salt to be transferred from a given amount of water due to shorter retention time in the membrane stack.

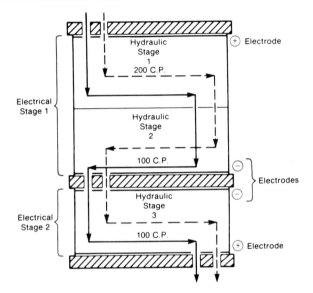

a) – Multistaging within a single stack.

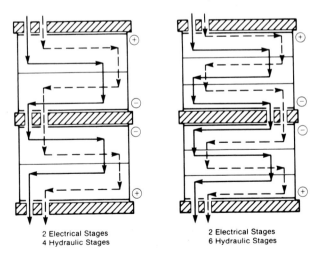

b) – Multistaging of flow and electrical current.

FIGURE 7 – Staging of electrodialysis flow stacks.

If three hydraulic stages were contained within one electrical stage, the current and hence the salt removed remains constant for every cell pair. To achieve the same percentage salt removal in each stage requires that the flowrate is reduced. For example with a 50% salt removal the flow rate per cell pair would have to be doubled for each successive stage. Since the total flow into each hydraulic stage is identical, the only way to increase the flow rate per cell pair in successive stages is to decrease the number of cell pairs in those stages.

However, by decreasing the number of cell pairs in a hydraulic stage, and increasing the flow rate per cell pair, the pressure drop through that stage is increased. At some point, the pressure drop through the entire stack will exceed the stack inlet hydraulic pressure limit of typically 50 psi (3.4 bars). If the limit of a 50 psi (3.4 bars) stack inlet pressure is exceeded, an additional type of staging, electrical staging, is used. Electrical staging is accomplished by inserting additional electrode pairs into a membrane stack. This gives flexibility in system design, providing maximum salt removal rates while avoiding polarisation and hydraulic pressure limitations. Each electrical stage allows the use of an independently controlled current to the cell pairs within that stage. If two electrical stages are used, there would now be two independent currents, (I_1) and (I_2), in which $I1>I2$. Thus more cell pairs can be used in hydraulic stage 3 in comparison to that possible with a single electrical stage.

Sheet Flow Spacer

Operation with sheet flow spacer is essentially similar to that with tortuous flow spacers. However solution velocities in sheet flow stacks are typically five times lower at $3-10$ cms^{-1}, to give similar solution residence times to achieve the typical approximate 50% desalting capacity per pass. The mass transfer in this unit is not as good as in the tortuous flow unit although hydraulic pressure losses, at $0.5-2$ bars, are significantly lower (c.f. tortuous flow $2-7$ bar).

The sheet flow spacer has a marginal gasket area with an open area containing a woven or non-woven plastic screen or expanded plastic mesh. The solution flow is from one edge (or corner) across the membrane face to the opposite edge (or diagonally opposite corner). These spacers are relatively more expensive than tortuous flow spacers, but give better support to thin membranes, which are typically used by Japanese membrane module producers. It is reported that specially designed screen type spacers are able to achieve greater than 90% desalting in a single pass.

Cell Stack Performance

In electrodialysis the main variable for controlling the quantities of ions transferred through the respective membranes is the cell current. The main use of the electrical energy supplied to the cells is to drive ions through a combined resistance barrier of the membranes and the respective electrolyte solutions. This resistance is determined largely by the ionic concentrations in the diluate and concentrate solutions. The higher the concentration that is used, generally the lower the resistance of the electrolyte, up to a limiting value at very high concentrations. For a given cell pair, the voltage required to overcome the single cell resistance, R_s, is largely determined by the relative ionic

INTRODUCTION TO MEMBRANE SEPARATIONS

concentrations in the diluate and concentrate streams and the degree of desalting. In a typical desalting operation a high ionic concentration in the concentrate, will render this resistance component negligible in comparison to the diluate. The voltage is given generally by Ohm's Law (see Table 2)

TABLE 2 – Performance Factors in Electrodialysis.

FARADAY'S LAW

Faraday's Law as related to the ED process states that the passage of 96,500 amperes of electric current for one second will transfer one gram equivalent of salt. The quantity, 96,500 ampere-seconds, is called a Faraday. This is equal to 26.8 amperes of current passing for one hour.

Faraday's Law therefore is the basis for calculating the amount of electric current needed in an ED system for transferring a specific quantity of salts. When put in a form for use in ED calculations, Faraday's Law is:

$$1 = \frac{F^x \times F_d \times \Delta N}{e \times N^x}$$

where:
- 1 = direct electric current in amperes
- F^x = Faraday's constant = 96,500 ampere seconds/equivalent
- ΔN = change in normality of demineralized stream between the inlet and outlet of the membrane stack.
- F_d = the flfow rate of the demineralized stream through the membrane stack (liters per second)
- e = current efficiency
- N^x = number of cell pairs

OHM'S LAW

Ohm's Law states that the potential (E) of an electrical system is equal to the product of current (1) and the system resistance (R).

$$E = 1 \times R$$

where:
- E   is expressed in volts
- 1   is expressed in amperes, and
- R   is expressed in ohms.

$$R_{cp} = R_{cm} + R_{am} + R_c + R_d$$

where:
- R_{cp} = Resistance, per unit area of one cell pair
- R_{cm} = Resistance per unit area of cation membrane
- R_{am} = Resistance per unit area of anion membrane
- R_c = Resistance per unit area of concentrate stream
- R_d = Resistance per unit area of demineralized stream

(dimensions of all resistance units are ohm-cm²)

CURRENT EFFICIENCY

Based on Faraday's Law, the efficiency of the current being used to transfer salts in the membrane stack can be calculated. Theoretically, for every 26.8 amphere-hours one gram equivalent of salt will be transferred in each cell pair. In such a case, the process would be 100 percent efficient.

The curreny efficiency (e) is calculated according to the following equation:

$$e = \frac{F^x \times F_d \times \Delta N \times 100}{1 \times N^*}$$

For an ideal operation of a cell stack with no electrical losses (E.g. current leakage, current efficiency losses) the passage of one Faraday of charge will transfer one gram equivalent of ions through the membranes of adjacent cells. For a stack with N cell pairs the ion transfer from one stream will be

$$j \, A \, N \, / F \quad \text{gm. equivalents/sec}$$

where j is the cell current density.

For a total flow, F_d, of one stream to the stack the change in ionic concentration, or salinity S (gm equiv. / m³), is given by

$$\Delta N \, F_d = j \, A \, N / F$$

For a given duty, at a specified current, the power required to operate the cell stack is

$$P = (F_d \, S \, F/N)^2 \, N \, R_s / A$$

Thus for a fixed amount of desalination the power required is inversely proportional to the installed membrane area, N A. This membrane area can, in principle, be installed as a either a large number of small cells or a small number of larger cells. This clearly ignores the power losses which occur in the electrode compartments. Now because the installed cost of electrodialysis equipment will increase with installed membrane area there will be an optimum membrane area which minimises the total cost of the electrodialysis operation. This optimum must also allow for other costs such as pumping and also allow for the effect of other phenomena, such as polarisation, on the system performance.

The major component of the resistance of any cell pair is that of the diluate stream. In a typical brackish water desalination the conductivity of the recirculated concentrate solution builds up to a value of around 20,000 ms. The diluate conductivity will vary between 500 to 5,000 ms. Typical commercial membranes have a resistance of approximately 5 ohm/cm² (combined). Thus on the basis of a 1 cm x 1 cm x 1 mm thick electrolyte the comparable resistances of the concentrate and the diluate are 5 and 20 to 200 ohm/cm².

Polarisation

During practical operation an accumulation or depletion of solute(s) takes place at the surface of a membrane because of the permselectivity of the membrane. Thus a concentration variation is set up in the vicinity of the membrane in which the diffusion of the solute is a major factor determining performance. This phenomena is known as polarisation. In general an increase in polarisation decreases the efficiency of separation of the process as the flux of the more permeable component decreases. In forced convection conditions, experienced in electrodialysis cells, there are boundary layers at the membrane surface in which the solution velocity is much lower than that of the free flowing solution. The concentration profile, as measured from the membrane surface, develops in a similar way to the velocity profile and is seen as a concentration boundary layer. This is much thinner than the velocity boundary layer.

The boundary layers in electrodialysis cells are influenced by the transport of ions through the membrane. Fig 8 serves to illustrate the phenomena of polarisation at an electrodialysis (cation) membrane situated between two electrodes in a salt solution. On

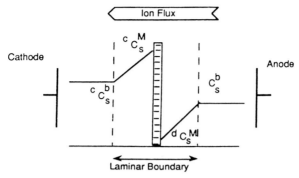

FIGURE 8 – Polarisation at the surface of ion exchange membranes.

application of an applied emf the cations will move from right to left across the membrane. Generally the flux of ions in solution is not equal to that in the membranes because the values of the transport numbers, t_m and t_s, in the respective membrane and solution phases are different. Thus, in the case of the cations in which $t_m > t_s$, the flux, J_m, due to electrical transport through the membrane, is greater than the flux, J, in the solution.

This difference in the values of the fluxes leads to a concentration gradient being established at the anode side of the membrane, in which the supply of ions required to maintain a steady state is by diffusion. A similar but opposite concentration gradient is established on the cathode side of the membrane. Thus under a steady state, the flux of ions through the membrane is equal to the combined electrical and diffusive fluxes through the boundary layers. The diffusive flux through the membrane is negligible in most applications. With an approximation of a linear concentration profile in a diffusion layer of thickness d, the current density is given by

$$j = D F (C_b - C_m)/ (t_m-t) d$$

where C_b and C_m are the bulk solution and membrane surface concentrations respectively.

If the applied emf is increased, to increase the rate of transport, the concentration at the membrane surface will fall to provide the required increase in driving force for diffusion. As the emf is increased the concentration of cations at the membrane surface will approach zero and a limiting current density will be reached. The polarisation in ED membranes is largely concerned with hydrodynamic characteristics of the cell. In a practical situation the effect of polarisation would be seen as an apparent increase in resistance of the stack, i.e. a small increase in current is achieved for a relatively large increase in voltage.

The effect of polarisation also instigates transport of other species, typically H^+ ions which, when the concentration of metal cation approaches zero, is readily transported in the membrane because of its high mobility. The supply of H^+ ions can be supplemented by the dissociation of water and thus the 'water splitting' phenomena is experienced. A similar effect due to polarisation is also experienced at anion exchange membranes due to the mobility of OH^- ions. The afore mentioned phenomena can lead to changes in the pH of the electrolyte solutions, because of the difference in mobilities of the anions and cations involved. In practice the lower mobility of the cation will cause the limiting current condition to be established first at the cation exchange membrane. Thus depletion of H^+ ions will cause an increase in pH and if the current density is increased further, such that limiting current conditions take place at the anion exchange membrane, then the pH of the solution will decrease and become quite acidic.

Bipolar membrane stacks

Membrane modules for bipolar membrane salt splitting are essentially identical to those for similar electrodialysis applications. Figures 9 and 10 give details of bipolar membrane stacks and operation.

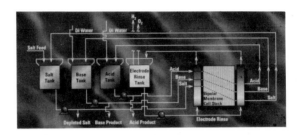

FIGURE 9 – Schematic of a typical 3-compartment water splitting system.

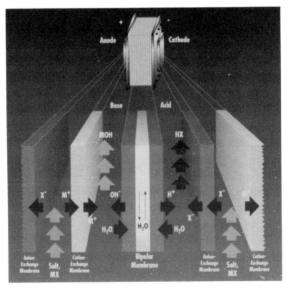

FIGURE 10 – Water splitting is an electrically driven membrane process, similar to electrodialysis in that conventional ion exchange membranes are used to separate ionic species in solution, but distinguished by the unique water splitting capabilities of a bipolar membrane. In simplest terms, a water splitter effeciently generates acid and base products from aqueous salts without chemical addition. It can also be used to directly acidify (or basify) a process stream, avoiding by-products and extensive purification steps.

SECTION 1.8 – LABORATORY EQUIPMENT

The rapid expansion in the adoption of membrane separations has resulted in a need for pilot scale and laboratory scale equipment to perform preliminary tests of the technology prior to scale-up. Equipment is often a scaled-down version of the larger industrial plant although several manufacturers supply equipment primarily intended for laboratory applications. A major area of supply is in crossflow filtration units and dead-end MF and UF systems.

Microfiltration/Ultrafiltration

The selection of a laboratory unit for microfiltration will generally depend on the aim of the study
- simple filtration of low volume dilute macromolecular solutions or suspensions
- filtration of more concentrated solutions where gelation or fouling are anticipated
- filtration of larger volumes leading to industrial applications

To satisfy these requirements laboratory units are available from several manufacturers for dead-end filtration and crossflow filtration which offer similar but not always the same features.

Dead-end microfiltration can be performed quite routinely in simple funnel type filters where the driving force is supplied by applying vacuum to the permeate. Stirred cells are used to give improved mass transport in ultrafiltration and microfiltration applications. Flat sheet disc membrane are supported in sealed cylindrical beaker containers and the "fluid" is mixed using magnetically driven stirrers. Filtration is carried out by applying pressure externally from a compressed gas.

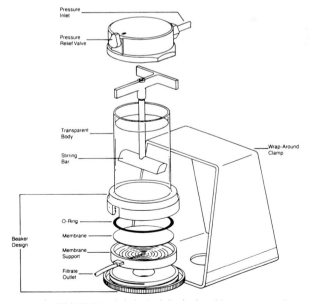

FIGURE 1 – Schematic of a typical stirred cell.

Stirred cells come in wide range of sizes to treat volumes as small as 0.6 cm^3 to 500 cm^3 and greater. Materials of construction can vary from acrylic, polycarbonate or glass bodies, with viton O-ring seals and PTFE stirring bars. Depending upon the material, pressures of 2 to 6 bar are available.

Stirred cells.

TABLE 1 – Types of laboratory membrane filtration units.

Type	Features	Applications
Dead End Filtration	No polarisation control	Dilute solutions
Stirred Cell	Improved polarisation control but not reproducable. Affected by product accumulation	Small volume, high concentration
Thin Channel Cross Flow	Good reproducible polarisation control, better usually than stirred cell	Any applicable concentration or volume
Cartridge	Good polarisation control with spiral wound or hollow fibre units	Any applicable concentration or volume

Cross Flow Filters

Laboratory scale cross flow microfilters and ultrafilters are supplied as ready to use modules to be fitted into suitable laboratory flow systems. They can be used for a broad range of applications (Table 2) such as concentration and diafiltration of proteins, cell harvesting, clarification and separation of emulsions. A typical unit comprises of two parts, a stainless steel flow distributing and module holding device and the filter unit or module. The module is a rectangular (137 x 127 mm) sealed 9 mm thick unit containing a stack of filters. The filters are layered parallel in pairs and located between each pair is

INTRODUCTION TO MEMBRANE SEPARATIONS

a layer of mesh. The feed flow is routed through this layer of mesh in the narrow channel formed between the filters. The mesh acts as turbulence promoters to counteract the effect of concentration polarisation during cross flow. Each filter "cassette" has as area for filtration of 0.1 m² and the unit can accommodate up to five modules i.e. 0.5 m² filtration area. Each module has a row of openings on the top and bottom which precisely match the ports of the flow distributing system integrated in the retentate and permeate plates of the module holding device. The fluid to be processed enters via four inlet ports on the cassette bottom and flows out as retentate through four corresponding ports at the top. The permeate flows through three openings located on the top and bottom of the permeate plate.

TABLE 2 – Application of laboratory cross flow filters.

Substance/Product	Application	Approximate class of module
Cells	Harvesting and washing	Microsart module 0.45 µm pore size
Difficult-to-filter liquids	Clarification	
Fermentation products	Clarification Concentration	Microsart module 0.2 µm pore size
Sera	Clarification	
Viruses	Concentration Ultrapurification	Microsart module 0.1 µm pore size
Pigments	Concentration	
Bacteriophages	Concentration Ultrapurification	Ultrasart module
Immunoglobulins	Fractionation	100,000 NMWCO*
High-molecular-weight solutes	Fractionation	Ultrasart module
Oil-water emulsions	Separation	20,000 NMWCO*
Biologicals	Diafiltration	
Pyrogens	Depyrogenation	
Proteins and enzymes	Concentration Ultrapurification	Ultrasart module 10,000 NMWCO*
Sera	Diafiltration	
Proteins	Diafiltration	Ultrasart module
Peptides	Concentration	5,000 NMWCO*

*NMWCO = nominal molecular weight cutoff

Laboratory cassette module.

Cassettes are supplied as microfilters with typically polyolefin and cellulose acetate membranes or with asymmetrical cellulose triacetate or polysulphone ultrafilters. Filtration capacities are in the range of 0.05 to 30 dm^3/h/m^2 depending upon applications.

Alternative designs of laboratory cross flow filters use either linear channels or slotted channels which are less obstructed and are more applicable to higher solid applications. Membranes can also be supplied as single pairs in a sealed sandwich of two membranes and a support plate (polystyrene). When testing of a wide range of membranes is required there are several module designs which allow the introduction of separate sheets as single or multiple membranes. These are offered as either rectangular sheet cross flow devices or flat disc membranes with radial flow or with spiral flow channels.

Spiral flow is achieved in a machined thin spiral channel in contact with the membrane.

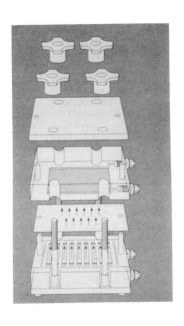

INTRODUCTION TO MEMBRANE SEPARATIONS 179

The secondary flows induced by the spiral results in much higher filtration performance in comparison to simple cross flow at similar velocities.

Many of the laboratory cross flow devices are supplied as complete systems, with membrane modules, pumps, reservoirs, valves, fitting, instruments and controllers.

Ceramic Filter Elements and Systems

For the purpose of testing mineral and ceramic UF and MF tubular membranes a series of "micro' tubular units are available. For example a ceramic single monolithic membrane with either 7 or 19 flow channels – gives an area of 370 or 560 cm^2 for filtration. The housing is a 316 L stainless steel or PVC. Alternatively ceramic filters are available with cylindrical elements with flow channels running coaxially along the full length. Each filter element is made of 99.6% alpha-alumina (Al_2O_3).

In the preparation of the membrane a porous substrate, which includes the flow channels, is extruded, and the ceramic membrane layer is deposited on the inside diameter of each flow channel. This membrane layer is sintered in place, making an extremely strong ceramic-to-ceramic bond.

Ceramic filters are compatible with high pressure operation, pH extremes, high temperatures, organic solvents, high viscosities, high solids, high chloride levels, and abrasive materials. In operation, the dilute fluid is fed into the flow channels. Different fluid pressure from inlet to outlet of the filter element forces fluid and small particles from the inside of the flow channel, through the membrane layer, into and out of the porous substrate. This leaves the larger solids concentrated at the device outlet. An elastomeric seal mounts the filter elements in specially designed housings. As clear, filtered process fluid leaves the substrate, it is captured by the housing and directed for further processing.

The ceramic systems can be automated, configured for batch or continuous operation, and cleaned or steamed-in-place. Systems are fully enclosed, and in-line diafiltration (solids washing) is simplified. Typical applications are
- fermentation broth clarification for antibiotic, steroid, and enzyme production
- catalyst recovery
- crystal concentration

A typical application guide and selection guide for ceramic filter elements is given in Table 3

Table 3 – Application and Selection Guide for Ceramic Filters.

Flow Channel Diameter	Applications
6 mm	Fluid viscosity >30 cp or solids smaller than 0.5 mm diameter
4 mm	Fluid viscosity up to 30 cp or solids smaller than 0.4 mm diameter
3 mm	Fluid viscosity up to 15 cp or solids smaller than 0.2 mm diameter
2 mm	Fluid viscosity up to 10 cp or solids smaller than 0.2 mm diameter
Pore size, µm	Applications
UF50 kDalton NMWL	Colloid removal and ultrapure clarification general clarification and 4-6 log bioburdent reduction
0.45 µm	Clarification where dissolved solutes are retained by 0.2 µm
1.0 µm	Large particle removal; crystal recovery

Table 4 – Specifications of Ceramic Laboratory Filter Units.

Dimensions*		Connection		Number of Elements	Filter Element Length	Maximum Filter Area (ft^2/m^2)
A	B	Inlet/Outlet	Permeate			
36"	4"	1 1/2" flange	1/2" flange	1	836 mm	1.5/0.1
36"	4"	3" flange	1/2" flange	3	836 mm	4.5/0.4
36"	4"	4" flange	1" flange	10	836 mm	15.0/1.4
36"	4"	6" flange	1 1/2" flange	22	836 mm	33.0/3.1
36"	4"	8" flange	1 1/2" flange	37	836 mm	55.5/5.2
11"	2 15/16"	1 1/2" TC	3/4" TC	1	209 mm	0.4/04
36"	2 7/8"	1 1/2" TC	1" TC	1	836 mm	1.5/0.1
36"	2 3/4"	3" TC	1" TC	3	836 mm	4.5/0.4
36"	3 1/16"	4" flange	1 1/2" TC	10	836 mm	15.0/1.4
36"	3 1/16"	6" flange	1 1/2" TC	22	836 mm	33.0/3.1
36"	3 1/16"	8" flange	1 1/2" TC	37	836 mm	55.5/5.2

Notes: TC denotes "Tri-Clamp"; flanges are back-up rings, 10 bar (150 psi) rated.

Ceramic filter units.

Cartridge Units

Scaled-down versions of spiral wound modules are also available with diameters as small as 25 mm with membrane areas of 0.2 m². Tubular membranes modules are also available as shorter versions of the large size units with a full compliment of membranes or with single pair membranes.

Hollow fibre modules are available in a wide range of sizes down to 0.03 m³ membranes area. The basic fibre membrane is the same diameter as in the large modules, scaling down is achieved by using fewer membranes and/or shorter fibres (more).

Reverse Osmosis

For evaluation of reverse osmosis is a laboratory scale there are stirred cells, thin channel cells and tubular modules as well as scaled down versions of commercially available equipment. A high pressure stainless steel stirred cell is used for filtering up to 400 ml of solution without a separate reservoir. The cell is designed primarily for reverse osmosis with a maximum pressure rating of 100 bar (1500 psi). The stirred cell is also suitable for UF where solvent incompatibilities exist with polymer UF cells. A height-adjustable magnetic stirring bar minimises concentration polarisation on the membrane surface. The cell may be operated by pressurising with either high pressure gas or an air driven pump. The cell is autoclavable.

For cross flow studies of reverse osmosis a Radial Flow Cell utilises 76 mm diameter membranes at pressures up to 1700 psi. Short and long term studies of retention, separation factors and flux as a function of pressure, flow rate, concentration and time are possible.

Typical laboratory spiral-wound reverse osmosis membrane cartridges designed for the concentration of aqueous solutions containing solutes smaller than 1500 Daltons molecular mass are available in 51 to 203 mm diameters with 0.3 to 27.9 m^2 membrane areas. The reverse osmosis membranes consist of a very thin (0.1-0.5 µm) semi-permeable membrane integrally bonded to a substrate that contributes strength and durability without affecting either product filtration rate or membrane retention characteristics. Each spiral wound cartridge is made by wrapping alternating layers of reverse osmosis membranes and plastic separator screens concentrically around a hollow core. Product enters one end of the cartridge under pressure, flowing tangentially down the cartridge axis. Permeate, i.e. water and species not rejected by the membranes, flows through the membrane into permeate channels and spirals to the central core. The retentate flows out of the opposite end of the cartridge. Operating pressure are determined by the membrane material but typical maximum inlet pressures are up to 70 bar.

For larger scale short-term evaluation of reverse osmosis cross flow separation using tubular membranes units constructed in 316 stainless steel are available. The module can be fitted with a range of 1.25 cm diameter, 30 cm long tube membranes. For evaluation at greater capacities equipment is available as stand alone units.

The basic unit contains one module, 1.2 metres in length, with a 316 stainless steel shroud. The membrane area within the module is 0.9 m^2. A total of 6 modules may typically be fitted to give up to 5.4 m^2 membrane area.

A heat exchanger is incorporated to maintain the temperature of the recycled liquor at the desired value. The heat exchanger is of similar construction to the module, is 0.6 metres in length and is constructed of 316 stainless steel. The recirculation pump is a triple plunger type with stainless steel contact points and high pressure cut-off switch. The unit will deliver either 22 l/m at up to 70 bar for the RO duty, or by changing the drive pulleys, 30 l/m at up to 12 bar for the UF duty. The permeation rate (flux) of the unit will depend upon the type of membrane fitted, the type of liquid being processed and the operating conditions, and will usually be in the range 15 l/m$_2$ hr to 60 l/m^2 hr.

The tubular construction means that the modules are particularly easy to clean by simple clean-up-place procedures at low pressures. Where necessary chemical cleaners can be recycled using the main pump which also cleans the main pipework and fittings.

TABLE 5 – Operating conditions and specifications of laboratory modules.

Operating conditions	
Typical permeate flow	5-50 ml/min
Typical recommended pressure	4000 kPa (40 Bar) for RO
	400 kPa (4 Bar) for UF
Typical recommended recycle flowrate	15 l/min (RO) $\equiv$ 2 m/s
	30 l/min (UF0 $\equiv$ 4 m/s
Pressure drop (water) at 2 m/s	15 kPa (2 psi)
Pressure drop (water) at 4 m/s	50 kPa (7 psi)

NOTE: Optimum operating conditions will depend upon the application and membrane type.

Specifications	
Description	2 off 1.25 cm dia x 30 cm tubes connected in series
Membrane area	240 cm^2 (0.024m^2)
Tubeside volume	75 ml
Permeate volume (running full)	approx 750 ml
Permeate volume (running empty)	approx 50 ml
Maximum operating pressure	5500 kPa (55 Bar) at 70°C
	7000kPa (70 Bar) at 20°C
Materials (excluding membranes)	316 stainless steel plus nitrile rubber seals. (Viton seals on request)

Tubular RO laboratory module.

FIGURE 2 – Laboratory scale RO unit.

Electrochemical Membrane Cells

These are available from several manufacturers of membrane electrochemical cells a wide range of scale-down versions of the commercial units. These are either available as single cell unit or as operating laboratory systems. In most respects the bench units and full scale units are identical, regarding materials of construction, electrodes, flow channel size etc. The selection of the membrane is largely up to the particular user to specify in terms of ion-exchange characteristics and chemical and mechanical stability. It is likely that the performance of the membrane is a major part of the evaluation studies at the bench scale and hence several membrane manufacturers products will be tested. As can be seen materials of construction are typically polymers of propylene, vinyl chloride etc. Membranes areas vary from tens of cm^2 to several thousands of cm^2.

Electrodialysis

For the range of applications from de-salting to production of acids and bases from corresponding salts a small scale electrodialysis unit consisting of 11 or 12 cells each with effective membrane surface areas of 58 cm^2 (or 37 cm^2) is available. The cell unit comes with three storage tank, three pumps an adjustable DC power supply and optional

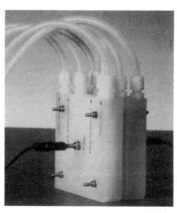

Electrodialysis laboratory equipment.

conductance gauge and timer. The three flow circuits in the standard unit are for batch recirculation of concentrate, diluate and electrode rinse solution.

Electrodialysis units and also diffusion dialysis units are available with individual membrane areas as small as 200 cm² which can be formed into multi-cell stacks. They are supplied as complete operating units ready for laboratory tests. Intermembrane gaps are between 0.4 mm and 2.0 mm formed with polyester or polyethylene spacers.

Three or four component electrodialyser module.

Gas Permeation and Membrane Reactor Ceramic Units

For the evaluation of ceramic membranes for high temperature gas separations and combined catalytic reaction and gas separation two systems are available, one based on flat membrane discs, the other on tubular ceramic membranes. The disc unit is most suitable for material development and evaluation of catalytic and permeation properties, whereas the tubular system in preferred for process evaluation and development. Both ceramic substrates and ceramic carrier tubes can be modified with specific selective top layers and/or catalysts. The catalyst can be distributed homogeneously inside the porous support or can be applied in the ceramic membranes top layer. Membrane units can be supplied as complete systems with ovens, temperature and mass control systems, safety devices, and computerised data acquisition and process control. Sealing is obtained by ductile inert metallic or graphite seals. Potential areas of application include:

- separation by Knudsen diffusion

- separation based on high selectivity of one component in a reaction mixture. For example dehydrogenation
- propane/prpoene, c-hexane/benzene, ethylbenzene/styrene
- catalytic processes combined with separation eg methane coupling, esterification, with separation of H_2O, CO_2
- catalytic process with stoichiometric dosage of reactants

SECTION 2

Membrane Materials, Preparation and Characterisation

INTRODUCTION

CHARACTERISATION OF MEMBRANES

ELECTRODIALYSIS AND
ION EXCHANGE MEMBRANES

Société des Céramiques Techniques

A subsidiary of
U.S. FILTER

BP 1
65460 Bazet
France

Tel: (+33) 62 38 95 95
Fax: (+33) 62 38 95 50

SCT, a subsidiary of U.S. FILTER, develops, manufactures and sells high efficiency ceramic membrane products for liquid crossflow micro-, ultra- and nanofiltration in the most demanding industries. An extensive experience in ceramic technology allied to specialist skills has made SCT a worldwide leader in this field. For many years, our MEMBRALOX® ceramic elements have earned us an international reputation for their durability and resistance to thermal and chemical attack.

SCT has developed its own support made of high purity alpha-alumina and a variety of membrane materials: alumina, zirconia and titania (surface modifications are also available). SCT also acquired CERAFLO® ceramic microfilter technology in 1994.

SCT is dedicated to a real partnership with its customers offering a wealth of technical assistance and advice for the development of applications, both through in-house lab testing and on-site pilot testing. Systems are supplied through an extensive worldwide network of equipment manufacturers or through U.S. FILTER specialist companies.

Typical applications using MEMBRALOX® ceramic membranes include: milk, beer, fruit juices, acids, solvents, wastewater, oily emulsions, fermentation broth....

MEMBRALOX®: ONLY THE BEST!

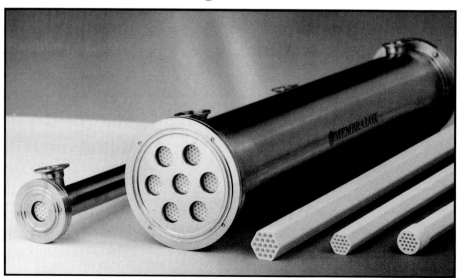

MEMBRANE MATERIALS, PREPARATION AND CHARACTERISATION

SECTION 2.1

A synthetic membrane is a permeable or semi-permeable phase, often a thin polymeric solid, which restricts the motion of certain species. This membrane is a barrier which controls the relative rates of transport of various species through itself and thus, as with all separations, gives one product depleted in certain components and a second product concentrated in these components. The performance of a membrane is defined in terms of two simple factors, flux and selectivity. Ideally a membrane with a high selectivity and permeability is required although typically attempts to maximise one factor are compromised by a reduction in the other.

Membranes are put to varied uses, the separation of mixtures of gases and vapours, miscible liquids (organic mixtures and aqueous/organic mixtures) and solid/liquid and liquid/liquid dispersions and dissolved solids and solutes from liquids.

Due to the very different nature of these separations the types of membrane used-materials, method of fabrication, differ quite significantly. In addition membrane material science has, and still is, developed rapidly over recent years to produce a wide range of materials of different structure and with different ways of functioning

Types and Materials

Table 1 summarises materials of some synthetic membranes of technical interest. Generally these materials can be classified into three types:-
- synthetic polymers; a vast source in theory although simple hydrocarbons (eg polypropylene) perfluoropolymers, elastomers, polyamides and polysulphones are prominent
- modified natural products; cellulose based
- Miscellaneous; include inorganic, carbon, ceramic, metals, dynamic and liquid membranes.

To be effective for separation membrane materials should ideally be chemical resistant (to both feed and cleaning fluids), mechanically stable, thermally stable, have high permeability, have high selectivity and generally be stable in operation. All these

TABLE 1 – The type, structure and preparation of synthetic membranes.

Membrane type	Membrane structure	Preparation	Applications
Asymmetric CA, PA, PS, PAN	Homogeneous or microporous, "skin" on a microporous substructure	Casting and precipitation	UF and RO (MF) GP, PV
Composite CA, PA, PS, PI	Homogeneous polymer film on a microporous substructure	Deposition on microporous substructure	RO, GP, PV
Homogeneous S	Homogeneous polymer film	Extrusion	GP
Ion exchange DVB, PTFE	Homogeneous or microporous co-polymer film with positively or negatively charged fixed ions	Immersion of ion exchange powder in polymer, or sulfonation and amination of homogeneous polymer film	ED
Microporous: Ceramic, metal Glass	0.05 to 20 m pore diameter 10 to 100 m pore diameter	Moulding and sintering Leaching from a two component glass mixture	GP F (molecular mixtures)
Microporous: Sintered polymer PTFE, PE, PP	0.1 to 20 m pore diameter	Moulding and sintering	F (suspensions, air filtration)
Microporous: Stretched polymer PTFE, PE	0.1 to 5 m diameter	Stretching a partial crystalline film	F (air, organic solvents)
Microporous: Track-etched PC, PEsT	0.02 to 20 m pore diameter	Irradiation and acid leaching	F (suspensions, sterile filtration)
Symmetric microporous phase inversion CA	0.1 to 10 m pore diameter	Casting and precipitation	Sterile filtration, water purification, dialysis

PTFE – polytetrafluoroethylene CA – cellulose esters PVC – polyvinylchloride PA – polyamide PE – polyethylene PS – polysulfone
PP – polypropylene S – silicon rubber PC – polycarbonate PEst – polyester PAN – polyacrylonitrile PI – polyimide
DVB – divinylbenzene UF – ultrafiltration RO – reverse osmosis GP – gas permeation MF – microfiltration ED – electrodialysis
F – filtration PV – pervaporation

GFP MEMBRANEN

Finding New Ways

... for Filtration

®NADIR and ®MOLSEP membranes for ultra/nano/microfiltration and pervaporation

spirally wound, cassette, hollow fibre and tubular modules

... and for Phase Contact

®LIQUI-CEL contactors for gassing and degassing

®CELGARD microporous membrane for blood oxygenators and battery separators

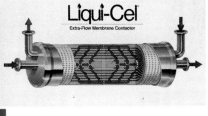

Let us help you find new ways to improve your separation process. Please contact us at any time under:

Hoechst AG
GFP Membranen
Rheingaustr. 190
65174 Wiesbaden
Phone +49-611-962-6237
Fax   +49-611-962-9237

SelRO™
Chemically Stable UF & Nanofiltration Membranes

Industrial Applications

Food
Recovery and reuse of caustic soda and acids in dairies, breweries, beverages, etc.

Pharmaceuticals
Concentration of intermediates and antibiotics in organic solvents or water/ solvents
Recovery of spent solvents from HPLC

Chemicals
Reduction of COD from effluents
Desalination of dyestuffs
Removal of heavy metals from acids
Recovery of catalysts from organic solvents

Metal
Acid purification

Dairy Caustic

Brewery Caustic

CuEdta Recovery

Metal Removal

Acid Decoloring

Separations in Organic Solvents

Membrane Products
KIRYAT WEIZMANN LTD.

P.O. Box 138 Rehovot 76101 Israel, Tel. +972-8-407557,
Fax. +972-8-407556, Internet address: mpw@netvision.net.il

Membranes available in Tubular and Spiral Wound

MEMBRANE MATERIALS, PREPARATION AND CHARACTERISATION

properties are obviously relative in terms of individual processes and the respective capital and operating costs. Chemical resistance relates more to the operating lifetime of the membrane. A gradual deterioration of the membrane can occur over months and years with perhaps only a relatively small loss of selectivity. The initial membrane costs and cost of refitting can determine the material chosen. For example the introduction of inorganic membranes nearly an order of magnitude more expensive than organic counterparts, has occurred because of the much improved operating lifetimes (5 years or more), greater tolerance to extreme conditions of operation, eg higher temperature, aggressive chemicals, and subsequent saving in maintenance.

Generally the intended application of the membrane will determine the mechanism of transport and how the membrane functions which will in turn define the membrane structure. Two types of structures are generally found in membranes (solid material) symmetric or asymmetric. Symmetric membranes are of three general types; with approximate cylindrical pores, porous and non-porous (homogeneous) (see Fig 1). Asymmetric membranes are characterised by a non uniform structure comprising an active top layer, or skin, supported by a porous support or sub-layer (see Fig 1). There are three general types porous, porous with top layer and composites. There are several methods for producing membranes, which come under headings of asymmetric and symmetric types.

Symmetric membranes which by definition are of a uniform structure, are generally produced by one of the following methods:

- *Sintering or stretching* – for the manufacture of microporous membranes
- *Casting* – for the manufacture of ion exchange membranes and membranes for pervaporation
- *Phase inversion and etching* – the manufactured materials function as pore membranes and are used in MF, UF and dialysis
- *Extrusion* – materials produced by this method function as diffusion membranes for gas permeation and pervaporation.

Microporous membranes are the simplest of all the symmetric membranes in terms of principle of operation. They are primarily used in filtration but have other applications in separations such as pertraction and liquid membranes. Microporous membranes have defined pores or holes and separation is achieved by a sieving action. Materials for fabrication include ceramics, metals, carbon and polymers. The simplest membranes, eg ceramic, are produced by moulding and sintering, although sintered polymeric membranes are available. An alternative relatively simple method for the manufacture of polymeric membranes is by extrusion and stretching (perpendicular to the direction of extrusion) which leads to partial fracture of the film.

The manufacture of microporous membranes with uniform cylindrical pores is achieved by track etching.

A number of inorganic materials such as microporous glass and ceramics come under the category of symmetric membranes. These materials can however also be coated to form composites to introduce specific properties, improved rejection, structural improvements, increased flux and selectivity.

190 HANDBOOK OF INDUSTRIAL MEMBRANES

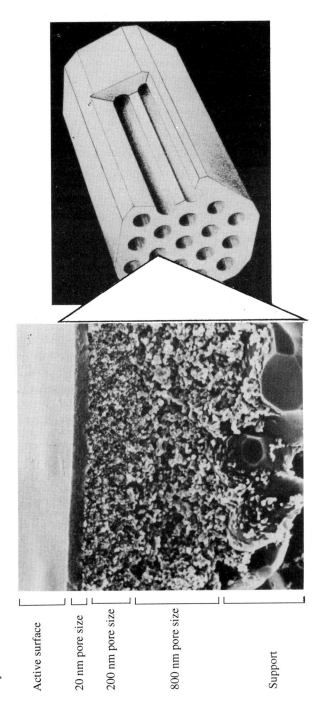

FIGURE 0 – Geometry and microstructure of a Membralox® ceramic membrane.

Layer structure:
- Active surface
- 20 nm pore size
- 200 nm pore size
- 800 nm pore size
- Support

MEMBRANE MATERIALS, PREPARATION AND CHARACTERISATION

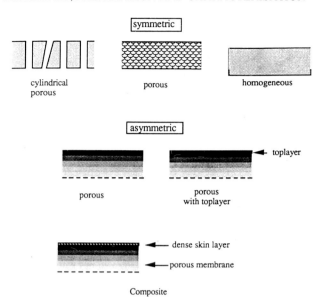

FIGURE 1 – Structure of different membranes; symmetric porous, non-porous, cylindrical pore, asymmetric membrane, composite membrane.

Asymmetric membranes are produced either by phase inversion from single polymers or as composite structures. Phase inversion incorporates porous structures which are formed by precipitation from a homogeneous polymer solution. They (see Fig 1) are made up of a relatively thick porous support layer (0.2-0.5 mm) with a dense active "skin" layer (< 1 micron). These are classed as pore membranes and are used in NF and UF. Phase inversion is also used for the manufacture of microporous symmetric membranes.

Composite membranes differ from those produced by phase inversion in that the skin and support are of different materials. This enables a certain amount of tailoring of membrane function for specific applications and thus gives potential improvements over phase inversion. Generally these are classed as diffusion membranes and are used in RO, GP and PV. There is however overlap between the areas of application of both types, ie composite membranes are used for UF (and MF). The composite is generally restricted to two layers, but can comprise a number of regions, or coatings, as for example found in modern RO membranes. These membranes are typically made on a polysulphone substrate, itself asymmetric, which is also supported on a fabric. An ultra-thin desalting layer is produced in situ on the active layer to which a protective membrane coating can be applied. The skin layers and coatings generally can be selected to exhibit certain characteristics, e.g. biocompatibility.

Polymer characteristics

In the case of synthetic polymers there are several basic features and characteristics which determine the physical and chemical properties which in turn have a significant bearing on the polymers application as a membrane. These features include:

1. The number and molecular weight of repeating units in a polymer.
2. Whether the material is a homopolymer, the repeating unit is the same (eg CH_2-CH_2 in ethylene), or a copolymer, the repeating units are different.
3. How the material monomers are linked, as random chains, in for example synthetic rubbers (nitrile-butadiene, acrylonitrile-butadiene-styrene etc).
   - A block copolymer where the chain is built up by linking blocks of the monomer eg styrene-isoprene-styrene (SIS).
   - Are grafted copolymers in which irregularities are attached in the side chain, the second monomer is attached by chemical means or radiation.
4. Whether the polymer is linear, branched or cross linked. Crosslinked polymers have chains connected by covalent bonding often by chemical reaction or by physical cross links as in for example semi-crystalline polymers. Cross linking has the effect of making the polymer insoluble.
5. Stereoisomerism (see Fig 2). In certain polymers types different side groups are in the repeating units (eg vinyl polymers). These side groups can be attached in different ways.
   - All lie on the same side of the chain, isotactic
   - Arranged randomly on either side or the chain, atactic
   - Arranged on alternative sides, syndiotactic

   This behaviour has a significant effect on the property of the polymer, a regular structure ie isotactic will produce a crystalline polymer and correspondingly atactic polymers are non-crystalline or amporophous. The crystallinity has a major effect on the permeability of a polymer membrane.

   When polymers contain a double bond in the main chain they can exhibit cis-trans isomersion in for example the polymerisation of 1, 3-isoprene, chloroprene (neoprene) and butadiene rubber. This cis-isomer is natural rubber which is used as a membrane material. The trans-isomer is stiff and leathery with thermoplastic properties.

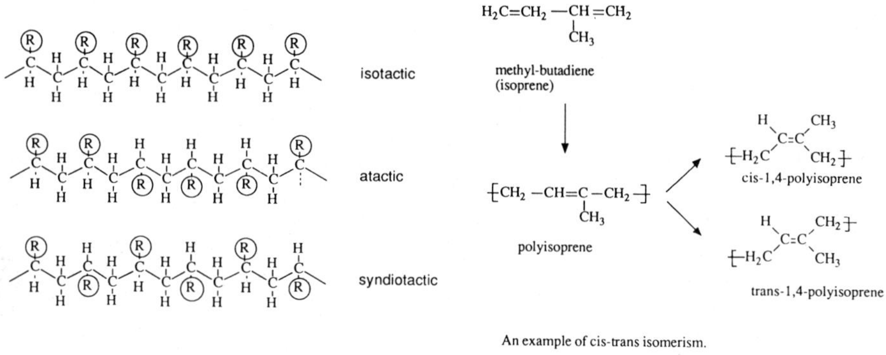

FIGURE 2 – Steroisomerism in polymer.

Membrane Technology
an international newsletter

Every month, in just 12 succinct pages, Membrane Technology newsletter brings you an up-to-date international digest to follow all the news and developments affecting Industrial membranes and membrane technology.

Your monthly snapshot of world wide news

Each issue is packed with essential information... from the latest news & views to case studies, and covering the entire range of membrane technologies – from micro filtration to reverse osmosis.

In every issue

- Latest news and views on the development and application of Industrial membranes
- Case studies
- New product launches
- The latest patents – designs and inventions
- Research
- Events

For more information contact:

Elsevier Advanced Technology
The Boulevard
Langford Lane
Kidlington
Oxford OX5 1GB

Tel: (+44) (0) 1865 843842
Fax: (+44) (0) 1865 843971

FILTERS & FILTRATION HANDBOOK 3RD EDITION

With over 3000 sales of the second edition, the Filters & Filtration Handbook has become the leading reference manual for the selection and application of filtration & Separation products.

The new edition has been extended and updated to incorporate all the latest developments in liquids, solids and air filtration and separation technology supplied by both manufacturers and users.

The Handbook is an essential reference tool for managers, engineers, designers, technicians, plant operators and consultants and those with responsibility for purchasing, planning, sales and marketing. It is directly relevant to numerous industries including water/waste treatment, fluid power, chemicals, pharmaceutical, food and beverages processing, general engineering, electronics and manufacturing.

OUT NOW!

OVER 750 PAGES,

1500 ILLUSTRATIONS

TABLES AND CHARTS!

SECTION 1 — BASIC PRINCIPLES:
Filters & Separators • Contaminants • Filter Ratings • Filter Tests • Surface and Depth Filtration

SECTION 2 — FILTER MEDIA
Absorbent and Absorbent Media • General Types of Media • Membrane Filters • Woven Wire, Expanded Sheet & Mesh

SECTION 3 — TYPES OF FILTERS:
Strainers • Screens • Cartridge Filters • Electrostatic Precipitators • Candle Filters • Magnetic Filters • Sintered Filters • Bag Filters • Precost Filters • Filter Presses • Rotary Drum Filters • Rotary Disc Filters • Belt Filters • Leaf Presses • Tipping Fan Filters

SECTION 4 — TYPES OF SEPARATORS:
Separators • Centrifuges • Hydrostatic Precipitators • Wet Scrubbers • Dynamic Precipitators • Coalescers • Mist Eliminators

SECTION 5 — LIQUIDS AND SOLIDS
Water Filters • Process Filters • Backfinishing Filters • Industrial Wastewater Treatment • Oil/Water Treatment • Metal Working Fluids • Dewatering and Fuel Treatment • Industrial Chemical Filters

SECTION 6 — AIR FILTRATION
Air Filters • Air Filter System • Dust Collectors • Fume and Hot Gas Filters • Machine Intake Filters • Compressed Air Filters • Respiratory Air Filters • Sterile Air and Gas Filters

SECTION 7 — OILS
Hydraulic Oil Filters • Engine Filters • Homgenizers. Oil Cleaning

SECTION 8 — FILTER SELECTION
Selection Data • Tables and Graphs • Buyers Guide

TO ORDER

☐ Please send me _____ copies of the Filters and Filtration Handbook (3rd Edition) at £102/$204
☐ Payment is enclosed (Please make cheques payable to Elsevier)
☐ Please invoice me
☐ Please charge my ☐ Access/Mastercard ☐ Visa/Barclaycard Eurocard ☐ American Express

Card No: .... Expiry Date: ....
Signature: .... Name: ....
Position: .... Organization: ....
Address: ....
.... State: ....
Post Code/Zip Code: .... Country: ....
Tel: .... Fax: ....
Nature of Business: ....

ELSEVIER ADVANCED TECHNOLOGY

FAX OR POST TO:
Elsevier Advanced Technology • PO Box 150 • Kidlington • Oxford • OX5 1AS • UK
Tel: +44 (0)1865 843848 • Toll Free From US: 1-800-85-0085 • Fax: (0)1865 843971

6. Chain Flexibility. This is a major structural characteristic determined by character of the main chain and the nature of the side chain or groups (if present).
7. Chain Interactions. The type of interaction between the polymer chains have a bearing on the physical property of the polymer. In linear and branched polymers secondary interaction forces (dispersion, dipole and hydrogen bonding) act. The strongest of these is hydrogen bonding when a hydrogen atom attached to an electronegative atom (eg oxygen) is attracted by the electrogative group in another chain eg O, N. In some cases the extent of bonding causes significant insolubility and can improve crystallinily.

The relative mechanical, chemical, thermal and permeation properties of a polymer membrane are influenced by the state of the polymer. The separation application of the polymer membrane, whether porous or non-porous media is required, and the chemical resistance and thermal stability has a major influence on the choice of polymer. The selection of polymers for porous membranes is less critical. Microfiltration membranes generally consist of crystalline polymers, which show high chemical resistance and thermal stability.

The two important parameters which directly influence the membrane performance of the polymer ar the crystallinity and glass transition temperature, Tg. The glass transition temperature divides the tensile modulus of the polymer into two regions, a glass state and a rubbery state. In the glassy state the tensile modulus is high and the mobility of the polymeric chain is very restricted. The tensile modulus in the rubbery state is typically several orders of magnitude lower than in the glassey state, and thus a high degree of chain mobility is seen. Glass transition temperature of several polymers are shown in Table 2.

TABLE 2 – Glass transition temperature of various polymers.

Polymer	Tg (°C)	Polymer	Tg (°C)
polydimethylsiloxane	−123	poly (vinyl alcohol)	85
polyethylene	−120	poly (vinyl chloride)	87
poly-(cis-1,4-butadiene)	−90	polymethylmethacrylate	110
poly-(cis-1,4-methylbutadiene)	−73	polyacrylonitrile	120
natural rubber	−72	polytetrafluoroethylene	126
butyl rubber	−65	polycarbonate	150
polychloroprene	−50	polyvinyltrimethylsilane	170
poly-(cis-1,4-proplylene)	−15	polysulfone	190
poly (vinyl acetate)	29	polytrimethylsilylpropyne =	200
polymethylpentene	30	poly (ether imide)	210
Nylon-6 (alif. polyamide)	50	poly (vinylidene fluoride)	210
cellulose nitrate	53	poly-(2,6-dimethylphenylene oxide)	210
polyethyleneterephtalate	69	poly (ether sulfone)	230
cellulose diacetate	80		

The degree of crystallinity of a polymer has a large influence on the mechanical and transport properties of a polymer. Some polymers are semi-crystalline, consisting of an amorphous and a crystalline fraction, eg polypropylene, various polyamides. Completely crystalline polymers exhibit little change in mechanical properties (tensile modulus) up until the melting point. Semi-crystalline polymers exhibit mechanical properties somewhere in between the completely crystalline and amorphous states ie the tensile modulus decreases with temperature after the glass transition temperature. Generally, although there are exceptions rubbery polymers (elastomers) exhibit high permeabilities and glassey polymers low permabilities for gases such as N_2 and O_2. However the choice of polymer is largely determined by the selectivity of transport of different gas. Therefore the state of polymer dictates the appropriate choice for a particular separation. Amorphous glassy polymers are generally used for the separation of permanent gases such as N_2 from air, whereas rubber polymers (often crosslinked) are employed for separations of organic vapours from air. Amorphous glassy polymer are generally used for ultrafiltration membranes as this is convenient for control of the small pore size required.

In general it is desirable to use a polymer which is chemically and thermally stable. Materials with high glass transition temperatures and high melting temperatures and those with high crystallinity are desirable. However as a polymer becomes more stable into becomes more difficult to process into a membrane as it is no longer soluble in convenient solvents. A selection of thermally and chemically stable polymers is presented in Fig 3.

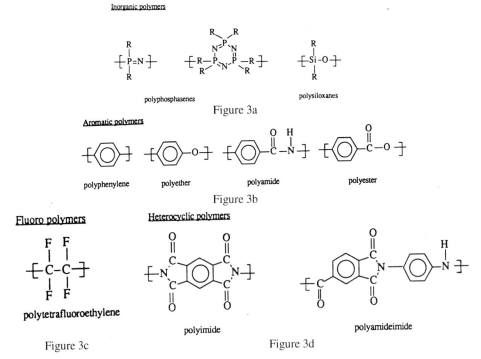

FIGURE 3 – A selection of thermally and chemically stable polymers.

TABLE 3 – Molecular structure and thermal characteistics of typical MF and UF membrane materials.

Polymer	Chemical Structure	$T_g/°C$	$T_m/°C$
Polyethylene (PE)	$-(CH_2)_n-$	−60 – −90	137-143.5
Polyvinylidenefluoride (PVDF)	$-(CH_2-CF_2)_n-$	−40	160-185
Polypropylene (PP)	$-(CH_2-CH(CH_3))_n-$	−10	167-170
Polycarbonate (PC)	(bisphenol A carbonate structure)	150-155	240
Teflon	$-(CF_2)_n-$	−113	327
Cellulose Acetate (CA)	(cellulose acetate structure, Ac : OCOCH$_3$)	—	230
Polyethersulfone (PES)	(aryl sulfone ether structure)	225	—
Polysulfone	(bisphenol A sulfone structure)	190	—
Polyvinylalcohol (PVA)	$-(CH_2-CH(OH))_n-$	65-85	228-256
Polyacrylonitrile (PAN)	$-(CH_2-CH(CN))_n-$	80-104	319
Polyphenylenesulfide (PPS)	$-(C_6H_4-S)_n-$	85	285

Crystalline polymers show a high chemical and thermal stability and are generally used for microfiltration membranes. Polymeric membranes with a high glass transition temperature are generally recommended for ultrafiltration membranes. Thus polysulphone is widely used for UF membranes, despite its poor resistance to solvents.

Mechanical properties of polymer membranes are generally only important when the material is not supported as in the case of hollow fibre and capillary membranes. Materials with high tensile moduli are therefore used, eg polyimide, as fibres. In addition to the value of the tensile modulus the toughness or brittleness is a factor in mechanical stability. The ability to withstand a large deformation under stress (ie toughness) is important. Cellulose esters and polycarbonate are examples of tough polymers.

Membrane Materials

In the use of particular polymers for membranes it is convenient to classify the membranes as either microporous, as applied in MF and UF, and non-porous or dense as applied in gas separations. For porous membranes the important factors in selecting a material are processibility and stability. For non-porous membranes, the flux and selectivity characteristics of the material are of immediate interest.

Porous Membranes

The objective in producing a microporous structure is to achieve a selectivity of separation based on dimensions of the pores. The method of processing of the polymer into a thin membrane is then the factor of importance. The material of the polymer can effect the selection of the method of processing in many cases. Other characteristics dictated by the material are the adsorption behaviour and wettability (hydrophilicity).

The majority of porous membranes have what is described as a microporous structure. Such structures comprise a matrix of a solid material, usually a polymer, and a network of interconnected pores. Microporous membranes (Fig 4(a)) can have a uniform distribu-

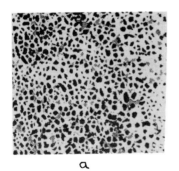

a

FIGURE 4 – Structure of microporous membranes.

TABLE 4

Nominal pore size (μm)	Membrane/ support	Manufacturing process	Module configuration	Mechanical thermal & chemical stability	Degree of asymmetry
0.1-5	polysulfone	immersion precipitation	hollow fiber	+	++
0.1-0.65	polypropylene	thermal precipitation	hollow fiber/tubular	++	-
0.1-0.4	nylon-6		hollow fiber/tubular	++	-
0.1	polysulfone	several phase inversion	hollow fiber/flat sheet	+	+
	cellulose acetate	techniques	flat sheet	—	+/-
	fluoropolymer		flat sheet	++	+/-
2-5	polyethylene	phase inversion technique	tubular	+	-
0.02-0.4	polypropylene	stretching	hollow fiber	++	-
0.1-5	polysulfone	immersion precipitation	flat sheet	+	++
0.1-5	fluoropolymer		flat sheet	++	++
0.2-10	nylon	evaporation precipitation	flat sheet cartridge	++	-
0.1-5	polysulfone?	immersion precipitation	spsiral wound/tubular	+	+
0.01-0.5	polyvinylalcohol	immersion precipitation	hollow fiber	-	
0.2	polysulfone/ fluoropolymer	phase inversion technique	flat sheet (rotary module)	++	
0.2	polyolefine	phrase inversion technique	hollow fiber	++	
0.1-0.65	cellulose nitrate	evaporation/immersion	flat sheet	-	+/-
	fluoropolymer	precipitation	flat sheet	++	+/-
0.1-1.2	polysulfone	immersion precipitation	spiral wound	+	
	fluoropolymer		spiral wound	++	
0.1-1.5	fluoropolymer	immersion precipitation	flat sheet	++	+
0.2-5	fluoropolymer	evaporation/immersion	flat sheet	++	+/-
0.6-0.8	nylon	precipitation	flat sheet	++	+/-
0.05-1	poly(ether)imide	immersion precipitation	hollow fiber/flat sheet	+	+
0.05-1	polysulfone		hollow fiber/flat sheet	+	+
0.2	polycarbonate	tract etching	flat sheet		
0.025-0.2	γ-A_12O_3	anodic oxidation	flat structure	+/-	-
0.2-5	α-A_12O_3	sintering/slip casting	tubular	+++	++
0.1	ZrO_2/α-Al_2O_3		tubular	+++	
0.2-3	glass	leaching of soluble phase	tubular		
0.1	ZrO2/C	dynamically formed from suspension	tubular	+	++
0.15-8	SiC	sintering/slip casting	tubular	+++	++
0.2-1	carbon	-	tubular	+++	++
0.2-1	α-A_12O_3	sintering/slip casting	tubular	+++	++
0.2	SiC	sintering/slip casting	tubular	+++	++
0.6-0.14	ZrO_2/C	dynamically formed from suspension	tubular	++	++
0.05-0.3	glass	leaching of soluble phase	hollow fiber	+/-	-

FILTERS & FILTRATION HANDBOOK 3RD EDITION

With over 3000 sales of the second edition, the Filters & Filtration Handbook has become the leading reference manual for the selection and application of filtration & Separation products.

The new edition has been extended and updated to incorporate all the latest developments in liquids, solids and air filtration and separation technology supplied by both manufacturers and users.

The Handbook is an essential reference tool for managers, engineers, designers, technicians, plant operators and consultants and those with responsibility for purchasing, planning, sales and marketing. It is directly relevant to numerous industries including water/waste treatment, fluid power, chemicals, pharmaceutical, food and beverages processing, general engineering, electronics and manufacturing.

OUT NOW!

OVER 750 PAGES,

1500 ILLUSTRATIONS

TABLES AND CHARTS!

SECTION 1 — BASIC PRINCIPLES:
Filters & Separators • Contaminants • Filter Ratings • Filter Tests • Surface and Depth Filtration

SECTION 2 — FILTER MEDIA
Absorbent and Absorbent Media • General Types of Media • Membrane Filters • Woven Wire, Expanded Sheet & Mesh

SECTION 3 — TYPES OF FILTERS:
Strainers • Screens • Cartridge Filters • Electrostatic Precipitators • Candle Filters • Magnetic Filters • Sintered Filters • Bag Filters • Precost Filters • Filter Presses • Rotary Drum Filters • Rotary Disc Filters • Belt Filters • Leaf Presses • Tipping Fan Filters

SECTION 4 — TYPES OF SEPARATORS:
Separators • Centrifuges • Hydrostatic Precipitators • Wet Scrubbers • Dynamic Precipitators • Coalescers • Mist Eliminators

SECTION 5 — LIQUIDS AND SOLIDS
Water Filters • Process Filters • Backfinishing Filters • Industrial Wastewater Treatment • Oil/Water Treatment • Metal Working Fluids • Dewatering and Fuel Treatment • Industrial Chemical Filters

SECTION 6 — AIR FILTRATION
Air Filters • Air Filter System • Dust Collectors • Fume and Hot Gas Filters • Machine Intake Filters • Compressed Air Filters • Respiratory Air Filters • Sterile Air and Gas Filters

SECTION 7 — OILS
Hydraulic Oil Filters • Engine Filters • Homgenizers. Oil Cleaning

SECTION 8 — FILTER SELECTION
Selection Data • Tables and Graphs • Buyers Guide

TO ORDER

☐ Please send me _____ copies of the Filters and Filtration Handbook (3rd Edition) at £102/$204
☐ Payment is enclosed (Please make cheques payable to Elsevier)
☐ Please invoice me
☐ Please charge my ☐ Access/Mastercard ☐ Visa/Barclaycard Eurocard ☐ American Express

ELSEVIER ADVANCED TECHNOLOGY

Card No:.. Expiry Date: ....
Signature:.. Name: ....
Position:.. Organization: ....
Address: ....
.... State: ....
Post Code/Zip Code:..................................... Country: ....
Tel: .... Fax: ....
Nature of Business: ....

FAX OR POST TO:
Elsevier Advanced Technology • PO Box 150 • Kidlington • Oxford • OX5 1AS • UK
Tel: +44 (0)1865 843848 • Toll Free From US: 1-800-85-0085 • Fax: (0)1865 843971

tion of pores throughout the film or a sharp degree of asymmetriy in the structure of the membrane. The asymmetric membrane, shown in Fig 4(b) has long finger-like pores that reach to one surface of the membrane, while towards the outer surface of the membrane the pores become much smaller and a thin skin layer can be detected. In contrast, the membrane shown in Fig 4(c), has no larger finger-like voids, but there is a strong gradation in the size of the pores from one surface to the other. Many microfiltration membranes are symmetric in structure.

Microfiltration

Microfiltration membranes (see Table 4) can be produced by several methods, sintering, stretching, phase inversion (solvent casting) and track etching.

Three hydrophobic materials are commonly used as MF membranes, PTFE, PVDF and polypropylene. These all exhibit excellent to good chemical stability. PTFE is insoluble in most common solvents and cannot be produced by solvent casting. PVDF is less stable than PTFE, and is soluble in aprotic solvent such as dimethylformamide and can be produced by solvent casting. Polypropylene is the least stable of the three and can be produced by stretching and phase inversion.

Many polymer membrane materials exhibit detrimental adsorption characteristics. Solute adsorption has the effect of reducing flux and can lead to difficulties in membrane cleaning. A range of hydrophobic membranes are consequently widely used owing to reduced adsorption behaviour. The best known materials are based on cellulose such as cellulose esters (acetate, triacetate, nitrate and mixed esters). Cellulose is a polysacharide, derived from plants, and is quite crystalline. The polymer is very hydrophilic but is not water soluble. It is limited in terms of chemical, thermal and mechanical stability. The pH range of operation is limited to 4 to 6.5. The polymer is also very susceptible to biological degradation. Despite these limitations cellulose and its derivatives are used in several membrane separations eg

- cellulose (regenerated) in dialysis
- cellulose nitrate and acetate in UF and MF
- cellulose triacetate in reverse osmosis

Polyamide is another important class of membrane material with good chemical, thermal and mechanical stability. The aliphatic polyamides such as nylon-6, nylon 6-6 and nylon 4-6 are widely used as microfiltration membranes. Aromatic polyamides are widely used for reverse osmosis.

For the use of ultrafiltration membranes the usual method of preparation is phase inversion. Three widely used materials are polysulphones, polyimides and polyacrylonitrile. The polysulphones and polyimides alone are chemically and thermally stable, hydrophobic materials. Polyacrylonitrile often has a co-monomer eg methylmethacrylate, added to increase its hydrophilicity.

Ultrafiltration membranes are frequently used as supports for the production of composite membrane for reverse osmosis and gas permeation.

TABLE 5 – Chemical compatibility of membrane materials.

	Cellulose Acetate	Hydrophobic PUDF	Hydrophilic PUDF	MF-Millipore Cellulose Esters	Fluoropore PTFE	Mitex PTFE	Isopore PC Polycarbonate	Isopore PET Polycarbonate	AN and PP Polypropylene
ACIDS									
Acetic acid, glacial	X	●	●	X	●	●	O	●	●
Acetic acid, 5%	●	●	●	●	●	●	●	●	●
Boric acid	●	●	●	●	●	●	●	—	●
Hydrochloric acid (conc.)	X	●	●	X	●	●	●	X	●
Hydrofluoric acid	X	●	●	X	●	●	●	X	O
Nitric acid (conc.)	X	O	X	X	O	●	●	X	O
Sulphyric acid (conc.)	X	●	X	X	O	●	X	X	●
BASES									
Ammonium hydroxide (6N)	X	●	X	X	●	●	X	O	●
Sodium hydroxide (conc.)	X	●	X	X	●	●	X	X	●
SOLVENTS									
Acetone	X	X	X	X	●	●	O	●	●
Acetonitrile	X	●	O	X	●	●	—	—	O
Amyl acetate	X	●	●	X	●	●	●	●	X
Amyl alcohol	X	●	●	X	●	●	●	●	●
Benzene	X	●	●	●	O	●	O	●	O
Benzyl alcohol (1%)	O	●	●	●	●	●	●	●	●
Brine (sea water)	●	●	●	●	●	●	●	●	●
Butyl alcohol	●	●	●	●	●	●	●	●	O
Carbon tetrachloride	X	●	●	●	O	●	O	●	X
Cellosolve (ethyl)	X	●	●	X	●	●	—	●	O
Chloroform	X	●	●	●	O	●	X	●	O
Cyclohexanone	X	●	●	X	●	●	O	●	X
Dimethylacetamide	X	X	X	X	●	●	—	—	O
Dimethylformamide	X	X	X	X	●	●	X	●	●
Dioxane	X	●	●	X	●	●	X	●	O
DMSO	X	●	●	X	●	●	X	●	●
Ethyl alcohol	●	●	●	O	●	●	●	●	●
Ethers	O	●	●	●	●	●	●	●	X
Ethyl acetate	X	●	●	X	●	●	O	●	●
Ethylene glycol	●	●	●	X	●	●	●	●	●
Formaldehyde	●	●	●	X	●	●	●	●	●
Freon TF or PCA	X	●	●	●	●	●	●	●	X
Gasoline	X	●	●	●	●	●	●	●	●
Glycerine (glycerol)	●	●	●	●	●	●	●	●	●

TABLE 5 – Chemical compatibility of membrane materials.

	Cellulose Acetate	Hydrophobic PUDF	Hydrophilic PUDF	MF-Millipore Cellulose Esters	Fluoropore PTFE	Mitex PTFE	Isopore PC Polycarbonate	Isopore PET Polycarbonate	AN and PP Polypropylene
Hexane	X	●	●	●	●	●	●	●	●
Hydrogen Peroxide (3%)	X	●	●	X	●	●	●	●	●
Hypo (photo)	—	●	●	●	●	●	—	●	●
Isobutyl alcohol	●	●	●	●	●	●	●	●	●
Isopropyl acetate	O	●	●	X	●	●	—	●	●
Isopropyl alcohol	●	●	●	X	●	●	●	●	●
Kerosene	X	●	●	●	●	●	●	●	●
Methyl alcohol	●	●	●	X	●	●	●	●	●
Methylene chloride	X	X	●	X	O	●	X	●	X
MEK	X	X	X	X	●	●	O	●	●
MIBK	X	●	X	X	●	●	—	—	●
Mineral spirits	X	●	●	●	●	●	●	●	●
Nitrobenzene	X	●	●	X	●	●	X	●	●
Paraldehyde	X	●	●	X	●	●	—	—	—
Ozone (10 ppm in water)	X	●	●	●	●	●	●	●	●
Pet base oils	X	●	●	●	●	●	●	—	—
Pentane	X	●	●	●	●	●	●	●	X
Perchloroethylene	X	●	●	●	O	●	●	●	X
Petroleum ether	X	●	●	●	●	●	●	●	X
Phenol (5.0%)	X	●	●	●	●	●	—	●	●
Pyridine	X	●	●	X	●	●	X	●	●
Silicone oils	X	●	●	●	●	●	●	●	●
Toluene	X	●	●	●	O	●	O	●	O
Trichloroethane	X	●	●	●	O	●	O	●	X
Trichloroethylene	X	●	●	●	O	●	X	●	X
TFA	X	●	O	X	●	●	—	—	●
THF	X	●	●	X	●	●	X	●	O
Xylene	X	●	●	●	O	●	●	●	O
GASES									
Helium	X	●	●	●	●	●	●	●	●
Hydrogen	X	●	●	●	●	●	●	●	●
Nitrogen	X	●	●	●	●	●	●	●	●
Ozone	X	X	X	X	X	●	●	O	O

Codes: ● - Recommended   O = Limited applications, testing prior to use is recommended
X = Not recommended
Recommendations are based upon static soak for 72 hours at 25°C and atmospheric pressure. Dynamic (operating) conditions at moderate (± 10%) fluctuation will not change the recommendations, but high liquid temperature may do so in some cases.

SYMMETRIC MICROFILTRATION MEMBRANES

| pore structure | straight pores | soap bubble-like (foam-like) | coral-like (tortuous) | stretched |
| production technique | track-etching/ anodising processes | casting + leaching/evaporation | | film-stretching |

ASYMMETRIC MICROFILTRATION MEMBRANES

| pore str. | finger-like substructure sieve-like toplayer | foam-like substr. nodular toplayer | double toplayer | sintered ceramic spheres |
| production technique | phase inversion | phase inversion | phase inversion | sintering/ slip casting |

FIGURE 6

Preparation Methods

Stretching

Thermoplastic polymers can be melted and extruded through a die to produce microfiltration membranes with a pore structure induced by stretching the material. The partially crystalline polymeric material is stretched perpendicular to the direction of extrusion so that crystalline regions are located parallel to the direction of extrusion. Under mechanical stress, small ruptures occur in the membrane which are slitlike, generally 0.2 m in length and 0.02 m in width. Pore sizes in the range of 0.1 m to 3 m are generally produced. The porosity of the membranes produced is high and can approach 90% is some instances. A widely used material is polypropylene.

Sintering

The production of membranes by sintering involves the pressing of a powder of a given size and then heating to an elevated temperature. With the correct temperature of sintering the interface between the particles disappears to produce a porous structure. The pore size produced depends on the particle size and size distribution, but is limited to sizes of 0.1 m. The porosity of the membrane is relatively low in the range of 10 to 20% for polymers. The method is also widely used to produce membranes from metal, ceramics, carbon and glass. In the case of metal membranes porosities up to 80% can be achieved.

Track-etching

An ideal microfiltration membrane will be a sieve with a uniform and regular hole size. Such membranes can be produced by track-etching. Track-etched membranes differ fundamentally in structure from other membranes. They are the closet analogues to a sieve -the membranes are essentially dense polymer films that are punctuated by cylindrical holes. Figure 6 shows the surface of a track-etched membrane, in which the holes are about 0.2 m in diameter. Membranes are manufactured by a process divided in two steps:

MEMBRANE MATERIALS, PREPARATION AND CHARACTERISATION

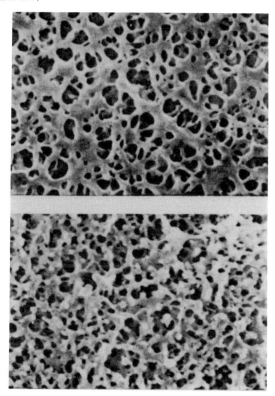

FIGURE 5 – Scanning electron micrograph of a polypropylene membrane produced by stretching.

tracking and etching. During the tracking phase, a thin polymer film is unrolled at high speed while exposed to a fast moving beam of accelerated Ar6+ ions (see Fig) The highly energetic ions pierce the polymer film and break the polymer chains. This leaves "tracks" in the membrane material. The activated film is then subjected to "etching". In this the "tracked" film is exposed in baths of aggressive chemical agents e.g. NaOH, which attack the polymer (Fig 7). The polymer film is rapidly attacked by the accelerated ions along the tracks, which are now converted into clean, cylindrical pores of a defined, uniform diameter. The pore density of the membrane is controlled during the tracking step, by modifying the speed of the film, an the open diameter is controlled during the etching step, by varying the immersion time in the etching baths. The use of a highly energetic Argon ion beam during the tracking step allows the piercing of thicker films than was possible in older processes.

This beam of Argon ions, accelerated in a cyclotron during the manufacturing, avoids radioactive contamination, and enables the etching to be performed immediately after the tracking step. This reduces manufacturing time and improving quality control on the final product.

In older processes, the tracking phase consisted of unrolling a polymer film above a

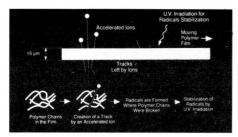 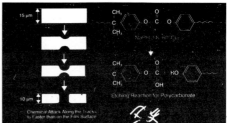

FIGURE 7 – Schematic representation of track-etching process for microfiltration membranes. Tracking and etching step.

source of fissile material inside a nuclear reactor. Accelerated neutrons coming from the heart of the reactor induced fission of the radioactive material. The fission products went through the polymer film leaving tracks of broken polymer chains. After waiting several months to reduce the radioactivity of the film to an acceptable level, the etching step was performed on the activated film.

It has only proved possible to manufacture membranes with pore sizes in the microfiltration range (0.03 – 8 m). Because track etched membranes are symmetric their resistance to the flow of water is proportional to the membrane thickness and they are therefore made thinner than asymmetric microporous membranes in order to have comparable flux.

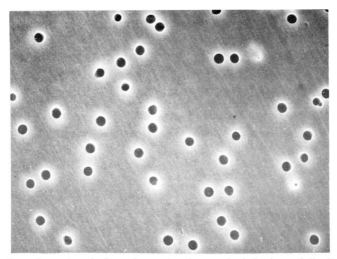

Surface of a track-etched membrane showing regular hole size and distribution.

A usual material which is track-etched is polycarbonate. Porosities of track etched membranes are of the order of 10% this being determined by the radiation time.

Leaching

A technique for preparing porous glass membranes is by leaching. The procedure takes a three component (eg $Na_2O - B_2O_3 - SiO_2$) homogeneous melt and cools it which causes separation into two phases. One phase consists mainly of insoluble SiO_2 while the other phase is soluble. The soluble phase is leached out by acid (or base) to produce a porous structure.

Phase Inversion

The majority of polymeric membranes can be produced by a method known as phase inversion. Phase inversion is a process whereby a polymer solution inverts into a swollen three-dimensional macromolecular complex or gel. Porous membranes are produced from a two or three (or four) component dope mixture containing, polymer solvent and non-solvent (and salt in some cases).

The usual method of phase inversion is by immersion precipitation in a gelatin solution.

The procedure is to first dissolve the polymer in the solvent solution, which may contain additives. This solution (casting solution) is spread directly onto a suitable support by using a casting knife. The support may be a glass plate, other inert support or a support for the membrane itself eg non-woven polyester. The casting thickness can typically vary from 50-500 μm. The cast film is then transferred to a non solvent (gelation) bath where exchange occurs between solvent and non-solvent (typically water) which leads to polymer precipitation. The use of an inert support during casting produces the free form of membrane.

The membrane performance characteristics (flux, selectivity) of phase inversion membranes depends upon many parameters

- polymer concentration
- evaporation time before immersion
- humidity
- temperature
- composition of casting solution
- coagulation bath composition and conditions.

Hollow fibre and capillary membrane are produced by a different procedure to that of flat sheet membranes, either wet spinning (or dry-wet spinning), melt spinning and dry spinning. The hollow fibres are self supporting and demixing occurs both outside and inside the fibre. The preparation method takes the viscous polymer casting solution and pumps it through a spinneret. The spinneret is a nozzle with a solid inner annular section. The annular section enables the polymer solution to be produced in a cylindrical form with an inside coagulation (or bore) fluid. This type of spinneret is used for the wet and wet-dry spinning. After spending some time in the air, or in a controlled atmosphere, the fibre is then immersed in a non-solvent coagulation bath.

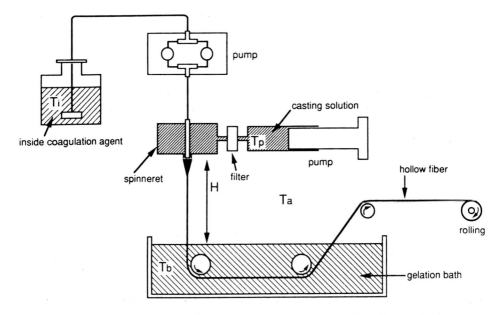

FIGURE 8 – Schematic representation of a wet spinning hollow fibre production.

The procedure for preparing tubular polymer membranes is different to that for the preparation of capillary and hollow fibre membranes, as the former are not self supporting. The casting has to be carried out on a supporting tubular material, eg non-woven polyester or carbon.

Other methods of phase inversion can be used for the preparation of polymer membranes, these are summarised in Table 6.

TABLE 6 – Phase Inversion Membrane Preparation.

Precipitation Method	Principle
Solvent evaporation	Evaporation on inert support or porous substrate in an inert atmosphere. Produce dense membranes (homogeneous).
Vapour phase	Casting of film into a vapour phase of solvent and non-solvent. Membrane formation is due to penetration of non-solvent into cast film, producing a porous membrane with no top layer.
Controlled evaporation	Polymer is dissolved in a solven/non-solvent mixture. Evaporation of solvent during evaporation shifts the composition to a higher non-solvent and polymer content. This leads to polymer precipitation and the formation of a skin on the membrane.
Thermal precipitation	A polymer and solvent solution is cooled to enable phase separation. Evaporation of solvent can allow the formation of a skinned membrane. Frequently used to prepare microfiltration membranes.
Immersion	A solution of polymer plus solvent is cast (on a support) and immersed in a coagulation bath. Precipitation occurs by the exchange of solvent and non-solvent in the coagulation bath.

A typical manufacturing process for cellulose acetate reverse osmosis and ultrafiltration membranes is by dissolving the polymer in a mixture of acetone, dioxan and formamide. The casting solution is spread as a thin film while exposed to the air and then plunged into an aqueous gelation bath. Generally the formulation of the casting solution affects the subsequent stages of the membrane manufacturing process. In this example both dioxan and acetone are solvents for cellulose acetate but acetone is the more volatile. Mixing both solvents enables the volatility of the solvent mix to vary, hence the rate at which solvent evaporates in air stage also is varied.. In this example the role of formamide in the casting solution is to reduce the affinity of the cellulose acetate for the mixture of solvents. Formamide is not a true solvent for cellulose acetate; when cellulose acetate is exposed to it the polymer swells so that during exposure to air the more volatile solvents evaporate from the cast film and the solution is brought to the point of polymer precipitation. The precipitation of a polymer phase from the casting solution is referred to as phase inversion.

After precipitation has occurred, a thin skin of polymer forms on the surface of the cast film. The resistance of this skin may limit the transport rate of solvents out of the casting solution. Further removal of solvents from the bulk of the film therefore takes place relatively slowly once the film has been immersed in the gelation bath.

In this gelation stage the microstructure of the bulk of the membrane is determined although the process of gelation is strongly influenced by the properties of the skin at the surface of the membrane exposed to the air. Three types of membranes are typically produced;

1. If the skin layer is dense, and has a very high resistance to the gelation medium and the outward transfer of the solvents, the rate of polymer precipitation can be limited by the rate of transfer across the skin, and a uniform porous structure produced.
2. If the resistance of the skin is such that the rate of diffusional transport across the bulk of the membrane is more important a gradation in pore size away from the skin surface is likely to occur. This gives an asymmetric sponge-like shown in figure 4.
3. Membranes with finger-like voids (figure 4), are formed if the precipitation of polymer occurs rapidly, i.e. where solvents with little affinity for the polymer are used or where the gelation bath has a high salinity. One proposed mechanism for the propagation of finger-like voids is rupturing of the skin layer in certain places, (due to rapid polymer precipitation), followed by the ingress of gelation medium.

Certain polymer membranes cannot be produced by conventional solvent casting because of the difficulty of identifying suitable organic solvents. In these cases heat treatment of the polymer may be used to induce the formation of a microporous structure. A thermally induced phase inversion process is used to produce polypropylene microporous membranes. The polymer is dissolved in a hot solvent and phase inversion is induced by thermally quenching the hot cast film.

Composite Membranes

The separation of a range of gases or liquid mixtures is effectively achieved by dense homogeneous films. However the permeation rates are very low due to the relative thick membrane used ie 20-200 µm. Thin membranes of the order of 0.1 to 1 µm are ideally required to achieve acceptable permeation rates. These membranes are mechanically weak and are in need of support. Asymmetric membranes with thin top layers such as cellulose acetate RO membranes prepared by phase inversion generally achieve the required objective. Solvent-casting techniques are successfull in the production of asymmetric membranes for reverse osmosis, ultrafiltration and microfiltration. However they do have certain limitations, certain polymers are not soluble in the preferred solvents, while for many RO the final membrane does not reproduce the known salt removal efficiency of the polymer. Hence, to overcome these problems composite membranes have been developed.

Composite membranes, are formed onto a porous, compaction-resistant, support layer, typically polysulphone ultrafiltration membranes, which possess most of the desired properties of a support layer. The critical stage is the formation of an ultrathin skin layer on top of the polysulphone support layer. The skin layer needs to be very thin, typically less than 0.1 m and is therefore very fragile. Composite RO membranes are produced by complicated routes in which the top desalting layer is formed in situ by some form of surface reaction. In this way the skin layer is bonded chemically to the support film and the physical stability is high. One of the first materials to be produced was skin layers of a polyurea, by first coating a polysulphone support with a second polymer (polyethylene imine) and then performing a surface crosslinking reaction to form the polyurea.

TABLE 7 – Methods of Composite Membrane Manufacture

Interfacial Polymerisation	Polymerisation reaction occurs between two very reactive monomers (or one pre-polymer) at the interface of two immiscible solvents. The support layer, impregnated with an aqueous solution of reaction monomer or pre-polymer, is immersed in a solvent containing the second reactive polymer. Often heat treatment completes the interfacial reaction and crosslinking.
Plasma Polymerisation	Plasma is formed by ionisation of gas by electrical discharge. Introduction of monomer into a reactor containing the membrane and the ionised gas produced radicals which react with each other. Eventually the molar mass of the polymer becomes too high and it precipitates on the membrane. Top layers of 50 nm can be produced which are highly cross linked.
Dip Coating	An asymmetric UF membrane is immersed in a dilute coating solution (typically <1%) of polymer prepolymer or monomer. After removal from the immersion bath, a thin adherent layer is left on the membrane. Exposure of the film to an oven evaporates solvent and causes crosslinking which fixes the thin layer to the sublayer.
Grafting	Polymer film is irradiated with electrons which generates radicals. Immersion of the film in a monomer bath enables the monomer to diffuse into the film. Polymerisation is initiated at the radical sites in the polymer substrate and a graft polymer is covalently bound to the basic polymer.

Commercial manufacturers produce a range of proprietary, composite reverse osmosis membranes typically polyamide skin layers, produced by crosslinking reaction.

Composite reverse osmosis membranes with a wide variety of desalting top layers are also produced by plasma polymerisation. In this process the desalting polymer layer is formed in situ by creation of an electric discharge in an atmosphere of an organic monomer. The process has been applied with some success to polysulphone substrates and a variety of vinyl monomers.

Composite membrane technology is also used to manufacture inorganic ultrafiltration membranes. Inorganic ultrafiltration membranes can be made by the deposition of a zirconium oxide/hydroxide layer on the surface of a porous carbon support tube, and then firing of the product. Composite ceramic ultrafiltration membranes are also prepared by the deposition of -alumina particles on top of a porous support layer of -alumina.

Generally the advantage of the composite membrane is that each layer can be chosen independently to give optimum performance of permeation, selectivity and stability. In addition composite membranes enable the use of elastomers which are difficult to use in phase inversion techniques. The techniques used to apply ultrathin top layers on supports (see Table 7) generally involve a polymerisation reaction which generates new polymers, an exception is by dip coating.

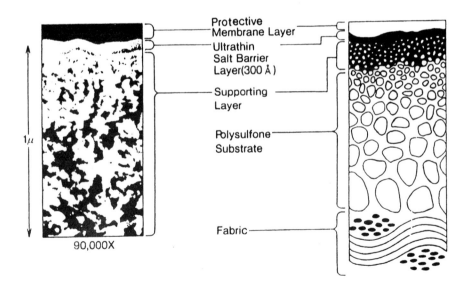

FIGURE 9 – Diagram of a composite membrane showing fabric support, substrate.

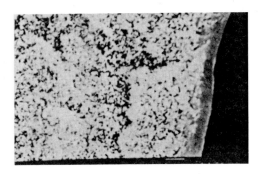

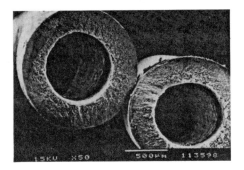

Hollow fibre thin film composite membranes.

In the production of active layers by immersion or spraying, the microporous support must be protected to prevent solution penetration into the pores.

Plasma Polymerisation

Plasma polymerisation is a method which can be used to produce polymer membranes by utilising high energy electrons, ions, atoms, radicals, exited molecules and other active seeds produced by electric discharge. The power to initiative these reactions is supplied by high energy electrons or ions at or above 10^4 K.

Plasma is partially ionised gas consisting of the energetic electrons, ions, photons etc. The plasma generates the chemically reactive species from otherwise inert molecular gas that combine with material to be polymerised or modified. Plasma polymerisation takes place in a low temperature plasma, provided by glow discharge operated in an organic

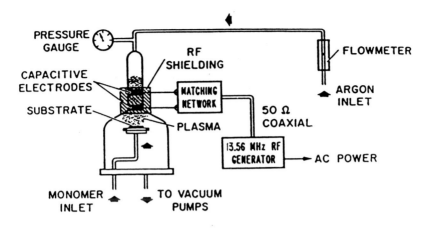

FIGURE 10 – Schematic of plasma polymerisation system.

vapour or gas at low pressure (133 Pa, 1 torr). The system utilises a radio-frequency power supply (13.5 MH$_z$) to generate plasma with electrons with energies in the range of 1-30 eV.

The method is an excellent way of forming surface coatings of polymers which are highly cross linked and which have excellent mechanical. chemical and optical properties. Coatings can be produced by introducing suitable monomer vapours in inorganic plasma gas or by creating the plasma from the organic vapour.

Plasma polymerised organic thin films have been applied in several technologies, optical, biomedical, electrical and in chemical industries. The material produced by plasma polymerisation have excellent properties of rejection of dissolved ions and organic solutes in application for RO and UF. Plasma polymerised organo silicic films eg hexamethyldisiloxane (PM$_2$) have high permeability and permselectivity for oxygen

Hollow fibre oxygen enricher made from plasma polymerisation of PM$_2$ onto hundreds of porous glass hollow fibres, 23 cm in length.

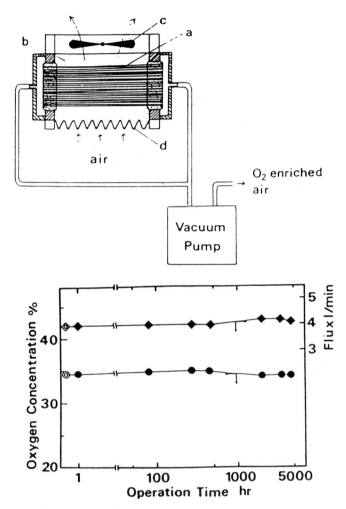

FIGURE 11 and 12 – Schematic diagram of oxygen enricher and membrane flux and permeability performance of PM_2 membrane

transfer form air. They have been used as an oxygen enricher in the medical treatment of patients with lung disease.

Examples of the use of plasma polymer surfaces in biomedical applications are given in Table 9.

If the support layer and active polymer layer are soluble in the same solvent then a intermediate layer is applied between them. Table 10, gives some examples of composite membranes produced by interfacial polymerisation.

TABLE 9 – Biomedical applications of plasma discharge treated polymer surfaces.

Gases (or monomers)	Polymers	Applications	Ref
NH_3 (or $N_2 + H_2$)	Polyproplene (PP) Poly (vinyl chloride) (PVC) Polytetrafluoroethylene (PTFE) Polycarbonate (PC) Polyurethane (PU) Poly (methyl methacrylate) (PMMA)	Heparin bonding for improved blood compatibility	21
Hexamethyldisiloxane (HMDS) $C_2H_4 + N_2$ Allene + N_2 + H_2O	Poly (ethylene terephthalate) (PET) Silastic (SR) Polysulfone (PS)	Improved blood compatibility (in some cases)	22
Hexamethyl-and octamethylcyclotetrasiloxane	PP	Improved membrane for blood oxygenator	23
C_2H_4, allene, styrene, acrylonitrile, C_2F_4, C_2H_3F, C_2F_3Cl, C_2H_3Cl	Polystyrene (PSt) SR	Improved tissue compatibility	24
C_2H_4, C_2F_3Cl styrene	SR	Improved tissue compatibility	25
$C_2H_2 + N_2 + H_2O$	PMMA	Modify corneal contact lens wettability by proteins	26
C_2H_4, Ar	PP, PET, PVC, SR Poly (methyl acrylate) (PMA)	Reduce leaching of small molecules from polymer into body	27
C_2F_4, Et_3SiH, pyridine	PVC (DOP plasticized)	Reduce leaching of DOP into blood	28
C_2H_4, C_2F_4, C_2H_6, Ar	Poly (2-hydroxyethyl methacrylate) [poly (HEMA)] or poly (HEMA-MA)	Control of pilocarpine release rate from hydrogel	29
C_2H_4, C_2F_4, C_2H_6, Ar	SR	Reduce progesterone release rate from SR	30

TABLE 10 – Composite membranes produced by phase interface polymerisation.

Membrane type	Aqueous reactant	Cross-linking agent	Flux $1\ m^{-2}h^{-1})$	R (%)
Polyester	Resorcinol	Isophthaloyl in hexane	75.0	40
Polyamide	Piperazine	Isophthaloyl chloride in hexane	45.8	98
Polyester	Sorbitol	Terephthaloyl chloride in hexane	37.5	97
Polyamide	Polyethyleneimine	Terephthaloyl chloride in hexane	50.0	96
Polyurea (NS 100)	Polyethyleneimine	Toluene diisocyanate in hexane	37.5	99.4
Polyamide type (FT 30)	m-Phenylenediamine	Trimesoyl chloride in hexane	75.0 62.8	98.0 18.06
Polyamide (PA 300)	Polyethyleneimine	Isophthaloyl chloride in hexane	50.0	99.4
Polyamine	1,3-Diaminobenzene	Formaldehyde vapour	75.0	97

Inorganic Membranes

There is growing interest in the use of inorganic membranes due to their superior chemical and thermal stability compared to polymer membranes. There are three general categories of inorganic membranes, ceramics, metals and glass membranes. Metallic membranes eg Ag are mainly made by sintering. Glass membranes are mainly prepared by leaching on demixed glasses. Ceramics are produced by sintering or sol-gel processes. They are generally a combination of a metal eg Al, Zr, with a metal oxide (nitride or carbide), with γ-Al_2O_3 and ZrO_2 prominent examples.

Potential types of inorganic membranes could include both passive and catalytically active types. Passive membrane separation is based on molecular size and generally include both MF and UF.

With polymeric membranes, it is assumed that most MF and UF separations take place on a size basis and RO and GP are assumed to take place mostly by a solution/diffusion mechanism but other mechanisms are also possible. The "inorganic membranes" are actually "super microporous sieves", separations is on the basis of a tortuous path and size than on true solution-diffusion membrane mechanism. However the inorganic separators are not simply sieves, because capillary action, adsorption phenomena, and surface charge play a role in the retention and separation. Tthe formation in-situ of a gel layer that may also initiate a solution/diffusion interaction.

Catalytically active membranes are passive type membranes with a catalyst layer. The catalyst layer can be organic such as an enzyme, or inorganic such as a metal. Solid electrolyte membranes would transfer specific ions through certain ceramic layers. Heterogeneous membranes would be large pore ceramics that incorporate media to

MEMBRANE MATERIALS, PREPARATION AND CHARACTERISATION

preferentially capture and migrate gases across a membrane. Hybrid or composite membranes would be an inorganic layer, probably a ceramic, with a thin permselective organic layer deposited on the top surface.

Inorganic membranes have good high temperature and solvent stability. Chemical resistance is obtained to a greater extent for inorganic membranes than is an overall robustness. Thermal stability is obtainable to a greater degree for inorganic membranes than for polymer membranes.

There are many companies which produce inorganic membrane materials, details of which are in the Appendix.

Not all of these companies or interests have fully commercial products. It can be seen from the list that ceramics are the dominant type of inorganic membrane material and that alumnia is the dominant type of ceramic at this time. In time, it may be that other materials such as zirconia, magnesia or silicon carbide will be equal in importance to alumnia. What seems more likely are the composites becoming the dominant type of ceramic configuration. Starting with a base substrate and coating it with another alumina or zirconia or titania is an evolving trend.

TABLE 11 – Range of available porous inorganic membranes.

Membrane material	Support material	Membrane pore diameter (nm)	Geometry of membrane element
Ni,Au		>500	tube
Ag,Pt			
Ag/Pd		0	tube
ZrO_2	C	4	tube
ZrO_2	C	4-14	tube
ZrO_2	metal	dynamic	tube
ZrO_2	Al_2O_3	10	tube
SiC	SiC	150-8000	tube
SiO_2 (glass)		4-120	tube capillary
Al_2O_3	Al_2O_3	4-5000	monolith/tube
Al_2O_3	Al_2O_3	200-1000	tube
Al_2O_3	Al_2O_3	200-5000	tube
Al_2O_3	Al_2O_3	200	tube
Al_2O_3	Al_2O_3	25-200	disk

Membrane Preparation

The types of porous inorganic membranesare either homogeneous, asymmetric or, composite membranes. The composite structure require the production of the supporting open structure. For ceramics these supports are typically tubes or monolithic elements with several channels, fabricated by ceramic shaping methods such as slip-casting extruding etc. Carbon supports are typically produced by pyrolysis of polymeric precursor or pressing of carbon materials. The supports typically have pore sizes in the range 5-15 mm and porosities of 40-50%, or greater for carbon supports.

Typical use of these supports is to produce microfiltration membranes, which have deposits layers 10-50 μm thick, with pores 0.2-1.0 μm in size and 40-50% porosity. The membranes are prepared by film coating the porous support with a suspension of the ceramic powder. The thickness of the coated layer is adjusted by changing the viscosity of the suspension, for example by changing the solids content of the suspension.

In preparation pinholes in the layer are avoided by suppressing the capillary force effect in the pores. The membrane layer is then formed by sintering at high temperatures, eg 1200 – 1450C for Al_2O_3, the temperature used depending upon material, powder particle size and required pore structure.

The suspensions used in the process are prepared by milling the powders or for finer suspensions by hydrolysis of salts or alkoxides (eg of Al(III), Zr(IV), Si(IV) etc) or other ions. Overall the method is used to apply layers having pore sizes from 50 nm to 500 nm, and in the case of ZrO_2 suspensions on C, 10 nm pore size.

The use of colloidal suspensions is applied in the "sol-gel" process, which uses the capillary forces in the support to improve adherence between top layer and support. The method produces a sharp pore size distribution with rapid production of layers to 10 m in thickness. In practice the support quality has to be very good and the pore size in the top of the support should be 1 m or less. Otherwise the capillary forces are too weak at the start of the cake filtration/slip casting process. The method can be used to produce membranes with multiple layers, the top layer being the thinnest and with the smaller pore size. Final pore sizes of the order of 3 nm can be produced (see Photo).

The production of crack-free membrane layers requires careful control of the drying, calcining and sintering stages. Drying control agents and organic additives, to adjust viscosity, are frequently employed to adjust the pore size distribution. During calcination and sintering the additives are burnt out. Membranes produced by these methods include

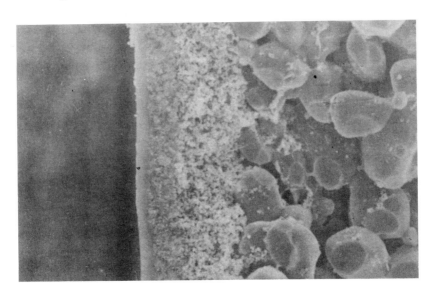

SEM photograph.

MEMBRANE MATERIALS, PREPARATION AND CHARACTERISATION 219

alumina and titania layers on oxides and glass, ZrO_2 on carbon and alumina, and silica on alumina.

Noninfiltrated ceramic membranes

Classical slip-casting of slurries or sols on ceramic supports is a common procedure for the preparation of commercial microfiltration and ultrafiltration materials. These materials are manufactured by association of various granular layers. Each ceramic layer is characterised by its thickness, porosity and mean pore diameter, and these parameter are controlled by the particle size an synthesis method. However the experimental hydraulic resistance of a inorganic composite membrane can be much larger than the resistance obtained by summing all the layers' resistances. This is explained by the existence of a transition boundary layer between two porous media of drastically different sizes.

A new route for membrane synthesis can minimise the infiltration effects.

The basic technique for membrane preparations is the classical slip-casting of a slurry. The supports used were tubular -alumna multilayered commercial supports ,inner and outer diameters 15 and 19 mm. The membrane is deposited on the inner part of the tubes; mean pore diameters were 0.8 and 12 µm depending on the required textural characteris-

Infiltrated (a) and non-infiltrated (b) membranes.

tics. An aqueous titanium (IV) oxide powder suspension is used made from a commercial paint pigment with a 0.2 µm mean particle size, a dispersant (a synthetic Na-based polyelectrolyte), to improve the powder dispersion, a binder and a plasticiser, to control the slurry viscosity and improve the film plasticity during drying and the first steps of the thermal treatment. By slip-casting the suspension inside the ceramic tubes, an infiltration of TiO_2 particles into the support is achieved. After firing treatment, a composite infiltrated material is obtained consisting of the support, the infiltrated boundary layer and the membrane itself.

To prepare noninfiltrated membranes, a new step is introduced, a pretreament of the support before slip-casting the slurry. An aqueous polymeric solution (methyl cellulose) is used to pre-impregnate the inner part of the support. After the polymer film is dried, the ceramic suspension is poured into the tube and evacuated 10 minutes. During this operation the polymeric dense film prevents any penetration into the pores of the support. During the firing treatment, the polymeric film is eliminated and leaves an inorganic noninfiltrated TiO_2 membrane. The final temperature and time of firing controls the mean pore size and the size distribution of the membranes.

The noninfiltrated membranes offers the following advantages:
- hydraulic resistance in the transition layer can be decreased
- noninfiltrated membranes of 0.2, 0.5 and 0.8 m mean pore diameters have water permeabilities increased by 30 to 70%;
- the noninfiltrated membranes are less sensitive to fouling;
- the microfiltration limiting flux is considerably increased.

Aluminium Asymmetric Membrane

An attractive structure for membranes is the asymmetric membranes formed from one single material. This type of structure is quite difficult to produce in general although an

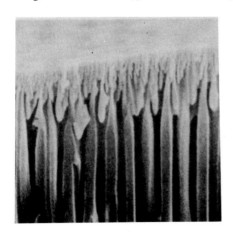
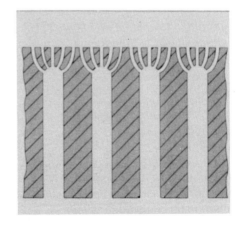

FIGURE 13 – Anodised alumina membrane with asymmetric structure.

anodised aluminium membrane is available (see Fig 13). The top side pores have a size of approximately 25 nm and the support pores are of the order of 200 nm. The material is available as a flat sheet membrane.

Anodised Aluminium

During anodic oxidation several metals develop coherent porous oxide coatings. In normal circumstances this porous layer adheres strongly to the metal substrate and is difficult to remove and thus limits the direct use as a porous membrane layer. Anodising in for example electrolytes of oxalic, phosphoric or sulphuric acid the porous structure is formed inwards from the outer surface only as for as an imperforate barrier layer. However by varying the voltage of the anodising cell, by for example reducing the starting value from 25 V to 0 V in steps of 0.5 V, the single pores which normally form branch into numerous small pores that weaken the film near the substrate metal. Collectively the branched pore system introduces a weakened stratum into the metal oxide film, enabling quite easy separation form the substrate. The barrier layer is very thin in this process and is generally left on the metal substrate, therefore the detached oxide film is porous on both sides of the film.

Before the porous layer is detached from the metal substrate, a perforated supporting layer can be attached to the film (Fig 2). The supporting material can be heat sealed or glued to the anodized film. The perforations of this support are large in comparison to the anodized film.

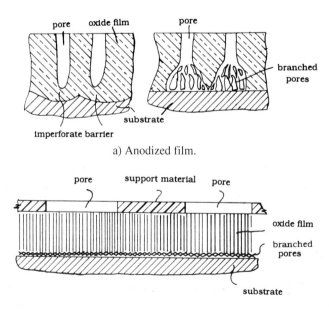

a) Anodized film.

b) Supported film before removal.

FIGURE 14 – Anodized films – a) normal film with imperforate layer, b) with branched pores.

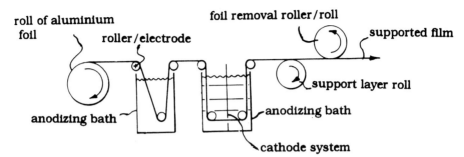

FIGURE 15 – Schematic of the membrane system and the process for production of anodized films.

Aluminium Etched Metallic Membrane

An inorganic membrane is made from aluminium foil by an etching process that generates a capillary pore structure with a pore size of 0.5-8 microns. The membrane is produced by etching recrystalised aluminium foil, either on both sides to produce a symmetrical pore structure, or on one side to produce an asymmetric structure.

Left: Double-sided etched foil.
Right: Cross-section of the same double-sided etched foil.

The membrane has also been made with a silicon rubber coating, and also with a finer pore size, down to 0.002 microns, produced by coating the pore walls with aluminium oxide. The coated version has been successfully demonstrated as a support layer for silicone in gas separation and pervaporation applications. Such membranes have been produced in both flat-sheet and spiral wound modules. An enrichment factor of 17 has been achieved with 4.3% vol CH_2Cl_2 in a nitrogen feed. Also in the pervaporation of aniline in water a selectivity factor between 8 and 23 is achieved.

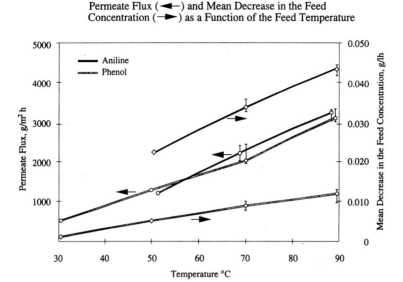

FIGURE 16 – Recovery of phenol or aniline from water using silicon polymer coated etchal membrane sheet.

Earlier tests, gave the following typical flow rates for the membrane

Air	up to 7000 m^3/m^2/h bar
Water	1,000 – 2,000 l/m^2/h bar
Methanol	2,000 – 3,000 l/m^2/h bar.

Recent pilot plant results using a newer, high cubicity foil, gave greatly improved flow rates – in the case of air, more than a thousand times better:

Air	1,000,000 m^3/m^2/h bar
Water	50,000 l/m^2/h bar

The aluminium foil membrane offers a number of advantages compared to polymeric or ceramic membrane types:
- the foil is easily formed; a laser-welded tubular format has been used for standard microfiltration test
- excellent resistance to organic solvents, even at elevated temperatures, and to radiation
- stable in aqueous solution and can withstand cleaning by bleaching with oxidising agents
- electrically conducting – a property which has been used to obtain flux enhancement and faster filtration in microfiltration, and also promises a route to easy cleaning
- tough – withstanding pressures up to 20 bar; this also allows increased filtration rates.

The main applications identified for these membranes are :
- basic etched foil -microfiltration
- coated etched foil – gas separation, pervaporation
- fine pore variant – ultrafiltration.

Major anticipated filtration uses are in the dairy, biotech, brewing and fruit juice sectors.

Thermoplastic Polymer Ultrafiltration Membranes

There is a range of semi-crystalline aromatic polyetherketones which are extremely useful high performance engineering polymers with a unique combination of mechanical toughness, high modulus, hydrolytic stability, resistance to oxidative degradation, and ability to withstand organic solvents and retention of physical properties at very high temperatures up to 250C. The structures of some available so called "thermoplastics" are shown in Fig 17.

The difficulty with the manufacture of membranes from these thermoplastics is their crystallinity makes them insoluble in conventional solvents at ambient temperatures.

Thermal phase-inversion is one of the standard methods of making microporous membranes, in which a polymer is dissolved at high temperature in a solvent, the solution is cooled to precipitate the polymer as a formed membrane, and the solvent is then removed. The polymeric materials made into membranes in this way include polyethylene, polypropylene, polycarbonate and PVC. Other materials, including some high-melting-point polyether ketone, have not previously been made into membranes via this route.

A thermal phase-inversion process which allows the manufacture of membranes for ultrafiltration or microfiltration, from an aromatic polymer containing in-chain ether or thioether and ketone linkages has recently been developed. The first step in the process is when the polymer is dissolved in a "latent" solvent. The polymer can be a homopolymer such as polyetherketone (PEK) or polyetheretherketone (PEEK), or a copolymer: PEK/

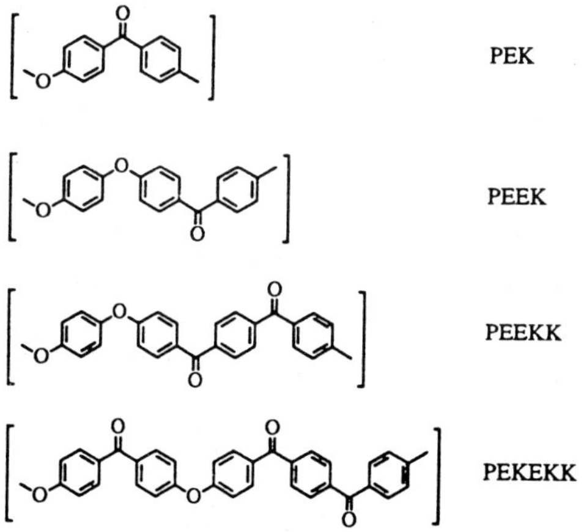

FIGURE 17 – Structure of poly(aryletherketones) thermoplastics
(effective membrane processes)

PEEK, PEEK/PES (polyethersulphone), PEK/PES are some of the combinations. The latent solvent is a compound in which the polymer is soluble at high temperatures, typically only 50C below the polymer melting point of 320-340C. At low temperatures, below 100C, the polymer is only poorly soluble in the chosen solvent. Examples of suitable solvents have a plurality of aromatic rings; tetraphenyls, hexaphenyls or polar polyaromatic compounds. The initial concentration of the polymer in the solvent is 10-50 wt%, preferably 20%. A pore-forming agent eg an inorganic salt or soluble polymer can be added to the solution.

The next step in the formation of the membrane is extrusion of the solution onto glass or another non-porous surface, if an unsupported membrane is wanted, or onto a PTFE, carbon fibre or stainless steel. After the solution cools and polymer is precipitated, the latent solvent is removed by washing the membrane with a suitable solvent for the latent solvent. The crystallinity of membranes produced in this way, typically 40%, can be increased by annealing or by treatment with a warm aprotic solvent such as dimethylformamide. The pore diameter range is $0.001 - 1.0$ microns, and both asymmetric and isotropic membranes can be manufactured.

A second method of formation uses certain very strong mineral acids of liquid hydrogen fluoride, trifluoromethane sulphonic acid, and sulphuric acid to dissolve the polymer. Casting solutions can be formed, containing for example 7-14% by weight of PEK in 98% sulphuric acid. Both flat sheet and hollow fibre membranes for ultrafiltration can be made:

- A reinforced flat sheet (on Ryton PPS paper) 250 m thick with a 0.15 m thick surface skin. Water fluxes of 0.2 m h^{-1} at 2 bar for a 15 kD MWCO membrane and 0.07 m h^{-1} for a 5 kD MWCO membrane are achieved.
- Hollow fibre membranes (see Fig 18) of internal diameter 0.65 mm and 0.15 mm wall thickness with nominal MWCO of 10,000. Pure water flux of 0.12 m h^{-1} at 2 bar pressure are a typical.

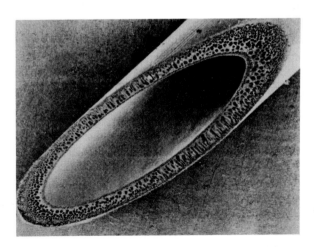

FIGURE 18 – SEM (diagonal cross-section, x 35) of a hollow-fibre PEK UF membrane.

Overall the membranes are steam sterlisable, resistant to solvents such as, acetone, toluene and perchloroethylene.

Microcapsule Membranes

Microcapsules are small containers, 1 to 500 mm in diameter made from synthetic or natural polymers and can be effectively considered as spherical ultrathin polymer membranes. They are used to protect encapsulated material form environmental conditions and to release this material into the surrounding medium is a sustained way. The microcapusle thus serves as a barrier against permeation of solute molecules.

Microcapsule membranes can be prepared by several methods which includes chemical methods (interfacial polymerisation, in situ polymerisation, insolubilisation), physico-chemical methods (aqueous and organic-phase separation, interfacial deposition, spray drying) and physical methods (electrostatic, vapour deposition, fluidised bed spray coating). The chemical methods of preparation are made suitable for preparation of membranes for medical and biochemical applications.

The majority of polymerisation reactions for preparing microcapsules are the condensation type, and consequently interfacial polymerisation is a suitable method for encapsulating liquids. The encapsulation process consists of three basic steps

  i) dispersion of an aqueous solution of water soluble reactant into an organic phase ie the formation of a water in oil emulsion
  ii) formation of the polymer membrane on the water droplet surfaces, initiated by the addition of an organic/oil soluble reactant
  iii) separation of microcapsules containing aqueous phase from the organic phase. This is followed by transfer of the capsules to an aqueous solution.

The production of microcapsule membranes by other chemical methods uses similar modifications of the emulsification process.

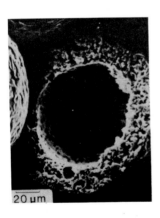

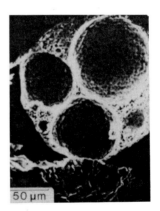

Scanning electron micrographs of microcapsule membranes.

SECTION 2.2 – CHARACTERISATION OF MEMBRANES

The characterisation of a membrane is important to enable a suitable material to be selected for a particular application. Characterisation is a means of relating structural and morphological properties to a class of membrane for a particular application. A problem is that membranes are put to a large number of different uses even within a particular separation category. Typically a membrane will be characterised in terms of its pore size, molecular weight cut-off, porosity, thickness, symmetry, crystallisation etc. These properties enable an indication of likely performance to be made, but often do not relate to the actual performance due to the interactions of the particles, or species, to be separated with the membrane media itself. Nevertheless the membrane or filter characterisation or, rating tests, are an important factor in system design.

Characterisation methods can be conveniently divided into two broad areas, for porous media and for non-porous media. Porous media have observable pores of a size approximately 2 nm and larger. These include membranes for microfiltration and ultrafiltration, dialysis and related applications where the actual 'pore' of the material are the major factor in characterisation. Non-porous membranes are of dense material, without pores and separation is dependent upon the material properties and morphology (crystallinity, glassy, rubbery etc), and conditions of operation e.g. temperature.

Porous Membranes

Characterisation methods for porous membranes can be divided into two areas; structure related parameters and permeation related parameters. Different methods are used for different applications but typically microfiltration and ultrafiltration membranes are of major interest.

Bubble Point Test

The bubble point test is a method for both characterising a particular membrane and also for testing the integrity of a membrane. The test is based on the fact that for a given fluid and membrane pore size and constant wetting the pressure required to force an air bubble through a pore is inversely proportional to the size of the pore or hole. The theoretical relationship between bubble point pressure and pore size is given by the Poiseulle's Law (or Laplace equation) for capillary tubes (see Fig 1).

The factor K is a shape correction term which allows for the fact that no practical membranes consists of symmetrical tubes. An immediate complication therefore arises in that the shape correction factor is dependent upon the method of material fabrication.

The test requires the measurement of the pressure required to blow air through a liquid filled membrane, i.e. when the first stream of bubbles are emitted from the upper surface of the membrane. A schematic diagram of a test system is shown in Fig 3. The membrane is flooded with the test liquid which may be water or frequently iso-propyl alcohol. The choice of fluid largely depends upon the anticipated pore size. The surface tension of a water/air interface is approximately $3^1/_2$ times that at a alcohol/water interface. At low pore size the pressures required in the bubble test for water will be quite high e.g. 145 bar with 0.01 m, and thus iso-propyl alcohol (i.p.a) is preferred. Although other liquids can be used

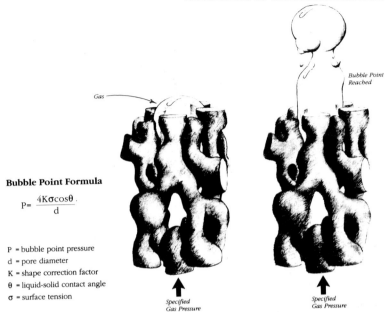

Bubble Point Formula

$$P = \frac{4K\sigma\cos\theta}{d}$$

P = bubble point pressure
d = pore diameter
K = shape correction factor
θ = liquid-solid contact angle
σ = surface tension

FIGURE 1 – Bubble point formula for bubble test.

these will give different value of pore size for the same membrane medium and thus a standard of i.p.a is preferred. The bubble point test is somewhat sensitive to the pore length and the rate of pressure increase. The test can also give an indication of the pore size distribution. The point at which the first stream of bubbles appear is the largest hole. A step increase in pressure will reveal an increasing number of holes. Eventually a pressure is reached where the entire surface is steaming with air bubbles i.e. the open bubble point ("boil-point"). This pressure is that corresponding to the mean pore size of the filter.

Integrity Test

The bubble point test is a non-destructive test; that does not contaminate the filter. It can be used to test the integrity of a membrane or filter so an alternative to destructive tests such as bacterial retention tests for sterilising membranes. The bubble point test detects minor filter defects and also out-of-size pores and correlates with the bacteria passage test (see Fig 2).

An in-process bubble point test can detect damaged membranes, ineffective seals, system leaks and distinguish filter pore size. For an in process bubble test, pressure from a nitrogen source (at 0.4 bar) (stage 1) forces liquid from a pressure vessel to fully wet the membrane. Once the liquid has passed through the membrane, gas is in contact with the filter surface while the pores and downstream piping are still filled with water, as the gas pressure is below the bubble-point pressure. Increasing the pressure to the bubble point pressure will eventually displace the liquid from the pores.

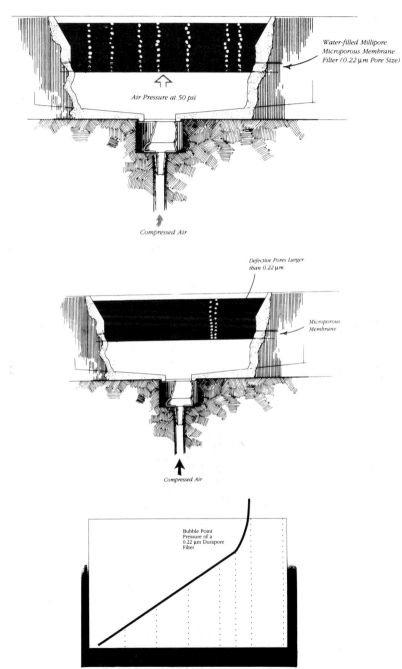

FIGURE 2 – Bubble point integrity text.

For an in-process bubble point test, pressure from a nitrogen source forces liquid from a pressure vessel to fully wet the membrane filter.

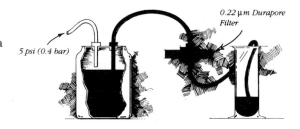

Once all the liquid has been passed through the membrane filter, gas is in contact with the filter surface. The pores and downstream tubing are still full of water.

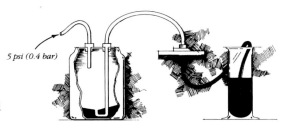

When the applied pressure reaches the bubble point pressure of the filter, liquid is displaced from the filter pores.

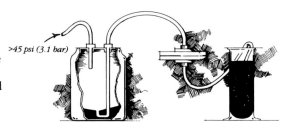

FIGURE 3 – In process bubble point testing.

Integrity test equipment.

MEMBRANE MATERIALS, PREPARATION AND CHARACTERISATION

Diffusion Testing

In high volume systems with final filter surface areas of 2 ft^2 or greater, a Diffusion Test is recommended. The diffusion test is based on the fact that gas will diffuse through the pores of a fully wetted filter. This diffusion rate is proportional to differential pressure and surface area. When the pressure begins to exceed the bubble point of the filter, bulk gas flow results. There are orders of magnitude of difference between diffusional flow and bulk flow.

The flow of gas is limited to diffusion through water-filled pores below the bubble point pressure of the filter under test. At the bubble point pressure, the water in the pores is forced out and bulk gas flows freely through the filter.

In the diffusion test, pressure is typically applied at 80% of the bubble point pressure of the filter under test. When there is liquid downstream of the filter, the volume of gas flow

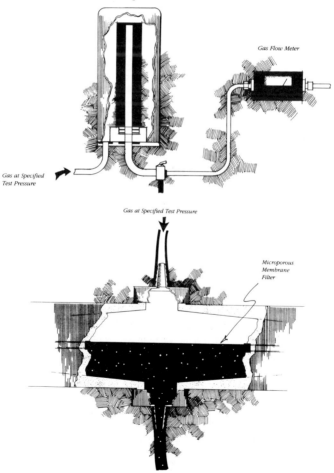

FIGURE 4 – Principle of diffusion testing.

is determined by measuring the flow rate of displaced water. The rate of diffusion can also be measured by a gas flow meter.

In industrial settings, the flow rate is often measured on the upstream side of the filter with an automated instrument. Upstream measurements do not require a tap into the sterile downstream side. The measurement technique used by many automated devices is pressure decay. After the gas on the upstream side is pressurised to the desired test pressure, the device isolates the filter from the gas source and measures the rate of pressure decay resulting from gas leaving the upstream side of the filter housing. This decay rate can be converted to the standard volumetric flow rate expression for the diffusion rate via the ideal gas law.

Mercury Intrusion Method

The mercury intrusion test is similar to the bubble point method relying upon the penetration of a fluid i.e. mercury into the membrane pores. In this case a dry membrane is exposed to mercury under pressure. The volume of mercury forced into the membrane is related to the pore size and pressure (the OLaplace equation), and can be measured very accurately. Thus at a particular pressure the pore size can be measured (the size is inversely proportional to the pressure) and also the fraction of pores of a particular size. As with the bubble point test a morphology or shape factor must be introduced.

The test involves exposing the dry membrane to a pool of mercury. The pressure of the mercury is gradually increased and at a certain lowest pressure, the largest pores will fill with mercury. Increasing pressure progressively fills the smaller and smaller pores until a maximum intrusion of mercury is achieved. At high pressure however, erroneous results may be obtained due to deformation or damage to the membrane material. In addition the method also measures dead-end pores, which are not active in filtration. The size range of the test covers 5 nm to 10 m pores meaning it covers microfiltration and some ultrafiltration membranes. Overall it gives both pore size and pore size distribution.

Water Integrity Test

Sterilising grade hydrophobic filters are utilised for the sterile filtration of air streams and gases in many pharmaceutical and biological applications. The integrity of these filters and their ability to retain bacteria are traditionally correlated to a non-destructive integrity test. e.g. the use of solvents (ethanol, isopropanol, freon) to wet the membranes in order to perform bubble point and diffusion integrity tests. The use of these solvents poses several problems.

In situ integrity testing using solvent wetted membranes requires elaborate bypass valve systems to prevent downstream contamination by solvent residuals. After testing great care must be taken to assure complete removal of the solvents. Solvents may reduce the active filtration area through particulate mediated membrane fouling, reducing filter performance and potentially leading to damage of the vented vessel. Here also, incomplete solvent removal during drying can cause the filter to rewet, rendering it hydrophilic and no longer capable of acting as a hydrophobic vent.

For these reasons solvent integrity is usually not performed in situ *post* sterilisation, but rather *pre*sterilisation. Then, sterility assurance can only be guaranteed if a filter integrity test is performed immediately prior to use but it proves extremely difficult to perform the required test and to install the filter assembly on the process equipment aseptically.

To overcome these problems an alternative integrity testing method based on water intrusion and penetration. The 'Water Pressure Integrity Test' (WPIT) may be performed in situ post sterilisation without any downstream manipulations and can be directly correlated to the retention or bacterial challenge tests. Other advantages include:
- applicability to any venting or gas filtration system
- filters can be tested after steaming or autoclaving in situ
- insensitivity to temperature fluctuations
- ability to test in hard to reach locations
- simplicity of testing routine

The water pressure integrity test is based on the same principles as the 'Mercury Intrusion' test. By definition, hydrophobic membranes resist wetting by water. The water repellent forces may be overcome through the use of sufficient pressure to 'wet out'. In practice this means that the upstream volume of the housing or filter must be completely flooded, then pressure applied by air on the water volume and the rate of water permeation determined.

Water permeation or intrusion is seen as an upstream pressure drop. The pressure drop can be converted into a diffusional flow which can be used to determine the integrity of

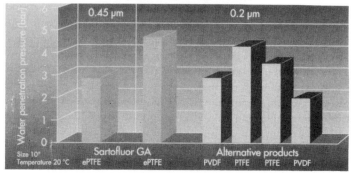

Water integrity test unit.

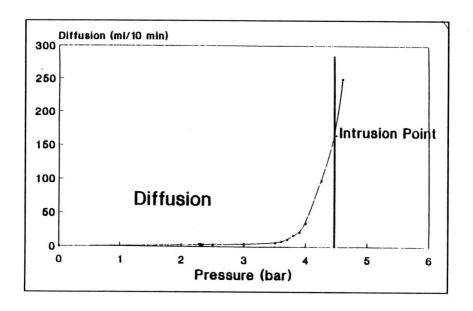

FIGURE 5 – Results of water intrusion test on 0.2 micron filter cartridge.

the membrane. The limited compressibility of water means that any permeation of water through the membrane – unlike air diffusion during the traditional pressure hold tests – will exhibit itself as a relatively large (volume dependent) pressure drop. This is the basis for the sensitivity and accuracy of the test method.

Tests are typically performed on 0.2 μm filter cartridges constructed with Gore expanded PTFE membranes. A dry cartridge is inserted into a housing and the upstream volume filled with water. An automatic filter integrity tester can be used to control and monitor the pressure test and programmed to measure water intrusion as a permeating flow over a range of pressures.

In the testing terminology, *water intrusion* is the point where water forcibly starts to enter and permeate the pore structure. The permeation of water during the intrusion stage initially shows a linear response to increasing pressure (Figure 1). Typically for a 0.2 μm filter, for example, the rate of water intrusion increases linearly with increasing pressure up to about 3.7 bar, when there is a rapid increase in flow and water penetration begins. The *water penetration* value (4.5 bar) is where the membrane becomes wetted and free water flow begins.

A series of comparative tests have established that WPIT is as sensitive and accurate as the current diffusion test method. All filters meeting the specifications for the diffusion test also met the specifications for WPIT. Furthermore, those filters outside the criteria for release for either test also failed the bacterial challenge.

Bacterial Challenge Test

The design and rating of filters was historically dictated by the pharmaceutical industry. The membrane filters is rated or categorised according to their ability to retain unwanted organisms and thereby could sterilise critical fluids. Because of the critical application, it is essential to have meaningful test procedures which are correlated with strict quality control.

A bacterial challenge test system for the evaluation of the effectiveness of high efficiency filters has been developed. The test system (see Fig 6) uses a *nebuliser* adapted from the original Micro Biological Research Establishment (MRE) design for high pressure operations.

The device uses two impinger type samplers in series upstream and a silt sampler downstream of the test filter. A minimum challenge of 3×10^8 spores is recommended for filters operating 300 days per year with average flows of 850 dm^3/min.

The Health Industry Manufacturers Association (HIMA) regulations and the FDA "Guidelines on Sterile Drug Products produced by Aseptic Processing" stipulate that when a sterilising filter when challenged with a minimum concentration of 10^7 Psuedomonas diminuta organisms per cm^2 of filter surface, must produce a sterile filtrate. Filters which produced sterile effluents according to this test were accepted as 0.2 µm in size. In fact

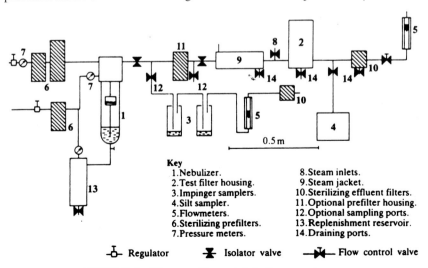

FIGURE 6 – Diagram of bacterial challenge testing apparatus.

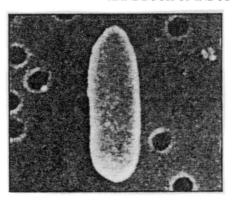

FIGURE 7 – Pseudomonas Diminuta and a 0.2 µm pore.

P.diminuta organism is much larger than 0.2 µm (see Fig 7) a minimum of 0.3 µm in diameter and 1 m in length. This means that filters with 'pore sizes' of 0.3 m and greater could retain the organism.

The bacteria challenge test is a destructive method and it therefore must be correlated with practical non destructive integrity test to ensure filtration reliability.

The FDA "Guideline on Sterile Drug Products Produced by Aseptic Processing" (June 1987) states:

"After a filtration process is properly validated for a given product, process, and filter, it is important to assure that identical filter replacements (membrane or cartridge) used in production runs will perform in the same manner. One way of achieving this is to correlate filter performance data with filter integrity testing data. Normally, integrity testing of the filter is performed after the filter unit is assembled and sterilised prior to use. More importantly, however, such testing should be conducted after the filter is used in order to detect any filter leaks or perforations that may have occurred during the filtration".

A typical method of correlation is to relate the results of the bacterial challenge test with the bubble point and diffusion methods.

It has been proposed that a way of overcoming the variability of filter testing methods and to establish realistic, high challenge tests, is to use the Log Reduction Value (L.R.V.) as recommended for liquid sterilisation filter tests. The L.R.V. is defined as the logarithm of the ratio of the total micro-organisms in the challenge to the micro-organisms in the filtered fluid.

The L.R.V. will depend upon the specified challenge and the challenge concentration and therefore a standard challenge is recommended. It has been suggested that such L.R.V. should be quoted for compressed air filters.

L.R.V. values for P. diminuta retention are generally related to bubble point of membranes. Generally hydrophilic filters, e.g. Durapore are totally retentive if they exhibit a bubble point of the order of 34.5 p.s.i, i.e. the L.R.V. is greater than 9. Similarly for hydrophobic filters a bubble point of 15 p.s.i. for Durapore measured with methanol implies a totally retentive filter. As the bubble point falls so does the value of the L.R.V. Such correlations are specific to particular membranes and filters.

TABLE 1 – Typical results.

Sartobran-PH SM 523 28 07 H1PH Lot no.	Air diffusion in ml/min. at a test pressure of 2.5 bar and at 20°C	Bubble point in bar (psi) at 20°C	Result
004/91- 6	5	≥3.2	Sterile
004/91- 5	5	≥3.2	Sterile
004/91- 9	5	≥3.2	Sterile
152/91-85	5	≥3.2	Sterile
277/91-59	5	≥3.2	Sterile
004/91- 4	6	≥3.2	Sterile
146/91-95	6	≥3.2	Sterile
149/91- 2	6	≥3.2	Sterile
149/91-40	6	≥3.2	Sterile
150/91-16	7	≥3.2	Sterile
151/91-14	7	≥3.2	Sterile
150/91-39	9	≥3.2	Sterile
150/91-40	9	≥3.2	Sterile
151/91- 2	9	≥3.2	Sterile
151/91- 3	10	≥3.2	Sterile
152/91- 0	10	≥3.2	Sterile
149/91-77	12	≥3.2	Sterile
277/91- 7	12	≥3.2	Sterile
151/91-84	14	≥3.2	Sterile
146/91-29	15	≥3.2	Sterile
149/91-57	15	≥3.2	Sterile
151/91-78	16	≥3.2	Sterile
146/91-24	18	≥3.2	Sterile
152/91-97	19	≥3.2	Sterile
237/91- 9	20	≥3.2	Sterile
152/91-75	23	<3.2	Non-sterile
346/91- 3	30	<3.2	Non-sterile

Latex Sphere Test

The organism challenge is a destructive test and obviously cannot be conducted on each filter manufactured. Instead, the organisms retention level is conveniently correlated to a non-destructive test-bubble point integrity. This simple, reproducible test confirms the functional integrity of membrane filters.

Filters which retain 100% of the challenge organism *P. diminuta* normally have water bubble point values of > 45 p.s.i. These filters are however labelled inaccurately, with the organism test as 0.2 µm absolute rated filters. Based on the bubble equation, finer pore-sized membrane filters with bubble point values twice those of the 0.2 µm rated membrane filters (or 90 p.s.i), are rated 0.1 m. This is a correct assumption if a 0.2 µm rated filter were really a 0.2 µm *pore sized* filter.

The pharmaceutical standard for rating filters is perfectly suited for the complete removal of *P. diminuta* under HIMA conditions. It does not, however, accurately address the retention capabilities of filters when challenged with particles other than this organism, nor does it address conditions other than those specified in the HIMA document.

Given the ubiquitous nature of microporous membranes it is surprising that the fundamental performance – that of separating particles from a fluid – is confusing. The performance of microporous filters is defined in many ways using terminology such as "Beta Ratio", "Nominal", "Absolute", "Rated", "Sterilising Grade" – in some cases without regard to the real performance of their products.

Another level of complexity is derived from the different manufacturing techniques that are employed in producing membranes. There are at least three major – and vastly different – processes used. Solution casting – used for Nylon 66, PVDF, Polysulfone etc., Mechanical stretching – used for PTFE, and Track Etching – used for Polycarbonate.

The availability of Polystyrene/Latex spheres with very precise diameters provided an opportunity to challenge filters with a very controlled and reproducible particle matched to the manufacturers rating of his product. Inconsistencies in filter ratings have been noted for several years, e.g. 0.2 μm filters retain 100% of the *P. diminuta* organism, but 100% retention is not achieved when challenged with 0.5 μm and even larger latex spherical particles. Latex spherical particles are nondeformable and readily available in a variety of diameters with very narrow size distributions. When monodispersed in a solution containing a surfactant, adsorption caused by particle and membrane charge interactions is eliminated or reduced. Absolute retention rating should ideally be independent of the filtration conditions. Flow rate, differential pressure, the pH and viscosity of the fluid, etc. can all drastically affect the small particle retention capabilities of membrane filters. These effects cannot be completely eliminated using latex spheres but can be controlled to simulate "worst case conditions" in order to maximise penetration.

In a typical test a suspension of latex spheres (Certified Nanospheres™ Size Standards from Duke Scientific Corp. Palo Alto CA) at a concentration of 2×10^9 per ml. in deionised water with surfactant (0.1% Triton X-100) is used. The Latex Spheres have a tendency to clump together and the use of the surfactant helped to ensure that they were monodispersed and to reduce the surface absorption tendency of some of the test membranes. To detect spheres that may pass through test filters a dual beam spectrophotometer is used. A sphere concentration versus absorbance calibration curve for each sphere size is prepared which is essentially linear.

The test membranes in the form of 47 mm discs are bubble point tested in accordance with the manufacturer's instructions to ensure that they were integral. Each test sample is then challenged with a total of 30 ml of the latex sphere suspension. Each challenge volume – having passed through the filter – was collected and measured for absorbance. Tests ae repeated using spheres of different diameters to get a retention profile of the filter and results recorded as retention efficiency versus sphere diameter.

Typical tests results showed that with the exception of the Track Etched Polycarbonate and the Nylon 66, the membranes tested all showed significant passage of spheres equal to or greater than the manufacturer's rated pore size. See Table A. The Nylon 66 filter exhibited an unusual characteristic in that it was more efficient at 0.220 μm challenge than with challenges with both larger and smaller spheres. This was explained by a surface charge effect and that the absorbed spheres could be removed by washing with further aliquots of the 0.1% Triton X-100.

TABLE 2A – Sphere Challenge in Triton X-100.

Membrane Type	Sphere Diameter μm vs. Percent Retention				
	0.149	0.198	0.220	0.300	0.398
PC	6	98	100	ND	ND
PTFE	12	33	62	100	ND
PVDF	ND	ND	16	24	92
PS	ND	16	13	21	90
N 66	ND	14	93	33	55

TABLE 2B – Sphere Challenge in SDS

Membrane Type	Sphere Diameter μm vs. Percent Retention				
	0.149	0.198	0.220	0.300	0.398
PC	9	99	100	ND	ND
PTFE	31	81	98	100	ND
PVDF	ND	ND	37	60	100
PS	ND	29	46	62	100
N 66	ND	50	41	59	84

The results show that many 0.2 m rated membranes actually passed spheres larger than 0.398 m. This despite the fact that they had been rated as sterilising grade 0.2 m "absolute") by the industry standard HIMA bacteria challenge using an organism with dimensions of 0.3 x 1.0 m. It is often assumed that the rating given by the manufacturer is an accurate representation of its performance and that the performance is independent of the challenge solution. The data generated in these tests shows that – with the exception of the Tracked Etched membrane – this is not the case. In many instances the "absolute" retention performance of the membrane is 2 or more times the rating on the box.

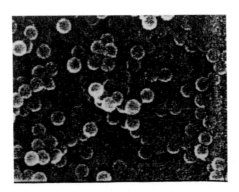

FIGURE 8 – SEM of latex spheres on a polycarbonate membrane.

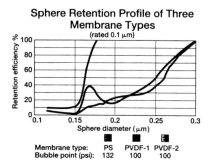
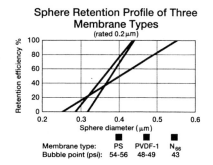

FIGURE 9 – Sphere retention profiles of membranes.

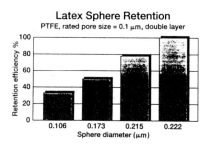
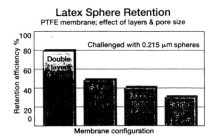

FIGURE 10 – Latex sphere retention.

Electron Microscope

The only direct way of producing pore statistics is by electron microscopy. Two techniques can be distinguished, scanning electron microscopy (SEM) and transmission electron microscopy (TEM). The SEM method is useful for studying the surface of microfiltration membranes as the resolution limit lies in the range of 0.01 mm. This however, is outside the pore size range of most ultrafiltration membranes. The SEM method enables an overall view of the membrane structure to be obtained, top and bottom surfaces and cross-section, which enables any asymmetry to be observed. Precautions in the use of SEM include prevention of damage or burning of the sample (by using thin gold layers) and prevention of drying of a wet sample.

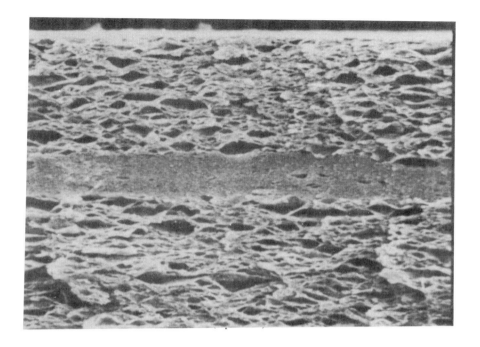

 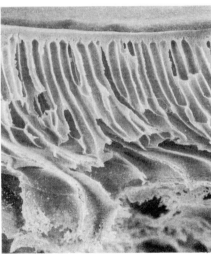

FIGURE 11 – Cross section and surface views of membranes by SEM.

The transmission electron microscope has a higher resolution than the SEM (3 – 4 A cf 50 A) but is more difficult to use with membranes, where sample perforation is critical. The sections of membranes would tend to evaporate under the beam, whereas thick samples would not reveal the porous structure. Therefore the 'replica' technique' is of potential use with membranes. Samples of the membrane are air dried, coated with platinum at 300 Hz, followed by a carbon coating using a freeze etch unit. The membrane is then dissolved from the replica using a series of acetone – dimethylformamide solutions to reveal the replicated surface. This replica is then thoroughly washed in sulphuric acid before viewing with the TEM.

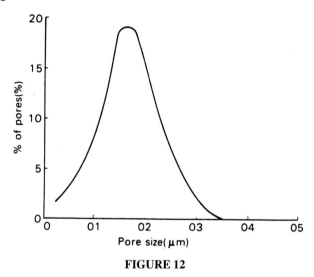

FIGURE 12

The additional attraction of electron microscopy is the ability to determine the pore size distribution. Typical pore size distributions are shown in Fig 12 which were obtained from micrographs. Generally there is good correlation between pore size, molecular weight cut-off and water flux through membranes. It is tempting to use the pore statistics of diameter and pore distribution, i.e. number per cm^2, to estimate the water flux using for example the Hagen-Poiseuille equation.

$$J = e\, dp^2\, \Delta P/32 \times \mu$$

e = porosity, dp = pore size, DDDP = pressure drop, x = membrane thickness, m = fluid viscosity.

Although giving an order of magnitude estimate, the change in the membrane pore tortuosity with pore size, the variation in the number of dead end pores and the chemical nature of material makes this a not altogether reliable procedure.

Permeability

The pore size of a membrane can be determined by measuring the relationship between pressure difference and flux of a fluid, typically water, through the membrane. This relationship is given by a form of the Hagen-Poiseuille equation which incorporates a membrane tortuosity factor. For an ideal cylindrical pore, once the membrane pore becomes wetted at a minimum breakthrough pressure, the flux and pressure difference are linearly related. This linearity will give the pore size (or mean pore size). For a real membrane in which a non-uniform distribution of pores exists, after the breakthrough pressure is reached, for the larger pores, more and more of the smaller pores become permeable. This results in a non-uniform variation of flux with pressure, up to a maximum pressure where all the pores are active. This flux pressure drop relationship is characteristic of the membrane pore distribution.

Pharmaceutical Filter and Filter Cartridge Tests

One of the most demanding applications of membrane filtration is in pharmaceutical industry where, final product quality is ensured by using filter cartridges and other devices. For example pharmaceutical products, such as injectable and infusion solutions and those which come in contact with open wounds, must conform to exactly defined quality standards. The desired quality of the final product can only be obtained when the entire production process is adequately safeguarded against contamination. Final product quality meeting the standards of the respective pharmacopoeias can be achieved by using membrane filter technology at critical points where particles or microbes could contaminate a product or must be separated from it.

Although heat-stable final products can be sterilised effectively by autoclaving. this process, does not remove particles or dead micro-organisms which may release pyrogens. Therefore, a prior membrane filtration run is required by CGMP regulations (Current Good Manufacturing Practice of the U.S. Food and Drug Administration) to ensure particle and microbe removal.

Solutions containing heat-labile products, such as antibiotics, can be cold sterilised by membrane filtration immediately before aseptic filling.

Microbe retentive filtration (bacteria retentive according to the European Pharmacopoeia 2 and DAB 9) or sterile filtration (sterilisation by filtration in conformance with the USP XXI), respectively, is an important process step in the manufacture of sterile pharmaceutical products. When sterilising filters are used in the manufacture of pharmaceuticals, the aseptic process must be validated, taking all aspects of the product and the production process into consideration. The term validation is defined by the F.I.P. guidelines as follows:

"Validation, as used in these guidelines, comprises the systematic testing of essential production steps and equipment in the R & D and production departments, including testing and inspection of pharmaceutical products with the goal of ensuring that the finished products can be manufactured reliably and reproducibly and in the desired quality in keeping with the established production and quality control procedures".

The manufacture of membrane filters must conform to current good manufacturing performance for quality assurance. Consistent high quality is assured by careful selection of raw materials validated production and test procedures (see Fig 13).

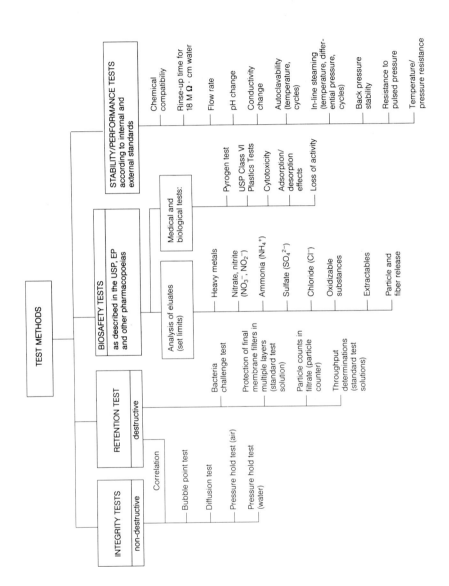

FIGURE 13 – Test methods for quality assurance of filter cartridges.

Multiple In-Line Steam Sterilisation Cycles (Thermal Stability)

The materials and construction of the Filter Cartridge allow several sterilisation cycles to be performed. Since multiple in-line stream sterilisation is a much-used sterilisation practice, the purpose behind this method is to examine the effects of thermo-mechanical stress, which is generated during multiple cycles and influences the integrity of the filter cartridge. This method also establishes guidelines as well as limits for multiple in-line sterilisation cycles. In the method a typical filter, with a final membrane pore size of 0.2 μm, and taken from several different production lots, is placed in a stainless steel cartridge housing and is sterilised using a continuous supply of saturated steam at 29 psi (2 bar). After a steam sterilisation temperature of 134°C is reached, 30 minutes is allowed for each sterilisation cycle. During this time, the differential pressure, which is held constant during sterilisation, must not exceed ~0.3 bar. After steam sterilisation the steam pressure is reduced to and then water is filtered for another 3 to 5 minutes (differential pressure of ~ 0.2 to 0.3 bar) to cool off the filtration system. Before beginning the test series and after the 1st, 5th, 10th, 15th, and 20th steam cycle, the cartridge integrity is checked by performing diffusion, bubble point and water flow rate tests. Once 20 steam cycles have been performed, the retention capability of the filter cartridges is tested using the bacteria challenge test according to the HIMA method.

To avoid damaging the membranes, water free of anticorrosive additives (which may give off alkaline vapours and vapours containing hydrazine) to produce the steam needed for in-line steam sterilisation is used.

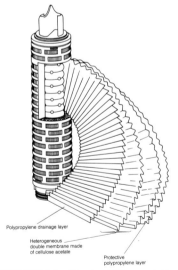

FIGURE 14 – Standard filter cartridge.

The actual number of multiple steam sterilisation cycles, which a filter cartridge can withstand, may vary according to the various process parameters, since the service life of the filter cartridges is influenced by process condition, e.g. the composition of the solutions to be filtered and any cleaning solutions or disinfectants used.

Particulates

Regulations require that the particulate matter in parenterals does not exceed defined limits i.e. the filtration system essentially must not allow any particulates or fibres to be released downstream. The particle release of a cartridge must be below the limits defined, for example, in the USP XXI "Large Volume Injections for Single-Dose Infusion."

In the test two cartridges each, are installed in a thoroughly cleaned filter housing and flushed with 20 litres of membrane-filtered (0.2 μm) de-ionised water. A differential pressure of ~ 0.3 bar is then applied to the housing, and samples (1st, 6th, 11th and 21st litres) are taken from the continuous filtrate stream for particulate analysis. Determination of the particle count and evaluation are done in conformance with USP XXI.

TABLE 3 – Typical results

Analysis	Limit according to the USP XXI	Rinse volume	Result
Total solids	≤1 mg/100 ml	1st l	Failed
		6th l	Passed
		11th l	Passed
		21st l	Passed
Oxidizable substances	Permanganate bleaching ≤ blank	1st l	Failed
		6th l	Failed
		11th l	Passed
		21st l	Passed
pH	between 5-7	1st l	Passed
		6th l	Passed
		11th l	Passed
		21st l	Passed
Heavy metals	≤1 ppm	1st l	Passed
		6th l	Passed
		11th l	Passed
		21st l	Passed
Sulfate	Turbidity ≤ blank	1st l	Passed
		6th l	Passed
		11th l	Passed
		21st l	Passed
Chloride	≤0.5 ppm	1st l	Passed
		6th l	Passed
		11th l	Passed
		21st l	Passed
Ammonia	≤0.3 ppm	1st l	Passed
		6th l	Passed
		11th l	Passed
		21st l	Passed

The table shows that after the cartridges are flushed with 11 litres of water, no extractables which would change the WFI quality are released by the Filter Cartridges. It is typically recommended that each filter cartridge installed is rinsed in the direction of filtration with 20 l of deionised water at a differential pressure of 0.3 bar. This volume of rinse water is sufficient to reliably prevent extractable substances from affecting or changing the medium during filtration.

Pyrogen (bacterial endotoxin) test

Many pharmaceutical products are required to be free of pyrogens and therefore filters must not contaminate the filtrate with bacterial endotoxings (pyrogens). The Limulus Amebocyte Lystate test (LAL test) is used to detect traces of bacterial endotoxins. Two test cartridges are shaken vigorously with 3.5 l of pyrogen-free water for one hour under non-pyrogenic conditions. The aqueous extract is tested with an LAL reagent as specified in the USP XXI for the "Bacterial Endotoxins Test".

Mycolasma Retention

Bacteria of the mycoplasma group (e.g. Mycoplasma, Acholeplasma and Spiroplasma genera) are some of the smallest self-propagating microoganisms. They are characterised by the absence of a cell wall and by their parasitic nature. Mycoplasmas are found in a variety of polymorphic shapes: coccal cells, filaments, or rosettes and their cell diameter is usually between 0.1-0.25 µm. The small size, combined with deformable characteristics due to the lack of a cell wall, exceed current retention capabilities of most 0.2 µm membrane filters. Membrane filtration using a pore size of 0.2 microns can usually only produce a log reduction value (LRV) of 5-6 thus liquids with a high mycoplasma concentration are filtered, eg serum, the filtrate will not be free of mycoplasmas.

A 0.1-micron membrane will not reliably retain mycroplasma. To solve the mycoplasma retention problem serial filtration using up to three 0.1-micron membranes can be performed. An even better method is to use a Sterilising Filter Cartridge with a pore size of 0.07 microns. The retention characteristics for mycoplasma are determined using a challenge test performed during a simulated actual filtration run in which factors common in biotechnology are taken into consideration, such as initial concentration, filtering pressure, and pulsation residence time.

Electronics Industry Filtration Test Methods

Characterisation of the effectiveness of filtration products for the electronics industry requires evaluation of both the membrane and filtration device. The following table briefly describes some of the tests used to characterise the membrane, filter cartridge, and integrity of the filter cartridge.

Other test procedures used to validate cartridge integrity in the pharmaceutical industry and other industries eg electronic, food include the air diffusion test, bubble point test, retention test, TOC test and resistivity test. These procedures are illustrated in the next set of captions.

TABLE 4 – Test Methods

Membrane Properties	Cartridge Properties
Strength The Mullen burst test. The membrane is placed over a rubber bladder and the bladder is filled with oil, stretching the membrane. The pressure at which the membrane ruptures or yields is recorded. The Mullen test may be replaced with a standard Instron tensile test. In this test, a strip of membrane is placed between two sets of jaws which are drawn apart slowly. The membrane elongates and eventually breaks. **Pore Size** The most direct is by use of a SEM image of the membrane surface and cross section. Initial bubble point gives an indication of the largest pore in the membrane. In this test, the membrane is wet with a solvent (generally alcohol). The test device allows the surface of the membrane to be visible under a thin layer of wetting solvent. Gas pressure is increased slowly. The pressure at which gas can be seen bubbling through the membrane is the bubble point. It can be related to the pore size of the membrane by the Young and Laplace equation. Alcohol porosymmetry gives an indication about the pore size distribution of the membrane. The air flow through a dry membrane is compared to that through an alcohol wet membrane. Below the membrane's initial bubble point, there is no air flow through the alcohol wet membrane. Above the bubble point of the smallest pore, the air flow through the dry and alcohol wet membrane are the same. The ratio of air flows over the pressure range gives an indication of the pore size distribution of the membrane. **Particle Retention** The particle retention capability of filter devices has classically been based on bacterial retention. A 0.2 m rated filter is capable of retaining a challenge of 10^7 Pseudonmonas diminuta per square centimetre of membrane area. The ability of the membrane to retain hard particles can be determined by challenging the membrane with a solution of monodiperesed latex beads. For beads larger than 0.1 m, the beads that pass through the membrane are collected on a track-etched filter and counted by use of a SEM image.	**Physical** The flow rate versus pressure drop of the device indicates the hydraulic resistance of the filter. For filters used in liquid service, the pressure drop test is generally done with water. For aerosol filters, the test is done with air. The graded failure test determines the physical strength of the filtration device. The device is put under increasing pressure differential and pulsing. The integrity (bubble point) is measured periodically to determine the point of failure. **Cleanliness** The cleanliness of the fabricated device is determined by measuring particle shedding in water for liquid-service filters and air for aerosol filters. Very clean test fluid is passed through the filter at steady flow and pulsed conditions. The particles downstream of the filter are detected using a laser particle counter. A condensation nucleus counter is also used to evaluate aerosol filtration products. **Extractables** The inorganic and organic materials that can be extracted from the liquid filtration cartridges are tested in a number of ways. Water of high resistivity and low TOC is pumped through the filter. Any degradation of the water quality is determined by an Anatel meter. Cartridges used for liquid chemicals are statistically extracted in water, alcohol, or 10% HCl for 24 hours. The materials added to the solutions are determined by use of a number of analytical techniques. For aerosol filters, the presences of moisture in the filtration device is determined by passing very dry nitrogen through the filter and detecting added moisture to the gas. The presence of organic extractables in aerosol filters is determined by drawing a high vacuum on the filter and using a residual gas analyser to determine the chemical components outgassing.

TABLE 4 (continued)

Integrity tests
Retention In addition to bacterial challenge testing, the liquid filtration device can be challenged with polydispersed latex beads. The concentration of beads in a number of size categories downstream of the filter is determined with a laser particle counter. The smallest bead that can be detected at this time is 0.1 μm. Aerosol filters are challenged with either a monodispersed or polydispersed aerosol particle challenge. The retention of the filter is quantified using particle counters. **Membrane Integrity Testing** An important feature of a filtration system is its ability to be integrity tested before and after each filtration. A filter integrity test should be performed prior to starting a filtration run. The test will detect a damaged membrane, ineffective seals, or a system leak. Two kinds of acceptable integrity tests, the bubble point test and the diffusion test, are described in this section. **Bubble Point Testing** A simple, non-destructive integrity test of performance is known as "bubble-point" testing. Membrane form one side to the other which can be thought of as fine capillaries. The bubble point test is based on the fact that liquid is held in these capillary tubes by surface tension and that the minimum pressure required to force liquid out of the tubes is a measure of the tube diameter. A bubble point test is performed by wetting the filter, increasing the pressure of air upstream of the filter, and watching for air bubbles downstream to indicate a damage membrane, ineffective seals or a system leak. A bubble point that meets specifications ensures the system is integral. *Diffusion Testing* In high volume systems, where a large volume of downstream water must be displaced before bubbles can be detected or high diffusive flow through the membrane strucutre can be mistaken for flow through defects, a diffusion test can be performed instead of the bubble point test. The diffusion test is based on the fact that in a wetted membrane filter, under a pressure, differential air flows through the water filter pores at differential pressures below the bubble point pressure of the filter by a diffusion process following Fick's Law. In a small area filter this flow of air is very slow. But, in the large area filter used in high volume system, it is significant and can be measured to perform a sensitive filter integrity test. In the diffusion test, pressure is applied at 80% of the bubble point pressure established for the particular filter system. The volume of air flow is determined by: (1) measuring the rate of flow of displaced water, (2) gas flow upstream or downstream of the filter or (3) the ability to hold pressure upstream of the filter. At 80% of the filter bubble point pressure, filter integrity is validated since there would be a dramatic increase in air (and water) flow above the diffusive flow if there was damaged membrane, wrong pore size filter, ineffective seals or system leaks.

250   HANDBOOK OF INDUSTRIAL MEMBRANES

Cartridge Retention Test

Air Diffusion Test

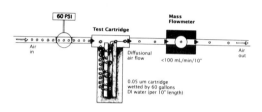

Bubble Point Test

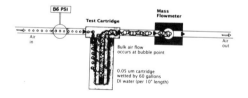

TOC Test

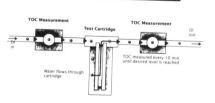

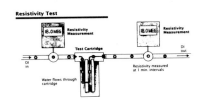

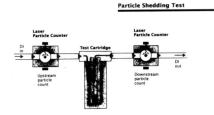

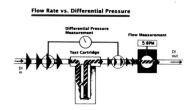

Ultrafiltration Membranes

The usual asymmetric structure of UF membranes in which the porous top layer in the size range of 20 – 1000 A is of interest means that many of the methods of characterisation of MF membranes cannot be applied. Bubble point and mercury intrusion methods require too high pressures which would damage or destroy the membranes structure. SEM is generally not possible and TEM is not always applicable.

The methods which can be used include the permeation experiments and methods such as, gas adsorption – desorption, thermoporometry, permporometry and rejection measurements.

The use of gas adsorption – desorption is frequently practised for measurement of pore size and size distribution of porous media. Typically nitrogen is used as the absorbing media and the method determines the quantity of gas absorbed (and desorbed) at a particular pressure up to the saturation pressure. A model is required which relates the pore geometry to the adsorption isotherms. The method is limited generally to more uniform structures. Ceramic membranes have been satisfactorily characterised by this method. The method unfortunately includes the contribution made to the membrane structure by dead-end pores.

Thermoporometry uses the calorific measurement of solid-liquid transition in a porous media. The method typically uses water as the medium and is based on the fact that the freezing temperature in the pores of a membrane (ie the top layer) depends upon the pore size. The extent of undercooling is inversely proportional to the pore diameter. The method also measures the dead-end pores in the membrane.

Permporometry is a method which only characterises the active pores in the membrane. It is based on the blockage of pores by a condensable gas, linked with the measurement of gas flux through the membrane. The pore blockage is based on the same principle of capillary condensation analogous to adsorption.

Solute Passage or Rejection

Ultrafiltration membranes are generally not characterised in terms of pore size but in terms of a molecular weight cut-off. Cut-off is defined on the basis of 90% rejection of a solute with a particular molecular weight. The method measures the permeability of selected solutes of different molecular sizes under controlled conditions. The solutes should cover the expected size range for 0% to 100% rejection and not interact with the membranes. Thus solutes such as sodium chlorides and glucose (MW 180) are used for the low end, ie 0% rejection with UF membranes and large proteins eg immunoglobulins (MW 7 900,000) or blue dextran are used for the upper size range as these have almost 100% rejection with most membranes. In all some 5 to 6 water soluble solutes are required to adequately characterise a membrane rejection (see Table 5).

There are no standard test conditions for solute rejection characteristics although recommendations do exist. These include, a pressure of 100 kPa (although this does vary) temperature of 25°C, 0.1% (w/v) of solution in a 1% saline solution, and a maximum possible degree of agitation. In addition only a relative small amount of solution should be filtered to avoid concentration effects. The membranes should be new, cleaned of preservative and conditioned using a series of soaking, washing and pressurisation until a stable and reproducible water flux is achieved.

The results of a series of rejection tests will be usually expressed as a plot of rejection versus MWCO (see Fig 15) with some membranes exhibiting sharp cut-offs, with a narrow range of MWCO, and other membranes exhibiting a diffuse cut-off, with a broad range of MWCO. Overall MWCO values of membranes are no more than a guide to a particular application. The abitrary setting of the cut-off value, the variability in test procedures and solutes and the effects of other operating parameters mean greater precision is not as yet possible.

TABLE 5 – Typical solute rejection data and solutes used to characterise UF membranes.

Solute	Molecular Weight	UM 05 pH 5	UM 05 pH 10	UM2	DM5	UM10	PM10	YM10	UM20	PM30	XM50	XM 100A[2]	XM 300[2]
D-Alanine	89	15	80	0	—	0	0	—	0	0	0	0	0
DL-Phenylalanine	165	20	90	0	—	0	0	—	0	0	0	0	0
Tryptophan	204	20	80	0	—	0	0	—	0	0	0	0	0
Sucrose	342	70	80	50	—	25	0	—	—	0	0	0	0
Raffinose	594	90		—	—	50	0	10	—	0	0	0	0
Inulin	5,000	—	—	80	70	60	—	45	5	0	0	0	0
PVP K15	10,000	—	—	90	85	65	35	75	—	—	—	—	—
Dextran T10	10,000	—	—	90	90	90	5	—	—	—	—	—	—
Myoglobin	17,800	>95	>95	>95	—	95	80	>90	60	35	20	—	—
α-Chymotrypsinogen	24,500	>95	>95	>98	—	>95	>95	—	90	75	85	25	0
Albumin	67,000	>98	>98	>98	—	>98	>98	>98	95	>90	>90	45	10
Aldolase	142,000	>98	>98	>98	—	>98	>98	—	>98	>98	>95	—	50
IgG	160,000	>98	>98	>98	—	>98	>98	—	>98	>98	>98	90	65
Apoferritin	480,000	>98	>98	>98	—	>98	>98	—	>98	>98	>98	>95	85
IgM	960,000	>98	>98	>98	—	>98	>98	—	>98	>98	>98	>98	>98

[1] Measured at 55 psi (3.8kg/cm2), except where noted
[2] 10 psi (0.7kg/cm2)

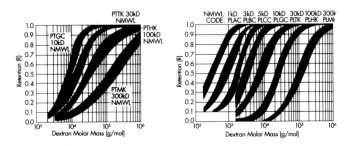

FIGURE 15 – Rejection data of typical ultrafiltration membranes.

Other factor which affect the rejection measurements are the membrane material, the shape and flexibility of the macromolecualr solute, its interaction with the membrane, the membrane configuration, interaction of the membrane and solute, concentration polarisation phenomena and interactions between different solutes or macromolecules. Secondary

TABLE 6 – Tests for Characterising Membranes

Integrity Tests	Medium	Characteristic
Microfiltration		
Air Diffusion	Air	Integrity
Bubble Point Test	Air	Pore size
Cartridge Rentention Test	Water	Filtration Efficiency
Flow rate vs Differential pressure	Water	
Particle Shedding Test	Water	
TOC Tests	Water	
Resistivity Test	Water	
Bacteria passage test	Pseudomonas dimiuta	Sterility
Mercury intrusion test	Hg	Pore size and pore distribution
Latex sphere test	Latex sphere dispersion	Integrity
Water penetration test	Water	Integrity
Electron Microscopy (SEM, TEM)	Pore size, shape and	distribution, pore density
Permeation Measurements	Water flux for pore size and distribution.	
Ultrafiltration		
Gas adsorption – desportion	N_2	Pore size and distribution
Thermoporometry	Water	Pore size and distribution
Permporometry	Gas	Pore size and distribution
Solute Rejection	Various solutes	MWCO

membranes can be formed on the membrane surface which impede the permeation of lower molecular weight solutes. The adsorption of solutes by the membrane can also result in reduction in the apparent rejection of a UF membrane. Overall therefore these membrane surface phenomena must ideally be taken into account in characterising the true (instrinsic) membrane properties. This generally involves indirectly measuring the concentration of solute at the membrane surface using equations describing boundary layer phenomena or alternatively eliminating these effects by experiments at low pressure driving forces and low feed concentrations.

Molecular weight cut off determination of UF membranes

Techniques for the characterisation of ultrafiltration (UF) membranes generally use marker compounds of a known molecular weight (for example, dextrans and proteins) to challenge the membrane under controlled conditions of temperature, flow rate, and pressure. This leads to a nominal value or rating for the membrane. Manufacturers of UF membranes tend to use different marker compounds and ratios to determine the ratings of their membranes. For a majority of end-user applications this technique is more than adequate. However where the application calls for the separation of molecules whose molecular weights are not greatly dissimilar, then the rating of a membrane from one supplier's batch to another, or between different suppliers of the same rated membrane, may have a significant effect on the performance of the membrane and its subsequent effect on product quality and/or yields

The basis of this technique employs a size exclusion chromatography (SEC) system which has been previously calibrated with polymer standards of known molecular weight. Accurate percentage mass data is generated by the SEC system software across part or all of the MWD of a sample. The MWD of a sample is divided into slice intervals (SI), which can be related to molecular weight intervals (MWI), with each SI being 250 daltons in size. This is shown in the table below,

SI	MWI (Daltons)
1	1-250
2	250-500
3	500-750
4	750-1000

The total number of SI is such that the expected MWCO value is spanned by sufficient SI to enable the data to identify the MWCO value. The test compound used is for example a commercially available maltodextrin with a broad MWD which extends above and below the expected MWCO value of the membrane under test. When the percentage mass in each SI is plotted against each respective SI then an intersection point (IP) is revealed which can be related to a specific SI/MWI, and hence MWCO, or perhaps more correctly the average molecular weight cut off (AMWCO).

The technique enables an accurate molecular weight value to be determined and assigned to a membrane, enabling the manufacture, development and subsequent use of a membrane to be critically assessed. The use of expensive marker compounds is

TABLE 8

Membrane type	Stated MWCO	Measured AMWCO
Polysulphone (Flat Sheet)	20000	16500
Polysulphone (Flat Sheet)	5000	3000
Polysulphone (Flat Sheet)	5000	2625
Polysulphone (Cassette)	3000	3750
Polysulphone (Cassette)	1000	3000
Polysulphone (Spiral)	1000	1500
Cellulosic (Spiral)	3000	2825
Cellulosic (Spiral)	1000	1500
Thin Film Composite (Flat Sheet)	1000	1600

eliminated, being replaced by a relatively cheap and easily obtainable test compound, where differences in the MWD from batch to batch are not critical to the characterisation of a membrane.

Non-Porous Membranes

The transport through non-porous membranes is at a molecular level by a solution – diffusion mechanism in which separation is achieved by differences in solubility and/or diffusivity of solutes. Characterisation of membranes is clearly not related to porosity measurement but to physical and chemical properties. Indeed tests for porosity ie pin holes, are part of membrane integrity tests. A number of physical methods are used to characterise the parameters which affect the permeability or solute transport through a membrane. Two major structural parameters are the glass transition temperature and the degree of crystallinity. The degree of crystallinity generally has a direct bearing on permeability as transport usually proceeds via amorphous regions in polymers. For most commercial polymers, the glass transition temperatures and the degree of crystallinity are known, for other materials they can be measured using methods listed in table. Polymer density is an important parameter in membrane characterisation and is directly related to the glass transition temperature and the degree of crystallinity. Density of a polymer decreases as the temperature increases, this decrease increase more rapidly after the glass transition temperature. Membranes prepared from high density polymers tend to have lower permeabilities.

The preparation of many non-porous membranes requires the introduction of specific surfaces or top layers or the modification of the surface to achieve desirable permeation properties. In many cases the chemical nature of this layer or surface is not precisely known and therefore it can be necessary to carry out surface analysis.

Overall the principal method for characterising a non-porous membrane is to determine its permeability towards particular liquids, solutes and gases. A wide variation in permebility for particular species is seen for particular membranes. Notably gas permeabilities can vary by over size orders of magnitude.

In the case of reverse osmosis the permeability is primarily concerned with the transport of water. The other important characteristic relates to the transport of solute through the

MEMBRANE MATERIALS, PREPARATION AND CHARACTERISATION

TABLE 7 – Characterisation Tests for Non Porous Membranes

Permeability	Measurement of the steady state flow of liquid or gas under an applied differential pressure or other driving force.
Physical Methods	Measurement of glass transition temperature and degree of crystallinity using differential scanning calorimetry (DSC) and differential thermal analysis (DTA). Density gradient measurement. X ray diffraction to measure size and shape of crystallites and the degree of crystallinity.
Plasma etching	To measure the thickness of the top layer in asymmetric and composite membranes.
Surface Analysis	Methods include electron spectroscopy for chemical analysis, ESCA, X ray photoelectron spectroscopy (XPS), secondary ion mass spectroscopy (AES). The general basis is excitation of the surface by radiation or particles bombardment and the emission of products which provide information about specific groups which are present.

membrane. This is characterised in terms of the retention or rejection coefficient for a particular solute in a particular membrane.

SECTION 2.3 – ELECTRODIALYSIS AND ION EXCHANGE MEMBRANES

Ion exchange membranes commercially manufactured are either homogeneous or heterogeneous. For general electrodialysis applications, where extremes of pH or strong oxidising media are not encountered, the much cheaper heterogeneous membrane can be employed. Commercial ion exchange membranes are effectively ion exchange resins in sheet form which have a cross linked structure. Cation exchange membranes are often sulphonated to attach SO_3^- groups to the polymer and anion exchange membranes have a similar structure with attached quaternary ammonium groups.

In water the attached groups are associated with a mobile counter ion which freely migrates under the applied potential field, accounting for the high electrical conductivity and selectivity of the membrane. The high selectivity of say a cation exchange membrane to cations is due to the Donnan exclusion effect. On immersing in an aqueous solution of salt for example the membranes swell producing a network of aqueous "porous" regions where the sulphonic acid groups dissociate to produce mobile cations and fixed negatively charged sulphonic groups. The mobile cations either transport under the application of an applied potential or are free to exchange with other cations. The fixed charge group essentially electrostatically repel the solution based anions which cannot contribute to the flow of current.

The membranes used in ED stacks must be easily handled in stack assembly and be resistant to osmotic swelling and be impermeable to water under hydraulic pressure. The membrane must also be resistant to fouling or poisoning and the adverse effects of

Ion exchange membrane sheet used in electrodialysis.

polarisation. Fouling can be caused by any material which settles on the membrane surface and de-activates the ion diffusion capability. This may be from organic macromolecules present in the water or by precipitation of colloids on the membrane surface. This is a particular problem in the case of anion-exchange membranes as colloids are typically negatively charged. Scaling of the membranes can occur due to the presence of Ca and Mg ions. The scaling is from the formation of crystalline inorganic salts usually on the concentrate side of the anion exchange membrane where the concentration of the cations is highest due to concentration polarisation. Poisoning is caused by the adsorption of strongly held ions. These ions may exchange with the original counter ions drastically increasing the membrane electrical resistance and selectivity.

The general problem of fouling of membranes is less severe at low current densities where the effect of polarisation is not as significant. Procedures which are adopted to minimise the effect of fouling are;

i   Pretreatment of the feed to remove material such as colloids.
ii   Reduction in the concentration of $CaSO_4$ to below the saturation level or the addition of salts which delay crystal formation.
iii Acidification of the concentrate stream to prevent carbonate precipitation.
iv   Electrodialysis reversal (EDR).

Heterogeneous Ion Exchange Membranes

Heterogeneous membranes consist of fine colloidal ion-exchange particles embedded in a suitable inert binder (phenolic resins, PVC etc.). The difficulty with the production of these membranes is combining suitable electrical conductivity with adequate mechanical

strength. The ion-exchange particles swell when immersed in water and thus to produce membranes with suitable strength to be used in large "thin" sheets, requires a sacrifice in ion-exchange capacities, which is typically 1-3 mequival/g dry membrane (cf 3-5 mequival/g resin). Procedures for the preparation of heterogeneous membranes include
- calendering ion-exchange particles into a polymer film
- dry moulding of film-forming polymers and ion-exchange particles and milling of this mold stock
- dispersion of ion exchange particles in a film-forming binder solution followed by evaporation of solvent
- dispersion of resin into partially polymerised binder polymer followed by more polymerisation.

The swelling in ion-exchange membranes is limited by the covalent cross linking between the resin and the polymer. Generally the membranes contain over 65% of cross-linked ion-exchange particles to have practical low electrical resistnaces.

Homogeneous Membranes

The procedures for preparation of homogeneous ion-exchange membranes are either by polymerisation of functional monomers, or by additional functionalisation of a polymer film.

In the production of homogeneous membranes by polymerisation or polycondensation of monomers, one or more of the monomers must contain a species that either is or can be made ionic. There are many procedures for preparation of membranes by polymerisation, two early developed examples are
i) polycondensation of phenol sulphonic acid in paraform with formaldehyde, which is cast into a film which polymerises at room temperature
ii) polymerisation of styrene and divinyl benzene followed by either sulfonation or amination for cationic and anionic exchange membranes.

Homogeneous membrane preparation by introducing ionic groups into a preformed film uses either monomers which contain a cross-linking agent (eg divinyl benzene) or using grafting by radiation methods. The usual starting materials are polyethylene or polystyrene although hydrophilic polymer may be used, e.g. poly vinyl alcohol, cellulose. An example of this type of membrane is in production of polyetheylene cation and anion exchange membranes, which exhibit excellent mechanical strength, with low electrical resistances and high permselectivity. The membrane is made by exposing the polymer film to a sulphur dioxide and chlorine gas mixture under UV light radiation at room temperature to form the sulfochlorinated polyethylene. The cation membrane is formed by removing the chloride with NaOH, the anion membrane is formed by amination of SO_2Cl group, followed by reaction with methyl bromide.

A third method of membrane manufacture is by the introduction of anionic moieties into a polymer chain followed by dissolving the polymer and casting into a film. One example is sulfonated polysulphone which is a reinforced membrane with good mechanical and electrochemical properties.

Figure 1a
Fluorocarbon cation exchange

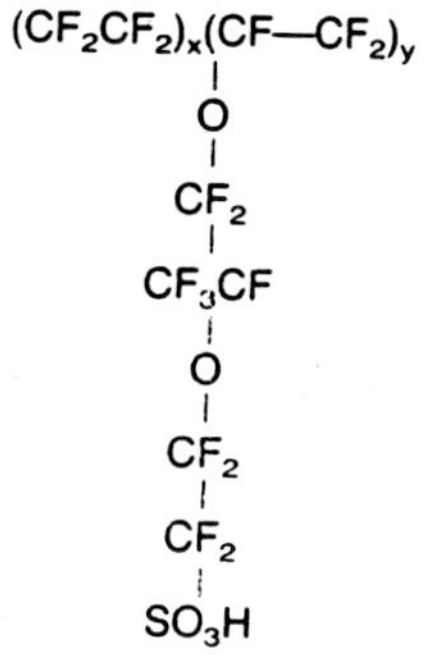

$(CF_2CF_2)_x(CF-CF_2)_y$
|
O
|
CF_2
|
CF_3CF
|
O
|
CF_2
|
CF_2
|
SO_3H

Figure 1b

$(CF_2CF_2)_x(CF-CF_2)_y$
|
O
|
CF_2
|
CF_2
|
SO_3H

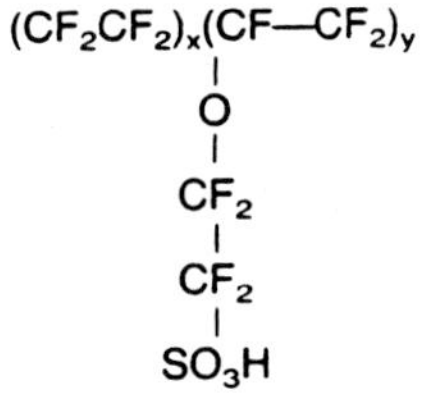

Figure 1c – Heterogeneous styrene
and divinyl benzene copolymer
Cationic   Anionic

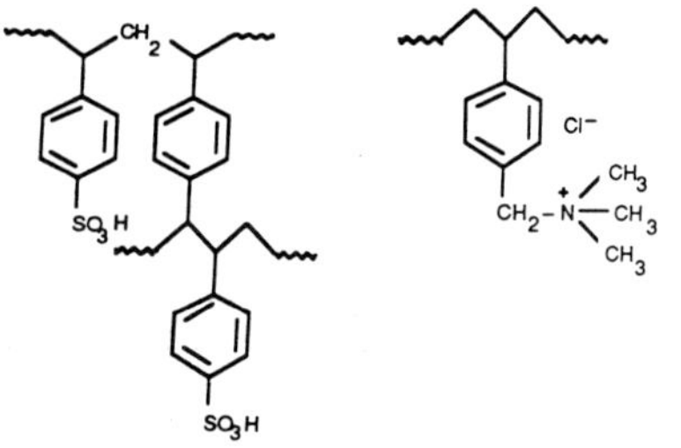

Figure 1d
Polyethylene membranes

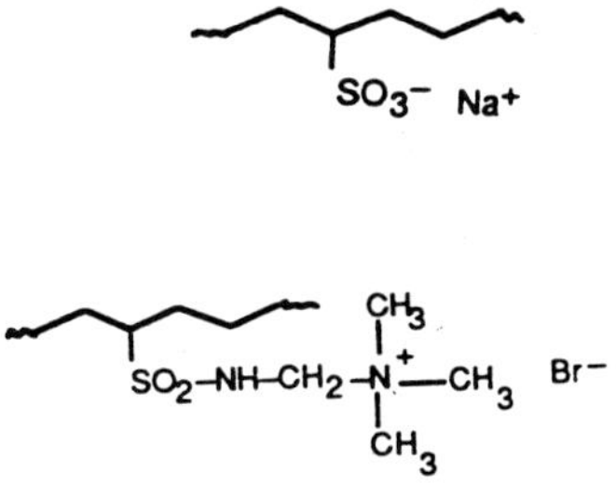

Figure 1e – Sulphonated polysulphone cation exchange membrane

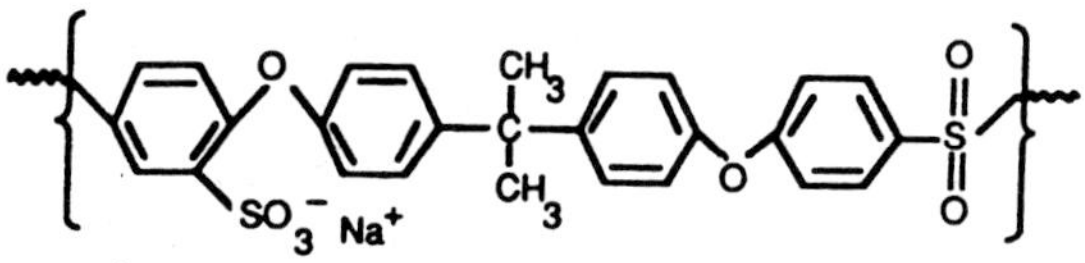

FIGURE 1 – Structure of ion exchange membranes.

Fluorocarbon Membranes

Fluorocarbon type ion-exchange membranes were developed primarily to overcome the limitations of hydrocarbon based ion-exchange when subjected to a degradation, oxidation environment. The major use is in chlor-alkali cells for the production of chlorine and sodium hydroxide in which they show excellent thermal and chemical stabilities. The membrane is produced in a several stage synthesis of an ionogenic perfluorovinyl ether and its copolymerisation with tetrafluorethylene. The resulting copolymer is extracted as thin film (100 μm) and then the inorganic moiety is converted to membranes that carry sulfone groups, ie the SO_2F groups are reacted with sodium hydroxide to -SO_3Na group. Similarly carboxylic acid membranes can be produced in various ways by converting the -CF_2 SO_2 F groups to -COOH groups.

There has been less success with the production of anion exchange fluorocarbon membranes with alkaline stability. However recent materials have been produced with similar structures to the cation exchange fluorocarbon membranes by replacing the sulphonic acid group with amine -NR_3^+ group. They are reported to exhibit good alkaline stability characteristics.

An alternative homogeneous anion exchange membrane with good alkaline stability has been manufactured by incorporating a quaternary ammonium ion into a polysulphone (or polyether sulphone) matrix. This can be done by chloromethylating the polysulphone and then adding the monohydroxide – quaternary salt of 1-benzyl-1-azonia-4-azabicyclo [2.2.2] octane hydroxide (DABCO). The cross linking density is adjusted by the ratio of DABCO to chloromethylated polysulfone. The membrane has good electrical characteristics. IEC = 1.2 meqg^{-1}, area resistance 1.05 OOO cm^2, 97.5% permselectivity and 8% water swelling.

Bipolar Membranes

Bipolar membranes offer an energy efficient alternative to electrolysis for the splitting of water into hydrogen ions and hydroxide ions. The major challenge has been the production of suitable compatible anion and cation exchange materials which can be combined in a single membrane with appropriate chemical and thermal stability and with good electrical characteristics. Bipolar membranes can be produced in several ways
- Laminating conventional anion and cation exchange membranes back to back. Special surface coating prior to lamination can lead to quite good water splitting membranes.
- Casting a cation selective layer on top of a previously prepared cross-linked anion-exchange membrane.

In bipolar membranes the chemical stability is determined primarily by both the properties of the charged anion-exchange moieties and the polymer matrix. To minimise the bipolar membrane electrical resistance the thickness of the interface between the two oppositely charged layers should be very small (< 10 nm). A recent membrane has been developed in which a very thin (10 nm) intermediate layer of an insoluble polyelectrolyte complex of poly (4-vinyl pyrindine) and poly acrylic acid is formed between an anion membrane and a cation membrane. The anion selective layer is the homogeneous polysulfone anion exchange membrane (Fig 1) and the cation layer is a cross linked sulfonated poyetherether ketone (Fig 1) with a strongly acid sulfonic acid group as fixed charge. Cross linkage is achieved during membrane formation via reaction of hexamethylene-diamine with partly modified sulphonic acid chloride. Bipolar membrane production is achieved by spray coating the dilute solution polyelectrolyte complex onto the anion-selective layer (30 μm thick), which after evaporation to produce the 5-10 nm "intermediate layer" has the cation layer (30-60 μm) cast on top. Reinforcement of the final membrane is achieved by an inert polymer screen. The electrical characteristics of the 60 μm thick cation layer are; IEC 1.7 meqg^{-1}, 1.31 cm^2 area resistance, 98.5 % permselectivity and 12.5% swelling.

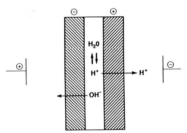

FIGURE 2 – Bipolar membrane operation.

The bipolar membrane typically operates with a voltage of 0.95 V at a current density of 1000 Am^2 in 1 mol dm^{-3} sodium sulphate solution.

Special Property Ion Exchange Membranes

The individual applications to which membranes are put often require specific properties and characteristics. Special properites are required for membranes used in the production of salt from seawater, in chlor-alkali process, as battery separators and as ion-selective electrodes. These membranes include
- Monovalent ion permselective
- Protein permselective cation exchange
- High proton retention anion-exchange
- Anti fouling anion exchange

These membranes are typically produced by applying an appropriate surface coating to the membrane. In the case of cationic membranes this is a thin cationic charged layer. Antifouling anion-exchange membranes are particularly important in ED where there is a risk of precipitates forming on the surface. Two methods which are used to improve the tolerance to organic (anion) fouling are i) adjustment of the degree of cross linkage and the chain length of the cross linker in the polymer and ii) surface coating with a thin layer of cation exchange groups which cause electrostatic repulsion of the organic molecules.

A list of the wide range of membranes available is given in Appendix X with typical properties of the membrane. There are many other membranes available from the different manufacturers with specific properties for specific applications. These membranes exhibit different chemical resistance and are recommended for specific applications.

Table 1 illustrates the chemical resistance characteristics and properties of one manufactures range of materials with details of recommended applications. These applications include electrolysis of organics, electrodialysis with resistance against organics at high temperatures and organic foulants, electrodialysis in concentrated acid and alkali (perfluorinated) to standard electrodialysis applications.

TABLE 1a – Properties of ion-exchange membranes.

Grade	CMS	ACS	AFN	ACLE-5P	CLE-E	C66-10F
Type	Strongly acidic cation permeable Na-form	Strongly basic anion permeable Cl-form	Strongly basic anion permeable Cl-form	Strongly basic anion permeable Cl-form	Strongly acidic cation permeable Na-form	Strongly acidic action permeable Na-form
Properties	Mono-cation permselective	Mono-Anion permselective	Diffusion-dialysis Resistant against organic fouling	Operable in high pH solution	Cation permselective Resistant against organic fouling Operable in high temperature solution	Application suggested for electrolysis of organics
Electric Resistance	1.5 - 2.5	2.0 - 2.5	0.4 - 1.5	10 - 20	15 - 25	5 - 8
Total Cation or Anion	0,98<	0,98<	0,98<	0,98<	0,98<	0,98<
Na$^+$ + K$^+$	0,90<	0,02>	0,02>	0,02>	0,98<	0,98<
Ca^{++} + Mg^{++}	0,10<					
Cl$^-$	0,02>	0,98<	0,98<	0,98<	0,02>	0,02>
SO$_4^-$		0,005>				
Burst Strength	3 - 4	3 - 5	2 - 5	8 - 10	8 - 10	6 - 8
Water Content	0.35 - 0.45	0.20 - 0.30	0.40 - 0.55	0.2 - 0.3	0.3 - 0.4	0.3 - 0.4
Exchange Capacity	2.0 - 2.5	1.4 - 2.0	2.0 - 3.5	1.3 - 2.0	1.3 - 1.8	1.7 - 2.2
Thickness	0.14 - 0.17	0.15 - 0.20	0.15 - 0.20	0.20 - 0.30	0.8 - 1.3	0.25 - 0.35
Reinforcing	yes	yes	yes	yes	yes	yes
Standard Size (m)	1.00x1.00	1.00x1.00	1.00x1.00	1.00x1.00	1.00x1.00	1.00x1.00

Measurement Basis
- Electric Resistance: Equilibrated with o.5N-NaCl solution, at 25°C (Ω cm^2)
- Transport Number: Measured by electrophoresis with sea water
  Current density: 2[A/dm^2], at 25°C
- Burst Strength: [kg/cm^2]
- Water Content: Equilibrated with 0.5N-NaCl solution
  [g.H$_2$O/g.Na-form dry membrane (or Cl-form)]
- Exchange Capacity: [meg./g.Na-form dry membrane (or Cl.-form)]
- Thickness: [m/m]

TABLE 1C – Properties.

Grade	CL-25T	CM-1	CM-2	ACH-45T	AM-1	AM-2	AM-3
Type	Strongly acidic cation permeable	Strongly acidic cation permeable	Strongly acidic cation permeable	Strongly basic anion permeable	Strongly basic anion permeable	Strongly basic anion permeable	Strongly basic anion permeable
		Low electric resistance	Little diffusion coefficient	Operable in high pH solution	Low electric resistance		Little diffusion coefficient
	Na-form	Na-form	Na-form	Cl-form	Cl-form	Cl-form	Cl-form
Electric Resistance	2.2 - 3.0	1.2 - 2.0	2.0 - 3.0	2.0 - 2.7	1.3 - 2.0	2.0 - 3.0	3.0 - 4.0
Total Cation or Anion	0,98<	0,98<	0,98<	0,98<	0,98<	0,98<	0,98<
$Na^+ + K^+$ $Ca^{++} + Mg^{++}$	0,70 0,28	0,70 0,28	0,70 0,28				
Cl^- SO_4^{--}	0,02>	0,02>	0,02>	0,02>	0,02>	0,02>	0,02>
				0,98<	0,98<	0,98<	0,98<
Burst Strength	3.0 - 4.0	2.5 - 3.0	3.0 - 3.5	3.5 - 4.5	3.0 - 3.5	3.0 - 3.5	3.0 - 3.5
Water Content	0.25 - 0.35	0.35 - 0.40	0.25 - 0.35	0.20 - 0.35	0.25 - 0.35	0.20 - 0.30	0.15 - 0.25
Exchange Capacity	1.5 - 1.8	2.0 - 2.5	1.6 - 2.2	1.3 - 2.0	1.8 - 2.2	1.6 - 2.0	1.3 - 2.0
Thickness	0.15 - 0.17	0.13 - 0.16	0.13 - 0.16	0.14 - 0.20	0.13 - 0.16	0.13 - 0.16	0.13 - 0.16
Reinforcing	yes	yes	yes	yes	yes	yes	yes
Standard Size	1.00x1.00	1.00x1.00	1.00x1.00	1.00x1.00	1.00x1.00	1.00x1.00	1.00x1.00

TABLE 1B – Chemical resistant properties.

Grade		Standard CL-25T	Standard CL-25T	Standard CL-25T	Standard CL-25T	Standard CL-25T	Standard CL-25T	Standard CL-25T	Standard CL-25T	Standard CL-25T	Standard CL-25T
Sodium Chloride	.3%	O	O	O	O	O	O	O	O	O	O
	.20%	O	O	O	O	O	O	O	O	O	O
Sulfuric Acid	.10%	O	O	O	O	O	O	O	O	O	O
	.40%	O	O	O	O	O	O	O	O	O	O
Hydrochloric Acid	.10%	O	O	O	O	O	O	O	O	O	O
Nitric Acid	.20%	O	O	O	O	O	O	O	O		O
Caustic Soda	.5%	O	O	▲	▲	O	▲	X	▲	O	O
Ammonia	.4%	O	O	▲	▲	O	▲	X	▲	O	O
Ethylene Glycol	.50%	O	O	O	O	O	O	O	O	O	O
Phenol	.7%	X	X	O	O	X	O	O	O	X	X
Acetone	.30%	▲	O	O	O	▲	O	▲	O	▲	▲
Dioxane	.30%	▲	▲	▲	▲	▲	▲	▲	▲	▲	▲
Methanol	.50%	O	O	O	O	O	O	O	O	O	O
Ethanol	.50%	O	O	O	O	O	O	O	O	O	O
Acetic Acid	.50%	O	O	O	O	O	O	O	O	O	O
Sodium Thiosulfate	.3%	O	O	O	O	O	O	O	O	O	O
Citric Acid	.50%	O	O	O	O	O	O	O	O	O	O
Strong Oxidizing Agent		X	X	X	X	X	X	X	X	X	X

O — Not attacked - ▲ — Slsightly attacked - X — attacked
• Tested by immersion in the respective solution at room temperature and subsequent equilibration with 0.5 N-NaCI solution.
• NE▲SEPTA-F® is chemically stable or resistant against most chemicals but may slightly swell in alcohols and ethers.

Ion-Exchange Membrane Characterisation

The characteristics of ion exchange membranes, typically listed in the Appendix, which are used to define their performance include electrical and mechanical properties, permselectivity and chemical stability.

Ion-Exchange Capacity (IEC)

Ion-exchange capacity is measured by titration of the fixed ions eg R_4N^+, $-SO_3$ in the membrane with either 1 M NaOH or HCl for cation or anion exchange membranes respectively. For example the membranes are first soaked in 1 M HCl, for cation and 1 M Na OH for anion membranes to achieve equilibrium. After rinsing the membranes free of chloride or sodium, the samples are then back-titrated with NaOH (cation) or HCl (anion).

Ion-Exchange Permselectivity

The permselectivity of an ion-exchange membrane measures the transfer of electric charges by specific counter ions in relation to the total charge transport through the membrane. This can be determined by two methods

  i) Measure the change in concentration of specific ions from the feed to the diluate during electrodialysis at a fixed current. The utilisation of the current over a given time interval will give the ion transport number in the membrane.

ii) Measure the potential difference across the ion-exchange membrane, with a pair of calomel reference electrodes. In chambers of the cell either side of the membrane ionic solutions of known but different composition eg 0.1 M and 0.5 M KCl are placed and well mixed. With assumptions of negligible osmotic flow across the membrane, constant ion mobiliites and small concentration gradients the theoretical potential difference $\Delta\phi_{the}$ for a 100% permselective membrane is a function of the ratio of solution activities, a, either side of the membrane.

$$\Delta\phi_{the} = - \frac{RT}{F} \ln\left(\frac{a_1^s}{a_2^s}\right)$$

The measured "membrane" potential difference is a function of the ion transport number tm in the membrane given approximately by

$$\Delta\phi_{the} = - (2t^m - 1) \frac{RT}{F} \ln\left(a_1^s / a_2^s\right)$$

Thus the ratio of measured to theoretical potential difference gives the transport number as

$$t^m = \frac{1}{2}\ 1 + \Delta\phi/\Delta\phi_{th}$$

Electrical Resistance

The electrical resistance of an ion-exchange membrane is determined by measuring the potential drop across the membrane when placed in a cell with the same solution situated on either side. The solution is typically 0.5 M NaCl, at 25C. The potential difference is measured (see Fig 3) by two colomel reference electrodes connected across the membrane via two lugging capillary probes, with their tips placed directly on the membrane surface. The electrodes in the cell are isolated from the solutions, typically by using membranes, to prevent interference from electrode reactions eg gas bubbles. In the experiment the potential is measured as a function of current density and the slope of the gradient of the plot of potential vs current density gives the area resistance in $\Omega cm2$. The use of area resistance is preferred as it gives a direct measure of the membrane potential drop from a known value of current density.

Mechanical and Chemical Properties

The chemical stability of a membrane is determined by exposing the material to various test solutions (acids, bases etc) or process solutions. Assessment of chemical stability is

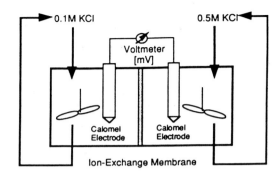

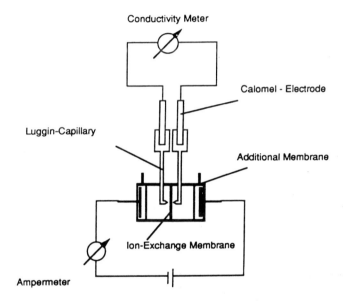

FIGURE 3 – Cell for electrical resistance measurement.

by comparison with new membranes, visually, and be determining electrical and mechanical properties. The mechanical properties of interest are thickness, dimensional stability, swelling, tensile strength and hydraulic permeability. These properties are determined by the materials used, the presence of reinforcement and the homogeneity or heterogeneity of the membrane. Membranes should be free of pin holes. Pin holes can be determined by exposing one side of the membrane to an appropriate indicator solution and looking for penetration of the dye through the other side.

Hydraulic permeability is determined by measuring the flow of de-ionised water through the membrane under a known hydrostatic pressure.

Membrane swelling is determined by equilibrating a sample of known weight in deionised water. The sample is then weighed wet after surface water is removed, dried and weighed again. The swelling is then the water uptake measured as a percentage of the weight of the dry membrane.

APPENDIX 1 – Summary inorganic membrane materials by company

Company	Ceramic	Al_2O_3	Porous metal or C/ZrO_2	SiC	C	Glass SiO_2	Metal Salts
Air Products							X
Agency Ind Sci/Japan	X						X
APV (distributor)	X	X					
Arco Chemical	X						
Asahi Glass							X
Corning (monolith)	X						
CeramMem	X	X		X			
Ciba-Geigy							X
Du Pont			X				
GFT						X	
General Motors	X	X					
Golden Technologies (Coors)	X	X		X			
Hoogovens	X	X					
Millipore	X	X					
Mitsubishi Heavy Ind	X	X					
Mitsubishi Jukogyo	X						
NGK	X	X					X
Negev Nuclear Research					X		
Nitto Electric	X						
NASA (US)			X				
Osmonics	X	X	X				
Pall			X				
Schott Glass						X	
Stichting Energioen (Neth)		X					
Sumitomo	X	X					
Swiss Aluminium		X		X			
TDK	X	X	X				
TechSep/Rhone Poulenc	X	X	X				
Teijin							X
Toray					X		
Toyota	X						
Union Carbide			X				
US Filter	X	X					
Weitzham Inst.	X						
Whatman/Anotec	X	X					
Carre			X				
Mott			X				
SFEC			X				
Asahi						X	
Fuji						X	

APPENDIX 2 – Typical properties of commercial ion-exchange membranes.

Membrane	Type	Structure Properties	IEC (Meq/g)	Backing	Thickness (mm)	Gel Water (%)	Area Resistance 0.5N NaCl, 25°C (Ωcm^2)	Perm-selectivity 1.0/0.5 N KCl (%)
Ashai Chemical Company Ltd								
K 101	Cation	Styrene/DV B	1.4	Yes	0.24	24	2.1	91
A 111	Anion	Styrene/DV B	1.2	Yes	0.21	31	2-3	45
Ashai Glass Company Ltd								
CMV	Cation	Styrene	2.4	PVC	0.15	25	2.9	95
AMV	Anion	Butadiene	1.9	PVC	0.14	19	2-4.5	92
ASV	Anion	Univalent	2.1		0.15	24	2.1	91
DMV	Cation	Dialysis			0.15		–	–
Flemion	Cation	Perfluorinated						
Ionac Chemical Company								
MC 3470	Cation		1.5	Tergal	0.6	35	6-10	68
MA 3475	Anion		1.4	Tergal	0.6	31	5-13	70
MC 3142	Cation		1.1		0.8		5-10	–
MA 3148	Anion		0.8	Tergal	0.8	18	12-70	85
Ionics Inc								
61AZL389	Cation		2.6	Modacrylic	1.2	48	–	–
61CZL386	Cation		2.7	Modacrylic	0.6	40	~9	–
103QZL386	Anion		2.1	Modacrylic	0.63	36	~6	–
103PZL386	Anion		1.6	Modacrylic	1.4	43	~21	–
Du Pont Company								
N 117	Cation	Perfluorinated	0.9	No	0.2	16	1.5	–
N 901	Cation	Perfluorinated	1.1	PTFE	0.4	5	3.8	96
N 324	Cation	Perfluorinated		PTFE	0.4	–	–	96
Pall RAI, Inc								
R-5010-L	Cation	LDPE	1.5	PE	0.24	40	2-4	85
R-5010-H	Cation	LDPE	0.9	PE	0.24	20	8-12	95
R-5030-L	Anion	LDPE	1.0	PE	0.24	30	4-7	83
R-1010	Cation	Perfluorinated	1.2	No	0.1	20	0.2-0.4	86
R-1030	Anion	Perfluorinated	1.0	No	0.1	10	0.7-1.5	81
Rhone-Poulence GmbH								
CRP	Cation		2.6	Tergal	0.6	40	6.3	65
ARP	Anion		1.8	Tergal	0.5	34	6.9	79
Tokuyama Soda Company Ltd/Eurodia								
CL-25T	Cation		2.0	PVC	0.18	31	2.9	81
ACH-45T	Anion		1.4	PVC	0.15	24	2.4	90
AMH	Anion	Chemical resistant	1.4	–	0.27	19	11-13	–
CMS	Cation	Univalent	>2.0	PVC	0.15	38	1.5-2.5	–
AFN	Anion	Antifouling	>3.5	PVC	0.15	45	0.4-1.5	–
AFX	Anion	Dialysis	1.5	PVC	0.14	25	1-1.5	–
ACLE-SP	Anion	High pH resistant	1.3		0.2	35	10-20	–
Neosepta-F		Perfluorinated						
Stantech								
STC-IT	Cation		2.0	Yes	0.13		1.2-2.0	
STA-IT	Anion		1.8	Yes	0.13		1.3-2.0	
Solvay								
	Cation	Fluorinated	2.6	–	0.25	–	0.5-1.5	0.7
	Anion	Fluorinated	0.9	–	0.16	–	1.5-4.5	0.93

SECTION 3

Gas Separations

GAS SEPARATION - GAS DEHYDRATION - GAS REMOVAL

Try the Innovators, try Aquilo

NITRO
On the spot nitrogen generators, capacities from 0,1 m³/hr to 100 m³/hr, high quality, low operation costs.

DRYPOINT
Membrane air dryer, always the right dewpoint, compact, simple, no noise, no maintenance.

PRODUCE
Various concepts for controlled atmosphere storage of fruit and vegetable.

Aquilo Gas Separation, Etten Leur, the Netherlands. A company committed to quality and performance in all aspects of vapour and gas treatment, with in-depth expertise in membrane technology. Aquilo supplies a wide range of standard systems and components. It is the willingness to custom design, engineer and manufacture your exact system that separates Aquilo from all the others. Standard solutions for end-users, custom engineered answers for OEM's. You name it, Aquilo produces it.

Aquilo, the name to watch!

Aquilo Gas Seperation b.v.
Oude Kerkstraat 4
4870 AG Etten Leur
The Netherlands
Phone: +31 (76) 508 53 00
Fax: +31 (76) 508 53 33

GAS SEPARATIONS

The separation of mixtures of gases is performed with the objective of obtaining one or more of the constituents in a pure form. There are many applications in both large scale and small scale processes. Separation can be achieved by several methods based on different physical and chemical properties of the species. There are four principal methods applied to the separation of gases; absorption, adsorption, cryogenics and membranes (see Table 1). Process economics will determine which of these methods is used for any particular application.

Absorption is a physical process which is frequently chemically enhanced by reacting the gas with the solvent or a component within the solvent. The transfer of gas into the liquid phase can be also enhanced by the application of high pressure. Absorption is typically applied in the selective removal of acid gases such as CO_2, H_2S and SO_2. Absorption requires regeneration of the solvent which for physical solvents is achieved by a reduction in pressure and for "chemical" solvents, by the action of thermal driving force or by chemical means.

Adsorption processes use a solid surface (absorbent) to preferentially remove one component in an analogues way to the solvent in absorption. The component is reversibly bound to the surface which is of high surface area, up to $1500 m^2/gm$ absorbent, due to a highly microporous structure of the material. Typical adsorbents are zeolites, carbons, molecular sieve, silica gel and alumina and are used in the form of fixed beds. Adsorption is a dynamic process in which "concentration waves" of absorbate are generated in the bed and eventually the absorbent will become "fully" loaded. Regeneration is then achieved by parametric sieving, which in gas separation is in one of two manifestations, thermal swing adsorption (TSA) or pressure swing adsorption (PSA). The former is the older of the two processes and uses a higher temperature to regenerate the bed by displacing the solid/gas equilibrium (isotherm) in favour of the gas phase. Pressure swing adsorption uses reduced pressure to release the absorbate from the surface. TSA is used in natural gas sweetening and in gas drying operations. PSA is used in recovery of carbon monoxide, carbon dioxide, hydrogen, methane, nitrogen, oxygen and other gases and also in air drying and hydrogen isotope separation. Cryogenic separation uses very low temperatures to separate gas mixtures e.g. for nitrogen, 77K and for helium, 4K. Liquid and vapour phases are produced and separation is achieved by distillation or analogous processes. The cryogenic distillation involves a sequence of vaporisations and condensations; high boiling species concentrate in the liquid phase flowing down the column and low boiling constituents concentrate in the vapour phase flowing up the column. The low temperature in cryogenic separation is produced through compression followed by cooling using for example cooling water, followed by refrigeration and Joule-Thomson expansion. Product from cryogenic separation may be a cryogenic liquid or a gas. Cryogenics is used in the separation of atmospheric gases, methane from nitrogen, ethane, ethylene and in hydrogen separations, etc.

Membrane separation based on "selective" gas permeation competes directly with the three above mentioned methods in many applications. Membranes offer versatility and simplicity in comparison to other methods, which must be balanced against limitations of medium purity and the need for recompression:

TABLE 1 – Comparison of Gas Separation Technologies.

Technology	Advantages	Disadvantages
1. Cryogenic without distillations	a) High recovery of products b) Moderate purity of light product (e.g. H2 up to 98%) c) Can operate at high pressures d) Low cost e) Low pressure loss of light product	Cannot achieve very high purity of light product products
2. Cryogenics (with distillation)	a) High recovery of products b) High purity of light product (e.g. H2 up to 99.5%+) when using hydrocarbon wash processes c) Can operate at high pressures d) Good purity of heavy products e) Low pressure loss of light product	a) High cost b) High energy consumption
3. Absorption	a) Simple process b) Low pressure loss of light product	Poor separation characteristics i.e. low purity light ends or low recovery of heavy ends
4. Adsorption (Pressure Swing)	a) Very high purity of light product (e.g. 99.99% H2) b) Simple process	a) Low recovery b) Operates most favourably at lower pressures (20-30 bar)
5. Adsorption (Thermal Swing)	Can remove minor components virtually completely	Expensive for bulk removal of impurities
6. Hydrides	a) High purity product b) Simple process	a) Limited experience on a commercial scale b) Low capacity

GAS SEPARATIONS

Membrane Separation of Acid Gases

Membrane separators are applied in the separation of acid gases in four general ways:
i) as a direct method of separation in the gas phase based on the difference in permeabilities of species through the membrane, usually a polymer
ii) as a gas/liquid contactor (absorption) offering a means of controlling the interfacial area
iii) as a method of regeneration of the solvent in chemically assisted absorption
iv) as a support for an appropriate "chemical" carrier, which may be electrochemically assisted.

The last of these methods using facilitated transport is still primarily at a developmental stage for separation of involving CO_2, CO, H_2S and SO_2 and NO_x· O_2 from N_2, olefins from hydrocarbons. Pilot scale operation is reported but no commercial installations are yet known. The functionality of the carriers can enable high selectivities to the achieved and separations which are not possible with polymer gas permeation membranes.

Membrane Gas Absorption

Absorption in many cases relies on providing a high interfacial area between the gas and the liquid phase to achieve high overall rates of gas mass transport. Many commercial devices based on packed columns, plate columns, spray columns, venturi scrubbers etc are available. In many cases equipment design can be restricted by limitations in the relative flows of the fluid streams. The use of a membrane, in the form of low cost hollow fibres (e.g. polypropylene), gives a contact surface area between the gas and liquid phases, which is independent of the gas and liquid flows. The small dimensions of the hollow fibre

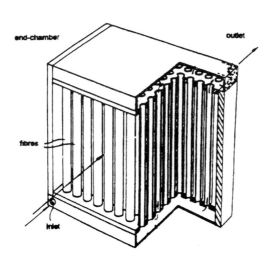

FIGURE 1 – Cross flow hollow fibre membrane gas absorber.

membranes give high specific surface areas which are significantly greater than most conventional absorbers. Commercially available hollow fibre modules used for filtration duties can be used as absorbers with gas flow through the fibres. These units however do not offer the best distribution of flow over the outer surface of the fibres and thus a more regular fibre packing configuration is advantageous (see Fig 1). In this design absorption liquid is fed through the fibre tubes and gas flows over the fibre bundle. A requirement in operation is that the gas and liquid phases do not "mix" to avoid entrainment of liquid in the gas stream.

For porous membranes the liquid should not enter the pores – which is controlled by pressure difference and the membrane pore structure and "wetting" characteristics. Aqueous based absorbents can be used in conjunction with hydrophobic membranes such as polypropylene. Applications of hollow fibre membrane absorbers are seen in Table 2.

TABLE 2 – Applications of hollow fibre absorbers

Flue gas	SO_2, NO_X, HCl, CO_2
Natural gas, landfall gas	H_2O, H_2S, CO_2
Biogas	
Indoor air	O_3, SO_2, NO_X, VOC
OFF gas	NH_3, H_2S, CO_2, VOC
Process Water Calculation	CO_2
Scrubbing of reactor vents	Various

TABLE 2A – Comparison of hollow fibre and packed column contactors

Benefits	Liqui-Cel® Contractors	Packed Columns
Surface Area/Volume, ft^2/ft^3	>1000 ft^2/ft^3	25-75 ft^2/ft^3
Space Efficiency	1	3
Capital Equipment Costs	2	2
Installation Costs	2	3
Instrumentation Costs	2	3
Independent Flow Control for Gas & Liquid Streams	1	4
Vapor Load per ft^2 Contacting Area for Vacuum Operation	6	5
Predictable Scale-up	1	3
Low Maintenance Contactor Design	2	2
Materials of Construction Approved for Food Contact	2	2
Stripping Gas/Liquid Feed Ratio	6	5
Expandable for Increased Process Capacity	1	4
Expandable for Increased Efficiency	1	4
Process Turn-down Capability	1	4
No Flooding and Loading Operating Flow Contraints	1	4

KEY
1 = Excellent   3 = Fair   5 = Low
2 = Good   4 = Poor   6 = High

GAS SEPARATIONS

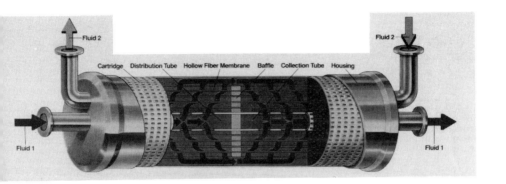

Gas transfer contactor.

These applications are not known to be practiced at a commercial level. Recent tests at a 120 Nm^3/h scale of SO_2 recovery from a flue gas resulting from combustion of biogas containing H_2S have demonstrated over 95% removal of SO_2.

Hollow fibre gas/liquid contactors offer several advantages over standard equipment such as packed columns (see Table 2A).

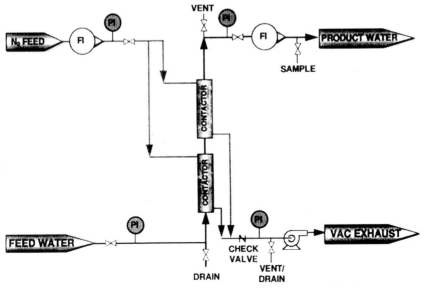

FIGURE 2 – Schematic using nitrogen and vacuum combination.

GAS PERMEATION

The process of gas permeation is a relatively simple process which has expanded in use rapidly since it was introduced commercially in 1979. Since that time approximately 20 or more companies have come to the market place with equipment (see Appendix). The range of applications (see Table 3) covers the supply of pure gases such as He, N_2 and O_2 from air, the separation of acid gases such as CO_2 and H_2S, the separation of H_2 in the petrochemical and chemical industries and a variety of smaller applications.

TABLE 3 – Applications of gas permeation membranes

Recovered Gas	
Hydrogen	Ammonia purge gas (H_2, N_2, NH_2, Ar)
Hydrogen	Methanol synthesis Syngas refineries purge (H_2, C_nH_m) Petrochemical (H_2, N_2 CH_4, CO) Syngas ratio adjustment Hydrogenation reactions
Carbon dioxide	Natural gas Elimination from fuel cell Hydrocarbon gases Control for food presservation and fermentation Landfill drainage gas (CO_2, CH_4 C_nH_m)
Hydrogen Sulphide	Hydrocarbon gases Sour gas treatment
Carbon monoxide and Hydrogen	Synthesis gas
Nitrogen	Preservation and inert blanketing Air separation (O_2, N_2)
Oxygen	Oxidation in chemical industry Air separation Combustion use Medical gases
Helium	Hydrocarbon gases
Helium	Diving gas (He/N_2, O_2, air)
Water vapour	General gas dehydration Well head gas Air dehumidification Natural gas dehydration
Oxygen enriched air	Use as combustion gas
Hydrocarbons/air	Pollution control HC recovery

Natural Gas Industry

The application of membranes in the gas industry is principally in the treatment of natural gas The composition of gases from different wells is variable and may require treatment

GAS SEPARATIONS

Treatment of natural gas at well head by gas permeation.

before entering the distribution system. Toxic materials such as hydrogen sulphide must be maintained at safe levels. The risk of corrosion in pipelines by CO_2, H_2S and water must also be minimised. The principle separations required in the treatment of natural gas (predominantly CH_4) are, CO_2/CH_4, H_2S/CH_4, H_2O/CH_4, N_2/CH_4, He/CH_4.

The established technology for removing CO_2 and H_2S from natural gas is amine absorption, which accounts for around 70% of natural gas treatment processes. It is an efficient method which adapts well to variations in gas composition. However there are several limitations which have enabled membrane gas permeation to impact in the market. These limitations include:-
- absorption units are large and heavy which present problems for use on offshore platforms
- a high degree of supervision and maintenance is needed
- the amine regenerator consumes moderate amounts of gas
- the amine regenerator presents a fire hazard on offshore platforms

Membrane units are more cost effective than absorption for smaller applications and they can treat gas at the well head, which reduces safety problems and corrosion problems. The fact that the well head gas is at high pressure (up to 130 bar) is a significant bonus for gas permeation membranes and additionally when the product gas methane is the permeating species on the high pressure side of the membrane.

CO_2 Separation

The content of carbon dioxide in natural gas and hydrocarbon gases is variable and can be

Dow H fibre unit.

as high as 40-60% in cases with predictions as high as 80% as hydrocarbon capacities decline.

These high content CO_2 gases can often arise from the practice of enhanced oil recovery (EOR). In miscible-flood EOR, the oil bearing strata is injected with CO_2 to stimulate oil recovery. Gas released by this process at the well head is therefore heavily laden with CO_2 and is in need of treatment.

All appropriate gas permeation membranes show much higher affinities for carbon dioxide than methane with permeability ratios greater than 20. This therefore enables effective CO_2 removal from hydrocarbons. Modules used for separation are hollow fibre designs with double-ended tube sheets. The feed gas is introduced to the outside of the fibre bundle with the CO_2 rich stream permeating to the inside of the fibres and the hydrocarbon rich steam flowing to the central core of the module where it is piped away.

The application of the membrane units depends upon the scale of the operation. Single-stage membrane units are suitable for low flow rate application but with higher flowrates multiple module designs with gas recycle are used to minimise losses of hydrocarbons. Typical designs are shown in Fig 3. In this operation the permeate gas (CO_2 rich) has sufficient calorific value to be used as a fuel.

Typical performance of modules, which are installed as trains, is to achieve permeate CO_2 compositions of 92-95% operating a feeds containing 55% to 65% of CO_2. Permeate flows per module are between 7,000-8,000 MSCFH an actual feed flows of between approximately 16,000 to 18,000 SCFH.

Enhanced oil recovery processes are often used to increase production from depleted fields. High pressure CO_2 is pumped into the ground at the periphery of the field and diffuses through the formation to drive residual oil towards the oil wells. Large quantities of CO_2 are required in the initial stages of the process. In time the CO_2 arrives at the well head and the concentration of CO_2 in the gas associated with the well will increase to values as high as 70% to 90%. The objective of the gas separation is to recover the natural gas

GAS SEPARATIONS

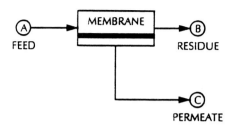

Composition (Mole %)	Stream		
	A	B	C
CH_4	93.0	98.0	63.4
CO_2	7.0	2.0	36.6
Flow rate (MMSCFD)	20.00	17.11	2.89
Pressure (PSIG)	850	835	10
Methane Recovery = 90.2%			

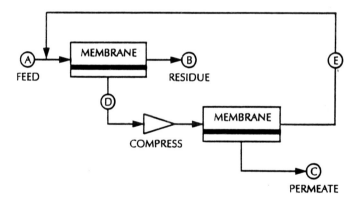

Composition (Mole %)	Stream				
	A	B	C	D	E
CH_4	93.0	98.0	18.9	63.4	93.0
CO_2	7.0	2.0	81.1	36.6	7.0
Flow rate (MMSCFD)	20.00	18.74	1.26	3.16	1.90
Pressure (PSIG)	850	835	10	10	850
Methane Recovery = 98.7%					

FIGURE 3 – Single stage and two stage membrane processes for natural gas sweetening.

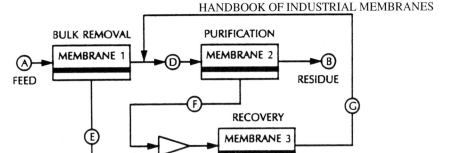

	Stream						
Composition (Mole %)	A	B	C	D	E	F	G
CH_4	93.0	98.0	49.2	96.1	56.1	72.1	93.0
CO_2	7.0	2.0	50.8	3.9	43.9	27.9	7.0
Flow rate (MMSCFD)	20.00	17.95	2.05	19.39	1.62	1.44	1.01
Pressure (PSIG)	850	835	10	840	10	10	850
Methane Recovery = 94.6%							

FIGURE 3 (contd.) – Multistage membrane processes for natural gas sweetening.

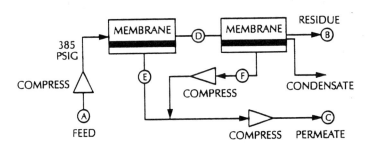

	Stream					
Composition (Mole %)	A	B	C	D	E	F
Methane	9.6	26.9	2.9	20.0	2.0	6.7
Carbon dioxide	70.0	10.0	93.4	35.4	95.2	86.2
Hydrogen sulfide	0.6	0.01	0.8	0.1	0.9	0.4
Ethane	6.3	10.0	1.0	14.0	0.7	2.4
Propane	5.6	18.4	0.6	12.9	0.4	1.5
I-Butane	2.5	8.6	–	6.0	–	0.2
N-Butane, C5+	1.3	4.3	–	3.0	–	0.1
Nitrogen	4.1	11.8	1.1	8.6	0.7	2.5
Flow rate (MMSCFD)	22.5	6.25	16.0	9.5	13.0	3.2
Pressure (PSIG)	235	365	290	375	80	15
Condensate flow = 81 gal/hr						

FIGURE 4 – Membrane process for CO_2 and natural gas recovery from EOR operation.

and also the CO_2. The CO_2 is recompressed and re-used in the process. The variable composition of CO_2 in the gas makes membrane separation more attractive than amine absorption, and especially when high CO_2 concentrations are encountered. The operation of the system (shown schematically in Fig 4), requires optimisation of the permeate pressure – high permeate pressure minimises recompression costs but decreases membrane efficiency.

There is a tendency in enhance oil recovery for membranes to be used in hybrid combinations particularly with amine absorption, due to a reduced cost in overall operation.

In the case of low CO_2 content hydrocarbon streams, technologies based on chemical or absorption technologies are typically used. Physical absorption based methods can be used on high CO_2 content feeds where lower purity requirements are acceptable. Typical chemical amine based technologies can remove over 90% of the CO_2 from the stream and result in only 1% hydrocarbon losses.

TABLE 3A – Typical solvents for absorption of CO_2

Chemical:	Monoethanolamine (MEA) Di-ethanolamine (DEA) Tri-ethanolamine (TEA) Hot Potassium Carbonate (Benfield) Methyldiethanolamine (MDEA)
Physical:	Methanol (Rectisol) N-methyl-2-pyrrolidone (Purisol) Polyethylene Glycol (Selexol) Propylene Carbonate (Fluor Solvent)
Physical/Chemical:	Sulfolane/Diisopropanolamine/Water (Sulfinol)

Improvements in module design using "small" fibre configuration with increased tube bundle sheet diameter offered increased capacity for gas membrane permeators to operate "competitively" in this CO_2 content range. Such single stage modules when processing gas containing 10% CO_2 to produce a pipeline specification product of 2% CO_2, permeate between 8% to 15% of the inlet hydrocarbon. This is significantly greater than with chemical absorption systems. However, with a permeate hydrocarbon recovery the overall hydrocarbon losses can be reduced to levels below absorption. The permeate hydrocarbon recovery units are composed of one or two additional membrane separation stages and recycle recompression. Overall there is not much difference between the performance of membrane separation and absorption.

Membrane carbon dioxide removal from liquid (adapted from the *Membrane Technology* Newsletter)

A membrane gas-processing technology for CO_2 removal has been developed that lowers operating costs by reducing membrane surface-area requirements using "sweep gas" technology. The process treats the separated NGL stream instead of the raw inlet-gas, and in place of 200 MMcfd of inlet gas, only 3000 b/d of liquid must be treated.

The gas plant extracts NGLs from sweet natural gas produced at various offshore Louisiana platforms. The plant uses cryogenics to extract up to 86% of the incoming C_2 and virtually all C_{3+} fractions. The 200 MMcfd inlet-gas stream contains approximately 1% CO_2, less than the 2.0% maximum concentration allowed for cryogenic processing and gas sales.

Because the boiling points of CO_2 and C_2 are similar, the CO_2 concentrates in the liquid ethane product so that, at optimum recovery, the liquid product becomes overly contaminated with CO_2 and therefore out of sales specification. As a result, the plant was operated with a warmer than desired bottom reboiler temperature to remove the CO_2 which also removed part of the ethane product liquid. If the reboiler temperature were lowered until the methane content of the liquid product was instead limited, C_2 recovery would be approximately 10% higher. At this optimum recovery level, however, CO_2 content is more than double the 5% maximum CO_2 specification.

Several possible solutions exist;

1. to remove the CO_2 from the gas as it came into the plant. This could be accomplished with an amine treating system. Although feasible, the system would be quite large and have significant operating and maintenance costs. It was also not environmentally desirable.
2. membrane gas permeation. At low CO_2 concentrations, such a membrane system can be made to operate effectively if the losses can be recompressed and recycled through a second set of membranes. The disadvantages to this method are the high capital cost and the maintenance and operating costs for the additional compressor. The recycle compressor required would also be a large unit because it must take suction at approximately atmospheric pressure and discharge at the inlet-gas pressure for recycling.
3. to remove the CO_2 from the liquid product. An amine treating system could be used but was not favoured because of its associated operating and maintenance costs and environmental liabilities. The higher CO_2 concentration would allow membranes to work more efficiently, however, with reduced hydrocarbon losses because a smaller membrane surface area would be required.

Membrane Separation

In a typical setup for semi-permeable membrane separation, the stream to be purified is introduced at high pressure on the exterior (shell side) of the membrane fibre. The higher permeable components like CO_2 travel through the fibre wall and are collected at low pressure in the centre bore as a gas. This gas in the bore then flows out both ends of the membrane unit at approximately atmospheric pressure in order to maximise the pressure difference across the membrane fibre to increase

GAS SEPARATIONS

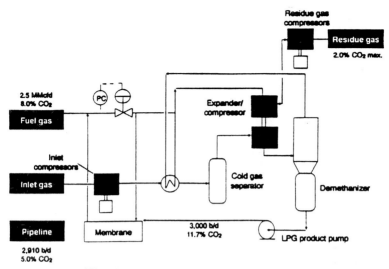

Flowsheet of carbon dioxide separation process.

the driving force across the fibre wall. The membrane unit was modified so that the low pressure permeate may be purged and diluted by a flow of fuel gas countercurrent to the flow of the high-pressure stream. Pressure regulators are arranged so that the flow of fuel through the membrane unit is maintained. The outlet of the permeate side of the membrane, which will contain a mixture of permeated impurities and fuel gas, is operated at a pressure just sufficient to enter the plant's fuel system serving plant boilers, engines or other fuel users. Despite the higher total back-pressure on the permeable membrane, the dilution effect actually reduces the partial pressure to the extent that overall driving force across the membrane is increased.

In the case of an NGL feed stream, the driving force for the CO_2 is increased by a greater proportion than that for the hydrocarbon, thereby improving membrane selectivity, or separation factor. In a fuel gas sweep of 15% of the molar flow rate of liquid is used, the driving force for CO_2 removal is increased by approximately 17%, whereas the driving force for product hydrocarbon loss by permeation remains relatively unchanged. As the ratio of molar flow-rate of fuel-gas sweep to molar flow-rate of liquid product is further increased, the partial pressure on the permeate side of the membrane is further reduced, causing an even greater driving force.

In the actual application, the liquid would go into a pipeline after being treated. The sweep gas was introduced on the tube side of the device and the permeate stream drawn off near the bottom of the case. The resulting permeate gas volume was small and was sent to the flare. In the actual application, this permeate gas would be sent back to the fuel-gas system. The membrane system operated at liquid feed pressures up to 1000 psig while the permeate pressure was at 100 psig. The

analysis of the permeate gas samples consistently showed negligible, if any, heavy hydrocarbon constituents, confirming that the membrane device was operating successfully.

Separation of Hydrogen Sulphide and Water Vapour

Natural gas is often saturated with water at the well head and this has important implications in subsequent pipeline transport.
-   Under high pressure, water reacts with methane to form solid methane hydrate which could lead to blocking of the pipeline if left unchecked.
-   In the presence of CO_2 and/or H_2S condensation of water in pipelines will lead to the formation of aqueous S^{2-} ions and carbonic acid which will react with and thus corrode pipelines.

The formation of methane hydrate can occur at temperatures below approximately 13C at typical pipeline pressures of 70 bar. Prevention of this requires removing as much water as possible and adding an inhibitor, e.g. ethylene glycol or methanol. Glycols are also useful dehydration agents and are used to remove water. This may also include the use of a silica gel or molecular sieve solid bed adsorption (dehydration) unit. Glycol used in dehydration in offshore platforms, requires regeneration in situ for re-use. This is normally achieved by heating at reduced pipeline pressure. A glycol dehydration unit can reduce the natural gas water content from a saturation level of approximately 1370 ppn (70 bar and 40C) to approximately 140 ppm. Hydrocarbon loss is around 0.5%. Gas permeation is a possible method of dehydration as most membranes are very permeable to water vapour with reported selectivities over CO_2 of greater than 100 (up to 1,000 for cellulose acetate).

GAS SEPARATIONS

The use of gas permeation has to overcome two major limitations.
- the driving force for water vapour permeation is low
- the waste vapour permeate may condense out in this stream unless a sweet gas is used

These factors are believed to have impeded the application of membrane separation in this area.

Hydrogen sulphide is corrosive to pipelines even in the dry state and thus it would be advantageous to remove it from natural gas at the well head. Any separation of H_2S would require the treatment of the gas due to its toxicity. Thus a membrane separation which could produce a more concentrated H_2S gas would have to precede a sulphur recovery unit e.g. the British Gas Stretford Process. The effect would therefore to reduce the size of the recovery unit, at the expense of the membrane separation unit. The relatively poor selectivities of current membranes for H_2S over CH_4, coupled with low driving forces (partial pressure of H_2S) have slowed down industrial adoption. A membrane unit with a capacity of 100 MMSCFD operating at a pressure of 83 bar is reported to reduce H_2S content to 10 ppm. Final polishing of the gas to 33.3 ppm is achieved with a zinc oxide bed.

Nitrogen is also found in well head gases in varying quantities depending upon the country of source. Nitrogen is also an option to CO_2 for use in enhanced oil recovery and would in this application be present in gases released in this operation. Nitrogen selective membranes are awaited before this can be considered a viable option to cryogenics.

In gas fields where helium is in a relatively high concentration (0.1%), membrane gas permeation can be used to recovery the helium. Operation in a multi-stage membrane plant in Canada is reported to produce a helium rich gas containing 60% by volume, with possible recovery up to 90% by volume.

Hydrocarbon Gas Dewpointing

The quantities of heavier hydrocarbons e.g. ethane occurring in natural gases can sometimes lead to condensation in the pipelines. The removal of these gases is routinely done by compression and cooling the gas streams to induce condensation of the heavier hydrocarbons. The disadvantage of this process is that all the gas steam is cooled. Selective permeation of heavier hydrocarbons over methane, with recycling and recompression to recover the HC's is reported to be near (or current) commercialisation.

Separation of CO_2 from Hydrocarbon Liquids

Processing of hydrocarbon gas streams for NGL products can result in the build up of CO_2 in several of the streams, typically in the de-methaniser bottoms and the de-ethanizer overhead or reflux streams. Implementation of gas permeation to remove CO_2 (at 6.4%) from de-methanizer bottoms to a level of 3% has resulted in commercial adoption. In this operation feed pressure was approximately 65 bar and permeate pressure was between approximately 7 to 14 bar.

Nitrogen from Air

Nitrogen is used in many industries (see table 4) as an inert gas for purging of equipment and blanketing of potentially dangerous processes. This is in addition to its use as a

TABLE 4 – Application of Nitrogen

In the chemical industry:	for purging tanks and vessels, blanketing storage tanks.
In oil and gas production	for inserting storage tanks and vessels, for purging pipelines, for blanketing rotating gaskets and de-ionised water, as well as separators etc
In the food industry and in the packaging industry:	for the packaging of food, the production of wine and the blanketing of fruit juices, edible oils and syrups, etc
In the heat treatment and production of metals:	including the tempering, hardening, carbonizing, extrusion and sintering of metals.
In the shipping trade:	for inerting void spaces and storage tanks, for the blanketing and purging of tanks in for example LNG and LPG tankers.
In the rubber industry:	for the production of all types of rubber products
In laboratories:	for purging gas in spectrometers, as a carrier gas where high purity is required, as an agitation gas, etc.
In aviation:	for the blanketing and inerting of fuel tanks, mobile Nitrogen systems for the ground crew.
In the storage of perishable products:	for example the storage of fruit at ultra low oxygen percentages (ULO).

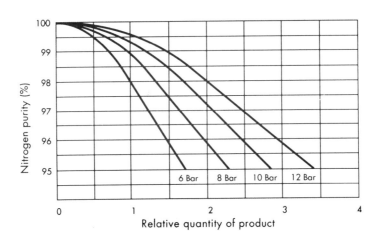

FIGURE 5 – Relative production rate as a function of nitrogen purity and pressure.

GAS SEPARATIONS

Hollow fibre gas permeator for N_2 generation.

cryogenic fluid in laboratories and as a bottled gas in analysis and producing oxygen free environments.

Most nitrogen (and oxygen) is produced by cryogenic air separation techniques in large industrial plants. Gases are then distributed by pipeline, road or rail tankers or in cylinders. This makes nitrogen very expensive in remote and offshore locations. Ideally on site air separation to produce N_2 is desirable, as there is no limitations in the supply of feedstock. Cryogenic separation cannot be considered for relatively small scale generation because of the high capital cost of small equipment. The on-site generation of N_2 has been satisfied by either pressure swing adsorption or gas permeation membranes. There is an increased use of gas permeation owing to developments in the technology and the introduction of hollow fibre thin film and coated membranes. Gas permeation is particularly simple with a basic system consisting of air filters and the membrane module(s). Complete units are available which include air compressors, monitors and controllers. Equipment is compact and lightweight, reliable and virtually maintenance free. There is no start-up time for the supply of nitrogen which can be produced at up to 99.9% purity. The performance of the unit depends upon the required purity of the N_2, and pressure (see Fig 5). In many applications for low pressure blanketing or purging applications 5% maximum oxygen content is acceptable, which has a major impact on unit cost. The production capacity can virtually double by switching to a 95% N_2 rather than 99% N_2 product.

Small Scale N_2 Generation for the brewing industry

The traditional method of dispensing keg beers is with pressurised carbon dioxide. The use

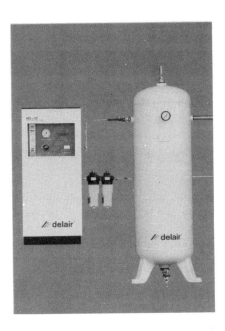

Small scale N_2 generator.

of high purity CO_2 results in slower dispense times, poor foam/froth retention and waste. Brewers therefore moved to the use of mixed gas, of nitrogen and carbon dioxide, for dispensing. The ratio of nitrogen and carbon dioxide used depends on the particular type of beer and can be in the range of 2.3 to 0.43. The mixed gas almost eliminates all the disadvantages of using pure CO_2, but can only be stored as gas in cylinders. This, in comparison to liquid CO_2 storage, reduced the dispensing capacity of each delivered cylinder of gas. In additional individual cellars will require several cylinders of mixed gases of different composition to accommodate the dispensing of a range of beers.

This problem has been resolved by using small capacity membrane nitrogen generators which operate in conjunction with liquid CO_2 storage. With appropriate gas mixers, a mixed gas of any specification is available in situ. This radically increases dispensing capacity and gives added flexibility in dispensing a wide range of beers.

Oxygen Enriched Air

High purity oxygen cannot currently be produced from air using membranes although oxygen enrichment is possible and is used.

Oxygen enriched air has applications in chemical processing industries, medical and healthcare use, metal and steel industries etc. Membrane modules are typically plate and frame modules based on thin film silicon polymers. The module configuration is preferred due to the requirement for low pressure operation to minimise air compression costs. In the process (see Fig 6) feed air is blown by a fan giving a pressure of around 1.03 bar and

GAS SEPARATIONS

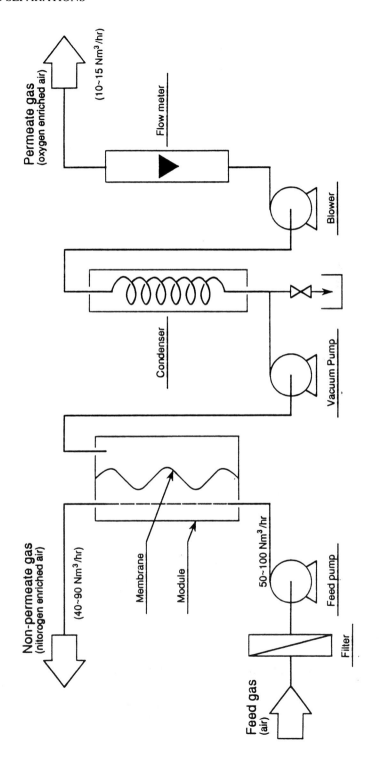

FIGURE 6 – Oxygen enrichment process.

290 HANDBOOK OF INDUSTRIAL MEMBRANES

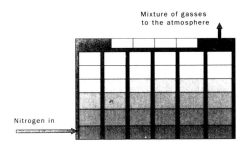

Conventional Nitrogen flushing (O_2 removal)

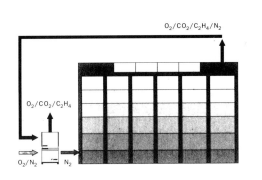

Recirculation principle of the Delair separator

FIGURE 7 – Recirculation principle of perishable goods storage.

oxygen enriched nitrogen permeate is sucked by a vacuum pump (0.2-0.3 bar) or alternatively by a suction blower. Typical product gas contains 27% to 35% O_2. Modules contain typically 100 membranes giving areas of up to 86m² for industrial use. High permeability (3.4 Nm³/m²/h/bar) low selectivity (2.1 O_2:N_2) are used in industrial application whereas high selectivity (4.0, O_2:N_2), low permeability membranes are used in medical applications. The use of a composite hollow fibre membrane has also been reported in this applications in a pilot scale operation. A major requirement for oxygen enriched air is in methane fired furnaces.

Medical applications of oxygen enriched air are for use of patients with disease of the respiratory organs. In this application a 40% oxygen rich air is produced in modules of 2.8 m² membrane area.

Perishable Food Storage

The storage of perishable products, typically foods, under modified gas conditions such as controlled atmosphere (CA) or ultra low oxygen (ULO) is important due to the variable market price of a high value commodity which means that a loss of quality during storage is no longer acceptable. Systems which provide flexible storage and marketing of high-quality products are required. Gas permeation units provide an effective means of storage, the principle of operation is based on gas recirculation through the storage cell [see Fig 7] and is based on the rapid removal of oxygen. As well as the removal of oxygen, gases such as CO_2, ethylene and aromatics are removed. The system also operates as a nitrogen generator, from the air, which recirculates through the gas permeation unit. This provides significant savings in energy in comparison to the washing or flushing principle of operations. Overall the benefits of oxygen recirculation separations are:
- maintenance of picking/harvesting quality
- lengthening of product storage time
- rapid oxygen removal enables partial marketing of stored produce
- positive pressure system maintains ULO condition
- low maintenance and fewer losses

Typical "pull-down" capacities for oxygen removal of 21% to 40%, based on a 400 m³ storage cell are approximately 31 hours for a 7.5 kW unit.

A combination of oxygen permeation and scrubbing in one unit is available which makes it possible to monitor and adjust the CO_2 and oxygen content independently.

The application of the membrane units depends upon the scale of the operation. Single-storage membrane units are suitable for low flow rate application but with higher flowrates multiple module designs with gas recycle are used to minimise losses of hydrocarbons. Typical designs are shown in Fig 8. In this operations the permeate gas (CO_2 rich) has sufficient calorific value to be used as a fuel.

Overall the production of oxygen enriched air by membranes competes with pressure swing adsorption and other supply methods. The effective areas are generally when lower flowrates are required. The relatively low purity levels achieved has not seen a large adoption of membranes in this area. One potential application is in the supply of 35% oxygen to fermentation processes.

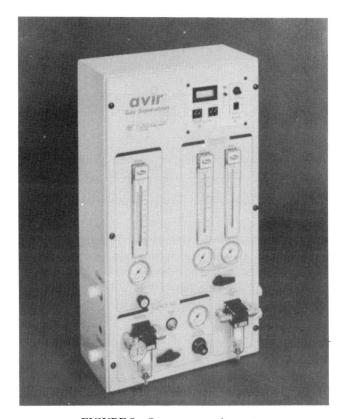

FIGURE 8 – Oxygen separation system.

Chemical Industry Applications

A principle use of gas permeation in the chemical industry is in H_2 and CO_2 recovery and separations.

Hydrogen Recovery

Ammonia is produced by the steam reforming process using natural gas and air as principle feedstock sources of N_2 and H_2 (synthesis gas). In ammonia synthesis (Fig 9) a purge gas is withdrawn from the ammonia synthesis loop to control the level of inert gases e.g. argon and methane. This purge steam also contains hydrogen (plus ammonia and nitrogen) which is separated out and fed back to the synthesis process loop.

For this separation two technologies can be used cryogenic separation or membrane gas permeation.

The membrane process uses several hollow fibre (or spiral wound) modules arranged in two sets. In the first set, hydrogen rich gas is produced as the permeate from the purge which is at a pressure of 90 bar. This hydrogen rich product is fed back to the synthesis gas intermediate stage compressor. The retentate from the first module set is then fed back

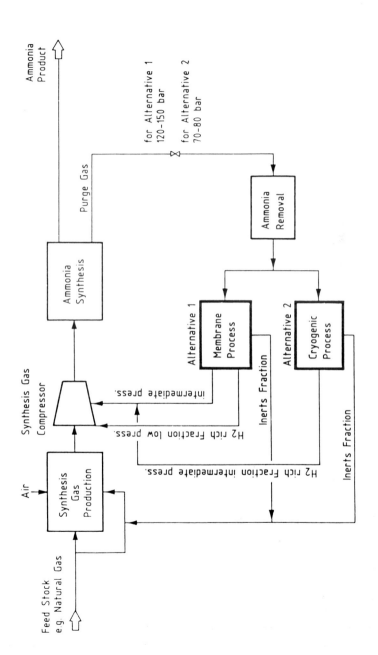

FIGURE 9 – Simplified flowsheet of ammonia synthesis.

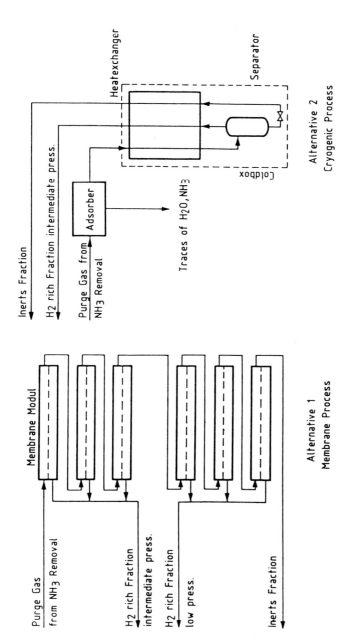

FIGURE 10 – Purge gas separation by gas permeation or cryogenics.

to the second set of modules to produce a second hydrogen rich permeate which is also fed back to synthesis gas compression (First Stage). A lower pressure permeate is provided here to obtain a high recovery rate of hydrogen of approximately 90-95%. The retentate gas from this stage, containing 70-80% inerts (argon and methane) is used as fuel for the steam reformer.

In cryogenic separation Fig 10 the purge gas is first treated to remove ammonia and then traces of water (and ammonia) are removed by absorption. The gas is then cooled in a plate film heat exchanger, to liquefy most of the inerts (80-95%) and part of the nitrogen. This liquefaction is against the cold hydrogen rich product and the evaporating inerts fraction. The hydrogen rich fraction is then fed to the intermediate stage of synthesis gas compression.

Overall performance is typically a 92% recovery of H_2 as a 92% pure product from a feed gas containing 61% H_2.

Hybrid Systems

Synthesis gas (syngas) is a mixture of hydrogen and carbon dioxide which has uses in petrochemical and metallurgical processes. It is either used as a mixture or separated into its two high purity components. Syngas is produced mainly from hydrocarbon by:
- steam reforming natural gas or light hydrocarbons
- partial oxidation of heavy oils
- gasification of coal or coke

The principle use of syngas is in the production of chemicals with carbon-carbon bonds (C_1 chemistry). There are many commodity chemicals which are produced by C_1 chemistry, ethylene glycol, acetic acid, oxoalcohols etc. (see Table 5) It can be appreciated in these processes that there is a significant variation in the required stoichiometric ratio of H_2/CO, from 0 to 2.0. For the benefits of these chemical synthesis it is desirable for the syngas to be introduced in the appropriate ratio. However syngas is usually available with H_2 : CO ratios greater than 2.0 and thus processes which can adjust the ratio accordingly without sacrificing the hydrogen are attractive. Fig 11 is a typical process flowsheet for

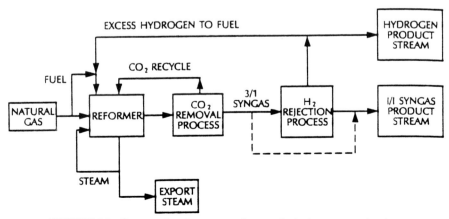

FIGURE 11 – Steam reforming process for oxoalcohol syngas production.

a steam reforming process which produces a 1:1 syngas ration for oxoalcohol production. Hydrogen has a significantly higher diffusion rate across polymer membrane than carbon monoxide and other gas such as methane and carbon dioxide.

The polysulphone hollow fibre membrane system was developed for this application to debottleneck a cryogenic separator in an oxysyngas application.

Cellulose acetate spiral wound membrane systems have also been quite successful in these separation e.g. production of oxoalcohol by syngas from natural gas reforming. The process requires the production of two streams 'pure' hydrogen (797%, C2% CO) and a 1 : 1 ratio of H_2 CO syngas. The process requires the use of two membrane separations to meet the specifications of product. After CO_2 has been removed from the feed gas, it is compressed and separated into the pure hydrogen product permeate, which is either compressed or utilised as a reformer fuel. The residual gas from the first membrane stage has a composition of 1.83 H_2 : CO ratio and is thus passed to the second membrane stage to reduce the hydrogen content down to the required ratio of 1 : 1. The hydrogen rich gas from the second stage is recycled to the first stage to maximise hydrogen use and enhance the separation in the first membrane stage.

TABLE 5 – Chemicals produced from synthesis gas.

Product	Reaction Stoichiometry	Feed/Product Weight Ratio	Required H2/CO/Ratio
Acetic acid	(1) $2H_2 + 2CO \rightarrow CH_3COOH$	1.0	1.0
	(2) $CH_3OH + CO \rightarrow CH_3COOH$	1.0	0
Methanol	$2H_2 + CO \rightarrow CH_3OH$	1.0	2.0
Ethylene glycol	$3H_2 + 2CO \rightarrow CH_2OH-CH_2OH$	1.0	1.5
Acetic anhydride	$CH_3COOCH_3 + CO \rightarrow CH_3COOCOCH_3$	1.0	0
Acetaldehyde	$3H_2 + 2CO \rightarrow CH_3CHO + H_2O$	1.4	1.5
Ethanol	$4H_2 + 2CO \rightarrow CH_3CH_2OH + H_2O$	1.4	2.0
Oxoalcohols	(a) $RCH=CH_2 + CO + H_2 \rightarrow RCH_2CH_2CHO^b$	Varies	2.0
	(b) $RCH_2CH_2CHO + H_2 \rightarrow RCH_2CH_2CH_2OH^b$		Overall

[a] From S.P. DiMartino et al. (3), reprinted with permission from Air Products and Chemicals, Inc., (c) 1988 APCI.
[b] Both normal and iso-isomers are formed.

Carbon Monoxide Production

In a similar way to the production of hydrogen rich gas from syngas, pure carbon monoxide can be produced by membrane separation. The porches is based on a two stage membrane separation with recycle, removing the permeate in two stages. Membrane permeation is used to adjust the ratio of hydrogen to carbon monoxide in synthesis gas, where it offers the advantage of precise adjustment coupled with a relatively low pressure drop.

Improvements in performance of the syngas separation can be achieved by combining it with the operation of pressure swing adsorption. Typically the membrane performs the bulk separation giving the required hydrogen to carbon monoxide ratio. The hydrogen permeate is then purified by PSA and the tail gas from the PSA system is recompressed and returned to the synthesis gas.

GAS SEPARATIONS

Alternative processes use membrane separation to purify the tail gases produced from PSA. The PSA can realise high purity product but recovery of gas is lower. Recompression of the gas and subsequent membrane separation enables the remaining desired component to be purified and recycled as feed to the PSA.

Cryogenic separation coupled with membrane separation is also used in two general forms
- the membrane performs the upstream bulk separation of the more permeable component prior to entering cryogenic separation. Typically hydrogen is removed and additionally water by the membrane stage. Applications are in de-bottlenecking cryogenic plant.
- the membrane purifies the overhead product from the cryogenic plant. An application is in production of both a pure CO_2 (> 97%) product and sales gas (<3% CO_2) from natural gas (64% CO_2, 29% HC's, 7% N_2). The cryogenic unit provides the pure carbon dioxide stream whilst giving a top product of 36 mol % CO_2 with the hydrocarbons. After heating this stream, membrane separation gives a permeate rich in CO_2 which is recycled to cryogenic separation and the product sales gas.

Hydrogenation Processes

For hydrogenation processes a hydrogen feed gas with a purity of at least 98% can be required. Two separations methods would generally be considered in this case, PSA and gas permeation. The selection of either depends both on the scale of the process and the initial source of the impure hydrogen feedstock, e.g. naptha cracker gas (90% H_2, 10% CH_4) and cryogenic product gas (92.3% H_2, 6.2% N_2). In the case of the cryogenic product gas, gas permeation is a preferred option especially at a relatively small scale of 1700 Nm^2/h.

Hydrogen/Hydrocarbon Separations

A significant and practical use of membrane gas separation is in the recovery and purification of hydrogen gas in refining operations. Hydrogen is an expensive chemical feedstock in these processes and maximum use of this material is clearly desirable. Applications of membranes exist in the catalytic reformer and in hydrotreater, hydrocracker, hydrogenerator and catalytic cracker purges (see Table 6)

TABLE 6 – Typical hydrogen membrane performance in refining applications.

	Hydrogen Membrane Recovery			
Process Stream	Primary Separation	Feed Purity (%)	Permeate Purity (%)	Recovery (%)
Catalytic reformer offgas	H2/CH4	70 to 80	90 to 97	75 to 95+
catalytic cracker offgas	H2/CH4	15 to 20	80 to 90	70 to 80
Hydroprocessing unit purge gas	H2/CH4	60 to 80	85 to 95	80 to 95
Pressure swing adsorption offgas	H2/CH4	50 to 60	80 to 90	65 to 85
Butamer process	H2/CH4	70	90	

298 HANDBOOK OF INDUSTRIAL MEMBRANES

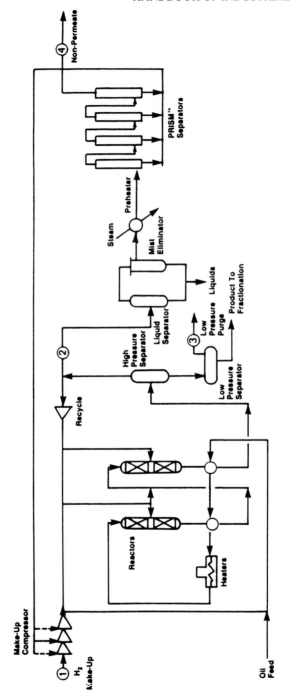

FIGURE 12 – Process flow diagram of hydrocracker operation with membrane separation.

GAS SEPARATIONS

The hydrogen content of these streams ranges from 15 to 80%. 90 to 95% hydrogen purity is required to recycle the hydrogen to a processing unit.

In the case of a hydrocracker, hydrogen and oil are introduced to the reactor, where hydrogen reacts with the oil to produce liquid hydrocarbons. Liquids and gas are disengaged in a high-pressure separation and the gas phase, contains hydrogen, is compressed and returned to the reactor.

A purge must be taken from the recycle to prevent the buildup of light hydrocarbons in the reactor loop, which reduce the efficiency of reactor by lowering the hydrogen partial pressure in the gas phase.

Traditionally the inerts are removed either by a direct high pressure purge or by absorption in oil to concentrate the inerts prior to purging them. The former process removes 4 mol of hydrogen from the process for every mole of hydrocarbon removed. The latter has a high capital cost and still loses about 1 mol of hydrogen per mol of hydrocarbon. Membranes can be used in lieu of the absorber system, at a reduced cost and better process efficiency.

In a typical process (Fig 12) a portion of reactor gases is treated in a hollow fibre membrane system to produce a hydrogen-rich permeate stream. This hydrogen-rich stream is then pressurized so that it can be introduced into the reactor. The retentate stream, enriched in hydrocarbons, is purged from the reactor loop and used as fuel gas. This purge maintains only about 20% hydrogen, compared to 50% from an oil absorber and 80% in a high-pressure

Commercial hydrogen separation system.

The applications of membrane gas separation for other purge gases is similar to that described above.

The general economic benefits of membranes over adsorption and cryogenic as shown in Table 7, for hydrogen recovery from refinery offgas is generally accepted.

TABLE 7 – Comparison of the separations for hydrogen recovery from refinery offgas.

	Membrane Process		Adsorption	Cryogenic
	80°C	120°C		
Hydrogen recovery (%)	87	91	73	90
Recovery H_2 purity (%)	97	96	98	96
Product gas flow rate (MMscfd)	2.76	2.86	2.24	2.86
Power (kW)	220	220	370	390
Steam (kg/h)	230	400	—	60
Cooling water (t/h)	38	38	64	79
Investment ($ millions)	1.12	0.91	2.03	2.66
Installation area (ft^2)	86	52	651	1292

A further development in refinery operation has been the use of membrane systems to recover the purge hydrocarbon gas in addition to the hydrogen. The hydrogen is removed from the purge gas, in a specially designed membrane unit, until hydrocarbon condensation occurs due to a reduction in dew point. The membrane unit recovers the condensed hydrocarbon but the membranes are not damaged by the condensed hydrocarbons.

Landfill and Digester Gas Methane Recovery

A relatively new source of methane is from gases from landfill and also digester gas. This biogas is formed in the anaerobic digestion of landfills, sewage, municipal waste and agricultural waste. Both gases have approximately 40-50% CO_2 and also in the case of landfill contain up to 2-3% air. H_2S, NH_3 and waste vapour. The treatment of the biogas to remove the CO_2 is a process which can be readily carried out by gas permeation, the capacities of operation are generally low 1,000 to 100,000 m^3/day. The membrane

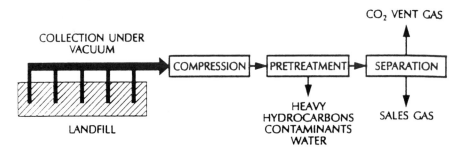

FIGURE 13 – Schematic diagram for landfill gas recovery.

separation has to cope with both the variable production and composition of the generated biogas and the natural variation which occurs between different land fill sites. Carbon dioxide removal by gas permeation with hollow fibre membranes upgrades the calorific value of the gas. Other so called 'fast gases' HS_1 and water permeate through the membrane. A schematic of a typical land fill biogas recovery is shown in Fig 13. The membrane separation is carried out in two stages of membrane permeation with recycle of the CO_2 rich permeate from the second stage to the feed compressor suction. The process can produce 9.5% methane with over 80% recovery over substantial periods of operation.

Helium Recovery

The rapidly developing importance of offshore technology in oil and gas fields has put greater demands on underwater operation. Depths of operation can be 450 m or greater which introduces significant hydrostatic pressures for manned operation, of 45 bart 450 m. At these depths there are severe problems of toxicity to man from oxygen and nitrogen. For long exposures the partial pressures of these gases must be low, in the case of oxygen below 0.5 bar to avoid poisoning and in the case of nitrogen, to avoid narcosis, less than 3 bar. The breathing mixtures for depths greater than 100 m, is a "trimix" of helium, oxygen and nitrogen (Table 8)

Welding operations which use argon as the shielding gas results in an atmosphere contaminated with argon which is toxic at high partial pressure. Argon has a relatively high melting point (-189.2C) and is a solid at the temperature of liquid nitrogen. This presents a potential problem of blocking in cryogenic separation.

TABLE 8 – Trimix gas compositions.

Depth	% He	% N_2	% O_2
100	90	5	5
200	92.5	5	2.5
400	93.8	5	1.2

It is estimated that in an offshore dive of 30 days, at a depth of 300 m, using six dives each working two hours per day requires a total of approximately 38,000 m^3 of helium. Thus helium recovery and purification is used to substantially reduce the associated large operating costs (Fig 14). The contaminated gas can be treated by gas permeation to separate out the helium as a 99.75% to 99.9% permeate. A two stage operation using polyetherimide flatsheet membranes operating at 60 bar pressure is used to restrict helium losses to less than 0.5%. The feed, contaminated helium, is stored at 200 bar typically, and reduced to a pressure of 60 bar for gas permeation. The first stage gives the required permeate product and the second stage recovers the remaining helium from the first stage, which is recycled back to stage 1. Membrane area requirements are quoted at 10 m^2, stage 1 and 16 m^2 stage for 100 Nm^2 h^{-1} helium production.

The contaminated helium gas contains a maximum of 1.5% (vol 1 CO_2, 10% Ar and 50 ppm CO).

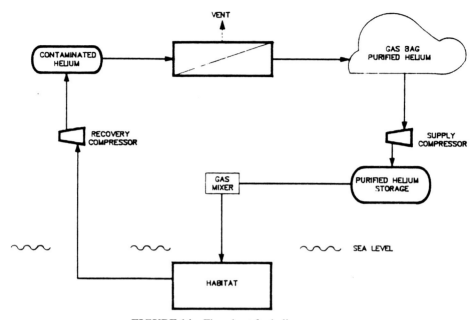

FIGURE 14 – Flowsheet for helium recovery.

Water Vapour Removal

There is significant potential for dehydration of a range of process streams and air. For example dryers are used commonly in final processing steps to remove moisture from final products. The heat from the dryer which is transferred to the air is generally lost to atmosphere. By using membranes with high permeabilities for hot water vapour, the vast amount of heat contained in the water (approximately 90%) can in principle be recovered.

Membranes are used to dehydrate process air streams as replacements for dessicant dryers or adsorption systems. The membranes used to produce dry air have extremely high water to air selectivities, otherwise nitrogen enrichment may occur, due to O_2 permeation with the water vapour. A system is available as a 'point of use' air dehumidifaction unit which is connected to a compressed air line. The unit operates much like a pervaporation unit with a dry gas sweep on the permeate side. This dry air is taken from part of the product dry gas. The loss of pressurised gas in the permeate is a major cost in operation of the unit. Typically a unit weighing 3.2 kg can treat an air feed of 27 000 scfd at a pressure of 10 bar.

The area of water vapour removal and the removal of organic vapours is discussed in more detail in Section6 covering Pervaporation and Vapour Permeation.

Electrochemical Membrane Gas Separations

There are several methods for the electrochemical membrane separation of gases, primarily used as a means of gas concentration. Processes have been developed for concentration of oxygen, carbon dioxide, acid gas such as chlorine, sulphur dioxide, nitrogen oxides and hydrogen sulphide. The latter three are proposed applications in the

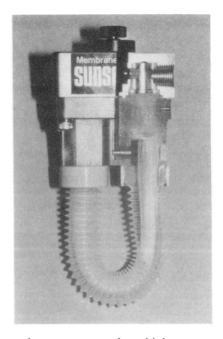

treatment of flue gases and waste gases and are high temperature processes based on eutectic membranes have been researched. The principle of operation are generally similar utilising the effects of an electric field generated by two electrodes across a suitable membrane. To illustrate the technique the case of oxygen separation is considered further.

The commercial processes for the separation of gases, such as oxygen, utilise a potential gradient to achieve the desired separation. The theoretical potential for a fuel cell which utilises oxygen at different partial pressures in the anode and the cathode side is given by the Nernst equation in the form,

$$E = (R\,T/4F)\ln(P_c/P_a)$$

where P_c and P_a are the oxygen partial pressures in the cathode and the anode sides of the membrane cell respectively. In practice the voltages achieved in such a cell would be extremely low unless large partial pressure differences were used. The reverse view of this equation is one representing the theoretical behaviour of an electrochemical cell for concentration, whereby on the application of a potential gradient oxygen is transferred from the cathode (oxygen reduction) to the anode (O_2 evolution). Thus the net result is that oxygen is removed from the cathode side and transferred to the anode side of the cell. There are several designs of electrochemical cells proposed for this application based on the use of alkaline electrolyte, polymer electrolyte and solid oxide electrolyte, and oxygen binding transition metal complexes.

The alkaline electrolyte cells are similar to those used in alkaline fuel cells utilising PTFE-bonded platinum black cathodes and PTFE-bonded $NiCo_2O_4$ anodes with a 32% KOH aqueous electrolyte. Operating voltages for systems to convert air at the cathode

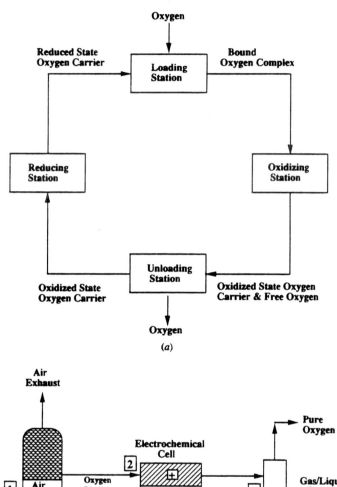

FIGURE 15 – Oxygen separation system.

to pure oxygen at the anode are typically greater than 1.0 V at current densities of 1000 A/m^2. The presence of carbon dioxide in the air poses problems with carbonation in this type of application. This can be overcome by the use of solid polymer electrolyes and using components for the cell similar to those in SPE water electrolysers (see Section 18), 99.5% oxygen can be produced from air under typical operating conditions of 1000 A/m^2, 0.75 V and 60°C.

Solid oxide electrolyte used for oxygen separation are typically stabilised zirconia doped with calcia or yttria. A variety of other materials have been tested e.g yttria doped thoria, bismiuth-erbium oxide $SrCeO_3$, $CaTiO_3$. The proposed applications include the production of oxygen for industrial processes such as steel manufacture, where waste heat is available to maintain the operating temperatures, typically around 800-900°C, required for these solid oxide cells. The design of the oxygen concentration cells models that of the analogous fuel cell technology, adopting either tubular, flat plate or disc shaped membranes. The electrodes in these cells are typically porous Pt, Ag or metal oxides.

The ability of metallo-organic ligands to bind oxygen led to the development of an electrochemical based separation technology for the removal of oxygen from fluids. The principle of the method is according to the following steps;

i) first the complex is formed with oxygen, for example by using a hollow fibre membrane to contact the oxygen containing stream with the ligand stream. The gas and ligand stream are on opposite sides of the membrane and transport occurs through the porous walls of the membrane.

ii) Anodic oxidation is then used to release the oxygen gas in the electrochemical cell, O_2 and ligand are then separated in a gas, liquid separator.

iii) The released cationic form of the ligand is then regenerated by cathodic reduction in the other compartment of the electrochemical cell.

To be successful the ligand must form a stable, reversible complex with O_2, undergo an electrochemical redox reaction, be soluble in aqueous solution and show a greater selectivity for oxygen than for other components, such as nitrogen, which may be present. Ligands which have been considered are cobalt complexes of porphyrins, amino acids and polyamines.

SECTION 4

Air and Gas Filtation and Cleaning

AIR AND GAS FILTRATION AND CLEANING

Sterile compressed air and gases are required in many industrial sectors including biotechnology, foods, pharmaceuticals, electronics, chemicals etc. Production of many of the products such as proteins, vaccines, vitamins etc, requires aseptic and sterile gas supplies throughout the production cycle. The manufacture of foods and beverages frequently requires the use of gases such as air, nitrogen, oxygen, carbon dioxide which must be sterile and free from bacteria and other particulates in use. Electronic components require the use of purified gases in manufacture to ensure the integrity and reliability of the final products.

In several manufacturing operations gases and air are exhausted or discharged which contain significantly large amounts of particulates and other species which can constitute a risk to the environment, health and safety. In addition the air or gas borne species represent a valuable product which ideally should be recovered from the gas prior to discharge or further treatment.

This section looks broadly at these areas whilst specific applications of gas filtration technology are considered in sections on microfiltration and the food industry.

Compressed Air Cleaning and Sterilisation

Compressed air is used as a motive force for the transport of materials, as a carrier medium for energy to a point of use where it acts as a driving force for pneumatic equipment, as a substrate in many industries and as a processing medium in mixing, packaging and pressurising etc. The production of compressed air requires several pieces of equipment, air in-take filters, appropriate compressors, aftercoolers, air receivers and associated pipework and fittings (see Fig 1). Aftercoolers are used to remove the water present in saturated air down to a temperature of 8 to 10°C.

The very production of this compressed air is the cause of many impurities and contamination in the gas. The initial source of the contamination is the air itself, in the form of atmospheric dust, smoke and fumes, bacteria and viruses, water vapour and unburnt

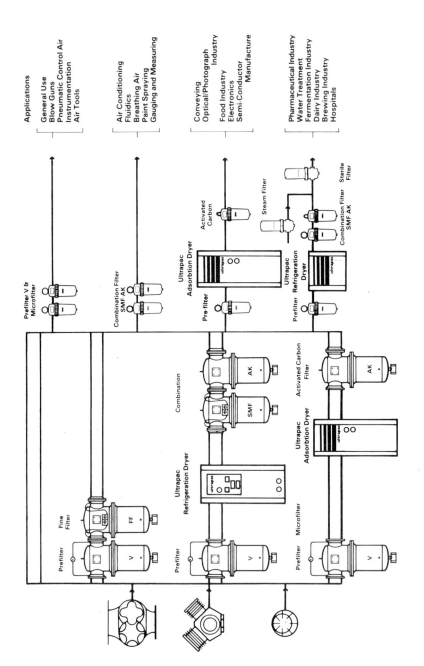

FIGURE 1a.

AIR AND GAS FILTRATION AND CLEANING

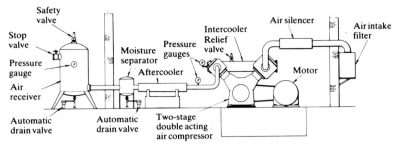

FIGURE 1b – A typical compressed air installation.

hydrocarbons used as fuels. It is estimated that compression of air to 8 bar at 20°C from a typical metropolitan environment amounts to a product containing 1120 million particles (dirt) per cubic meter. As much as 80% of this pollution is less than 2 μm in size and will pass through the intake filters of most compressors. These filters are used to protect the compressors and not the compressed air. Micro-organisms including bacteria and bacteriophage, or virus particles which are not removed by the intake filter, are not resident in the compressor for a time sufficient to be killed by the heat of compression. In addition to environmental pollution in the air the other sources of contamination are water vapour from the air, gases generated in the compressor, oil carried over from the compressor and solid contaminants within the system including wear particles and carbon. This source includes all equipment, compressors, pipelines, valves and air receivers and will even be present with stainless steel in the form of installation dirt.

TABLE 1 – Compressed Air Quality Classes.

Class	Solid Contaminants		Water Pressure Dew Point	Total Oil Content
	Maximum Particle Size (m)	Concentration (mg/m3)*	°C	mg/m^3
1	0.1	0.1	-40	0.01
2	1	1	-20	0.1
3	3	5	+2	1.0
4	40	10	+10	5
5				25

* 1 bar absolute process, 120C and a relative vapour pressure of 0.6

The extent of treatment required for air from a compression system depends largely on the application. Compressed air quality classes for solids, water and oils are given in Table 1 General industrial applications utilising mains compressed air supplies may simply require partial water removal, by aftercooling, followed by filtration to remove solid contaminants above a certain size. Air quality specifications may dictate high purity requiring several stages of filtration, drying and oil removal. Applications of compressed air have been put in four general categories

1. General use. Blow gas. Pneumatic control air. Instrumentation. Air tools.
2. Air conditioning. Fluidics. Breathing Air. Paint spraying. Gauging and measuring.
3. Conveying. Optical/photograph industry. Food industry. Electronics. Semi-conductor manufacture.
4. Pharmaceutical industry. Water treatment. Fermentation industry. Dairy industry. Brewing industry. Hospitals.

The different air quality requirements are met by different combinations of filters and other air contaminant removers. This include
- Humidity control – to lower the dew point to below operating/ambient temperatures

TABLE 2 – Quality Class Recommendations For Some Typical Applications.

Application	Typical quality classes		
	Solids	Water	Oil
Air agitation	3	5	3
Air bearings	2	2	3
Air gauging	2	3	3
Air motors, heavy	4	4-1	5
Air motors, miniature	3	3-1	3
Air turbines	2	2	3
Boot and shoe machines	4	4	5
Brick and glass machines	4	4	5
Cleaning of machine parts	4	4	4
Construction	4	5	5
Conveying, granular products	3	4	3
Conveying, powder products	2	3	2
Fluidics, power circuits	4	4	4
Fluidics, sensors	2	2-1	2
Foundary machines	4	4	5
Handling of food, beverages	2	3	1
Industrial hand tools	4	5-4	5-4
Machine tools	4	3	5
Mining	4	5	5
Packaging and textile machines	4	3	3
Photographic film processing	1	1	1
Pneumatic cylinders	3	3	5
Precision pressure regulators	3	2	3
Process control instruments	2	2	3
Rock drills	4	5-2	5
Sand blasting	—	3	3
Spray painting	3	3-2	3
Welding machines	4	4	5
Workshop air, general	4	4	5

Note: The above values are indicative only. For certain applications more than one class may be considered. The ambient conditions will influence the selection, especially as regards dewpoint.

AIR AND GAS FILTRATION AND CLEANING

- Prefilters – installed downstream of compressed air receivers, they consist of regenerable porous filter elements with pore size of 5 to 25 µm. They removal heavy contamination, but not oil, water and finer particles.
- Sub-micrometer filter – this is required to produce an oil (and water) free compressed air. Typical filters are of the coalescing depth type.
- Activated carbon adsorbers – these "filters" remove hydrocarbon vapours and odours which must be removed in food, dairy, brewing and pharmaceutical industries.

In the use of distributed air lines, standard filters which removal particles down to 40 to 50 µm in size, or medium filters, (5 to 40 µm), are used. Further filtration to 1 to 5 µm, or ultrafine filtration down to 0.1 µm or better is usually installed at the take-off point to individual supplies. Thus generally these filters follow coarse filters, which reduces the burden on the second stage, or third stage, filters. The recommended filter quality class for solids, water and oil for specific applications is given in Table 2.

Further details of filtration for compressed air supplies is given in the Filters and Filtration handbook 3rd Edition, Elsevier.

Air and Gas Sterilisation

There are several industrial sectors where sterile compressed air and other gases is essential.

- The screw compression of air will generally produce aerosols in the 0.1 µm to 0.5 µm size range and a significant fraction of aerosols will be under 0.1 µm in size. In many technical uses of compressed air a virtual oil free material is demanded requiring a degree of retention of 10^6 and greater.
- In the various branches of microbiology (genetics, biotechnology, medicine and pharmaceuticals) absolute filtration of air for sterility is important.
- In the electronics industry where filtration down to 0.01 µ is generally required.
- the production and packaging of food and dairy products eg beer, and cheeses.

The direct use of air into preparatory chemicals demands that it be as sterile as the chemical medium itself. The removal of solid particulate, liquids (oil and water) odours and micro-organism is thus essential. Air when used in the energy mode for mixing of powders, batching of materials and instrumentation should also be free of particulate.

Micro-organisms, ie bacteria in the size range 0.2 µm to 4 µm, viruses and bacteriophage (0.3 µm down to 0.04 µm), are a serious problem in many industries because of their ability to "grow" and multiply as living organisms under suitable conditions.

A system for the sterilisation of compressed air (or gas) must possess the following attributes

1. Remove permanently all micro-organisms potentially dangerous to the process
2. Reliable over long operating periods
3. Inert and not support growth of bacteria
4. Economic to install and maintain
5. Easy to install, use and maintain

6. Steam sterilisable "in-situ"
7. Able to be integrity tested "in-situ"

There are a number of alternative methods to achieve sterile air
1. Dry heating in an oven at a temperature of at least 165°C for 2 hours
2. Irradiation, using high intensity ultraviolet, X-ray or gamma ray sources
3. Physical filtration

Dry heating is very expensive and only possible with small flowrates. Irradiation, although less costly than heating, is impractical for other than specialist systems with small flowrates. Additionally shadowing of micro-organisms by particles may also cause incomplete sterilisation.

Filtration is generally accepted as the only practical and effective method to sterilise both small and large volumes of air and gases. Historically, packed deep bed designs were used for air sterilisation. These however were frequently unsatisfactory due to, channelling and by-passing of gas and thus poor filtration, possibility of bacteriological growth in the filter with incomplete steam sterilisation and carry over of bed filter/fibre material into the "sterile" air. These have thus been replaced by modern air filtration cartridges. There are two types of cartridge filtration in common use. The first older type is usually referred to as depth filtration and there is wide acceptance of binder free micro fibre filters which can offer 99.999999% efficiency with particles of 0.01 m in size. The second type are based on membrane filters with the most commonly utilised material being polytetrafluorerethylene.

The removal of particles and micro-organisms in filter cartridges is by a combination of three mechanisms, direct interception, inertial impaction and diffusion, ie Brownian motion. Direct interception generally applies to comparatively large particles which can't penetrate the filter. Particles in the size range of 0.3 to 1 μm have sufficient mass such that they cannot follow the gas streamlines through the filter and thus impact onto the filter material. Submicron particles are affected by intermolecular forces which cause them to wonder, by Bromnian motion, and thus increase the probability of collision with the filter.

There is widespread use of air sterilisation filter cartridges which are constructed using an amalgamation of filter media. The cartridges which are constructed have a relatively coarse outer layer (2 μm filtration rating) to trap dirt particles before the gas enters the micro fibre web. This web (borosilicate) removes particles down to the 0.01μ m level and removes mists, converting the aerosols back into liquid. Liquid is then forced through the filter by gas flow and gravity enabling it to be removed from the filter. A second internal layer of 2 μm filter medium is used to hold the micro fibre filter, within the cartridge. For the removal of bacteria, the filter medium should not contain a binder material which would acts a nutrient for growth.

Typical membrane sterilisation cartridge filters use expanded PTFE membranes. The membranes are formed into a high surface area cylinder pleat pack held between two polypropylene support screens. These screens are permanently heat bonded to polypropylene end caps. Stainless steel inner cores are also utilised to give robustness.

Membrane cartridge filters based on a multi-layer of membrane filter medium, support and irrigation mesh, are also widely used. Cartridges are fabricated from a pleated filter

AIR AND GAS FILTRATION AND CLEANING

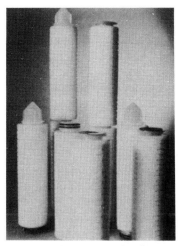

Air sterilisation cartridge filters.

pack comprising of a polypropylene support net one or two nylon membranes (0.2 μm or 0.4 μm cartridge) and a very fine polypropylene fibre pre-filter layer. The filter media is a thin (110 μm) homogeneous cast microporous nylon membrane which cannot itself release fibres and contaminate the filtered area. Sealing of the filter into the cartridge is by melt sealing in polyproylene. The section of nylon and polypropylene materials alone allows the cartridge to be autoclaved.

Hollow fibres

Novel hollow fibre membrane cartridge filters (see Fig 2), available in hydrophobic polypropylene or hydrophilic polyamide, give the required high performance for absolute sterility of compressed air supply. They offer a very large surface area for filtration and a considerable amount of thermo-mechanical resistance. The rated retention sizes are 0.1,

FIGURE 2 – Hollow fibre membrane cartridge filter.

0.2 and 0.45 µm and with a narrower pore size distribution than conventional membrane filters, the filtration capability, measured as the number of pores per unit area is much greater.

The filter cartridge is fitted into a pressure filter housing and sealed using single or double sliding O-rings. This allows for expansion during in-situ steam sterilisation.

Gas Filtration

Several manufacturers produce capsules and cartridge filters for the membrane filtration of a range of other gases. The chemical resistance of PTFE makes this material a frequent choice for the membrane material. Applications of these filters include:
- point of use filter for
    1. reactive and corrosive gases in applications such as dry etching and doping
    2. atmospheric gases to process tools
- final filter before gas distribution to process delivery systems
- integral filtration in semi conductor processing equipment

A typical unit will be 100% PTFE filter cartridge in a 316 L stainless steel capsule, with 0.01 m absolute gas particulate removal rating. Membrane areas of 300 cm^2 are used for gas flow rating of 1.2 scfm per psid at 30 psig (2.07 bar).

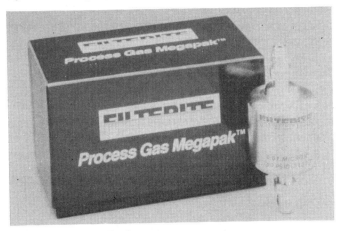

Also available are filter assemblies which are designed to remove submicron particles ($\geq$ 0.003 µm) from ultra-high purity point-of-use gas filtration applications. The filter design allows a high flow capacity. The filter assembly has typically a diameter of 25 mm which makes it easy to install. It is 100% integrity and helium leak tested, and contains a fluorocarbon element, which is constructed of a PTFE medium, and a PFA 440HP grade Teflon core. The 316 L housing, constructed of electropolished stainless steel, exceeds VIM/VAR specifications, and the O-ring consists of Teflon encapsulated Viton. It is manufactured and packaged in a cleanroom environment. The filter assembly is offered in a wide variety of fitting options including a 0.25" (6 mm) gasket seal (VCR compatible) and 0.25" (6 mm) butt weld.

AIR AND GAS FILTRATION AND CLEANING

Filter Design and Selection

The filter housings are designed with air flow from outside to inside, with housing's body seal, vent connection, housing seal and drain connection on the "dirty" side of the process. This eliminates these components as sources of contamination leaving only one critical seal, between cartridge and housing.

Sterile filters are generally sized to flow rate versus pressure drop information. Initial pressure losses in pre-filters are approximately 1 psi and 2 psi if liquid is being removed. Coalescing prefilters are designed to withstand high pressure losses and are replaced when these are between 5 to 10 psi. Air sterilisation filters are sized to an initial pressure drop of approximately 1.5 psi. Replacement of these filters is generally not based on pressure loss increase but after a prescribed number of steam sterilisation have been performed. Steam sterilisation is a routine requirement in any system. Replacement of cartridge filters is generally after 1 year service life, but this is not a critical factor as costs are low and replacement easy. Integrity of the system sterilisation is of prime concern. It is essential that any filter used has been integrity tested to ensure satisfactory duty. Integrity tests are described in section 2.

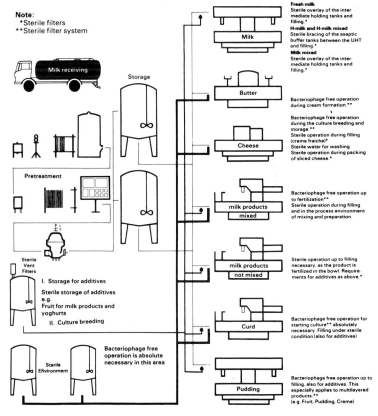

Filter applications in the dairy industry.

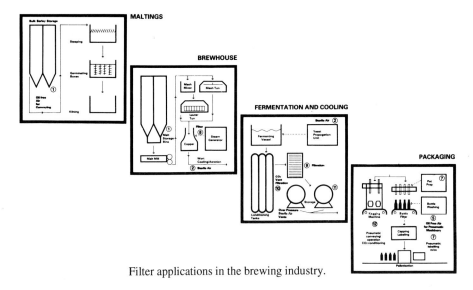

Filter applications in the brewing industry.

Compressed Air Dryer

Hollow fibre membrane – composite based on polyethersulfone support with hydrophobic coating are used in the cross flow mode for drying of air. Wet air flows though the hollow fibre membranes which allow only water molecules to pass, as permeate, thus drying the air. A small flow of dry air is applied on the permeate side to purge the water molecules. A condensate drain is not required as water leaves in the gaseous state. A drain is required on the pre-filters. The hollow fibre module is suitable for both screw and piston compressors and can be integrated into the compressor or can be stand-alone or point of use. Dew points below 0°C (-20°C possible) are reached. The attraction of the system are:-
- maintenance free – no moving parts
- no refrigerant (CFC's)
- no electrical connections
- oil resistant
- suitable for installation between compressor and receiver
- no noise

Air Cleaning by Membrane Gas Absorption

Environmental tobacco smoke (ETS) is a nuisance, irritation and health hazard to non-smoking people who involuntarily are exposed. ETS contains several hundreds of chemical compounds, both particulate and gaseous. It is not generally practical to install mechanical ventilation systems to remove these components to provide the required air quality. An alternative is to use "portable" air cleaners which circulate air through a filter system, which can consist of a particulate filter (electrostatic, cloth) and also a charcoal filter. Many components generated by tobacco burning are not removed in these systems. Recent tests using lab scale hollow fibre membrane absorbers (see section 8.) with water

AIR AND GAS FILTRATION AND CLEANING

TABLE 3 – Membrane air cleaner removal efficiencies.

Component	Removal Efficiency [%]
Acetone	96.6
Styrene	15
Formaldehyde	98.4
Nicotine	99.1
Ammonia	95
Odour (threshold)	54
Odour (decipol)	49

as the absorbing solvent (see Table 3.) have demonstrated efficient removal of water soluble components with low pressure loss operation.

The removal of hydrocarbons will require the use of appropriate solvents.

Gas Cleaning

It is generally accepted that, in the removal of dust and particulate from air and gas streams that barrier filters give the potentially superior performance to other types of separators.

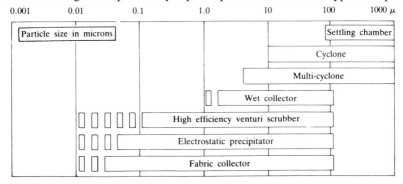

FIGURE 3 – Performance of various gas cleaning equipment.

For operation at relatively low temperature fabric collectors are employed. Such filters operated in the dead-end mode with particulate laden gas flowing through the filter element – envelope or tube and particulate capture is by a sieving action, where particles are trapped by the smaller interstices of the fabric. This results in the build up of a particulate cake, which acts as a highly efficient filter for smaller sub-micro particles, the fabric essentially then acts as the support for the filter cake. The operation of fabric filters requires the removal of the particulate cake at regular intervals to maintain reasonable gas flow without excessive pressure losses. This is achieved by mechanical shaking of the filter element, low pressure -reverse air or by high pressure-reverse jet pulsing. The choice of cleaning depends on the fabric of the filter, woven fabrics are adequately cleaned by low pressure reverse-air, or mechanically, while non-woven (felted) fabrics generally require high pressure pulsing. The efficiency of separation of new filters improves after use.

The performance of all filters varies with particulate size, at above approximately

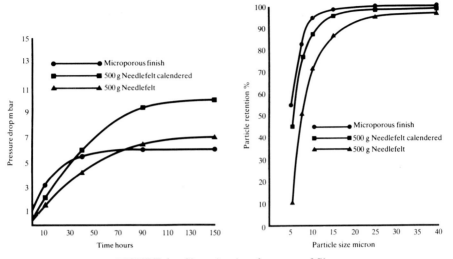

FIGURE 4 – Size related performance of filters.

10 μm almost 100% removal can be achieved, while at the micron and sub-micron level efficiency falls off rapidly with decreasing size – see Fig. 4

The efficiency of any fabric can be improved by using smaller fibre diameters and closer weaving felted fabrics have smaller open areas than woven fabric and are thus more efficient. This clearly is at the expense of reduced permeability (gas flow) and also poorer cleaning ability. There is a wide selection of materials for fabric filters although many have limited physical and chemical resistance and also restricted range of operating tempera-

Needle felt cartridge and pleated cartridge filters.

TABLE 4 – Characteristics of some fabrics for filters.

Generic name	Max temp °C (°F) Continuous	Intermittent	Physical resistance					Chemical resistance				
			Dry heat	Moist heat	Abrasion	Shaking	Flexing	Mineral acid	Organic acid	Alkalis	Oxidising	Solvents
Cotton	80 (180)	—	G	G	F	G	G	P	G	F	F	E
Polyester	135 (275)	—	G	F	G	E	E	G	G	F	G	E
Acrylic	235 (275)	140 (285)	G	G	G	G	E	G	G	F	G	E
Modacrylic	70 (160)	—	F	F	F	P-F	G	G	G	G	G	G
Nylon (polyamide)	115 (240)	—	G	G	E	E	E	P	F	G	F	E
Nomex†	205 (400)	230 (450)	E	E	E	E	E	P-F	E	G	G	E
Poly-propylene	95 (200)	120 (250)	G	F	E	E	G	E	E	E	G	G
PTFE (fluorocarbon)	260 (500)	290 (550)	E	E	P-F	G	G	E	E	E	E	E
Fluorocarbon	230 (150)	—	E	E	P-F	G	G	E	E	E	E	E
Vinyon	175 (350)	—	F	F	F	G	G	E	E	G	G	P
Glass	290 (550)	315 (600)	E	E	P	P	F	E	E	F	E	E
Wool	100 (215)	120 (250)	F	F	G	F	G	F	F	P	P	F

†Du Pont de Nemours International S.A. trademark Key: E = Excellent. F = Fair. G = Good. P = Poor

tures – see Table 4. The best all round materials are fluoropolymers, with operating temperature in excess of 200°C.

Fabric filters are produced as tubes, flat envelopes or pleated cartridges.

Improvements in the efficiency of the fabric filters can be achieved by the application of a membrane to the surface either

i) bonded to the fabric, typically needle felt to form a laminate
or
ii) as a micro-porous polymer coating – microporous finish fabrics

The membrane restricts filtration to its surface and prevents particle penetration into the fabric, which provides the mechanical strength to withstand wear.

The use of membranes is limited because of cost and is restricted to areas where severe blinding of the fabric occurs or where particulate release is a problem. Microporous finish fabrics formed by specialised coating method from a polymer or emulsion are a significant advanced in technology of gas fabric filters and have found applications in the filtration of PVC dust and fly ash.

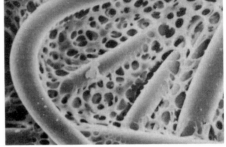

FIGURE 5 – Needle felt fabric and micro-porous membrane surface.

A typical gas filter membrane medium is microporous expanded polytetrafluroethylene. In filtration the particulate material is collected exclusively on the membrane, without penetration into the substrate. The non stick character for the membrane ensures that the accumulated cake dust is removed totally after each cleaning. In practical operations the membrane is supported (laminated) on to a range of fabric packings, eg polyester, polypropylene, glass fibre, as appropriate. The membrane is capable of functioning at up to temperatures of 288°C.

The filter is formed in bags ranging in size from 6 ft long by 4.5 in diameter for pulse jet filters to large 9 m (36 ft) long, 0.3 m inch diameters for reverse air bag houses. The filter bags are meeting emission rates for less than 2 mg/Nm^3 in application such as

- metal smelters, scrap lead and battery market
- chemical works, lead oxide removal in battery manufacture
- power stations
- incinerators, of hazardous, clinical wastes, industrially polluted soils, tyres
- cement works.

The PTFE membrane bag filters can be rendered anti-static to overcome any problems of sparking when used in electrostatic precipitators.

The use of PTFE membrane/glass fibre bag filters offers a chemical and thermal resistance material. One cited application uses the bags to filter the flue gas from incineration of hazardous waste. A novel feature of this process is the addition of activated carbon and lime into the gas to adsorb toxic substances (eg dioxin) and remove acids. The gas loading to the bag house is 7 g/m^3 which is reduced to 10 mg/m^3 or as low as 2 mg/m^3 if required.

High Temperature Gas Cleaning

There are many industrial processes which generate hot gases contaminated with solids, liquids and gaseous pollutants. (see Table 5). These are typically from the burning of waste and slurries, non ferrous metal plants, heavy fuel oil-fired systems etc. With the introduction of more stringent emission standards many branches of industry are compelled to assess the environmental performance of their processes and introduce the most cost effective means to meet emission targets. Existing processes must comply with standards, set for example by the UK Environmental Protection Act, over prescribed timescales whilst newly installed processes must meet emission targets at the specified time.

The pollutants to be controlled can be put into two groups, particulate and gaseous. Gaseous pollutants are frequently colourless and difficult to see and present their own particular environment problems. The particulate which are visible are a high priority in environmental clean up, due to adverse publicity form the media and pressure groups. The removal of pollutants from hot gases can present considerable problems and options available are limited.

Traditionally particulate removal has been achieved using bag filter technology using fabric filters, cyclones, wet gas scrubbers and electrostatic precipitators. The use of bag filters and wet gas scrubbers generally required the cooling of gas, which can be costly and impose its own particular problems. The waste heat cannot always be economically recovered using heat exchangers which may be susceptible to blockage and fouling by particulate. Chemical corrosion may also result at lower temperatures due to the presence of water and acid. Wet gas scrubbers are used when both acid gases, eg HCl, and dust are present, to remove both simultaneously. The dust removal efficiency is not sufficient to meet the higher performance levels dictated by legislation.

TABLE 5 – Sources of hot gas effluents.

Sub-micron Metal Oxide
Chimney Stack Emissions
Boiler Flue Gases
Micro-fine Mineral Particles
Foundry Emissions
Secondary Aluminium Production
Secondary Aluminium Smelting
Dry Scrubbing on Clinical Waste Incinerator Flue Gas
Boiler Flue Gas From Smokeless Fuel Process
Energy From Waste Gasification Process

Cyclones, although inexpensive, do not offer the required performance levels and are typically employed for coarse particulate removal prior to filtration. Electrostatic precipitators, although offering high removal efficiencies, are expensive and large and under increasingly tighter controls will become uneconomic. Overall the developments in particulate filtration technology for hot gases are towards barrier type filters, which can respond to increasing stringent limits and are not venerable to temperature fluctuations.

For application with operating temperature of around 280°C and less there are a range of modern treated needle felts which can be used.

PTFE is used when corrosive chemicals are encountered. At higher temperatures of operation metal fibres are manufactured to withstand heavy duty load and high air flows. At extreme temperatures, up to 900°C, ceramic filter elements can be used to give high filtration efficiency.

Titanium Dioxide Recovery in Membrane Coated Bag Filters

As one of the last stages in the production of the titanium dioxides the material is ground in fluid energy mills under an atmosphere of steam at 180°C. The exhaust from the mills is a steam flow of 38,600 m^3/h containing a high concentration of titanium dioxide particles, (95% below 1 μm in size). An application of membrane bag gas filtration is in the recovery of the product titanium tioxide from an exhaust gas stream. This system replaces two venturi scrubbers originally installed for this application.

In the plant the exhaust stream is drawn into two custom-designed membrane bag dust collectors by a heavy, corrosion-resistant fan. The filter housings are fabricated from 316 L stainless steel to maintain the high purity of the titanium dioxide and because of the corrosive nature of the conveying gas. The tubeplate in the top of each filter housing

AIR AND GAS FILTRATION AND CLEANING

carriers 442 membrane-coated glass fibre filter bags (3 m long) supported by specially-designed stainless steel cages with integral venturis. Bag changing is quick and easy and the bags are cleaned by low-pressure pulses of compressed air. At start-up a pre-heat system raises the temperature of the filters to 150C to prevent condensation.

The filtered titanium dioxide is discharged through a rotary valve into stainless steel screw conveyors for return to the mills. The cleaned conveying steam passes through a condenser (two venturi scrubbers originally used to remove titanium dioxide) to remove the steam. The remaining non-condensable gases are then discharged to the atmosphere. The overall recovery of titanium dioxide has increased from 90% to 99.99% with this filtration system.

Gas cleaning installation for TiO_2 recovery.

Ceramic Filters

Ceramic membrane filters are designed to achieve cake, not depth, filtration. Particles are collected on the surface of the membrane and as filtration continues the layer of deposited particles thickens forming a cake. This increases the pressure drop across the filter which at some time becomes too high and then the cake must be removed for efficient filtration. In nearly all cases this involves reverse pulse cleaning, where air (or nitrogen) is applied to the clean side of the filtration medium detaching the cake to. In time the pulse removes a higher and higher fraction of cake until an equilibrium is eventually reached and a standard thickness of cake remains. The time for this equilibrium to be reached is known as the conditioning period which is when a stable layer is formed which actually performs the separation.

Generally there are two different types of ceramic media available: high-density media comprising bonded ceramic granules of silicon carbide, alumina, alumina silicate, etc. and low-density media use bonded ceramic fibres of alumina or alumina silicate. In both cases the bonds are silicaceous but the porosity of the low density media is much higher – 0.8-0.9 compared to 0.3-0.4 for the high density media.

Although the ceramic filters are inert, there are certain conditions that can cause failure. Chemical attack can occur, with alkalis, fluxes and reducing agents may attack the silicaceous bond. Although the speed and severity of the attack is difficult to predict, high temperatures and high concentration of chemicals will increase the rate. Embrittlement and loss of strength in the material are symptoms of chemical attack. High alumina concentrations in the binder phase will reduce the rate of attack.

Thermal shock is a potential problem, which can affect high-density ceramics more seriously. Premature failure can occur with frequent and rapid changes in process gas temperature.

A new generation of gas filter medium based on low density alumino-silicate ceramic is available for high temperature duties up to 900°C. It is a highly porous (88%) medium formed as a candle closed at one end and open at the other. Candles are mounted on a header plate separating clean and dirty gas. The dirty gas impinges on the outer surface of the candle tube and clean gas is carried away down the hollow centre. The filter offers high efficiency (>99.9%) at low pressure drop and high filtration velocity which can lead to smaller and more compact filtration units. The ceramic filter can be used for lower temperature operation, although in the presence of liquid or if liquid are formed the filtration will be hindered.

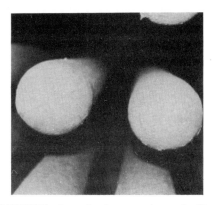

FIGURE 7 – Low density ceramic candle filter.

This ceramic is tolerant to attack by fire and sparks and can operate at temperatures up to 900°C, which is a major advantage when dealing with hot, particulate-laden exhaust gases. By using a low-density ceramic, the problems associated with brittleness and thermal shock are reduced. A typical system will have over 36 elements (candles), some have as many as 2400 – means that if one element fails, any drop in performance won't be dramatic. Another advantage is that you only have to replace the failed elements not the whole system.

The ceramic filters operate with low pressure drop, which combined with high filtration velocity and high operating temperature lead to smaller, more compact filtration units.

The filter candles may also be used to support lime, or other alkalis, in dry scrubbing processes to remove acid gases. In sulphur dioxide removal, operating at the elevated

AIR AND GAS FILTRATION AND CLEANING

temperatures increase the efficiency of the neutralisation reaction. The advantage of the dry scrubbing process is that the dry waste product is more easily disposed of than a solution or wet slurry effluent.

Applications of the filter include, waste incineration, non ferrous, cement, coal and ferrous industries. Because the filter has good temperature stability it is useful at temperatures where bag filter are about at their limit ie 250°C.

A recent example of the use is filtration of furnace exhaust gas from an aluminium smelting works, to remove particulate consisting of magnesium chloride, sulphur, alumina, silica, sodium and potassium chlorides and organics, all within the size range of 0.2 – 5 µm. Filtration down to 0.5 mg/Nm^3 is achieved. The ceramic candle elements are mounted within a conventional bag filter frame enabling retrofitting of exhausting element (see Fig 8).

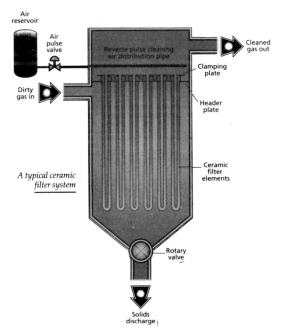

FIGURE 8 – A typical ceramic gas filtration unit.

36th year of publication: all previous editions sold out

PUMPING MANUAL
9th Edition
By Christopher Dickenson

Widely recognised as the first source of reference on all aspects of pump technology and applications the Pumping Manual will enable you to...

Specify the right pump for the task
Design cost-effective pump systems
Understand the terminology
Ensure effective installation, operation and maintenance of all your pumping equipment
and much more!

Section One: Introduction
SI Units
Pump Evolution
Pump Classification
Pump Trends

Section Two: Pump Performance and Characteristics
Fluid Characteristics
Pump Performance
Calculations, Type Number and Efficiency
Area Ratio
Pipework Calculations
Computer Aided Pump Selection

Section Three: Types of Pumps
Centrifugal Pumps
Axial and Mixed Flow Pumps
Submersible Pumps
Seal-less Pumps
Disk Pumps
Positive Displacement Pumps (General)
Rotary Pumps (General)
Rotary Lobe Pumps
Gear Pumps
Screw Pumps
Eccentric Screw Pumps
Peristaltic Pumps
Metering and Proportioning Pumps
Vane Pumps
Flexible Impeller Pumps
Liquid Ring Pumps
Reciprocating Pumps (General)
Diaphragm Pumps
Piston, Plunger Pumps
Self Priming Pumps
Vacuum Pumps

Section Four: Pump Materials and Construction
Metallic Pumps
Non Metallic Pumps
Coatings and Linings

Section Five: Pump Ancillaries
Engines
Electric Motors and Controls
Magnetic Drives
Seals and Packaging
Bearings
Gears and Couplings
Control and Measurement

Section Six: Pump Operation
Pump Installation
Pump Start-up
Cavitation and Recirculation
Pump Noise
Vibration and Critical Speed
Condition Monitoring and Maintenance
Pipework Installation

Section Seven: Pump Applications
Water Pumps
Building Services
Sewage and Sludge
Solids Handling
Irrigation and Drainage
Mine Drainage
Pulp and Paper
Oil and Gas
Refinery and Petrochemical Pumps
Chemical and Process
Dosing Pumps
Power Generation
Food and Beverages
Viscous Products
Fire Pumps
High Pressure Pumps

Section Eight: User Information
Standards and Data
Buyers Guide
Editorial Index
Advertisers Index

800 pages
1500 figures and tables
ISBN: 1 85617 215 5

For further details post or fax a copy of this advert complete with your business card or address details to:
Elsevier Advanced Technology, PO Box 150, Kidlington, Oxford OX5 1AS, UK
Tel: +44 (0) 1865 843848 Fax: +44 (0) 1865 843971

CIP4699

ELSEVIER
ADVANCED
TECHNOLOGY

SECTION 5

Separation of Liquid Mixtures/ Pervaporation

New Edition

SEALS & SEALING HANDBOOK
4th Edition
by Mel Brown

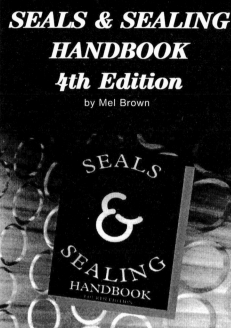

Packed full of useful reference information and data this book is the essential reference guide for all those involved in sealing technology.

Find out how to...

- Cut down on time consuming selection procedures

- Maximise seal life and reduce maintenance costs

- Ensure compatibility of seal materials

- Comply with environmental guidelines and legislation

and much more!

Includes full coverage in all these areas:

- Mechanics of sealing
- Materials
- Static seals
- Dynamic seals
- Mechanical seals
- Fluid power seals
- Special seal types
- Seal selection
- Engineering data

600 + pages
2000 diagrams, charts, tables and illustrations

ISBN: 1 85617 232 5

Publication date:
July 1995
Price: £115/US$184

price: £115 US$185

Reworked, updated and extended to include the latest developments in sealing technology

---

PRE-PUBLICATION ORDER FORM (valid until 30.6.95)

☐ Please send me _____ copy/ies of the Seals & Sealing Handbook - 4th Edition @ £115 US$185* [PIS12 + B1]

Payment
EC orders (not UK) please add VAT at your national rate or supply your VAT registration no:..
☐ Please find payment enclosed (please make cheques payable to Elsevier)
☐ Please charge my Access/Mastercard/Visa/Barclaycard/Eurocard/Am Ex
(delete where applicable)
Card No. ....Card Name ....
Expiry Date................................... Signature ....
Date...........
☐ Please invoice me. Company Purchase Order No.............................

Delivery Address
Name .... Position.....................................[JT:]
Organization ....
Address ....
Town.................................... State ....
Post/Zip Code............................ Country ....
Tel.. Fax ....
Nature of Business ....[SIC:]
(CIS1299)

*Prices may be subject to exchange rate fluctuations but orders with prepayment will be honoured at the advertised rate. Postage will be added at cost to all orders received without payment.

ORDERS & ENQUIRIES TO

Elsevier Advanced Technology
P O Box 150, Kidlington,
Oxon OX5 1AS, UK
Tel: +44 (0)1865 843848
Fax: +44 (0)1865 843971
e-mail: eatsales@elsevier.co.uk

OR

In North America
Elsevier Advanced Technology
660 White Plains Road, Tarrytown,
NY 10591-5153, USA
Tel: (914) 524 9200
Fax: (914) 333 2444

REFUND GUARANTEE
Remember our money back guarantee. Should you wish to cancel your purchase, simply return the book within 28 days and we will refund your payment.
sealhdk.3d

ELSEVIER ADVANCED TECHNOLOGY

SEPARATION OF LIQUID MIXTURES /PERVAPORATION

Applications of Pervaporation

Pervaporation is unique among membrane separations, involving a change of phase to achieve separation. It offers a means of separating miscible liquids of similar molar mass and is an alternative method to distillation. However the relative cost of membrane units in comparison to distillation equipment means that in most separations, where distillation performs efficiently, pervaporation is not a viable alternative. In addition pervaporation will not generally be economically viable as a multi-stage separation process. Where pervaporation can prove to be a useful method is in the separation, or removal, of small amounts of one liquid from a liquid mixture. Pervaporation also then becomes attractive in the separation of liquid mixtures which form azeotropes, ie where vapour and liquid have the same composition in equilibrium, as standard distillation cannot achieve this separation. There are many mixtures of organic solvents and organic solvents with water which exhibit azeotropic compositions. The composition frequently has a low percentage of one component, for example of water, in azeotropes of alcohols (see Table 1).

Included in Table 1 are typical permeation rates and selectivites achieved in pervaporation with a grafted (poly vinyl pyrrolidine onto PTFE) membrane.

There are potentially a large number of separations amenable to pervaporation which can be classified as either mixtures with water (aqueous) or non-aqueous mixtures (see Table 2).

This range of mixtures occurs predominately in the chemical industry, although other applications are in the food and pharmaceutical industries to process heat-sensitive products, in waste water applications to remove trace quantities of volatile organic (chlorinated hydrocarbons) and in analytical procedure to concentrate components for detection.

Pervaporation is used to overcome the difficult stage of a separation, ie to overcome azeotrope limitations in distillations. It would rarely be used as an isolated method rather it would be part of a hybrid separation with for example distillation or even reverse

TABLE 1 – Liquid mixtures with azeotropic composition.

Typical positive azeotropic mixtures (characterised by a minimum boiling temperature) which can be separated by pervaporation through a permselective membrane obtained by grafting polyvinylpyrolidone onto a thin polytetrafluoethylene film [1]

A-B Azeotropes (A=Fast component)	T_b (°C)	Azeotrope characteristics		Selectivity		Permeate flux (kg/h m²)
		Tb(°C)	c(%)	α	β	
A: Chloroform B: n–Hexane	61.2 69.0	60.0	72.0 28.0	3.9	1.25	2.65
A: Ethanol B: Cyclohexane	78.5 81.4	64.9	30.5 69.5	16.8	2.89	1.10
A: Butanol–1 B: Cyclohexane	117.4 81.4	78.0	10.0 90.0	23.5	7.23	0.30
A: Water B: Ethanol	100.0 82.8	78.2	4.4 95.6	2.9	2.68	2.20
A: Water B: t–Buthanol	100.0 82.8	79.9	11.8 88.2	41.0	7.17	0.35
A: Water B: Tetrahydrofurane	100 65.5	63.8	5.7 94.3	19.1	9.24	0.94
A: Water B: Dioxane	100.0 101.3	87.8	18.4 81.6	18.1	4.36	1.33
A: Ethanol B: Ethylacetate	78.5 77.2	71.8	31.0 69.0	2.4	1.67	0.95
A: Methanol B: Acetone	64.7 56.2	55.7	12.0 88.0	2.9	2.36	0.65
A: Ethanol B: Benzone	78.5 80.1	67.8	32.4 67.6	1.3	1.18	2.90

The selectivities α and β are defined by the following ratios, respectively:

$$a = \frac{(c'_A/c'_B)}{(c_A/c_B)} = \frac{c'(1-c)}{c(1-c')}, \quad b = \frac{c'}{c}$$

Where c and c' are the weight concentrations of the faster permeant (A) in the feed (c) and the permeate (c'), respectively. T_b denotes the normal boiling point of an organic compound or of an azeotropic mixture (at 1 atm). Pervaporation temperature: $T = 25°C$.

TABLE 1 (continued) – Liquid mixtures with azeotropic composition.
Fractionation of negative azeotrpic mixtures (characterised by maximum boiling temperature) by pervaporation through a membrane obtained by grafting polyvinylpyrolidone onto a thin polytetrafluoroethylene film [1]

A-B Azeotropes (A=Fast component)	T_b (°C)	Azeotrope characteristics		Selectivity		Permeate flux (kg/h m²)
		Tb(°C)	c(%)	α	β	
A: Chloroform B: Acetone	61.2 56.2	64.7	80.0 20.00	1.8	1.10	0.85
A: Chloroform B: M.E.K.	61.2 79.6	79.9	17.0 83.0	1.0	1.0	1.50
A: Butanol-1 B: Pyridine	117.7 115.3	118.7	71.0 29.0	1.4	1.09	1.25
A: Water B: Formic acid	100.0 100.7	107.1	22.5 77.5	1.0	1.0	2.74
A: Acetic acid B: Dioxane	118.1 101.3	119.5	77.0 23.0	2.7	1.17	0.27
A: Acetic acid B: D.M.F.	118.1 153.0	159.0	26.0 74.0	1.2	1.14	0.04
A: Formic acid B: Pyridine	100.7 115.3	150.0	53.5 36.5	2.8	1.31	0.22
A: Acetic acid B: Pyridine	118.1 115.3	139.7	35.0 65.0	1.0	1.0	0.09
A: Propionic acid B: Pyridine	140.9 115.3		74.0 26.0			

A = Faster permeant; T = 25°C
Parameters α, β, c and c' and T_b are the same as in Table 5.1
M.E.K. = Methylethylketone; D.M.F. = N,N-dimethylformamide.

TABLE 2 – Classification of mixtures for pervaporation.

Aqueous	Dehydration Trace organics	water/ethanol (ethanol/water, aromatics/water)
Non Aqueous	Polar/non polar Aromatics/aliphatics Saturatd/unsaturated Isomers	alcohols/aromatics (methanol/toluene) alcohols/aliphatics (ethanol/hexane) (cyclohexane/benzene) (butane/butene) (C8 isomers)

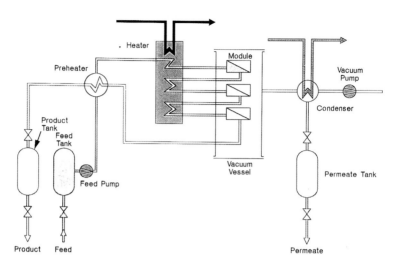

FIGURE 1 – A process flow diagram of pervaporation.

osmosis. Thus for example with an azeotrope with a high water content (water/pyridine) pervaporation would only be used once to produce the two non-azeotrope fractions for further processing.

The most important applications of pervaporation are
- removal of water from organics
- removal of organics from water(and gases)
- separation of organic mixtures
- concentration of aqueous solutions

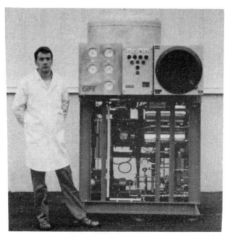

Isopropyl alcohol dehydration plant.

TABLE 3 – Applications of pervaporation.

MAIN APPLICATIONS
Dehydration of organic solvents and mixtures • alcohols : ethanol, propanol, butanol, etc., • ketones : acetone, MEK, MIBK, etc., • esters : ethyl, butyl, propyl, acetates, etc., • others : THF, dioxane, MTBE, glycol ether, acetonitrile, etc., • acetic acid, • organic amines, pyridine. The PV process can easily dehydrate solvent mixtures. The binary or ternary azeotropes can be dehydrated without the use of entrainers. In some cases, where distillation is not possible, prevaporation is the only alternative to costly incineration of waste solvent streams. • Debottlenecking of existing entrainer distillation plants is also an effective application.
Removal of organics from aqueous streams Several applications are covered by PERVAP® organophilic membranes. • *Waste Water purification* A wide range of organic solvents can be extracted: • hydrocarbons, chlorinated hydrocarbons, • esters, ketones, ethers, • alcohols. The PERVAP® systems can reduce COD requirement of water water streams going to biological treatment units, or preconcentrate organic wastes to incinerators. • *Wine and beer dealcoholisation* The main advantage of PERVAP® systems in this case is a far better recovery of aromas than with any other process. • *Aroma recovery and concentration* PERVAP® membranes allow recovery and concentration of many compounds in the food industry.

The main industrial applications are in the dehydration of organic solvent mixtures and the removal of organics from aqueous streams (see Table 3).

Dehydration

The dehydration of organic/water mixtures has rapidly become one of the main areas of application of PV. The splitting of water/organic azeotropes by pervaporation is common, either used as the sole dehydration step or used in conjunction with distillation for final dehydration. There are many organic streams in industry which become contaminated with, or contain low concentrations, of water (<10%). Methods to achieve dehydration include distillation and adsorption. Distillation will become very expensive when used to remove small amounts of water. Adsorption with desiccants (alumina, zeolities), although having relatively low capital outlay, requires regeneration. This regeneration requires high energy costs. Adsorbent replacement costs, disposal and generation of hazardous gaseous effluents are also other factors against adsorption. Table 4 gives a list of organics currently dehydrated by pervaporation. Pervaporation is generally economic with water contents of approximately 10% and less, with final product water content of hundreds of ppm to 10 ppm attainable. To go much below these water contents requires significantly greater installed membrane area and possibly a greater reduced pressure on the permeate side.

TABLE 4 – Organic solvents dehydrated by pervaporation.

Solvent	Water Content Feed (wt.%)	Product (ppm)	Solvent	Water Content Feed (wt.%)	Product (ppm)
1-Butanol	8.4	135	Ethanol/MeOH	2.9	780
n-ButanoL	5.4	800	Ethanol/benzene	14.1	320
t-Butanol	10.4	581	Allylalcohol	4.85	620
THF	0.4	220	Trichlene	0.01	8
Xylene	0.1	140	MEK	3.8	220
Methanol	7.1	1650	Methylene chloride	0.20	140
Methanol/IPA	0.21	300	Ethylene dichloride	0.22	10
Caprolactam	10.3	671	Chlorothene	0.0617	12
Ethanol/IPA	0.6	610			

For water contents above 10%, extraction and other methods are generally more economic, while below 0.1% content of feed, adsorption is generally preferable.

Many of the applications cited in Table 4 are for the small scale recycling of solvents (acetone, isopropyl alcohol) in pharmaceutical and speciality chemical companies where water contamination arises from a reaction or from other separations. Other applications are for bulk chemical processing (ethylene glycol, methyl ethyl ketone) where the objective is generally debottling and improvement in economics of existing distillations and perhaps adsorption.

Anhydrous Alcohol Production

Ethanol production is either based on fermentation or on synthesis methods such as the sulphuric acid process and direct catalytic hydration (of ethene).

Fermented ethanol product is typically 8% to 12% by volume, which after several stages of distillation to rectify and purify is produced as a near azeotropic mixture. Anhydrous ethanol for chemical and fuel use is obtained typically by azeotropic distillation with benzene trichoroethene, etc. The direct hydration route uses extractive distillation, with water, to free the ethanol of impurities and is thus purified by distillation in a similar manner to the fermented product.

Azeotropic distillation is a relatively expensive procedure and in addition there is some concern on environmental and health grounds over the use of some of the dehydrating agents. Pervaporation is considered to be an appropriate and competitive replacement for azeotropic distillation in the production of anhydrous ethanol (see Fig 2). A product of 99.5% ethanol is produced and a permeate, containing a relatively high percentage of

SEPARATION OF LIQUID MIXTURES/PERVAPORATION

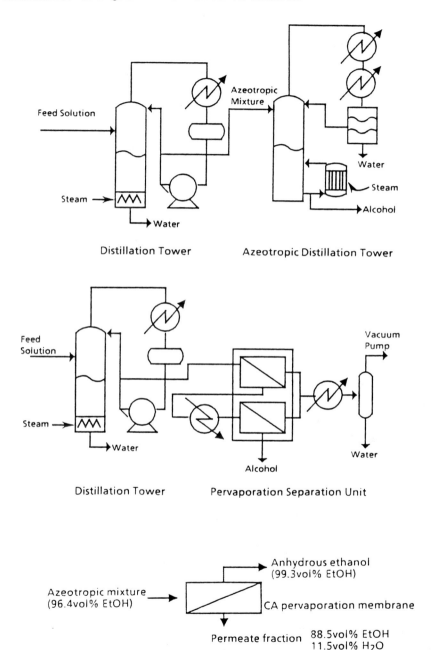

FIGURE 2 – Dehydration of alcohol by pervaporation and distillation.

ethanol, which is recycled back to distillation. It is particularly acceptable method of ethanol production for medical use.

The installation of pervaporation plants has slowly increased over the last ten years (see Table 5). These units are of varying capacity, with the largest designed to produce 150,000 dm^3 per clay of anhydrous ethanol (0.2%) water from a pre-distilled feed of 93% ethanol. The membrane area required is 2,200 m^2, which is the form of the flat sheets of a composite of a PVA active layer/PAN backing layer/PET supporting layer in several plate and frame modules.

TABLE 5 – Pervaporation unit installations.

Prevaporation operation	*No. of plants*
Ethanol dehydration	
Béthéniville sugar refinery, France (150,000 l d^{-1})	1
Provins sugar refinery, France (30,000 l d^{-1})	1
Smaller plants (1,000-12,000 l d^{-1})	11
Isopropanol dehydration	
Production capacity ranging from 5,000 to 15,000 l d^{-1}	5
Dehydration of ethylacetate (1,000-6,000 l d^{-1})	3
Dehydration of ethers (tetrahydrofurane, dimethoxyethane)	
Production capacity ranging from 2,000 to 6,000 l d^{-1}	2
Dehydration of ketones (6,000 l d^{-1})	1
Dehydration of other organic solvents	
Production capacity ranging from 750 to 15,000 l d^{-1}	6
Multipurpose plants (integrated systems)	3
Total number of operational units	33
+ 25 pilot plants (4m^2 surface area membrane each) installed to test the applicability of the technique to potential fractionation problems	

SEPARATION OF LIQUID MIXTURES/PERVAPORATION

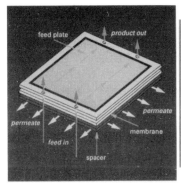 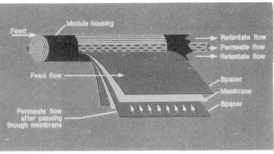

Plate and frame module. Spiral wound module.

Solvent	Ethanol	Ethanol	Ethanol	Ethanol
Concentration				
Inlet	85.7%	93.9%	85.7%	93.9%
Outlet	99.8%	99.8%	99.95%	99.95%
Flow rate	1,195 kg/h	1,500 kg/h	840 kg/h	970 kg/h
Utility requirements				
Steam	195 kg/h	110 kg/h	145 kg/h	83 kg/h
Power	85 kW	85 kW	85 kW	85 kW
Membrane area	480 m2			
Height	3,000 mm			
Length	7,500 mm			
Width	2,000 mm			

Ethanol dehydration plant.

Plate and frame pervaporation module.

Solvent Recycling

Electronics and pharmaceutical industries require good quality solvents for drying and cleaning purposes. During use many solvents become contaminated with water and thus need to be refined before recycling. Azeotropic distillation is commonly used in these applications. However pervaporation can economically remove this water and reduce the concentration to a few tens of ppm (see Table 6).

TABLE 6 – Relative economics of pervaporation and azeotropic distillation for isopropyl alcohol refining.

		Prevaporation	Azeotropic distillation
I	Utility consumption		
	Steam (kg/h – IPA)	0.3	1.6
	Electricity (kwh/h – IPA)	0.03	0.01
	Cooling water (t/h – IPA)	0.055	0.01
II	Running cost (Yen/kg – IPA)		
	Steam	0.9	4.8
	Electricity	0.6	0.2
	Cooling water	0.55	1.0
	Entrainer (Benzene)	—	0.03
	Membrane	1.9	—
	Total	3.95	6.03

where:
Steam : 3,000 yen/t
Electricity : 20 yen/kw – h
Cooling water : 10 yen/t
Membrane renewal : once/3 years

Solvent vapours such as chlorinated hydrocarbons are generated in several industries, eg electronic, pharmaceutical, and for environmental reasons are recovered and not released to atmosphere. Adsorption on activated carbon is effective, but this on regeneration with steam, produces a mixture of condensed water and solvent. After phase separation of this mixture a solvent phase, containing a low percentage of water, is produced. Dehydration of this solvent by pervaporation is effective.

Examples

In many cases where water contaminated solvents occur the concentration of the water in the permeate will give a mixture which will readily phase separate on cooling. This phase separation allows much greater recovery of the organics. A typical example is the dehydration of dichloroethene (shown in Fig 3.) or ethlene dichloride (EDC). In this plant saturated EDC (0.2% wt H_2O) form a condenser is preheated before entering the PV module equipped with PVA membranes. The pervaporation produces a purified EDC (< 10 ppm water) in one pass, with a permeate containing approximately 50% water. This

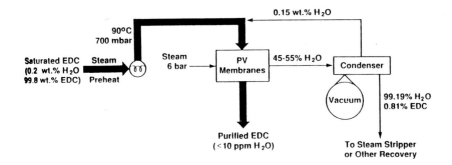

FIGURE 3 – Pervaporation of dichlorethene for solvent purification.

permeate is condensed and phase separation occurs to produce an EDC rich layer (0.15 wt % H_2O), which is recycled and a water rich layer containing 0.81% EDC.

Other commercial examples of this technique include the processing of jet engine fuels and halogenated refrigerants.

Another reported application is for dewatering isopropyl alcohol (IPA). This is a debottle-necking application in which the IPA at 85% is taken to 95% by PV prior to introduction into an extractive distillation column.

Dehydration of cleaning agent

Pervaporation systems are installed for use at an LCD manufacturing plants. Traditionally, microchip washing has been performed by chlorinated hydrocarbons but with the steady demise of CFCs, isopropanol is becoming a preferred cleaning agent. The alcohol picks up water during the process and eventually is either discarded or regenerated. This expensive prospect can be avoided by the use of a compact pervaporation unit, which allows the straightforward dehydration of isopropanol and similar solvents, whilst also offering the convenience of on-site recycling.

The Pervaporation system typically dehydrates 6 m^3 of isopropanol per day, which is used as a cleaning agent. The IPA final concentration is typically 98.9%.

Biotechnology

Pervaporation has several features which are attractive for separations in biotechnology; i) low temperature, ii) low pressures, iii) high cross-flow velocities are not needed and iv) additional chemicals are not required. There are four suggested classes of bioseparation for pervaporation, which require membranes with different characteristics.
- Direct bioproduct recovery
- Volatile by-product removal
- Concentration of sensitive bioproducts
- Dehydration of low molecular organics

The selective removal of organic byproducts is not a particularly well developed area, requiring highly selective membranes. The concentration of sensitive bioproducts, requires primarily high water flux membranes, as the products are usually high molecular species, eg amino acids, enzymes, which are non-permeable to pervaporation membranes. Dehydration of low molecular bioproducts by pervaporation will only be in the last stage of downstream processing.

Direct Product Recovery

Fermentation can produce many organic components, from a range of feedstocks, notably ethanol but also butanol, isopropanol, acetone, 2-3 butanediol, glycol and acetic acid. These products inhibit the fermentation process as they increase in concentration, eventually resulting in termination of fermentation. The volatile nature of the products makes it attractive to continuously remove these species during fermentation and maintain long-term fermentation at an optimum rate. The separation must retain the fermentation media, cells, nutrients etc, whilst removing fermentation product and by-product solvents. Practised procedures for extraction of ethanol from fermentations include

i) Vacuum evaporation – ethanol is removed at a temperature of 30-35°C
ii) Stripping – ethanol is removed in a stripping gas
iii) Extraction – liquid extraction with suitable water miscible organic solvents.

The PV membranes for this application must exhibit high flux and good selectivity for the organic solvent products. Economics of the process will likely require hollow fibre membranes, e.g. composite hollow fibre, consisting of a polysulphone porous support with a thin inner coating or organophilic polydimethly siloxane. However an economically competitive membrane for this application is still awaited.

Removal of Organics from Water

The applications of organic pervaporation from aqueous solution are generally targeted at pollution control, solvent recovery, organic concentration for disposal and speciality processes.

One suggested application is in the treatment of wash waters used to remove organics from solvent laden airstreams, ie air scrubbing. The dilute aqueous solution is treated by pervaporation to remove the solvent in the permeate and to produce water, with a minor amount of solvent, to recycle to the air scrubbing unit.

The contamination of water by small amounts of organics requires appropriate means of pollution control. There are many possible alternative methods for this problem, which include air stripping, carbon adsorption, biological treatment, steam stripping and incineration. If the aqueous stream contains a few percent of organic, then recovery of the solvent adds favourably to the overall process economics. Pervaporation is an economically viable alternative over a wide range of organic concentrations; approximately 100 ppm to 60,000 ppm (see Fig 4). The process of pervaporation can quite selectively recover the organic species. The membranes in these applications are rubbery polymers, such as silicone rubber, polybutadiene, natural rubber, polyether copolymers etc.

The selectivity of the separation depends significantly on the type of organic compound (see Fig 5. For example silicon rubber selectivity decreases as the hydrophobicity of the

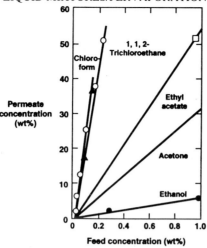

FIGURE 4 – Competitive range of pervaporation for organic in water separation.

organic decreases. Thus high selectivities are achieved separating benzene, 1, 1, 2 trichloroethene and toluene, good selectivities with ethylacetate and acetone, but only modest values with ethanol and acetic acid.

Furthermore different elastonomers (Table 7) give significantly different values of selectivity for individual compounds. Ethene-propene terpolymer is much more selective for toluene and trichloretlylene than silicone rubber, exhibiting a separation factor is excess of 30,000.

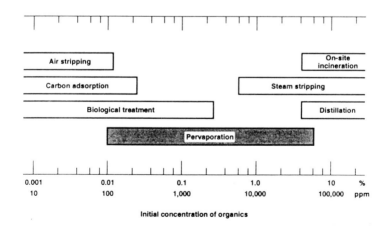

FIGURE 5 – Separation of organic compounds from water by pervaporation using a silicone rubber composite membrane.

TABLE 7 – Elastomeric membranes for removal of organics from water by pervaporation.

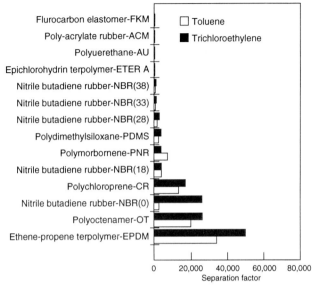

The process of pervaporation with rubbery membranes is also suitable for recycling dilute solutions (1% to 2%) of organic solvents such as ethyl acetate. A process has been developed to treat approximately 90 m^3/day of a 2% ethyl acetate solution using silicone rubber membranes. 90% of the ethyl acetate is recovered as a 96.7% product. The process is similar to that depicted in Fig6 ...

The separation of polar solvents such as ethanol, acetic acid and formic acid from water can not generally be achieved with a separation factor much greater than 5 to 10, with most rubber membranes. At the moment this has prevented commercial applications in solvent recoveries and fermentations. Progress however is being made in this area with modified rubbery membranes, where for example silicone rubber containing dispersed zeolite has shown selectivity greater than 40 for ethanol/water.

Pollution Control

An example of pollution control is the treatment of groundswater contaminated with 0.1% of 1, 1, 2-trichloroethane. As shown in Fig 6, the process, with a membrane unit having a separation factor of 200, removes 99% of the organic compound and produces a permeate of 4.1% organic. The permeate rapidly phase separates into an organic rich (>99%) layer and aqueous stream containing 0.4% organic, which is recycled to the incoming groundwater for re-treatment.

Organic/Organic Separations

The dominant separation method for organic mixtures in the petroleum and chemical processing industries is distillation. It is an energy intensive process estimated that around 40% of the total energy consumed by the chemical processing industries is in distillation.

SEPARATION OF LIQUID MIXTURES/PERVAPORATION

When the organic components have similar boiling points and relative volatilties, the separation is difficult and the energy cost is particularly high. Separation of azeotropes is not possible by simple distillation and thus entrainers are added to increase relative volatility of the major components and thus break the azeotrope. Several important separations in petroleum and chemical industries rely on the use of entrainers (see Table 8). These processes require additional separations to recover the entrainer.

TABLE 8 – Binary hydrocarbon azeotrope mixtures and possible entrainers.

Hydrocarbon system	Relative volatility	Best entrainer	Relative volatility
2,2-Dimethylbutane/cyclopentane	1.006	n-Propylamine	0.987
Cyclohexane/2,4-dimethylpentane	1.006	Acetone	1.025
Methyl cyclohexane/ 2,4-dimethylpentane	1.046	Ethanol	1.085
3-Methylpentane/1-hexane	1.009	Methylene chloride	1.159
3-Methylpentane/2-ethyl, 1-butene	1.037	Ethyl formate	1.156
2-Ethyl, 1-butene/n-hexane	1.056	Chloroform	1.094
2,2,4-Trimethylpentane/ 2-2,4-trimethylpentene-1	1.040	Isopropyl acetate	1.129
n-Heptane/2,2,4-trimethylpentene-1	1.045	Isopropyl acetate	1.129
Ethyl benzene/p-xylene	1.035	2-Methyl butanol	1.079
p-Xylene/m-xylene	1.020	2-Methyl butanol	1.029
m-Xylene/o-xylene	1.105	Methyl isobutanol carbinol	1.150

In certain cases a suitable entrainer is not available and then alternative separation methods must be considered.

For a good overall separation factor for pervaporation, ideally both the membrane separation factor and the evaporation separation factor (relative volatility) should be large. With mixtures of low relative volatility, good separation is reliant upon a reasonable separation factor for the pervaporation membrane. For example with a benzene/cyclohexane mixture which has an azeotrope composition of approximately 50% benzene, pervaporation with a 20 μm thick crosslinked membrane (polymeric/alloy of polyphosphonate and acetyl cellulose) will produce a permeate with more than 90% benzene (see Fig 7).

Further purification of this benzene permeate, and the cyclohexane rich residue, by pervaporation would not be economic. Rather standard distillation of both these streams would be used to produce the respective pure components. Both distillations would realise azeotropic (or near azeotropic) mixtures as second products, which would then be recycled for further purification.

A critical factor in the development of organic/organic separations is the availability of membranes, to withstand the continuous long-term exposure to the organic compounds; frequently at elevated temperatures.

An interesting example of pervaporation applied to organic/organic separations is in the production of methyltertiary butyl ether from methanol and isobutene (C_4) (Fig 8). This

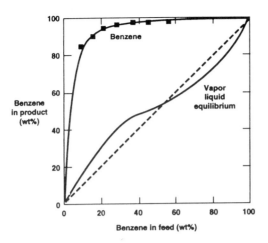

FIGURE 7 – Vapour product vs feed composition for pervaporation of a benzene/cyclohexane mixture.

process produces a reactor product mixture of all three components of which both the methanol and ether and methanol and C_4 form azeotropes. A process has been developed in which pervaporation is integrated in the system to separate out the methanol and recycle it back to the reactor. The membrane used is made from cellulose acetate. Cellulose acetate membrane has a separation factor for methanol from MTBE of over 1000, because the material is hydrophilic and methanol is more polar than MTBE or the isobutene.

Recovery of Volatile Bioproducts

The considerable variety of low molecular weight bioproducts can be put into the distinguishable groups of biosynthetic chemicals, aroma compounds and essential oils and "concrete's".

TABLE 9 – Oxychemicals produced from fermentation.

Low Boilers	High (or Non) Boilers
Ethanol	Ethylene Glycol
– Ethylene	Adipic Acid
– Butadiene	Acetic Acid
– Octane Enhancer	Acrylic Acid
– Industrial	Glycerol
	1,4-Butanediol
Isopropanol	Propylene Glycol
Acetone	n-Butanol
Methylethylketone	Citric Acid
	Sorbitol
	Propionic Acid
	Fumaric Acid

SEPARATION OF LIQUID MIXTURES/PERVAPORATION

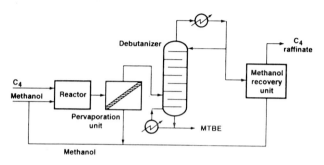

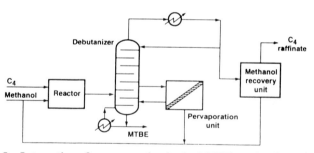

FIGURE 8 – **Integration of pervaporation in the MTBE production process.**

Biosynthestics are oxychemicals which can be produced by fermentation of renewable biomass on a large scale (see Table 9). Several of these have been produced in commercial quantities (ethanol, acetic acid, isopropanol, acetone, glycol and n-butanol) and are presently produced from petrochemcial feedstocks. Fermentation products are mainly constituents of the aqueous solutions and pervaporation is required to selectivity transfer the product or byproducts, through the polymeric membrane. Thus organophilic membranes such as silicone rubber (polydimethylsiloxcane, PDMS) in various configurations are used to selectively remove the organic solute. Much of the research on pervaporation membranes has focused on impeding transport of water. The enrichment of organic solute achieved by pervaporation is determined by the "pervapouration coefficient" which is essentially the product of activity coefficient and the ratio of pure component vapour pressure to permeate pressure. Thus although pervaporation of low boilers such as ethanol is relatively easy, the fact that, under pervaporation conditions, activity coefficients of organic solutes are high and increase with dilution enables high boiling solutes to be enriched.

Oxy Chemicals

Fermentation of biomass by specific yeast or bacterium typically produces several metabolites with the primary bioproduct. For example in the case of fermentation of glucose to ethanol by Sacc-horomyces cerevisiae (yeast) (see Table 10). These metabolites inhibit microbial activity as their concentrations increase and thus removal of these species as well as the main product during fermentation is of interest.

A typical study is in the fermentation of ethanol by S cerevisiae using a PDMS (1 m coating) on polysulphone membrane (see Table10).

TABLE 10 – Pervaporation of bioproducts form S cerevisiae using PDMS on PSU membrane.

Component	Fermenter (w-%)	Permeate (w-%)	Organic Flux (g/m²h)
Ethanol	4.8	26.1	63.9
Acetaldehyde	0.026	0.25	0.61
Ethylacetate	0	0.05	0.1
Methanol	0	0.02	0.06
Isobutanol	0.01	0.11	0.28
Methyl butanol	0.006	0.05	0.13
Acetic acid	0.007	0	0

A problem with the use of pervaporation in this mode is that certain inhibiting metabolites, such as acetic acid, do not pervaporate.

Acetone, Butanol-Ethanol

Fermentation of sugars by microganisms of the clostridium group yield mainly n-butanol, isopropanol, ethanol and acetone. The production of acetone, butanol and ethanol (ABE fermentation) using C. acetobutylicum Weizmann has seen commercial operation, although it is not in current use today. The fermentation is inhibited by butanol at concentrations of 1% and byproducts, acetic and butyric acids and thus accumulation of solvents in the fermentation broth is low. Recovery of n-butanol (B.pt 118C) by distillation from water at low concentrations (1%) is not economic. Thus processes based on membranes are alternatives for product recovery. These include, pervaporation, pertraction, membrane distillation and reverse osmosis. Pervaporation with PDMS membranes (Hollow fibre) of butanol gives enrichment factors of 20 and greater. Overall recovery of solvent products will be by an integrated procedure involving other separation such as stripping and three phase distillation.

AROMA Compounds

Aroma compounds (odours and fragrances) are volatile metabolites released from cultures of micro-organisms and fungi. These compounds are alcohols, aldehydes, aliphatic esters, lactones and terpenes and are correlated with the micro-organism and odours (see Table 11). The organic solute species are typically high boiling compounds and are in dilute aqueous solutions and thus pervaporation with organophilic membranes is ideally suited to product recovery, at least to achieve an immediate reduction in volume of feedstock.

TABLE 11 – Aroma compounds.

Microorganism	Odour	Constituents
Bacillus subtilis	Soybean	Tetramethylpyrazine
Ceratocystis moniliformis	Fruity, banana, peach, pear, rose	3-Methylbutyl acetate, δ-and γ-decalactone, geraniol, citronellol, nerol, linalool, α-terpineol
Ceratocystis variospora	Fragrant, geranium	Citronellol, citronellyl acetate, geranial, neral, geraniol, linolool, geranyl acetate
Ceratocystis virescens	Rose, fruity	6-Methyl-5-hepten-2-ol acetate, citronellol, linalool, geraniol, geranyl acetate
Corynebacterium glutamicum	Soybean	Tetramethylpyrazine
Daedalea quercina	Apples	
Inocybe corydalina	Fruity, jasmine	Cinnamic acid methyl ester
Kluyveromyces lactis	Fruity, rose	Citronellol, linalool, geraniol
Lentinus cochleatus	Anisaldehyde	
Lenzites sepiaria	Slightly spicy	
Mycoacia uda	Fruity, grassy, almond	p-Methylacetophenone, *p*-tolyl-1-ethanol, *p*-tolylaldehyde
Penicillium decumbens	Pine, rose, apple, mushroom	Thujopsene, 3-octanone, 1-octen-3-ol, nerolidol, β-Phenylethyl alcohol
Pholiota adiposa	Earthy	
Polyporus croceus	Narcissus	
Polyporus obtusus	Jasmine	
Poria xantha	Lemon	
Pseudomonas perolens	Musty, potato	2-Methoxy-3-isopropylpyrazine
Pseudomonas taetrolens	Musty, Potato	2-Methoxy-3-isopropylpyrazine
Stereum murrayi	Vanilla	
Stereum rugosum	Fruity, banana	
Streptomyces odorifer	Earthy, camphor	*trans*-1,10-Dimethyl-*trans*-9-decalol, 2-*exo*-hydroxy-2-methylbornane
Trametes odorata	Honey, rose, fruity, anise	Methyl phenylacetate, geraniol, nerol, citronellol
Trametes suaveolens	Anisaldehyde	
Trichoderma viride	Coconut	6-Pentyl-2-pyrone

Pervaporation of several aroma compounds has been achieved with silicone rubber (including chemically modified) and polyether-polyamide block copolymer
- apple (and fruit) juice aroma compounds
- gamma – decalactone (peach) produced by fermentation of castor oil
- 6-pentyl-2-pyrone, a natural aroma compound of coconut fragrance, from trichoderma viride culture medium

As a rule pervaporation enrichment is higher with esters, lower with aldehydes and lower again with alcohols.

Essential Oils

Essential oils are natural products of various parts of plants which can be recovered by steam distillation or solvent extraction. The characteristic and important fragrance is due to a combination of the many constituents (several hundred in many cases) and thus isolation of these constituents is rarely required and could detract from essential product quality. Pervaporation of essential oils, which are normally volatile with steam, is thus possible and has been demonstrated in the recovery of juniper oil.

Enhanced esterification by pervaporation

Esterification reactions generate water as part of the production of the ester. Esterification reactions are typically reversible processes and the degree of conversion is limited by equilibrium conditions. When water formed as a byproduct in a reaction is continuously removed from the reaction mixture, the formation of the wanted product can be shifted beyond the thermodynamic equilibrium and full conversion of the reactants can be obtained. The main advantages observed are:
- reaction (batch) times are reduced;
- space-time yield of a given reactor can be increases by a factor of 2 to 3;
- downstream separation and purification of the wanted product is facilities or even eliminated;
- energy costs are reduced; and
- when one of the reactants is used at a slight excess, 100% yield of the wanted product is possible.

Pervaporation has been applied to the continuous removal of water from esterification reaction mixtures and can extract water either directly or from a side loop of the reactor, as the membranes used are fully stable against the reactants and the acidic catalyst. When at least one of the reactants has a volatility sufficiently higher than that of the product, an evaporated stream can be treated in the vapour phase and returned to the reactor, thus water can be removed at the reaction temperature or at any freely chosen temperature.

SEPARATION OF LIQUID MIXTURES/PERVAPORATION

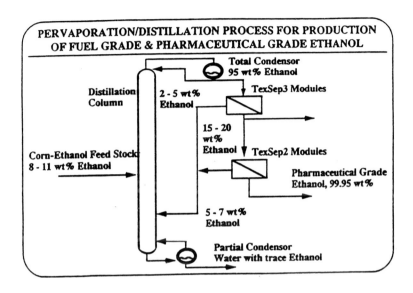

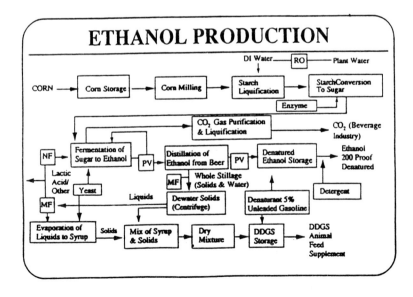

SECTION 6

Separation of Organic Vapour/Air Mixtures

SEPARATION OF ORGANIC VAPOUR/AIR MIXTURES

Volatile organic compounds (VOC) and other hazardous organic compounds are common pollutants emitted from many chemical and pharmaceutical processes. The scale of the problem is large with estimates in excess of 0.25 million tonnes of organic pollutants in 5 million cubic metres of air per year in the chemical industry. Clean air and environmental legislation requires elimination or control of a large percentage of these emissions. The technologies generally considered in process chemical recovery and end of pipe control of organic vapour emisions are either incineration, adsorption, absorption and condensation. A new membrane based technology with several applications for organic vapour emission control is based on vapour permeation.

The selection of a control technology for VOC emissions depends on several factors, including the type and concentration of the organic compound, the capacity required, the removal or recovery efficiency, the value of the organic and the specific site characteristics. Each particular application must be separately evaluated and there are general guidelines (see Table 1) which may be used to aid the decision. In general terms the choice of recovery process depends upon the concentration of the VOC and the capacity of the air flow. Fig 1 indicates the domain ranges for separations based on condensation, adsorption and membranes. As can be seen one process alone may not be able to meet the requirements of a particular recovery operation and consequently a combination of methods may prove to be the best option.

Recovery of Organic Vapours from Air by Vapour Permeation

Organic contaminants in air and in other permanent gases can be recovered using vapour permeation through an appropriate membrane. The membranes used are similar to those used for pervaporation ie non-porous. The mechanism of transport is similar to pervaporation, driven by differential pressure induced by vacuum on the permeate side. The membrane are relatively impermeable to air and permanent gases and are usually made from rubbery polymers. Table 2, shows selectivities (relative to N_2) for a range of

TABLE 1 – General guidelines for the selection of VOC recovery processes.

1.	Carbon adsorption is most commonly applied to dilute mixtures of VOC's and air. Inlet concentrations greater than 5,000 ppmv usually require dilution prior to adsorption. The cost of carbon adsorption is dependent on the total amount of VOC removed from the air, in addition to the flow rate. Thus, carbon adsorption is typically most economical for dilute concentrations. Carbon adsorption is not effective for the most volatile materials and is rarely used for ketones and other highly reactive compounds because of the danger of igniting the bed.
2.	Steam or hot gas may be used to recover VOCs from the adsorption medium. Large streams warrant the installation of dual adsorption beds, so that one bed may be regenerated on-site while the other is adsorbing. Inlet flow rates less than approximately 100scfm are typically handled by off-site regeneration of the carbon. Regeneration with steam produces a liquid waste stream containing VOCs that may not be easily recycled or may need to be handled as a hazardous waste.
3.	Condensation generally requires high inlet concentrations (above 5%) to achieve any VOC recovery. Typically, condensation cannot meet stringent environmental requirements without the use of very low temperatures or high pressure, both of which drive the cost up considerably.
4.	Membrane systems are best suited for treating streams containing more than 5,000 ppmv organic vapor. Membrane system costs increase in proportion to the flow rate of the inlet stream, but are relatively independent of the organic vapor concentration.

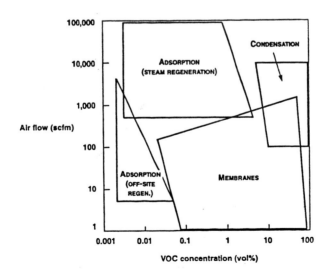

FIGURE 1 – The application range of VOC recovery processes.

TABLE 2 – Selectivities of vapour permeation of organic compounds, relative to nitrogen.

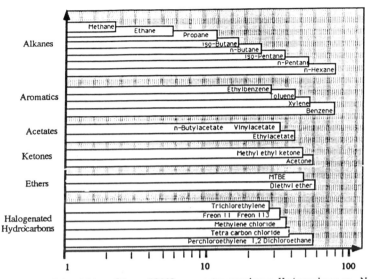

Selectivities of 1 μm PDMS-composite membrane Hydrocarbons vs. N2

alkanes, aromatics, ketones and halogenated hydrocarbons for a 1 m thin silicone rubber (PDMS) composite membrane.

The permeability of many organic vapours of low volatility decreases with an increase in temperature, as the solubility in the membrane decreases with an increase in temperature. This is an opposite trend to many permanent gases and volatile organics.

Off-gas Treatment in Gasoline Tank Farms

The storage, handling and distribution of gasoline results in off gases which, for environmental reasons and to comply with relevant clean air legislation, must be subjected to appropriate vapour recovery. The organic compounds of gasoline vapours, eg propane, butane or pentane and also carcinogenic substances, eg benzene must be kept to low levels. Membrane modules used for vapour permeation use flat sheet membranes in plate and frame configurations, some 0.5 m long and 0.32 m diameter. Gasoline recovery from off gases takes place in two stages of membrane separation (see Fig 2) with the permeate phase undergoing post treatment in a catalytic incinerator or in gas engines. The residual concentration of the membrane permeate is between 50-100 gm HC/m^3.

An alternative procedure is to adopt a hybrid system of a vapour permeation module with pressure swing adsorption (PSA) (see Fig 3).

Off gas from the tank form is held in gasometers. The gas is first compressed and scrubbed (clean absorption) with gasoline before undergoing vapour permeation. The retentate is then passed to a pressure swing adsorption (activated carbon or molecular sieve) unit consisting of two parallel units operating alternatively. While one unit carries

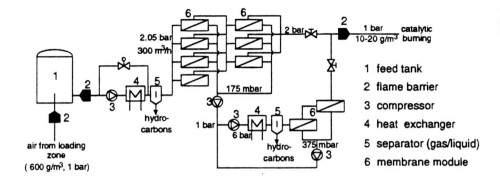

FIGURE 2 – Vapour recovery unit using two stage vapour permeation modules.

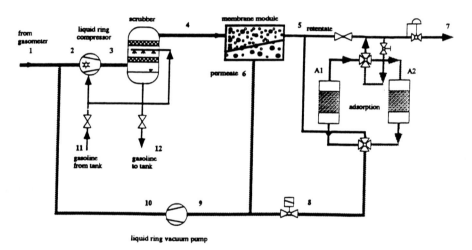

FIGURE 3 – Vapour recovery hybrid unit of vapour permeation and pressure swing adsorption.

out the adsorption function the other undergoes desorption and regeneration using a by-pass stream of the clean stream. The vacuum pump supports both the vapour permeation and the desorption of the PSA column.

Gasoline Station Vapour Return

Regulations on clean air in various countries have set guidelines for vapour return systems at gasoline stations. Systems on the market include

i) balance system – consisting of a bellow which covers the filling pipe and in which the filling point is sealed to the atmosphere

ii) open vacuum assisted system – consisting of a nozzle with a vapour spout and internal vapour channel.

SEPARATION OF ORGANIC VAPOUR/AIR MIXTURES

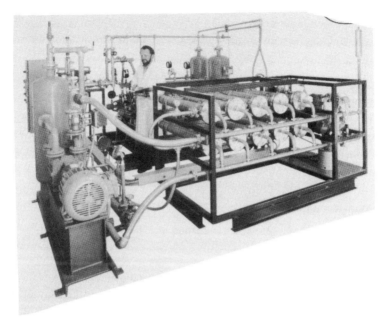

There are limitations to the efficiency of the vacuum recovery systems, which are typically 65 – 70%. The ratio of recycled air/gasoline vapour volume must be equal to the pumped liquid volume to prevent an emission transfer from the vehicle fuel tank filling point to the storage tank. Improved efficiency in vapour return requires suction of surplus air/vapour. Incorporation of a membrane, vapour permeation module in the system enables this surplus to be treated. The permeate, a hydrocarbon rich stream is returned to storage at the pumped fuel volume while the vapour lean surplus retentate is vented at atmosphere (Fig 4).

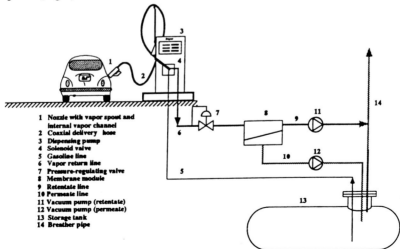

1 Nozzle with vapor spout and internal vapor channel
2 Coaxial delivery hose
3 Dispensing pump
4 Solenoid valve
5 Gasoline line
6 Vapor return line
7 Pressure-regulating valve
8 Membrane module
9 Retentate line
10 Permeate line
11 Vacuum pump (retentate)
12 Vacuum pump (permeate)
13 Storage tank
14 Breather pipe

FIGURE 4 – Vapour permeation applied to fuel return system.

The use of vapour permeation allows vapour recovery efficiencies greater than 90% at the cost of installing the membrane module and an additional vacuum pump.

End of Pipe Solvent Recovery

The recovery of solvents such as methylethyl ketone, hexane, toluene and 1, 2 dichloroethane by vapour permeation is applied in the chemical and pharmaceutical industries. Capacities of these plants are of the order of 1 to 20 m^3h^{-1} to recover solvent vapours of around 1 to 2% by volume in concentration, for example in the case of 1, 2 dichloroethane. The membrane module operates with a feed consisting of the original feed and a recycle of the permeate which has the solvent partially recovered. The retentate has a 0.0075 vol % solvent concentration giving an overall recovery of greater than 99.5%

Alcohol Dehydration

Dehydration of alcohols can also be carried out by vapour permeation as well as pervaporation. Under identical isothermal operating conditions, (feed concentration, temperature, feed pressure, permeate pressure and specific feed mass flowrate) the permeate flux and selectivity of vapour permeation and pervaporation are identical.

Dehydration of alchohols, notably ethanol is performed using a three layer of composite utilising polyvinylalcohol as the active layer. Membrane module design is similar to that of a pervaporator, e.g. plate and frame modules with flatsheet (rectangular or disc membranes). The basis of design is a commercial plate-type heat-exchanger constructed from stainless steel.

This design concept allows ready operation at high temperatures which allows the permeate flux to be maximised, without loss in selectivity, within the operating limits of

Industrial size vapour permeation modules.

the membrane material. Higher temperatures can mean that the permeate side pressure can be increased enabling permeate condensation to be carried out by relatively cheap water cooling rather than by refrigeration. Obviously the vacuum must be well below the dew point to avoid condensation on the membrane, and within the membrane structure, which would inhibit permeation.

Vapour permeation has advantages over pervaporation in technical applications:

1) Isothermal operation is possible. In pervaporation there is a temperature drop between feed and retentate and permeate as the heat required for the evaporating component is removed locally in the unit.

2) At the same mass flowrate, vapour permeation operation has much higher volumetric flowrates and thus linear velocities across the membrane. This reduces the impact of any concentration polarisation, particularly at the upper feed concentration range. Thus the use of VP offers effectively higher permeate flux or lower membrane area requirements than pervaporation.

An industrial vapour permeation plant for ethanol dehydration consists of a three stage vapour permeation train operating on a saturated vapour of 94% ethanol at 2.2 bar pressure and 100°C (see Fig 5). The feed from one module is the feed to the next. The alcohol feed is preheated with the dehydrated alcohol vapour leaving the membrane permeation train. Each membrane stage experiences an approximate 0.5 bar pressure loss, feed to permeate, which is compensated for by recompression between modules. The final product of 99.9% ethanol is condensed by water cooling. The permeate, water rich phase, is returned to a rectification column.

An alternative operation is to attach the vapour permeation to the vapour product from a normal pressure alcohol rectification column to dehydrate the vapour. Vapour permeate from the membrane modules is recycled to the column after condensation.

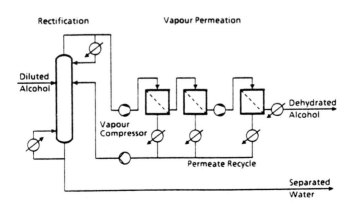

FIGURE 5 – Vapour permeation train for dehydration of ethanol.

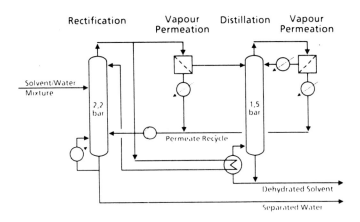

FIGURE 6 – A combination of vapour permeation with distillation and rectification.

Dehydration of Isopropanol

Composite PVA membranes give almost the same permeation flux versus water content characteristics for IPA as for ethanol. As IPA has a lower alcohol azeotrope concentration (88% out) the specific membrane fluxes are higher than for the ethanol azeotrope. These fluxes are also higher than for pervaporation under similar conditions.

The result of this is that in a typical dehydration of 83% wt IPA solution to 99.9% wt PA, only six vapour permeation modules are required in comparison to thirteen for pervaporation. In addition membrane area requirements for vapour permeation are approximately half those for pervaporation. In both cases however, the required membrane area increases steeply with the required purity of IPA. If very high purity IPA is needed, say 99.995% wt, a combination of vapour permeation with distillation in a hybrid process may offer advantages over membrane processes alone. In one system shown in Fig 6, rectification produces a subazeotropic IPA vapour (83%) at a pressure of 2.2 bar, which after a single vapour permeation stage is concentrated to above its azeotrope composition at a pressure of 1.5 bar. This effectively reverses the vapour/liquid equilibrium characteristics of the mixture enabling a distillation (stripping) of the vapour to occur which produces a dehydrated bottom product of pure IPA. The second vapour permeation dehydrates the top product vapour from the distillation which is returned to the column as reflux after condensation. The permeates from both membrane units are returned to rectification.

VOC Recovery by Compression-Condensation and Vapour Permeation

Condensation is widely used to minimise organic vapour losses to atmosphere. The gas stream containing vapour is chilled to a temperature at which a substantial fraction of the condensable content of the stream liquifies. The amount of volatile material that can be recovered in this way depends on the concentration of the VOC in the stream, the boiling point of the VOC, and the temperature and pressure of the stream. Chilling only affords efficient recovery for organic compounds with relatively high boiling points and therefore, compression is often combined with chilling to raise the dew-point temperature of the condensable organic component. The typical removal afforded by condensation alone and by condensation with compression of the gas stream is illustrated for several examples in Table 1.

Condensation (see Table 3) can only recover relatively non-volatile compounds, e.g. perchloroethylene, with reasonable efficiency at temperatures above 0°C. For example, if the stream contains 10% perchloroethylene, simple cooling to 10-15°C will reduce the emissions by 88% (leaving approximately 1% perchloroethylene). Moderately volatile compounds e.g. benzene, are not easily condensed unless a combination of compression with condensation is used. For benzene, cooling to 10 – 15°C reduces the emissions by only 24%. However in the case of highly volatile organic compounds, a combination of cooling and compression will not produce streams containing acceptably low concentrations of VOCs. Effective recovery of many VOCs can only be accomplished by very low-temperature chilling, but low-temperature chilling has several disadvantages:

- Most streams contain water vapour; at condensing temperatures below 0°C, ice forms in the condenser, necessitating periodic defrosting of the condenser or dehydration of the stream.
- If temperatures lower than –40°C are necessary, a two-stage chiller is required, which dramatically increases the energy costs.

TABLE 3 – Compression-condensation recovery efficiencies for representative VOCs

Solvent	Boiling point	Condenser vent gas composition (vol % VOC)		
		15°C, 14 psia	5°C, 14 psia	5°C, 150 psia
Vinyl chloride	—13.8%	up to 100%	up to 100%	19.9
CFC-11	24°C	71.4	48.2	4.8
Methylene chloride	40°C	36.7	23.5	2.3
Benzene	80°C	7.6	4.5	0.5
Perchloroethylene	121°C	1.3	0.7	0.1

A recent system for the recovery of VOCs from air streams is based on a combination of vapour compression-condensation and vapour permeation.

The vapour separation process, shown in Fig 7, typically consists of two steps: compression-condensation and membrane vapour separation. The compression-condensation step is conventional. A vapour/air mixture is first compressed to 3-13 bar and then cooled in a condenser where a fraction of the organic vapour condenses, and is directed to a solvent storage tank for recycling or reuse. The non-condensed portion of the vapour/air mixture flows across the composite membrane surface that is much more permeable to organic vapours than to air (10-100 times more per permeable to organic compounds

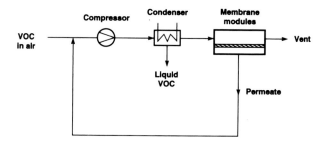

FIGURE 7 – Flow diagram of compression/condensation and membrane separation.

than to air). The membrane separates the gas into two streams: a permeate stream containing most of the remaining solvent vapour from the condenser, and a solvent-depleted stream essentially stripped of organic vapour. The solvent-depleted air is vented from the system. The permeate stream is recycled to the inlet of the compressor.

Vapour transport through the membranes is by virtue of a lower vapour pressure on the permeate side of the membrane than on the feed side. This pressure difference is either

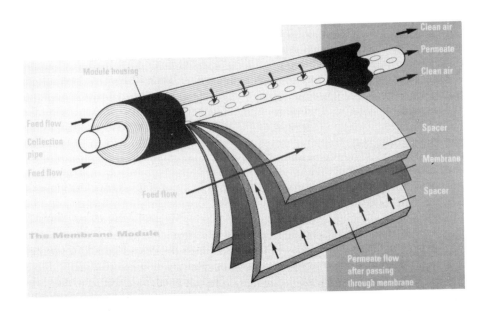

achieved by compressing the feed stream, as shown in Figure 7, or by applying a vacuum on the permeate side of the membrane. In some cases, a combination of vacuum on the permeate stream and compression on the feed stream is used.

Air and organic vapour permeate rates are determined by their relative permeabilities and the pressure difference across the membrane. Depending on the system design between 90% and 99.99% of the organic vapour is removed from the feed air stream.

Overall, to achieve an effective and economic separation, the membrane has good selectivity for organic vapours from air, is formed into high-flux, defect-free materials and are formed into space-efficient, low-cost membrane modules. The modules are based on composite membranes and spiral-wound modules.

Generally, condensation and membrane separation can be used separately for recovering VOCs from air streams. The feature of the membrane vapour recovery system design is that it combines the advantages of compression-condensation and membrane separation to create an optimised overall process with a higher efficiency, than either method alone.

TABLE 4 – Important Air Pollutants Recoverable by the VaporSep™ Membrane Process

Acetaldehyde	HFC-134a
Acetone	Hexane
Acetonitrile	Methanol
Benzene	Methyl bromide
Carbon tetrachloride	Methyl chloroform
CFC-11	Methyl isobutyl ketone
CFC-12	Methylene chloride
CFC-113	Propylene oxide
Chlorine	Styrene
Chloroform	Toluene
Ethylene dichloride	Trichloroethylene
Ethylene oxide	Vinyl chloride
HCFC-123	Xylenes

Advantages of the system include;
– Addition of the membrane separation step eliminates the requirement for multi-stage cooling, and often allows condensation to take place at temperatures above the freezing point of water. In many applications, readily available cooling water can be used.
– The membrane separation step also benefits from the driving force provided by compression prior to condensation.
– In many cases, compression of the feed eliminates the need for a vacuum pump on the permeate side of the process.

With combined compression-condensation and membrane separation, it is possible to optimise the three key parameters that affect recovery performance; membrane area, condenser temperature, and the pressure ratio across the membrane. The impact of these

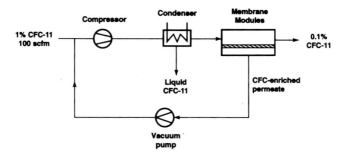

Example of a membrane vapour separation system designed to recover CFC-11 from a 100 scfm air stream containing 1 vol% CFE-11. The process conditions used as the basis of this calculation are as follows: Feed pressure = 72.5 psia, Permeate pressure = 3.6 psia, Pressure ration = 20, Condenser temperature = 20°C, and Membrane area = $80m^2$.

FIGURE 8.

process parameters on the overall recovery of a typical system (Figure 7) is shown in Figure 8. The concentration of CFC-11 in the vented air stream can be lowered by increasing the membrane area, reducing the temperature of the condenser, and by increasing the pressure ratio across the membrane.

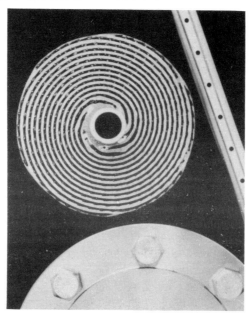

Applications

The membrane VaporSep process can be used to treat a wide range of important industrial vapour streams (Table 15). Ideal applications are those that can be accomplished upstream, at a point where the VOC concentration is maximised and the total air flow is minimised.

Promising applications include the recovery of chlorinated hydrocarbons, chlorofluorocarbons (CFCs) and/or hydrochlorofluorocarbons (HCFCs). These solvents are typically expensive (around US$2 – 8/lb) and difficult to recover using conventional processes, such as carbon adsorption and compression-condensation.

Three representative applications for the technology are illustrated below.

Application	Objective
Recovery of HCFC-123 from a film drying process	Replacement of volatile organics CFCs and III-trichlorethane
Recovery of CFC-12 (R-12) refrigerant	Purge air stream from coolant line in compressor
Recovery of ethylene oxide with CFC-12	Replacement to catalytic oxidation or scrubbing for sterilant effluent

Overall the adoption of the above method is expanding with many thousands of potential applications in hospitals estimated. The advantages of the method include:
- the membrane process is continuous unlike alternative adsorption.
- there is no secondary waste produced.
- low efficiencies – associated with low concentration and low boiling point of condensable components – are avoided.

- regular defrosting of the condenser is not required.
- reduction in costs associated with using low temperatures of condensation.

Polyolefin resin degassing

A pair of closed-loop, membrane systems for separating monomers and other hydrocarbons from nitrogen during polyolefin resin degassing offer resin producers an entirely new approach to dealing with the vent from degassing operations. By recovering valuable monomer, hydrocarbons such as hexane, and nitrogen that would otherwise be lost from the vent, resin producers can save on the cost of raw materials.

An important source of waste in polyolefin manufacture occurs when the solid polymer resin is purged with nitrogen to remove unreacted monomer and other hydrocarbons, such as hexane. The vent from this degassing operation is typically flared to eliminate emissions of hazardous air pollutants. Vapor Separation systems for polyolefin resin degassing, however, provide a recovery alternative to high-cost flaring. In one design option, the system recovers 95-99% of the hexane from a purge stream consisting primarily of hexane and nitrogen. The system offers several benefits to resin producers; recovering the nitrogen and recycling it to the degassing operation; capturing the hexane, which is a hazardous air pollutant; and conserving natural gas by avoiding the need to flare the purge vent.

Based on the value of recovered raw materials, it is said that the system payback period is just over a year of continuous operation.

In another design option, the Vapor Separation system for polyolefin resin degassing recovers valuable monomer (propylene or ethylene) from nitrogen. The residue from the first membrane vapour separation unit, which is 90-95% nitrogen, is recycled to the degassing operation. The permeate from the second membrane vapour separation unit, which is 90–95% monomer, is sent for purification and reuse as feedstock in the polymer

production process. In addition to recovering both monomer and nitrogen, this system also conserves natural gas by eliminating the need for flaring.

The following gives brief details of other applications of VOC recovery using vapour permeation and compression/condensation

Many industrial process chillers use CFC-12 (R-12) as a refrigerant fluid. Since the low-presure side of the compressor operates under a vacuum, air leaks into the chiller. Many low-temperature chillers are equipped with a mechanism that allows the air tro be periodically purged from the coolant line. The CFC-12 lost during purging can be effectively recovered by a VaporSep™ unit, as shown below. The purged CFC-12/air mixture, containing up to 95% CFC-12, is drawn under pressure from the purge vent into the membrane system. The mixture first passes through a dryer to remove water vapor and then into a condenser that cools the vapor/air mixture down to 3°C. Some of the CFC-12 condenses inside the condenser and is returned to the suction side of the existing refrigeration unit compressor. The non-condensed solvent vapor and air mixture passes acros the surface of the membrane, where it is separated into two streams: a permeate stream containing most of the remaining CFC-12 vapor from the condenser, and a vent stream containing a very low concentration of CFC-12. The VaporSep™ system can be designed to recover up to 99.99% of the CFC-12 from the purge vent.

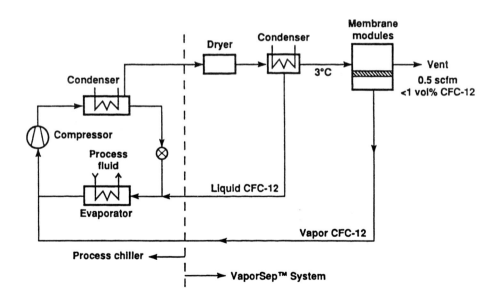

FIGURE 9 – Flow diagram of a membrane VaporSep™ system designed to recover CFC-12 from an industrial chiller purge stream.

Ethylene oxide is a widely used sterilant in both hospitals and industrial applications. Ethylene oxide is flammable and toxic, and it is commonly diluted with a carrrier gas such as CFC-12 or HCFC-124 to reduce the flammability hazard. The most common sterilant gas mixture consists of 12 wt% ethylene oxide and 88 wt% CFC-12. The ethylene oxide vented from the sterilization chamber is currently controlled using catalytic oxidation or a scrubbing technique. Due to their ozone depleting properties, CFC-12 and HCFC-124 will also require control technology. California and Wisconsin already have regulations in place, and other states are expected to follow their lead.

MTR Vapor Sep™ technology c an be used to recover the ethylene oxide/CFC-12 mixture, as shown below. The membrane system is able to recover more than 95% of the sterilant gas mixture, which can then be returned to the supplier for purification and re-blending. Residual ethylene oxide not recovered by the membrane system is catalytically c onverted to carbon dioxide and water.

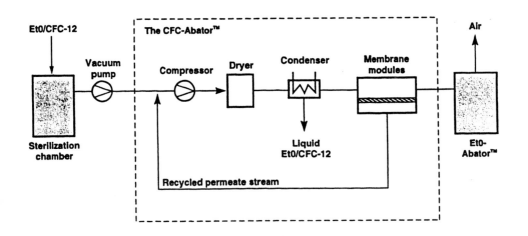

FIGURE 10 – The combination of MTR's membrane process and the Donaldson ethylene oxide abator addresses the recovery requirements of hospital and industrial sterilization applications.

SEPARATION OF ORGANIC VAPOUR/AIR MIXTURES

Industrial coating operations, including film formation and pill coating, rely heavily on the use of volatile organic solvents, such a CFCs (chlorofluorocarbons), methylene chloride and 1, 1, 1-trichlorethane. The production of many of these solvents will be phased out during this decade. The available replacement solvents, such as HCFC-123, are very expensive and subject to stringent environmental emission limits. Therefore, viable conversion of an existing process to the new materials also requires the availability of efficient recovery technology. The MTR membrane vapor recovery process has been demonstrated to be effective for recovery of HCFCs (hydrocholorofluorocarbons), HFCs (hydrofluorocarbons) and perfluorinated compounds.

A VaporSep™ system recovers HCFC-123 vapors from the drying chamber of the film-coating operation. A flow diagram and photograph of the system are shown below. HCFC-123 vapors exit the drying chamber at a concentration of 6.3% in air. The vapor stream is compressed to 125 psig and cooled to $-15°C$. A dryer removes water vapor from the stream to prevent ice formation in the condenser. The condenser vent is routed to the membrane modules. The concentration of HCFC-123 in the resulting vent stream never exceeds 100 ppm. The unit recovers 33 kg (70 pounds) of HCFC-123 per hour.

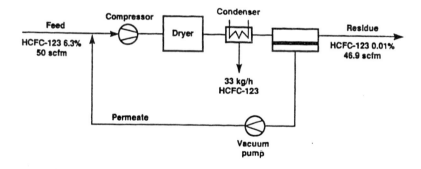

FIGURE 11 – Flow diagram of a VaporSep™ membrane system to recover 99.99% of the HCFC-123 lost during coating and drying operations.

Removal of VOC from Air by Membrane Absorption

Removal of VOC from air through, for example, cross-linked polydimethylsiloxane (PDMS) gas separation membranes is limited to relatively high concentrations (70.5% VOC), because at lower concentrations compressor and vacuum pump costs become prohibitive. The use of selective membranes in conjunction with a absorbing liquid with a high affinity for the VOC, e.g. non-polar hydrocarbon oils or silicon oil, can result in recovery of VOC's to low levels (0.6 mg/m^3). The membrane is a supported PDMS hollow fibre which selectively removes the VOC's whilst retaining the organic phase, which would pass through a porous hydrophobic membrane. The VOC's are concentrated in the absorbing liquid which is regenerated by air stripping. The VOC's are then recovered by condensation and stripping air is recycled to remove residual VOC's.

SECTION 7

Microfiltration

In Industry, a lot of money goes down the drain

Doesn't it?

On the other hand, economic material recovery and multiple usage of water means considerable profit.

Optimum recycling solutions are no longer mere dreams.

Microdyn specialists in modern crossflow microfiltration have the perfect system for the economic, ecological processing of water. using the proven Microdyn modules, both solids and emulsified substances are separated from water and other liquids. In combination with biological processes in fact, even dissolved heavy metals can be reliably segregated.

The advantages are plain to see; less waste and effluent costs; better use of water and materials; ease of compliance with limit restrictions; and no emissions from the closed compact installations. Process controls can be automated.
Is this something you could profit from?
Then why not ring us right now and let us advise you on reducing costs. Because micro-fine filtering pays!

MICRODYN
MODULBAU GMBH & CO KG
Ohder Straße 28 42289 Wuppertal
P.O. Box 24 02 62 42232 Wuppertal
Germany
Tel: 202/60 20 92
Fax: 202/60 30 87

Fairey

ADVANCED CROSSFLOW CERAMIC MEMBRANE TECHNOLOGY

- LOW ENERGY INPUT
- PORE SIZE 0.2-1.2 MICRON
- STAINLESS STEEL MODULES 0.06-6.48 M2

TYPICAL APPLICATIONS

- FERMENTATION BROTH
- OILY WATER SEPARATION
- CHEMICAL WASTE
- MINERAL PROCESSING
- MILK/BEER PROCESSING
- WATER RECOVERY
- LEACHATE TREATMENT

FAIREY INDUSTRIAL CERAMICS
FilleyBrooks, Stone, Staffordshire, ST15 0PU, England
Telephone: +44 1785 813241; Facsimile: +44 1785 818733

MICROFILTRATION

Microfiltration is widely used in industry in both dead end and crossflow modes. Dead-end MF is applied extensively in clarification, sterilisation, in laboratories etc., while the use of cross flow is increasing rapidly. This section details applications in pharmaceutical, food and semi-conductor industries and in waste water treatment. Other applications of microfiltration are discussed in other sections of this book.

Applications of Microfiltration

Pharmaceuticals/Cosmetics – Sterile Filtration: diagnostics, ophthalmics, paraenterals, vaccines Air vents, solvent recovery systems, vaccines, antiobotics, perfumes, vaccines/blood products	***Bulk Chemicals/Petrochemicals*** – Sterile Filtration: solvents and reagènts, inorganic solutions, caustic, aggressive solvents, fatty acids, waxes, polymer fibres and films
Biotechnology/Fine Chemicals – Sterile Filtration: antibiotics, growth media, solvents, acids, tank venting, sterile compressed air, fermented liquid Steam filtration, acid bases, pyrogen removal, chromatography column protection	***Paints/Coatings*** – Painting solutions, hydraulic fluid inks, foam, plating solutions, waste waters ***General Industrial*** – Process water, dyes, treatment of grinding and polishing waters, radwaste dewatering, finely dispersed suspension Retention of activated carbons, recovery of catalysts, recovery of process waters, filtration of process chemicals, solvents, hot or concentrated acids, treatment of aqueous cleaners, Removal of heavy metals as hydroxides, removal of lignin, removal of oil water effluents
Electronics/Nuclear – Sterile Filtration: demineralised water, phot-resits, gas, process solvents, high pressure cryogenics	
Food/Beverage – Sterile Filtration: wine, beer, bottled water, amino acids, brine broths, gases, minerals (soft drinks) sugar solutions, edible oil, syrups, cheese whey clarification, whisky, brandy, vinegar Clarification of cheese whey, defatting and reducing microbial load of milk Clarification of wine, beer, fruit juicesvinegar, bottled water, beet and sugar cane solutions, purification of dextrose streams	
	Beverages – Cold sterile bottling of wine, soft drink manufacutring, cross flow filtration of wine, fruit juices, beer products, sugar solutions
	Concentration – Dyes, pharmacuetical pre-products, finely dispersed suspensions

Membrane Microfiltration – Dead-end

Dead-end microfiltration is primarily concerned with the separation of suspended matter, in the 0.1-10 μm size range, from gases and liquids. The separation is primarily based on size, although adsorption can play a role in some applications. In operation a pressure differential drives the fluid through the membrane and the particles or suspended matter are trapped by the medium and accumulate on the surface. In time the accumulation, either in the pores or on the surface of the membrane, is sufficient to stop or severely restrict the flow such that the medium must be replaced (or cleaned).

The application of dead end filtration are generally in the areas of purification, clarification, sterilisation and analysis. Materials other than membranes are also used in many of these applications, as either surface or depth filters. These materials include paper, fiberglass, nonwoven materials, fabric screens, and other pressure formed particles. In many cases the materials are for coarser filtration of larger particles although in many applications the filter medium perform a similar function and duty to membrane filters. Filter media are discussed in detail in the Handbook of Filters and Filtration.

Dead-end Microfiltration Cartridges

Conventional dead-end filtration centres around the pleated flat membrane in filter cartridges.

The nature of operation results in the build up of filtered suspended solid on the surface as a cake which causes a continuous reduction in flow. It is important to prevent rapid blockage of the membrane filter by using an effective prefiltration – typically using depth filters. the latter filters, in comparison to membrane filters, have a large solid holding

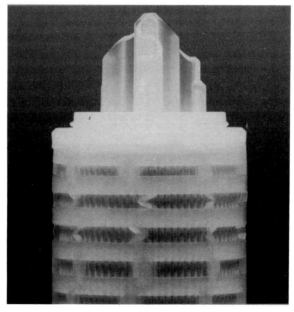

Membrane filter cartridge.

MICROFILTRATION

capacity due to relatively high thickness and internal surface area. It is essential to match the retention characteristics of both types of filter in any application.

Cartridge filter technologies can be classified as either monolayer or multi-layer units depending upon the number of membranes, pore sizes and pore geometries used.

Monolayer Technology

In this cartridge, only a single membrane is used with particles predominately retained on the surface. This therefore results in quite rapid blocking of the membrane pores which generally limits the service life. Additionally the absolute reliability cannot be guaranteed

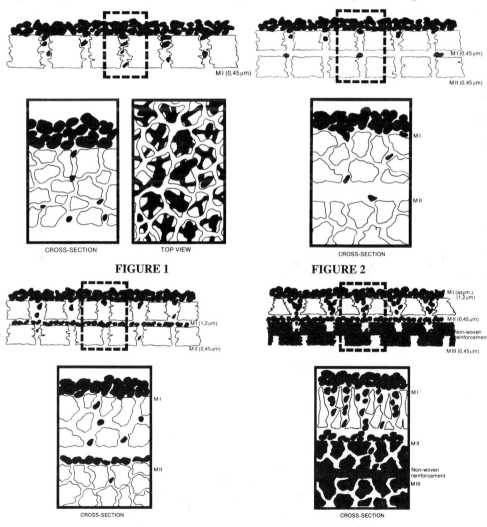

FIGURE 1   **FIGURE 2**

FIGURE 3   **FIGURE 4**

Filter cartridge technologies.

and thus the risk of micro-organism permeation limits applications to non-critical microbial areas.

Multi Layer Technology

More than one membrane layer is used in these cartridges. Constructions vary depending upon types of membrane, pore sizes and pore geometries.
- Cartridges with homogeneous double layers
- Cartridges with heterogeneous double layers
- Cartridges with three layers (hybrids)

Homogeneous double layer cartridges contain two membranes of similar pore size and symmetrical structure normally laminated together. One of the membranes is usually reinforced with non-woven material to render mechanical stability. This style of cartridge gives high microbial retention reliability but have a limited service life due to primary surface retaining most of the particulate matter. Heterogeneous cartridge filters use two membranes of different pore size and symmetrical pore structure. The largest pore size membrane faces the feed and thus the design gives fractional retention of particles and turbid matter on two membrane surfaces. This increases the material retention capacity of the membrane whilst making it less effective than the homogeneous cartridge filter in the retention of micro-organisms, when exposed to very high microbial challenges.

Hybrid cartridge designs contain three membranes. The upper most (feed side) membrane is coarser than the other two and is asymmetric. The membrane is more open to the feed and thus the pronounced asymmetry opens up the internal surface area of the membrane to particulate. Retention of turbid matter is therefore achieved in the depth of the filter and is not confirmed to the surface. This imparts a very high dirt holding capacity to this configuration. The following two membranes are finer, symmetric and of the same pore size and chemical characteristic. The first of these two membrane is primarily responsible for retaining the particulate material penetrating to top asymmetric membrane. The combination of the two membranes gives the high filtration reliability. Suitable selection of materials for the symmetric and asymmetric can give appropriate filtration requirements. Overall the hybrid cartridge is as reliable as the homogeneous cartridge and has a higher capacity for particulate. A comparative example of the hybrid cartridge (0.8 μm + 2 x 0.45 μm) and homogeneous cartridge (2 x 0.45 μm) filtering a 2.5% crude sugar solution demonstrates the increased capacity of the former. At a differential operating pressure of 2.5 bar and an initial flow of 400 dm^{-3} h^{-1}, the double membrane produced 65-100 dm^3 m^{-2} and the hybrid produced 155-200 dm^3 m^{-2} filtration volume in the same time internal of filtration.

The selection of appropriate microfiltration systems depend on the particular applications and the final ultimate goal required by the user. Because of the variety of application to which microfiltration is used a range of microfiltration systems exist. The choice of system depends on which is the desired product, clarified liquid, recovered solid or both, to what degree and tolerance is particulate solid to be removed as specified by regulatory, economic, health, environmental and other factors. To what extent other competing techniques might be used should be considered as well as using a combination of

MICROFILTRATION

technologies, including membranes. The design and selection of a microfiltration system depends on several process parameters (see Chapter 2) relating to the fluid to be filtered, the suspended material to be retained and various system considerations eg pumping requirements. Membrane and module material compatibility are clearly important.

Pharmaceutical Industry

In the pharmaceutical industry a range of filtration and membrane separations are employed to maintain product purity and product quality (See Table 2). For most high column filtration applications depth, surface and membrane filters have complementary characteristics. Depth filters allow the removal of the bulk of particles whilst surface filtration comines relatively high dirt-holding capacity with a defined retention behaviour.

TABLE 1 – Characteristics of Different Cartridge Filters

	Depth Filter	*Surface Filter*	*Screen/Membrane Filter*
Structure	A fibrous, grandular or sintered matrix that produces a random porous structure. Particles become trapped in the tortuous network of flow channels. The principle retention mechanisms are random adsorption and mechanical entrapment throughout the depth of the matrix.	A multilayer filter medium constructed of glass or polymeric microfibers. Particles larger than the spaces within the filter matrix are retained. primarily on the surface. Smaller particles tend to be trapped within the matrix, giving the surface filter properties of both a screen and a depth filter.	Screen filters may be thought of as a geometrically regular porous matrix. Particles are retained on the surface primarily by a sieving or size exclusion machanism. All particles and micro-organisms larger than the pore size will be retained.
Construction	Filtration media may be wound cotton, polypropylene, rayon cellulose, fibre-glass, sintered metal, porcelain or diatomaceous earth.	Typically, surface filters are constructed of polypropylene, cellulose/resin bonded paper, or fibre glass/paper.	For critical submicron and macromolecule separations, screen filters are constructed of cast polymeric membranes.
Performance	Depth filters characteristically exhibit high dirt-holding capacity and will also retain a large percentage of contaminants smaller than their pore size rating. They are generally less expensive than surface or screen filters.	Due to their multiple layers of pleated media, surface filters exhibit a high dirt-holding capacity. As a result of their "controlled" pore structure, they provide more predictable filtration than depth filters. They are generally less expensive than screen (membrane filters).	Predetermined, controllable pore size limits the largest particle than can pass through a screen filter. Retention efficiency is independent of flow rate and pressure differential. Membrane filters exhibit low hold-up volume and are non-fibre releasing.

The membrane filters give complete removal of particles (including microorganisms) to a pre-established size. By suitably combining either a depth filter or a surface filter upstream of a membrane filter, for final filtration, an optimum performance of retention efficiency, high dirt capacity and process economics is achievable.

TABLE 2 – Filtration Applications in the Pharmaceutical Industry

Clarification/Prefiltration	Sterile/Final Filtration
Antibiotics	Blood fractions
Blood fractions	Facilities water
Compressed gas	Facilities gas
Diluents	Fermentor exhaust
DI Pre/Post treatment	Fermentor feed air
Facilities water	Filling machine filters
LVP	LVPs
Make-up water	Ophthalmics
Ophthalmic	Serum
Orals	Solvent filtration
Reagents, buffers	SVPs
Resin traps	Sterile products tank vent
Rinse fluids	Tissue culture media
Serum fractions	Veterinary pharmaceuticals
SVP	WFI tank
Synthesis	
Tissue culture media	
Topicals	
Vaccines	
Vial Washers	

The range of processes include antibiotic processing, production of biologically-based diagnostic or therapeutic proteins, facilities filtration of various fluids, fermentations, large volume and small volume parenterals, ophthalmics preparations.

Antibiotics Processing

Antiobiotics are produced by fermentation processes, after which they are generally purified by crystallisation in solvents. Antibiotics processing may include a variety of subprocessing steps (see Fig 5). The common process is the fermentation and its subsequent bulk contamination removal steps. Filtration of solvents involved in the primary clarification steps takes place prior to further purification.

After clarification, the process may involve one or several of the following subprocesses: solvent extraction, activated carbon, ion exchange, column separations and high pressure liquid chromatography. Typically, both the solvents employed in the extraction phases and the solvent/antibiotic mix itself are subjected to prefiltration steps at approximately 2.5 µm, 1.2 µm or 0.6 µm levels to further reduce contamination.

Solvent prefilters must remove large amounts of particulate and micro-conntamination from a process stream requiring broad chemical compatibility. Low pressure drops and high flow rates couple with moderately high retention efficiencies enable the contamination to be retained to maximise cartridge service life.

Millipore S.A.
39, Route Industrielle de la Hardt
B.P. 116
67124 Molsheim
France

Tel: [+33] 88 38 90 00

Fax: [+33] 88 38 91 93

Millipore Corporation, a multinational company leading the worldwide separation industry, develops, manufactures and sells filters, filter housings and turn-key custom filtration systems in a wide range of industries, life science research and health care laboratories.

Millipore Corporation's BioProcess and Microelectronics Divisions supply the Pharmaceutical, Food and Beverage and Microelectronics industries need for sophisticated filtration technology products and validation support for sterility assurance in aseptic processing, tank venting, fermentor air filtration, virus removal, economic fluid clarification, protein concentration, cell harvesting, chemical filtration, gas filtration and purification and photochemical dispense, and other process separations applications.

International Contacts:

Austria: Tel (01) 877 8926 Fax: (01) 877 1654

Belgium - Luxemburg: Tel: (02) 726 8840 Fax: (02) 726 9884

Denmark: Tel: 46-59 00 23 Fax: 46-59 13 14

Finland; Baltic Republics: Tel: 90-804 51 10 Fax: 90-859 66 16

France: Tel: (1) 3012 7000 Fax: (1) 3012 7180

Germany: Tel: (06196) 494-0 Fax: (06196) 482237

Italy: Tel: (02) 25078-1 Fax: (02) 265032-4

Norway: Tel: 02-67 8253 Fax: 02-68 5315

Spain: Tel: (91) 729 0300 Fax: (91) 729 2909

Sweden: Tel: 08-628 69 60 Fax: 08-628 64 57

Switzerland: Tel: (01) 9083060 Fax: (01) 9083080

The Netherlands: Tel: (07650) 22000 Fax: (07650) 22436

UK/Ireland: Tel: 01923 816 375 Fax: 01923 818 297

USA + R.O.W: Tel: 617 275 9200 Fax: 617 533 8630

domnick hunter
PROCESS DIVISION

Your Partner in Critical Filtration

domnick hunter limited	domnick hunter	domnick hunter	domnick hunter gmbh	domnick hunter Iberica	domnick hunter inc
England	Scandinavia	France S.A.	Germany	Spain	USA
Tel: (44) 0191 410 5121	Tel: (45) 47 107775	Tel: France 74 62 34 51	Tel: (49) 2 11583660	Tel: (34) 3351 4807	Tel: (1) 704 568 8788
Telefax: (44) 0191 410 5312	Telefax: (45) 42 173332	Telefax: 74 62 35 44	Telefax: (49) 2 15139 5779	Telefax: (34) 3351 7102	Telefax: (1) 704 568 8787

MICROFILTRATION

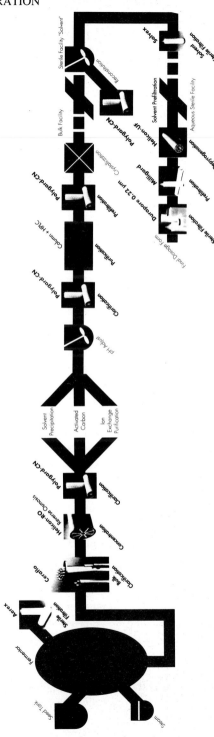

FIGURE 5 – Typical flow diagram of antiobiotic production

Solvent sterile filters must withstand exposure to a variety of different solvents for long periods of time. Solvent sterile filters are pre-extracted to eliminate contribution of extractables to the process stream. Qualified, validated removal of organisms at 0.22 µm in the presence of these solvents is critical. The filters are generally made from hydrophobic materials such as polypropylene, PVDF, PTFE.

The processing of the bulk final solution in the solvent based process involves the use of sterilising membrane filters (0.22 µm). Other filter applications include sterile hydrophobic tank vents (0.22 µm) and stainless steel steam filters (2.0 µm). *Sterilising tank vents* must provide low pressure drops and high flow rates during process conditions ranging from low pressures/ambient temperatures to high pressure/elevated temperatures. Particulate and micro contamination must be consistently and effectively removed.

Preparation of the aqueous final dosage product involves sterile filtration (0.22 µm). Prefiltration is important because it reduces the risk of premature plugging of the final filter.

Biologicals

The production of biologically-based diagnostic or therapeutic proteins from mammalian cells is a relatively new technology. The wide range of proteins that can be produced requires a variety of separation methods. Figure 6 shows a general processing scheme for the production of biologicals, highlighting characteristics of production that are widely shared in many processes. To support proper growth all cells have important nutrient requirements and thus all fluids which come in contact with cells in a reactor or fermentation vessel must be sterile and free from any potentially toxic agents or pyrogens.

In addition to liquid nutrients, growing cells require control of pH and O_2 concentrations within the reactor vessel. Thus, all gases entering the system must also be sterile. To separate the biomass from the desired product clarification techniques including filtration and/or centrifugation are used. In some cases, the desired product is intracellular and cell disruption techniques must first be employed. Isolation techniques such as extraction and adsorption are commonly employed prior to purification procedures. Large-scale purification procedures generally rely on precipitation, chromatography, and filtration.

The last step in the process is contamination removal. Dedicated contaminant removal steps are utilised as well as previous purification steps which may give some removal as well. Sterility of biological parenterals is crucial. Removal of bacteria and other biological associated contaminants such as viruses is necessary to meet stringent world wide regulatory requirements.

The range of application include:

Sterilising grade filters must provide low protein binding, low extractables and low particle shedding, as well as high flow rates to insure a sterile fluid stream. Sterile filters are utilised both upstream to ensure fermentation tank/cell culture sterility, and prior to the final filling point as a final bacterial removal step.

Sterilising tank vents must provide high flow rates, ensure protection of the environment from pathogens, prevent bacterial growth through and maintain aerobic integrity. These filters must be capable of sterilising gases in both the forward and reverse direction.

Prefilters must be capable of providing high retention efficiency and adsorptive

MICROFILTRATION

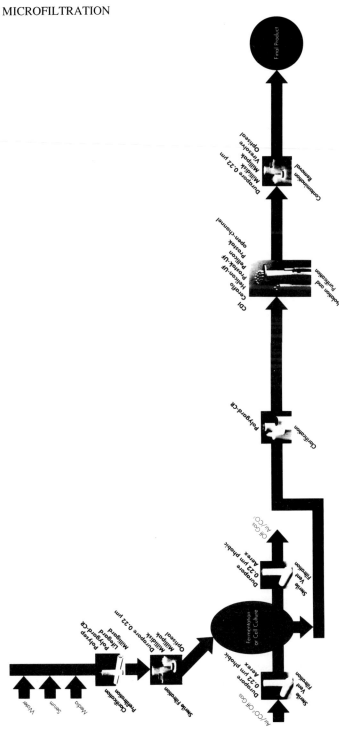

FIGURE 6 – Processing schemes in the production of biologicals.

capacity. They should operate at low differential pressures, while achieving maximum protection of final filters.

Tangential flow filtration is also utilised for a number of operations during the purification of biologicals and include microporous systems, ultrafiltration, continuous de-ionisation and virus removal. Microporous systems provide better yield and purity of filtrates than centrifuges in an enclosed system that require no handling of solids or pastes. Applications of microporous systems include: cell harvest, perfusion of continuous cell culture vessels, clarification of biological, and pre-column clarification.

Ultrafiltration (UF) systems improve yield and preserve activity while improving productivity of allied purification technology like chromatography. Applications for UF systems include: concentration of biological, desalting and buffer exchange, an depyrogenation of media and buffers.

Process continuous de ionisation (CDI) systems remove salts to extremely low levels without chemical regeneration of dilution, and are less expensive to operate than electrodialysis or mixed-bed ion exchange. Applications for CDI systems include: chaotrope (eg urea) desalting and peptide desalting.

Virus removal filters must be able to achieve predictable, high resolution removal of viral contaminants. The system must be able to achieve high protein recoveries with concomitant high viral log reduction values. The mechanism or removal must utilise a size exclusion technology in order to properly validate mammalian viruses removal.

Facilities Filtration

The need for high purity gases and water is critical in many industries. The supply of these "facilities" free of particulates, bacteria and virses relies heavily on microfiltration in conjunction with other processes. Facilities filtration may include a variety of fluids, but commonly refers to the filtration of processes for water, compressed air, and nitrogen. The filtration of these fluids takes place in two distinct areas, the central distribution systems and points of use.

In the central distribution systems microfiltration is one of several unit operations which typically includes gravitational settling, ion exchange, and chlorine removal (in carbon beds). A central air system will include an intake filter, compressor, receiver, dryer, and an oil removal step (coalescing filter). Point of use filters, for both water and air/nitrogen, are typically 0.22 μm sterilising-grade filters, although they are sometimes preceded by prefilters if contamination risk exist between the central distribution system and points of use. Filters in a facilities gas application are expected to provide high flow rates (low pressure drops) and a long service life, while being subjected to high pressures 6.9 bar and, sometimes, high temperatures. Aerosolised bacteria, viruses, and particulate must be retained within the microfiltration system, sometimes for several weeks, without allowing grow-through to occur. Filters must also retain their hydrophobicity and not wet out.

A typical example where stringent water quality requirements must be met is the supply of de-ionised water in the electronics industry. The electronic components, wafers are rinsed many times with copious quantities of water is critical and must conform to the current semiconductor equipment manufacturing institute (SEMI) guidelines. The production of the de-ionised water in the electronics industry requires many stages of treatment. These are generally put into three sections; primary, secondary and tertiary. The primary and secondary stages provide the deionised water for use in the recirculating system of the deionised water supply system.

The function of the tertiary treatment is to remove all the contaminants from the recirculating water. This system includes, mixed-bed ion exchangers, UV sterilisation and sanitisation with hydrogen peroxide or ozone. Two stage microfiltration is incorporated to remove, any break down products of ion-exchange resins, bacteria still resident in equipment and pyrogens produced by decomposition of bacteria in sanitisers and any particulates generated in the process equipment. The microfilters must be virtually absolutely stable in de-ionised water and not shed particulate material. The POU filters associated with the electronic component de-ionised water recirculation loop are used to guarantee final water quality. These filters must be stable and not shed particulates under variable and pulsed conditions of flow associated with there operation.

Filters used in the electronics industry (and others) de-ionised water supplies are typically cartridge configuration available from several manufacturers. The material components vary from one manufacturer to the other (see Table 4) but all tend to be of a standard size approximately 10 in high and 2.8 inch diameter. For other pore water filtration applications some variation in materials will occur. Alternative stacked disk modules are available from certain manufacturers.

Filters in the central water systems may be exposed to a variety of contaminants. The filters must reduce the bioburden and contaminant load, while maintain a low pressure drop for extended periods of time. There also may be 0.22 μm or 0.45 μm membrane filters in the recirculation loops of deionised and distilled water systems. Filters in the deionised water system must maintain low pressure drops throughout long throughput cycles. They may be exposed to high particulate loads from the resin beds.

As the quality of incoming facilities fluids varies widely, the good design of the microfiltration system is critical for both efficiency and economics. Attention should be paid particulary to clarification and prefiltration to avoid costly oversized final filters. The housing design can play an important role in maintaining minimal pressure loses.

In the case of a Water for Injection (WFI) systems maintained at elevated temperatures, up to 80°C, oxidation and hydrolysis are of particular concern, not only for the filter media, but also for the support materials and cartridge components.

Fermentation

Fermentation is used in the production of antibiotics, hormones, enzymes, amino acid, blood substitutes and alcohol in the range of 10-litre seed fermenters to as large as 400,000 litres in industrial fermentation. During the fermentation process, air is in contact with the fermentation fluid for agitation, liquid transferences or sparging. Thus the fermentation inlet and outlet air is sterile-filtered to eliminate airborne microorganic contaminants from the fermentor or prevent fermented organisms exiting to the environment.

Good sizing of a fermentation air system is extremely important to be cost effective – a typical allowable pressure drop is < 140 mbar. In gas streams, system sizing is based on flow rate and differential pressure, not blockage, and the lower the pressure drop, the better. The inlet and outlet of the fermenter are both sterile filtered. Many manufacturers are more concerned with phase outbreaks in the manufacturing area or growth-through of microorganisms if the cartridges inadvertently wet out during the fermentation cycle.

Filtration applications include:

Bioreactors, fermentors used in biotechnology, are usually smaller than their industrial counterparts and have great available air pressure. Due to regulations (eg FDA and EPA), manufacturers must insure that microogranism do not exit to the environment. A liquid bacterial retention claim on the sterilising cartridge is necessary. The cartridge may be reused, but it is generally changed after each batch. Steaming-in-place is performed under controlled conditions.

The larger industrial fermentors require high air flow rates and low pressure drops. Cartridges will be installed in the housing, steamed-in-place up to 100 times and integrity tested after six months to one year.

The hydrophobic cartridges require high oxidative stability since temperatures may reach 80°C for extended periods. Phase retention is critical since an outbreak will shut the facility down and therefore, cartridges tested for aerosolised bacteriophage retention are recommended. A bacterial grow-through claim is desirable. A prefilter may be used to reduce steam contaminats prior to final filtration.

Injectable Drugs (Parenterals)

The application of MF in the preparation and processing of parenterals (injectable drugs) and their constituents is for three primary aims:

- Final sterilisation – total removal of bacterial from heat labile products
- Reduction of bacterial burden prior to final sterilisation by autoclave
- Removal of particles (organic and inogranic), from parental solutions, their consitutent feedstocks and any aerosols generated as part of their processing.

The sterilisation of drugs or the product of sterile drugs is a prime requirement in the pharmaceutical industry. Sterile drugs are processed by either aspectic processing or terminal sterilisation. Aseptic sterilisation is used for those drugs which cannot be subjected to terminal sterilisation using steam or an autocalve. Aseptic processing requires sterilisation in all steps of the drug processing:-

- sterilisation of the containers
- sterilisation of drugs
- filling the containers
- closing with presterilised closures

The only accpetable and practical method of production and processing of the sterile drugs is by filtration.

In terminal sterilisation filtration is used to reduce the bioburden, and for pyrogen management, upstream of the container filling and final sterilisation of the product.

TABLE 5 – Pharmaceutical applications of microfiltration.

Definition	Process Fluid	Product Quality	Operating Constraints
WFI Water for injection	Water	Low pyrogen Low particles	High discharge flow rates Recirculation loops: frequent changeout
Vents Storage tank "breathe"	Gas Water vapor	Sterile product tanks Aerosol retentive filters	Pressure equilibrium, post steaming: Post steam during filing and emptying, environmental fluctuations. High gas flow rate
LVPs Parenterals: 100- to 1000- mL quantities	Large volume dextrose – particles – variable Amino acids Emulsions: – high viscosity – low viscosity	Low particle load Low pyrogen load Low bacteria count (<1/mL)	Time ≤8-16 h High throughput: continuous operation No downtime low-value added products Operations done in sterile area Variable pressure feed pump Filling machine pressure derived from centrifugal pump
SVPs Parenterals: <100 mL	Small volume Some: quarantined sugar particles	Low bacteria Low pyrogen Some: sterile	Time ≤8 h High throughout: batch operation Only final sterilization and filling performed in sterile area Batch operation
Fermentor Vent	Offgas	Sterile	High air flow rates. Micro-organism containment. Warm temperatures
Air/oxygen	Gas	Sterile Particle free	High flow rates Presterilized lines
Nutrient feed	Aqueous Particles Serum	Sterile Particle free	Continuous feed up to weeks
Liquid product	Nutrient broth Multicomponent Cell debris Emulsions Antifoam agents	Sterile Particle free Low pyrogen	Time: rapid processing
Antibiotics Viral/bacterial antigens	High solids Aqueous in fermentor, crystallized in solvents	Sterile Low pyrogen Low particles	Crystal particles Aggressive solvents High pyrogen adsorption. Manufacture in nonsterile area. Fast to minimize pyrogens
Recombinant proteins Therapeutic proteins	Aqueous Highly purified	Sterile Low pyrogen Low particles High product activity recovery	Low protein binding; low losses High throughput Nonsterile area
Serum Animal/human serum	Aqueous Multicomponent Colloids High particles Proteinaceous gels Bacteria Cell debris Mycoplasma Highly variable	Sterile In cell culture applications: high cell propogation Low pyrogens Low particles No mycoplasma (FCS)[a]	Low pressure Cool: 10 to 15°C Constant volume flow operation Low protein binding Low losses FCS[a] = fetal calf serum

LVP Filtration

The ingredients in typical LVPs are water, dextrose (5% to 50%), amino acids, salts and occasionally viscous, Total Parenteral Nutrition (TPN) components. Final products include IV suspension bags and Continuous Ambulatory Peritoneal Dialysis (CAPD) exchange solutions. Large Volume Parenterals (LVPs) are different to other pharmaceutical products in that 0.45 µm filtration is used for particle removal prior to terminal heat sterilisation. There is a trend toward filtration to 0.22 µm in the LVP industry. LVPs are single unit doses of volumes greater than 100 cc and as large as 3-5 litres. Batch sizes are in tens of thousands of gallons, multiple filling lines may work in parallel and thus LVP plants operate with very large volumes per day.

In LVP manufacture the materials are formulated and prefiltration and final filtration reduce the particle load. Prefiltration removes colloidal or hard particle from the dextrose in solution. The final filter is a 0.45 µm cartridge, with or without a prefilter. Occasionally two final filters in series will be used. LVP filters are expected to provide high consistent flow rates to insure that all product containers are filled accurately and consistently. LVP manufactures operate at extremely high differential pressures and push the limits of the hydraulic strength of a pleated cartridge. A Start-up of greater than 5.5 bar is common, and pressure peaks to 6.9 bar occur. Depending on the percent of dextrose in a solution, the cartridge may contain a cellulose ester prefilter to extend the throughput of the cartridge. A cartridge without a prefilter will decrease the differential pressure across the cartridge and increase the flow rate with cleaner solutions.

Since LVPs are not viewed as high-value-added products in the market, manufacturing practices strict cost containment. After terminal sterilisation in an autoclave, product is tested and released. A product bubble point integrity test may be substituted for a water bubble point. This eliminates the breakdown of the filter housing for an in-process integrity test and the need for water flushing, and saves operator time. The product bubble point can be determined experimentally, but depends upon matrix criteria, such as solutes, surface active ingredients and the formulation of the batch.

SVP Filtration

Small Volume Parenterals include a wide variety of drug products, both traditional and bio-engineered. They are typically less than 20 ml and packaged as small vials, prefilled syringes, amplouels or lyophilized powders. Many SVPs are heat labile and therefore packaged using aseptic processing,

There are three basic areas for filtration in SVP processing: sterile filtration, prefiltration and facilities filtration. Facilities include production of WFI, sterile filtration of air or nitrogen.

The compounded SVP is pumped by peristaltic pumps typically, to the holding tank via two prefilters and a final filter. In this tank the product can be quarantined while the quality is assessed.

Prefiltration of raw materials is used to reduce bioburden and to remove particles which can prematurely plug final filters. Prefilters are available in a wide range of pore sizes to efficiently filter any solution. For solutions with high particle or colloid content, more than one prefilter in series is often used. The purpose of prefilteration is to protect the more

MICROFILTRATION

costly final filter and to prevent plugging, which can affect fill rates or can cause a shutdown to replace filters. Because each fluid is different, a prefiltration system must be designed for each fluids stream. Some of the fluid characteristics to consider when designing a prefiltration train are: particle size distribution, particle load, chemical compatibility and viscosity

Prefilter characteristics important in the SVP process are:
- flow rates
- batch sizes
- low extractables
- steam sterilisable
- high retention
- non-pyrogenic

Typically, raw materials used to produce SVPs are of high quality, therefore making high retention efficiency more important than dirt-holding capacity.

Sterile filtration is typically with 0.22 µm rating and can be employed after compounding or at the filling head. For increased sterility assurance, 0.22 µm filters can be used at both locations. Sterile filtration of air or nitrogen is used for tank vents, lyophilizer vents, and for drying in vial and stopper washes. The purpose of the sterilise filtration is to remove all bacteria without altering the efficacy or concentration of the drug product. It is also important that the filter holdup of the product is a minimum, when it is of high value as many biotech products are. Characterisitcs of sterilising filters are:
- retention of *pseudonomas diminuta*
- steam sterilisable
- low extractables
- non-shedding
- integrity testable
- low protein binding
- high flow rates
- non-pyrogenic

Alternative strategies may use autoclaving for terminal sterilisation followed by filtration with a 0.45 µm filter.

Filling machine filters are specially designed to fit on a filling head to remove bacteria and particles just before the product is dispensed into the container. A good filling machine filter will have all the characterisitcs of any sterilising filter plus a rigid support system to prevent filter media flexing. Filling machines subject the filter to pulsed flow which can cause the filter to flex, releasing particles and causing after drip and dispense volume problems.

Ophthalmic Filtration

Ophthalmic preparations for use in or on the eye must meet general regulatory requirements for sterility, isotonicity and neutral pH. Applications of ophthalmic products range from contact lens cleaning and storage solutions to washes used during eye surgey and products for the treatment of eye disease or inflammation. The major component of most ophthalmics is sterile water, salt buffers such as sodium chloride, petrolatum for the production of ointments and thixotropic polymeric substances used as lubricating agents are present. Heat-sensitive ophthalmic components are sterilised using total aseptic processing via sterile 0.22 μm filtration or by the addition of sterile componets to a bulk solution. A key ingredient in ophthalmic ointments is petrolatum, which is extremely viscous and must be filtered at 70-90C in order to pass through a sterilising filter. Benzylalkonium chloride is a common preservative for ophthalmic preparations. Since BAK binds to the sterilising membrane, various alternatives to minimise binding such as preflushing, smaller surface area or recirculation must be taken.

Product batch sizes range from 200 litres to 40,000 litres, depending on the type of container to be filled. The production of ophthalmics operates at differential pressures of between 2.07-2.76 bar, sometimes across a single ten-inch cartridge. The filling operation vary from one-day fills to multi-day compounding.

Water for Injection (WFI) is compounded with water-soluble ingredients and filtered through a 0.22 μm filter with a hydrophoic 0.22 μm tank vent. Some producers use a 0.45 μm prefilter to acheive a log reduction in microorganisms.

Filtration of Semiconductor Process Chemicals

Several procedures used to manufacture semiconductor microcircuits are very sensitive to contamination by particulate leading to a loss in product yield. Sources of particulate are varied and include chemicals used in wafer cleaning. The removal of particulates from these chemicals can be achieved by filtration with small pore size membranes. Operation requires that the filters experience minimal hydraulic shock due to large flow rate changes, which would result in particulate shedding and the release of previously captured particles. In practice filters can be operated with a stabilised flow.

The chemicals used to clean wafers are typically aggressive and filters are usually made from PTFE. A particularly difficult semiconductor process fluid to filter is 96% sulphuric acid. Typical PTFE membranes used in this application are made by stretching sintered PTFE particles. Alternative membranes are dual asymmetric materials which has a smaller size layer sandwiched between two open layers. The effective pore size of the thin layer is 0.05 μm, but the asymmetry means that pressure drop characteristics approach conventional stretched 0.1 μm filters. Asymmetric filters are claimed to provide lower particle concentrations than conventional filters in both steady flow and during conditions of fluctuating flow.

The semi conductor industry uses chemicals for several purposes to clean the wafer surfaces, to etch circuits, to remove unwanted materials in processing and also as photoresits. The photoresits are radiation sensitive polymer which are applied to wafer surfaces to impart the desired circuit features with the appropriate processing. The materials typically have high viscosities in the range of 10 to 100 cp but can be as high as 1000 cp with certain high molecular weight polymer solutions. These high viscosities make photoresits more difficult to filter than other chemicals.

In general filtration of semi conductor chemicals is carried out both by the chemical manufacturers to appropriate semi conductor manufacturer specifications (SEMI) and by the semi conductor industry after storage and at point of use.

For critical filtration of aggressive chemicals in the semiconductor industry cartridge

MICROFILTRATION

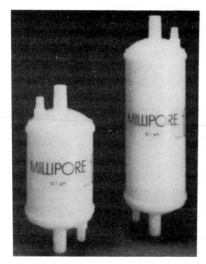

FIGURE 9 – Disposal filters for a chemical filtration.

filters with Teflon or PVDF microporous membranes are recommended. The use of PFA (Perfluoralkoxy – housing makes such filters ideal for many aggressive chemicals, ozone, solvents, hot and concentrated acids. Typical cartridges are 4 inch and 10 inch in length, 2.75 OD offering 0.84 m^2 of filtration area. Membrane pore size ratings vary from 0.05, 0.1 and 0.2 μm which are supported on a PTFE non woven binder free material.

For point of use filtration either pleated or stacked disc membrane cartridges are used. In the case of all the filters the type of chemicals to be filtered will dictate the material(s) of construction to the all the components. Some semi conductor manufacturers have a preference for the use of fully disposable filters which includes the housing. This filter is made from either ethylene-chlorotrifluoroethylene polymer (ECTFE) with ECTFE stacked discs and a PTFE (0.2 μm) filters of form all PTFE.

For the filtration of photoresits, pleated cartridge filters, with nylon of PTFE membranes and stainless stell housings are used at the source of manufacture. The POU filtration of photoresits is generally performed with all PTFE filters. An all Teflon pump and filter assembly (consisting of stacked disc membranes is available for the POU supply of photoresits see Fig 10).

Sludge Dewatering with Tubular Membrane Pressing

The treatment of sewage and the chemical coagulation of surface waters are two processes which result in sludges. Satisfactory disposal of such sludges is clearly of world wide concern and the scale of the problem is continuously increasing. Estimates of sewage sludge disposal requirements in the UK alone are greater than 30 million tonnes, equivalent to 1.2 million tonnes of dry solid. Although 30% of this sludge is disposed of at sea, this practice is to be phased out. Water supply works produce vast amounts of liquid sludge effluent which is part disposed of as solid waste or in many cases dewatered in lagoons, prior to discharge of supernatant to local water courses. This is often inadequate to achieve required discharge levels.

FIGURE 10

The common use of polyelectrolytes in sludge treatments makes the recovered water unsuitable for drinking and thus recycling. Recycling of waters also leads to concerns over recycling of *Cryptosporidium* occysts. Overall therefore with increasing demand, increasing environmental pressure and legislation coupled with increasing costs of landfill, efficient methods of dewatering sewage or water works sludge are of great importance.

Traditional methods of sludge dewatering rely on barrier filtration using, for example, large plate and frame filter presses. Alternatively centrifugation is used, although it is an expensive and relatively complex procedure.

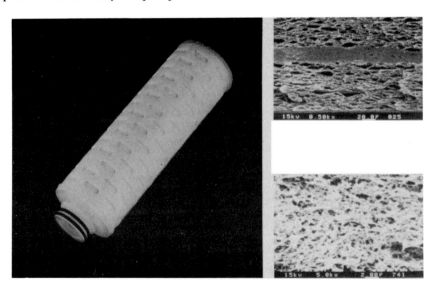

FIGURE 11 – Cartridge and membrane filters.

MICROFILTRATION

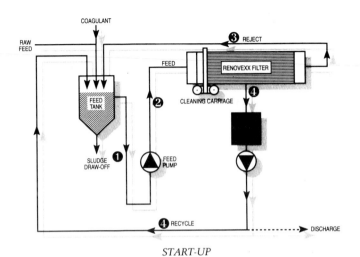

START-UP

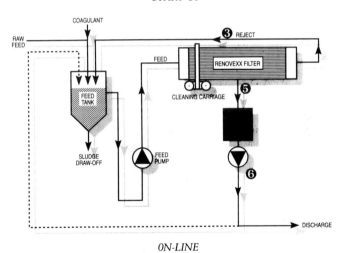

ON-LINE

FIGURE 12 - Tube processing dewatering process.

An effective membrane separation procedure for dewatering sludges uses an array of flexible woven polyester tubes, typically 25 mm in diameter. The array can contain up to thirty tubes woven into a single fabric which is suspended vertically, with the tubes horizontally aligned. Sludge is pumped to an inlet manifold through the inside of the tubes and out through the outlet manifold (see Fig 12). An initial period of cross-flow filtration takes place and a layer of solid forms on the inside of the tubes. Retentate is returned back to the feed tank. After a short time the outlet valve is closed and the tubes are operated in a dead-end mode. The solid thus builds up on the inside of the membrane as the permeate filters through. The permeate is collected underneath the tubular array and returned to the works or disposed of suitably, eg by discharge to surface waters.

After a suitable time, when a relatively thick annular cake has formed on the inside of the tube a decaking cycle is commenced. The outlet valve is opened and simultaneously a flushing fluid is pumped along the tubes and a pair of cleaning rollers, mounted externally on a carriage, is traversed along the array of tubes. These rollers gently squeeze the tube which is sufficient to break the sludge cake off the membrane wall. The compression and flow produces a venturi effect which through the increased velocity scours flakes of solid from the membrane. The flushing fluid is usually the untreated sludge. The resulting mixture of solid and flushing fluid is phase separated, on a wire screen or open-texture conveyer, excess fluid being returned to the sludge feed tank.

Typical dewatering cycle times are between 20 to 40 minutes for waterwork sludges and operating pressures are in the range of 200 – 400 kPa. Feed solid content is typically 1-2% solids and cake solid content is up to 17% w/w before drying. The method performs well on a variety of sludges form dissolved air flotation and from sedimentation processes used for drinking water treatment.

There are several applications of this technology which include the removal of precipitated metals (hydrated metal oxides, sulphides) from industrial effluents; processing of fruit juices and removal of oil and metals fines from industrial waste water recycle.

MICROFILTRATION

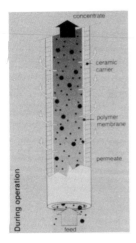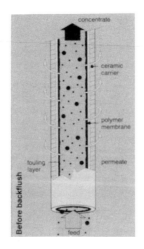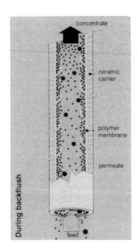

Membrane pack.

Removal of Heavy Metals from Industrial Waste Waters by Microfiltration

A wide range of waste waters, containing dissolved metals are produced in manufacturing processes. Metalliferous waste waters are generated from the production of a wide range of articles such as central heating boilers, washing machines, electronic components, bathroom and kitchen taps etc., and from zinc phosphating, battery and mercury recycling, thermal galvanising etc.

A process which combines a pre-precipitation stage, for formation of metal hydroxide flakes, with a cross-flow membrane filtration can reach discharge levels approaching 0.05 mg dm^{-3} of dissolved solid. The metals to be removed are generally (partially) dissolved in the waste water and by adding caustic solution (lime or sodium carbonate) they are converted into metal hydroxide flakes in a flake formation chamber in approximately 5 minutes. The flake suspension is then pumped under pressure through a cross-flow membrane module consisting of tubular membranes. Feed flows inside the tubes at

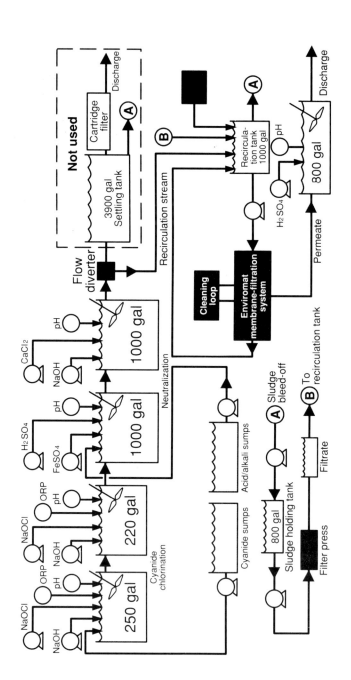

FIGURE 13– Plating – waste water treatment and microfiltration system.

velocities of 4 m s^{-1}, which maintains the flakes in suspension preventing a layer forming on the membrane. A typical membrane pack consists of 37 tubes 1.8 m in length and 14.5 mm inside diameter giving an area of 3 m^2/module. The membrane is a polymer attached to the inner surface of ceramic membrane support tubes. Sealing is by standard EPDM rubber.

Membrane filtration is performed with permeate back flushing which enables flux rates of 0.5 m^3/m^2/h to be achieved depending on the membrane and the medium (presence of colloids, oils, silicates etc). The use of built-in recirculation pumps in each module adds to energy efficiency by minimising pressure losses in external pipes and bends. Modules can be mounted in series on a single common manifold/base plate to enable a gradual concentration of the suspension.

Effluents generated from printed circuit board facilities contain metals such as Cu, Pb, Ni and Sn. Treatment of the solutions is by precipitation of the metal from chemically pre-treated solutions (see Fig 13). Cross-flow filtration of the metal precipitate suspensions is a viable alternative to other filtrations based on pre-coat filters. The process concentrates the precipitate into a slurry which is drawn off from the recirculation loop and processed in a filter press.

Microfiltration in the Food Industry

Filtration processes are indispensable techniques in the food and beverage industry, in many applications where high product quality and long shelf life is required. It is estimated that over 1.3 10^5 m^2 of membranes area for MF and UF is in use world wide in the beverage industry. The two most important filtration processes in the beverage industry are

- dynamic or cross-flow filtration
- static or dead-end filtration

Static filtration processes generally use kiselgur filtration, sheet filtration and membrane filtration. Membrane filters are either multi-layer membranes or uniform or heterogeneous porosities.

As a general guide static filtration processes are broken down into the following areas

- High turbid matter loads (> 10 EBC) – kieselguhr or sheet filtration for coarse filtration to achieve for example pre-clarification of the product (> 20 m size)
- Moderate turbid matter loads (2 > EBC > 0.3) – fine filtration using all three mentioned filtration method can be considered
- Low turbidity (< 0.4 EBC) membrane microfiltration

There is considerable overlap in the areas due to the wide variation of colloidal material and other properties of the various products. Various grades of the three filter medium can also be selected.

The use of microfiltration is generally for the retention of cellular components, micro-organism and other solids from alcoholic or non-alcoholic beverages ie clarification and sterilisation. A wide range of materials are processed including milk, beer, wine, whiskies, soft drinks, potable water, syrups, edible oils, and vinegar.

Microfiltration of Milk

The main applications of microfiltration of milk are for fat separation, bacteria removal and concentration of caseinate. Membranes with a narrow size distribution which are uniform in the range of 0.3 to 1.0 μm will separate fat and bacteria from the other milk components. Membranes are susceptible to fouling by proteins and/or minerals which has generally ruled out polymeric materials. The retained polarised particles form a secondary or "dynamic" membrane with ultrafiltration characteristics and thus also separate out casein and whey proteins.

Microfiltration with inorganic/organic membranes is widely adopted. Protein rejection is typically less than 3% and bacterial retention around 99%. Permeate is therefore a virtually bacteria free skimmed milk. Furthur details of this process can be found in section on Food industry.

Cheese Whey

In the making of cheese every 100 kg of milk used results in 10-20 kg of cheese and 80-90 kg of whey, the liquid fraction which is drained from the curd. The whey is a valuable source of materials eg proteins and lactose, although of limited direct use. The ratio of protein to lactose is too low and ultrafiltration is used to increase this ratio. However the whey usually contains small amounts of fat, as small globules 0.2 to 1 μm in size, and casein as fine particulate (5-100 μm). These components can have detrimental effects on the functional properties of the protein concentrations if not removed. Centrifugal separation does not effectively remove the casein and fat. Microfiltration using the HFLF, constant pressure filtration process is effective, reducing the fat/protein ratio from 0.07-0.25 to 0.001-0.003. An added benefit is the removal of some of the precipitated salts and a reduction in bacterial load.

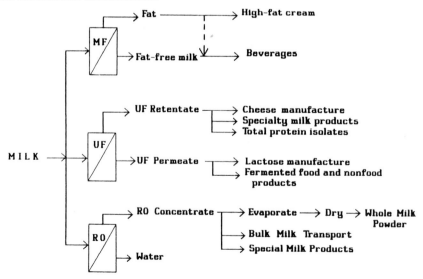

FIGURE 14 – Fractionation of milk by membrane separation.

MICROFILTRATION

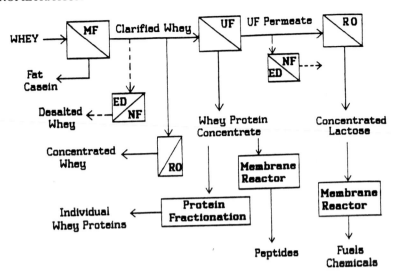

FIGURE 15 – Processing of whey with membranes

Clarification by Microfiltration

Cross-flow microfiltration has the attraction of being an effective method of clarification of liquids. In the food industry it is becoming more widely adopted as an alternative to for example rotary vacuum filtration using filter aids such as diatomaceous earths. It is generally an easier operation, avoiding the cost and disposal problems of the filter aids.

Clarification of apple juice by microfiltration has been practised for several years as a means of producing a sparkling, clear beverage. More recently the cross-flow microfiltration of highly coloured fruit juices, eg grape, cranberry and cherry has been introduced.

Gelatine

Gelatine is the proteinaceous solution derived from selective hydrolysis of collagen, the major constituent of the connective tissue is skin and bone. Gelatine has use in the pharmaceutical and photographic industries as well as in foods. It has the properties of high viscosity in aqueous solution, ability to reversibly change from gel to solution and is effective as a protective colloid.

The raw product produced by hydrolysis required filtration to remove dirt, coagulated protein, fats and other particles prior to concentration. Cross-flow microfiltration gives a product of consistent quality, is a cleaner and easier process than vacuum filtration. Effective membrane units use spiral wound or tubular asymmetric membranes of polyether sulphone cast onto a polyolefin non-woven backing.

Syrups and Sweeteners

Glucose (dextrose) solutions as produced from starch are required to be free of suspended solids, etc ie clear and haze free. The percentage sugar content (brix) is high (30-40) in comparison to fruit juices and thus solutions have much higher viscosity. The type of suspended solid and loading are however different, less fibrous, and thus spiral wound modules can be used to make for more compact module design.

Applications in Wine Production

Both dead and cross-flow membrane filtration are applied in the production and bottling of wine. The reliable bottling of wine by membrane filtration has almost become accepted practice due to several factors.

- *Microbiology reliability.* The functioning of the filter system prior to operation can be checked using Integrity Tests (see Chapter 2)
- *Product quality assurance* – change over from one product to another without mixing of the products is achieved by emptying the filter system (and prefilter) with carbon dioxide or compressed air
- *Economics.* Average filtration outputs are between 50 to 100 m^3h^{-1} (67-124 thousand bottles)

Cartridges come in a range of standard sizes constructed of polypropylene with silicone, EPDM or Viton gaskets, etc. All components are thermoplastically welded. Membranes are for example of hydrophilic polysulphone with pore sizes from 0.2, 0.45, 0.65 to 0.8 µm.

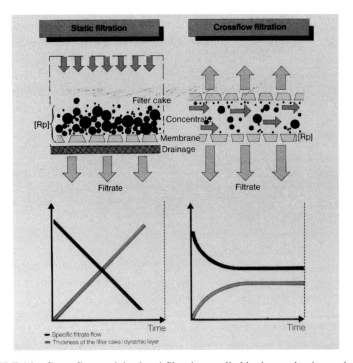

FIGURE 16 – Cross-flow and dead end filtration applied in the production and storage of wine.

MICROFILTRATION

There are different membrane types and geometries used in the cross-flow microfiltration of wine. Popular systems are capillary and flat sheet modules utilising 0.2 μm MF membranes and also 10^6 Dalton cut off (approximately 0.1 μm) UF membranes are used. In this application the filter performance varies widely and is affected by several product specific and membrane specific parameters. Typical filtration capabilities are in the range of 30 to 80 dm^3 m^{-2} h^{-1}, depending upon the wine variety, vintage and grape quality at picking. This is largely affected by the extent of wine pretreatment.

Cross-Flow Filters

Membrane filters are supplied as modularised plant utilising for examplepolypropylene or polysulphone capillary membrane, 1.8 and 1.5 mm i.d. respectively. The remaining plant and filter housing is stainless steel. Plant capacity can vary from 600 dm^3/h size to 3600 dm^3/h size.

The plant contains a twin pumping system to accommodate requirements of high cross-flow velocities (2-3 ms^{-1}) and high pressure 3 bar. A feed pump introduces the unfiltered product into the system at high pressure at a low volumetric rate and a circulation pump operates at low pressure but with a large flow.

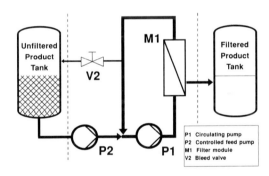

FIGURE 17 – Batch cross-flow microfiltration.

The system operates with continuous bleed of a small volume of product (concentrated) from the recirculation loop to enable slow continuous concentration of solid up to the final desired solid concentration. Typically in wine production the concentrate residue is below 0.3% of the original product volume. This concentrate can then be processed via a lees filter or a rotating vacuum filter.

Membrane filters are supplied as modularised plant utilising polypropylene or polysulphone capillary membrane, 1.8 and 1.5 mm i.d. respectively. The remaining plant and filter housing is stainless steel. Plant capacity can vary from 600 dm^3/h size to 3600 dm^3/h size.

The plant contains a twin pumping system to accommodate requirements of high cross-flow velocities (2 – 3 ms^{-1}) and high pressure 3 bar. A feed pump introduces the unfiltered product into the system at high pressure at a low volumetric rate and a circulation pump operates at low pressure but with a large flow.

The system operates with continuous bleed of a small volume of product (concentrated) from the recirculation loop to enable slow continuous concentration of solid up to the final desired solid concentration. Typically in wine production the concentrate residue is below 0.3% of the original product volume. This concentrate can then be processed via a lees filter or a rotating vacuum filter.

The use of cross-flow microfiltration (CMF) offers several options as a replacement to conventional processing steps in the production of wine. there are several advantages to the production of wine in the use of cross flow membrane filtration:

- reduced use of SO_2 due to reduced concentration of microbes
- reduced requirements for expendables and lower costs for diposing of diatomaceous earth and filter pads at waste disposal facilites
- reduced need for final safety filters for cold sterilisation during bottling
- improved tartrate stabilisation by removing colloids that retard tartrate precipita tion, which reduces energy costs compared to traditional tartrate stabilisation by cooling.
- improved quality/taste of wine
- high pigment yield or colour intensity of red wines compared to those passed through pad filters
- improved brilliance
- faster preparation of base wines used for making sparkling wines and young wines which are rapidly marketed
- no undesireable changes in the wine constitutents

The following example illustrates the many filtration steps used in wine making

Filtration Processes in Wine Making

The separation of yeasts and other microogranisms immediately after fermentation is desirable to prevent changes in the wine caused by microbes. However, filtration is not only concerned with the wine itself but also in other process media involved in wine making such as, water, steam, air and unfermented must – used for sweeting and blending – a common practice in Germany.

The following is an example of the many microfiltration steps which are concerned in the production of bottled wines.

1 – Filtration of wine before storage (clarification)

The cross flow membrane system filters wine after fining and preclarification and so can eliminate all the steps normally performed after these procedures in conventional wine making. The gentle cross flow technology incorporated in the modular design with periodic back flushing with filtrate, in-process hot water sanitisation, feed reverse cycles

MICROFILTRATION

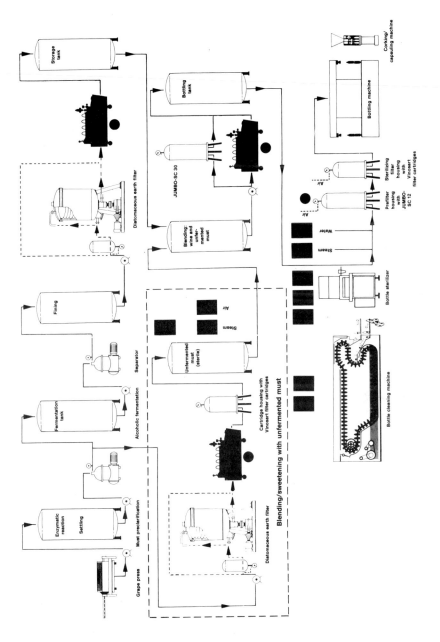

FIGURE 20

and computer-controlled monitoring of the process data, can produce wines of a higher quality eg according to the 5-point scoring procedure established by the German Society of Agriculture, than those conventionally filtered through diatomaceous earth and pad filters. Stopping malo-lactic fermentation whenever suitable reduces requirements for expendables and eliminates the cost for disposal of diatomaceous earth and pad filters at waste disposal facilities.

2 – Cold sterilisation of unfermented must

Cross flow filtration can also be used for pre-treatment of cold sterlised, unfermented must to replace a diatomaceous earth filter and a filter pad directly upstream of membrane filter cartridges in stainless steel cartridge housings for cold sterilisation of unfermented must. The membrane filter cartridges typically contain pleated filter media, comprising a polyester prefilter and two gauze-reinforced cellulose acctate membrane filters (heterogeneous double membrane), the first one has a coarser pore size rating than the second one. The cartridges are completely suitable for food contact use since all components are FDA listed. They are available in a selection of pore sizes, a 0.65 µm pore size retains yests, lactobacilli and pediococci. With 0.45 µm, it holds back yeasts and wine-spoilage microbes and has been additonally validated for microbial removal with average challenges of *Leuconostoc oenos*.

3 – Sterile venting of the unfermented must tank and sterile filtration of compressed air for removal of unfermented must

For the sterile venting of unfermented must tanks a unit referred to as the Tank Guard Capsule which is a complete sterile air filter with a polyester prefilter and a PFTE membrane filter in a polypropylene housing can be used. The tank outlet is attached to a pipe fitting assembly designed to withstand short-term heating. A fermentation tube is connected to the inlet of the Guard for visual inspection during sterile storage. Water containing a bacteriostatic agent for microbiological stabilisation is added to the tube. This elimiates the need for sulfuric acid and formaldehyde because the Capsule does the job of sterlising the incoming air.

The sterile removal of unfermented must portions from a tank typically requires preparation of the filter by steam sterilisation. The Tank Guard eliminate this step for sterile removal of a partial quantity of must for sweetening or blending with other wines. The compressed gas or air tubing is attached to the inlet of the Guard which is designed to withstand pressures up to 400 kPa. Slight pressure is built up to remove a partial quantity of the must.

4 – Filtration of steam and air

For the filtration of steam for tank sterilisation and sterile filtration of air for subsequent cooling of the tank a Tank Sterlising Assembly is available. This is an integrated steam and air filtration system that saves time for steam sterilising large tanks and cooling them with sterile air to prevent vacuum caused by condensation. A portable assembly consist of a steam filtration sytem, with a 50 cm stainless steel filter cartridge for reliable removal of particles from steam, and a system for sterile filtration of air with a membrane filter cartridge. The stainless steel filter cartridge has a double-layered, sintered stainless steel fiber web which is reinforced by sintered-on wire mesh. It filters steam up to a temperature of 180°C. The membrane filter cartridge contains a polyester prefilter and a PTFE membrane that prevents both breakthrough and growthrough of microbes even when pressure surges occur or high operating pressures are used.

5 – Last cellar filtration of wine clarified by CFMF

After clarified wines have been aged in storage tanks and, depending on the wine making procedure, blended with unfermented must to sweeten them, two different systems are used in the cellar filtration:

If a short time (about 6 months) has elapsed between clarification and the last cellar filtration run, cartridge filtration is used. Either standard cartridges or large JUMBO cartridges can be used in this application. The large area cartridges 36 m^2 is 98 cm (39") high, have a circumference of 82 cm (~ 32") and contains an efficient filter combination. This combination consists of a staged double layer of charged depth filter material, and a gauze-reinforced, asymmetrical membrane filter with a 3 µm pore size. A 90 mm (3.5") filtration depth ensures long service lives, excellent flow rate characterisitics and filtration indices. JUMBO cartridges are installed in stainless steel housings with low hold-up volumes, and therefore constitute closed filtration systems. With the JUMBO systems, there will be no loss of product due to the dripping and no wine mix zones. Since the last filter is a membrane, the system can be drained with compressed air without adversely affecting its flow rate performance and breakthrough is prevented when the cartridge is spent or when the product feed rate exceeds the depth filters' flow rate specifications. The JUMBO cartridges replace pad filters and module systems, therefore saving work and time during filter change-outs.

If more than six months have elapsed between clarification and the last cellar filtration run, the cross flow system is used again in cellar filtration. The previously filtered wine is practically sterile after the last cellar filtration run, and so the membrane filter upstream of the bottling machine serves as a final saftey filter. They trap any microbes picked up in the piping instead of having to retain a complete load. This greatly extends the in-service life and reduction in particulate is especially advantageous when filtering vintages that have a high colloid content which frequently causes blockage in conventional filtration systems. Wines filtered by cross flow membranes typically have an unprecedented level of brilliance; and treated red wines have a better red pigment yield or colour intensity than do conventionally filtered wines.

Last cellar filtration without previous use of CFMF

If wine is clarified, for example, by a separator and by a coarse-grade diatomaceous earth filter, the last cellar filtration run can be preformed with a combination offine-grade diatomaceous earth filter, standard cartidges with a nominal retention rating of 1 µm, and a large cartridge (JUMBO). The cartidges are depth filters that have a 12 mm thick layer of polypropylene fibers with decreasing diameters and therefore progressively higher retention efficiencies.

6 – Cold sterilisation of wine immediately before bottling

When CFMF is used in the last cellar filtration run, the wine available for bottling is almost sterile. Therefore, the only remaining function of the membrane filter cartridge is to stop any microbes, which may have been picked up in the piping, from entering the bottling machine. As a result, the membrane filter cartridges with the following pore sizes of 0.65 µm (retains yeasts lactobacilli and pediococci), 0.45 µm (additionally validated for a microbial retention rating with average challenges of *Leuconostoc oenos*) or with maximum challenges of *Leuconostoc oenos*) have very high throughputs for this process. These beverage-grade membrane filter cartridges are installed in stainless steel cartridge housings which have been designed for the beverage industry.

If the last cellar filtration run is performed using pad filters, a greater number of filters and equipment will be needed directly upstream of the bottling machine or filter. Alternatively use either use a JUMBO cartridge with a 1.2 µm membrane as a prefilter upstream of the final filter cartridges or a standard cartridge types (70 mm diameter) as a prefilter for 0.45 µm membrane filter.

9 – Pre- and final filtration of water for the bottle cleaning machine

In addition to filtration of wine and must filtration of water for bottle cleaning is required depending upon the quality of the water, it may be sufficient to use a 0.45 µm membrane filtration upstream of the bottle cleaning machine or to add a 0.2 µm cartridge filters for prefiltration.

JUMBO Cartridge Filtration

For larger flow application, large (JUMBO) cartridges have been developed. The standard JUMBO depth cartidge filter with its height of 1,000 mm and diameter of 265 mm offers a total filtration are of 36 m^2. With this area a single JUMBO cartridge offers the capacity of twenty 3-high (30") standard cartridges. A variety of pleated filters are used in the cartridges in the 90 mm space between the outer support and core. These diverse filters are matched precisely to meet specific filtration requirements in terms of area, retention rating and filter material. The JUMBO cartridge filter is a combination of a polypropylene/glass fibre prefilter and a heterogeneous, multiple combination of aysmmetrical, reinforced cellulose acetate membrane filters, with the finest pore size being 0.65 µm. The recommended flow rate of each cartridge is 100 hl/h with a maximum differential pressure of 2.5 bar. The combination of depth and membrane filters form a closed, integrity testable system and guarantees reliable retention of microorganisms, particles and haze, long service live and low filtration cost.

The JUMBO cartridges can bring filtration costs down to an acceptable level in many applications where 70 mm standard cartridges are not economical, without compromising filtration efficiency.

Advantages the JUMBO cartridges include:

The combination of depth and membrane filters in a closed system makes it possible to fully utilise the features of both types together for optimum preformance. The depth filters

MICROFILTRATION

ensure long in-service life while the downstream membrane filters provide microbiological security.

The closed system allows the use of pressure for maximum recovery of product, and also keeps mix zones to a minimum.

The shorter installation time and the extended intervals between individual filter change-outs reduces the amount of labour involved.

A lower cost than a standard cartridge system designed for the same flow rate.

Applications of JUMBO cartridges include filtration of

i) **Beer**: sterile filtration downstream of sheet prefilters.
Reliable removal of beer-spoilage microorganisms.
Cost containment within a given filtration cost budget.
Retention of the full flavour of beer without adversely affecting quality.
Integrity testable filtration system.

ii) **Water**: pre- and sterile filtration of process water.
Reliable removal of haze and microorganisms.
Replacement for chemical treatment and other alternative methods.
High hourly flow rate performance, long in-service life and high economy.

iii) **Wine**: cold sterilisation immediately prior to bottling.
Yeast-free filtrate – closed filtration system – drainable by piping in compressed air – minimial wine-water mix zones – better protection against blockage of the cartridge system in the bottling area – cost reduction.

The JUMBO cartridge filter is a combination of a polypropylene/glass fibre prefilter and a heterogeneous, multiple combination of aysmmetrical, reinforced cellulose acetate membrane filters, with the finest pore size being 0.65 µm. The recommended flow rate of each cartridge is 100 hl/h with a maximum differential pressure of 2.5 bar.

FIGURE 21 – JUMBO cartridge filter.

Fruit Juice

Conventional filtration/clarification of beverages containing haze particles operate on the principle of deep bed filtration or on the principle of surface sieve (membrane) filters in the dead end mode. In both cases the build-up of filtrate increases the resistance to flow. Cross-flow filtration can be effectively used to produce beverages (fruit juices) of guaranteed turbidity, colour and flavour. In general cross-flow filtration of fruit juices is applied in the same way as that for wine, although operation must allow for the higher turbid matter and higher pectin contents of the juices. This essentially requires pretreatment with pectinase and fining with gelatine (or gelatine/kieselsol) and also suitable module geometry selection with adequate spacing or capillary diameters. Filtration of juices can generally be performed at higher temperature than for wines and thus higher fluxes are possible.

Typically, capillary modules are used which can be applied in two stages
- Small diameter eg 1 mm, membranes are used to produce the major part of clarified juice in a main recirculation loop.
- In the second stage the concentrate from the first stage is further concentrated in wider diameters, eg 3 mm, membranes to a limit of flow ability.

The chemical sludge is then processed to a compact filter cake using a membrane cloth filter.

Beer

Beer as a natural product has a high colloidal load which makes it more difficult to filter than wine and juices. Thus the filtration rates per unit membrane area are lower which has limited cross-flow techniques to a fewer applications such as filtration of yeast beer and sterile filtration of rest and return beer. Rest beer is generally returned keg beer.

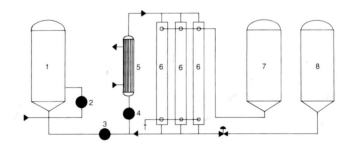

Scheme of a TFF-Plant

1 Yeast tank
2 Recirculation pump
3 Feed pump TFF-Plant
4 Recirculation pump for concentrate
5 Cooler concentrate circulation
6 Ceramic modules, number acc. to capacity
7 Filtrate tank
8 Concentrate tank

FIGURE 22 – Filtersysteme flow circuit.

Cross-flow filtration of beer requires very high cross-flow velocities (2-3 ms^{-1}). This is particularly the case for yeast lees where initial dry solid content can be 12%. With suitable module design the yeast content can be concentrated to over a 20% (dry solids) suspension. The product is generally suitable for direct blending with finished beer.

Typical recovery of beer from yeast is achieved with ceramic membranes. Mutlichannel modules consisting of a single monolith with 19 flow channels of 6 mm diameters each are used. The support body is a 15-20 µm diameter porous α-Al$_2$O$_3$ block with and internal thin microporous membrane of α-Al$_2$O$_3$, pore sizes of 0.2, 0.5, 0.8 and 1.0 µm. Typical performance on discharging a 20% yeast retentate (dry solid basis) is 10-17 dm^3 h^{-1} for each module which is equivalent to a flux of 30-50 dm^3 m^{-2} h^{-1}, depending upon yeast quality.

Crossflow microfiltration system for cider production

The traditional British cider apple used to make cider is small, hard and inedible, as well as being fibrous and bitter – qualities which produce the characteristic strength of flavour and aroma. The apples are first reduced to a pulp by a rotary mill; the juice is then extracted on a continuous belt press where the pulp is fed between two stainless steel mesh belts and persuade by rollers. Further juice extraction is carried out using a traditional hydraulic press.

The juice is pumped on to the fermentation vessels – oak, and ranging from 4,000 to 42,000 gallons capacity – where it is treated with SO$_2$ to destroy unwanted organisms before yeast is added to start the fermentation. After fermentation has ceased the cider is racked off the yeast deposit into more oak vats where it is matured for at least six months to allow its full character to develop. When the cider is judged by tasting to be ready, it is centrifuged, filtered and clarified. Blending with ciders from other vats is carried out in the acidity and gravity are adjusted. The cider is then further filtered and sterilised before being bottled or kegged.

Previous filtration systems was labour intensive – in operation, cleaning and repair – and resulted in high losses of cider at a time when much of the rest of the operation had been made more efficient. The seven-stage filtration process was undertaken in two parts: after the maturation process it involved a pulp filter, a coarse grade Nellie filter, two sheet filters and a polishing grade Nellie filter; after the blending operation there was another sheet filter, a cartridge filter and a nylon membrane filter.

The membrane system is based on membrane modules of 1 m^2 membrane surface area – and was supplied complete with feed tank, pumps, control system, self-cleaning gas backwash regime and integrity test equipment The heart of the CMF system is the filter module, which contains approximately 3,000 hollow fibres manufactured from permeable polypropylene to a 0.2 µm pore specification. The fibres are capable of filtering out and separating all suspended solids, bacteria and yeast colloids.

A low pressure feed pump is used to force the cider into the membrane module through the feed port and into the body of the module surrounding the hollow fibre permeable membranes. The cider continues along the shell and exits the return port. Contaminants are continuously being swept away by the crossflow stream, and some of the cider filters through the fibre walls to exit down the centre of the fibres as pure filtrate.

To remove built-up contaminants a gas backwashing technique is used. The gas pulse provides total removal of contaminants whin seconds and is repeated at a set interval which was determined during the earlier trials. An additional feature of the gas backwash is that it uses no filtrate.

A particular feature of the plant is its automatic clean-in-place system which is designed to chemically clean and sanitise the CMD either at the end of every batch, or, in some cases, when the transmembrane pressure reaches a predetermined point. The process is controlled by a PLC which automates the filtration operation and manages the gas backwashing. For the cider applications the membranes are guaranteed for one year but from experience they are expected to have an average life of two to three years.

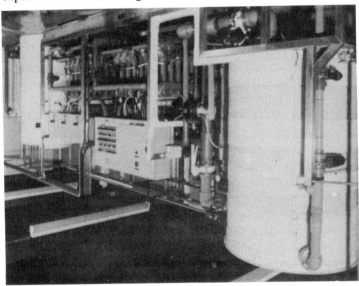

Filtration and Carbon Membranes

In high temperatures processes with hot aggressive fluids filtration requirements can only be met with a limited number of materials. Carbon is one such material which is well known for its corrosion resistance. Membranes are typically a composite structure of a very thin layer of porous carbon 910 µm thick) applied to the internal surface of a narrow diameter carbon support tube.

The membrane tubes are assembled in monolithic joint free bundles using carbon composite tube sheet. The bundles are then mounted in a shell made from an appropriate compatible material such as PTFE lined steel or stainless steel. The carbon tubes are 6 mm internal diameter and 1.5 mm thick giving good mechanical resistance, ie bursting pressures of 40 bar.

The module has a great tolerance to back flushing cycles typically 300,000 at up to 8-12 bar. Sealing of the bundles in the end plates ensures no by-pass problems. Temperature resistance is high up to 165°C and thus enables steam sterilisation.

The membranes have cut off sizes of 0.8 and 0.2 µm with 0.1 µm the lowest achieved.

MICROFILTRATION

FIGURE 23 – Carbon fibre composite membrane.

FIGURE 24 – Carbon fibre cross-flow filtration module.

TABLE 8 – Applications of carbon fibre composite membranes

Surface treatment baths: acid pickling, alkaline degreasing, hot solvent
Recycling of acids: separation of precipitates or catalyst
Concentrated salt solutions, hydrocarbons
Clarification and sterilisation of water and beverages
Separation and recycling of living cells
Separation of colloids and proteins
Separation of high value precipitates in the pharmaceutical industry
Pre-post treatment separation
Waste recovery, drinking or industrial water treatment

The membranes are tolerant to strong acids at all concentrations, hot organic solvents and alkaline baths but not strong oxidising agents. Carbon is a fully biocompatible material recommended for alimentary fluids and biological fluids.

Applications of the membranes include filtration of corrosive and aggressive fluids, separation, clarification and sterilisation of biological fluids and difficult separations of mineral particles (see Table 8).

Membranes have successfully filtered titanium dioxide suspensions in sulphuric acid, salts crystals, fat and proteins form hydrochloric acid and radioactive dust form nuclear industry laundry effluents.

Metal Finishing Industry

An important application is in metal finishing industry in the treatment of effluent from degreasing baths.

Aqueous degreasing baths use either acid or alkaline solutions containing a variety of additives and chemicals to remove oily hydrocarbons from surfaces. Progressively baths become contaminated with oil and loose their effectiveness, meaning the contents must be eventually removed and disposed. In many cases this occurs on a daily basis otherwise problems of product quality arise. Treatment of the degreasing solution by cross-flow filtration to remove the oily contaminants can extent the effectiveness of the degreasing solution to up to 5 weeks. The carbon membranes are particularly unaffected by frequent shut-down and start up and give fluxrates of approximately 300 dm^3 h^{-1} m^{-2}.

The cross-flow filtration process, shown in Fig 25, can remove oil from either spray or immersion type baths. In operation some of the bath solution is purged and pretreated to remove free oil and potentially damaging solids, then passing to a concentrate tank. Solution from the concentrate tank is continuously passed through the membrane to remove the oil which is returned to the concentrate tank in the retetentate. The oil free permeate is returned to the degreasing bath at the same rate it is purged from the bath. At an appropriate time the concentrate tank is emptied to dispose of the "concentrated" oil

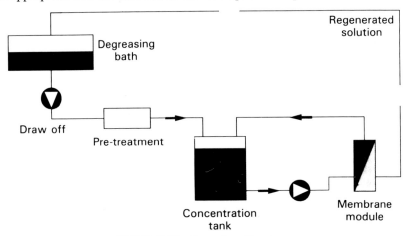

FIGURE 25 – Cross flow filtration process.

degreasing solution. The process radically reduces the volume of spent oil/solution required for disposal by a factor of several hundreds. The payback time from a reduction in disposal costs, on installation of cross-flow carbon filters in less than one year. Additionally saving in manpower, chemicals, heating and in continuity of production are made.

Microfiltration of Adsorbates: Stainless Steel Filters

The separation of poorly biodegradable substances from suspensions or emulsions is often readily and economically achieved by adsorption. Preliminary flocculation may also be necessary. Absorbents usually used in liquids include

Activated carbon. Separation of hydrophobic substances and dissolved substances.

Activated bentonite. Hydrophobic and oleophilic versions for oil/water and emulsion separation.

Synthetic resin absorbers. Adsorption of organic molecules eg surfactants.

Molecular sieves. Hydrophilic zeolities.

Filters acids (with or without absorbents) eg kieselgurs, silica gels, aluminas.

Selective adsorption by flotation with surfactants or zeolities.

In all the above cases, the characteristic property of adsorption, sieving, flotation or catalytic capacity is enhanced by a small structure. This however, may necessitate rapid and complete separation of the fine particles which may be submicron in size. Stainless steel microfiltration membrane modules are suitable for these applications.

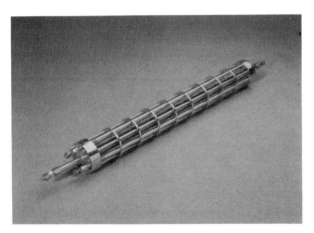

Stainless steel microfilter.

Suitable membranes are tubular (see photo) connected in parallel in tubular modules of stainless steel or polypropylene. A feature of one device is the ability to adjust manually and steplessly the membrane pore size, from 0.05 µm to 100 µm, without dismantling the module.

In operation the finely dispersed adsorbents etc bridge over the pores of the stainless steel membrane producing a modified top layer which is continuously regenerated by

Adjustable pore size microfiltration membrane.

cross-flow. The characteristics of the stainless steel membrane filters in these adsorption applications are
- relatively small amounts of adsorbates are needed
- other auxiliaries may be added to precipiting agents or flocculants
- simple to clean (or sterilise) with steam, hot water, acid and alkali cleaning agents
- high filtration pressures possible, up to 35 bar
- backwash pressures up to 10 bar

The stainless steel membranes have been used in several applications.
- waste water from grid or sand trap of sewage plant
- dumping ground seepage water
- condensate containing emulsion
- brewery waste water
- polishing effluents from metal industries
- wash water
- supernatant liquid from oil separators
- paint pigment suspension
- chemical process water

Ceramic Microfiltration

Microfiltration is being increasingly used to treat various waste waters. There has been substantial development of ceramic membranes, and the specific features of this type of membrane include: heat-resistance, a high degree of resistance to organic solvents, good cleaning postential, high mechanical strength, applicable in a wide pH range, a long life cycle and a good price/performance ratio. These features mean that this type of membrane has many potential applications in separation processes, for example, in Paint/water mixtures, Oil/water emulsions, Compressor condensate, Waste water in olive oil processing, degreasing baths, etc. (see Table 10).

Ceramic membrane units are produced as complete, mobile systems ready for implementation into the treatment process. Systems can incorporate automatic back pulsing for membrane cleaning to cope with the dirtiest of waste waters. Modules are supplied in a wide range of sizes to meet most applications of microfiltration with a pore size of 0.1 μm until 0.5 μm. The filtration systems are constructed as cylinders equipped with ceramic filtration tubes.
- the strong ceramic tube allows high cross-flow velocity of the feed, resulting in a

turbulent flow. This turbulent flow prevents the formation of the dirt layer and guarantees a high permeate flux;
- the secure connection between the membrane layer and the supporting tube allows frequent back pulsing – not back-flushing with permeate. On the permeate side a regular pressure boost is generated (once every 3 to 10 minutes), with a pressure of 2 to 3 times the transmembrane pressure. Back pulsing removes most of the dirt layer which builds up on the membrane despite the turbulent flow and maintains an acceptable permeate flux. The amount of dirt determined the frequency of back pulse interval.

Ceramics are good materials for the production of membrane filtration tubes. Advantages of ceramic membranes compared to polymer membranes include:
- resistance to high temperatures up to 280°C (in specially developed modules and systems up to approximately 700°C);
- good corrosion resistance: resistant to organic solvents and a wide pH-range;
- ceramic membranes are suitable for cleaning and steam sterilisation;
- large mechanical strength: the possibility to back pulse, resulting in an efficient removal of the dirt layer and the possibility of treating fluids with a high viscosity;
- chemically inert: a wide range of application possible in the chemical industry;
- long operational life;
- high membrane flux; from the composite structure.

The disadvantages of ceramic membranes compared to polymeric membranes:
- brittle and so they must be handled with care;
- the surface area/volume ratio is low so systems have larger dimensions;
- the investment in ceramic membranes is high.

Ceramic membranes can be constructed in several layers: a ceramic support tube of high porosity and one or several membrane coatings. The top layer is the real membrane layer and is responsible for the separation. The top layer is very thin (20 µm) to reach a high flux. Sizes start from membrane areas of 0.05 m^2 with typical filtration flows of 3-9 dm^3 h^{-1} up to 4 m^2 membranes areas at filtration flows of 175-500 dm^3 h^{-1}.

Alternative ceramic membranes are available including a high performance ceramic membrane made from silicon carbide base material and ceramic layers.

The elements are either single channel or 19 channel design. The latter gives a 0.2 m^2 filtration surface (internal) for a 900 mm long, 25 mm diameter unit. The operating range of pH is 1-11 (70°C) and 1-14 (25°C) and operating temperatures of several hundred degrees. Membrane pore size/cut off sizes vary from 0.05 µm to 10 µm. Typical water fluxes are

$$1.38 \text{ dm}^3 \text{ h}^{-1} \text{ m}^{-2} \text{ bar}^{-1} - 0.05 \text{ mm}$$
$$2.2 \text{ dm}^3 \text{ h}^{-1} \text{ m}^{-2} \text{ bar} -1 - 0.4 \text{ mm}$$

Maximum operating pressure is 10 bar.

The following table and flowsheets illustrate typical applications and performance of ceramic microfiltration.

TABLE 10 – Application of Ceramic Membrane Filters

paint water mixtures	compressor condensate
emulsion treatment	compressor
emulsion treatment (pilot plant)	waste water form enamelling plant
degreasing baths	bilgewater, on board
degreasing baths (pilot plant)	restbeer recovery
paper mill (pilot system)	soil cleaning
car wash water recycling	

TABLE 11 – Microfiltration with Ceramic Membranes. Application and Performance.

Application	Medium and Requirement	System Parameters	Performance and Cost
Detergents/Water/Oil/Fats Laundry	In industrial cleaning of laundry, water and detergents are used. This solution becomes fouled with fats, proteins, metals and detergents and has a high chemical oxygen demand (COD) and metal content. Production of water of a quality to enable the producer to reuse it or discharge it in the municipal waste water system.	pore size of membrane: 0.2 micron average membrane flux: 150 $l/m^2/hr$ average cleaning interval: 1 week, in line temperature of medium: 50-70°C operating time: 24 hours/day	suspended solids in permeate: < 10 ppm hydrocarbons in permeate: < 10 ppm concentration factor: > 250 total operating costs: 5 ECU/m^3
Oil/Water Emulsion Compressor Condensate	In an oil lubricated compressor a condensate is formed. This condensate is an oil in water emulsion with a typical oil content of 0.5%. Removal of oil content to a dischargeable level, (< 10 ppm) and concentration of oil to the highest possible level (> 90%).	pore size of membranes: 0.2 micron average membrane flux: 150 litre/m2/hr average cleaning interval: 3 weeks size of static separator: 100 litre	oil content of the permeate: < 10 ppm oil content of the concentrate: > 90% concentration factor: > 180 energy: 40 KWh/m^3 cleaning chemicals: 0.05 ECU/m^3 total operating costs: 20 ECU/m^3 pay back time: < 2 years

TABLE 11 (continued).

Application	Medium and Requirement	System Parameters	Performance and Cost
Waste water treatment Waste water from olive oil production and from washing olives	Waste water polluted with solids and a high chemical oxygen demand (COD). Production of water of a quality to enable the producer to discharge it in the municipal waste water plant.	pore size of membranes: 0.5 micron average membrane flux: 120 litre/m^2/hr temperature of medium: 30-50°C R.O. modules: high temperature modules	permeate of MF clear with high COD permeate of RO clear with COD < 2000 total operating costs: 8.5 ECU/m^3
Detergents/Water/ Oils Caustic Degreasing Baths	In industrial cleaning and degreasing a caustic solution of detergents in water is used. This solution becomes fouled with oils, fats and metals. Removal of oils, fats and metals to such a level, that after addition of detergents the bath can be used, in other words, extension of the lifetime of the bath, resulting in lower water use and less discharge.	pore size of membranes: 0.2 micron average membrane flux: 250 litre/m^2/hr average cleaning interval: 1 week temperature of medium: 40-70°C pH of medium: 9-11	oil content of the permeate: < 100 ppm solids content of the permeate: < 100 ppm life time extension: > 5 energy: 12 KWh/m^3 cleaning chemicals: 0.10 ECU/m^3 total operating costs: 10 ECU/m^3 pay back time: < 2 years due to the constant quality of the bath, the degreasing process improves

TABLE 11 (continued).

Application	Medium and Requirement	System Parameters	Performance and Cost
Beer Beer recovery from yeast	In fermenting cellar and storage cellar yeast, ~2-3% of the annual output of breweries is lost, and has to be taken care of. This so called yeast consists of 90% beer & 10% solids. Recovery of beer from yeast with high yield (~60%), which can be returned to the fermenting or larger cellar without further treatment.	pore size of membranes: 0.5 micron average membrane flux: 40 1/m2/hr average cleaning interval: 8 hours temperature of medium: 15°C operating time: 18 hours/day	total operating costs: 3 ECU/hl Comment:
Oil/Water emulsions Emulsion treatment	All different types of oil in water emulsions, as are generated in industry. Concentration of oil to a sufficient level for further treatment or refinery. Production of waste water which can be discharged directly, or treated in a biological way.	pore size of membranes: 0.2 micron average membrane flux: 125 litre/m2/hr average cleaning interval: 1 week temperature of medium: 40-70°C	oil content of the permeate: < 10 ppm concentration factor per step: ~10 energy: 12 KWh/m3 cleaning chemicals: 0.10 ECU/m3 total operating costs: 10 ECU/m3 pay back time: < 3 years

Ceramic filtration module.

MICROFILTRATION

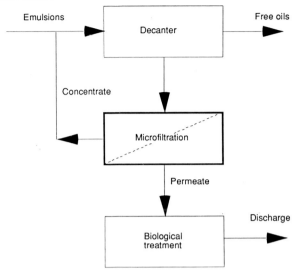

FIGURE 26 – Treatment of oil/water emulsions.

After coarse filtration and decantation the remaining oil/water emulsion is concentrated in the ceramic microfiltration system. The concentrate is returned into the decanter, and after removal of the free oils again microfiltered, until all oils are removed. The extracted water can be fed into a biological treatment plant or discharged directly, dependent on the composition of the starting emulsions.

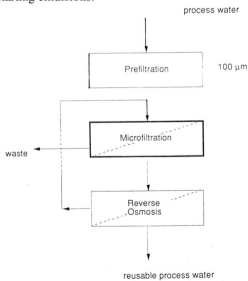

FIGURE 27 – Treatment of waste water from laundry cleaning.

Prefiltration with 100 μm, microfiltration with ceramic membranes to reduce the amount of suspended solids, fats, oils and metals. Final treatment with reverse osmosis.

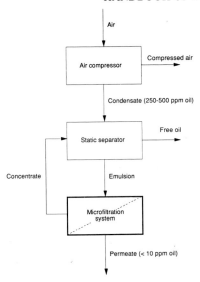

FIGURE 28 – Treatment of air compressor condensate.

The emulsion is treated in a combination of a static separator and a ceramic membrane; the membrane extracts water out of the emulsion, leaving a more concentrated retenate and subsequently the static separator extracts pure oil out of this retentate. The residue of the static separator is filtered again in the membrane unit. In this way there remain two effluents: water and oil.

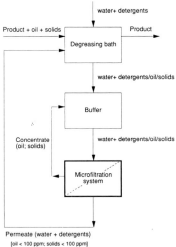

FIGURE 29 – Lifetime extension of a degreasing bath.

The bath is cleaned continuously by a ceramic microfiltration system, operating in by-pass mode. The cleaned liquid is returned to the bath, the pollution is concentrated into a separate container. The chemical contents of the bath need to be monitored and adjusted.

MICROFILTRATION

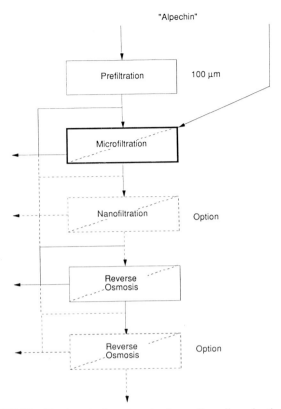

FIGURE 30 – Treatment of waste water from olive oil production.

Prefiltration with 100 μm filter, microfiltration with 0.5 μm and reverse osmosis to reach a COD that is accepted by the local authorities.

Paint/Water Separation by Ceramic Microfiltration

Paint producing and paint applying industries use water in three different applications:
- Cleaning components in the production processes for solvent based paints
- As a solvent for water based paints
- To catch over-spray from water based paints in spray booths

In the first application the waste water contains heavy metals as well as pigments and solvents. Treatment prior to disposal is essential and has three main purposes:
- Reduction of the quantity of waste water and thus reuse the water
- Reuse of the pigments
- Easier disposal of the waste water fraction

These three applications can be met by using either electroflotation followed by filter pressing or by ceramic microfiltration followed by treatment of the retentate in a filter press. For microfiltration, tubular carmic membrane (0.2 μm pore size) modules can produce a retentate of 12% dry matter, mainly pigments. This retentate can be directly processed in the filter press and the product can be re-used in the production process. Although the permeate is colourless and contains no pigments, it contains solvents, COD and heavy metals which necessitates post-treatment before disposal. Typically cross flow operation is with a velocity of 6.5 ms^{-1} and a differential membrane pressure of 2.0 bar which gives a flux of 100-250 dm^3 m^{-2} h^{-1}. Operation required frequent membrane pulsing to remove membrane surface deposits which cause a rapid decrease in flux rate.

In the second and third applications above, increasing demand for solvent-free paints has led to the introduction of water-based paints. Both the production and the application of this type of paint often result in a water/paint waste mixture. Two examples are leftovers diluted with (cleaning) water, and water used in water curtains in spray booths, which becomes increasingly contaminated with paint (fillers, organics, pigments). Treatment can be either by organic membrane ultrafiltration followed by reverse osmosis or by ceramic membrane microfiltration followed by reverse osmosis. The ceramic membrane is attractive in its ability to treat very viscous flows, ie high solids content and its effective back-pulse procedure combined with chemical cleaning. The ceramic membrane can produce solids concentration of up to 65% which depends upon the paint quality and the ability to recirculate the mixture at high velocities across the membrane. The permeate produced as effluent can be reused or, depending on the paint in question, directly discharged into open waters, or further treated with reverse osmosis membranes before discharge. The permeate flux through the membranes varies from 50 to 250 dm^3/m^{2}h, depending mainly on paint type and concentration. The specific overall treatment costs are in the same range as in the previous example (about NLG 30-35/m$^{3)}$.

The permeate contains some organics and is slightly coloured but contains no pigments. It must be treated, by reverse osmosis, before disposal.

The operating costs and pay back times of ceramic membranes in these applications is attractive. Microfiltration is typically used in conjunction with other separations, membranes and bio-treatment.

1 – Dewatering of Radioactive Waste Sludges

Proposals for the encapsulation of wet intermediate level waste sludges (ILW) and resins required that such material should be dewatered. Among the possible candidates for such a process are
- Centrifuge scroll decanter
- Lamella settler with filter
- Rotary drum vacuum filter
- Membrane filter

The dewatering typically requires the treatment of 5 w/o Magnox sludge waste (magnesium hydroxide with slate dust) with particle sizes in the range of 1-100 µm up to product concentrations of 30% to 50%.

All the above mentioned devices can achieve the required de-watering but considerations of maintenance, cleaning/decontamination, complexity fault, conditions and process compatibility favoured cross-flow filtration. Cross-flow filtration using tubular stainless

MICROFILTRATION

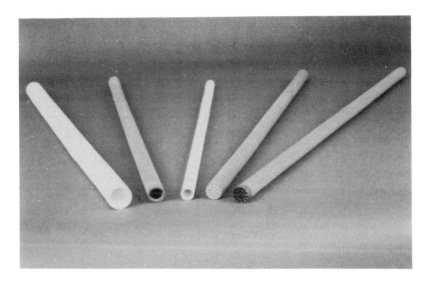

steel sintered membranes or ceramic cylindrical channel block membranes are appropriate. Permeate clarity of 1 ppm solid is achieved with average flux rates of between 150 to 180 x 10^{-6} m^3 m^{-2} s^{-1} at 2 bar (g) inlet pressure.

TABLE 12 – Wastewater components in the paint producing industry

pH	6 — 8
Temperature	10 — 12°C
Dispersed pigments	0.8%
Specific gravity	1.5 — 2
Particle size	30 — 8400 nm
Monoethylene glycol	12%
Surface-active matter	present
Hydrocarbons (TOC)	600 — 1400 mg/l

TABLE 13 – Comparison of operating costs for ceramic microfiltration and electrofiltration. Basis 8m^3/d at 20 hours per day treatment time. 4000h/yr operation with a 2.7 m^2 membrane and flux of 150 $dm^3/m^2/h$

Component	Microfiltration, NLG/m^3	Electroflotation, NLG/m^3
Interest and depreciation	18.50	14.00
Energy and chemicals	2.70	8.00
Maintenance/operation	6.00	6.00
Spare parts	5.00	16.70
Total	32.20	44.70

TABLE 14 – Applications of Paint/Water Mixtures

Characteristics of Mixture	Treatment Method	Operational Characteristics	Test Results	Membrane Performance
Solvent based paint producing industry. In the solvent-based paint producing industry, water is used to clean components in the production process. As well as pigments and solvents, the waste water contains heavy metals, so that disposal without treatment is impossible. Table 1 gives several characteristics of the waste water	Treatment of the waste water has three purposes: • Reduction of the quantity of waste water (i.e. reuse of the water fraction). • Ease of disposal of the water fraction. • Reuse of the pigments as a raw material (concentration of the dry matter). These combined requirements can in general be achieved in two ways, by using either: • Electroflotation followed by filter pressing; or • Ceramic microfiltration in combination with a downstream filter press for treatment of the retentate.	The waste water is treated batch by batch. About 8 m³ is released every day (1500 m³/year), which must be treated automatically in about 20 hours. The water is recirculated across the microfiltration unit until a sufficient concentration is reached in order to make treatment in the filter press possible. During the tests up to 12% dry matter in the retentate appeared to be attainable. This makes possible direct processing in the filter press. The dehydration product contains mainly pigments, and can be reused in the production process.	Feeding pressure 2.5 bar Pressure difference across membrane 2.0 bar Pore size 0.2 m Specific flux 100 - 250 l/m²h Temperature 45C Crossflow rate 6.5 m/s	The permeate is colourless and contains no pigments. However, there are still solvents, COD and heavy metals present in the permeate, so that post-treatment is necessary before disposal is possible. The permeate flux decreases as a result of deposition on the membrane surface. By using a back-pulse procedure that is specific to ceramic membranes, contaminants can be removed promptly from the membrane surface. The procedure entails that every 5 minutes the permeate is put under pressure for 1.5 s, so that the flow direction is reversed and the contaminants removed with the retentate flow. Once a week a standard purification procedure is applied using biodegradable chemicals. In this way a high specific flux can be guaranteed.

TABLE 14 continued

Characteristics of Mixture	Treatment Method	Operational Characteristics	Test Results	Membrane Performance
Water based paint producing industry. In this industry, waste water originates both from leftovers and from cleaning procedures. In addition to pigments and water, the waste mixture contains fillers and some organic components, so that treatment is necessary.	Treatment of the waste serves three purposes: • Reduction of quantity through reuse of the water fraction. • Disposal of the water fraction after secondary treatment. • Reuse of pigments as raw material after concentration to a high dry matter content. These can be achieved by using: • Organic membrane ultrafiltration followed by reverse osmosis. • Ceramic membrane microfiltration followed by reverse osmosis. There are two main advantages in using ceramic membranes instead of organic membranes: • The possibilities offered by ceramic membranes in treating very viscous flows (high solids contents). • The effective cleaning procedures available for ceramic membranes.	The waste mixture is treated batch-wise. The water is recirculated through the filtration unit until the concentration in the storage tank prohibits further extraction of water. The permeate is further treated by reverse osmosis membranes and discharged, or used as cleaning water. The solids concentration can reach values up to 65%, depending on the paint quality. The permeate still contains some organic materials and is slightly coloured, but it contains no pigments. Because of the organic materials, final treatment - e.g. using reverse osmosis membranes - is necessary before disposal is allowed.	Feeding pressure 2.5 bar Pressure difference across membrane 2.0 bar Pore size 0.1 m Specific flux 50 - 250 l/m^2 Temperature 55C Crossflow rate 6.5 m/s	The permeate flux depends on the viscosity of the mixture. This viscosity increases with increasing concentration of paint in the waste water. The final concentration of paint in the retentate varies between 35% and 65%. This difference in potential concentration is caused by the flow behaviour of the mixture: as long as the mixture can be recirculated at high velocities over the membranes, and the chemical bonds between water and paint are still weak, separation is possible. The back-pulse mechanism, combined with an effective chemical cleaning step, ensures high specific fluxes. The membrane flux can always be restored to the original level, even after severe fouling or complete blocking of the pores.

TABLE 14 continued

Characteristics of Mixture	Treatment Method	Operational Characteristics	Test Results	Membrane Performance
Water based paint applying industry. In spray-booths a water curtain is used to collect the so-called 'over-spray'. The water is recirculated over a storage tank (typically 2 - 5 m^3) and refreshed every week. Refreshment is necessary as a result of the increasing paint content in the water, ranging from 2% to 10%.	When the water is demineralised before use, the bath contains only water and paint. Removal of the paint from the water not only increases the lifetime of the bath (saving on the use of demineralised water), but also cuts the disposal costs. Furthermore, if the paint used is manufactured to be 'ultrafilterable', the concentrated paint can be reused. The paint removal can be achieved by organic membrane ultrafiltration and by ceramic microfiltration. The operational advantages of ceramic filtration - high overall flux, possibility of high concentration, very little operational personnel involvement and relative inertness to contamination of the bath - combine to make ceramic microfiltration the better choice.	The bath is cleaned continuously in bypass. This means that a certain fraction of the bath (say 10%) is cleaned hourly. In this way an equilibrium state is reached in the bath at a sufficiently low level of contamination. During stand-stills the retentate of the filtration system, which is kept apart in a separate storage tank during the normal operating cycle, is concentrated to the desired level.	Feeding pressure 2.5 bar Pressure difference across membrane 2.0 bar Pore size 0.1 m Specific flux in operating cycle 150 l/m^{2}h Specific flux in concentrating cycle 75 l/m^{2}h Temperature 20 - 50C Crossflow rate 6.5 m/s	

Vibratory Shear Enhanced Filtration (VSEP)

The traditional method of reducing the effect of fouling in membrane systems is to operate with cross-flow of feed over the membrane. There are economic limits in module designs, to the cross-flow velocity which can be used (typically 10,000-15,000 s^{-1} shear rate). Thus cross-flow membranes will still plug and foul because the flow cannot remove solids and particulate retained within the turbulent boundary layer.

An alternative method of creating increased shear rates on the face of the membrane is to move the membrane itself. One technique referred to as vibratory shear enhanced processing (VSEP) moves the membrane (leaf) elements in a vigorous vibratory motion tangential to the face of the membrane. The feed slurry moves at an almost leisurely pace between parallel membrane leaf elements. The shear waves induced by vibration of the membranes repels solids and foulants from the surface giving free access for liquid to the membrane pores.

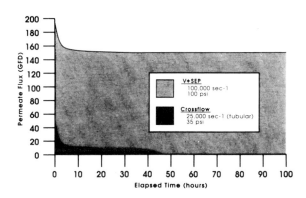

FIGURE 31 – Effect of VSEP and cross-flow on particulate retention at a membrane surface.

The commercial module for the VSEP consists of an array of parallel leaf membrane discs separated by gaskets. The stack of disc is spun at high speed in a torsional oscillation with a rim peak displacement of 1.5 inch at 60 Hz creating a shear rate of approximately 150,000 s^{-1}. Unlike cross-flow filtration, almost 99% of the energy input is converted into shear at the membrane surface. With a transmembrane pressure difference of 6 – 10 bar the attainable filtration rates is some five times higher than with cross-flow. In addition the process has the ability to concentrate to a much higher end solid content and the slurry can become viscous. The unit essentially extrudes the final product from between the vibrating membranes. The disc stack can be operated in a single pass configuration. The VSEP machines are normally constructed of either kynar or polycarbonate. A variety of membranes can be used eg polypropylene, polyetrafluorethylene, polycarbonate, stainless steel, nylon, acrylic, ranging in pore sizes from reverse osmosis, ultrafiltration to microfiltration sizes and even woven screens (1 µm and above).

The separator unit is supplied as a compact skid mounted system with available membrane area up to 300 ft^2 and a filtration removal capacity up to 150,000 gal/day.

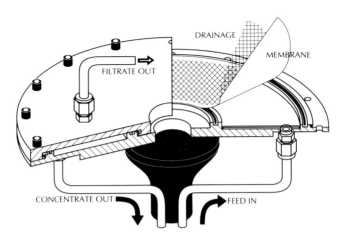

VSEP schematic layout.

Applications

The VSEP approach can be put to a wide range of industrial applications (Table VSEP).

Filtration Manufacture of Kaolin Clay

The normal method of filtration is accomplished with the feed material in a flocculated condition. A problem with flocculated feed is that the final percent solid content of product is typically 50% or less. The product leaving a rotary vacuum filter is generally not pumpable and has to be transported with solids handling systems.

An example of this is in Kaolin clay production. Clay is flocculated with acid and vacuum filtered on a rotary drum to give a cake of 55% solids (w/w). The application of VSEP can produce a 70% by wt solid dispersion, which is still pumpable.

Generally the upper operating limit of solid concentration with VSEP is determined by the pumpability of the dispersion. The technology can be used as a pre-concentrator eg 10% to 20% prior to further concentration in for example a filter press or a spray drier. This greatly reduces the overall energy requirements.

Other applications include the concentration of finely divided colloids up to the gel concentration.

Latex Polymer (UF)

Chemical plants producing latex typically use cross-flow membrane separation to diafilter and concentrate a range of latex products. Ultrafiltration enables the size, and thus, cost of evaporation systems to be reduced and also the associated energy consumption's.

Latex wash waters are also reclaimed using UF. This both reduces disposal costs and enables recycling back into the process.

For some latex streams ultrafiltration can result in severe membrane fouling problems, necessitating daily cleaning and shutdown. The high shear VSEP system is reported to perform well in these applications (eg latex polymerised PVC and ABS) giving more

MICROFILTRATION

stable fluxrates, and a higher end solid content. Additionally, significant energy savings are made due to reduced pumping costs.

Pharmaceutical – Antibiotic Concentrations

For the concentration of a 2% solid content antibiotic solution to 36% solids prior to spray drying. The capacity of one machine is 60 m^3.day of solution. the use of a 200 molecular weight cut off membrane in the VSEP gives a permeate with a low TOC, COD, BOD, which does not require further processing before disposal.

Bioreactor Effluent (UF)

Waste product from an automobile manufacturer is treated in a bioreactor which produces a low solids output hazardous waste. A VSEP ultrafiltration unit is reported to concentrate this waste to a 50% solids sludge. The sludge is then used as fuel in a boiler. The VSEP unit produces 80 gal/ft^2/day under stable flux conditions with a reported pay back time on disposal costs alone, of only 4 months.

Pigments

Most pigments, organic and inorganic, are currently washed and dewatered in large pressure filters. This operation involves continuous operator supervision and expensive maintenance, in removing filter cloths. The VSEP system can wash and concentrate pigment up to 40% solids at typical flux rates of 300 gal/ft^2/day.

Waste Water Treatment

- Removal of metal hydroxides.
- Reduction of COD, TOC and BOD in pulp mill effluents.
- Treatment of oil/water wastes.

Membrane Belt Filter

The Membrane Belt Filter incorporates membrane technology with a belt filterand combines the advantages of the fine particle removal using a microporous membrane with the suspended solids handling capabilities of a belt filter.

In the process liquid feed containing suspended matter is collected in a feed tank and pressurised by a low-head pump. From here the pressurised feed continuously enters a pressure feeder through two entry ports, and is directed almost parallel to the membrane-covered belt at a much higher velocity than the belt is moving through the feeder. The system's membrane pore sizes range from 0.01 to 16 μm, and in contrast with depth filters, there are no large columns of filter backwash, which cause further treatment and disposal problems required by additional concentration of waste.

The flow cross-sectional area inside the filter's pressure chamber continually decreases in the flow direction. This helps to maintain the high cross-flow velocity, since filtered water is rapidly leaving the chamber through the membrane belt. A downstream port on the pressure feeder is used to recycle water back to the feed tank. The pressure feeder system is capable of operating at up to 50 psig. In microfiltration, inlet pressures rarely exceed 10 psig, but higher pressures are necessary when using ultrafiltration membranes.

Table 15 – Some Applications of VSEP Filtration

Chemical Process Industry	Waste & Pollution
Latex Concentrating	Bio-Sludge Concentration
Catalyst Washing and Concentration	White Water Treatment
Titanium Dioxide Filtration	Oil/Water Separation
Colloidal Silica Filtration	Metal Hydroxide Filtration
Acid Clarification	BOD/COD Reduction
Calcium Carbonate Concentration	Sludge Dewatering
Othalic Acid Catalyst Fines	Scrubber Effluent
NaOH Recovery in Bayer Process	
Boiler Water Treatment	**Mining**
Calcium Chloride Clarification	Mineral Clay Dewatering
	Red Mud Recovery
Food & Beverage	Mine Tailing Processing
Wine Clarification	Size Classification
Beer Clarification	
Beer Tank Bottoms	**Municipal Water/Waste Treatment**
Wine Lees Recovery	Drinking Water
Sugar Processing	Reclaimed Water
Vinegar Clarification	Activated Sludges
Juice Clarification	
Juice Colour Removal	**Oil Production/Processing**
Juice Extended Recovery	Produced Water
Starch Processing	Drilling Muds
Whey Processing	Completion Fluids
Soy Protein Concentration	Injection Water
Modified Starch Recovery	Hydraulic Cutting Fluids
Steeps Water Purification	Extraction Brine Recovery
Paints & Pigments	**Biotech/Pharmaceutical**
Organic & Inorganic Pigment Washing	Drug Washing and Concentration
Organic & Inorganic Pigment Concentrating	Cell Lysate Removal
Latex Paint Recovery	Drug Purification
Electrolytic Paint Recovery	Cell Harvesting
Dye Recovery	
Pulp & Paper	
Process Water Filtration	
Spray Head Water Filtration	
Lignin Removal (Plant Effluent Treatment)	

The outer edges of the pressure feeder system are sealed against the membrane surface with a proprietary method that minimise wear on the membrane surface. Flow through the membrane is driven by pressure above the vacuum beneath the belt.

Feed water, which has become highly concentrated in suspended matter within the pressure chamber, is directed onto the membrane belt surface immediately downstream of the pressure chamber for further dewatering by vacuum applied beneath the belt. After

MICROFILTRATION

vacuum dewatering, the solids content of the filter cake typically ranges from 20-80%, depending upon the nature of the solids.

Dewatered solids are removed from the belt by a doctor blade. A soft brush and a fine liquid spray clean the remaining solids from the membrane surface during the return cycle of the belt. Also, as the membrane belt passes over the tension roller during the return cycle, filtered liquid entrained in the fabric portion of the belt backflows through the membrane. The backwash removes small particles that are trapped in the membrane pores.

A vacuum chamber is located immediately below the membrane belt as it travels through the pressure chamber and subsequent filter cake dewatering sections. Liquid filtrate is drawn from the underside of the membrane belt into the vacuum chamber through a perforated steel plate. The vacuum enhances filtrate flow through the membrane belt and aids in drying solids in the downstream section of the belt.

SECTION 8

Analytical Application of Membranes

FOR UPDATES ON A REGULAR BASIS READ:

The international magazine for the Filtration & Separation Industry

Filtration & Separation is read by thousands of engineers, specifiers, designers and consultants across a range of industries:
- process engineering ● chemical engineering ● food and beverage ● petrochemical ● biotechnology ● water supply & treatment

and they rely on it to keep them up to date with developments in their industry

Coverage includes:
● Industrial Pollution Control ● Pressure Filters: Belt, Vacuum and Drum ● Centrifuges and Cyclones ● Municipal Water & Waste Treatment ● Utilities ● Filter Presses ● Metal Wire Cloth ● Screens & Plates ● Air Filters ● Contamination ● Dust Control ● Clean Rooms ● Filtration Media (& Fabrics) ● Overview of the World Filtration Industry ● Environmental Protection ● Toxic Waste ● Settling, Flotation and Gravity Filter Systems ● Food & Beverage Filter Processes ● Membranes ● Separation processes in chemical & allied Industries ● Cartridge & Bag Filters

Plus a World Directory of Filtration & Separation Equipment

Filtration & Separation is published 10 times a year, and is available from Elsevier Advanced Technology — see address on order form.

Because we are so confident of the value of **Filtration & Separation**, we offer a no-nonsense guarantee — if you are not satisfied, we will refund the cost of all unmailed issues without question!.

☐ Please bill me for _____ subscription(s) to
Filtration & Separation @ £75/US$115

Name: .... Position: ....
Organization: ....
Address: ....
....
.... Post/Zip: ....
Country: ....
Tel: .... Fax: ....

Return to:
Elsevier Advanced Technology ● PO Box 150 ● Kidlington ● Oxford OX5 1AS ● UK
Tel: +44 (0)1865 843841; Fax: +44 (0)1865 843971

ANALYTICAL APPLICATION OF MEMBRANES

Microfiltration membranes offer a general way of removing particulate material from fluid streams and are thus routinely used in a range of analytical procedures to determine particulate contamination in a wide range of gases and liquids. The procedures include the detection of microorganisms in a variety of waters and process fluids (foods, beverages, pharmaceuticals) where the membrane traps the microorganism and is subsequently used as the culture medium, passive cell growth studies and in so-called blotting applications.

Detection and Analysis of Particulate Contamination

The identification and monitoring of suspended particles in a range of environments has become increasingly important for reasons of health, hygiene and product quality and precision. The increased precision and lower tolerances for electronic and mechanical components used for example in aircraft, machine tools, hydraulic systems and nuclear power plants, and the decrease in line widths of integrated circuits demand that particulate contamination be controlled down to smaller particle sizes and lower numbers of particles. Particle monitoring is required for fuels, hydraulic oils, lubricants, water, chemicals, precision components, medical devices, pharmaceutical products and clean room environments.

Prolonged workplace and environmental exposure to airborne contaminants has been linked to a number of occupational diseases. Various national and international regulatory agencies have established standard methods of analysis and set threshold limits for a large number of airborne contaminants.

Overall membrane filtration is widely used as an analytical tool for collecting, identifying, and measuring particles in industrial hygiene and for the sampling, detection, and analysis or particulate contamination in liquids and gases by filtration through microporous membranes. In general, these methods involve passing a representative sample of a liquid or gas through a suitable filter disc. All particulate matter which exceeds the membrane pore size are retained on the surface where the contaminants may then be analysed.

Sampling methods for the determination of particulate contamination in gases includes area sampling for clean rooms and industrial hygiene and as personal monitoring for worker safety.

Adequate attention must be given to proper sampling techniques to ensure valid, reproducible results. Sampling variables can be categorised as contamination, sample adequacy, sample representativeness and time.

Sampling equipment, containers, analytical apparatus and membrane filters must be clean if subsequent measurements are to be a valid index of the system being tested. The number of samples, the sample volume, and sampling time depend upon both the level of system contamination and the type of measurement being employed. In generally, sufficient material must be sampled so that the collected contaminant will be clearly measurable at "dangerous" levels of contamination. To be meaningful and reproducible, samples should be as representative as possible of the entire fluid system when the system is operating normally. Airborne particulate concentration can vary widely in industrial hygiene situations depending on the operations being carried out.

There are numerous technique of analysis performed on microfiltration membranes to monitor over 120 workplace airborne contaminants. Table 1 illustrates the wide range of

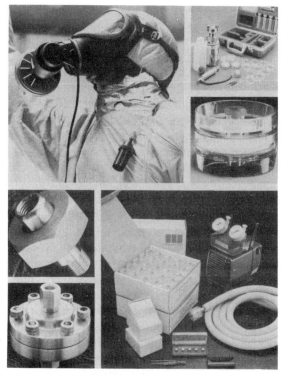

Top left – Personal monitoring. Top right – Patch test kit
Middle left – Gas line filter holder. Middle right – Fluid contamination analysis monitor.
Bottom left – High pressure stainless steel filter holder. Bottom right – Aerosol monitoring kit.

ANALYTICAL APPLICATION OF MEMBRANES

TABLE 1 – Air Monitoring Filter Selection Guide.

This Guide is a handy reference tool procided to help you monitor any of the over 120 workplace airborne particulates listed. The Guide summarizes most of the relevant information you need, including filter types, NIOSH Analytical Methods and OSHA Exposure Limits. See Explanatory Notes for further details.

NIOSH/OSHA ANALYTICAL METHOD: These methods are referenced in NIOSH and OSHA literature. For some contaminants, more than one analytical method is presented.

NIOSH REFERENCE: Most numbered analytical methods are described in the NIOSH Manual of Analytical Methods, third (Aug. 1987) edition. Methods that have not been revised for this edition are described in the second edition, Vols. 1-3 (Apr. 1977), and Vol. 4 (Aug. 1978).

OSHA PERMISSIBLE EXPOSURE LIMIT: Exposure limits are promulgated and enforced by the Occupational Safety and Health Administration. The above values are referenced in the Code of Federal Regulations, Title 29, Part 1910 (July 1, 1988). These references should be consulted to establish 8-hour TWA's (Time Weighted Averages) and Ceiling Values. A blank in the table indicates that dno standard has been published.

CONTAMINATION ANALYSIS
    air sampling   water pollution
    hydraulic oil monitoring   beverage analysis
    fuel testing

CONTAMINATION CONTROL
    particle removal in HPLC camples   tissue culture media sterilization
    solvent clarification   air and gas filtration
    water purification

Contaminant	Filter Type	NIOSH/OSHA Analytical method	NIOSH Ref.	OSHA Permissible Exposure Limit	ACGIH Threshold Limit Values
Aldrin	GF	Gas Chromatography	5502	0.25 mg/m3	0.25 mg/m^3
Alkaline Dusts	PTFE 1 μm	Titration	7401	2 mg/m3 (NaOH)	
Alumium	MCE 0.8 μm	Atomic Absorption	7013		10 mg/m^3 (metal oxide) 5 mg/m^3 (pyro powders, welding fumes) 2 mg/m^3 (soluble salts; alkyls)
4-Aminobiphenyl	GF	Gas Chromatography	269	(car.)	
Antimony and compounds, as Sb	MCE 0.8 μm	Atomic Absorption Anodic Stripping Valtammetry	S2,261 189	0.5 mg/m^3	0.5 mg/m^3
ANTU (alpha naphtyl thiourea)	MCE 0.8 μm	Gas Chromatography	S276	0.3 mg/m^3	0.3 mg/m^3
Arsenic, as As	MCE 0.8 μm	Atomic Absorption ICP	7900 7300	0.5 mg/m^3 (c)	0.2mg/m^3
Arsenic, organo	PTFE 1 μm	Ion Chromatography/ Hydride Atomic Absorption	5022	0.5 mg/m^3	0.2 mg/m^3 as As

TABLE 1 (continued).

Contaminant	Filter Type	NIOSH/OSHA Analytical method	NIOSH Ref.	OSHA Permissible Exposure Limit	ACGIH Threshold Limit Values
Arsenic Trioxide as As	MCE 0.8 µm	Atomic Absorption	7901	0.01 mg/m^3 (As)	
Asbestos	MCE 0.8-1.2 µm	Phase Contrast Microscopy TEM	7400 7402	footnote (5)	footnote(5)
Barium, soluble compounds	MCE 0.8 µm	Atomic Absorption	7056	0.5 mg/m^3	0.5 mg/m^3
Benzidine and 3,3-Dichlorobenzdine	GF	HPLC	5509	(car.)	
Benzoyl Peroxide	MCE 0.8 µm	HPLC	5009	5 mg/m^3	5 mg/m^3
Beryllium and compounds	MCE 0.8 µm	Atomic Absorption ICP	7102 7300	2 µg/m^3	2 µg/m^3
Bromoxynil, and Bromoxynil Octanoate	PTFE 1-2 µm	HPLC	5010		
Cadmium, dust	MCE 0.8 µm	Atomic Absorption ICP	7048 7300	0.2 mg/m^3	0.5 mg/m^3
Cadmium, fume	MCE 0.8 µm	Atomic Absorption ICP	7048 7300	0.1 mg/m^3	0.5 mg/m^3
Calcium	MCE 0.8 µm	Atomic Absorption ICP	7020 7300	5mg/m^3 (CaO)	2 mg/m^3
Carbaryl (Sevin®)	GF	CVolorimetric	5006	5 mg/m^3	5 mg/m^3
Carbon Black	MCE Matched Weight 0.8 µm	Gravimetric	5000	3.5 mg/m^3	3.5 mg/m^3
Chlorinated Camphene (Toxaphene)	MCE 0.8 µm	Gas Chromatography	S67	0.5 mg/m^3	0.5 mg/m^3
Chlorinated Diphenyl Oxide (Ether)	MCE 0.8 µm	Gas Chromatography	5025	0.5 mg/m^3	0.5 mg/m^3
Chlorinated Terphenyl (60% chlorine)	GF	Gas Chromatography	5014		
Chromium, as Cr	MCE 0.8 µm	Atomic Absorption ICP	7024 7300	1 mg/m^3	0.5 mg/m^3

ANALYTICAL APPLICATION OF MEMBRANES

TABLE 1 (continued).

Contaminant	Filter Type	NIOSH/OSHA Analytical method	NIOSH Ref.	OSHA Permissible Exposure Limit	ACGIH Threshold Limit Values
Chromium, soluble salts, as Cr	MCE 0.8 μm	Atomic Absorption ICP	7024 7300	0.5 mg/m^3	0.05 mg/m^3
Coal Tar Pitch Volatiles, (benzene solubles)	PTFE 1, 2 & 0.5 μm	Gravimetric (benzene extraction)	5023	0.2 mg/m^3	0.2 mg/m^3
Colbalt and compounds, as Co	MCE 0.8μm	Atomic Absorption ICP	7027 7300	0.1 mg/m^3	0.05 mg/m^3
Copper, dust and fume	MCE 0.8 μm & 5 μm	Atomic Absorption ICP	7029 7300	1 mg/m^3	1 mg/m^3 (dust) 0.2 mg/m^3 (fume)
Cyanides, aerosol and gas	MCE 0.8 μm	Ion Specific Electrode	7904	5 mg/m^3 (as CN) 11 mg/m^3 (as HCN)	5 mg/m^3
DDT	GF	Gas Chromatography	S274	1 mg/m^3	1 mg/m^3
2,4-D and 2,4,5-T	GF & PTFE 5 μm	HPLC	5001	10 mg/m^3	10 mg/m^3
Demeton	MCE 0.8 μm	Gas Chromatography	5514	0.1 mg/m^3	0.1 mg/m^3
Diborane	PTFE 1 μm & 0.5 μm MCE 0.8 μm	Plasma Emission Spectrometry ICP	6006 7300	0.1 ppm	0.1 mg/m^3
Dibutyl Phosphate	PTFE 1 μm	Gas Chromatography	5017	5 mg/m^3	5 mg/m^3
Dibutyl Phthalate & Di (2-Ethylhexyl) Phthalate	MCE 0.8 μm	Gas Chromatography	5020	5 mg/m^3	5 mg/m^3
Dieldrin	GF	Gas Chromatography	S283	0.25 mg/m^3	0.25 mg/m^3
4-Dimethylamino-azobenzene	GF	Gas Chromatography	284		10 mg/m^3
Dinitrobenzene	MCE 0.8 μm	HPLC	S214	1 mg/m^3	1 mg/m^3
Dinitrotoluene	MCE 0.8 μm	HPLC	S215	1.5 mg/m^3	1.5 mg/m^3

TABLE 1 (continued).

Contaminant	Filter Type	NIOSH/OSHA Analytical method	NIOSH Ref.	OSHA Permissible Exposure Limit	ACGIH Threshold Limit Values
Dyes, Benzidine-o-Tolidine, o-	PTFE 5 μm	HPLC	5013		
EPN	GF	Gas Chromatography	5012	0.5 mg/m^3	0.5 mg/m^3
Ethylene Glycol	GF	Gas Chromatography	5500		125 mg/m^3 (c)
Ethylene Thiourea (2-imidazolidine-thione)	MCE 0.8 μm	Spectrophotometric	5011	(car.)	
Fibers, asbestos and fibrous glass	MCE 0.8 to 1.2 μm	Phase Contrast Microscopy	7400	footnote (5)	footnote (5)
Fluorides, aerosol and gas	MCE 0.8 μm	Ion Specific Electrode Ion Chromatography	7902 7903	2.5 mg/m^3	2.5 mg/m^3 (as F)
Formaldehyde	PTFE 1 μm	Visible Absorption Spectrophotometry	3500	3 ppm	1.5 mg/m^3
Graphite	MCE Matched Weight Monitor)	Gravimetric	0600 0500		2.5 mg/m^3 (natural resp. 10 mg/m^3 (synthetic) total dust
Hafnium	MCE 0.8 μm	Atomic Absorption	ID-121 (OSHA)	0.5 mg/m^3	0.5 mg/m^3
Hexachloronaphth-alene	MCE	Gas Chromatography	S100	0.2 mg/m^3	0.2 mg/m^3
Hexamethylene-tetramine	MCE 0.8 μm	Spectrophotometric	263		
Hydroquinone	MCE 0.8 μm	HPLC	5004	2 mg/m^3	2 mg/m^3
Iron Oxide fume	MCE 0.8 μm	ICP	7300	10 mg/m^3	5 mg/m^3
Kepone	MCE 0.8 μm	Gas Chromatography	5508		
Lead, metal, fume, & other aerosols	MCE 0.8 μm	Atomic Absorption ICP	7082 7300	50 μg/m^3	0.15 mg/m^3
Lindane	GF	Gas Chromatography	5502	0.5 mg/m^3	0.5 mg/m^3
Lithium Hydride	MCE 0.8 μm	ICP	7300	25 μg/m^3	0.025 mg/m^3

TABLE 1 (continued).

Contaminant	Filter Type	NIOSH/OSHA Analytical method	NIOSH Ref.	OSHA Permissible Exposure Limit	ACGIH Threshold Limit Values
Magnesium Oxide fume	MCE 0.8 μm	ICP	7300	15 mg/m^3	10 mg/m^3
Malathion	GF	Gas Chromatography	5012	15 mg/m^3	10 mg/m^3
Manganese	MCE 0.8 μm	ICP	7300	5 mg/m^3(c)	5 mg/m^3(c) 1 mg/m^3 (fume)
Mercury (particulate) (non-particulate)	GF or MCE 0.8 μm	Atomic Absorption	6000	0.1 mg/m^3	0.1 mg/m^3 (inorganic)
Methoxychlor	GF	Gas Chromatography	S371	15 mg/m^3	10 mg/m^3
4,4'-Methylenebis-(2-chloroaniline) (MOCA®)	GF	HPLC	236	(car.)	
Mineral Oil Mist	MCE 0.8 or 5 μm or GF	I.R. Spectrophotometry	5026	5 mg/m^3	5 mg/m^3
Molybdenum, insoluble compounds	MCE 0.8 μm	ICP	7300	15 mg/m^3	10 mg/m^3
Molybdenum, soluble compounds	MCE 0.8 μm	ICP	7300	5 mg/m^3	5 mg/m^3
Naphthylamines	GF	Gas Chromatography	5518	(car.)	(car.)
Nikel, metal and soluble compounds, as Ni	MCE 0.8 μm	ICP	7300	1 mg/m^3	1 mg/m^3 (metal) 0.1 mg/m^3 (soluble compounds as Ni)
4-Nitrobiphenyl	GF	Gas Chromatography	273	(car.)	(car.)
Nuisance dust	MCE Matched Weight 0.8 μm	Gravimetric	0600 0500	5 mg/m^3 (respirable fraction) 15 mg/m^3 (total dust)	10 mg/m^3
Octachloronaphthalene	MCE 0.8 μm	Gas Chromatography	S97	0.1 mg/m^3	0.1 mg/m^3
Oil mist, mineral	MCE 0.8 or 5 μm or GF	Infrared Spectroscopy	5026	5 mg/m^3	5 mg/m^3

TABLE 1 (continued).

Contaminant	Filter Type	NIOSH/OSHA Analytical method	NIOSH Ref.	OSHA Permissible Exposure Limit	ACGIH Threshold Limit Values
Organo (alkyl) mercury	MCE 0.8 μm	Atomic Absorption	6000	0.01 mg/m^3	0.01 mg/m^3
Paraquat	PTFE 1 μm	HPLC	5003	0.5 mg/m^3	0.1 mg/m^3 resp.
Parathion	GF	Gas Chromatography	5012	0.1 mg/m^3	0.1 mg/m^3
Pentachloro-naphthalene	GF	Gas Chromatography	S96	0.5 mg/m^3	0.5 mg/m^3
Pentachlorophenol	MCE 0.8 μm	HPLC	S297	0.5 mg/m^3	0.5 mg/m^3
Phosphoric acid	MCE 0.8 μm	Ion Chromatography	ID-111 (OSHA)	1 mg/m^3	1 mg/m^3
Phosphorus	MCE 0.8 μm	ICP	7300	1 mg/m^3	0.1 mg/m^3 (yellow)
Phthalic Anhydride	MCE 0.8 μm	HPLC	S179	12 mg/m^3	6 mg/m^3
Picric Acid	MCE 0.8 μm	HPLC	S228	0.1 mg/m^3	0.1 mg/m^3
Platinum, soluble salts, at Pt	MCE 0.8 μm	ICP	7300	2 μg/m^3	1 mg/m^3 (metal) 0.002 mg/m^3 (soluble salts as Pt)
Polychlorobenzenes	PTFE 5 μm	Gas Chromatography	5517		
Polychloro-biphenyls	GF	Gas Chromatography	5503	1 mg/m^3 (42%Cl) 0.5 mg/m^3 (54%Cl)	1 mg/m^3 0.5 mg/m^3
Polynuclear Aromatic Hydrocarbons	PTFE 1-2 & 0.5 μm GF & AG 0.8 μm	HPLC Gas Chromatography Spectrophotometric (column chromatography)	5506 5515 184	0.2 μg/m^3 (benzapyrene)	
Pyrethrum	GF	HPLC	5008	5 mg/m^3	5 mg/m^3
Rhodium, metal, fume & dust, as Rh	MCE 0.8 μm	Atomic Absorption	S188	0.1 mg/m^3	1 mg/m^3

TABLE 1 (continued).

Contaminant	Filter Type	NIOSH/OSHA Analytical method	NIOSH Ref.	OSHA Permissible Exposure Limit	ACGIH Threshold Limit Values
Rhodium, soluble salts	MCE 0.8 μm	Atomic Absorption	S189	1 μg/m^3	0.01 mg/m^3
Ronnel	GF	Gas Chromatography	S299	15 mg/m^3	10 mg/m^3
Rotenone	PTFE 1 μm & 1μm (13 mm)	HPLC	5007		5 mg/m^3
Selenium compounds (as Se)	MCE 0.8 μm	ICP	7300	0.2 mg/m^3	0.2 mg/m^3
Silica, crystalline	MCE 0.8 & 0.45 μm MCE 0.8 μm	Visible Absorption Spectrophotometry Infrared Absorption Spectrophotometry	7601 7602	footnote (4)	0.1 mg/m^3 (Quartz) fee nootnote (4)
Silver, metal and soluble compounds	MCE 0.8 μm	ICP	7300	0.01 mg/m^3	0.1 mg/m^3 (metal) 0.01 mg/m^3 (soluble compounds as AG)
Strychnine	GF	HPLC	5016	0.15 mg/m^3	0.15 mg/m^3
Sulphur Dioxide	MCE 0.8 & 0.45 μm	Ion Chromatography	6004	5 ppm	5 mg/m^3
Tellurium	MCE 0.8 μm	ICP	7300	0.1 mg/m^3	0.1 mg/m^3
O-Terphenyl	PTFE 1-2 μm	Gas Chromatography	5021	9 mg/m^3	5 mg/m^3(c)
Tetrachloro-anphthalene	GF	Gas Chromatography	S130	2 mg/m^3	2 mg/m^3
Tetryl	MCE 0.8 μm	Colorimetric	S225	1.5 mg/m^3	1.5 mg/m^3
Thallium, soluble compounds, as T1	MCE 0.8 μm	ICP	7300	0.1 mg/m^3	0.1 mg/m^3
Thiram	PTFE 1 μm & 5 μm	HPLC	5005	5 mg/m^3	5 mg/m^3
Tin, inorganic compounds except oxides	MCE 0.8 μm	ICP	7300	2 mg/m^3	2 mg/m^3

TABLE 1 (continued).

Contaminant	Filter Type	NIOSH/OSHA Analytical method	NIOSH Ref.	OSHA Permissible Exposure Limit	ACGIH Threshold Limit Values
Tin, organic (organotin compounds)	GF	Atomic Absorption	5504	0.1 mg/m^3	0.1 mg/m^3
Titanium Dioxide	MCE 0.8 μm	Atomic Absorption	S385	15 mg/m^3	10 mg/m^3
Tributyl Phosphate	MCE 0.8 μm	Gas Chromatography	S208	5 mg/m^3	2.5 mg/m^3
Trichloro-naphthalene	GF	Gas Chromatography	S128	5 mg/m^3	5 mg/m^3
2,4,7, Trinitro fluoren-9-one	PTFE 0.5 μm	HPLC	5018		
Triorthocresyl Phosphate	MCE 0.8 μm	Gas Chromatography	S209	0.1 mg/m^3	0.1 mg/m^3
Triphenyl Phosphate	MCE 0.8 μm	Gas Chromatography	S210	3 mg/m^3	3 mg/m^3
Tungsten, soluble and insoluble	MCE 0.8 & 0.45 μm	Atomic Absorption ICP	7074 7300		5 mg/m^3 (insoluble) 1 mg/m^3 (soluble)
Vanadium, total V	MCE 0.8 μm	ICP Atomic Absorption	7300 7101	0.5 mg/m^3 (c)	0.05 mg/m^3 as V$_2$O$_5$
Warfarin	PTFE 1 μm	HPLC	5002	0.1 mg/m^3	0.1 mg/m^3
Welding and Brazing Fume (Ag, Cd, Cr, Cu, Fe, Ni, Mn, Zn-metals & oxides)	MCE 0.8 μm	X-Ray Fluorescence ICP	7200 7300	5 mg/m^3	5 mg/m^3
Ytrium	MCE 0.8 μm	ICP	7300	1 mg/m^3	1 mg/m^3
Zinc, and compounds, as Zn	MCE 0.8 μm	Atomic Absorption ICP	7030 7300	5 mg/m^3 (as oxide)	5 mg/m^3 (as oxide)
Zirconium compounds, as Zr	MCE 0.8 μm	ICP	7300	5 mg/m^3	5 mg/m^3

FOOTNOTES TO TABLE 1
1. **Abbreviations**
   HPLC - High Performance Liquid Chromatography
   ICP - Inductively Coupled Argon Plasma, Atomic Emissions Spectroscopy
   TEM - Transmission Electron Microscopy
   m^3 - cubic meter
   c - ceiling limit
   Car - carinogen
   resp. - respirable
2. **Filter Type**
   Millipore filters (especially mixed cellulose esters, Type AA 0.8 µm) are referenced as the standard in many NIOSH methods including asbestos, arsenic, and lead.
3. **Gravimetric Analyses**
   For most methods that reference the use of a PVC-type filter, the Matched Weight Monitor is an alternative. With these preassembled monitors, you eliminate the need for preweighing filters, and the potential for contamination in assembly.
4. **OSHA Permissible Exposure Limits for Crystalline Silica**
   Quartz respirable $\dfrac{10}{\%SiO_2 + 2}$ mg/m^3

   Cristobalite and Tridymite: use $^1/_2$ the value calculated from the formulae for quartz.
   ACGIH: Please consult the "Threshold Limit Values" handbook, second printing, 1987-87
5. **Fibers (Method 7400)**
   It is virtually impossible to distinguish between glass and asbestos fibers using Phase Contrast Microscopy. Therefore, the limits specify level of fibers, or asbestos fibers. These limits are as follows:
   OSHA: 0.2 asbestos fibers (>5 µm long) mL.
   ACGIH: 0.5 amosite; 2 chrysolite; 0.2 crocidolite; 2 other forms, filter/mL.
   Asbestos (Method 7402)
   Using TEM, this method is definitive for asbestos fibers. The limits are as follows:
   OSHA: The same as for Fibers (see above)
   ACGIH: 0.5 amosite; 2 chrysolite; 0.2 crocidolite; 2 other forms, filters/mL.

EXPLANATORY NOTES
For some contaminants represent most workplace airborne particulates referenced in NIOSH/OSHA literature. These airborne contaminants require a filter collection step in the sampling method.

FILTER TYPE
MCE: mixed cellulose esters (membrane filter)
PTFE: polytetrafluorethylene (membrane filter)
GF: glass fiber, binderless (depth filter)
AG: silver (metal membrane filter)

procedures which vary from simple gravimetric methods to more sophisticated procedures of ICP and HPLC etc. The sampling procedure should use a filter/membrane of the correct type in terms pore size, particle capture mechanism (surface or depth) and of a suitable material of construction. Many factors have to be considered, such as is the sampling from a gas, liquid (or surface), is the membrane to be used as a substrate for chemical analysis, is the membrane to be used in a microscopic analysis etc?

A range of different types of filter medium are used (see Table 2). These include

i) Mixed cellulose esters. Biologically inert mixtures of for example cellulose acetate and cellulose nitrate. Suitable for a wide range of analytical procedures including gravimetric analysis by the ashing technique and light microscopy.

ii) Track etched polycarbonate. Recommended for scanning and transmission electron microscopy.

iii) PTFE either backed with high density polyethylene or polypropylene or unbacked. For applications with gases and non-aqueous fluids with acids and alkalis and higher temperature operation.

TABLE 2 – Membrane Filter Selection for Analysis.

Filter Type	Mean Pore Size(μm)	Typical Flow Rate[1] Water[2]	Air[3]	Typical Porosity (%)	Typical Refractive Index	Minimum Bubble Point[4] bar	psi
MF-Millipore (mixed cellulose acetate and nitrate)							
SC	8.0	630	65	84	1.515	0.42	6
SM	5.0	400	32	84	1.495	0.42	6
SS	3.0	296	30	83	1.495	0.70	10
RA	1.2	222	20	82	1.512	0.77	11
AA	0.80	157	16	82	1.510	0.98	14
(black)		157	16	82	N.A.	1.12	16
DA	0.65	111	9	81	1.510	1.20	17
HA	0.45	38.5	4	79	1.510	2.11	30
(black)		38.5	4	79	N.A.	2.32	33
PH	0.30	29.6	3	77	1.510	2.46	35
GS	0.22	15.6	2	75	1.510	3.52	50
VC	0.10	1.5	0.4	74	1.500	14.1	200
VM	0.05	0.74	0.2	72	1.500	17.6	250
VS	0.025	0.15	0.15	70	1.500	21.1	300
Fluoropore (PTFE)							
FS	3.0	286	20	85	See Note (5)	0.05	07
FA	1.0	90	16	85		0.21	3
FH	0.5	40	8	85		0.49	7
FG	0.2	15	3	70		0.91	13
Mitex (PTFE)							
LC	10.0	126	14	68	N.A.	0.04	0.5
LS	5.0	51.9	9	60	N.A.	0.06	0.9
Durapore (polyvinylidene difluoride)							
SVLP	5.0	288	N.A.	70	N.A.	0.21	3
DVPP	0.65	69	N.A.	70	N.A.	0.98	14
HVHP	0.45	35	6	75	N.A.	0.56	8
HVLP	0.45	29	N.A.	70	N.A.	1.5	22
GVHP	0.22	15	3	75	N.A.	1.20	17
GVWP	0.22	6.9	N.A.	70	N.A.	2.46	35
VVLP	0.10	2.5	N.A.	70	N.A.	4.9	70

1. Flow rates listed are based on measurements with clean water and air, and represent initial flow rates for a liquid of 1 centipoise viscosity at the start of filtration, before filter plugging is detectable. Actual initial flow rates may vary from the average values given here.
2. Water flow rates are mililiters per minute per cm² of filtration area, at 20°C with a differential pressure of 0.7 bar (10 psi). Flow rates for Fluoropore, Durapore hydrophobic, and Mitex filters are based on methanol instead of water.
3. Air flow rates are liters per minute per cm² of filtration area, at 20°C with a differential pressure of 0.7 bar (10 psi) and exit pressure of 1 atmosphere (14.7 psia).
4. Bubble-point pressure is the differential pressure required to force air through the pores of a water-wet filter (except methanol-wet for Fluoropore, hydrophobic Durapore, and Mitex filters).
5. Crystalline and amorphous regions of Fluoropore filters have differing refractive indices, and it is therefore not possible to obtain uniform clearing.

TABLE 3 – Standard Procedures for Analysis of Particulate Contaminants.

Literally hundreds of "in house" variations of published procedures have been developed by individual companies and government agencies. Major published procedures are listed below.
NIOSH 7400 Asbestos Method (most recent revision available through OSHA, Salt Lake City, Analytical Laboratory Microscopy Branch, Salt Lake City, Utah) adopted by OSHA and EPA with some modifications, requires use of a 25 mm Monitor and mixed cellulose ester (MCE) membranes for collection and preparation of asbestos samples prior to analysis with phase contrast microscopy.
OSHA Reference Method effective July 26, 1986 (29 CFR Parts 1910 and 1926, referenced in Federal Register of June 20, 1986), Occupational Exposure to Asbestos, Tremolite, Anthophyllite, Actinolite Final Rules, requires use of a 25 mm monitor with conductive cowl and 0.8 µm or 1.2 µm pore size MCE filter and phase contrast microscopy.
OSHA ID 160, available from OSHA, Salk Lake City, Utah (tel. number 801-524-4270) adoption of the NIOSH 7400 Method for asbestos requires use of a conductive 25 mm monitor with extension cowl and 0.8 µm or 1.2 µm pore size MCE filter and phase contrast microscopy.
EPA rule regarding Asbestos Containing Materials in Schools under section 203 of title II of TSCA referenced in Federal Register of October 30, 1987 (40 CFR part 263, P. 41826-41905) requires use of a 25 mm monitor with center ring or conductive cowl and 0.45 µm mixed cellulose ester filter (MCE) with 5.0µm pore size MCE backup filter and phase contrast microscopy. Sample preparation and analysis with transmission electron microscopy (TEM) is also covered in the October 1987 Federal Register (40 CFR part 63).
SAE ARP-598A Procedure for the Determination of Particulate Contamination of Hydraulic Fluids by the Particle Count Method which uses oblique lighting of the Millipore filter surface. White or black Millipore filters may be used. Particles are counted in five size ranges (fibers are reported separately).
Federal Test Method Standard No. 791a, Method 3009-T, Lubricants, Liquid Fuels, and Related Products; Methods of Testing which is cirtually identical to SAE ARP-598 above.
SAE ARP-743A Procedure for the Determination of Particulate Contamination of Air in Dust Controlled Spaces by the Particle Count Method which is similar to ARP-598, specifically for air sampling, but which counts only two particle size ranges and fibers.
ASTM F25-68 Standard Method for Sizing and Counting Airbourne Particulate Contamination in Clean Rooms and Other Dust-Controlled Areas Designated for Electronic and Similar Applications equivalent to ARP-743 except only two particle size ranges only are counted.
ASTM F24-65 Standard Method for Measuring and Counting Particulate Contamination on Surfaces counts fibers and three particle size ranges either directly on, or washed from, the surface of components.
ASTM F51-68 Standard Method for Sizing and Counting Particulate Contaminants in and on Clean Room Garments measures detachable particulate contamination of 5µm and larger collected from garment fabrics.
F31-78 Processing Aerospace Liquid Samples for Particulate Contamination Analysis using membrane filters.
F312-69 Microscopical Sizing and Counting Particles from Aerospace Fluids on membrane filters.
F318-78 Sampling Airbourne Particulate Contamination in Clean Rooms for Handling Aerospace Fluids.
D2276-83 Particulate Contamination in Aviation Turbine Fuels.
ARP 575 A Filter Patch Testing Procedures for Aerospace Hydraulic Pumps and Motors.

iv) Silver. Ideal collection media for analysis of crystalline silica by X-ray diffusion and for the analysis of organics.

v) PVDF. Suitable for aqueous or organic samples.

There are several different types of sampling devices for the wide range of applications:-

    Aerosol monitors
    Clean room monitoring
    Garment monitoring
    Personal monitoring
    Gas line sampling
    Fluids sampling
    Fuel samplers
    Laboratory filters

Microbiological Assay

Microporous membranes are important for the detection of microorganisms in foods, beverages, pharmaceutical products and potable water sources. The technique involves the filtration of the samples through a microfiltration membrane to trap the microorganism, then culturing the microorganism on the membrane, and then counting the grown colonies.

Testing of Foods, Beverages and Pharmaceuticals

The steadily growing requirements for the quality and the longer shelf life of foods and beverages means that manufacturers cannot limit quality assurance to inspection of the final product alone, such as a bottled beverage or a prepared food product. Continuous inspection of raw materials and in-process quality control tests throughout production are requiredif he wants to avoid later losses and customer complaints. Microbiological and aseptic testing play a significant role in such quality assurance.

In the soft drink industry, the microbiological and hygienic quality including the biological stability of the products are important criteria for their assessment as only a few microbes are often all it takes to spoil large quantities of a beverage.

Although technological development has reduced the risk of contamination by spoilage microbes, the issue of shelf life is now more critical as a result of the enormous production output now possible. Quality control of bottling and filling, in terms of chemical, and above all, biological stability, is now paramount.

The requirements for a practical microbiological test method that permit quantitative and reproducible detection of trace contamination and that can be performed efficiently and economically under routine conditions are fulfilled optimally by the membrane filter method. The principle of this method is based on the concentration of microorganisms from relatively large samples on the surface of the membrane filter and on culturing these microbes on a nutrient pad or an agar culture medium.

The basic membrane filter method takes a membrane filter of the appropriate pore size placed in a filter holder,through which the sample is filtered and microorganisms in the test sample are retained on the filter surface by the screening action of the membrane filter. Growth inhibitors can be removed by flushing the holder with sterile water following filtration. Afterwards, the membrane filter is placed on a culture medium and incubated. Nutrients and metabolites are exchanged through the pore system of the membrane filter. Colonies, which have developed on the membrane filter surface during incubation, are counted and related to the sample volume.

The method offers the following advantages:

Compared with the direct method, considerably large sample volumes can be tested. This concentration effect increases the accuracy of microbial detection.

The membrane filter with colony growth can be filed as a permanent record of the test.

The visible colonies can be related directly to the sample volume and thus give quantitative results.

The microorganisms can be detected by different methods. Methods involving culturing techniques and the microscope are used to detect microbes, whereas biochemical and serological techniques are commonly applied to differenitate among such organisms. For detecting microorganisms in cultures, liquid and solid culture media are employed. Microoranisms are concentrated by growth in or on these culture media. Quantitative detection is only possible with solid culture media because the individually developing colonies can be evaluated and counted on the surface.

The following culture media can be used for microbiological testing:

-Nutrient pad sets. These sets offer the most convenient way to use the membrane filter method.

-Absorbent pads to be wetted with culture media.

-Culture media with agar or gelatine as the solidifying agent; nutrient broth.

FIGURE 1 – Vacuum filter holders made of stainless steel (sterilise by flaming) or glass are primarily used.

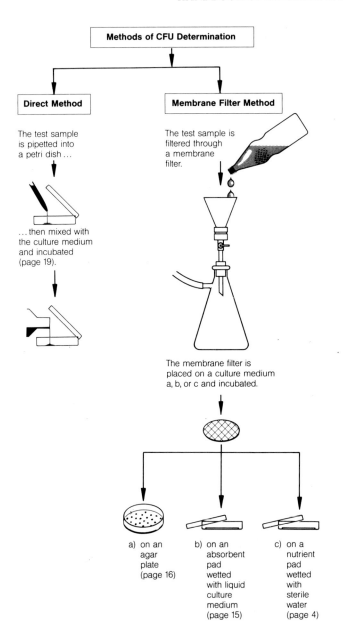

FIGURE 2 – Methods for culture medium determination.

Typical application of the culture medium in microbiological testing are given in Table 4a. The principle membrane used is cellulose nitrate or (acetate) with a 0.45 μm pore size.

TABLE 4a – Typical application examples.

Product:	Detection/determination of:	Nutrient pad type
Beer	Pediococci and lactobacilli Wild yeasts Yeasts and molds	VLB-S7-S Lysine Wort
Foods	Colony count Enterobacteria Mesophilic bacteria and thermophilic spore formers Pseudomonas aeruginosa Salmonellae Staphylococci Streptococci Yeasts and molds	Standard TTC, Standard, Caso Endo, Teepol, M-FC or Tergitol TTC Glucose-Tryptone Cetrimide Bismuth-Sulfite Chapman Azide Wort or Sabouraud
Milk	E. coli and coliforms Lactose-positive and lactose-negative bacteria Salmonellae Streptococci	Endo China Blue Bismuth-Sulfite Azide
Pharma- ceuticals and cosmetics	Candida albicans Colony count Fecal streptococci Mesophilic bacteria and thermophilic spore formers Pseudomonas aeruginosa Staphylococcus aureus	Wort or Sabouraud Standard TTC, Standard, Caso Azide Glucose-Tryptone Cetrimide Chapman
Soft drinks	Acid-tolerant microbes Colony count (CFU) Lactic-acid bacteria Slime-forming bacteria (Leuconostoc) Yeasts and molds	Orange Serum Standard VLB-S7-S Weman Wort or Schaufus-Pottinger
Sugar	Mesophilic bacteria and thermophilic spore formers Slime-forming bacteria (Leuconostoc) Yeasts and molds	Glucose-Tryptone Weman Wort or Schaufus-Pottinger
Water	Colony count E. coli and coliforms Fecal streptococci Pseudomonas aeruginosa	Standard TTC or Standard Endo, Tergitol TTC, Teepol or M-FC Azide Cetrimide
Wine	Acetobacter Lactic-acid bacteria Lactic-acid bacteria (especially Leuconostoc oenos) Yeasts	Wort, wetted with 3-5% ethanol Orange Serum Tomato Juice "Jus de Tomate" Wort or Schaufus-Pottinger

TABLE 4b – Typical applications for ready-to-use agar culture media.

Detection/determination:	Culture medium
CFU count (colony count) Hemolysis Resistance	DST Blood Agar
CFU count	Standard Agar Nutrient Agar acc. to DEV
Aerobic microbes	TS Agar (Caso Agar)
Pediococci and lactobacilli	VLB-S7-S Agar
Lactobacilli and yeasts	Tomato Juice Agar
Yeasts and molds	Sabouraud Agar Malt Extract Agar Wort Agar
Wild Yeasts	Lysine Agar
E. coli and coliforms	Endo Agar
Acid-tolerant microbes	Orange Serum Agar

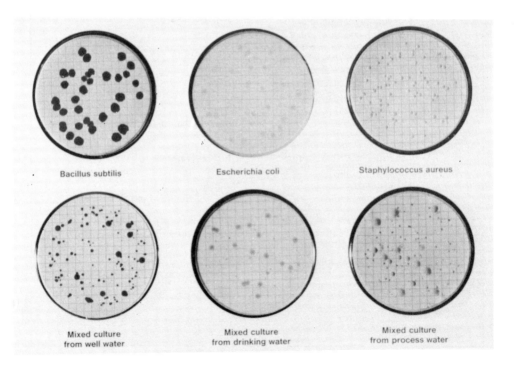

Bacillus subtilis   Escherichia coli   Staphylococcus aureus

Mixed culture from well water   Mixed culture from drinking water   Mixed culture from process water

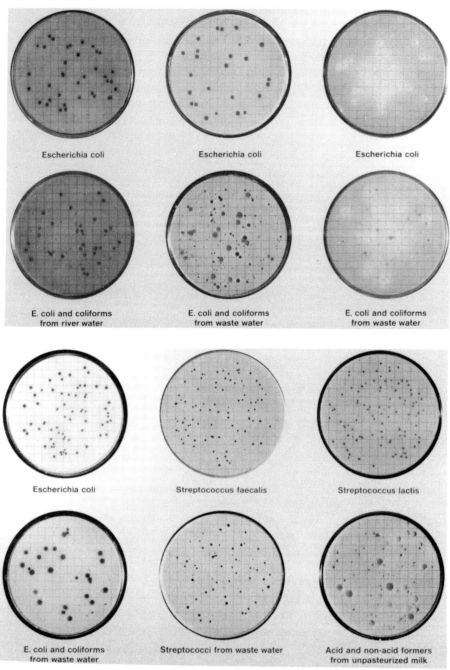

FIGURE 3 – Nutrient pad sets.

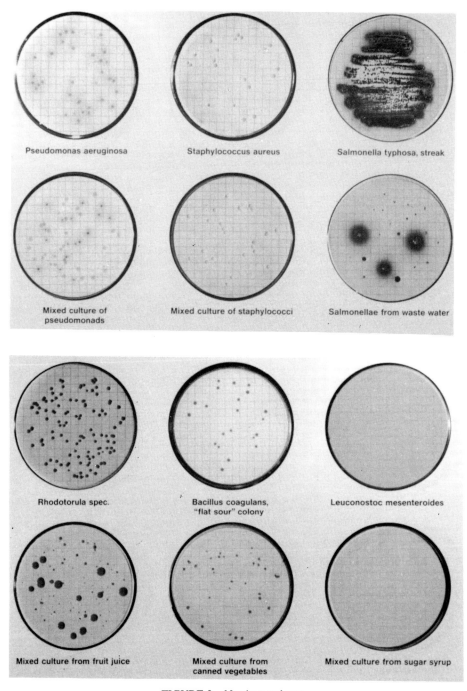

FIGURE 3 – Nutrient pad sets.

ANALYTICAL APPLICATION OF MEMBRANES

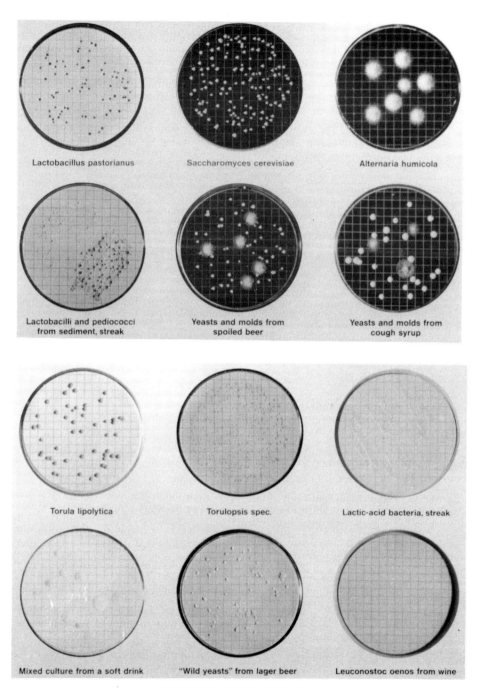

FIGURE 3 – Nutrient pad sets.

Water Microbiology

Coliform are a group of gram-negative . nonspore-forming, rod-shaped bacteria that ferment lactose at 35°C in 24 to 48 h. They are widely distributed in nature and many are native to the gut of warm-blooded animals and man.The presence of coliform organisms which are relatively harmful themselves is a usefull indicator of the presence of enteric pathogens. Coliform organisms are almost always present in water containing enteric pathogens. They are relatively easy to isolate and they survive longer than the disease-producing organisms. Thus coliforms are a useful indicator of the possible presence of enteric pathogenic bacteria and viruses and generally water that is free of total coliforms is free of disease-producing bacteria.

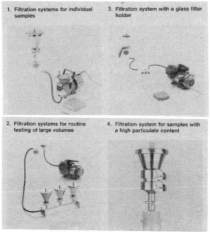

Filtration systems.

The classical test for the examination of coliform density in water supplies and water disposal systems has been the tube fermentation or MPN (most probable number) methods. This test is based on the ability of the coliform group of organisms to ferment lactose at 35°C in the appropriate growth medium, producing CO_2 in a 46-h period. The coliform group are known to be always present in water when enteric pathogens are present and thus coliforms satisfy the major criteria of the ideal indicator.

A major limitation of this method is that the overallprocedure requires four to five and thus alternative faster methods were developed based on the membrane filter technique.

The membrane filter technique uses a 45 µm pore size polymer membrane as a collection device utilising the capillary action of its the membrane pores to draw a selective medium for coliform growth onto its surface. When coliform organisms ferment lactose aldehyde is produced as one of the intermediate materials and is the indicator for coliform growth. Aldehyde formation is identified by the appearance of a metallic sheen on the surface of the coliform colonies caused by an aldehyde complex formed from the interaction with basic fuchsin and sodium sulphite present in the growth medium. Incubation periods of about 18 to 24 hours at temperature of 35°C, are required to complete the procedure, much less than the MPN system.

ANALYTICAL APPLICATION OF MEMBRANES

The membrane filter method is widely used for total coliform and also for faecal coliform, total bacteria and other bacterial tests. An advantage of the membrane filter is its ability to concentrate and localise bacteria from large samples thus increasing the sensitivity of quantitative bacteriology below one organism per milliliter. The membrane method is the easier and cheaper to run than the MPN method. The major limitations of the membrane filter method are associated with elevated turbidity, injured coliforms, high numbers of noncoliform bacteria, and that the membrane filter type may severely influence the sensitivity of the procedure.

The membranes for quantitative bacteriology must have pores that are small enough to retain bacteria but are open enough so not to interfere with growth of the organism. They also must not cause aberrant colony morphology. In general, gridded (to ease colony counting) membranes made from cellulose esters with average pore sizes of either 0.45 or 0.7 μm are used. The small pored membrane is commonly used for total coliform analyses, and the larger one for faecal coliform analyses.

There are several specialised system configurations to facilitate these microbiological analyses from various membrane suppliers. The samples can be vacuum filtered through 47 mm diameter membranes in nondisposable glass or stainless steel filter holders. If

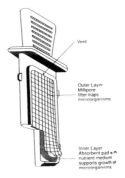

Sampling device.

several analyses are to be performed, it becomes more effective to use presterilised, self-contamined disposable holders; a 100 mL funnel with a 0.45 μm membrane. The system offers several features, such as:

1. No time-consuming preparation
2. Simplicity of use
3. Faster filtration
4. Consistent recoveries
5. Simplified colony counting
6. Choice of media
7. Compliance with worldwide standards.

Alternatively, integrated devices have been developed to facilitate the filtration of occasional samples. For field sampling, integrated samplers are available that can integrate a 0.45 μm membrane filter, the growth medium, the incubating dish, and sample filtration in one device.

TABLE 5 – Applications of polymer membrane substrates.

Pore size	Membrane Materials	Applications	
3.0 μm	Polycarbonate nontissue culture treated	Coculture Suspension culture Cell motility Cell invasion Transport studies	Tumor cell metastasis Chemotaxis Cell/cell interactions Plant cells
0.4 μm	Polycarbonate nontissue culture treated	Transport studies Plant cells Suspension culture	Cell/cell interactions Coculture
3.0 μm	Polycarbonate tissue culture treated	Enhanced differentiation Electropysiology Virus infection In vitro toxicology Permeability Endothelial cell penetration	Polarised functions Transport studies Cell invasion Tumor cell metastasis Plant cells Chemotaxis
0.4 μm	Polycarbonate tissue culture treated	Enhanced differentiation Electrophsiology Polarised functions In vitro toxicology	Transport studies Virus infection Permeability Plant cells
0.45 μm	Triton-free mixed esters of cellulose nitrate and acetate	Enhanced differentiation Electrophysiology Polarise functions In vitro toxicology	Transport studies Virus infection Permeability Plant cells
0.4 μm	Hydrophilised PTFE	Enhanced differentiation Electrophysiology Polarise functions In vitro toxicology Requires ECM coating Microscopically transparent	Transport studies Virus infection Permeability Plant cells

Other Microbiology

The membrane filtration method has found a wide range of applications in analysis for the presence, quantity and variety of microorganisms. All tests generally use mixed cellulose ester (or single) ester membranes with 0.45 μm (or 0.7 μm) pore size. Variations in the test procedures are largely found in the choice of culture medium and conditions for culture growth. Application areas include

1 Porous streams
2 De-ionised water
3 Airborne microorganisms
4 Soil
5 The effect of antibiotics
6 Detection of yeasts
7 Detection of alcoholic fermentation
8 Examination of solid foods

Cell Growth

Microfiltration membranes are used as passive substrates to which cells can attack and grow. Physiologically and anatomically, a microporous membrane is an ideal artificial surface for growing mammalian cells. This is particularly true for epithelial cells, which constitute more than 60% of all the recognised cell types in the human body.

Previously cells gown *in vitro* have typically been grown on impermeable "tissue culture" (plasma-discharge) treated polystyrene. The solid, impermeable nature of the polymer ensures that the cells attaches, grows, obtain nutrients, discharge metabolic wastes, and be exposed to fluid on only one side. The first commercial introduction of presterilised cell culture inserts occurred in 1985 using cell culture inserts, which contain a surfactant-free, mixed cellulose ester (acetate and nitrate), 0.45 μm microporous membrane. The convenience and numerous advantages of growing cells on microporous membrane is increasing.

The idealised properties of a membrane for cell culture are given in Table 6. A variety of polymeric membranes are currently used for cell culture depending upon the application. These include mixed cellulose ester (MF) membranes, hydrophilised PTFE membranes and track-etched polycarbonate membranes.

The relevant characteristics of these membranes are listed below:

Mixed cellulose   - biocompatibility but opaque and bind proteins
Hydrophilised PTFE   - microscopically transparent when wet

TABLE 6 – Characteristic proerties of a membrane required for cell culture.

1.	Biocompatibility; no extractables; inert-ness; nontoxic
2.	High porosity
3.	Wide range of pore sizes (0.2 to 30 μm)
4.	Superior cell adhesion characteristics
5.	No protein binding
6.	No autofluoresence
7.	Gamma sterilisable
8.	Reproducible (on a micron scale)
9.	Extremley thin for rapid diffusion
10.	Microscopically transparent
11.	Thin sections for both light and electron microscopy
12.	Mechanically strong and stable
13.	Easy to handle
14.	Inexpensive to use

Track-etched polycarbonate — negligible protein binding and background fluorescent - very thin (10 μm) but low porosity and microscopic capacity

TABLE 7 – Application areas of membrane cell culture.

1.	*Toxicology* Toxicity testing Animal testing alternatives Artificial tissue or organ equivalents
2.	*Cancer* Metastatic inhibition Chemotherapy susceptibility testing Gene alterations
3.	*Hormone receptors*
4.	*Pharmacology*
5.	*Drug discovery and mechanisms*
6.	*Cellular transport mechanisms*
7.	*Inflammation* Chemotaxis
8.	*Cellular aging*
9.	*Immunology*
10.	*Neurology*
11.	*Cell/cell interactions*
12.	*Mechanisms of virus infection and propagation* (eg HIV, influenza, etc)

The thinness of polycarbonate makes for rapid diffusion and hence rapid times to equilibrate. Some of the applications of these substrates are given in table and general areas of application in table .

Blotting of Macromolecules

Blotting generally refers to a process where macromolecules are transferred from an electrophoresis gel to a microporous membrane without disturbing their pattern. The macromolecule is typically immobilised by adsorption or by a covalent linkage to reactive groups on the solid phase. Once transferred to the membrane, macromolecules are more easily accessed by stains and probes and, in some cases, can be reprobed with different reagents. Consequently the "blotting" process can facilitate a range of analytical procedures and be particular useful for solid phase protein immunoassays or hybridization analysis with nucleic acids. There are several techniques for blotting as can be seen in Table 8.

ANALYTICAL APPLICATION OF MEMBRANES

TABLE 8 – Binding and Transfer Media Application.

Application	Immobilisation										Nucleic acid purification		Protein binding Glass fibre filter
	On cellulose nitrate membrane filter				Selex pp	On Nytran-N		Cova-lent DPT	Ionic				
	0.1 um CN	0.2 um CN*	0.45 um CN*	0.6 um CN	0.2 um SX	0.2 um N	0.45 um N		NA-49	NA-45	Elutip-d	Elutip-r	
DNA capillary bolts		b	B			S	S	S		S			
Electro-		S	S			B	B	S		S			
Dot-bolts			B	B			S	S		S			
RNA capillary bolts		B	B			S	S	S		S			
Electro-						B	B	B		B			
Dot-bolts		B	B	B			S	S		S			
LMW nucleic acids	S	S				S	S	B					
Immobilisation and recovery of:													
RNA										B		B	
mRNA										B		B	
DNA										B	B		
Frequent rehy-bridising of:													
DNA		S	S			B	B	B					
RNA		S	S			B	B	B					
Proteins:													
Capillary bolts		B	B			S	S	S		S			
Electro-bolts		B	B		B	S	S			S			
Dot-bolts		B	B	B	S	S	S			B			
Immuno-bolts		B	B			S	S			S			
Stained		B	B		B					S			
LMW	B	B	B		B	S		B		S			
Basic		S	S						B				
Colony hybridising		B	B	B		S	B						
Plaque-lifts			B	B	B		S	B					
TCA preci-pitation				S	S			S					B
Receptor binding studies				S									B
Hybrid selec-tion studies		S	S			S	S	B					
Protein micro-sequencing					B								B

LMW = low molecular weight
S = suitable
B = best medium for application
* = fleece-reinforced and non-reinforced   PP = polyproplene   CN = cellulose nitrate
N = polyamide

Materials used for the blotting techniques include membranes of cellulose nitrate, nylon 6 and 66 (and charge modified), PVDF, PVDF with surface cationic groups, polyamide (amphoteric and positively charged), as well as activated cellulose papers (eg DEAE) and activated glass fibre filters.

Nitrocellulsoe membranes are the commonly usedl matrices for binding proteins and nucleic acids. Their physical and chemical properties make them universal media for all transfer and immobilisation procedures. Nnitrocellulose membranes are used more often than any other binding matrix – for southern, northern and western blots, for colony and plaque transfers and dot blots. Nitrocellulose Membranes are available in three pore sizes; 0.45 μm pore are used preferably for nucleic acid transfers and colony/plaque transfers, 0.2 μm porosity have become the standard material for protein blotting. Their higher binding capacity is especially usefull in electrotransfer of low molecular weight proteins and 0.1 μm pore size are used for binding of very small proteins or peptides and for other special techniques. The are compatible with all important detection procedures; hybridisation with radioactive probes and autoradiography,

liquid scintillation counting,
hybridisation with non-radioactive chromogenci probes,
reaction with anionic dyes and India ink,
and binding of chromogenic enzyme-conjugated antibodies.

The nitrocellulose Membranes used consist of 100% pure nitrocellulose, contain extremely low extractable and are 100% triton-free. Supported Nitrocellulose membranes contain a synthetic precision tissue which is coated with nitrocellulose on both sides. Surface properties and binding capacity conform to those of unsupported nitrocellulose membranes and show the same low background binding. The supporting tissue are mechanical stable and are suitable for use with multiple reprobing and for repeated colony transfers.

Polyamide membranes are preferably used for transfer and immobilisation of nucleic acids. They exhibit better binding of double-strand DNA, small DNA fragments and RNA compared to that of nitrocellulose membranes and they have higher mechanical and thermal stability. This make them extremely well-suited for multiple reprobing and repeated colony transfers. The hydrophilic membrane structure ensure immediate wetting and well-defined behaviour during blocking.

Polyamide Membranes, because of their intermediate positive surface charge, exhibit optimal ratio of signal to background binding and allow covalent binding of single-strand DNA by alkaline transfer.

Hyrophobic PVDF based substrate have increased in use as an alternative to nitrocellulose in many protein blotting applications. This is due to its high binding capacity and physical robustness, together with its chemical stability, major attributes of this new substrate. Molecules which are problematic with other membranes appear to be well retained on its surface by a combination of hydrophobic and electrostatic interactions and the strong bipolar character of the molecules.

Overall applications of these membranes in blotting can be seen in Table 9

ANALYTICAL APPLICATION OF MEMBRANES

TABLE 9 – Suitable Immobilon Membranes.

	NC Pure	NC HATF	NC HAHY	P	P^{SQ}	CD	N	S
Applications								
• Southern Blotting (DNA)	1		4				3	2
• Northern Blotting (RNA)	1		4				3	2
• DNA Blotting and Reprobing	2							1
• Western (Protein) Blotting (general applications)	2		3	1				
• Protein Blotting prior to:								
– HIV detection				1				
– N-term,inal sequencing				2	1	3		
– Internal protein sequencing						1		
– High yield protein sequencing				2	1			
– Elution and recovery				2		1		
– AA analysis				1	2			
• Glycoprotein Visualization				1	1			
• Peptide Mapping				1		2		
• Phospho Peptide Sequencing				2			1	
• Dot/Slot Blotting (DNA, RNA, Proteins)	2	3		1				
• Binding Assays (Proteins)	2		4	1	3			
• Lipopolysaccharide Analysis				1				
• Colony Hybridization		1						2
• Cell Culture		1						
Detection Methods								
• Chromogenic	x		x	x	x			
• Radioactive	x		x	x	x		x	x
• Fluorescent	x		x	x	x			
• Chemiluminescent								x
Compatible Stains								
• Coomassie brilliant blue				x	x			
• Amido black	x		x	x	x			
• India ink	x		x	x	x			
• Ponceau S red	x		x	x	x			
• Colloidal gold	x		x	x	x			
• CPTS	x		x	x	x			
• Immobilon-CD stain	x	x	x	x	x			
• Transillumination	x		x	x	x			

NC = Nitrocellulose   P = PUDF

Biochemistry – Molecular Biology Sample Preparation

One of the most important steps in biochemistry and molecular biology research is the correct preparation of samples and reagents for analysis. Procedures such as concentrating, desalinating, deproteinising, clarifying and precipitating of the samples can effect the yield, activity and purity of the material. The effectiveness of these procedures is critical to obtaining reliable results. Membranes (UF and MF) perform valuable functions in many procedures from removing unicorproated nucleotides from labelled probes to elimination

TABLE 10 – Applications overview.

Applications	Sample/Method
DNA isolation	Elution from agarose gel
Nucleic acid clean up	PCR-amplified preps Nick-translated DNA, End-labeled DNA Random-primer labeled DNA Fluorescein-labeled oligos Genomic DNA drop dialysis
Cell separation	Single samples Multiple samples
Dialysis or diafiltration	Ultrafiltration Drop dialysis
Protein/DNA concentration	Ultrafiltration Ion-exchange chromatography
Serum/plasma concentration	Ultrafiltration
Minicolumn bead- or resin-based purification	Microfiltration for bead or resin retention
Particulate removal	General purification, clarification
Protein removal/separation	Ultrafiltration Ion-exchange chromatography Adsorption
Free vs. bound studies	Complex mixtures Single samples Multiple samples
Precipitate removal or solid phase study	TCA ppt. Individual samples Multiple samples Filtrate Recovery
Sterilization	Sterilizing filtration and mycoplasma removal
Precolumn purification for HPLC, FPLC®	Ion-exchange chromatography
IgG isolation	Ion-exchange

of primers form PCR preparations, from simple reagent clarification to purification techniques. Typical applications are illustrated in Table 10.

Blotting and analysis.

The procedures used can largely be achieved with a centrifugal filter unit for one step, small volume sample preparation, giving excellent recovery. Applications include both MF and UF although sterilising filtration and clarifying filtration and multiscreen filtration systems are used.

Centrifugal filtration unit.

MEDICAL APPLICATIONS
Filtration in Hospital Pharmacies

Microbiological contaminants include bacteria, mould and fungi that are endemic to the hospital environment. In the pharmacy, they can be transferred to critical solution by touch contamination, by the equipment used in preparing the solution, or by the air. There is also a chance that microbial contamination is in the solution itself. Clinically, microorganisms are a problem alive or dead. Alive they can multiply at logarithmic rates and cause nosocomial infections in compromised patients. There are a number of reported incidences of pharmacy-related nosocomial outbreaks. Dead microorganisms are a source of pyrogens which can cause pyretic reactions.

Mirobiological contamination rates for pharmacy controlled admixture programs have been reported from 1 to 26%. The number of additives made to a solution in the pharmacy increases the risk of such contamination.

Particulate contaminants also have clinical consequences. These contaminants include metal bits from needles, rubber corings, undissolved drugs, debris from the wall of containers, glass particles from ampoules, airborne fibres and soil, hair, dandruff, and so on. As with microorganisms the general sources of these contaminants include people, the environment, the equipment, and the solutions.

Studies have linked particulate contamination to phlebitis, granulomas, embolisms, silicosis, and other pathologic conditions. Several studies have indicated that particles below 5 μm may be linked to these conditions and that particles below 5 μm are present in hospital solutions.

The pharmacy distributes solutions throughout the hospital: hyperalimentation solutions, IV solutions, antibiotic solutions, extemporaneous solutions, irrigation solutions – all are compounded from bulk materials, reconstituted or added to in the pharmacy. Because of the widespread distribution of pharmacy solutions, it is especially critical to

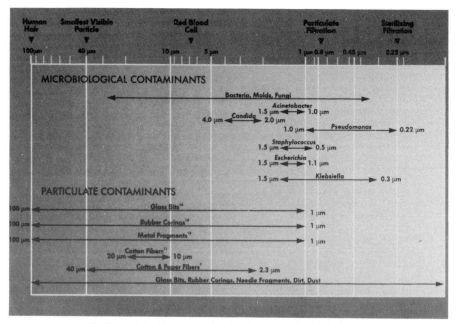

FIGURE 4 – Contamination in pharmeceutical solutions.

maintain the sterility and cleanliness of solutions and solution additives, as well as the technique, facilities and equipment used to prepare those solutions. The use of filters and filter-based systems simplifies this task.

ANALYTICAL APPLICATION OF MEMBRANES

For media that are difficult to filter, ie with high particulate load or colloids, depth filters with a high retention capacity can be used upstream of the membrane filters to prevent premature blocking.

A range of drugs to which membrane sterilisation and filtration is applied is listed in the Appendix.

Direct Patient Care

The protection against contamination from particulate and biological species is critical in the direct care of patients. Typical applications where procedures are susceptible to contamination are in irrigation solutions used in ophthalmic surgery, epidural anaesthesia, neonatal IV therapy solutions and general IV therapy. In all cases (see Table 11) sterilisation of solutions is achieved using 0.22 μm filters, typically with mixed cellulose membranes in a PVC housing. For IV therapy the filter unit can consist of two filters, a sterilising hydrophobic membrane disc that vents any trapped air and a hydrophilic cylindrical membrane that filters the intravenous solution. A cylindrical design is one way of offering a larger membrane are compatible to patient use.

TABLE 11 – Direct Patient Care Filter Application.

Application	Typical Solutions	Pore Size	Degree of Protection
Intraocular	Acetylcholine	0.22 μm	Sterilization
Injections	Air Alphachymotrypsin Atropine Buffered Salt Solutions Epineprine Fluorescein Pilocarpine Saline Tetracaine		Particle removal over 0.22 μm
Intrathecal Injections	Hydrocortisone Streptomycin Penicillin	0.22 μm	Sterilization Particle removal over 0.22 μm
Epidural Anesthesia Spinal Anesthesia	Bupiracaine Lidocaine Pontocaine Procaine Tetracaine	0.22 μm	Sterilization Particle removal over 0.22 μm
Neonatal I.V. Therapy	Hyperalimentation Solutions	0.22 μm	Sterilization Particle removal over 0.22 μm

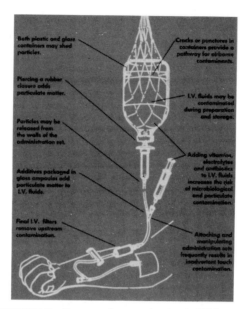

FIGURE 5 – Sources of contamination in IV therapy.

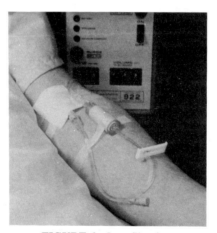

FIGURE 6 – Ivex filter in use.

Laboratory Filtration and Membrane Applications

Membrane filters are very thin, highly porous media composed of foamed and/or stretched polymeric compounds. Because of their homogeneous structure they cannot contaminate the filtrate with fibres or particles from the membrane matrix.

Microfiltration membranes retain particles according to the size of the pore, and the affinity of the filter materials for the solute. While materials are held largely on the surface of the membrane, they can also retain within the matrix itself, as in depth filtration. Unlike depth filters, however, which bind solutes of nominal size ranges only, membrane filters retain particles with absolute accuracy due to their controlled, predetermined pore size. Some membranes also demonstrate retention of specific substances as for example in the case nitrocellulose membranes which bind proteins and nucleic acids in high concentration. This feature may be an advantage for some applications where exclusion based on chemical properties rather than porosity alone is required.

Membrane filters for microfiltration are used primarily for separating particulate materials or micro-organisms, larger than the rated pore size, from gases or liquids. On filtration of gases particles are also retained which are smaller than the pore size rated. Also they show a low particle capacity per unit area as compared to a depth filter. For some filtration applications it may be advisable to prefilter through a depth filter, eg, a glass fiber filter, prior to microporous filtration to prevent clogging. This may be necessary, because the solution is highly viscous, contain molecules which may precipitate in the membrane filter or are heavily contaminated by microorganisms and particulates.

This separation is carried out either for the purpose of cleaning to obtain highly pure or sterile products or recovering (concentrating) these materials in order to carry out further chemical, microscopic, microbiological or other analyses on the separated sample.

Typical applications include
- Sterile filtration
- Dialysis
- Fluid clarification/purification
- Gas filtration/particle control
- Microbiological investigations
- HPLC solvent filtration and sample preparation

In the selection of a filtration unit considerations such as ease-of-use, reliability and short filtration times are important. The materials for both the membrane itself and housing and support structures must be carefully selected. Assembly and testing procedures should also be considered. For sterilisation applications the membrane must be validated as sterile, using the LRV value (usually > 8.56) which is related to bubble point in non-destructive tests.

Sterilising filtration is generally performed under pressure to minimise contamination risks and to decrease foaming problems when filtering protein solutions. Operation above 1 bar pressure (ie above the vacuum limit) enables higher flowrates to be achieved.

To select the optimal micro-filtration membrane for an application the following characteristics of the different microporous filters, should be addressed;
- Pore size of the filter, relative to the size of the particle to be filtered. This criterion controls particle exclusion.
- Hydrophilicity or hydrophobicity of membrane. Wetting properties and chemical compatibility of the different membranes vary.
- Chemical properties of the fluid or gas to be filtered.
- Temperature of the filtrate.
- Surface colour or grid pattern, depending on the application requirements. Gridded membranes are frequently used in bacterial or water analysis.

TABLE 12 – Overview of applications of membrane filters.

Pore Size	Type	Material	Examples of Application
12 μm	AE 100	cellulose nitrate	**Pre-filtration** Leucocyte chemotaxis Air filtration
8 μm	AE 99	cellulose nitrate	Pre-filtration Infrared spectrographic analysis of asbestos Monitoring of high viscosity oils **Dust measurement**
5 μm	AE 98	cellulose nitrate	Cytological investigations Cleaning of viscous fluids Monitoring of lubricants Gravimetric dust analysis
	TE 38	PTFE	Steam filtration Measurement of radioactivity in the air, final cold ashing
3 μm	ME 29	mixed asters of cellulose	General cleaning and microfiltration **Air monitoring, e.g. for radio-activity** In-line filtration of solutions and gases for particle remove
	RC 61	reg. cellulose	Clarification of organic solvents
1.2 μm	ME 28	mixed esters of cellulose	Removing yeasts from wine **Air monitoring**, particle removal Heavy metal contamination analysis Gravimetric analysis of **lubricants**
	ST 69	cellulose acetate	Plankton investigations Isolation of impurities in hot gases
1.0 μm	RC 60 TE 37	reg. cellulose PTFE	Clarification of organic or aqueous solutions Separation of emulsified drops of water from organic solvents Cell labelling
0.8 μm	ME 27	mixed esters of cellulose	Clarification of pharmaceutical or biological solutions, X-ray fluorescent analysis **Microscopic analysis** of fibrous particles, e.g. open atmospheres Heavy metal analysis in air monitoring Stabilizing of wines **Analysis of yeasts and moulds**
0.8 μm	ST 68	cellulose acetate	Fine filtration of alcohols Sterilization of gases **Particle control** in hot gases Filtration of hydraulic fluids
0.6 μm	ME 26 BA 90	mixed esters of cellulose cellulose nitrate }	**Microscopic investigations** Microbiological control, e.g. wild yeasts Stabilization and analysis of **drinks** Filtration of milk products Fine filtration of dispersed sediments
	RC 59	reg. cellulose	Fine filtration of organic solutions and solvents
0.45	ME 25	mixed esters of cellulose	Analysis of materials which can be filtered off **Sludge investigations** Microbiological analyses of coliform bacteria Total bacterial count

ANALYTICAL APPLICATION OF MEMBRANES

- Flow rate required. Higher flow rates may be required for large volumes.
- Pressure tolerances of filter, wet-strength and mechanical properties. These properties are described for each individual filter.
- Retention properties of the membrane with regard to the filtrate eg protein binding.
- Holders and housing for individual filters are incorporated into either disposable filtration devices or reusable holders for membrane filters.
- The refractive index of the membrane may be an important criterion for signal detection.

In some applications cost may be a factor which can decide the choice of membrane or housing type.

TABLE 13 – Membranes and Characteristics for Sterile Filtration.

Mixed ester membranes with a low acetate content, constant weight and an insignificant amount of extractable. They are ideal for gravimetric analyses and microbiological contamination controls. Excellent flow rates and ideal handling properties.
Nitrocellulose membranes. They are very versatile filters made of 100% pure cellulose nitrate, used for aerosol sampling, for the filtration of aqueous solutions and the, in bacteriophage and virus filtrations.
Cellulose acetate membranes are highly moisture resistant and bind very few proteins. The selective exclusion properties make them the filter of choice for basic sterilisation/clarification procedures. They are used for filtering aqueous, alcoholic and oily solutions, even under high temperature conditions (up to 180C).
Regenerated cellulose membranes are made of pure cellulose. They exhibit excellent chemical stability against organic solvents – comparable to PTFE – and can be used for purifying and filtering of both aqueous and organic solutions. It is usefull for the preparation of HPLC samples and eluent buffer filtration.
Polyamide membranes are universal filters, which are naturally hydrophilic and resistance to most solvents. They are ideal for sterilisation filtrations, for purifying alkaline solutions and for the production of high-purity solvents. Excellent handling is the attractive feature of this material.
Polytetrafluoethylene (PTFE) membranes (with polyester supporting fabric) are hydrophobia and are notable for their extremely high chemical and temperature resistance. They are a preferred choice for air and gas filtration, eg for venting sterile containers, fermenters and other tanks, and are also suitable for filtering strong acids, bases, aggressive solvents etc.
Polycarbonate membranes filters are specially developed membranes for the AOX (absorbable organic halogens) determination. This type of filter is a radiation track filter. PC membranes have exceptionally low halogens blank value.
Polypropylene – are designed for sterilising filtration of antibiotics solutions in aggressive organic solvents.
Polyvinyldifluoride (hydrophilic) – for sterilising filtration made from of PVDF coated with hydrophilic molecules attached by covalent binding. This membrane offers important benefits to the user: Boarder chemical compatibility, Superior mechanical resistance, Higher resistance to temperature, Lower non-specific protein binding, Extremely low extractables)

Mixed cellulose esters 0.22 μm, which have been specially treated to minimise the quantity of wetting agents (Triton-Free) have been used fore more than 30 years for the reliable sterilisation of aqueous solutions such as buffers, drugs and vitamin solutions.

PVDF can be used to sterile filter not only aqueous solutions, but also alcoholic solutions, tetrahydrofuran, concentrated acids and bases. Cellulosic membranes can easily be torn apart, and may break during autoclaving procedures. This risk is eliminated with PVDF membranes. PVDF can filter solutions up to 85°C (versus 75°C for a standard mixed cellulose ester membrane) and can be autoclaved up to 126°C (versus 121°C for mixed cellulose). The hydrophilic membranes bind only 4 μm lgG per cm^2, versus 150 μg/cm^2 for mixed cellulose. The molecules making the membrane hydrophilic are covalenty bound to the PVDF backbone. Polyvinylfluoride membranes when wetted with methanol and water exhibits excellent protein affinity. The proteins are mainly bound through strong hydrohopic interactions. They have high mechanical strength and do not shrink or distort even in the presence of methanol.

Further details of membranes can be found in the Appendix.

Range of filtration devices.

In the use of membrane filters appropriate procedures in their care and handling should be observed. Membranes are sandwiched between Patapar, a smooth blue paper that eliminates static and aids handling of the individual membranes. They are available in 13, 25, 476, 50, 90, 142, 293, 304 mm disc sizes from many manufacturers.

Membrane should be kept cool, preferably not above 15 – 20°C and kept away from sunlight and chemical vapours. Improper storage may destroy pore structure of the membranes or interfere with wettability. Membranes have limited stability and should be used within 12 months of receipt.

Most membrane filters may be autoclaved at 121°C and 15 psi for 15 minutes or sterilised with ethylene oxide gas. Sterile pack membranes are sold presterilised with ethylene oxide gas. Membrane filters should be handled carefully with for example flat-tipped forceps or rubber gloves only. Care should be taken to minimise pressure applied to the membrane surface as excessive pressure can damage the membrane pore structure.

One of the benefits of sterilising ready to use filtration devices is that they eliminate some of the limitations of classical sterilising filtration systems:

The relatively high investment required for hardware (filter holders).

The time needed for installation, autoclaving, testing and cleaning of the filtration system.

The handling of a sometimes heavy equipment (a 293 mm filter holder weighs 22 kg).

The autoclaving procedure itself and the associated requirements, limitations and risks init use.

A major benefit of using sterilising filtration devices in laboratories is the increased quality and reliability associated with a product. This is made and controlled in production facilities operating according to pharmaceutical GMP (Good Manufacturing Practices) and, more recently, the need to adhere to ISO (International Standards Organisation) requirements.

The sterilising filtration devices should pass a number of tests including, among others: visual examination, integrity check, control of downstream particulate contamination, sterility testing and control of apyrogenicity. Sterilising filtration devices are also designed to minimise the "dead volume", ie the volume of liquid remaining in the device at the end of filtration to reduce product loss. Units are usually delivered ready-to-use, sterile and pyrogen-free.

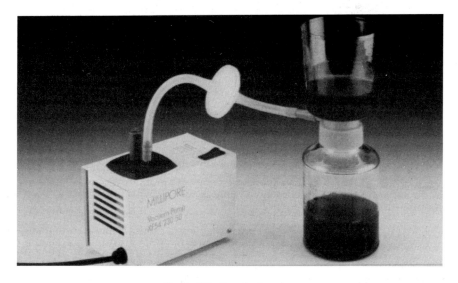

Sterile filtration device.

Applications of Sterile Filtration

For fast, simple and reliable filtration of small to medium volumes, complete ready to connect filter units are finding great popularity in laboratories. These device eliminate time consuming preparatory work, such as changing the filter and cleaning the filtration system. These units come in a wide array of different types for a variety of applications, including
- sterile filtration, ultrafiltration, clarification/prefiltration of aqueous media
- ultracleaning of small volume samples containing solvents of HPLC and GLC analysis
- sterile filtration of gases
- sterile filtration of difficult to filter aqueous solutions
- preclarification of liquids heavily laden with coarse particles

The hold-up volumes of the filters vary from 1 l to 150 µl with samples volumes typically varying from 0.1 to 50 cm^3.

Mycoplasma Removal

Mycoplasma is the trivial name assigned to microorganisms in the class mollicutes which comprises three families: Mycoplasmataceae, Acholeplashmataceae and Spiraoplasmataceae. Unlike other prokaryotes, mollicutes lack a cell wall and they are bound by a trilaminar plasma membrane and, therefore, are deformable. This characteristic, together with their size (between 0.05 and 0.8 µm in spherical form), means they are able to pass through a 0.22 µm membrane, which makes them a threat to any experimentation involving tissue culture. Introducing serum with undetected low levels of mycoplasma contamination (< 10 colony-forming units per 100 ml) to a cell culture can result in mycoplama concentrations which are 2 to 3 logs higher than the cell concentration!

Mycoplasma infections can have a devastating effect on cell cultures. For instance, certain strains of mycroplasma (e.g. M. *orale* and M. *Argini*) hydrolyse arginine. Chromosomal damage can be induced in mammalian cells by arginie depletion in culture media via mycroplasma metabolism. Mycoplasma infections can also cause cell culture cytopathology and decrease cell growth rate.

Studies on the extent of cell culture contamination by mycroplasam shown that 17 to 93% of continuous cell lines are contaminated. Antibiotic treatment of infected cultures is unreliable and the best means of removing the infection is by destroying the culture.

Membrane filters are the only suitable method to remove mycoplasma from tissue culture media with for example PVDF hydrophilic 0.1 m achieving LVR value superior to 8. The membranes are available as discs in different diameters to fit into suitable filter holders. Filtration membrane are also available sterile and pyrogen-free:

Solvent Filtration

Solvents for liquid chromatography must be filtered primarily to remove particulate matter That otherwise block the frits downstream and accelerate wear in the moving parts of the system (eg pump and injector). Particulate contamination in liquid chromatography mobile phases can arise from many sources: particulates from the distillation system used to produce the solvents, glass or metal fragments from solvent containers, plastic particles from the bottle closures, undissolved substances, precipitates formed during the mixing of buffers with organic solvents, dust and airborne microorganisms, etc.

The filtration of mobile phases to remove microorganisms is also important, especially from buffers which promote bacterial growth, such as those used in biochromatography. Filtration is traditionally performed under vacuum to simultaneously remove particulate contamination and degas the solvent. Solvent degassing is important because when the pressurised mobile phase leaves the chromatography column and enters the detection chamber, it undergoes fast decompression. Any disolved gases (mainly nitrogen and oxygen), present in solution under the higher pressure conditions in the chromatography column, suddenly form bubbles which strongly interfere with the detection signal, making the chromatogram difficult to interpret.

Solvents generally used in HPLC applications do not promote the growth of microorganisms and the main concern is presence of particulate contamination. Usually a 0.5 µm pore size membrane filter provides adequate clarification of solvents and the choice of the membrane is dictated by its compatilbity with the solution to be filtered. The membrane not contribute to the contamination of the filtered solution, i.e. filters should not release extractables (organic substances or ions) or dissolved membrane polymers into the filtrate.

Membrane filters recommended for HPLC solvent filtration are mixed cellulose ester membrane, 0.45 µm, low extractables, PVDF hydrophilic membrane, 0.45 µm, PTFE unlaminated membrane, 0.5 µm hydrophilised PTFE.

When performing buffer filtration for biochromatography applications, the use of 0.22 m membranes is often recommended as buffers form an environment in which microorgansims could develop. A suitable filter is the low extractable PVDF hydrophilic membrane validated for microorganisms retention and is compatible with buffers normally used in biochromatography e.g. phosphate, citrate, urea, etc).

Sample Filtration

The level of contamination in samples for instrumental analyses is very often much higher than in solvents.:. a solution may contain as many as 100,000 bacteria per millilitre and still shown no signs of turbidity. The particulate matter originating from the sample can damage the injection valve, block the in-line filter or column inlet frit or connecting tubing of HPLC systems. Particulate contamination which is retained in the upper part of a column packing can act as a secondary phase, distributing the separation patterns. Microorganisms can easily grow in biochromatography columns using the sugar based polymer matrix and the buffer as a nutrient. To prevent operational problems all samples should ideally be filtered prior to chromatographic separations, even in particulate contamination seems to be unlikely:.

It is also important to ensure that sample matrices are compatible with mobile phase composition and will not cause the formation of precipitates. This can be verified by simple turbidity tests.

The ideal sample filtration device should easily and safely remove all unwanted particulate contamination from the sample (100% retention), without adsorbing the analyte and without contaminating the filtrate with extractables. HPLC sample filtration is usually performed with 0.45 μm (or 0.2 μm) membranes of hydrophobic PTFE for organics, cellulose acetate for aqueous solutions or a hydrophilic PTFE for both.

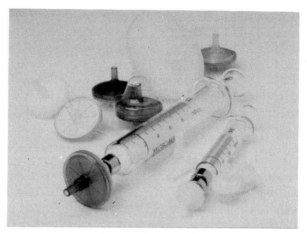

Sample filtration units.

TABLE 14 – Selection Guide for Pharmaceutical Filters.

Application (contaminant)	Objective	Membrane Pore Size
Clarification and visible particulate	Visual/optical clarity	5 μm depth or surface filter
Large organism/particle	Yeast and mould removal	0.65 μm final filter
Bacteria reduction	Pyrogen management	0.45 μm final filter
Bacteria retention	Sterilisation	0.22 μm final filter
Small organism removal	Mycoplasma removal	0.1 μm final filter

Centrifugal Ultrafiltration Concentration

For concentration and purification of macromolecular solutions in the laboratory a fast, efficient useful procedure is the combination of ultrafiltration with centrifugal forces. The application of the centrifugal force drives liquid through the low-binding ultrafilter while solutes with a size greater than the nominal molecular weight cut-off (MWCO) of the membrane are retained. For dilute solutions (1 mg/ml or less), flow rates are directly proportional to g-force, while at higher concentrations, increased viscosity and a polarising layer at the membrane surface reduces the flow. Membrane polarisation is reduced in the units when they are spun at a fixed angle There are several applications of centrifugal ultrafilters.

In the use of the centrifugal separators it is important to chose a suitable cut-off for the membrane. Generally it is best to choose a device with cut-off at about half of the MW of the protein to be concentrated to maximise protein recovery and minimises the filtration

ANALYTICAL APPLICATION OF MEMBRANES

time. For example, if bovine serum albumin (MW = 67,000) were the protein of interest, a 30,000 MWCO (rather than 50,000 MWCO) device would result in the most efficient concentration and recovery of the protein in the retentate.

For nucleic acids the cut-offs (based on the number of bases or base pairs in a fragment of DNA or RNA) which correspond to the MWCO of typical low binding membranes have been determined (see Table 15). The nucleotide cut-off (NCO) indicates the fragment length of single- or double-stranded DNA or RNA that is expected to be recovered at 90% efficiency with a unit of the named NCO. As with cut-off selection for proteins, it is best to chose the NCO at about half the length of the fragment of interest. For example for nucleic acids above 500 bases (or base pairs), a device with 100,000 MWCO membrane is suggested.

TABLE 15 – MWCO of low binding membranes for DNA and RNA recovery.

Porosity (μm) or MW cutoff (D)	Microseparation							
	Micro-pure	Micro-con	Centri-con	Centri-prep	Centri-free	MPS	Centriflo	Minicon
0.45 μm	•							
0.22 μm	•							
100,000		•	•	•				
50,000		•	•	•			•	
30,000		•	•	•	•	•		
25,000							•	•
15,000								•
10,000		•	•	•		•		
3,000		•	•	•		•		
1,000						•		
500						•		

The following tabulates several applications of centrifugal ultrafiltration devices.

Desalting and Buffer Exchange

Centrifugal concentrators are a fast, convenient, high-recovery alternative to dialysis and ethanol precipitation. Sample dilution, often associated with spin columns, is not a problem. Salt transfer across the membrane is efficient and independent of microsolute concentration or size. When desalting with a centrifugal ultrafiltration device, one should note that there is no change in buffer composition as a result of ultrafiltration. Rediluting the retentate with water and spinning again (diafiltration) effectively decreases the salt concentration of the sample by the concentration factor of the ultrafiltration. For greater salt removal, multiple concentration and redilution spins are required. For most samples, three concentration/reconstitution cycles will remove about 99% of the initial salt content.

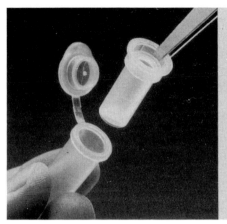
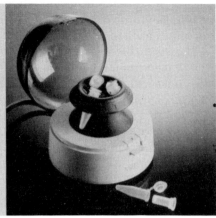

Centrifugal ultra filtration system.

Removal of Small Molecules

Centrifugal concentration is effective in removing primers, liners and unincorporated label from DNA or protein solutions. However, ultrafiltration membranes do not function like size exclusion columns as fully membrane-permeating molecules pass at the same rate as salt because their transport through the membrane is independent of their individual molecular size. Proteins above 3,000 MW may be separated from each other and from smaller molecule if they differ by a factor of about ten in molecular size. Small volumes should be prediluted to assist in purification.

Ultrafiltration is not a fractionation technique. It can only separate molecules which differ by at least an order of magnitude in size. Molecules of similar size can not be separated by ultrafiltration.

Detergent Removal

The chemical nature of most detergents causes micelle formation above a critical concentration limit, the Critical Micell Concentration (CMC) and the detergent forms aggregates leading to gross changes in molecular structure. This affects the amount of the detergent which can be removed from solution with UF membranes of specific cut-off. For example, the monomer of Triton X-100 (500-600 MW) should pass readily through a 3,000 MWCO membrane. However, at concentrations above its CMC of 0.2 mM, Triton X-100 forms micelles composed of approximately 140 monomeric units. During ultrafiltration, the micelles behave like 70,000-90,000 MW globular proteins. Therefore, above the CMC of Triton X-100, a 100,000 MWCO membrane would be required to removed the detergent effectively.

Concentrating and Desalting Nucleic Acids

Centrifugal concentrators are also useful as alternatives to ethanol precipitation of nucleic acids. DNA and RNA samples with starting concentrations as low as 5 ng/ml can be routinely concentrated in minutes with 99% recovery of starting material, without use of coprecipitatnts. Ultrafiltration can also be used to change solvents by diafiltration. In this process, the sample is concentrated, then diluted to the original volume with the desired buffer and concentrated again. This, in effect, "washes out" the original solvent. Recovered samples are immediately available for subsequent procedures or analysis.

Polymerase Chain Reaction (PCR)

For successful PCR amplification, inhibitory substances that might co-purify with DNA must be eliminated. Diafiltration can remove impurities, such as heme, heparin and EDTA, often found in blood-derived DNA. Ultrafiltration is also efficient in the separation of PCR-amplified DNA from other components of a reaction mixture as excess primers, dNTPs and buffer salts. For this application, devices with 100,000 or 50,000 MWCO are recommended.

Improving Protein Recovery

Centrifugal devices can concentrate and desalt ng/ml levels of nucleic acids, with excellent recovery. However for very dilute protein solutions (generally below 25 g/ml starting concentration), concentrate recovery is often not quantitative. This decrease in recovered protein is caused by non-specific binding of the protein to exposed binding sites on the plastic of the concentrated device. The extent of non-specific binding varies with the relative hydrophobicity of individual protein conformations. Pretreating (passivating) the plastic by blocking the available binding sites before concentration can often improve the recovery yield from dilute protein solutions.

Concentration of solutions with high solid content

Centrifugal separators are ideal for the separation of high solid content solutions. They can achieve high flowrates without the need for prefiltration. Centrifugal UF is suitable for separation of antibodies from culture broths or for the desalting of bacteria.

Recovery of Gel-purified samples

There are four basic methods for recovery of proteins or nucleic acids from gel slices; electroelution, elution by diffusion, gel dissolution and extrusion by gel compression. The choice of technique depends upon several factors. The micro-electroeluter allows rapid and high yield recovery of small amounts of proteins or nucleic acids that have been purified by gel electrophoresis. Eluted molecules are concentrated to a small volume and can be further purified by diafiltration.

The micro-electroeluter consists of two chambers, each filled with a buffer. The bottom of the upper chambers has circular openings into which up to nine concentrators fit smoothly, partly in the upper and partly in the lower chamber. Electrodes in both chambers provide a uniform electric field when the apparatus is plugged into a standard power supply.

In operation, gel slices containing the protein or nucleic acids are inserted into perforated sample tubes which allow buffer and target molecules to pass. The gel matrix is retained. Under applied voltage, the macromolecules in the gel slices are eluted and migrate downward until they are retained by the ultrafiltration membrane at the base of the devices. A suitable cut-off (3,000, 10,000, 30,000, 50,000 or 100,000 molecular weight) is selected in advance. After elution, the protein or nucleic acids may be concentrated by spinning the Cnetricon units in a laboratory centrifuge to remove or exchange the buffer. The final sample, in concentrated form, is recovered in a sample cup by spinning the devices once more upside down, according to standard procedures. Typically, over 90% of the original sample may be recovered in as little as 25 l. Membrane compatibility enables the use of standard electroelution buffers. If desired, the buffer may be exchanged by dilution and reconcentration in the same unit after electroelution.

An alternative method for a rapid and easy method to extract proteins and DNA from acrylamide and agarose gels is microfiltration inserts, combined with ultrafiltration microconcentrators, provide The gel is loaded into the insert, which is then centrifuged. The DNA or extracted protein permeates the microporous membrane and is subsequent retained above the ultrafiltration membrane in Microcon where the sample can be diafiltered and concentrated. Although typical recoveries are lower than by electroelution, the procedure is faster and simple when 100% recovery is not required.

Sample Preparation for HPLC Analysis

Centrifugal micropartition is a fast, easy method to remove protein from biological samples prior to HPLC. The resulting ultrafiltrate, containing only low-Mw analytes, is ready for injection. Ultrafiltration does not dilute the sample nor add chromatographic interferences, cause baseline disturbances or shift retention due to pH changes. The low-adsorptive membrane assures high analyte recovery.

Microfiltration separators can also remove suspended solids before analytical separations such as HPLC and can also be used in conjunction with microconcentraters for simple, one-step particle removal and protein or nucleic acid concentrations.

Micro-electroelutor.

Separation of Free From Protein-Bound Microsolute

In the past, free drug levels in serum or plasma samples were not widely measured, partly for want of a convenient means of separating free from protein-bound drug. The technique of choice was generally equilibrium dialysis, a time-consuming procedure, subject to effects of dilution and buffers. It does not directly indicate the free drug concentration in the sample. Other techniques such as ultracentrifugation and gel filtration arealso time-consuming and there in inadequate standardisation of results. An alternative for free/

TABLE 16 – Application of Centrifugal Ultrafiltration.

Device	MW Cut-Off Choice	Process Volume (ml) Max.	Min.*	Avg. Process Time	Major Applications
MICROCON	3,000 10,000 30,000 50,000 100,000	0.5 0.5 0.5 0.5 0.5	0.005 0.005 0.005 0.005 0.005	95 35 12 6 15	• Concentration/desalting of proteins or nucleic acids • Removal of primers from PCR**-amplified DNA
MICROPURE	0.22 0.45	0.35 0.35	— —	0.5 0.5	**Micropure separators:** • Clarification of samples before HPLC **Micropure insert and Microcon:** • DNA recovery from agarose gels • Oligonucleotide recovery from acrylamide gels • RNA recovery from acrylamide gels • Protein recovery from acrylamide gels • Purification and concentration of antibodies from hybridoma cells
CENTRICON	3,000† 10,000 30,000 50,000 100,000	2.0 2.0 2.0 2.0 2.0	0.025 0.025 0.025 0.025 0.025	120 60 30 15 30	• Concentration/desalting of proteins, enzymes, DNA, monoclonal antibodies, immunoglobulins • Concentration of urine, spinal fluid, serum • Removal of primers from PCR**-amplified DNA • Recovery of protein and DNA from electrophoresis gels, with Centrilutor micro-electroeluter • Removal of labeled amino acids and nucleotides
CENTRIPREP	3,000 10,000 30,000 50,000 100,000	15 15 15 15 15	0.5 0.5 0.5 0.5 0.5	140 55 30 20 80	• Concentration/desalting of proteins • Purification of antibiotics, hormones, drugs from biological fluids, fermentation broths • Recovery of biomolecules from cell culture supernatants, lysates
CENTRIFREE MPS	30,000 —	1.0 1.0	0.15 0.15	25 25	• Free drug, hormone assays • HPLC sample preparation • Amino acid sample preparation
CENTRIFLO	25,000 50,000	7.0 7.0	— —	30 30	• Rapid concentration of dilute protein • Deproteinization of samples
MINICON B15 CS15 A25	 15,000 15,000 25,000	 5.0 2.5 0.75	 0.05 0.03 0.04	 300 130 120	• Fast, simple and reliable one-step enrichment of biological solutions (urine, cerebrospinal fluid) for electrophoresis or immunoelectrophoresis.

* At temperature of 25°C. ** PCR (polymerase chain reaction) is covered by U.S. Patents issued to Hoffmann-LaRoche, Inc. † For higher solvent resistance, Centricon-SR3

bound separation is centrifugal ultrafiltration in a standard laboratory centrifuge (preferably angle-head). Free drugs readily pass the membrane for collection and analysis. The sample is not diluted in the process.

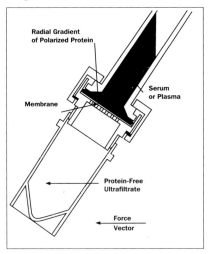

FIGURE 9 – Polarisation control in a fixed angle rotor.

Vacuum Filtration

It is routine practice in laboratories to utilise vacuum filtration in applications of analysis. Vacuum can frequently be supplied by using simple water jet pumps connected to taps or alternatively by using laboratory vacuum pumps. Membrane filter discs (of standard sizes) are clamped between two flanges position between the retentate chamber and the filtrate collection vessel. The pre-cut discs are mounted on appropriate porous supports. The housing are typically made from glass, stainless steel and polystyrene. The range of applications are many and include gravimetric analysis, biological investigations, bacte-

Vacuum laboratory filtration system.

riological control of water, foodstuffs and beverages, pharmaceutical and cosmetic products, general clarification, ultra-purification and sterilisation of all types of fluid.

Sensors and Monitors

There are a range of devices for sensing and monitoring specific species in gases or liquids which utilise the electrochemical response initiated by the detected species as the measurement signal. In the devices segregation of the external environment from the internal detector (and reference electrodes) requires the use of an appropriate membrane, which can be polymeric, solid state, ion-selective or a liquid membrane. The most familiar device is the glass pH electrode utilising a glass membrane. A range of other devices are available for the detection of specific ions in solutions, specific components in the gas phase (H_2S, CO_2, SO_2, H_2 etc), oxygen in the gas or liquid phase (notably blood and other biomedical applications). A range of other analytical devices are currently under development.

Ion-Selective Electrodes

Ion-selective electrodes have many applications in water analysis and environmental analysis. Electrodes include those for the determination of pH, F^-, CN^-, NH_3 and total hardness ($Ca^{2+} + Mg^{2+}$). The principle of operation of the electrode is shown in Fig 10. The measurement of the potential between the ion-selective electrode and the reference

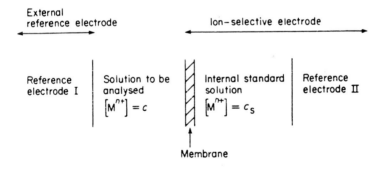

FIGURE 10 – Basic electrode principle for determination of a ion M_n+ with an ion-selective electrode.

electrode allows the determination of the ion M^{n+} between the analysed solution and an internal standard solution. In most cases the ion-selective electrode and reference electrode are mounted in a single (combination) probe (see Fig 10).

The analysis with an ion-selective electrode is almost completely dependent on the membrane properties. The potential measured by the ion-selective electrode is a function of the membrane potential

$$E_m = (2.3\,RT/nF)\log C/C_s$$

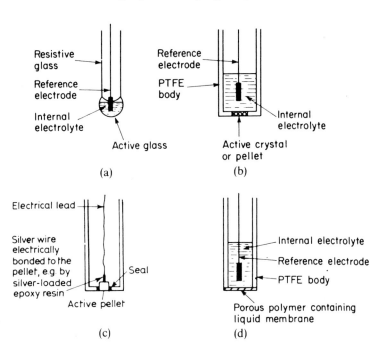

FIGURE 11 – Common construction of ion-selective electrodes.

As, in principle, all factors associated with the use of the electrode, other than the ion concentration to be measured, remain constant (eg C_s), the measurement of the cell potential is directly proportional to the logarithm of the ion concentration.

In practice the response of the membrane should be independent of other species in solution and reach a stable potential rapidly. In many cases this is difficult to achieve and reflects the largely empirical way of selection and design of the membrane. Overall a range of different types of membranes are used in ion-selective electrodes; including glass, solid state, heterogeneous and liquid membranes. Table X gives a selection of commercial ion-selective electrodes in which these four types of membranes are used.

Glass membranes are generally based on Na_2O-Al_2O_3-SiO_2 mixtures. Membranes selective to H^+ ions are rich in SiO_2, eg 72% SiO_2 0% Al_2O_3 and low in Al_2O_3. Membranes selective to alkaline metal ion have a higher content of Al_2O_3. In the latter cases there are major interferences from other alkali metal ions.

A variety of different solid state membranes are used for different ion detection. Many of these use a combination of $Ag_2S + AgX$ ($X = Cl^-$, Br^- or SCN^-) in the form of a pressed disc and the electrode responds to X^-. For the determination of cations (M^+) the membrane

TABLE 17 – Typical membranes and commercial ion-selective electrodes.

Ion electrode	Membrane	Concentration range/mol dm^{-3}	Major interferences
H^+	Glass	10^{-14}–1	None
K^+	Valinomycin	10^{-6}–1	Cs^+, NH_4^+
Na^+	Glass	10^{-6}–sat.	Ag^+, H^+, Li^+
F^-	LaF_3	10^{-6}–sat.	OH^-, H^+
Cl^-	$Ag_2S/AgCl$	10^{-5}–1	Br^-, I^-, CN^-, S^{2-}
Br^-	$Ag_2S/AgBr$	10^{-6}–1	I^-, CN^-, S^{2-}
I^-	Ag_2S/AgI	10^{-7}–1	CN^-, S^{2-}
CN^-	Ag_2S/AgI	10^{-6}–10^{-2}	I^-, S^{2-}
S^{2-}	Ag_2S	10^{-7}–sat.	Hg^{2+}
Ag^+	Ag_2S	10^{-7}–1	Hg^{2+}
Cd^{2+}	CdS/Ag_2S	10^{-7}–1	Ag^+, Hg^{2+}, Cu^{2+}
Pb^{2+}	PbS/Ag_2S	10^{-7}–1	Ag^+, Hg^{2+}, Cu^{2+}
Cu^{2+}	CuS/Ag_2S	10^{-8}–sat.	Ag^+, Hg^{2+}, S^{2-}
Ca^{2+}	$(RO)_2PO_2^-/(RO)_3PO$	10^{-5}–10^{-1}	Zn^{2+}, Fe^{2+}, Pb^{2+}, Cu^{2+}
$Ca^{2+}+Mg^{2+}$ (hardness)	$(RO)_2PO_2^-/ROH$	10^{-7}–1	Cu^{2+}, Zn^{2+}, Fe^{2+}, Ni^{2+}, Pb^{2+}
NO_3^-	R_4N^+/ether	10^{-5}–1	ClO_4^-, ClO_3^-, I^-, Br^-

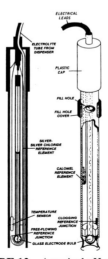

FIGURE 12 – A typical pH electrode.

material is a mixture of $Ag_2S + M_{n/2} S$.

Heterogeneous membranes use similar active components as solid state devices but instead the active material is deposited into the pores of an inert support, eg silicone rubber.

Ion-selective electrodes based on liquid membranes immobilise an active species, an organic molecule (dissolved in a solvent), in the pores of an inert polymer. The typical active species are phosphate diesters for calcium ion detection, metal complexes for anion detection and neutral macrocyclic crown ethers for alkali metal detection.

New developments in ion-selective electrode are in the combination of ion-selective membranes with metal oxide semi-conductor field-effect transistors (MOSFET). The so-called ion-selective field-effect transistor (ISFET) is an electronic device to measure a selected ion concentration in a solution using an exposed gate insulator or membrane. the membrane use plasma polymerised materials, eg tetrafluorethylene, chlorobenzene, doped with ions eg iodide.

TABLE 18 – Construction and performance of gas sensing ion-selective electrodes.

Gas	Ion-selective electrode	Electrolyte	Membrane	Concentration range/mol dm^{-3}	Serious interferences
H_2S	S^{2-}	Citrate buffer pH 5	Polypropylene	10^{-6}-10^{-2}	None
NH_3	pH	0.1 mol dm^{-3} NH_4Cl	PTFE	10^{-1}-1	Volatile amines
CO_2	pH	0.01 mol dm^{-3} $NaHCO_3$ +0.1 mol dm^{-3} NaCl	PTFE	10^{-2}-10^{-4}	Volatile weak acids
NO_2	pH	0.1 mol dm^{-3} $NaNO_2$ +0.1 mol dm^{-3} NaCl	PTFE or polypropylene	10^{-6}-10^{-2}	Volatile weak acids
SO_2	pH	0.1 mol dm^{-3} $K_2S_2O_5$ +0.1 mol dm^{-3} NaCl	PTFE or silicone rubber	10^{-6}-10^{-2}	Volatile weak acids
HCN	Ag^+	10^{-2} mol dm^{-3} $KAg(CN)_2$	Polypropylene	10^{-7}-10^{-1}	Sulphide

Gas Sensing Electrodes

A number of sensors for measuring specific species concentration in gases are based on the electrochemical response of an electrochemical cell. Many of these gas-sensing electrodes position a gas permeable membrane of for example, PTFE, polyethylene, silicone rubber, between the solution to be tested and the sensing element, a thin electrolyte and an ion selective electrode. A range of different devices is listed in Table . The membranes typically have pore sizes of several microns.

The monitoring of oxygen in for example blood is also based on similar devices using PTFE or polypropylene membranes (6 – 12 m pore size). Potable devices are also available for detection of gaseous oxygen and also hydrogen gas.

TABLE 19 – Some industrial applications of electrochemical gas sensors.

	O_2	CO	H_2O	SO_2	NO	NO_2	Cl_2	H_2	HCN	HC	$NH3$	CH_4
Air quality		•										
Auto exhausts	•				•							
Battery rooms								•				
Chemical industry	•		•	•		•	•	•	•	•	•	•
Chimney stacks	•	•		•	•	•				•	•	•
Construction sites	•		•									
Domestic boilers	•	•		•	•	•						•
Fertiliser plants											•	
Fire detection		•										
Food processing	•			•							•	
Grain storage	•					•						
Mining	•	•			•	•						•
Nuclear industry	•							•				
Oil rigs	•		•					•				•
Paper mills				•			•					
PCB manufacture				•						•		
Refrigeration											•	
Semiconductors								•	•	•	•	•
Sewage industry	•		•									•
Steel works		•		•								•
Tunnels		•	•									•
Water treatment				•			•					

SECTION 9

Water Desalination

WATER DESALINATION

Desalination is defined as the removal of dissolved salts from various waters; brackish, sea, etc. Desalination can be performed by several processes including two membrane processes, reverse osmosis and electrodialysis (see Table 1). The appropriate choice of separation process, either membrane or non-membrane based eg ion-exchange, evaporation depends on local economic (and occasionally political) factors. The relative economics of the processes depend to a large extent on the concentration of dissolved solids.

TABLE 1 – Processes for Water Desalination.

	With a Phase Change	Without a Phase Change
Evaporation	Multistage flash Multiple effect Evaporation with Vapour Recompression	
Crystallisation	Freezing Hydrate	
Membranes	Membrane distillation	Reverse Osmosis Electrodialysis
Others		Ion-exchange

Water Desalination Processes

Ion exchange (IX) is established as an important desalination processes. Ion exchange uses a polymer resin containing fixed charged chemical groups that interact with the anions or the cations in the water. Undesirable ions from the water are exchanged for more desirable ions in the resin and held within the resin bed. Typical desalination is accomplished by exchanging cations, such as sodium calcium, magnesium for hydrogen ions; and exchanging anions such as chloride, sulfate, or phosphate for hydroxide. Both cation and anion exchange operate together and thus the hydrogen and hydroxide ions then combine to form water. Because the salts are removed from the water and fixed in the resin bed, exhaustion of the resin sites eventually results. The resins can be regenerated for another cycle by contacting the cation resin with acid, and the anion resin with sodium hydroxide. This is the reverse process of the original desalination step.

Although ion exchange itself does not directly consume energy, apart from that required in pumping solutions, there are associated operating costs in supplying regenerating chemicals and water. The regeneration frequency in IX is directly proportional to the total dissolved solids (TDS) of the feed water and thus the amount of product water produced per pound of regenerate chemical consumed decreases rapidly with increasing TDS. In addition each regeneration consumes substantial amounts of water, so at high TDS levels the amount of waste water can approach and even exceed the quantity of product water. Therefore IX is not normally used to remove salts from water containing more than 1,000 ppm of dissolved solids.

A typical IX system is shown schematically as a simple two-bed system consisting of a cation exchanger and an anion exchanger. Each exchanger has a regenerate preparation system in which the concentrated chemical regenerates are diluted prior to contact with the exchanger resin. The regeneration sequence involves a backwash to remove accumulated suspended solids, and purge the bed of resin fines and then chemical addition, followed by a rinse.

Dissolved solids and contaminants can be removed from water by two processes involving a phase change; evaporation and crytallisation. In evaporation the water is vaporised by heating and because the solubility of the dissolved salts in the vapour stream, is much less than in water, the separation factor, defined as the concentration of the salts in the liquid phase divided by the concentration of the salts in the vapour phase, is typically greater than 1000. Several evaporation processes are practiced; multiple effect evaporation, vapour recompression and flash evaporation.

Multiple effect evaporation involves several stages of water conversion to vapour, and then recondensation. In multiple-effect evaporation the feed water to all but on eunit is the brine from an adjacent unit. The source of heat is product vapour received from the adjacent unit on the opposite side. Figure 10 shows a flow diagram of a multiple-effect evaporator. Each effect operates at a different temperature; the lowest is the last in the series. The use of product vapour as the heat source forall but one of the units achieves high energy efficiency; the efficiency of a multiple-effect system is proportional to the number of effects. For example a five-effect system would produce slightly less than 5 kg of product per kg of steam supplied, compared to only 1 kg of product for a single effect.

Vapour recompression is a process in which water is evaporated and then compressed, causing it to condense at a higher temperature, thus providing a source of latent heat for further evaporation (see Fig 1b). The difference in temperature between the vaopur and the liquid is relatively small to reduce compression requirements although a relatively large heat-exchange surface area is consequently required. Vapour recompression is suitable for producing relatively small amounts of purified water where thermal energy is not readily available, and electrical energy is.

Flash evaporation is a process in which the water is heated under pressure and then the pressure is suddenly reduced, resulting in the evaporation of only a percentage of the water. Flash evaporation is generally accomplished in several stages each stage involving relatively small temperature and pressure differences. A method to improve the economy of a basic multistage flash plant is to reuse a portion of the brine waste as feed and thus reduce the amount of feed water to be pretreated. The concepts of multistage flash and multiple-effect evaporation can be combined to achieve greater utilisation of heat energy and reduced construction costs by utilising the product of one plant as the heat source for a second plant.

The method of crystallisation uses freezing for separation. The crystal formation excludes foreign materials from the solid lattice and this generates a two-phase system of brine and pure ice. The two phases can then be physically separated, and purified water recovered by melting the ice. The method is not a common desalination technique due to difficulties in solids handling and separating ice crystals form the liqiud. The maximum water purity obtained is controlled primarily by the extent to which the ice crystals can be separated from the brine. A vapour recompression unit, removes sufficient vapour from the refrigeration unit to both cool the feed water and provide latent heat for melting the final product.

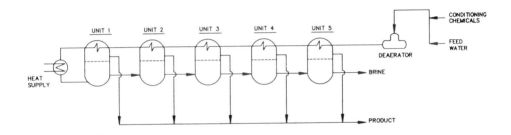

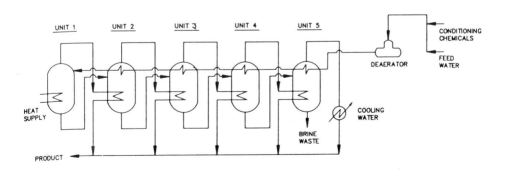

FIGURE 1 – (a) Typical crystallising desalination system.
(b) Multiple effect evaporator.

Membrane Distillation

The concept of membrane distillation has been proposed as a method of water desalination. The process utilises porous membrane for the permeation of water as vapour and not as liquid. The method is analogous to mulitstage flash (MSF) desalination with direct condensation and also an infinite number of stages. A module design (shown in Fig 2) has been proposed, based on a spiral wound concept, for desalination in which feed is the coolant for condensation of the vapour. A major cost of this process is the large heat exchanger surface which is required and which consequently is unlikely to see the method adopted, at least for large scale desalination.

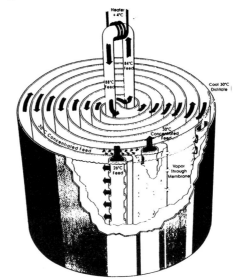

FIGURE 2 – Module design for membrane distillation.

Membrane Processes

Electrodialysis and reverse osmosis compete in many water desalination applications. Although the theoretical energy requirement is the same for both processes, there are considerably different energy requirements in practical applications due to the irreversible energy dissipation. In RO this energy dissipation is associated with the applied transmembrane pressure difference, which is much greater than the theoretical osmotic pressure difference, that is required to overcome friction experienced by water molecules as they pass through the membrane. In ED the energy loss is associated with the friction of ions as they move through the membrane. Thus in ED, energy requirements are directly related to the concentration of dissolved ions whereas in RO it is largely independent of the concentration of dissolved ions. Thus overall it is expected that the energy requirements of ED are smaller than those of RO at low dissolved ion concentrations. At a particular dissolved ion concentration the costs of ED will begin to increase above those of RO and RO will then begin to become favoured. In the comparison of separation process costs clearly the cost of non-membrane based separations must also be included.

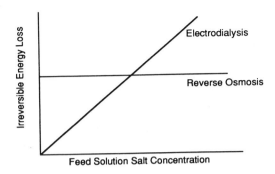

FIGURE 3 – Comparison of the cost of reverse osmosis and electrodialysis for desalination.

Reverse Osmosis

Commercial reverse osmosis membranes can eliminate up to 99% of the dissolved solid content of a water in one stage. The relative small variation in treatment costs with increased TDS makes RO economically attractive for desalination of brackish water with TDS concentrations of 100 ppm and for seawater supplies at 45,000 ppm. It has been estimated that RO accounts for 31% of the total world desalination capacity, with the greatest proportion (73%) for the desalination of brackish water, while only 7% is for seawater applications (see Table 2). The growth area appears to be in process water applications. The capacities of the large plants are between 20,000 to 300,000 m^3/day. There appears to be a growing trend towards the use of spiral wound composite modules, away form hollow fibre modules.

The module arrangements in reverse osmosis plant depend upon the application but are typical staged on either the permeate (water) side or on the retentate (brine) side. Typical arrangements of flows in RO plants are shown in Fig 4.

In the case of brackish water treatment the individual membrane rejection performance is sufficient to require only one stage to achieve the required reduction in salt content. The second, and possible third stage are used to further concentrate the retentate and recover additional treated water. In some brackish water plants the recovery can be as high as 95% depending upon requirements of feed pretreatment and the presence of dissolved impurities.

For typical seawater desalination plant there are two possible configuration
i) One pass desalination using retentate (brine) side staging to improve recovery, typically around 50%. High salt rejection membranes are used in this procedure.
ii) Two pass desalination. The first bank of membranes produces a brackish water, which is desalinated further in a second bank and possibly back blended with feed. Retained brine from the second stage is recycled back to the feed of the first stage.

TABLE 2 – Large Seawater Reverse Osmosis Plants.

Location	Nominal capacity '000 m³/day	Feed	Membranes	Overall energy kWh/m³	% Recovery	Pressure MPa
Large Seawater Reverse Osmosis Plants						
Jeddah [47] Saudi Arabia 1989	57	43,300 mg/l TDS	Toyobo CTA hollow fibers	8.4	35	6
Bahrain [48] 1984	46	12,000-30,000 mg/l TDS	DuPont B10 hollow fibers	5.0	61	6.5
Malta [49] 1983	20	39,200 mg/l TDS	DuPont B10 hollow fibers	5.9	35	8.1
Las Palmos [46] Spain 1989	24	Seawater	Filmtec spirals			
Large Brackish Water Reverse Osmosis Plants						
Yuma [3] Arizona USA 1990	274	Brackish 3,000 mg/l TDS	69% Fluid Systems/UOP Cellulose Acetate Blend 31% Hydranautics Cellulose Acetate Blend			70
Daesan [46] Korea 1990	95	Brackish	Toray Spirals			
Iraq [46] 1983-86	64	Brackish	DuPont Hollow Fibers			
Riyadh [50] Saudi Arabia 1984	60	Brackish	DuPont Hollow Fibers Brine staged 4:2:1		90	2.8
Salboukh [46] Saudi Arabia 1979	60	Brackish 1610 mg/l TDS	DuPont Hollow Fibers		90	
Unayzah [51] Saudi Arabia 1989	52	Brackish 1500 mg/l TDS	Envirogenics Spirals	1.8	90	
Ras Abu Jarjur [52] Bahrain 1984	46	Brackish 19000 mg/l TDS	DuPont B10 Hollow Fibers	5.0	61	5.8
Cape Carol [53] Florida, USA 1976	46	Brackish 1600 mg/l TDS	42% DOW TCA Fibers 58% Hydranautics Spirals	0.53	75 85	1.4 1.7
Fort Myers [46] Florida, USA 1989	46	Brackish 400-500 mg/l TDS	Hydranautics/ Nitto Denko Spirals		90	0.7
Bayswater [54] Australia 1985	36	CW Blowdown 2500 mg/l TDS	Hydranautics Spirals		82	3.0

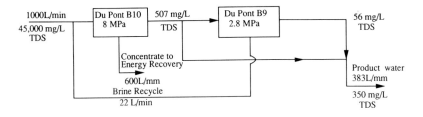

FIGURE 4 – Flow arrangements for a typical seawater installation.

This second method has the attraction of improved utilisation of the pressure energy in the concentrate. In large desalination plants it is usual to recover the energy from the brine which can result in power savings of up to 30% in the case of seawater feed.

A typical reverse osmosis plant, shown in Fig 5, consists of appropriate feed pretreatment, high pressure pumps, the array of membranes modules, pipework and valves and appropriate process control systems to maintain required performance. A cartridge filter is always used to remove both particles which might otherwise block the relatively small spaces between the membrane layer and also submicron particulates which may foul the membrane surface. The particular membrane module configuration in Fig 5, is referred to as a 3-2-1 array and is frequently used in practice as it enables a relatively high water recovery, while maintaining brine flows within the minimum and maximum specified by membrane manufacturers.

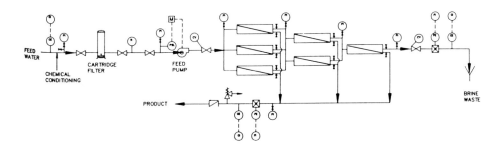

FIGURE 5 – A typical reverse osmosis system.

Plant Operation

The key to the successful operation of reverse osmosis plant is pretreatment of the feed coupled with appropriate process control and monitoring. This is essential because currently available reverse osmosis membranes are sensitive to chlorine and oxidants, have limited pH tolerance and are susceptiblel to fouling by submicron particles, microorganisms and by precipitation of salts at the membrane surface as the feed is concentrated. Particular attention must be paid to calcium carbonate and calcium sulphate scaling and the salt density index.

The prevention of calcium carbonate scaling requires measurement of the calcium ion concentration, pH and temperature of the feed. It is usual to keep the pH of the feed at approximately 5.5 by the addition of sulphuric acid (see Fig 6). The potential for calcium sulphate scaling can be estimated from measurement of the calcium ion and sulphate ion content of the feed which enables the solubility product of calcium sulphate in the concentrate to be determined. Suppression of calcium sulphate precipitation is achieved by the addition of sodium hexametaphosphate (SHMP) or alternatively proprietary sequestering agents (e.g. Flocon 100 from Pfizer).

The quantity of submicron particles present in the feed is defined by the silt density index (SDI). The SDI is determined by monitoring the declined in flux when the feed water is filtered through a 0.45 μm membrane for 15 minutes at a transmembrane pressure differential of 207 kPa

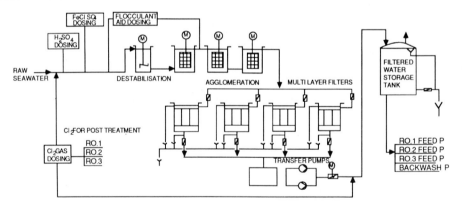

FIGURE 6 – Feed pretreatment scheme for a seawater reverse osmosis plant.

A typical pretreatment scheme used at the Doha reverse osmosis facility in Kuwait is shown in Fig 6. In this scheme the raw seawater feed and the treated water are chlorinated to minimise biological activity. Before use in the desalination plant, the chlorine content of the water is lowered using sodium bisulphite. The SDI for the seawater is reduced, to a value less than 3, using flocculation in conjunction with filtration. If there is a possibility of trace organic chemicals in the feed an activated carbon filter is used.

The pretreatment plant performance is monitored by a range of regular tests to ensure that the feed water to reverse osmosis is of the required specification. The RO plant itself is monitored and controlled through measurement of key parameters; permeate conduc-

WATER DESALINATION

tivity, pressure, flowrate, temperature and pH. Reverse osmosis is applied in several water desalination and purification applications which are described in Section 10.

A typical application of reverse osmosis applied to a municipal wastewater process is illustrated in Fig 7, showing a two step automated process.

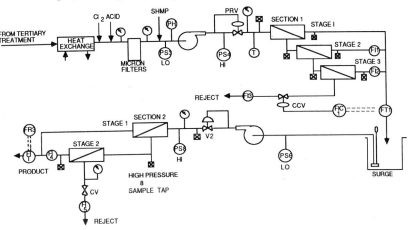

FIGURE 7 – Application of reverse osmosis to municipal wastewater treatment.

Reverse osmosis plant.

Electrodialysis for Water Desalination

Electrodialysis involves the removal of ions from water by transport through ion-permeable membranes under an applied electrical field. Solution flow is ideally across the membrane surface and not through and across the membranes, as in most other membrane processes, and partly eliminates membrane fouling common to RO, although salts can still precipitate at the surface of the membranes. The efficiency of the process is determined largely by the effectiveness of the membranes in conducting electricity with ions of one charge but preventing the passage of oppositely charged ions. The transport of ions consumes electrical energy in proportion to the amount of ions (salts) transported from feed to concentrate. This is governed by Ohm's Law and Faraday's Law. Thus ED finds applications in desalination of water with relatively low salt content, where in terms of the absolute number of ED installations, the most important industrial application is in the production of potable water from brackish water.

Generally in virtually all cases when brackish water is converted to potable water either RO or ED is used. Rapid developments in membrane technology and in particular in RO have improved the energy consumption and economics and simplified the operational procedures to levels far above those of alternatives such as evaporation. Potable water production costs in electrodialysis are a strong function of feed water concentration, and at low feed water concentrations (< 3,000 to 5,000 ppm) electrodialysis is generally more economic than reverse osmosis (or multi-stage flash evaporation). Clearly the relative economics of electrodialysis vary from country to country depending on several factors, such as electrical energy costs.

The salinity of brackish water can vary from approximately 2 to 10 kg/m^3 and this factor, in particular, affects the economics of ED (or RO). Typically in ED, a brackish water feed solution containing 1,000 to 8,000 ppm of dissolved salts is divided into a potable water product stream, containing less than 500 ppm (world health organisation regulation) of salts, and a concentrated brine stream, containing 20,000 to 50,000 ppm of salts. The concentration of salts to this and higher level is a significant feature of electrodialysis. The degree of desalination achieved in one stage of electrodialysis is around 50% and when high levels of de-ionisation are required several stages in series are utilised.

The design of any desalination plant must accommodate the individual requirements of pretreatment of the brackish water before the membrane separation. The pretreatment, which includes cooling, softening, coagulation and acidification's, is partly required to maintain high operating lifetimes of membrane which would fall drastically at high temperatures and if precipitation of silica and other species on the membrane were to occur. Typical operating lifetimes of membranes are between 2 and 3 years.

One of the first brackish water desalination plants was built in the USA in Webster, South Dakota and Figure 8 shows a flow diagram of the demineralisation plant. The brackish water drawn from a city well is first treated to remove manganese, filtered, and then passed through a 4-stage electrodialyser to be demineralised. The demineralisation process is divided into 4 stages as the product water of 350 ppm as TDS must be separated from the raw water 1,500 to 1,800 ppm. Such a mutli-stage system is more economical with a high capacity of 950 m^3/day. Each stage of this process uses approximately 220 pairs of ion-exchange membranes, limped together into a stack, and the 4 stages are set on a filter press.

The main ionic constituent of the raw water in this plant was $CaSO_4$ which is slightly soluble. This plus the low temperature of the water, at 9C, required a relatively large membrane area in comparison to other plants operating with a higher salinity of water, but mainly as NaCl (see Table 3).

WATER DESALINATION

TABLE 3 – Performance of an Electrodialysis Plant.

Outline of the desalination plants			
Item	Name of the plt. Webster USA	Shikine Japan	Oshima Japan
Raw water	Well water	Well water	Well water
Capacity of the plant M^3/d)	950	200	1000
Salinity of the raw water (ppm)	1500-1800	1145	2992
Salinity of the product water (ppm)	350	500	500
Temp. of the raw water (°C)	915-25	25	
Main ionic component of the raw water	$CaSO_4$	NaCl	NaCl
Operation type	Continuous	Batchwise	Continuous
Operation control	Automatic	Automatic no-man	Automatic no-man
Construction completion	Sept. 1961	June 1970	April 1972
Size of membrane (mm)	1115 x 1115	1150 x 575	1115 x 1115
No. of memb. cell pairs per stack	216	150	250
No. of stage	4	1	4
Operating data			
Temp. of the raw water (°C)	9.0	11.0	23.0
Total electric power consumption (Kwh/m^3)	1.40	1.31	1.00
H_2SO_4 consumption (g/m^3)	280	4	4
Quantity of the raw water (m^3d)	1660	227	1540
Quantity of the desalinated water (m^3d)	954	200	1340
Desalting efficiency of the raw water (%)	57.5	88.1	87.0
Salinity of the raw water (ppm)	1450	1084	2083
Salinity of the desalinated water (ppm)	356	484	425

The electrical power consumption of these plant is 1.0 to 1.4 kWh m^{-3} of product water. It is reported that there are currently over 2000 plant installed worldwide for the desalination of brackish water with a capacity in excess of 10^6 m^3/day utilising membrane area greater than 1.5 10^6 m^2.

During the operation of electrodialysis salts may be formed on the surface of the membrane as a result of concentration polarisation, which can result in scaling of the membrane. The size of this problem depends on the quality of the water supply and on the

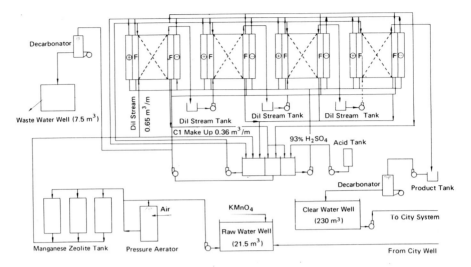

FIGURE 8 – Flow diagram of an electrodialysis water desalination plant.

extent of water pretreatment before electrodialysis. In many cases a modified ED operation is used, referred to as "polarity reversal", or electrodialysis reversal (EDR), to eleviate the problem of scaling. In EDR the electrical potential of the terminals alternates approximately every 15 minutes, which cycles the function of each compartment from demineralisation to brine concentration, thus eliminating solid salt build up. A drawback of EDR is that after each reversal the unit must be flushed to ensure concentrated brine is removed from the diluate stream to be produced.

Electrodialyser plant.

Nitrate Removal from Drinking Water

Nitrate levels exceed EC guide-line levels at a number of pumping stations in Europe. One method to combat this problem is blending low nitrate water in order to achieve acceptable nitrate levels in the water supply. This alternative is not always available or often proves

very expensive. As an alternative, water treatment technologies aimed at removing nitrates from drinking water have been developed;
- reverse osmosis
- ion exchange processes
- anaerobic biological digesters
- electrodialysis

A drawback of RO processes is that the filtrate or permeate contains too few ions, ie it approaches the quality of distilled water. Therefore, it has to be mixed with untreated water in order to achieve a drinkable composition. This fact limits the applications to nitrate reductions of approx 50%. The treated water will then have roughly 50% of the original concentration of ions. An attraction of reverse osmosis on the water composition is its ability to also reduce or totally remove organic matter with molecular weights higher than approx 100 g/mol. This means that pesticides and other organic impurities, which often pollute the water along with nitrate, will also be reduced.

In ion exchange processes for nitrate removal, nitrate may be either exchanged against hydroxyl, lowering the salinity of the water or against either chloride or hydrogen carbonate. The last two exchange reactions lead to either a higher chloride content or raised water aggressively, thus posing certain problems regarding water quality parameters. Also, because of the dynamics of the exchange process, the water quality of the treated water will not be the same as the resin is progressively depleted of its exchange capacity. This poses some problems for the post-treatment stages. Also, such variations in water quality are not desired from a quality point of view. Even though ion exchange processes have reached a stage where they are considered state-of-the-art for nitrate removal from drinking water, their future use in this area is questioned, largely due to the environmental concerns related to discharging the brine stream into rivers and estuaries.

Anaerobic biological digesters

Certain naturally occurring bacteria have the ability to digest nitrates under favourable conditions. These conditions comprise an anoxic or anaerobic environment (the bacteria are thus forced to use nitrate as their source for oxygen) and the presence of a source of carbon as well as an energy source. Two types of mechanisms for the digestion are know. The first, the autotrophic denitrification, uses hydrogen carbonate as a carbon source and either hydrogen or sulphur as an energy source. More popular and simpler to implement is the hetreotrophic denitrification, where the bacteria digest an organic compound, typically methanol, ethanol or acetic acid which at the same time is used as carbon and an energy source. All of these processes ultimately lead to the total reduction of nitrates to nitrogen gas. However, the reaction mechanism invariably leads through the intermediate product of nitrite. This fact poses certain health and safety risks for the precise, since nitrate is the true source of toxicity for the human body. If too little carbon and/or energy substrate is provided for the reaction mechanism, nitrite will invariably be found in the treated water. On the other hand, the substrates should not be provided in excess, otherwise they will not be digested completely and contaminate the drinking water.

Since the bacteria are living organisms, they reproduce and die. Therefore, a certain loss of the bacteria cultures is unavoidable, either as excess growth biomass or as dead biomass.

This means that steps have to be implemented to ensure that the produced drinking water is freed of these again. This is generally achieved by a slow sand filtration, an active carbon filtration and a disinfection step. This entails an added investment as well as more intensive supervision of the plant. Furthermore, the drinking water is depleted of oxygen by the bacteria, making an oxygenation stage necessary. A more critical problem is the fact that the bacteria adapt very slowly to changing conditions, ie flow, temperature, nitrate load changes. This means that such a treatment plant has to be operated in a continuous, constant mode. and its use will most likely be limited to very large water works.

The Electrodialysis Process

With the development of nitrate selective membranes a number of key advantages are introduced to the application of electrodialysis for nitrate removal from drinking water:-
- low overall water desalination;
- no chemical dosing;
- high recoveries;
- low energy requirements;
- low TDS in waste water.

The selectivity of a membrane is governed by a number of physical and chemical effects. The transport limitations imposed by concentration polarisation can render a selective membrane useless if the operating conditions (flow, current density) are also not suitably defined. Basically, a potentially selective membrane has to be operated at low current densities and at high flow velocities. The water produced by ED contains all the original components desired and required by the human "organism" and fulfils the limits set for nitrate concentrations by the authorities. The ED process can produce drinking water with less than 25 mg/l of nitrates, even when nitrate concentrations in the drinking water increase, as is expected for the near future. Along with the removal of nitrates a certain softening effect is inherent to the process, a welcome side-effect.

The design of a plant (fig 3) consists basically of three streams: the diluate or drinking water, the concentrate or brine and a third small stream used to rinse the electrode chambers at the end of each stack. The drinking water is passed through a prefiltration prior to feeding it to the stack to avoid any blockage of the stack by fine particulates. After this prefiltration a small portion of raw water is diverted to the brine loop (this portion eventually becomes the waste waters) and an even smaller portion is diverted to the electrode rinse loop. The remaining raw water is passed through the dilute chambers of the stacks (if more than one stack is required they each treat a portion of production capacity) were the water is treated.

In the electrode chambers water is electrolysed, thereby producing oxygen at one electrode and hydrogen at the second. This leads to small but significant changes in pH in these chambers, which could potentially lead to a scale precipitation at the electrodes, provided that the water has enough hardness to do so. For this reason, a small softening unit is installed to treat the electrode rinse make-up in order to remove hardness. The amount of softened make-up required generally is approx 0.5% of plant production capacity or less and is reused as brine make-up.

WATER DESALINATION

Table 3a outlines some of the changes in water composition that an ED plant will produce. An important factor in nay such treatment technology that it does not require any chemicals or additives. A deciding factor in the choice of nitrate removal processes will always be the specific production costs for the treated water. Table 3b shows the specific treatment costs for a number of processes and for the ED process at two plant sizes typical of drinking water installations. The ED process is a technically and economically attractive alternative for denitrification of drinking water. A key advantage in this respect is the quality and composition of the treated water, which in most cases does not have to be subjected to any kind of post-treatment. Also, the fact that the separation can be done without using any chemicals makes this process specially attractive. Operation and maintenance of an electrodialysis plant is very simple and fully automated, making the process applicable to pumping stations of any size.

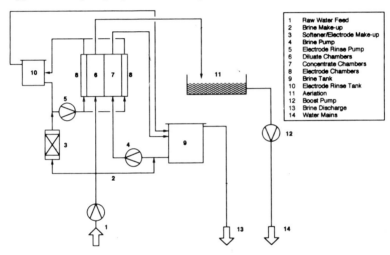

FIGURE 8a – Schematic diagram of the electrodialysis nitrate removal process.

TABLE 3a – Typical performance of the Ed nitrate removal.

Temperature	basically unchanged
pH	slight drop by 0.1-0.2 units
Calcium	removal of approx. 10-20%
Magnesium	removal of approx. 10-20%
Sodium	very little change
Potassium	very little change
Chloride	removal of approx. 50%
Nitrate	removal to levels below 25 mg/l
Sulphate	almost no change, only slight removal
Carbonate	Reduction by about 10%
Organic contents	no change
Trace metals	no change
Bacteria counts	no change

TABLE 3b – Comparison of Costs of Different Nitrate Removal Processes.

Process (substrate for bacteria)	Plant 50 m³/h Specific Treatment Costs [DM/m³]	Plant 500 m³/h Specific Treatment Costs [DM/m³]
Reverse Osmosis	0.68	0.56
Conventional Electrodialysis	0.62	0.44
Cl^- and HCO_3^- - Exchange	0.90	0.57
Demineralization by Ion Exch.	0.74	0.44
CARIX-Process (HCO_3^- Exch.)	1.02	0.47
Anaerobic denitrification (EtOH substrate)	1.16	0.49
Anoxic denitrification (acetic acid substrate)	0.58	0.43
Autotropic denitrification (H_2 substrate)	1.59	0.66
HITREM Process	0.45	0.37

Sea Water Desalination

There has been several ED plants constructed for the production of water from seawater, often in regions or locations where the underwater source is brackish. These include plants of between 2.7 – 7.5 m³/day capacity in the Arabian Gulf, USA and USSR. Smaller scale units have been developed for the production of portable water on shipping vessels and for inclusion in survival kits [2 dm³/day capacity] see Photo. A typical unit for on-board ship, producing 10 m³/day consumes 20 kW power in a 200 cell pair stack (see Table 4). The typical quality of water produced is shown in Table 5. However, apart from specialist applications, the economics of ED in seawater desalination are not generally attractive.

TABLE 4 – Specification of Sea Water Electrodialyers.

	UX-10	UX-25	UX-50	UX-75
Capacity (M³/D)	10	25	50	75
Operation	Full automatic			
Power Source	AC/3-Phase 50/60 Hz 200, 440V			
Consumption of Electric Power (KVA)	20	50	100	130
Type of Cell	TS-10	TS-10	TS-25	TS-25
Number of Membrane-pairs	200	500	400	600
Space Requirement (M x M)	2.2x1.1	2.8x1.3	3.3x1.5	4.3x2.3
Weight of Equipment (Kg)	1,600	2,400	3,900	6,500

Small scale (on board ship) electrodialyser.

TABLE 5 – Quality of Product Water.

	Sea Water	Product Water
Appearance	Colorless, transparent	Colorless, transparent
Taste	Saltish	None
pH	8.2	6.7
Ammoniacal Nitrogen (ppm)	Not detected	Not detected
Nitrous-acidic Nitrogen (ppm)	Trace	Not detected
Nitric-acidic Nitrogen (ppm)	Not detected	Not detected
Chlorine ions (ppm)	18,000	184.5
Consumption of Permanganate (ppm)	3.5	2.5
Hardness (ppm)	1,264	68.0
Total Irons (ppm)	Not detected	Not detected
Number of Bacteria	Below 10	Below 10
Colon Bacilli	Negative	Negative

Reverse osmosis is more suited to the production of fresh water from sea water. This has been brought about by developments in membrane materials for these applications which now withstand temperatures up to 30 °C and pressures up to 10 bar with improved efficiencies. A typical plant will perform the desalination in two RO stages; the first at high pressure, to overcome the high osmotic pressure and the second at lower pressure when the salt concentration is typically around 0.8 kg/m^3. In some applications ED can be used to polish the product of a seawater RO plant. In other applications where disposal of large quantities of concentrated brines is difficult a further concentration using ED is desirable.

High Purity Water

In most industrial application water is usually de-salted to between 100 to 200 kg m^{-3} total dissolved solids. When purer water is required the ED is followed by ion-exchange. To reduce the burden on the ion-exchange system, the use of polarity reversal in ED has been applied giving a dissolved solid content of 3 g m^{-3} in the water. The production of high purity water by ED (and RO) is described in detail in section 10 on water purification.

Electrodialysis Reversal

Since classical electrodialysis was commercialised in 1952 for desalting brockish well water in the Arabian desert, the expansion in use has been massive. The application was always plagued by the problem of concentration polorisation, organic fouling, films, mineral scale, slime and other deposits on the concentrate side. Similar problems also plague the operation of RO plants. The practice of chemical addition (acids and polyphosphates) to control precipitation of C CO$_3$ and CaSO$_4$ salts on the brine side of the membrane became a standard practice. Even with finely tuned maintenance programs, operational shutdowns frequently occur for chemical cleaning of the membranes. The cleaning regimes require stack disassembly, chemical treatments, spacer scrubbing and re-assembly. These problems created interest in electrodialysis reversal (EDR) as a means of minimising carbonate and sulphate (calcium) scaling, and colloidal fouling. . The EDR system uses electrical polarity reversal to continually control membrane scaling and fouling

The EDR reversal system is designed to continuously produce demineralised water without constant chemical addition during normal operation, thus eliminating the major problems encountered in unidirectional systems. The process of EDR is shown schematically in Fig 9.

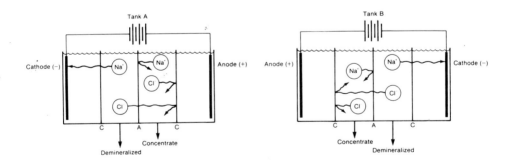

FIGURE 9 – Principle of electrodialysis reversal in which the polarity of the electrodes in tank 'A' is opposite that of tank 'B'.

Electrodialysis transfers ionic species from the water being desalted through the ion exchange membranes to a concentrate waste stream under the application of a direct current. Electrodialysis reversal is the same process, with the exception that the polarity of the electrodes is reversed 3 to 4 times each hour. This reverses the direction of ion movement within the membrane stack (as shown in Fig 9), thus controlling film and scale formations.

Reversal occurs approximately every 15 minutes automatically. Figure 10 shows that upon polarity reversal the streams which formerly occupied demineralising compartments become concentrate streams, and conversely the streams which formerly occupied concentrate compartments become demineralised streams. Therefore, at polarity reversal, automatically operated valves are required to switch the two inlet and outlet streams. The incoming feed water now flows into the new demineralising compartments and the recycled concentrate stream now flows into the new concentrating compartments. An adverse effect of reversal is that the concentrate stream remaining in the stack, whose salinity is higher than the feed water, must now be desalted. This means that in a short time interval the demineralised stream (product water) salinity is higher than the specified level and water is *off-spec product*.

Polarity reversal, ensures that flow compartment in the stack are exposed to high solution concentrations for only 15 to 20 minutes at a time. Any build-up of precipitated salts is quickly dissolved and carried away when the cycle reverses.

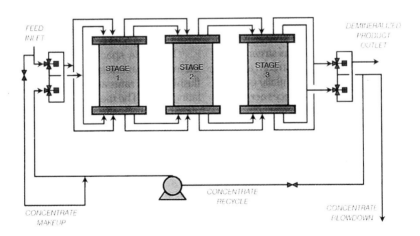

FIGURE 10 – A typical EDR flow diagram.

When polarity reversal takes place, the chemical reactions within the electrode compartments are also reversed. EDR requires the use of platinum coated titanium electrodes to withstand the polarity reversal and the alternate function as anodes. The alkaline environment at the new cathode now occurs at the anode of the previous polarity and the acidic environment at the new anode now occurs at the cathode of the previous polarity. The acidic environment generated at an anode helps prevent scale formation and is now used at both electrodes in cycles. For maximum benefit from anodic acid generation, the anode stream may be operated without flow which allows the concentration of H⁺ ions to increase, the pH falls to about 2 to 3. However, chlorine and oxygen gases generated at the anode tend to accumulate on the electrode to form a layer of gas (blanket) which increases the resistance at the electrode. To reduce this problem, the anode stream flows for a brief period to flush out the gases. However the cathode flows continuously to minimise OH⁻ concentration build upsince an alkaline environment increases scaling tendency. Solenoid valves in the electrode streams automatically control the flows in the anode and cathode compartments.

EDR reversal unit.

Colloidal Fouling

Membrane processes have driving forces that tend to deposit colloidal material on the membrane surface. Colloidal particulates interact with water to form an aggregate charge at the surface of their bound water layer. This charge is usually negative in natural and most waste waters. The DC electric field applied to electrodialysis stacks moves colloids toward the anion membrane. At the membrane surface, the electric field and electrostatic attraction to the positively-charged ion exchange sites in the anion exchange membranes, tend to hold the colloid deposit in places.

Figure 11 illustrates electrostatic colloidal deposition in the and the effect of EDR process. When the DC field is reversed in the EDR process, the electric driving force is reversed, and tends to displace deposited colloids from the anion membrane surface into the brine stream.

The mechanisms for depositing and attraction of colloids on the membrane surface depend on the degree of charge on the particles. The limits for turbidity (or SDI) for fouling cannot be absolutely established, although colloidal fouling is never experienced below a feedwater five-minute SDI of 12, while fouling is likely if the five-minute SDI is above 16. These values generally correspond to turbidities of 0.25 to 0.50 NTU.

The tortuous flow spacer design also helps reduce colloidal fouling in EDR systems. Tortuous path EDR spacers provide uniform turbulent flows essential to optimum operation which minimise the potential for solids entrapment and the effects of deposits on system pressure drop.

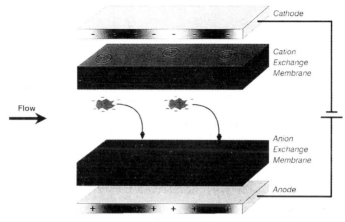

FIGURE 11 – Colloidal deposition in the EDR process.

Overall the EDR process gives five major benefits:
- breaks up polarisation films 3 to 4 times each hour, thus preventing polarisation scale
- breaks up freshly precipitated scale or seeds of scale and flushes them to waste before they can grow or cause damage
- reduces slime or similar formations on membrane surfaces by electrically reversing the direction of colloidal particle movement
- eliminates the complex practical problems associated with continuous feed of acids or complexing chemicals
- automatically cleans the electrodes with acid formed during anodic operation

High Recovery System for Water Desalination

Obtaining an adequate supply of quality water is becoming an increasingly difficult problem in many parts of the world and great care must be taken in utilising the available supplies. Water recovery is now an important consideration when designing and specifying desalination systems. Water recovery is defined as the amount (as a percentage) of product water produced from the total amount of water used: The factors which contribute to requirements of high water recovery are; reserves of raw water, high cost of waste disposal, well pumping and raw water treatment systems

There are several systems and schemes for improving water recovery; concentrate recycle, offspec product recycle, electrode stream recycle and chemical addition.

Concentrate Recycle

Concentrate recycle is simplt the recirculation of the concentrate stream is called. In ED the flow rates of concentrate and demineralised water through the membrane stacks are essentially equal – there must be little pressure difference between the two streams (differential pressure is 0.5 – 1.0 psi). In the simplest case, this would lead to a recovery of demineralised water equal to only half of the saline water treated. However, in almost all ED (and EDR) plants much of the effluent concentrate is recycled to the concentrate to stream, to conserve water. the fraction of concentrate recycled is limited by the solubility of the least soluble mineral in the concentrate stream, i.e. until the least soluble mineral begins to precipitate. This concentration level is controlled by purging a fraction of the concentrate stream to waste and adding an equal volume of new feed water to this recycle stream.

Off-Spec Product Recycle (OSPR) in EDR

During EDR the demineralised and concentrate flows are interchanged in the membrane stack and thus the salinity of the original concentrate compartment changes from a value of, say 10,000 ppm, to a product value of say 250 ppm. The off-spec period (high salinity product) is the time required for feed water entering the original concentrate inlet at the time of polarity reversal to pass through the entire membrane system. Therefore, the more hydraulic stages in the system, the longer the off-spec time will be.

In a standard EDR system, all diversion valves and stack polarities reverse simultaneously. The salinity profile of the product water during this period is shown in Figure 12 for a typical 3 stage system with a 30 second per hydraulic stage contact time. The product water exiting the system in this time interval (0 and 90 seconds) is diverted to waste since its salinity is too high.

However, by delaying the reversal of the outlet diversion valves for a controlled time period, the product salinity profile changes to that shown in Figure 12. This lowers the average salinity of the off-spec water to the point where, in most cases, it is lower than the feed water salinity and can be recycled back to the system feed and is no longer a source of waste. Table compares the recoveries of two typical 3 hydraulic stage systems operating under identical conditions, with and without an Off-Spec Product Recycle system.

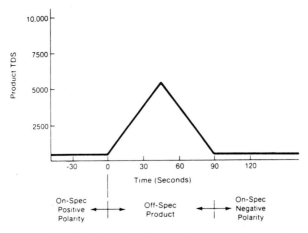

FIGURE 12 – The effect of off-spec product recycle.

TABLE 6 – The Performance of High Product Recovery Systems.

	Base Case	Phased Reversal	Phased Reversal and OSPR	With OSPR*	30 Minute Cycle	30 Minute Cycle OSPR, & Phased Reversal
Gross Product	333,333	333,333	333,333	333,333	333,333	333,333
Off-Spec	33,333	11,111	11,111	33,333	16,667	5,556
Net Product	300,000	322,222	322,222	300,000	316,666	327,777
Electrode Waste	3,000	3,000	3,000	3,000	3,000	3,000
Concentrate Blowdown	72,000	72,000	72,000	72,000	72,000	72,000
Off-Spec	33,333	11,111	—	—	16,667	—
Total Waste	108,333	86,111	75,000	75,000	91,667	75,000
Total Feed	408,333	409,333	397,222	375,000	408,333	402,777
Percent Recovery	73.5	78.9	81.1	80.0	77.6	81.4
Power Consumption KWH/KGal	6.90	6.43	6.43		6.54	6.32
KWH/M³	1.82	1.70	1.70		1.73	1.67

Phased reversal

Phased reversal EDR is based on independently controlling the reversal of the inlet valves, the electrical polarity of each stage and the outlet diversion valves. The sequence starts with the reversal of the inlet valves, which creates a slug of reversal water (A mixture of concentrate and demineralised streams). As this slug proceeds through each stage the stage reverses electrical polarity. Finally the outlet valves are reversed. Thus overall the offspec

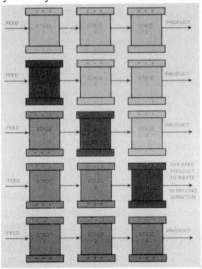

Diagram of phased reversal in EDR systems.

product is reduced to that required for the water to pass through one stage regardless of the number of stages in operation. Table 6 compares the effect of phased reversal with other high recovery systems.

Electrode Stream Recycle

When the cathode and anode streams are combined during ED, they chemically neutralise one another with the resulting pH and salinity being approximately that of the feed water. The liquid after separation of gases is normally sent to waste. However, this liquid is essentially the same as the feed water and can be sent back to the feed water tank along with the off-spec product and thus increase system recovery while reducing the overall amount of waste water.. In smaller systems, the electrode flow represents a significant portion of the treated water. In larger systems, the electrode stream flow represents a relatively small fraction of the total amount of water being treated.

Chemical Addition

In the case of EDR systems which do not use chemical dosing to prevent precipitation, the maximum allowable concentrate salinity is determined by limiting values for $CaSO_4$ and $CaCO_3$ saturation. If a higher recovery of feed water is desired, this can be obtained by increasing the cycles of concentration of the concentrate streams with addition of sodium hexametaphosphate (SHMP) and/or acid into the concentrate recycle stream. This enables a higher concentrate stream salinity, and $CaSO_4$ and $CaCO_3$ levels which are higher than the established limits.

In addition to being able to obtain higher recoveries (90 plus percent) on most feed water sources, the combination of chemical dosing and EDR has two particular advantages; very low SHMP consumption per unit of product and low acid consumption per unit of product water, a fraction of what would otherwise be required.

Membrane Properties

The membranes are important to the successful application of the EDR process in surface and wastewater desalination. Membrane performance has significantly improved since both cation and anion membranes were first based on cross-linked styrene-divinyl benzene polymers. Newer anion membranes of aliphatic composition exhibit exceptional resistance to harsh chemical environments and organic fouling. EDR membranes are unaffected by exposure to the range of pH 0 – 10. This allows the use of strong acid solutions (5% hydrochloric acid) to remove scales and metal hydroxide deposits.

Membrane desalination systems are likely to support bacterial growths if a disinfectant residual is not maintained, especially when treating organic-laden waters. Acrylic-based anion exchange membranes also have excellent resistance to strong disinfectants such as hypochlorite, chloramines, and peroxides EDR systems typically operate with 0.3 to 0.5 mg/l of free chlorine residual in the feedwater for bacterial control with no membrane degradation. Higher disinfectant residuals, in the range of 25 mg/l free chlorine, can be used for shorter periods for sterilisation or other purposes. Typical membrane lifetimes are; anion:ten years, cation membranes for up to 17 years, without loss of efficiency or chemical stability,.

Applications of EDR in Water Demineralisation

Overall EDR plants are used in their thousands producing in excess of 200 million gal/day of desalted water for municiple fresh water supplies and industrial use. EDR applications are widening, because of the reduced scaling problems, into more exacting applications such as the desalination concentration of highly fouling and scaling water such as cooling tower waste, refinery wastewater and RO waste effluent. Ion exchange demineralisers now serve well over a thousand industrial and utility plant sites in the United States alone. However membrane demineralisers (electrodialysis reversal and reverse osmosis) are finding increased use to supplement or replace ion exchange demineralisers when expansions and upgrades are required. Membrane demineralisers are attractive because they eliminate or minimise regenerant chemical use and disposal, supervisory labour, and service interruptions of ion exchange demineralisers. In addition, EDR and RO reduce levels of total organic carbon (TOC) as required by semiconductor, nuclear, and other industries that require high purity water.

Much of the industrial water is from surface sources which contain particulate, colloidal, and organic contaminants, which result in a high silt density index (SDI) measurement and which tend to foul reverse osmosis (RO) membranes. The SDI and hence the membrane fouling tendency of surface waters can be reduced by conventional pretreatment: clarification with lime, alum, ferric salts, and/or polyelectrolytes, followed by media filtration. Operation of such plants is chemical-and labour-intensive and generates sludges for disposal. Under typical industrial conditions, however, conventional pretreatment is frequently inadequate or inconsistent for sensitive down-stream processes or uses.

Historically, typical industrial pretreatment has been adequate for ion exchange demineralisers, which can be backwashed if particulate matter breaks through and fouls the beds. Such pretreatment is also usually adequate for EDR systems, which have high SDI tolerance and tend to be self-cleaning. However for optimum performance of RO systems on surface waters, there is a need for more reliable pretreatment which combine several operations. The following are some examples of what are referred to as Triple Membrane Demineralisation Systems.

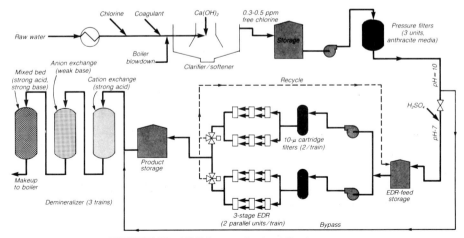

FIGURE 13 – EDR in a predemineralisation system.

Triple Membrane Demineralisation at Nuclear Power Stations

A typical triple membrane demineralisation system is shown in the system flow sheet of Figure 14. Raw lake water is obtained at the plant's cooling water intake. The water is pumped through a 200 – 300 mesh screen and fed to the pretreatment trailer at approximately 80 – 100 psi. In the pretreatment trailer the feed is split into three separate lines each of which feed a 54 inch multimedia filter. The filters contain five layers of media including a gravel support layer, two layers of different size garnet, a layer of sand and a top layer of anthracite. The filters are semi-automatic and require manual initiation of the backwashing procedures. Chlorine of 0.3 – 0.5 ppm is fed directly upstream of each filter. A product stream from each of the filters is fed directly to one of the three identical triple membrane trailers (TMTs). These trailers utilise the three membrane processes of ultrafiltration (UF), electrodialysis reversal (EDR) and reverse osmosis (RO) in series to produce a consistent high quality product water regardless of feedwater conditions. Each of the three TMTs in this system is a 120 gpm unit which operates independently.

The water entering the TMTs first flows into the ultrafiltration system. There are 14 vessels in the system arranged in a parallel, single pass design. Each vessel contains 4-8 inch x 40 inch spiral wound, polysulfone membranes with a nominal molecular weight cut-off of 50,000. The membranes used have a nominal filter rating of 0.012 microns and will reject bacteria and high molecular weight organics. The bulk of the reject stream in the UF is recycled to the suction of the UF feed pump. This maintains adequate flow rates in the reject stream, and maintains a low element recovery rate and allows high overall UF recovery rates. The permeate from the UF is sent into the EDR feed tank.

In the EDR system the incoming water is desalted. Typically 90% salt removal is achieved in the EDR stacks with approximate power consumption of 0.1 KWh/1000 gallons. In addition a large amount of organic matter is removed from the feedwater.

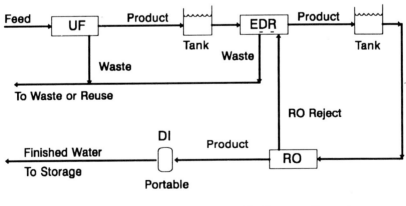

FIGURE 14 – The basic triple membrane flow diagram.

Product water from the EDR is sent into the RO feed tank. In the RO system thin film composite membranes are used to obtain maximum rejection of salts, organics and silica. If the membranes are not chlorine tolerant sodium bisulfite is injected in the EDR product line upstream from the RO tank to remove any residual chlorine in the water. In addition caustic is added to this stream to raise pH and thus convert carbon dioxide, which would pass through the RO membrane, into bicarbonate, which is rejected by the RO membranes.

In the triple membrane unit the series arrangement of UF followed by EDR and then utilises the best features of each individual process while providing optimum conditions for the downstream process. Polysulfone spiral wound UF membranes with a nominal 50,000 molecular weight (MW) cut-off are used for pretreatment and eliminate the need for filter cartridges for the EDR and RO steps. UF performance is maintained with periodic cleaning by chlorine or peroxide solutions.

The EDR step removes up to 90 percent of the dissolved solids including some TOC components below the 100 MW size. The EDR process results in a product that has lower scaling indices and thus ideal as feed to the reverse osmosis stage. The thin film composite RO membranes provide maximum overall removal of TOC, silica, dissolved solids and particulate. Following the TMT treatment the water is then treated by ion exchange. 18 megohm resistivity and less than 1 ppb individual ion levels are achieved with small (3.6 cu. ft.) mixed bed ion exchange vessels. These polishers, due to their extremely low loading, can be economically regenerated off site under strict quality control conditions.

	ANO	GGNS	NAPS	FNP
CONDUCTIVITY - mmhos/cm	0.08	0.10	0.08	0.056
TOC - ppb	100	50	100	50 (20)
SILICA - ppb (SiO_2)	3	5	5	5 (3)
SODIUM - ppb (Na)	1	1	1	1
SULFATE - ppb (SO_4)	1	1	NS	1
DISSOLVED OXYGEN - ppb (O_2)	NA	NA	100	10
CHLORIDE - ppb (Cl)	1	1	NS	1
FLUORIDE - ppb (F)	1	1	NS	1

NA - Not Applicable   NS - Not Specified

FIGURE 15 – Water quality specifications from triple membrane systems.

Triple membrane systems are also finding applications in the electronics industry as replacement to standard RO/IX plant. The system improves the reliability of the RO step and reduces by one or two orders of magnitude the frequency of regeneration of ion-exchange demineraliser.

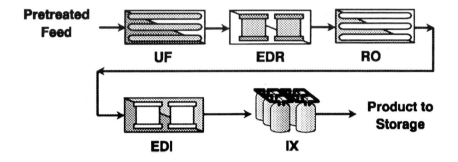

FIGURE 16 – Triple membrane and EDI de-ionisation process.

	EDI Feed (ppb)	EDI Product (ppb)	Rejection (%)
Na^+	595	4.4	99.26
K^+	80	0.31	99.61
Cl^-	40	0.85	97.88
SO_4^{--}	11	< 0.1	> 99.09
NO_3^-	32	< 0.1	> 99.69

FIGURE 17 – EDI feed and product resistivity.

Electrodialysis cell stack.

In further efforts to reduce the amount of regeneration of the ion-exchangers, the use of elctrodeionisation has been practised on at least one power plant in the USA. The EDI process is operated between the reverse osmosis and ion-exchange stages (see Fig 16) to enable significantly improved reduction in dissolved solid and ion content. (see Fig 17) The techniques reduced ion-exchange regeneration by a factor of 17.

SECTION 10

Water Purification

WATER PURIFICATION

Water Treatment

It is estimated that the world's use of water including domestic, industrial and agricultural, is 250 cubic meters per person per year. This use varies widely from country to country being as high as 1500 m^3 in the USA. The production of tap water from sources such as rivers and underground reservoirs, water companies must ensure compliance with national health standards and produce water with acceptable taste, odour and visual appearances. The impurities that are present in water can be classified by size as:

Solutes. Small molecules both ionised (acids, bases and salts) and non-ionised (sugars, chlorinated hydrocarbons etc) and macromolecules including proteins, peptides, humic acid etc. These species impart colour to the water.

Colloidal suspensions (referred to as micellar or pseudo-solutions). These are generally two-phase heterogeneous systems, less than 100 m in size such as clays, hydroxides. Colloids are defined as particles that, due to their negative surface charge which is a barrier to aggregation, cannot settle naturally. Colloids appear as cloudiness in the water.

Particulate. These are the suspended solids visible under a microscope, eg sand, algae, bacteria etc.

Certain of these compounds are necessary for health but such compounds must be below maximum admissible concentrations. In most cases concentrations above these limits are due to human activities although in some cases the composition of the soil where the water is stored is a cause eg for Fe, H_2S, F, Mn.

Conventional water treatment plants to supply drinking water can employ several unit operating in different combinations. This depends on the source of water (see Table 1) which in the case of clean, unpolluted water may require only disinfection to meet the microbiological quality. On the other hand water containing a wide range of the three categories of pollutants may require oxidation, adsorption, coagulation, filtration and disinfection to achieve the required quality.

TABLE 1 – Conventional Treatment Processes for Drinking Water

Water Resource	Treatment Stages
1 Clean, unpolluted	→ Disinfection
2 Unpolluted but suspended solids	→ Filtration → Disinfection
3 Colloids, (importing colour) in small quantities	→ Coagulation → Filtration → Disinfection
4 Colloids in high concentation	→ Coagulation → Floc Settling → Filtration → Disinfection
5 Organic pollutants	→ Coagulation → Settling → Filtration → Disinfection ↑ ↑ Adsorption Adsorption ↑ ↑ ↑ Oxidation Oxidation Oxidation

With a high colloid concentration, the coagulation of the floc may require settling to avoid clogging of the filter. Flocculation can also remove certain dissolved organic pollutants by adsorption. Higher concentrations of these pollutants may demand the use of oxidation and adsorption with one or more of the clarification steps.

The traditional water treatment process requires the use of coagulants and flocculants to destabilise and agglomerate colloids. These reagents are inorganic (aluminium or iron) salts, natural polymers (alginates) and synthetic polymers (polyacrylamides) and have two major drawn backs

- the sludges formed contain these reagents which is now a greater disposal problem
- treated water may contain residual coagulants if the coagulation is not correctly controlled or operated.

Thus overall the demand for improved quality and improved quantities of water and the pressures for nonpolluting processes make it more difficult to employ chemical aids in clarification plants. Membrane technologies offer alternative treatment methods to overcome limitations of conventional treatment. In principle membranes, of the correctly selected cut off, will give absolute filtration of pollutants without using chemicals. The resultant waste (sludge) will only contain pollutants present in the resource.

The pressure driven membrane processes can solve the more usual problem:

i)   MF   clarification

ii)   UF   provides an absolute barrier for suspended solids and all microogranisms – clarification and disinfection

iii)   NF   removes all colloids and several small organics and bivalent salts. Can be used for removing colour and softening.

WATER PURIFICATION

iv) RO removes monovalent ions and most small organics. Used for desalination and for micropollutant removal.

A major area for membrane development for municipal drinking water supplies is in desalination. Electrodialysis and reverse osmosis compete with distillation processes such as multistage – flash evaporation (MSF), multiple – effect evaporation (ME) and vapour compression (VE). It is estimated that there are over 8,900 desalting units with an installed capacity of 15.6 million m^3/day. RO and ED account for 35% and 6% of this capacity respectively and these percentages are growing.

Membrane plant are typically of a smaller capacity than for example MSF facilities – on average 990 m^3/day for RO and 660 m^3/day for ED.

Nanofiltration is the second largest membrane process in water treatment, for softening and more recently the removal of disinfection by-product precursors. It is estimated there are over 150 plants with 6.10^5 m^3/day capacity. Ultrafiltration is used to remove particulate, microoganisms and colloidal material from potable water and thus replace conventional clarification and disinfection. Plants are commercially available which use cellulose ester hollow fibre membranes. These membranes (see Table 2) have been engineered to achieve appropriate mechanical strength necessary for operation where back flushing is used to remove foulants from the membrane to restore membrane flowrate.

TABLE 2 – Specification of hollow fibre UF membrane for water treatment.

Membrane	Anisotropic hollow fiber – Skin on the inside		
Material	Cellulose esters		
Inner Diameter	0.93 mm		
Outer Diameter	1.67 mm		
Cut-off	Solute Dextran Bovine serum albumen	MW 180 KD 67 KD	Retention 90% 100%
Maximum pore size	10 - 20 mm		
Permeability (clean water)	$7.3 \cdot 10^{-10}$ $mPa^{-1}s^{-1}$ (260 1 h^{-1} m^{-2} bar^{-1})		
Burst pressure	2.3 MPa (23 bars)		
Elastic modules	140 MPa		
Tensile stress at yield	3.1 MPa (470 g/element)		
Strain at yield	2.2%		
Tensile stress at break	6.5 MPa (980 g/element)		
Elongation at break	33%		

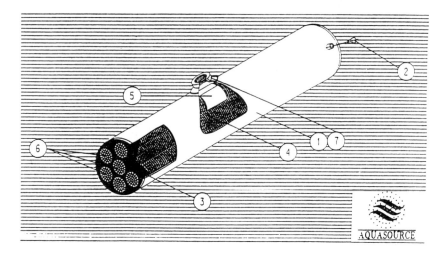

1 – Cylindrical envelope. 2 – Drain. 3 – Tube plate. 4 – Perforated sleeve. 5 – Baffle
6 – Elementary hollow fibre bundle. 7 – Permeate outlet.

FIGURE 1 – Hollow fibre UF multi bundle module.

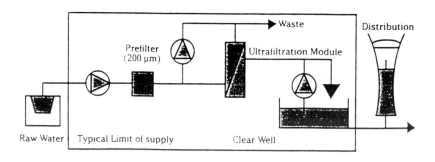

FIGURE 2 – Single stage continuous UF system for water purification.

The hollow fibre assemblies and tube end plates in one module are mounted in cylindrical envelopes (rigid perforated sleeve) of diameter 0.1 m. Larger diameter of fibre bundles decreased the pressure resistance of the tube plates and increased the pressure drop outside the fibres due to the longer tortuous flow path between the fibres. The membrane module is then an assembly of 7 fibre bundles, the ends of the sleeves and bundles are embedded in resin (see Fig 1.). A typical module has a filtration area of 50 m^2.

In operation the membrane modules are in single stage continuous flow systems (Fig

Ultrafiltration unit.

2) fitted with a 200 m prefilter to remove larger particles. When the pressure drop across the membrane reaches a pre-set value, the back flushing pump located in the permeate line is brought into use to restore the flux rate. This is typically 45 s every 60 min. Membrane plants are available with capacities ranging from 3 to 100 m^3 water/hr and greater.

The UF plants can treat various surface and groundwaters with occasional turbuidity and low load of organic matter, to produce potable water. For more polluted waters UF is combined with operations such as oxidation (for iron, manganese, colour) or/and adsorption on powdered activated carbon (for COT). For waters polluted with higher organic levels, eg surface waters, irreversible fouling of the membrane may occur requiring chemical cleaning. Surface waters are characterised by large numbers of bacteria, viruses and micro-organisms a high turbidity, high concentrations of organic matter, high concentrations of pesticides and taste and odour problems. MF and UF alone will not remove most organic matter and NF has limited production due to membrane fouling and module blocking. A combination of UF with oxidation and adsorption (PAC) is more efficient than NF (and UF) in terms of the treated water quality and hydraulic yield (see Table 3).

Table 3 – Comparison of UF and NF Processes Applied to Clarified Surface Water

Ultrafiltration and nanofiltration processes applied on clarified surface water. Organic matter removal. TOC: Total Organic Carbon; THMFP: Trihalomethane Formation Potential; AOC: Assimilable Organic Carbon; T and O: Taste and Odour (afte C. Anselme, V. Mandra, I. Baudin, J. Mallevialle. Paper submitted to AIDE Symposium, Budapest 1993).

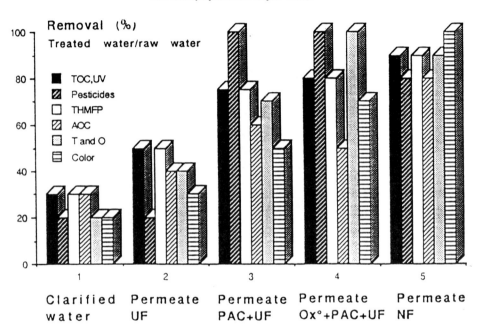

Water Treatment

Water Purification

A common requirement of water purification is to reduce ionic materials and both distillation and (IX) are suited to this task. Ion exchange (IX) is more popular in industry than distillation because it achieves much greater levels of separation, especially with the use of the mixed bed ion exchange. The operating cost of IX is almost directly proportional to the concentration of dissolved solids in the feed-water. The operating cost of RO, however, is relatively insensitive to the dissolved solids concentration up to levels of 1,000 or 2,000 ppm. Thus RO first appeared as a commercial water treatment process, as an alternative to IX for source waters containing high concentrations of dissolved solids. Generally, as the dissolved solids concentration in the feed water increased, the economics of purification mean conversion from an IX system to a RO system polished by IX. It is suggested that at 350 or 400 ppm of dissolved solids, RO is economically attractive.

The range of applications of Reverse Osmosis are given in Table 4.

WATER PURIFICATION

TABLE 4 – Typical Applications

- Pretreatment to Deionization systems
- Electronics – provides rinse waters low in colloidal, organic, and ionic impurities
- Manufacturing – for process, make-up, and high purity rinse waters
- Water Jet Cutting – improves operating efficiency and extends orifice life
- Printing – consistent water quality for fountain solution make-up
- Food & Beverage – provides water low in sodium and organics for product formulation
- Restaurants – spot-free glassware, coffee savings, and all drinking water
- Boilers – provides water with reduced scaling potential which improves energy efficiency
- Cosmetics – provides water for product formulation or rinsing applications
- Laboratories – provides water for reagent make-up and glassware washing
- Plating – spot-free rinse and solution make-up water
- Drinking water – reduces minerals and improves taste
- Humidification – reduces scaling and dusting
- Ice Production – improves taste and clarity
- Vehicle Wash – for spot-free rinses
- Horticulture – reduces leaf spotting and mineral build-up in planting soils

Water quality is a function of a variety of contaminants ionic materials, organics, silica and particulate including bacteria. Thus the method of water purification depends on the desired quality.

IX has the ability to remove a portion of the particulate and organic material present in the feed water. RO is intrinsically a broad-spectrum process, effective for the removal of virtually all particulate matter, and one of the most effective means of organic compound removal.

Ionic Removal

RO is widely used solely for removal of dissolved salts, such as sodium chloride and calcium bicarbonate. This reduces the ionic load on polishing IX equipment and, in some cases, depending on the water purity required, eliminates IX completely.

Organics Removal

The presence of organics can have several detrimental effects on particular processes. In electronics, organic material in the water tends to migrate toward the interface between the water and the product, reducing the quality of the surface. Organics present in water systems increase the food supply for bacteria, commonly residing in the piping distribution systems and thus increase biological growth rates. This generates more biological particulate material in the high purity water. Organics, although non-ionic and non interactive with IX systems, generally contain some sodium, chlorine, sulphur and nitrogen. Under the conditions present in a steam generator these organics can break down to liberate the associated ions and seriously degrade water quality. Organic removal is thus necessary to indirectly control ionic levels.

Silica Removal

Silica is non-ionic under normal conditions and is found at appreciable concentrations in most water supplies as discrete molecules or polymerised into colloids which have a range of molecular weights. Under certain conditions of concentration and pH, silica will precipitate to form discrete particles and thus IX is largely ineffective in its removal. However, RO is effective for both particulate silica removal and some dissolved silica removal.

Particles and Bacteria

Where particulate matter must be removed RO is effective for removal of virtually 100 percent of all discrete particle matter, including bacteria, silt and colloids.

A major problem with RO process operation is that of potential membrane fouling.

Fouling

The undesirable formation of deposits on surfaces is termed *fouling*. This can occur at the surface of a membrane when rejected solids (dissolved salts, suspended solids and microorganisms) are not transported from the surface into the bulk stream. Fouling can occur by the following processes

- Inorganic deposits (scaling)
- Organic molecules adsorption (organic fouling)
- Particulate deposition (colloidal fouling)
- Microbial adhesion and growth (biofouling)

The different types of fouling often occur simultaneously and all tend to decrease the performance of the permeator.

Reverse Osmosis

Power stations which utilise heavy fuel oil. The fuel oil is typically very dense, needing preheating to 130C to reach a required fluidity for use, and of poor quality with many impurities – mainly sulphur. The fuel oil must be purified before use in the boiler by washing with a high quality water: RO is appropriate to supply the required quality water (raw water composition in brackets)

Total hardness	$\leq 1F$	(77F)
Sodium	≤ 17 ppm as N	(550)
Potassium	≤ 0.9 ppm as K	(20)
Chlorides	≤ 27 ppm as C	(810)
TDS	≤ 50 ppm as CaCO	(2000)

The water had a final conductivity of 8 S/cm.

Car Wash Operations

In the US alone over 60% of the approximate 30,000 car wash operations use water softening systems to

WATER PURIFICATION

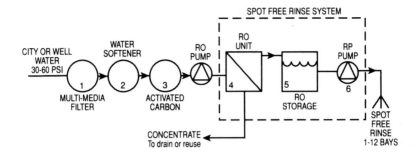

FIGURE 4 – System for Spot Free Rinse Water Production.

- enhance soap performance
- prevent scale build up in equipment
- provide a degree of filtering of large particles

Of these operations some 11,000 installed a purification system to provide spot free rinsing. Water spotting is caused by mineral salts such as calcium and magnesium and occurs when TDS are approximately 50 ppm and greater. This is generally above the average TDS of water (250 – 350 ppm).

Two purification systems are commonly in use, ion-exchange and reverse osmosis, with the latter having the greater market share, 65%. Nearly all new car wash water purification systems are now based on RO because of a much reduced cost, e.g. 20$ for RO, 300$ for IX per 1000 gal. A single stage RO system, with a 90% or greater rejection, will achieve required TDS levels for spot free washing. The design of a reverse osmosis based water purification system requires knowledge of the expected TDS of feedwater, chlorine content, temperature, pH, mineral analysis. A typical system is shown in Fig 4., which incorporates, media filters, water softener, activated carbon filters prior to RO. The design varies depending on the type of car wash and the manufacturer/design.

The most frequently used membranes are thin-film composites, rather than cellulose based, due to lower operating costs (lower operating pressures), higher removal of TDS and increased durability (double the lifetime of operation) of the former. Car wash operators require membrane elements, which last a long time, are easy to operate, require minimal maintenance, and readily adapt to changing feedwater conditions.

Tap Water Reverse Osmosis Systems

In the selection of RO membrane systems for home tapwater the primary factor is system performance; ability to provide water of high purity, the productivity of water and the durability of the membrane. In most respects thin film composite membranes are superior to cellulose acetate membranes except regarding their tolerance to chlorine in the raw water. The chlorine is added, in concentrations ranging from 0.5 to 1.0 mg/l, as a disinfectant which quickly inactivates microorganisms which pose a health hazard and

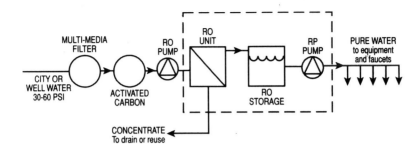

FIGURE 5 – Carbon filters in water treatment systems.

impart unpleasant tastes and odours to water. However, the chlorine tolerance of thin-film composites varies widely, thin film polyamide constructed spiral would membranes can operate up to 2000 hrs at 1 ppm chlorine levels before a noticeable loss in salt rejection. This tolerance is highly dependent on the amount of iron oxide contaminants which coat the membrane, ie more iron oxide, reduced tolerance. Chlorine attack on membranes depends on water pH and whether ammonia is present in the water to "blind" the chlorine in the form of chloramine, NH_2Cl (monochloramine) $NHCl_2$, NCl_3. Bound chlorine has a much reduced effect on thin film membrane and may result in negligible damage. This largely depends upon the amount of fee chlorine present with the bound chlorine and thus typically the value of pH.

Uncertainty over the effect of chlorine on thin film membranes can be removed by dechlorinating the water supply before the water reaches the membrane. This can be achieved by the use of carbon filters (see Fig5)

In most cases the installation of carbon filters in municipal tap water RO systems is not an added expense, whether the system uses cellulose acetate or thin film composites. With cellulose acetate RO elements, activated carbon filtration removes certain organic and inorganic contaminants, which impart unpleasant aesthetic characteristics, that the membrane is unable to remove effectively. In a thin film composite membrane system, the carbon filters primarily remove chlorine. Some supplementary removal of organic contaminants occurs although the membrane is the primary component for removing organic contaminants.

The location of the activated carbon filters with both membrane filters is different (see Fig 6.). With the thin film composite it is positioned before the membrane unit to avoid chlorine contact with the membrane. With cellulose membranes the carbon filter is after the membrane to collect organics which pass through the membrane. Cellulosic membranes generally require chlorine in the feedwater to reduce the potential for bacterial degradation of the membrane. Of course with either of these water treatment systems, the carbon filters have to be regularly replaced to prevent bacterial growth which can be a health hazard.

WATER PURIFICATION

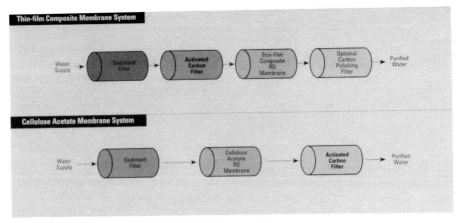

FIGURE 6

Commercial Drinking Water System

There is a large and diverse market for the use of water purification in commercial drinking water applications. Reverse osmosis systems can be used in several sectors, office buildings, manufacturing/retail facilities, health and educational institutions, water vending machines and bottled water. The driving factor behind this market use is identical to that in the home tap water market;- water which is safe, healthy and with good taste.

The application of RO water purification systems is in general as a large capacity central water system with multiple delivery points, as this is more cost effective. Clearly in certain instances point of use (POU) are preferred where for example there may be constructional constraints or other factors limiting centralised delivery. The main competition to RO systems in office buildings is simple water delivery. The attraction of RO lies in

- an on demand permanent system
- eliminating storage of bulky water bottles
- eliminates the inconvenience of delivery and the general wear and tear on furnishings and fittings
- much reduced cost of the water (eg $0.05 per gal compared to $0.9 – 1.3 per gal for bottled water)

In health and educational institutions there are opportunities for reverse osmosis systems both as central installations and POU systems supplying purified water to water coolers, drinking fountains, ice machines and coffee machines. In health and educational institutions purified water is frequently in use in kitchens for food and beverages preparation and spot free dishwashing and rinsing. Thus RO systems which both serve kitchens as well as drinking requirements is an additional attraction.

There are many thousand of bottled water vending machines, particularly in the USA, producing purified water. The most popular design uses a combination of carbon filtration, reverse osmosis, ion-exchange and ultraviolet sterilisation. Thin film composite membranes are the most popular due to reliability and ruggedness. Systems are either fitted with one or two elements with replacement required every six months approximately. These

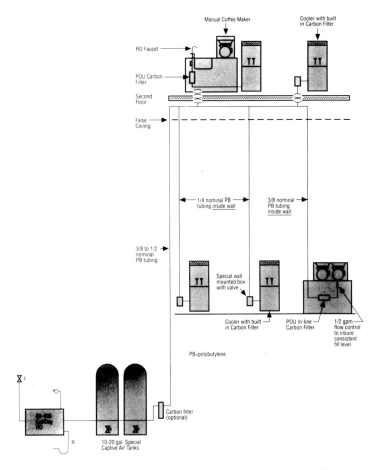

FIGURE 7 – Central RO system serving a small commercial building.

elements come in a range of sizes from two inch to four inch diameters for tap water, brackish water and other applications.

Food Service Applications

Food service operations are a potentially enormous marked for RO water purification systems

- Hospitals
- Nursing Homes
- Hotels
- Motels
- Resorts
- Clubs
- Schools
- Colleges
- Universities
- Military Installations
- Government Facilities
- Commercial Facilities

In the USA alone the total number of individual locations in this diverse market is over

WATER PURIFICATION

a million (include some 630,000 restaurants and fast food outlets). Estimates of the use of RO systems are currently in the range of 15% to 40% and include applications for drinking water, ice, food and beverage recipes, steamers and rinsing crockery and cutlery. There are several factors in favour of the use of water purification systems:
- reduction in detergent and rinse agent use
- eliminates scaling which can damage water heaters, coffee makers, steamers and other equipment
- economises the use of beverage ingredients, for example purified water reduces the amount of coffee, and also syrup in carbonated drinks

Performance specifications

	Nominal Surface Area (ft^2)	Product Water Flow Rate gpd (m^2/D)	Salt Rejection Cl" (%)
BW30-400	400	10,500 (40)	99
BW30-345	345	9,000 (34)	99
BW30-330	330	7500 (28)	99

- Permeate flow and salt rejection based on the following standard conditions: 2000 ppm NaCl, 225 psi (1.6 MPa), 77°F (25°C), pH 8, and 15% recovery.
- Flow rates for individual elements may vary but will be no more than 7% below the value shown for the BW30-400 and BW30-345 elements and no more than 15% below the value shown for the BW30-330 element.
- Minimum salt rejection for individual elements is 98.0% for the BW30-400 and BW30-345 elements and the 96.0% element.
- The BWE30-330 element is FilmTec's standard 8" element previously designated the BW30-8040.

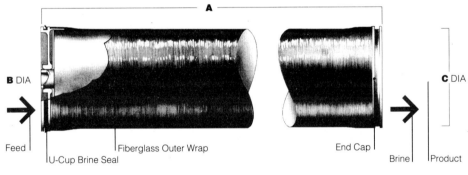

Operating Limits
Membrane Type   Thin-Film Composite
Maximum Operating Pressure   600 psi *(4.1 MPa)
Maximum Operating Temperature   113°F (45°C)
Maximum Feed Turbidity   1 NTU
Free Chlorine Tolerance   <0.1 ppm
pH Range:
     Continuous operation   2-11
     Short-term (30 min.), cleaning   1-12
Maximum Feed Flow   70 gpm (265 lpm)
Maximum Feed Silt Density Index   SDI 5

FilmTec RO element.

- improved taste
- spot free tableware
- benefits to health

The main competition to RO in food service applications are water softening and carbon filtration. Water softening removes calcium, magnesium and other dissolved solids and generally satisfies requirements of spot free rinsing and reduced maintenance on equipment such as coffee and ice makers. However scaling in steam generators is not prevented and can be increased due to the increased sodium level in the water. The taste of water is also not improved. Carbon adsorption does improve the drinking quality, by removing chlorine and sediments, but does not prevent scaling in steam generation equipment. Reverse osmosis is the most effective and efficient option for the full range of applications in food services.

Dishwashing

In the cleaning of food, debris and insoluble deposits – resulting from interaction of washing components and hard water – must be removed. While dish washing components usually remove food debris effectively they fail to eliminate hard water curd.

Soap used in commercial dish washers will react with water hardness to form a sticky, insoluble calcium and magnesium curd which streaks and spots dishes, glassware and utensils as well as leaving scale deposits in the dish washing machine.

Spots on dishes resulting from soap curd are a source of great concern in dishwashing, not only because they detract from the appearance, but also because hard water soap film protects bacteria and presents an environment for continued growth.

Washing dishes in completely soft water eliminates the formation of calcium and magnesium soap films, and thus improves dishwashing, in terms of both cleanliness and appearance.

Today, manufacturers of commercial dish washing equipment strongly recommend a water treatment mix combining ion exchange softeners for the washing cycle with demineralised or reverse osmosis water for the rinse cycle.

Potable Water Treatment

Reverse osmosis is commonly used as a method of desalination and purification of water. This is discussed in Section 9. A large application for producing potable water in Florida is at the Island Water Association Inc. This plant consists of six parallel trains of 600,000 gal/day capacity each for a total of 1.8 million gal/d (see Fig 8).

Each train contains 20 vessels in a 14 ft by 6 ft two-stage array, operating at a pressure of approximately 19 bar and 80% recovery. Each vessel houses six elements giving a total of 120 elements per train. This particular system is fitted with thin film composite membranes which give a production capacity of 400 gpm per train. The thin film membranes replaced the original cellulose acetate membranes which started to deteriorate significantly in performance after approximately 4 years of use. The thin film membranes gave significant improvements in performance over CA

- membranes increases flux rate, approximately 40 gpm per train
- reduced operating pressure, approximately 8 bar lower

WATER PURIFICATION

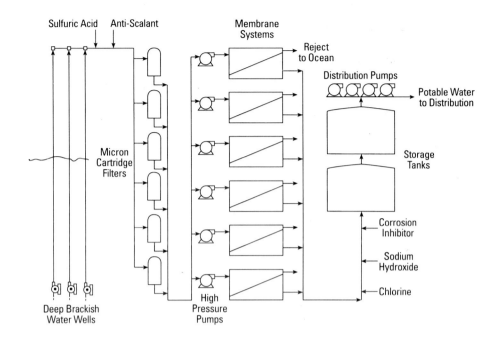

FIGURE 8 – Reverse osmosis plant in Florida.

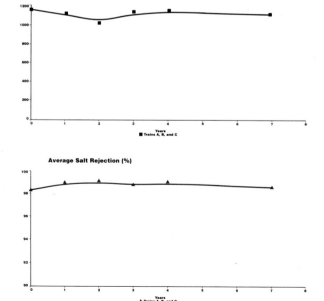

FIGURE 9 – The performance of a RO plant for potable water treatment.

- higher rates of salt rejection; gives improved permeate quality (80 ppm)
- greater tolerance to temperature (up to 45C)
- greater tolerance to pH (from 2 to 11)
- synthetic membrane composition makes it impervious to bacterial attack

TABLE 6 – A feed, concentrate and potable water analysis.

	Feed mg/l	Product mg/l	Brine mg/l	Rejection* %	Recovery %
Cations, mg/l					
Aluminum	0.25	<0.15	1.03		
Barium	<0.01	<0.01	0.06		
Calcium	98	0.27	460	99.9	78.7
Copper	<0.01	<0.01	<0.01		
Iron	<0.01	<0.01	0.02		
Magnesium	111	0.29	510	99.9	78.3
Manganese	<0.01	<0.01	<0.01		
Potassium	34	<4	148		
Sodium	730	26	3,100	98.2	77.1
Strontium	8.9	0.03	44	99.8	79.8
Zinc	<0.03	<0.03	0.06		
Anions, mg/l					
Alkalinity, as $CaCO_3$	1657	738	97.9	78.4	
Alkalinity, as HCO_3	201	9	900	97.9	78.4
Chloride	1,290	36	5,800	98.6	78.2
Nitrate	<1	<0.5	<5		
Sulfate	450	6	2,090	99.3	78.7
TOTAL IONS, mg/l	2,923	77	13,053	98.7	78.1
Other					
pH	7.5	6.1	7.8		
Conductivity, μmhos	4,.740	160	19,540		
TOC, mg/l	2	<0.1	7	>99.0	71.4
Silicon, mg/l as Si	10	0.3	54	99.1	82.0

*Salt rejection is calculated using the log mean average of the feed side to brine side concentration.

The operating cost with the thin film membrane (1992) was $0.26 per 1000 gallons of water. this compared with $0.60 per 1000 gallons with the CA membranes. Plant operation, total flow and salt rejection (see Fig 9) have remained stable for over 7 years of production.

Desalting Seawater

On the island of Lanzarote a RO plant is part of a self sufficient resort complex. The resort complex was originally equipped with three vapour decompression plants producing 1500 m³/d of drinking water from sea water wells. To satisfy expansion a RO system was selected, in preference to further vapour recompression, with a capacity of 1000 m³/d. The

WATER PURIFICATION

feed water to the RO system has a TDS of approximately 38,000. This water required pretreatment; sand pressure filters, cartridge filters and anti scalant and sodium bisulphate injection. The RO unit consists of a single stage, two pass system using thin film composite membranes with an operating pressure of about 62 bar. The rejection of the membrane was 99% giving a TDS content of 325 ppm. By using an energy recovery turbine, energy consumption was less than 6 kWh/m^3 of permeate. This was substantially less than vapour recompression and led to further expansion (1000 m^3/d) with the vapour recompression restricted to peak demand operation.

Reverse Osmosis Water Purity Unit

There is a need within the military to purify water from any available source for use by land based forces. The source may be fresh, brackish or seawater and it may be turbid, highly polluted or even contaminated by nuclear, biological or chemical (NBC) warfare agents. The major requirement is that the unit must not be site specific, ie must be capable of handling all waters from surface or subsequent sources with total dissolved solid content varying from 1.5 g/l and less to greater than 15 g/l. In addition to the water requirements the unit must meet stringent requirements on weight and size, mobility, maintenance, start-up and shut down times and straight forward operation.

In the early 1970's, the US military after evaluating several types of water treatment equipment, proceeded with the development of units based on reverse osmosis. Two standard Reverse Osmosis Water Purification Units (ROWPU) are currently in use, one rated at 600 gallons per hour (gph), used in processing fresh water and the second at 150,000 gallons per day for seawater (see Table 7). A system with a nominal rating of 3,000 gph for use with fresh water is under production.

TABLE 7 – Summary of ROWPUs.

	600-gph ROWPU	3,000-gph ROWPU	150,000-gpd ROWPU
Design accepted	1979	1987	1981
production rate (gpd):			
Fresh water	12,000	60,000	—
Seawater	8,000	40,000	125,000
Crew size	3	3	7
Elements used:			
Size	6-inch	8-inch	6-inch
Quentity	8	12	80
Weight (pounds)	16,975[1]	—	—
	11,380[2]	39,800[3]	—
Size (feet)	19 x 8 x 8[1]	30 x 13 x 8[4]	—
	10 x 8 x 8[2]	—	
Planned quantity	1,600	400	28
Total installed capacity			
(seawater)	12.8 mgd	16.0 mgd	3.5 mgd

[1] Trailer-mounted version. [2] Skid-mounted version. [3] ROWPU container only
[4] Includes trailer

TABLE 8 – Raw water characteristics at ROWPU.

	Pacific Ocean	James River, Virginia	River, Overseas	Panama City, Florida
pH (pH units)	7.8	7.8	7.6	8.4
Terperature (°F)	64.0	58.0	—	86.0
TDS	31,950.0	13,000.0	200.0	18,000.0
TSS	—	33.0	—	—
Turbidity (NTU)	—	—	19.0	0.4
TOC	—	4.1	3.9	4.9
Tot Alk (as $CaCO_3$)	—	66.0	13.0	67.0
Hardness (as $CaCO_3$)	2,920.0	2,600.0	109.0	2,830.0
Calcium	250.0	205.0	27.3	180.0
Magnesium	560.0	508.0	10.0	580.0
Sodium	11,000.0	4,110.0	22.1	5,100.0
Chloride	17,745.0	7,500.0	10.0	9,600.0
Sulfate	2,577.0	910.0	17.0	1,400.0
Silica	3.0	—	—	—

Units are mg/l unless otherwise specified.
TDS = total dissolved solids　　TOC = total organic content
TSS = total suspended solids　　Tot Alk = total alkalinity

TABLE 9 – Operational characteristics of the 600 ghp ROWPU.

Flow rates (at 77°F)	Feed	32 gpm
	Permeate	
	Fresh/brackish	10 gpm
	Seawater	8 gpm
Normal operating pressures	RO Feed	
	Fresh/brackish	≤500 psi
	Seawater	≤960 psi
Operator actions	Cartridge filter change	ΔP > 20 psi
	Multimedia backwash	ΔP > 10 psi
	Element chemical clean	ΔP > 100 psi
Cartridge filter	8 each, 40-in long, 5 μm (nominal)	
Multimedia filter	Flow rate	6.5 gpm/sq ft
RO vessels and elements	4 pressure vessels x 2 elements (6-inch)	
Chemicals	Coagulant Antiscalant Calcium hyphochlorite	
Operating temperature range	–25 to 110°F air temperature + 33 to 110°F water temp.	

WATER PURIFICATION

The ROWPU's can be used to treat water contaminated with NBC warfare agents. There are operating limits on the water to be treated, the temperature must be between 0C and 110F and the raw water turbidity must be less than 150 NTU (Naphelometric Turbidity Units). Typical raw water characteristics at some ROWPU sites are shown in Table 8.

The 600 gph ROWPU was designed for use in forward combat areas, behind front-line forces. The system (see Fig 10a and 10b) consists of several pretreatment steps: coagulation by addition of a cationic polymer followed by filtration through a multimedia filter and cartridge filters. Chemicals are added to the RO feed to reduce scale formation. The reverse osmosis consists of eight 6 inch elements in series to provide the final purification. Operating characteristics of the unit are given in Table 9.

After reverse osmosis, calcium hypochlorite is added to the permeate to render it potable. When operating in an NBC environment, although pretreatment and RO removes the bulk of the contaminant, additional post-treatment, with activated charcoal and ion exchange filters, is included to remove residual contamination.

The operation of the 600 gph ROWPU unit allows for chemical cleaning of the RO units using either citric acid or surfactant. This is initiated when there is a 20 % increase in the initial differential pressure or when there is a reduction in permeate flow or quality and when the maximum operating pressure is exceeded.

The 3000 gph ROWPU unit is employed in the rear portions of combat areas and was designed to survive in a nuclear battle field as well as to withstand contamination by NBC warfare agents. The system is enclosed in a 8" x 8" x 20" container mounted on a standard 22_ ton army trailer. The components of this unit are similar to the 600 gph unit. Additional features of pretreatment processes are a cyclone separator, which removes larger suspended particles such as sand and heavy dirt from the raw waste, and a basket strainer to remove large particles that may damage the media filter. The reverse osmosis is performed by two parallel sets of 6, 8 inch elements connected in series.

The ROWPU systems use standardised sea water RO spiral wound elements with thin film composite membranes. In military use the expected lifetime of the element is only 2,000 hours due to the greater demands of operation – including incomplete removal of suspended particles, biofouling, exposure to extremes of temperature and variability in raw water source.

In addition to the 600 gph, 3000 gph and (150,000 gpd) units a smaller water treatment

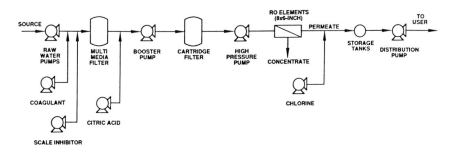

FIGURE 10a – Flow diagram of 600 gph ROWPUs.

unit for 50 – 150 personnel is under development which is expected to have use in disaster relief and other civic action operations in addition to military.

High Purity Water

Water is the most popular industrial fluid for a range of applications including the washing and rinsing of manufactured goods, driving turbines in power plants and diluting personal care products and a range of household chemicals. The required purity level of the water

TABLE 10 – Specification of pure water in various industries.

Pure water quality at 25°C	Cosmetics	Pharmaceuticals (non-injected drugs)	Microelectronics Conventional	New
Conductivity (µS cm^{-1})	10	3	0.055	0.055
Electrical resistance (MΩ cm)	0.1	0.3	18	18
Particle number (ml^{-1})	n.s.*	n.s	100	10
Filter grade (µm)	n.s.	0.5	0.5	0.2
TOC (ppb)	3000	1200	300	50
Bacteria count (ml^{-1})	1	1	1	0.1
Blocking index (VI)	0	0	0	0
Silicon (ppb)	n.s.	n.s.	20	5
Sodium (ppb)	n.s.	n.s.	n.s.	1
Chloride (ppb)	n.s.	n.s.	n.s.	2

*n.s., not specified.

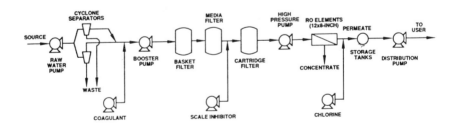

FIGURE 10b – Flow diagram of 3000 gph ROWPUs.

depends on the application (see Table 10.) but increasingly demands are for a product with parts per billion (ppb) of contaminants with expectations of parts per trillion to be more common in the future. The major industrial areas for high purity water are
- Boiler feed water
- Electronics
- Metal finishing
- Medical and pharmaceutical
- Packaging

There are several technologies which can be used to produce pure water for these applications. These are effective to different degrees depending upon the application and in many cases more than one method is employed to achieve the final desired quality. The classical method to prepare pure water is by distillation but this is not effective for removal of organic compounds with relatively high boiling points. The water purification processes of distillation, de-ionisation, carbon adsorption, filtration, UV oxidation, reverse osmosis and ultrafiltration vary in there effectiveness.

Boiler feed

High purity water is used for boiler feed to generate steam for production of electricity and for process heating. In these applications, water is fed to a steam generator (boiler), which converts the water to vapour (steam) through the use of fuel/oil, coal, natural gas or nuclear energy. The steam generated at pressures of between thermal 600 to 3,000 lb, is fed to a turbine, which removes thermal energy in the steam, converting it to mechanical energy as a rotating shaft. Mechanical energy is then converted to electrical energy in an electric generator. Exhaust steam from the turbine is condensed and repumped to the steam generator. The efficiency of the power generation cycle increases at higher steam temperatures and pressures.

Higher temperatures and pressures are more chemically and physically demanding, on the boiler and the turbine. Small amounts of contaminants, eg sodium, chloride, or silica, in the steam can cause corrosion or fouling of turbine blades. The higher pressures and temperature of the steam increased the volatility of salts in the feed water, especially silicon dioxide, increasing the level of these contaminants in the steam phase. Under conditions of higher temperature and pressure, the vaporisation is not as effective at separating out feed-water contaminants from the steam. Therefore the contaminants must be removed upstream of the steam generator.

Overall therefore, the quality of the water that must be maintained in the steam generator is determined by the operating pressure. The generator essentially concentrates the contaminants in the feed water because the steam has a lower level of contaminants than the water. A portion of the water in the steam generator must therefore be blown down or purged to maintain the desired generator water quality. To ensure that blowdown is only a small fraction of the total feed-water stream, the feed-water contaminant level must be much lower than the contaminant level tolerated in the steam generator. For example, in boilers with pressures of 600 pounds, it is common to have a blowdown stream of 1% of the feed-water rate, requiring feed-water qualities 100 times superior to water maintained

Table 12 – Steam Generator Water Quality (ASME).

Drum Pressure lb/bar	Silica (ppm SiO2)	Total Alkalinity (ppm CaCO3)	Specific Conductance (micromhos/cm) microseimens
0 - 300	150	350	3,500
301 - 450	90	300	3,000
451 - 650	40	250	2,500
651 - 750	30	200	2,000
751 - 900	20	150	1,500
901 - 1,000	8	100	1,000
1,001 - 1,500	2	0	150
1,501 - 2,000	1	0	100

in the steam generator. Table 12 gives typical guidelines for the maximum boiler water impurities necessary for operation of modern steam generators at different pressures. The particular features of each system boiler design, presence of a turbine etc will also greatly determine the quality limits that are appropriately maintained.

Steam generators used in chemical process plants, oil refineries, pulp and paper production plants, or food processing systems, are often used to produce steam only for heating. Precipitation of silica and corrosion of turbine blades is not an issue, the required steam quality is different than in power generation.

Boiler Feedwater Pretreatment

Boiler feedwater is traditionally treated by ion exchange demineralisers (both cation and anion) to remove ionic constituents which could cause scaling. There are several problems with the use of ion-exchange alone which include
- loss of strong-base capacity from frequent anion replacement
- regeneration requires large acid and caustic dosages
- the low throughput capacity necessitates frequent regenerations

These factors all contribute to high operating costs which can be reduced by the installation of RO prior to ion-exchange. The RO significantly reduces the total dissolved solid, e.g. from 480 ppm to 6 ppm (see Table 13). This causes a large reduction in the load on the ion exchange system with a much reduced need for resin regeneration, e.g. from three times a week to three times per year. In addition the volume of water produced between regenerations is dramatically increased (eg by over 6,000%). A typical utility, of 25 million gallon capacity, reduced the cost of deionisied water by some 6.28/1000 gal (ie $ 150,000), mainly as a result of reduced chemical usage.

The production of high quality water for industrial use such as boiler feed water from surface water sources becomes increasingly difficult because of the high salt content of the raw water. For feed water salt concentration in excess of 200 ppm ion-exchange becomes quite costly and furthermore burdens the environmental with additional discharge of chemicals during the regeneration process. Therefore, electrodialysis is often

TABLE 13 – Reverse osmosis/dioniser system profile water analysis.

Constituent	Raw Water	RO Permeate	% Rejection	Mixed Bed Effluent
Ca	47	0.11	99.8	<0.01
Mg	15	0.04	99.7	<0.01
Na	60	1.0	98.3	<0.01
K	4	<0.5	—	<0.01
Fe	<0.01	<0.01	—	—
Sr	0.88	<0.01	>98	<0.01
Cl	61	0.9	98.5	<0.01
SO_4	105	1.3	98.8	<0.01
HCO_3	190	4.9	97.4	—
SiO_2	28	0.45	98.4	<0.002
pH	7.8	7.5	—	6 to 8
Conductivity	705	10.7	98.5	0.1

Att constituents except conductivity and pH are expressed as mg/l.
Conductivity is expressed as micro S/cm @ 25°C, pH is expressed as standard units.

used to reduce the salt content of the feed water to less than 20 ppm prior to the ion-exchange deionising procedure.

The bulk of the dissolved solids can be removed either by ED or RO. These two processes are competitive and offer the advantage that they do not contribute additional water pollutants.

ED might appear to be less competitive than RO for this application, because RO removes particulates and substantial amounts of organic solutes form the water. Accumulation of these contaminants on RO membrane surfaces can cause premature failure. Therefore, it is common practice to remove organics and particulates before the rough desalting. With such pretreatment, ED stacks operate trouble-free for years with little maintenance. A typical system is shown schematically in Figure 11 for removing salts, principally $CaSO_4$ from water for use in rinsing integrated circuits.

Electrodialysis is also used for the production of ultrapure water in the electronics industry.

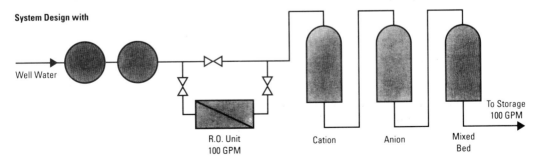

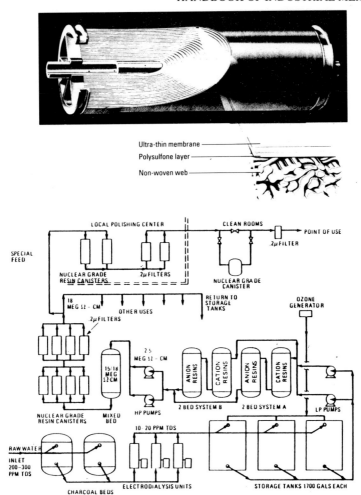

FIGURE 11 – Integrated circuit process water system.

TABLE 12 – Steam Generator Water Quality

Drum Pressure lb/bar	Silica (ppm SiO_2)	Total Alkalinity (ppm $CaCO_3$)	Specific Conductance (micromhos/cm) microseimens
0 - 300	150	350	3,500
301 - 450	90	300	3,000
451 - 650	40	250	2,500
651 - 750	30	200	2,000
751 - 900	20	150	1,500
901 - 1,000	8	100	1,000
1,001 - 1,500	2	0	150
1,501 - 2,000	1	0	100

WATER PURIFICATION

What is more important is to achieve low pH corrosion of the condensate returns lines which is caused by dissociation of bicarbonate and carbonate in the steam generator and the resultant carryover of carbon dioxide into the condensate lines. This is dealt with by removal of bicarbonate and carbonate alkalinity from the feed water.

Electronics Industry

The electronics industry produces a large range of products from cathode ray tubes, to discrete devices, such as diodes and transistors, to sophisticated integrated circuits. Electronics manufacturing processes typically involve surface washing. The features on an integrated circuit are so small that minute amounts of impurities in the water can render the circuit inactive. Ions such as sodium and chloride can adsorb into certain layers of the circuit, altering the electrical characteristics and thus the properties of the end product. Organic materials in the water tend to be surface active, and thus migrate toward, and become associated with, surfaces and disrupt placement of the following layer. Particles such as bacteria can disrupt additional layers or create electrical shorts between adjacent circuits.

Integrated circuit production requires water of the highest purity of any industry or application necessitating attention to all water contaminants, ionic, organic, particulate and silica.

Ultrapure water is generally used in the electronics industry as a cleansing rinse for surfaces. During production of integrated circuits, bare silicon undergoes many treatment (30 or 40), in which layers of conductive or insulating materials are added to the surface. Prior to addition of each layer a portion of the surface is etched away by chemicals, such

TABLE 14 – Ultrapure water quality in the electronics industry.

Integrate (bit)	256K	1 M	4 M	16 M	measurement method
Design rule (µm)	1.5-2.0	1.0-1.2	0.8	0.5	—
Resistivity (MΩ cm) 25°C	17 18<	17.5 18<	18.0<	18.1<	Specific resistance meter
Particle Size (µm) Number (pcs/ml)	0.2 0.1 30–50 50>	0.1 20> 10>	0.1 0.08 10> 5>	0.05 10>	
Live bacteria (CFU/ml)	0.2> 0.05>	0.05> 0.01>	0.05> 0.01>	0.01> 0.005>	M-TEG culture method
TOC (µg/l)	100> 50>	50> 30>	30> 10>	10> 5>	Wet oxidation TOC method
Dissolved oxygen (µg/l)	100>	100> 50>	50>	50> 10>	Dissolved oxygen meter
Silica (µgSiO2/l)	10>	10> 5>	5> 3>	1>	Ion chromatography

as sulphuric acid or hydrofluoric acid. Ultrapure water is used between all of the chemical etching steps to rinse and removal all of the chemical from the water surface.

A widely accepted specification for ultrapure water for manufacturing of 1 megabit DRAm product is shown in Table 14., although the appropriate water quality will vary with the exact nature of the product and the nominal feature size of the individual devices.

Metal Finishing

Products frequently encountered in daily life are often manufactured with bright and colourful finishes to make them more edurable and attractive. This typically involves metals such as gold, copper, cadmium, chrome etc. High purity water ranging in quality (measured conductivity) from 1 megohm to 10 megohms is customarily used to

i) rinse the starting surface of any interfering dirt or chemicals prior to plating to ensure a better, more uniform bond and more uniform application of metal

ii) to rinse off excess plating solution before drying to ensure the luster of the finish and to prevent spotting on the surface.

Packaging

High purity water is needed in the preparation of common household products (eg for dilution) because ions in most municipal supplies may precipitate with other ingredients in the product, thus rendering either performance or appearance unsatisfactory. Typical products include fabric softeners, window cleaners, and shampoos. This industry is collectively referred to as "packaging", because the purchasers of the high purity water are often facilities that concentrate on packaging not product development or marketing. The water purity specifications for most users depends on the product; however, quality comparable to a single-distilled water (less than 5 ppm of dissolved solids) is generally satisfactory.

Medical and Pharmaceutical Applications

The solvent of choice for most medical preparations, including pharmaceutical prescriptions, lotions, cleaning solutions and creams, is water. Water is the principal constituent to intravenous fluids for replacing natural body fluids in patients who have undergone serious illness or injury. The presence of any contaminants in these formulations can lead to unwanted side effects of, and interfere with, chemical characteristic of the medication, or may be directly harmful to the patient. These products require high purity. Small amounts of high purity water are also used in various medical laboratory testing procedures.

An area with a requirement for a significant large quantity of water is hemodialysis. In hemodialysis, the patient's blood flow is brought into close proximity with water, separated only by a semipermeable membrane. Contaminants in the blood diffuse through the membrane into the water, and are thus purged from the patient's body. The water used in this process, normally from the municipal supply, is carefully treated to remove most ionic materials that could interfere with the dialysis. Of special concern are chloramines, which can readily diffuse through the dialysis membrane into the blood and cause illness.

There are several standard specifications for water quality in medical and pharmaceutical uses, depending on the particular application. There are designations such as "sterile water for injection" and "purified water" and also various grades of sterile water. Water for injection is the most demanding to produce because of extreme quality control requirements. It must be purified by either distillation or RO to ensure reliable production of nonpyrogenic water. Purified water, for a variety of other applications including preparation of optical and oral medications, may be produced by ion exchange (IX) in addition to distillation and RO.

Water Treatment in Hospitals

Purified water of different quality are required in several applications in hospitals including, laboratories, hemodialysis, boiler water treatment, air conditioning, laundering and dish washing. Laboratories have traditionally depended upon distillation or ion-exchange to provide high quality water for a variety of chemical and analytical uses. The requirements of new medical techniques and more advanced analytical procedures has placed an ever increasing demand for high quality reagent grade water. Reverse osmosis offers a mean of providing reagent grade quality water in a simple, inexpensive near maintenance free manner.

A serious problem encountered in steam generation is the formation of scale and sludge deposits on boiler heat surfaces. Scale formation is the accumulation of deposits of solids within boilers. Scaling is caused principally by the breakdown of calcium bicarbonate to calcium carbonate at elevated temperature and pressure conditions within the boiler. Scale formation in boilers affects operation in several ways.

1. Reduced efficiency due to the insulation properties of scale.
2. Economic loss due to boilers out of service during cleaning periods.
3. Increased maintenance due to cleaning requirements.
4. Depreciation of equipment caused by scale formation and removal.

It is important that proper steps be taken to remove deposits from the boiler heating surfaces. Removal of scale deposits may be effected by one or combined water treatment methods, depending on the particular nature of the deposit and the technical characteristics of the boiler.

Air conditioning is the process of treating air, to control simultaneously its temperature, humidity, cleanliness and distribution to meet the requirements of the space to be conditioned. The temperature of air is controlled by either passing the air over chilled, or heated coils, or by exposing the air to a spray of water of controlled temperature.

The primary purpose of water for air conditioning (cooling water) is as a heat transfer medium ie waste heat is picked up at one site in the cooling cycle and disposed of in another site.

There are three major problems arising from the water for air conditioning systems and cooling towers; scale corrosion and algae and slime growth.

Scale can be eliminated by softening and chemical treatment.

Corrosion can be eliminated by chemical treatment.

Algae and slime growth are normally a problem in cooling towers. Both algae and slime clog air conditioning systems. Algaecides and chlorine are used to control such growths.

Water for laundering should ideally be of zero hardness and free from turbidity, colour, iron, manganese, and hydrogen sulphide. Water constitutes the most important individual reagent employed in the laundry process. The condition of the water used determines to a considerable extent the efficiency of soil and stain removal and the retention of the original appearance of a fabric during the wash or laundry cycle.

Using soft water will improve the efficiency and economy of a hospital's laundry operation in five different ways:

1. *Cleaner wash*. All fabrics come clearer when washed with the correct soap and soft water come cleaner.
2. *Water savings*. Softened water, with reduced hardness and less soap curd requires less rinsing.
3. *Reduced detergents*. With softened water, less laundry bleach is needed, chemical softeners are not required. Quoted savings in total supply costs can be as much as 50 per cent.
4. *Reduced linen replacement cost*. Linen and clothing replacement cost are a major concern. Two factors cause fabric to wear out – mechanical action of washing and chemical action. With soft water, linen replacement can be reduced by up to 70 per cent.
5. *Reduced repair and maintenance costs*. Water using appliances in hospitals can suffer from hard water damage. Hard water scale clog pipes and nozzles, damages valves, impairs machine efficiency. Soft water eliminates these problems and greatly increases the operating life of appliances.

Reverse Osmosis for Hemodialysis

The manufacturers of kidney machines need water with a high degree of purity for two purposes
- operation tests of the kidney machines
- production of concentrated solution for hemodialysis.

The testing of the kidney machines, which simulates their hemodialysis function, mixes the water with the concentrated solution, which reaches high levels of salinity up to 10,000 ppm. This solution itself presents an effluent disposal problem – the material cannot be simply discharged to watercourses and sewers. The application of a RO desalination process which consist of two desalinators in series (see Fig 12), ie bi-osmosis, produces an effluent with a relatively low salt content from the primary desalination. This flow can be used to dilute the highly saline drain from the kidney machine testing room. In this bi-osmosis treatment the effluent from the secondary desalinator has a very low salinity and can be recycled to the upstream of the treatment and thus reduce the water consumption.

The water presented to the desalinators goes through several essential phases of pretreatment.
- *Initial chlorination*
- *Filtration in selective multi-layer filters*

WATER PURIFICATION

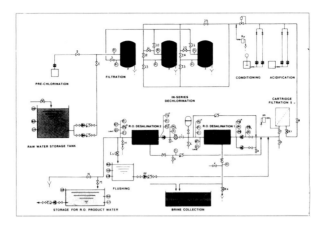

FIGURE 12 – Bi-osmosis water purification.

- *Activated carbon filters to prevent leakages and release of micropollutants*
- *Chemical treatment (injection of acid and sodium hexametaphosphate).* This can replace normal water softening processes to avoid increasing chloride content in the effluent which would be inevitable if ion exchange resins regenerated with sodium chloride were used.

Water for Hemodialysis

Hemodialysis is a long-term treatment designed to remove toxic substances from the blood of uraemic patients. Haemodialysis replaces the function of the kidney which has been destroyed by disease. The treatment is typically given three times a week and each session last between 4 and 5 hours. The patient's blood is passed through a semipermeable membrane, the "artificial kidney", which is in contact with a solution similar to plasma, the "dialysate".

The substances dissolved are transferred from the side of the membrane on which they are present in high concentrations to the other side on which they are absent or are present in only low concentrations. The process of exchange in the "artificial kidney", exploits the hydrostatic gradient on either side of the membrane, enables excess water and salt to be removed which the diseased or useless kidney cannot eliminate.

The equipment used for dialysis consists of the dialyser – also known as the "artificial kidney" – and the dialysis machine. The dialysis machine is used to prepare the dialysis solution, known as the "dialysate". The water is mixed by pumps with a concentrated salt solution composed of various electrolytes. Control devices permit the composition of the solution, the circulation of the blood, and the ultrafiltration process to be monitored. A permanent and reliable access to the blood stream is necessary for connection to the machine.

There are several possible contaminants which may occur in the dialysis solution. The dialysis solution is non-sterile solution in water, with a composition resembling plasma in terms of its electrolyte content. The solutions which have to be eliminated from the blood (urea, creatine and other waste products of nitrogen metabolism) are not present in the solution. The average electrolyte composition is calculated in order to correct any inbalance arising between two dialysis treatments. Table 15 lists the electrolytes and glucose which are usually present in dialysis solutions.

Possible contamination present in the water are given in Table 16. There can be severe potential medical problems with many metal ions commonly occurring in water. Aluminium sulphate is used as a flocculating agent in the purification of drinking water. Many symptoms, such as eg osteopathy, myopathy and anaemia, are attributed to aluminium poisoning. The maximum permissible level in a dialysis solution for use in haemodialysis is typically 0.3 µmol/l.

High zinc content can be due to galvanised components in the water supply system and causes fever and anaemia. Copper may occur either in water added to tap water to destroy algae, or originating from the water source itself, from the pipework or from the dialysis machine. Copper causes nausea, abdominal pain, headache and hepatitis (inflammation of the liver).

The iron content of the water is often the result of efforts at municipal water works to avoid using aluminium in drinking water purification. Large quantities of iron can also cause serious problems in patients and severely damage the dialysis machine.

Water from various natural sources can be very salty. When water with a high calcium content is "softened" a high sodium concentration can come about due to the exchange of calcium for sodium. A high sodium content in the dialysis solution can lead to elevated blood pressure and pulmonary oedema.

If the calcium and magnesium content of the dialysis solution are too high a condition known as hard water syndrome can occur with the symptoms of nausea, severe headache and arterial hypertension. Water which is completely free of calcium and magnesium must therefore be used with a specified quantity of ions added for dialysis purposes.

Medical problems can also arise from the presence of several anions. Chlorination is frequently used to kill bacteria in the purification of tap water. Free chlorine in the dialysis solution produces acute anaemia in the patient.

Fluorine is added to the water to prevent dental caries. However, some supplies have

TABLE 15 – Composition of the Dialysis Solution for Haemodialysis

Substance	Usual concentration mmol/dm^3	In exceptional cases mmol/dm^3
Sodium	10 - 145	150 - 170
Potassium	0 - 2	3
Calcium	1.25 - 2	0
Magnesium	0.25 - 0.75	0
Chlorine	100 - 120	–
Acetates or	35 - 45	–
Bicarbonate	33 - 45	5 (if bicarbonate)
Dextrose	0	100 - 150

TABLE 16 – Contamination of the Water and Possible Consequence for the Dialysis Patient.

Contaminant	Lowest Concentration At Which Poisoning Occurs	Symptoms Of Poisoning	Possible Means Of Removing The Contaminant
Aluminium	2 µmol/l	Encephalopahty Osteodystrophy Anaemia	Reverse Osmosis or Complete Softening
Calcium	2.2 µmol/l	Hard water syndrome	Softening or Reverse Osmosis
Magnesium	3.6 µmol/l	Hard water syndrome	or Complete Softening
Chloramine	5 µmol/l	Haemolytic Anemia	Activated Carbon Filter
Copper	8 µmol/l	Nausea, Vomiting Fever Hepatitis Haemolysis	Reverse Osmosis or Complete Softening
Fluorine	53 µmol/l	Disturbed Mineral Balance of the Bones	Reverse Osmosis or Complete Softening
Nitrates, Nitrites	32 µmol/l	Methaemoglobinaemia	Reverse Osmosis or Complete Softening
Sodium	15 µmol/l	Hypernatramia	Reverse Osmosis or Complete Softening
Sulphates	2 µmol/l	Nausea Vomiting	Reverse Osmosis or Complete Softening
Zinc	3 µmol/l	Haemolytic Anaemia	Reverse Osmosis or Complete Softening
Iron	36 µmol/l	Damage to Dialysis Unit	Filter
Manganese	36 µmol/l	Damage to Dialysis Unit	Filter
Formol	Traces	Haematuria Respiratory Depression	Flushing the system
Pyrogens		Shivering Attacks	Reverse Osmosis or Sterile Filter

Nephrosis = degenerative disease of the renal tubules. Myopathy = general term for any disease of the skeletal muscles. Haemolytic anaemia = cause: the life-time of red blood cells is reduced from the normal life-time of 120 days. Hepatitis = inflammation of the liver due to viral infection or propagation of inflammation of the bile ducts or due to the introduction of the virus into the blood stream by residues of human serum. Methaemoglobinaemia = elevated levels of methaemoglobin in the blood. Chelating agent = chemical compound which enters into complex compounds with metal cations, incorporating the metal into a ring (chelate). Widely proven for the removal of incorporated radionuclides.

a very high concentration which may disturb the mineral balance of the bones.

The NO_3 EC standard for drinking water is 50 mg/l of nitrates. However, in dialysis this concentration may cause anaemia, together with high blood pressure, and nausea. A level below 10 mg/l in the form of NO_3 is therefore recommended.

If the dialysis solution has too high a concentration of sulphate a condition with nausea and vomiting is observed in a patient.

Reactions to pyrogens are frequently observed in dialysis centres. These reactions are caused by microbial contamination in the water purification system.

The diluted solution for haemodialysis (called the dialysis solution) is subject to appropriate national standards (see Table 17), called the dialysis solution, is subject to medical prescription and therefore is the responsibility of the doctor treating the kidney patient. In each treatment session, the blood comes into contact with 120 – 1500 litres of dialysis fluid, making a total of approximately 23,000 litres per year.

The design of water purification facilities in dialysis centres must fulfil a number of criteria in order to supply adequate quantities of the right quality of water for the preparation of dialysis solution under all conditions. The choice of methods and the design of the unit are therefore important. Tap water supplied to dialysis centres varies considerably in quality and may be seasonally variable. Differences in water quality may thus arise depending on the changing application of purification methods or measures to safeguard the distribution system.

Since the process involves the innermost contact of the patients blood with a large quantify of water per year (around 25 m^3), the quality of the water to be used in the dialysate must be pure and controlled to a stringent level. There is clearly a high risk for the patient in using tap water for hemodialysis because although chemically and bacteriological safe for drinking, the water contains organic substances, non pathogenic bacteria, pyrogens and inorganic compounds (usually salts) at levels which greatly exceed prescribed limits of medical associations eg AAMI (Association of Advancement of Medical Instrumentation) and ASAIO (American Society for Artificial Organs).

The water used for dialysates must be suitably treated to the required level of contaminants. There are various treatments which are commonly used for such treatment which include softening, de ionisation and reverse osmosis.

Softening exploits the capacity of cationic ion exchange resin to retain the calcium and magnesium in the water, replacing these with sodium. This process does not alter the saline content of the water, does not ensure the removal of heavy metals eg Al, Mn, Cd, and does not provide control of the water sterility. De-ionisation is based on the capacity of the ion exchange resin, both cationic and anionic, to remove all dissolved salts from the water. However, its operation uses dangerous, corrosive, toxic chemicals eg HCl, NaOH, to regenerate the resin. This therefore calls for special precautions in the equipment room to prevent corrosion and requires neutralisation of drain flows to prevent damage to water courses and sewers and to comply with discharge regulations. Like softening it does not provide control of the water sterility.

Reverse osmosis is the only process which can guarantee the production of water with the require quality for hemodialysis ie low chemical and biological content. The reverse osmosis can remove 99% of bacteria, pyrogens and organic substances (molar masses

TABLE 17 – German DIN standards for Extracorporeal circulation Haemodialysis.

Parameters	Max. Concentration in mg/dm3
Aluminium (Al^+)	0.01
Arsenic (As^+)	0.005
Barium (Ba^+)	0.1
Lead (Pb^+)	0.005
Cadmium (Cd^+)	0.001
Calcium (Ca^+)	5
Free Chlorine (Cl^-)	0.1
Chloramine	0.1
Chromium (Cr^+)	0.014
Fluoride (F^-)	0.2
Potassium (K^+)	8
Copper (Cu^+)	0.1
Magnesium (Mg^+)	4
Sodium (Na^+)	70
Nitrate (N^-)	9
Mercury (Hg^+)	0.0002
Selenium (Se^+)	0.009
Silver (Ag^+)	0.005
Sulphate (SO_4)	100
Zinc (Zn^+)	0.1

Physical Chemical Properties	Standard for Drinking water mg/l	Standard for Haemodialysis mg/l
Calcium	-	2
Magnesium	125	2
Iron	0.3	–
Sulphates	250	50
Nitrates	44	0.2 - 10
Fluoride	1	0.5
Ammonium	0.08	0.2
Phosphates	–	5
Sodium	–	50
Potassium	–	2
Chlorides	250	50
Heavy Metals	-	0.1
Zinc	5	0.05
Tin	–	0.1
Mercury	–	0.004
Copper	1	0.1
Lead	0.1	–
Aluminium	–	0.03

> 200), 95% of polyvalent ions and 90% monovalent ions. Reverse osmosis offers the means of combating variability in contaminant levels in water supplies.

Reverse osmosis has been the established water purification process in the field of dialysis for many years. This is because the process is very reliable due to modern membrane materials, is simple to operate, is conducted completely without chemicals and the water is free of micro-organisms and pyrogens. The reverse osmosis process is suitable for the treatment of all types of water although various species contained in the water can damage the membranes, certain preliminary treatments have to be carried out on the untreated water, but this in turn increases the lifetime of the module and reduces operating costs.

A water purification system for the dialysis purposes is composed of the following basic elements:

Fine filter for untreated water   softening unit
                                     possibly filtration through activated carbon
Fine filter for softened water   reverse osmosis
Distributing system

The expedient method of removing solid impurities from the untreated water is with appropriate fine filters. These filters should be changed monthly to prevent build-up of microbial contamination in the filters. This sort of microbial contamination can have an adverse effect on the subsequent parts of the system, eg the water softening units. The water softening units have two functions;

i)   to protect the reverse osmosis unit form precipitation of lime-scale (risk of blockage)
     and
ii) to ensure an emergency supply of soft water to the dialysis department in the event of a breakdown in the reverse osmosis units.

The softening units are cation exchangers and are regenerated with sodium chloride. As a matter of principle twin water softening units are recommended in dialysis departments. With twin softening units there are three arrangements available: i) to operate the units in rotation, ii) to install the units in series and iii) to operate the units in parrallel. Water softening units are monitored automatically.

A final stage of very fine filtration must be installed before the reverse osmosis unit. The size of the filter must not be greater than 5 in order to protect the membranes from the smallest solid impurities. Filter cartridges are changed for reasons of hygiene at least once a month.

Depending on the elements present in the water, which must be established by analysis, the water may have to undergo further preliminary treatment before being supplied to the reverse osmosis units. This may be necessary if iron and manganese, a high level of organic substances, or excessive free chlorine are present in the water

Reverse Osmosis Unit

Water which has been softened should be used to prepare the dialysate, made from water

WATER PURIFICATION

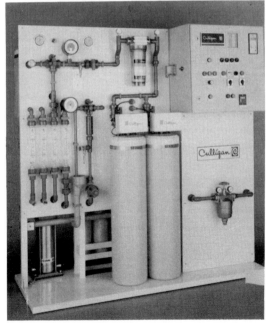

Culligan hermodialysis machines.

and a dialysis concentrate. Two different systems may be used today to prepare deionised water, the ion exchange process and the reverse osmosis process.

In recent years the reverse osmosis process has become increasingly favoured, because it produces a water for the preparation of a dialysate which is almost completely free of pyrogens and micro-organisms. In addition, this process ensures a constant pH.

The pore size of a reverse osmosis membrane with a deionisation rate of 97% NaCl is 5 angstroms, equivalent to a diameter of 0.0005 microns. This membrane prevents the passage of all organic substances with a molecular weight over 200 and, depending on their morphology, a certain percentage of those with a lower molecular weight.

Two types of membranes are mainly used today:
- spiral-wound in polysulphone,
- hollow-fibre modules in polyamide.

Polysulphone membranes are more suitable for dialysis because they can be disinfected if necessary with peracetic acid. Formalin is no longer required for the disinfection of these membranes.

The units developed for dialysis are available in different sizes, for outputs of between 150 and 200 dm^3/hour at operating pressure between 14 and 20 bar, with water feed at temperature of 20°C.

The dialysis techniques employed today are either acetate dialysis or bicarbonate dialysis, the method employed is basically a decision for the doctor, depending on the patients's condition.

If the acetate dialysis technique is used, dialysis can continue with soft water if the reverse osmosis unit breaks down. However, a different dialysis concentrate may have be required depending on the sodium content of the soft water.

With the use of bicarbonate dialysis, the emergency use of soft water, in the event of breakdown of the reverse osmosis units is not possible because the dialysis machine becomes completely blocked almost immediately and there will also be serious problems for the dialysis patient. If an emergency supply is available when bicarbonate dialysis is being used, then this must be a second reverse osmosis unit.

The use of reverse osmosis units in dialysis has proven to be a great benefit to patients. The feared hard water syndrome and pyrogenic reactions which accompany fever and other infections which may occur due to microbial contamination of the water are largely avoided.

Modular RO units are available to meet a range of requirements, from home use with single artificial kidney machines to dialysis centres with 36 dialysis machines.

The machines utilise from 1 to 6 RO membrane modules and are equipped with monitors to show the quality and quantity of water produced by each single membrane, automatic flushing systems, periodic sterilisation.

Pharmaceutical/Biotechnology

Ultrapure water is needed in the biotechnology industry for several applications including, tissue culture media, bacteriological media, buffer solution and analytical solvents, reagents and standards.

WATER PURIFICATION

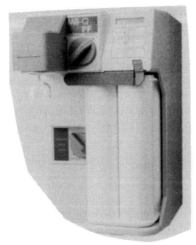

FIGURE 13 – Laboratory system for pyrogen free water.

A potential problem in water supplies is the presence of pyrogen, substances which cause a rise in temperature when injected. Most pyrogens are lipopolysaccharides from bacterial cell walls, with sizes ranging from 20 kD to 0.1 m aggregates. Pyrogens can be effectively removed by ultrafiltration, which is important for tissue culture research.

In this system carbon adsorption removes organics, mixed bed ion exchange removes inorganic species and UF with 10,000 MWCO membranes reduced pyrogen levels to below 0.05 ng/cm^3.

For the production of pure water at a larger scale a range of system designs have been employed depending upon the purpose of use and the country of origin. There is a strict distinction between purified water and water for injection (WFI) and in some countries only distillation and/or RO are the permitted method.

System Design

The water purification system needs to purge the incoming raw water of its contaminants to a degree determined by the process water specification; and to ensure that the high-purity water is delivered to the user points without significant loss of quality. An additional constraint – which applies particularly in the pharmaceutical industry – is the need to satisfy the requirements of the pharmacopoeias and regulatory bodies such as the US Food and Drug Administration (FDA).

The first objective is to produce process water of the requisite quality by choosing an appropriate combination of purification technologies. In some industries, fine chemicals for instance, the purity of the process water is governed by the demands of the manufacturing process or the product specification In others it is controlled by a regulatory body e.g. in the manufacture of pharmaceuticals, where both the quality of the process water and the methods of purifying it are specified in a pharmacopoeia. The second objective of maintaining water quality at the user points is met by meticulous design of the distribution system coupled with stringent validation, monitoring and maintenance regimes.

558 HANDBOOK OF INDUSTRIAL MEMBRANES

In the system design the distribution pipework must be free from "dead legs", and crevices in which water could stagnate and allow bacteria to proliferate. Ideally, the water should be continuously recirculated – at a velocity of 1-2 m/s – around a ring-main that includes a pressure-sustaining valve to ensure even flow from all the take-off points. Holding tanks should be sealed and fitted with sub-micron vent filters to prevent the ingress of dust, airborne chemicals or bacteria. Threaded joints and other non-sanitary fittings which could harbour micro-organisms should be avoided in the construction of the distribution loop. In addition, pipe runs must be sloped and tanks should have a conical base to facilitate draining the system during maintenance periods.

The selection of pipework material is an important factor in preventing impurities form leaching into the recirculating water. In many water purification systems, the tanks and pipework are made entirely from plastic materials. Acrylonitrile-butadiene-styrene (ABS) is widely used for constructing distribution loops, while polyvinylidene fluoride (PVDF), although expensive, has exceptionally low levels of leachable substances and can

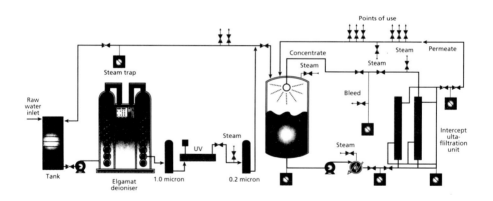

FIGURE 14 – A system designed for deionised and a pyrogenic free water.

WATER PURIFICATION

withstand high temperatures. In pharmaceutical systems, tanks and pipework requiring steam sterilisation are made from 316 grade stainless steel.

In cases where multiple grades of process water are required, they can be produced by incorporating suitable purification units into the distribution loops. Figure 1 shows a system which complies with cGMP principles and provides both deionised and apyrogenic process water. In the first section, a deioniser removes almost all the ionic impurities, while an ultraviolet lamp coupled with microfilters destroys most of the bacteria: residual bacteria and pyrogens are removed by the ultrafilter in the second loop.

After installation and commissioning, the water purification system must be validated to ensure that the purified water quality meets the specification. In the pharmaceutical industry – where the production plant must have regulatory approval – the water purification system is validated as part of the overall validation the manufacturing facility.

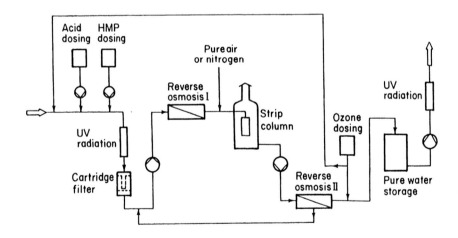

FIGURE 16 – Flow diagram of a two stage reverse osmosis process for pure water production in the pharmaceutical industry.

Alternative water purification systems for the pharmaceutical industry are shown in Fig 1 was designed to produce 12 m³/h for pyrogen free water at the French National Blood Centre (Orsay). The system consists of four parts; prefiltration, ultrafiltration, demineralisation and polishing filtration. Ultrafiltration is carried out with hollow fibre UF modules (Romicon) with 196 m² installed area. UF preceeds de-ionisation to increase the ion-exchange resin lifetime.

Pure water for the pharmaceutical industry can be produced using reverse osmosis units arranged as a two-stage cascade. After CO_2 stripping with sterilised air or nitrogen the permeate from the first stage is fed into the second stage. Pure water of sufficiently high quality is produced without the need for ion exchangers, which generally pose a problem with respect to bacteria growth, ionogenous germs and particles.

Operating Problems in High Purity Water Systems

There are several factors that can arise in RO systems for the production of high purity water which can have a negative affect on water quality which are a function of equipment manufacture and system operation.

1. Recontamination

This recontamination in normal circumstances is due to the release of construction fragments from the RO module, or to bacteria sloughage. Materials of manufacture should be maintained in the highest level of cleanliness throughout processing. Prolonged shutdown of RO modules should be avoided to prevent significant levels of bacteria growth in permeate channels.

2. Membrane Bypass

Membrane bypass results from two conditions; a loss of integrity of the seam seals between sheets of membrane allowing water to flow from feed side to permeate side, bypassing the membrane. This is generally not a common occurrence. The most common cause of membrane bypass is 0-ring seal leakage. These seal function between the membrane element and the connectors.

3. Differential passage of silica and carbon dioxide

In most applications requiring high purity water reverse osmosis alone does not provide adequate ionic removal. This is the case even with double pass reverse osmosis where RO is carried out with two modules in series. Therefore it is common for RO purified water to be polished with IX to give much higher ionic removal. The fact that RO membranes readily pass carbon dioxide and poorly reject silicon leads to much greater loads on the anion resin than on the cation resin, usually present as mixed beds. This, coupled with the fact anion exchange resins degrade faster and have less capacity than cation exchange resins, as well as being more expensive to purchase and regenerate, causes an in balance in IX polishing. Systems will as a result tend to be anion limiting, with silica being the first anion to pass through the resin bed.

There are some application for high purity water, e.g. medical, packaged products, human consumption, where purification with a two pass RO system is more desirable than

WATER PURIFICATION

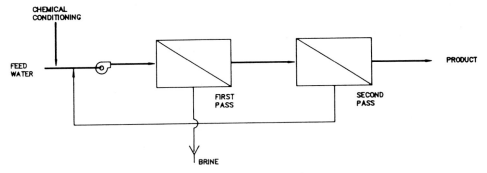

Double pass reverse osmosis.

IX or a combined RO and IX system. This arises from the constant risk that the chemical supply of IX could be contaminated and thus may add harmful materials to be finished water. In addition continuous RO is much simpler to operate than batch IX and requires much lower operating labour costs.

In virtually every ultrapure water system a polishing device is included to reduce the residual concentration of ions. Often there will be two treatment steps, i). *makeup ion exchange*, ii) *polishing ion exchange*. The makeup IX unit usually consists of a mixed bed system, but separate bed cation-anion units are sometimes used. In cases with concentrated feed-water supplied when the dissolved solids level exceeds approximately 50 ppm, separate cation and anion exchangers are generally used for the makeup system. The second IX system is always a mixed bed type, to obtain the lowest possible level of ionic impurities. IX units can remove substantial amounts or organic and particulate matter in addition to ions. Some organics, notably carboxylic acids and amines, are exchanged in the same way as inorganics ions. However, non-ionic organic materials can bind to the resin polymer backbone and, in this manner, can be removed from the water. Organics that are ionically exchanged also interact with the polymer structure and thus generally tend to bond more tightly with the resin. However difficulties arise on regeneration because the acid or caustic typically used for cation and anion resins is of a little benefit in removing adsorbed organic compounds. The IX resin may then have a reduced effectiveness for removal of additional organic after many cycles.

The chemical characteristics of IX resin that cause the removal of ions also develop an electrical surface charge which attracts fine particles. IX resins can remove micron-size particles by 20 to 90 percent, depending on the influent level. The extent to which IX resins can remove ionic materials is controlled factors such as resin chemistry, concentration of the regenerate, concentration of the impurities in the regenerate, pH of the water. Concentrations of cations and anions below 0.1 part per billion are theoretically achievable in IX systems.

RO has also been used in ionic polishing. There are plants, in the United States, in which RO is used to polish the effluent from polishing mixed bed IX units, thereby improving the ionic purity of the water.

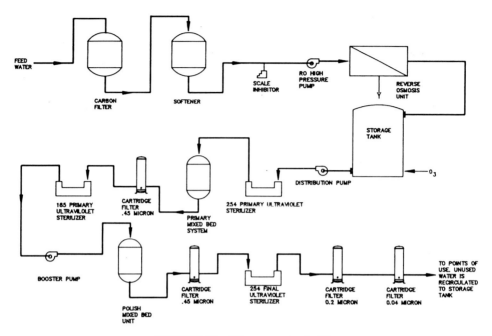

FIGURE 17 – Ultrapure water facility.

A flow diagram for an ultrapure water system to service a facility for fabrication of fine-geometry integrated circuit products is shown in Fig 17. The pretreatment system prior to the RO unit includes a carbon filter, a softener, and a feed of a chemical scale inhibitor. This pretreatment is tailored to thin film composite, polyamide RO membranes which are intolerant to chlorine but can operate at a wide pH range. The carbon filter is the preferred method of dechlorination of the feed water because of its reliability. The carbon filter is sometimes criticised as a source of bacterial particulate.

The presence of the softener tends to enhance the particulate quality of the water – the slight charge on the surface of the softener resin improves the softener's activity as a filter, in comparison to granular media beds having the same particle size. The softener pretreatment thus tends to reduce particulate fouling of the RO unit, reversing the particle generation by the carbon filter. Other polishing processes are used in high purity water installations. The most effective techniques are based on the principle that organics can be oxidised to inorganic carbon dioxide, or to carboxylic acids, which are ionised and interacts with IX. Methods include irradiation with 185 nanometer ultraviolet light, with and without the addition of agents such as hydrogen peroxide, to stimulate production of oxidants. Direct feed of ozone into water, followed by irradiation with 254 nanometer light is also very effective in reducing organic carbon levels to less than 5 ppb. Generally the method used for oxidation of the organic material only slightly changes the TOC. The primary effect is conversion of organics level to carboxylic acids, which are then removed with IX.

Hot Ultrapure Water Systems (Electronics Industry)

There has been an increase in demand for hot ultrapure water (HUPW) in the electronics industry to replace to UPW at 20C. HUPW facilities offer specific advantages of reduced cost and maintaining water quality, in comparison to UPW. The temperature of HUPW water is required at between 70C and 80C for best performance. Reaction kinetics are faster in HUPW and the quality of water must be as least as high as for UPW, to prevent defects arising from contaminants in the water. The supply of HUPW presented several problems of material selection, wetted materials have accelerated elution rates at 80C in comparison to 20C, many polymer suffer strength loss (stainless steel suffers from leaching of iron).

Traditional HUPW facilities used "spot" heating- a local heater and filtration system for each process tool. When not in use these systems remain at low flowrate until called on. During demands flowrates increase rapidly making temperature control difficult. In addition the low flowrates also caused problems with contaminant elution.

As a result centralised HUPW has been designed to supply 10 new spray tools and 2 automatic wet stations in an AT and T facility in Orlando. In each station five individually controlled 105 kW units raise the temperature of the existing UPW supply before filter housing. The cartridges are 0.05 m polycarbonate membranes with polypropylene support and hardware. All internal housing surfaces (except seals) are PVDF. The same polycarbonate membranes were used in local heater installations.

In the electronics industry there is distinct advantages in using hot water (95C) sterilisation and cleaning. This can be applied to sterilise the complete pipe system and avoids complication in the use of chemicals. The membrane is cleaned while it is sterilised

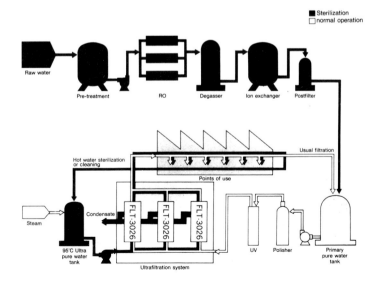

FIGURE 19 – In situ hot water sterilisation system.

and can filter hot ultrapure water. Hollow fibre polysulfone membranes in polysulfone housing have been designed with this capability, giving flux rates of 0.5 mh^{-1} bar^{-1} at 20°C.

Water for Air Conditioning

The water used for air humidiciation must have a low salt content and must be odourless, be non-corrosive and be sterile. Conventional processes such as phosphatisation, softening by ion exchange, acid dosing and decarbonation by ion exchange are not always satisfactory.

Reverse osmosis has the attraction of simultaneously deionising the water and removing bacteria and organic components. In addition, the requirements for service and maintenance are small and, with the exception of feed water conditioning, no chemical additives are required., The accumulation of bacteria in the air humidifier/washer can be avoided by continuous operation.

The integration of RO into the water circulation of an air washer is shown in Fig 19. After softening by polyphosphate dosing or anion exchange, the feed water is pumped into the external water loop of the air humidifer/washer. The RO unit is integrated into this external water loop. The flow-rate in the recirucltion loop depends on the rate of contamination of the wash water from washed out substances (fine particles, etc) of the air. The feed flow-rate is controlled by the loss of the water from the system (evaporation losses, discharged RO retentate).

The RO permeate is recycled into the water sump of the humidifer/washer unit and the retentate is discharged to the drainage system. In the humidifer/washer unit, the water is continuously circulated to prevent bacteria growth.

Reported capacities of hollow fibre RO unit are about 35 m^{3}d^{-1} of permeate in air-conditioning systems.

In the pharmaceutical industry UF is a suitable method for the removal of endotoxins. Modules based on hollow fibre membrane are routinely used.

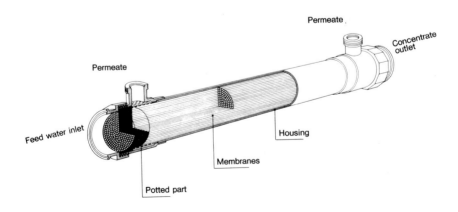

UF system.

WATER PURIFICATION

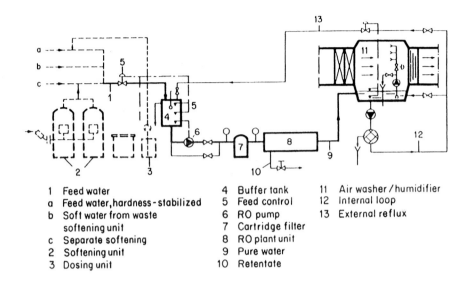

1	Feed water	4	Buffer tank	11	Air washer / humidifier
a	Feed water, hardness-stabilized	5	Feed control	12	Internal loop
b	Soft water from waste softening unit	6	RO pump	13	External reflux
c	Separate softening	7	Cartridge filter		
2	Softening unit	8	RO plant unit		
3	Dosing unit	9	Pure water		
		10	Retentate		

FIGURE 20 – Purification of water for air conditioning.

Continuous De-ionisation

The use of a combined electrodialysis and ion exchange system has been introduced on a commercial level as a means of producing high purity water form potable water sources. The process, continuous deionisation (CDI) removes ions from process streams by a unique combination of resins, membranes and electrical current. CDI has the removal efficiency of mixed-bed deionisation (MBDI) without the need for chemical regeneration cycles, resulting in minimal product loss. Liquid hold-up is also much less using CDI technology. The CDI module draws ions from the feed solution into the resin bed (Fig 11). The ions are then pumped out of the resin bed and into a separate salt concentrate stream through the action of an electrical field. Ion exchange membranes prevent ions from re-entering the deionised product.

The use of ion-exchange resins in the feed channel makes ion transfer possible even in solutions where salt concentration is less that 1 ppm. The use of ion exchange resins also prevents localised pH shifts even at low solution conductivities. CDI allows continuous operation, requires minimal power consumption and reduced product loss resulting in low operating costs. CDI systems can be used to remove ions form 10,000 ppm down to 1 ppm from streams containing 5% to 50% product and typically remove ions between 35 and 400 molecular weight, while retaining non-ionised product from 60 to 1500 molecular weight.

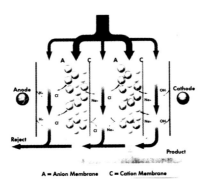

FIGURE 21 – Principle of operation of continuous de-ionisation.

A wide range of systems are available from 0.25 m² lab system to 200 m² production system.

Typical applications include (also see section 14):
- Deionisation/desalting of aqueous solutions of non-ionic solutes
- Urea recovery for recycle
- Urea desalting
- Contrast agent desalting

Overall the economics of the process are particularly attractive at feed salt concentrations of 50 ppm and above with costs of $0.01 per dm³ de-ionised water.

10.1 – LABORATORY WATER PURIFICATION

The quality of water used in laboratories is specified by several professional organisations concerned with the accuracy and reproducibility of laboratory tests. Summarised in Table X are the specifications published by the College of American Pathologists (CAP) and the American Society for Testing and Materials (ASTM). More recently, specifications and test methods for water for analytical laboratory use have been summarised in the international standard ISO 3696. The ASTM and CAP standards have been in place now for some time and with today's highly sensitive analytical techniques, which can detect very low levels of organic compounds, the water quality stated as "Type 1" is often not high enough. Water purification systems have therefore, developed to produce ultrapure water with a total organic content of less than 10 ppb, or even less than 5 ppb, which results in a water quality above "Type 1".

The water purification systems are available to supply water to the various grades and with different capacities. Depending upon the application the units will use appropriate combination of RO (or UF) membrane purification in conjunction with carbon adsorption, ion-exchange and UV radiation.

Laboratory Grade Water

For laboratory grade water a reserve osmosis systems produce purified water for general

TABLE 1 – Specifications of Water for Laboratory Use.

Type of Water	Electrical Conductivity (μS/cm at 25°C)	Electrical Resistivity (Megohms x cm at 25°C)	Silicate (mg/l)	Heavy Metals (mg/l)	$KMnO_4$ Reduction (min)	Sodium (mg/l)	Hardness $CaCO_3$ (mg/l)	Ammonium (mg/l)	Bacteria (col/ml)	pH (at 25°C)	Increasing Water Quality
Tap Water (example)	240	0.004	1	1	10	65	35	1	>10	—	Types III, IV 'Laboratory' Grade Water
CAP Type III ASTM Types III/IV	10	0.1	1	0.01	60	0.1	—	0.1	—	5-8	▶
Single Distilled	5-1 / 10-2	0.2-1 / 0.1-0.5	— / 1-0.5	— / 1-0.5	10 / 30	— / 5-2	— / 3-1	— / 0.01	— / <10	5-8 / 5-7.5	Type II 'Analytical' Grade Water
Lab Grade RO Plus	25-10*	0.04-0.1*	0.1	<0.04	30	6.5	1.6	0.4	<10	6	▶
CAP Type III ASTM Type III Double-Distilled	0.5 / 1	2 / 1	0.1 / —	0.01 / —	60 / 60	0.1 / —	— / —	0.1 / —	10^4 / —	— / —	Type 1 'Reagent' Grade Water
	2-1	0.5-1	0.7-0.1	0.8-0.1	60	1-0.5	0.3-0.1	0.01	<10	5-7.5	▶
RO Analytical/Grade 2	<1	>1	<0.1	<0.1	60	<0.1	—	0.01	<10	6-7	Above Type 1
CAP Type I	0.1	10	0.05	0.01	60	0.1	—	0.1	10	—	
ASTM Type I	0.06	16.6	—	—	60—	—	—	—	—	—	
Type I "Plus"	0.056	18.2	<0.01	<0.01	N/A	<0.01	—	<0.01	<1	—	
Above Reagent Grade	0.056	18.2	<0.01	<0.01	10 ppb TOC	<0.01	—	<0.01	<1	—	

*This value is a function of pressure, type of RO cartridge and feed water quality.

laboratory applications such as glassware washing and feed water for reagent grade water purification systems. Reverse osmosis is a broad based primary water purification technique to remove a high percentage of each of the four classes of contaminants present in potable tap water: inorganic ions, dissolved organics, particles and microorganisms. Protection of the RO membrane is typically provided by a pretreatment pack, which contains an antiscaling compound, activated carbon and a prefilter. The pack can be easily changed without tools.

The typical systems are equipped with alphanumeric displays which indicate function, performance and autodiagnostic parameters and a unique rinsing function that assures only good quality product water is delivered to the reservoir.

Storage is one of the biggest problems in pure water production as stand-by periods degrade the water quality of any purification system upon start-up. Reagent grade water systems recirculate the water through the purification elements is preset intervals when the system is not used. At each start-up, product water is automatically diverted to the drain for a short period until the specified ionic rejection level has been reached. Thus only high quality purified water will be taken from the system.

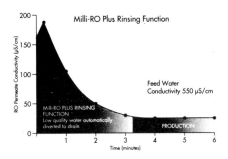

FIGURE 1 – Automatic rinsing function performance.

Analytical Grade Water

Analytical Grade Water (Grade 2) is used for the preparation of buffers, chemical/biochemical reagents and microbiological media. It is used as feed water to clinical analysers and instrumentation, such as weathering testing equipment and humidifiers.

A purification systems produces Grade 2 water from potable top water through a combination of a series of purification steps:

Stage 1: Pretreatment as with lab grade water

Stage 2: Primary purification, using a polyamide reverse osmosis module

Stage 3: Secondary purification

A choice of secondary purification is available based on ion-exchange or continuous electrodeionsiation. Systems are ideal for the production of purified water, according to European and US pharmacopoeia specifications.

Reagent Grade Water

Reagent grade water is used for many analytical applications such as atomic absorption, AA spectrophotometry, HPLC and ion chromatography.

To provide the additional purification level for water the systems are supplied with purification packs (all polypropylene) which contain activated carbon, nuclear grade ion exchange resins and an organic scavenger mixture. Pretreatment of water before reverse osmosis is either by a combination of prefiltration, anti-scaling and activated carbon or by ultrafiltration.

Depending upon the application, even minute amounts of organic carbon in ultrapure water can be disruptive. Therefore, systems are available which provides water with TOC levels below 5 ppb. This purity is attained by including ultraviolet photooxidation as an additional step in the multiple stage water purification process. Ultraviolet radiation oxidises dissolved organic compounds in water via a photochemical reaction involving both 185 nm and 254 nm wavelengths. This photochemical chain reaction requires both wavelengths to generate highly reactive hydroxyl free radicals which take part in the oxidation process.

Typical applications, where extremely low TOC levels are required, include environmental monitoring and critical instrumental analyses such as GC, GC/MS, HPLC and IC.

Pyrogen Free Water

Pyrogen-free water is used for critical biological applications such as cell culture, in vitro fertilisation and monoclonal antibody production. This quality of water is produced in

TABLE 2 – Range of Water Purification Systems.

Purification System	Purified Water Type	Water Quality	Flow Rates
RO Plus Pretreatment	Laboratory Grade Water	Removes: 94-99% Inorganic Ions >99% Dissolved Organics >99% Particles >99% Microorganisms	3 - 90 litres/hour
RO Plus pre and post treatment	Analytical Grade Water, Grade 2 - ISO3696/BS3978	Resistivity: 5-15 MΩ-cm T.O.C.: <50 ppb	6-75 litres/hour
UF + RO Plus pre treatment, RO and IX, Carbon adsorption & organic scavenging	Reagent Grade Water	Resistivity: 18.2 MΩ-cm T.O.C.: $\leq$ 10 ppb Particle-free: >0.2 µm Bacteria: <1 cfu/ml	0.5-10 l/min
Plus hollow Fibre UF	Pyrogen-Free Water	Resistivity: 18.2 MΩ-cm T.O.C: $\leq$ 10 ppb Particle-free: >0.2 µm Bacteria: <1 cfu/ml Pyrogens: 5 log reduction	1.3 l/min
Plus UV	Beyond Reagent Grade Water	Resistivity: 18.2 MΩ-cm T.O.C.: $\leq$ 5 ppb Particle-free: >0.2 µm	1.5 l/min

hollow-fibre ultrafiltration cartridges which dramatically reduces endotoxin that can cause pyrogenic reactions in cells. This provides up to 5 log reduction with feed water containing >2,000 EU/ml, resulting in a pyrogen level of <0.02 EU/ml. The other purification elements in this system are those used in the reagent grade water system.

For the production of larger quantities of reagent grade water, capacities of up to 10 dm^3/min system are available which are equipped with 3 or 4 individual treatment cartridges or bowls offering a range of purification options which can be tailored to meet a variety of water quality requirements and to suit the feed water quality.

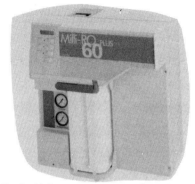

Typical laboratory water supply system.

Water Storage and Distribution System

For the supply of water in a distribution network a system is available designed to store purified water in a compact reservoir and to distribute it, under pressure, around an external distribution loop. A highly retentive vent filter in combination with a sanitary overflow device prevents the ingression of airborne microorganisms. The reservoir has a conical bottom to allow complete drainage for effective cleaning and sanitisation. A built-in quietly operating stainless steel distribution pump can operate continuously during a 10 hour period at 10 l/min and 3 bar pressure. The pump operating mode can be prepro-

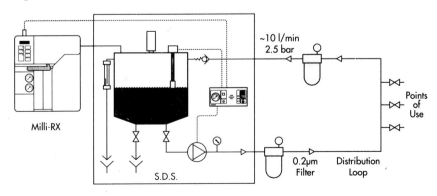

FIGURE 2 – Typical installation of the water storage and distribution system.

WATER PURIFICATION   571

Water storage and distribution system.

grammed for a seven day cycle (individual settings are possible for every day) resulting in a fully automatic operation of the system.

For applications requiring water for clinical analyses a further purification step can be introduced which removes dissolved gases. If not removed these gases could cause bubbles on surfaces or in the clinical analyser fluid lines. This dissolved gas removal is achieved using a hollow fibre unit.

A range of compact purification units are available to provide water suitable for most laboratory, pharmaceutical and small volume industrial applications. Typical units combine reserve osmosis (RO) with pretreatment and storage and can produce up to 140

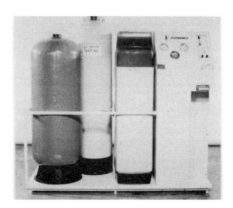

Water purification unit.

l/h. Overall the units treat water with activated carbon to remove chlorine and trihalomethanes, softens to reduce hardness, and removes 98% of feedwater dissolved solids and organics through reverse osmosis.

The unit includes an FRP activated carbon filter, water softener tanks, spiral-wound PA separators (membrane elements) contained in stainless steel housings, 5 µm cartridge prefiteers, and panel mounted instrumentation including a water quality monitor and pressure gauges. The unit can store up to 76 l of purified water for fluctuating demands.

SECTION 11

Industrial Waste Water and Effluent Treatment

INDUSTRIAL WASTE WATER AND EFFLUENT TREATMENT

Waste water and effluents vary considerably from one industry to another. The two phases which are of main concern and which membranes can play an important part in treatment are liquids, principally water based and air. In the case of air the principle pollutants generated by industry, combustions, power generation, traffic agriculture and municipal waste are:
- particulates
- volatile organics, CFC's, aromatics etc
- combustion acid gases, SO_2, NO_x, CO_2
- inorganics, NH_3, H_2S
- biogas products, CH_4, CO_2
- metals

There are generally a choice of processes which can be used to remove these components from the air, some of which achieve recovery and potential recycling. A range of filters can be used to remove particles including bag filters, electrostatic precipitators, cyclones and scrubbers. Acid gases can typically and efficiently be removed by gas

TABLE 1 – Membrane Processes for Gaseous Effluents.

Particulates	Membrane filters (polymer, ceramic)
Organic vapours	Vapour permeation
Combustion products (SO_2, NO_x, CO_2)	Membrane contactors (absorbers) Carrier membranes. Membrane reactors
NH_3	Membrane contactors, carrier membranes
H_2S (natural gas)	Gas separation, carrier membranes, membrane contactors, membrane reactors
CH_4, CO_2 (biogas)	Gas separation

scrubbing (absorption) processes as can other inorganics and metal species. Volatile organics can be removed by absorption, adsorption, thermal incineration, thermal oxidation, catalytic incineration etc.

Membrane processes are finding increasing applications in the treatment of gaseous (mainly air) effluents as shown in Table 1. These applications are not considered in this chapter but are described in more detail in sections covering Air Filtration, Gas Permeation, Pervaporation and Membrane Contactors.

In the treatment of liquid effluents, primarily waste waters, the type of pollutant or pollutants has a major bearing on the selection of treatment technologies. Table 2 gives a summary of the classes of pollutants which may be present in waters.

Some available technologies which can be used for pollution control are listed in Table 3, together with their effectiveness for various pollutants.

These processes include a diverse range of procedures: biological processes, chemical processes of oxidation, reduction, incineration and physical separations of filtration, distillation and stripping.

The existing technologies for the recovery of raw material can be classified into three general categories (see Table 4).

   i) Physical separations. These methods include gravity settling, filtration, flotation, flocculation and centrifugation. They are specifically used for the separation of fine and suspended solids and dispersed liquids from liquid.

TABLE 2 – Pollutants Present in Waste Waters.

Particulates, Suspended solids and micro-organism	Oil in globular form (150 µm circa) also settable solids Oil contamination in emulsified form, Settable solids Suspended solids (perhaps with settleable solid) Large solids, fibres and waste materials Fine particles Colloids, bacteria, virus
Inorganics (dissolved)	Heavy metals (Cd, Hg, Pb, Cr) Salts (cyanide, ammonia, nitrate etc) Nuclear waste (cesium - 137, strontium - 90) acids or alkalis
Volatile organics	Aromatics (benzene, toluene, xylene) Aliphatics (hexane, heptane) Alcohols (methanol, ethanol) Ketones (acetone, methyl ethyl ketone) Halogentated hydrocarbon (chloroform, ethylene chloride)
Non-volatile organics	Phenolics Polyaromatic compounds Microsolutes (pesticides, insecticides, herbicides) Surfactants and dyes

TABLE 3 – Technologies for Pollution Control of Liquid Streams.

	Suspended Colloidal Solids Removal	Dissolved Organic Removal	Dissolved Inorganic Removal	Micro-organism Removal
Biological Processes				
Activated sludge	X	X	—	X
Anaerobic digestion	X	X	—	—
Bio-filters	—	X	—	—
Extended aeration				
Bio-denitrification	—	L[a]—	—	
Bio-nitrification	X	X	—	—
Pasveer oxidation ditch	X	X	—	X
Chemical Processes				
Chemical oxidation				
Catalytic oxidation	X	X	—	X
Chlorination	X	X	—	X
Ozonation	—	L	—	X
Wet oxidation	X	X	—	X
Chemical precipitation	—	—	X	—
Chemical reduction	—	—	X	—
Coagulation				
Inorganic chemicals	X	X	—	X
Polyelectrolytes	X	X	—	X
Disinfection	—	—	—	X
Electrolytic processes				
ELECTRODIALYSIS	—	—	X	—
Electrolysis	—	—	X	—
Extractions				
Ion exchange	—	—	X	—
Liquid-liquid (solvent)	—	—	X	—
Incineration				
Fluidized-bed	X	X	—	X
Physical Processes				
Carbon adsorption				
Granular activated	X	X	—	—
Powdered	X	X	—	X
Distillation	X	X	X	X
Filtration				
Diatomaceous-earth filtration	X	—	—	X
Dual-media filtration	X	—	—	X
Micro-screening	X	—	—	X
Sand filtration	X	±	—	X
Flocculation-sedimentation	X	—	—	X
Foam separation	X	—	X	—
Freezing	—	X	X	—
MEMBRANE PROCESSING				
MICROFILTRATION	X	—	—	X
ULTRAFILTRATION	X	X	—	X
REVERSE OSMOSIS	X	X	X	X
Stripping (air or steam)	X	X	—	—

L = Under certain conditions there will be limited effectiveness.

TABLE 4 – Description of Technologies for Recovery of Materials.

Technology and description	Type of waste streams	Separation efficiency*	Typical industrial applications
Physical separation: Gravity Settling: Tanks, ponds provide hold-up time allowing solids to settle	Slurries with separate phase solids, such as hydroxide	Limited to solids (large particles) that settle quickly (less than 2 hours)	Industrial wastewater treatment first step
Filtration: Liquid passes through and solids are retained on porous media	Aqueous solutions with finely divided solids; gelatinous sludge	Good for relatively large particles	Various Tannery water
Flotation: Air bubbled through liquid to collect finely divided solids that rise to the surface with the bubbles	Aqueous solutions with finely divided solids	Good for finely divided solids	Refinery (oil/water mixtures); paper waste; mineral industry
Flocculation: Agent added to aggregate solids together which are easily settled	Aqueous solutions with finely divided solids	Good for finely divided solids	Refinery; paper waste; mine industry
Centrifugation: Centrifugal force causes separation by different densities	Liquid/liquid or liquid/solid separation, ie oil/water resins; pigments from lacquers	Fairly high (90%)	Paints
Component separation: Distillation: Boiling off materials at different temperatures (based on different boiling points)	Organic liquids	V. high separations achievable (99+% concentrations) of several components	Solvent separations: chemical and petroleum industry
Evaporation: solvent recovery by boiling off the solvent	Organic/inorganic aqueous streams, slurries, sludges, ie caustic soda	V. high separations of single, evaporated component achievable	Rinse waters from metal plating waste
Stripping: (air or steam)	Dissolved sorganics	Good	Solvent separations
Ion exchange: Waste stream passed through resin bed, ionic materials selectively removed. Ionic exchange materials must be regenerated	Heavy metals aqueous solutions cyanide removed	Fairly high	Metal-plating solutions
Ultrafiltration: Separation of molecules by size using membrane	Macromolecules. Heavy metal aqueous solutions	Fairly high	Metal-coating applications

TABLE 4 (continued).

Technology and description	Type of waste streams	Separation efficiency*	Typical industrial applications
Reverse osmsis: Separation of dissolved materials from liquid through a membrane	Heavy metals, organics; inorganic aqueous solutions	Good for concentrations less than 300 ppm	Secondary treatment process such as metal-plating pharmaceuticals
Carbon/resin adsorption: Dissolved materials selectivity adsorbed in carbon or resins. Adsorbents must be regenerated	Organics/inorganics from aqueous solutions with low concentration ie phenols	Good, overall effectiveness dependent on regeneration method	Phenolics
Solvent extraction: Solvent used to selectively dissolve solid or extract liquid from waste	Organic liquids, phenols, acids	Fairly high loss of solvent may contribute to hazardous waste problem	Recovery of dyes
Chemical transformation: Precipitation: Chemical reaction causes formation of solids which settle	Lime slurries	Good	Metal-plating wastewater treatment
Electrodialysis: Separation based on differential rates of diffusion through membranes. Electrical current applied to enhance ionic movement	Separation/ concentration of ions from aqueous streams; application of chromium recovery	Fairly high	Separation of acids and metallic solutions
Electrolysis: Separation of postively/ negatively charged materials by application of electric current	Heavy metals: ions from aqueous solutions; copper recovery	Good	Metal-plating
Reduction: Oxidative state of chemical changed through chemical reaction	Metals, mercury in dilute streams	Good	Chrome-plating solutions and tanning operations
Chemical dechlorination: Reagents selectively attack carbon-chlorine bonds	PCB-contaminated oils	High	Transformer oils
Thermal oxidation: Thermal conversion of components	Chlorinated organic liquids; silver	Fairly high	Recovery of sulphur, HCl
Chemical oxidation: Chlorination, ozonation	Dissolved organics, inorganics	High	Metal-plating

*Good implies 50 to 80 percent efficiency, fairly high implies 90 percent.

ii) Component separations. These technologies distinguish between constituents by virtue of difference in some physical property, electrical charge, boiling point, miscibility etc. Methods include evaporation distillation, solvent extraction, absorption, ion exchange and reverse osmosis.
iii) Chemical transformations. These methods require a chemical reaction to remove specific constituents and examples include, precipitation, electrolysis, electrodialysis and oxidation and reduction reactions.

An area which is not included in this list is biological processes. The conventional biological processes have been used in industrial waste treatment for many years and utilise either aerobic or anaerobic bacteria (see Table 3). They are effectively used for treatment of dissolved organics and sludges and involve destruction of the contaminant by microbial catabolism. There are continued developments in new microbial strains for improved degradation of recalcitrant compounds, with improved tolerance to more severe operating conditions and increased rates of degradation.

The biological process are generally limited to maximum BOD concentrations of around 10 gdm^{-3} and can be sensitive to heavy metals and certain organic (and inorganic) species. However strains are under development for such specific applications.

TABLE 5 – Comparison of Membrane Separation Technologies in Waste Treatment.

Feature	Micro filtration	Ultra filtration	Reverse Osmosis	Electro dialysis
Suspended solids removal	yes	yes	yes	no
Dissolved organic removal	no	yes	yes	no
Dissolved inorganic removal	no	no	yes	yes
Concentration capabilities	high	high	moderate	high
Permeate purity	high	high	high	moderate
Energy usage	low	low	moderate	moderate
membrane stability	high	high	moderate	high
Operating cost (S/1000 gallons feed rate)	0.50-1.00	0.50-1.00	1.00	1.00

The development of new microbial strains can be used to improve the degradation of recalcitrant compounds, achieve effective mutlicomponent destruction, improve rates of degradation and the ability to concentrate nondegradable constituents and improve the tolerance to frequently changing or severe operating conditions. There are several industries with experience in applying biological treatment process to waste management. Many of these processes could benefit from the incorporation of membrane separation within the overall operation, examples of which are discussed in this chapter.

Membrane separation has been used to achieve filtration, concentration, and purification. Large-scale applications, in pollution control have been in the past inhibited by two factors:
1. replacement costs associated with membrane use and
2. technical difficulties inherent in producing large uniform surface areas of uniform quality.

TABLE 6 – Wastewater Treatment Processes for Particulates.

Contaminants and/or problem	Suggested treatment
Oil in globular form *circa* 150 μm also setteable solids	Gravity type oil/water separator with inter-capacity to store oil. Depending on the permitted oil content of the treated waste water, flow velocity in separator may need to be limited and oil skimming incorporated
Oil contamination in emulsified form, no appreciable solids	Coalescing type oil/water separator
Setteable solids only	Gravity type separator. Performance may be enhanced by chemical treatment to provide agglomeration
Settleable and suspended solids only	Bulk clarifier
Large solids, fibres and waste material; no fines	Rotating screen
Sludge	Gravity separator or gravity filter. For large volumes, vacuum drum filter, belt filter etc.
Waste water requires pH adjustment	Chemical treatment. Other treatment necessary to deal with other contaminants present

Because of the inherent advantages of membrane separation over separation techniques such as distillation or evaporation, further development of membrane separation for large scale commercial applications is attractive. These advantages include lower energy requirements and a simpler, more compact system that generally leads to reduced capital costs. Commercial applications exist for most membrane processes but coupled transport designs are still mainly at the pilot stage. Microfiltration, ultrafiltration, reverse osmosis, and electrodialysis processes have more immediate application. Dialysis has been used on only a relatively small scale. The development of new materials for both membranes and supporting fabrics and the use of new layering techniques (eg composite membranes) have led to improved permeability and selectivity, higher fluxes, better stability, and a reduced need for prefiltering and staged separations.

Generally the most important factor in advancement of membrane separations technology has been improved reliability. New types of membranes have demonstrated improved performance, notably thin-film composites that can be used in reverse osmosis, coupled transport, and electrolytic membranes which have direct application to the recovery and reduction of hazardous materials and effluents from processing streams.

A major cost in a membrane separation system is the engineering and development work required to apply the system to a particular process. Equipment costs are secondary;

membranes generally account for only 10 percent of system costs. However, membranes must be replaced periodically and replacement membranes represent a running cost. Currently the largest profit items are for high-volume flow situations (eg water purification) or for high value product applications (eg pharmaceutical productions).

TABLE 7 – Selected Application of Membrane Separations in Waste Water Treatment.

Volatile organic compounds	Pervaporation. Membrane contactors
Salts (nitrate, cyanide)	Electrodialysis. Electrohydrolysis. Membrane contactors. Reverse osmosis. Liquid (carrier) membranes. Diffusion dialysis.
Heavy metals (Cd, Ni)	Electrodialysis, membrane contactors, liquid (carrier membranes). Reverse osmosis.
Microsolutes (pesticides)	Liquid (carrier) membrane contactors.
Heavy metal hydroxide ppts	Microfiltration
Oil-water effluents	Microfiltration
Oil-water and industrial emulsions	Ultrafiltration, reverse osmosis
Latex, dyestuff	Ultrafiltration
Electropaint	Ultrafiltration

Table 7 illustrates the range of waste water treatment processes which are amenable to membrane separation either as single treatment processes or as hybrid processes with many of the separation/recovery processes listed in table 4. Examples of hybrid processes include biological treatment with ultrafiltration (reverse osmosis) for landfill leachates, heavy metal precipitation with microfiltration, reverse osmosis with evaporation, for concentration of wastes, distillation with pervaporation, adsorption with vapour permeation in solvent treatment.

For the treatment of particulates (oils, suspended solids) gravity settling devices, coalescing type separators and clarifiers, are frequently used. The food industry is a particular large generator of waste water, estimates put 33 tonnes of water used for every ton of food processed. Although there are variations in contaminant and flowrates in this industry established system designs are based on skimming flotation, clarification and sludge disposal (see Fig 1). In many cases the effluent is unsuitable for direct disposal to the sewer and a biological treatment process is often used. Membrane processes are having some impact in waste water from the food industry and a range of other industries. A second example highlighted in this book is the treatment of metal finishing rinse water where traditional processes adopt ion-exchange for demineralisation which is frequently operated in a two stage operation incorporating activated carbon filtration. The carbon filter removes suspended solids and adsorbs organics (predominantly in surfactants and brighteners) prior to demineralisation. 95% water recovery for recycling can be achieved.

A membrane separation which finds major applications in wastewater treatment is UF. The primary uses of ultrafiltration in industry are as a method of fractionation, the separation of a stream into two fractions on the basis of particulate size or molecular weight cut-off. The ability to separate soluble macromolecules from other soluble species and solvents is the major reason for the use of UF in many industries and as a method of effluent treatment.

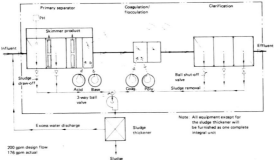

FIGURE 1 – Food processing waste water treatment process.

Two stage carbon filtration and ion-exchange treatment.

Treatment of Landfill Leachate

The landfilling of waste leads to the production of leachate and landfill gas, both usually derived from the decomposition of organic matter. Landfill leachate is the seepage water from solid waste landfill sites. It consists of rainwater containing pollutants washed out of the waste mass together with products of biological degradation. This highly polluting effluent is usually pre-treated chemically or biologically prior to concentration by reverse osmosis. A typical leachate treatment process is shown in figure 2. When the leachate is dilute it may be fed directly to a reverse osmosis plant, following simple grit screening; with no other pre-treatment being required.

The leachate usually has a high BOD and COD together with nitrogen, inorganics and organic halides. The quantity and composition of the leachate is determined by many factors
- waste types
- disposal methods
- construction and age of the landfill
- climate and seasonal effects

Leachate must be treated prior to discharge into surface waters. The selection of the most appropriate treatment system depends upon
1. the composition of the leachate at source
2. the required discharge standard required by the regulating authority
3. the anticipated flowrate that will require treatment.

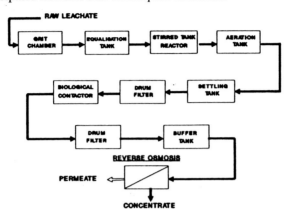

FIGURE 2 – Overview of leachate treatment process.

It is estimated that approximately twenty principal technologies can be applied in various combination for leachate treatment. The final solution is engineered to achieve the optimum system based on cost and quality of final discharge. A typical landfill leachate process may comprise of several of the following operations, biological pretreatment, ultrafiltration, reverse osmosis, ammonia stripping, activated carbon filtration, evaporation and drying/granulation.

FIGURE 3 – Schematic of leachate treatment technologies.

Biological Pretreatment

Biological treatment breaks down organic compounds and removes nitrogen and some inorganic species through flocculation. These are several biological treatment technologies which may be applied, anaerobic, aerobic and rotating biological contactors. In some cases, eg dilute or methanogenic phase leachate, a biological pretreatment may not be necessary.

INDUSTRIAL WASTE WATER AND EFFLUENT TREATMENT

Aerobic Treatment

There are two types of bacteria involved in the aerobic treatment of leachate. These are heterotrophes and autotrophes. Heterotrophic bacteria will oxidise organic compounds to produce carbon dioxide and water. They will also carry through a reaction between carbon and ammonia in the absence of oxygen (anoxic) to produce nitrogen, carbon dioxide and water. A species of autrotrophic bacteria (nitrosomas sp.) will convert ammonia to nitrite (NO_2) and another species (nitrobacter sp.) will convert to nitrate (NO_3) by combining it with oxygen. These two latter reactions are generally referred to as nitrification and denitirification.

It is, therefore, clear that the conversion of both organic carbon and ammonia to substances that can be released into the environment without harm depends very much upon the availability of oxygen. Oxygen can be input into the process either by pumping it into the base of the reactor, or by lifting the liquids into the air. There are, however, many other critical aspects to the use of this technology, such as sludge removal rates, that will require to be addressed in the operational design of such systems.

Anaerobic Treatment

Anaerobic treatment reactors carry out the same process on the leachate as takes places on a much larger and more diffuse scale within the landfill site itself. Organic compounds are degraded to volatile fatty acids and these in turn are degraded to form carbon dioxide and methane. The methanogenic bacteria which carry out the second stage are slow-growing bacteria which are ideally suited to treating high-concentrated early-stage leachate. Sulphates and nitrates are reduced to sulphide ions and nitrogen and heavy metals are removed.

Organic nitrogen, however, is broken down to ammonia, which is not removed. The ammonia levels in the effluent may, therefore, be higher than those in the leachate itself. For this latter reason it is generally not possible to obtain a discharge quality effluent with anaerobic treatment alone. A post-treatment stage is usually required, in particular to reduce ammonia levels.

The optimum temperature for the process is 35°C, although some activity will take place at temperatures as low as 5°C. Insulation and heating thus form an integral part of an optimised anaerobic digester. A further requirement is to ensure effective mixing of the reactor contents in order to maximise the mass transfer processes that are taking place. This may be accomplished by means of a simple paddle mixer or the specialised arrangement of the reactor internals, as encountered in the Upflow Anaerobic Sludge Blanket (UASB) or the Anaerobic Baffled Reactor (ABR).

Ultrafiltration and Bioreactors

Membrane bioreactors are biological treatment processes wherein membranes (UF) concentrate biomass, enhancing the effectiveness of the biological processes, clarifying bioreactor effluent, and virtually eliminating the production of sludge. The effluent can be further processed if required with reverse osmosis or activated carbon.

The system comprised denitrifaction tank, activated sludge tanks (nitrification), and an ultrafiltration (UF) system which separates the pretreated leachated from the activated

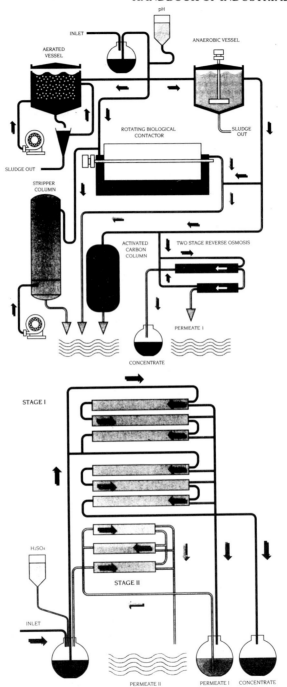

FIGURE 4 – a) a pilot scale treatment unit b) reverse osmosis system.

sludge, the latter being recycled. UF replaces the drum filter, biological contactor and settlement tank in the traditional biological pretreatment processes, and has a number of advantages:-
- UF is easier to control and operate, and thus requires lower manpower levels and lower maintenance.
- UF allows a higher recovery of biomass, and the denitirification/nitrification tanks are thus reduced to about one third of their normal size.
- These factors combined, result in significant savings in space.

The UF system is modular, comprising a pumping system to generate pressure and velocity, the membranes (housed in modules), and a flow control valve to regulate concentration. Design criteria involve correct membrane selection, optimisation of pressure, temperature, fluid velocity, and cleaning procedure, all of which depend on leachate type and composition.

Ultrafiltration stack for biomass treatment.

Reverse Osmosis

RO effectively removes recalcitrant materials, producing water fit for discharge to aquifer. RO removes all suspended and colloidal materials, and most dissolved solids, reducing COD, BOD, AOX, ammoniacal nitrogen, and heavy metals, which cannot be easily removed by other available technologies.

Membranes are thin barriers which allow small molecules (such as water) to pass through, while larger molecules are retained and concentrated. The process is driven by elevated pressure (>40 bar), while flow tangential to the membrane surface prevents the membrane from fouling or clogging.

Residual inorganic and non-degradable organic compounds are removed by RO. A high ammonia rejection rate is ensured through acidification of the RO feed. 75-80% of the RO feed is recovered as water fit for discharge to surface waters. The RO-concentrated leachate can be recycled to the landfill or further treated by evaporation and drying. The condensate from evaporation and drying stages is also treated in the RO process.

Depending upon the composition of the leachate, the RO process may comprise 1, 2 or 3 purification stages in order to generate permeate fit for discharge to surface waters. The first stage uses open tubular membranes which can handle large quantities of suspended material without blockage or damage; later stages use spiral-wound membranes., which economically treat the solids-free leachate (permeate). Thin-film composite membranes are used throughout, with the benefits of high pollutant retention, long membrane lifetime, and ease of cleaning.

The first stage usually uses RO membranes, but in the treatment of some leachates with high salt levels, nanofiltration (NF) membranes may be used. NF membranes allow salt to pass into the permeate. Since salt content limits the achievable concentration factor NF allows a high concentration factor to be achieved in the first stage. However, the NF permeate needs more comprehensive post-treatment than RO permeate.

Reverse osmosis unit for leachate treatment.

Rotating Biological Contactor

The rotating biological contactor is a method developed for achieving high rate aerobic treatment without the need to use forced air. This results in a significant reduction in the energy requirement to achieve complete aeration. A drum with a high internal surface area rotates through the liquid. Aerobic bacteria are grown onto the surface which are alternately immersed into the liquid and then pass through air. The process is particular suitable for the removal of ammonia and organic carbon. One factor that must be borne in mind is that sludge will be produced and must be either removed from site or returned to the landfill. The quantities of sludge produced will be dependent upon influent concentrations and flow rates.

Ammonia Stripping

As previously mentioned in anaerobic leachate treatment ammonia is not oxidised to nitrate and thus an additional treatment step may be required. The first step is to adjust the pH, by adding lime or sodium hydroxide. This addition not only raises the pH but also causes the precipitation of certain salts and larger organic molecules, thus decreasing residual COD. A plate separator can be used to remove suspended solids, thus keeping the column internals free from blocking and channelling.

From the plate separator the leachate is passed to the packed bed stripper column where the partial pressure of ammonia in water is used to extract it from the leachate. It may be necessary to have a multi-stage stripping process to achieve very low levels of residual ammonia. If required, the exhaust air from the column may be bubbled through a sulphuric acid absorber in order to remove the ammonia. The treated leachate is finally neutralised with acid before discharge.

Activated Carbon Filtration

Granular activated carbon is a highly porous and crude form of graphite with a wide range of pore sizes, from visible fissure sand cracks to ones of molecular dimensions. Adsorption is the process by which molecules with particular characteristics of size and molecular polarity are attracted and held to the absorbing surface. More than 130 different types of organics have been identified on spent carbon from leachate treatment plants. Carbon usage is minimised by staging container vessels in a lead-polish configuration. After the lead vessel is saturated it is recharged and then used as the polishing vessel until the new lead is saturated. The carbon needs to be re-activated in special furnaces and for this reason each fill will have a duration dictated by the volume of the chamber, the leachate flow rate and the concentration of absorbable components.

Evaporation and Granulation

If the RO concentrate cannot legally be recycled to the landfill, evaporation and drying of the concentrate may be necessary. Evaporation normally achieves a 10-fold concentration. The evaporator condensate is recycled to the first RO stage for purification. The evaporator has two or more effects; increasing the number of effects increases the achievable volume reduction while reducing specific energy consumption. Drying and granulation of the evaporator concentrate produces a granulate comprising less than 1% of the total raw leachate volume, with a dry matter content of at least 95%. It is free of dust and therefore easy to handle.

Two to four stages are usually employed in order to achieve cost effective operation. Each stage is identical in design to the others and is equipped with a circulation pump, a fluidised bed heat exchanger and an evaporator vessel. The fluidised bed heat exchanger is used to prevent deposits from building up on the heat exchanger surfaces. The evaporator is designed to allow the stream and concentrate to flow in parallel along the pressure gradient towards the vacuum pump downstream of the final stage, thus increasing the plant performance.

Dewatering of the concentrate proceeds continuously from stage to stage in the evaporator unit. The concentrate from the final stage can be drawn off and fed to a drying system.

The first stage is heated with primary steam from a steam generator. Subsequent stages are heated by the waste vapours from the proceeding stage. If the electrical conductivity of the concentrate from the final stage is too high the condensate is automatically recycled to the receiver vessel of the circulating evaporator.

Drying of leachate concentrates, produced by reverse osmosis, is the final state in treating leachate. granulation is a process that has been taken from the chemical process industries for the production of a dust-free, free-flowing, stable product of high bulk density, thus minimising residue volume.

The process consists of a fluidised bed drier with a classifier discharge system. RO or evaporator concentrate is sprayed into a fluidised bed. Solids in the liquid, accumulate on the fluidised particles and the water evaporates. Once the individual agglomerated grains have reached a critical size they fall through the classifier discharge. Distillate recovered from the water-cooled condenser is returned to the feed tank of the treatment process.

Gaseous emissions from non-condensable compounds are drawn off and burned in a muffled furnace.

Example of Landfill Site Treatment

At Damsdorf in Germany a new site is being constructed in 2 sections, each of 6 hectares, with a total refuse capacity of 3 million cubic metres. This new site will have bottom and top-sealing systems, and a leachate treatment system incorporating biological pretreatment with ultrafiltration, reverse osmosis, evaporation and granulation.

In order to minimise the time take to achieve methanogenesis in the new site, a special filling and leachate collection method is planned. Each of the main sections of the landfill will be subdivided into a number of compartments. The first compartment will be layered with compacted and loose refuse comprising materials from the old site mixed with new. This will reduce the time taken to achieve methanogenesis from the normal 2 years to about 9 months. When compartment 1 is full, the other compartment will be used. Leachate from all compartments is taken through compartment 1 (which is in the methanogenic phase) before collection. This ensures that the quality of the leachate changes only gradually after the first year of operation. The leachate treatment system will be installed in 2 stages, the second stage to come on-line after 2 years.

Biological pretreatment used the Activated Sludge Process, which comprises prefiltration, denitirification in a 125 m^3 stirred tank, and two 125 m^3 aeration tanks. Tubular ultrafiltration (UF) recovers the biomass post-nitrification. A 10-fold concentration of biomass is achieved, and the concentrated biomass is recycled to the denitirification stage. Approximately 3% of the UF permeate (biologically-pretreated leachate) is recycled to the aeration tanks for foam control.

FIGURE 5 – Landfill treatment site.

TABLE 10 – Design Capacities of Leachate Treatment System.

Design criteria	Initial installation	With extension
ROI	187 m² 2.5 m³/h	374 m² 5 m³/h
ROII	78 m² 2.7 m³/h	97.5 m² 3.4 m³/h
Evaporator	2-effect 5 m³/h	4-effect 10 m³/h
Drier	0.5 m³/h	1.0 m³/h

The UF permeate is concentrated by 2-stage RO. The first stage uses tubular membranes and achieves a concentration factor of 5. The permeate from the first stage is then concentrated in the second stage, which uses spirally-wound membranes and achieves a concentration factor of 4. Both stages operate at 40-50 Bar, and ambient temperature, and

TABLE 11.

Parameter		Raw Leachate	Biologically Pre-treated	RO Permeate	
				1st plant	2nd plant
COD	mg/l	5000	1500	<40	<15
BOD	mg/l	350	110	<10	<2
TKN	mg/l	2000	100	<10	<1
NH4-N	mg/l	1800	<10	<1	trace
NOX-N	mg/l	0	400	<20	<1
AOX	mg/l	4000	2000	100	<5
Conductivity (millisiemens/cm)		16-20	10-12	<1	<0.05

are chemcially-cleaned on a weekly basis. The cleaning solution is returned to the biological pretreatment process. The concentrate from the second stage is returned to the first stage.

The initial installation comprises a 2-effect evaporator to achieve a 10-fold concentration. A planned extension allows for a doubling in capacity by adding a further 2 effects.

The capacities for the individual units is indicated in Table 10 and the final leachate quality is indicated in Table 11.

The energy requirements of a leachate treatment process comprising biological pretreatment, reverse osmosis and evaporation, can usually be adequately satisfied by the energy recoverable from landfill gas. Landfill gas extraction and utilisation equipment for a fully integrated landfill system can also be supplied.

TABLE 12 – Characteristics of Landfill Leachate Treatment Technologies.

Technology	Application	Attributes	Cost
Aerobic Treatment	Aerobic treatment is generally employed, to reduce BOD and ammonia, where a robust treatment method is required and where a variation in effluent characteristics is acceptable. In order to mitigate the inherent unreliability of biological systems working on a substrate as variable in nature as leachate, it is often the case that some form of post-treatment is required. This may vary from a membrane to activated carbon and reed-beds.	Robust performance, proven in many varied applications. Low plant cost and simple plant operation.	One of the dominating costs involved in the application of this technology is the energy required to aerate the leachate. Associated liquid pumping costs are minimal when compared to that required for aeration. Each application will have its own particular requirements, such as the need to add nutrients bacterial seed or chemicals for pH adjustment.
Anaerobic Treatment	Anaerobic digestion will be particularly useful in treating high-strength leachates found in the early stages of landfill's decay processes. It will provide a compact and effective method of making a significant reduction to enable subsequent treatment methods, such as reverse osmosis or activated carbon, to be applied in an optimum manner.	Methane produced as by-product may be used to enhance the process kinetics. High organic removal rates mean that comparatively small vessels can be used to obtain a meaningful reduction in the organic loading of a leachate. Little or no excess sludge is produced. This compares most favourably to aerobic treatment technologies where greater levels of sludge production are encountered.	The cost of an anaerobic treatment system will in general be less than that of an aerobic system. It will involve the reactor vessel itself and the power consumption associated with pumping and agitating the leachate. Where highly acidic leachates are encountered it may be necessary to dose the influent with sodium hydroxide to bring the pH up above 6. Below this level anaerobic digestion will not take place.
Rotating Biological	The RBC is suitable for situations where biological activity may be supported by the organic carbon content of the leachate. It is also application for achieving significant reductions in ammonia levels where influent concentrations are adequate to support biological activity.	Low power requirement during operating. Easily transported. High ammonia and organic carbon removal rates. Low operator supervision requirement. Easily instrumented.	Costs associated with the use of the RBC involve capital cost (likely to be greater than that of equivalent aerated reactor) the chemical and nutrient additions that may be required, energy for heating the process (where this is determined to be advantageous) and power requirement of the motor turning the rotating drum.

INDUSTRIAL WASTE WATER AND EFFLUENT TREATMENT

Ultrafiltration	Both aerobic and anaerobic digestion processes have been applied to treat leachate, where metabolisation of BOD and/or reduction in ammoniacal nitrogen is required. Practically, aerobic treatment is far more widely-used, for operational reasons. However, aerobic treatment suffers from high energy costs, high capital investment costs, and a large footprint. UF achieves more than 75% reduction in the size of the bioreactor, and obviates the need for settling tanks, resulting in significant savings in investment and operating costs, and in land requirements. The UF process can also be used in conjunction with anaerobic processes.	Recues overall costs of the biological treatment processes. Processing efficiency is not greatly affected by changes in feed flow and composition. Robust, reliable technology, proven in many aggressive applications. Large reduction in the footprint of the biological treatment process. Easy to automate, and low maintenance requirement. Increased stability of the biological	The most significant cost involved in the use of UF is the pumping energy requirement. The use of UF increases the total energy requirement, but savings in overall capital cost offset this in many costs. Furthermore, savings in land usage may dominate other costs and may allow the use of biological treatment where land availability is restructed.
Ammonia Stripping	Ammonia stripping has its place in the range of techniques applied to landfill leachate treatment where ammonia levels must be reduced but other parameters, such as COD and heavy metals, are within the limits of a discharge consent. The use of an ammonia stripper should be specifically geared towards meeting such limits, rather than exceeding them by many orders of magnitude, as can be the case if this objective is not borne in mind.	A low cost method of treating one particular aspect of a leachate. This will be applicable where the only parameter falling outside of the discharge requirement is ammonia. Compact plant operating on a physical treatment method and hence more reliable than biological treatment systems. Options to instrument the plant for remote monitoring and control.	the cost of such a system will in general be low compared with other options that may be considered. The power requirement for air circulation is less than that for an equivalent aerated system and water pressure heads required will be a function of the height of the stripper column. When operated with in-line pre-filtration, such units can operate for many years with very little maintenance.
Carbon Filtration	Granular activated carbon is primarily used to remove AOX and COD, both of which will not be the primary focus of biological treatment systems and may, therefore, be found above discharge consent levels from such systems. With particularly dilute	Reliable performance within defined parameters. Low plant capital cost, with the facility to lease.	The cost of such a system is comprised of the capital cost of the adsorption equipment, operation and maintenance of this equipment and the transportation and reactivation costs that are inherent in the use of activated carbon that cannot be

TABLE 12 (continued).

Technology	Application	Attributes	Cost
	leachates it may be operated with only a plate separator or pressurised sand filter removing suspended solids from the flow, in order to ensure that the carbon filter is not blocked with solids. It is necessary to ensure that there are no substances in the leachate which will damage the carbon prior to selecting such a system. Activated carbon has beend used as a final polishing step after biological treatment, UV-oxidation, sedimentation and other physio-chemical treatment methods.	Low power requirement for operation.	regenerated or reactivated in-situ. It has been estimated that in 50% of installations less than 1 kg of carbon per cubic metre of leachate is required. Spent carbon is returned for reactivation and can then be reused. Activated carbon can be regenerated normally between 5 and 10 times., after which it must be replaced. The costs of this option are, therefore, critically dependent upon the cost of transport and reactivation and the rate of replacement.
Reverse Osmosis	RO is normally used in the following cases: Reduction in COD, BOD, ammonia, heavy metals, and other contaminants to very low levels, producing water fit for discharge to aquifer. Treatment of leachates with high inorganic loading, and/or with low volumetric flow rates (RO is more cost-effective at low capacities than biological treatment systems). Treatment of leachates from landfill sites without sewer connection, where the alternative option is costly tankering and off-site treatment. To meet the most stringent environmental discharge standards.	Energy efficient dewatering process. Robust process, proven over many years in the treatment of landfill leachate. Small footprint. Easy to automate, and low maintenance requirement. Processing efficiency is not greatly affected by changes in feed flow and composition.	Developments in membrane technology in recent years have allowed it to compete effectively with other processes. The initial capital cost, the cost of membrane replacement, and the pumping energy requirement are significant, but these are far outweighed by the cost of disposal of RO concentrate. If concentrate can be recycled to the landfill, RO is a very cost-effective process. At medium capacities, RO offers significantly lower specific costs than most other treatment processes, as well as superior degree of treatment, higher security in meeting consents, and lower operational requirements. Since tubular RO membranes are used, which are highly resistant to fouling and clogging, pre-treatment requirements are minimised.

INDUSTRIAL WASTE WATER AND EFFLUENT TREATMENT

			The pre-treatment normally comprises simple coarse filtration, and acidification with sulphuric or hydrochloric acid, which assists the retention of ammonia.
Evaporation	Evaporation drying can be a very expensive process in terms of heat required to facilitate the removal of water from solids. For this reason evaporation is often reserved for the concentrate from a reverse osmosis plant, where a significant amount of water has already been removed. Direct drying of RO concentrate is possible, but the use of evaporation reduces the total drying costs.	Reduces load on a granulator where a solid residue is required. May be used to reduce volume by operating as a stand alone system where leachate volumes are not high and landfill gas is available to power the steam generator.	Evaporation is not a complete treatment process in itself & normally requires influent pretreatment and exhaust air cleanup equipment, to prevent the release of noxious substances into the atmosphere. These capital items will, therefore, be a key component in evaluating the overall cost of evaporation systems. It is normal to pass the exhaust air form the process to a high-temperature muffled flare for complete destruction. Costs will include the power required for heating the evaporator and the exhaust air clean-up. In many cases, evaporation condensate must be recycled for further treatment, eg the reverse osmosis process.
Granulation	Granulation would be considered where it is necessary to remove the concentrate left from an evaporation or reverse osmosis treatment system, ie where the concentrate cannot be recycled to the landfill. The product is a solid granular material which must be disposed of as a hazardous substance.	Production of dust-free, free-flowing granular material of predetermined particle size, thus minimising further disposal costs. No airborne emissions: pollutants are concentrated in the granular material. Granules have a high solid and bulk density and, therefore, take up less space. Reduction of problems with materials of construction, as a result of low levels of corrosion and abrasion achieved with dry operation of the fluidised bed.	the cost of a granulator is based around the initial capital cost and the heat required to achieve final evaporation of leachate concentrate. This is not a low cost option for the final concentration of leachate residues and should be considered where no other option is available. The product from the system is a granular material that may be handled with mechanical tools. Disposal costs will also need to be clearly defined from the outset.

Anaerobic Digestion and Ultrafiltration of Wastewaters

There are many industries which need to dispose of organic effluents with COD levels ranging from 1500 mg/dm^3 (low load) to 100,000 mg dm^{-3} (high load). An alternative to conventional systems is to use an anaerobic digestion and ultrafiltration (ADUF) process, which is claimed to effectively eliminate sludge concentration and retention problems, prone to occur in standard systems. The process shown schematically in Fig 6 is a high rate, robust and reliable method which has applications in many industries: paper and pulp, breweries and distilleries, yeast and starch manufacturers, dairy, sugar refineries, vegeta-

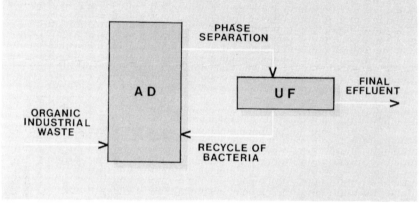

FIGURE 6 – Schematic of the ADUF process.

TABLE 13 – Performance characteristics of ADUF on various waste anaerobic digester effluents.

Source of industrial wastes	Avrge COD of waste to anaerobic digester (mg.l^{-1})	Measured COD after UF of digester effluents (mg.l^{-1})	Ultrafiltration permeate flux of digester effluents (litres.m^{-2}.d^{-1})
Tap Water	—	<15	
Beer Brewery	3 100	26	
Paper Mill	2 900	608	
Petrochemical	17 000	2 295	
Maize Processing	7 200	43	
Maize Processing	4 000	458	
Abattoir	3 900	59	
Wine Distillery	40 000	1 000	Stabilised flux
Yeast (Molasses)	35 000	3 577	Initial flux

UF Flux Standardised at 35°C, 500 kPa Inlet Pressure, 2 m.s^{-1} Tube Velocity
Suspended Soilds in Digester Effluents Varied in the Range 0,1 - 20,0 kg.m^{-3}

INDUSTRIAL WASTE WATER AND EFFLUENT TREATMENT

ble and fruit processing, abattoirs etc. The typical performance of the process in these industries is shown in Table , in terms of the UF permeate flux and COD level after UF, utilising tubular membrane modules. Typical cross flow velocities are in the range of 1-3 ms^{-1}.

The membranes used in the process are made from polyether sulfone with MWCO between 20,000 and 80,000. These membranes can operate at temperatures up to 60°C and with pH values between 1 to 10. The ability to operate at temperatures up to 60°C is an attraction in the process as it enables higher flux rates to be achieved and enables a choice to be made between a mesophilic or a thermophilic system. The overall benefits of the ADUF process include

- great flexibility in integrating the biological and the physical functions of the digester and the UF unit
- ultrafiltration and anaerobic digestion are complementary processes: anaerobic digestion decomposes organics which would otherwise foul the filter membranes, while the membranes serve to retain biomass which would otherwise be lost in the digester effluent
- the process achieves the desirable effects of longer sludge retention times and shorter hydraulic retention times
- at digester space load rates greater than 10 kg COD m^{-3} d^{-1} the process yields high quality effluent free of suspended solids
- the process reduces the organic loading of most biological wastes by as much as 95%
- in contrast to most types of purification processes which consume energy, anaerobic digestion actually produces energy. Consumption of 1 ton of COD liberates 500 m^3 of biogas equivalent to 500 kg coal
- anaerobic digestion yields a minimum of sludge as compared to aerobic (oxygen consuming) processes
- the process is completely enclosed – no undesirable odours will be detectable in the vicinity of the plant. This is especially important in sensitive areas, eg around food processing and beverage plants
- on-site anaerobic treatment offers significant savings in terms of effluent tariffs, effluent reuse capability and energy recovery, with a plant investment payback period of 1-2 years.

As is typical with this type of process selection of digester capacity and membrane area for processing a given waste requires pilot scale studies which can be done on mobile pilot plants.

Domestic and Industrial Wastewater *(MF & activated sludge)*

It is estimated that the average large chemical company in the UK pays £250,000 annually to the local water plc for supply and double (or treble) this for treatment before discharge. The scope for water treatment on plant is therefore enormous, however traditional treatment plants are generally large and difficult to place an exisiting sites. A submerged membrane system may well provide the answer to this dilemma. The process is based on an active sludge treatment with built in microfiltration membranes.

In the system (see Fig 7) the membrane module is integrated within the aeration basin. The module consists of a number of sheet membranes sitting in the activated sludge. Air is introduced at the bottom of the system to provide the motive force to move the sludge up through the gaps between the plates and to provide oxygen for the activated sludge and biomass respiration. In addition the air bubbles scour the surface of the membrane, thereby cleaning it. In the process the sludge moves up the gap and liquid (permeate) is drawn through the membranes due to an applied vacuum pressure and is then pumped off. The biomass and bacteria cannot pass through the membrane and are retained in the sludge. Due to the self-cleaning action of the air, the system does not clog or require any backwashing.

Domestic wastewater treatment plant.

The process is in operation in ten plants in Japan, the largest handles 130 m^3/d. The effluent quality is superior to that of conventional plant at comparable capital cost but lower operating costs. Trials on industrial effluents are underway in several countries including treatment of effluent in a Quorn plant.

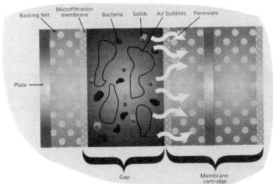

FIGURE 7 – Schematic of activated sludge membrane process.

Domestic Sewage Treatment

There has been interest in the use of RO in various stages of domestic sewage treatment. In the first stage of conventional sewage treatment the raw sewage is subjected to coarse straining. In the next stage the primary sewage is subjected to biological treatment which involves the use of an activated sludge reactor or a percolating filter. The secondary sewage effluent from this process has a greatly reduced organic content but contains a high concentration of bacteria and an increased loading of inorganic solids resulting from the biological degradation of the organics.

The processing of primary sewage by reverse osmosis is adversely affected by membrane fouling. Overall the cost of reverse osmosis is much greater than for the conventional process thus exploitation of this process is not realised.

However secondary sewage can, after further pretreatment, be processed by desalination. The pretreatment to reduce membrane fouling includes lime addition and sedimentation, ammonia stripping, recarbonation, break-point chlorination, mixed-media filtration and adsorption of activated carbon. A large reverse osmosis plant has been in operation on treated secondary sewage (Organic County, California). This plant desalts the treated sewage so that it can be injected into the aquifier, and protect the ground-water supply from sea-water intrusion. The plant is reported to use spiral wound cellulose acetate modules and has a capacity of 19,000 m^3 d^{-1}. The plant operates at a water recovery of 85%.

Microfiltration

The Dwr Cymru Welsh Water environmental policy is to disinfect all sewage discharges to coastal locations in the future and provide effective disinfection and solids removal and limit odour generation. Although this can be achieved by conventional primary, secondary and disinfection techniques membrane technology is seen as an effective alternative for the treatment of sewage for populations of up to 10,000 offering high level treatment combined with low environmental impact by using a compact and enclosed plant.

The process plant comprises preliminary treatment, inlet flow balancing, advanced primary treatment and microfiltration. Crude sewage is pumped to the works inlet chamber from pumping stations and the sewage passes through a packaged preliminary treatment plant which uses screening at a 600 m aperture to achieve grit and grease removal. The sewage then passes through a balancing tank and on to downstream treatment. The preliminary treatment and downstream processes provide full treatment capacity to 4 DWF (dry weather flow). Flows in excess of this are buffered firstly in the on-site balance tank and secondly in underground tanks and piepwork to provide up to 16 hours storage at average dry weather flow. Bypass flows pass through a manual 5 mm screen for direct discharge from the works.

Sewage flow from preliminary treatment passes to a packaged primary treatment process which combines coagulation, flocculation, lamella separation and sludge thickening. The coagulation stage combines ferric dosing with alkali pH correction. Two lamella separators are provided at 2 x 7-% summer flow. The lamella underflow is thickened to an expected sludge concentration of 3 to 4%, and sludge storage is provided to limit off-site transport. Odour control is applied to the sludge storage.

The primary treated sewage passes under gravity to 2 x 100% safety screens (500 m) and then to the microfiltration feed tank. The sewage feed is pumped to microfiltration arrays (3 x 50%). Filtered water is then discharged as final effluent under gravity through an outfall to mean low water mark. The filtered water is also utilised as plant service water.

The continuous microfiltration process employs microporous hollow fibre membranes which provide barrier filtration down to 0.2 μm, removing particulate BOD, oils and greases, bacteria an dviral organisms. The feed flow passes from the outside to the inside of the hollow fibre to maximise the working filtration surface are, and the process operates in direct flow filtration to obtain low energy consumption. Filtration rates are maintained by the combination of a gas backwash sequence and a chemical clean-in-place (CIP) system. The backwash uses a pulse of compressed air which is forced into the centre and through the walls of the fibres, lifting accumulated debris off the surface. The membranes are cleaned chemically on a routine basis with a biodegradable alkali solution which can be reused to minimise consumption. The use of microporous membranes permits online integrity test and diagnosis to validate disinfection capability. Backwash waters from the microfiltration process are returned to the works inlet to co-settle with the raw sewage feed. The coagulation and flocculation steps prior to sedimentation are included to control both the crude sewate settling and also the returned colloidal solids from the microfilter. The plant incorporates fully automatic process control and monitoring.

Crossflow Microfiltration Combined with Electrocoagulation and Flotation

Municipal waste water consists of a suspension of a very wide range of particulate, colloidal, soluble minerals and soluble organics. All treatment methods depend on a preliminary conversion of the colloidal and soluble pollutant into particulate pollutant prior to a final clarification process. This final stage is not always reliable unless crossflow microfiltration is used and would also achieve virtually total disinfection. Industrial development of the process, however, still depends on the particular flow of the permeate produced and regeneration of the porous membrane and thus pretreatment is necessary. Electrochemical destabilisation of municipal waste water for crossflow microfiltration is one attractive method.

The families of particles, based on different fractionation techniques which describe the pollution fractions contained in municipal waste water are shown in Table 15.

About 35% of pollution in untreated water is easily settleable, 35% is colloidal or

TABLE 15 – Definitions of the four families of particles in municipal waste water.

Soluble fraction	< 0.001 μm
Colloidal fraction	0.001 - 1 μm
Supracolloidal fraction	1 - 100 μm
Settleable fraction	> 100 μm

supracolloidal, and 30% is soluble. Using micro- or ultrafiltration can result in a reduction of at least 70% in the pollutants but this operation is only possible, if the medium's permeability can be controlled by the operator. Although external deposits can in theory be limited by tangential circulation flow, the colloidal fraction often causes an irreversible fouling of the medium. Microfiltration is then a only viable after the suspension has been conditioned. A large part of this colloidal fraction can be converted into a settleable particulate fraction through a pretreatment combining electrocoagulation and aeroflotation as shown in Figure 8. The treated water is mixed with recirculated water, pressurised, saturated with air, and then introduced into a separation chamber. In this chamber decompression releases fine gas bubbles which adhere to the particles and cause them to float. The electrocoagulation device placed at the entrance to the chamber, enhances the process by the release of gas (oxygen and hydrogen) caused by electrolysis of the water, and principally by the generation of flocculates of aluminium hydroxides (formed from Al anodes) which trap the colloidal and supracolloidal pollution.

The effluents are microfiltered with a mineral membrane with a porosity of 0.1 μm using a crossflow microfiltration process with an adjustable transmembrane pressure D P = 0.8 – 2.5 bar and an adjustable recirculation rate $V = 2 - 5$ m/s.

Permeation flow with a 0.1 μm microporous membrane increases from 0.02 m^3/m^{2}h for untreated water to a value approaching 0.35 m^3/m^{2}h for pretreated water.

The effect of electrocoagulation on the suspension's granulometric classification produces:
- A major reduction in the pollution's non-decantable fraction, either in the fraction smaller than 0.2 μm or in the fraction marked 'ad 2'.
- A corresponding increase in the settleable fraction.

Electrocoagulation thus causes a shift in the granulometric distribution towards larger diameters. Pretreatment can thus be used to reduce the waste water's fouling potential.

Although direct treatment of municipal waste water by microfiltration is generally considered uneconomic, it is possible to obtain increased permeate flows, approaching 0.35 m^3/m^{2}h, by combining this filtration with a preliminary electrochemical destabilisation and thus improving economics significantly. The crossflow microfiltration then becomes a potential treatment process, with virtual total disinfection of the effluent achieved with a membrane cut off of 0.1 μm. Only a part of the soluble fraction remains, which makes agricultural use of this water possible, especially as this water has a very limited fouling potential.

Farm Waste

Reverse osmosis using high retention polyamide thin film composite membrane is also being successfully applied to both the concentration of raw, settled pig slurry and to biologically pre-treated pig slurry. Up to 4 times volume reduction with the raw slurry and up to 6.5 times volume reduction with the pre-treated slurry are possible.

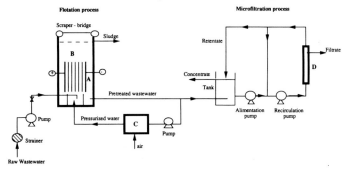

FIGURE 8 – Electrocoagulation - Flotation - Crossflow - Microfiltration.

PULP, PAPER AND TEXTILE INDUSTRIES
Pulp and Paper Industry

The pulp and paper industry consume a very large amount of water, with estimates of around 200 m³/tonne of cellulose produced and 100 – 1,000 dm³/tonne of paper (depending upon type) in addition to the water used in cooling. The water is the solvent for the pulping process and bleaching stages and is also used for washing, transport of fibres and purification. Consequently large volumes of effluent water are produced which is virtually non-bio-degradable because of the dissolved wood components. In addition efluents are hot highly coloured and are at extremes of pH. Cellulose production is based on two processes either the older traditional sulphite method or the sulphate method. Figure shows a typical process flowsheet of a sulphite pulping process. The decorticated and cut wood is treated (cooked) with a suitable solution to separate the cellulose fibres from lignin and hemicellulose. Sulphite liquors are evaporated to a solids content of approximately 50 – 55%, this concentrate can be utilised as fuel in the steam boiler of the plant. Chemicals are recovered from the combustion of magnesium hydrogen sulphite liquor but recovery is not possible in the case of calcium hydrogen sulphite solutions.

The cellulose fibres are washed and any non-decomposed parts are removed. This is typically followed by bleaching to improve the colour of the cellulose, i.e. to remove the characteristic dark brown lignin. A standard bleaching process consists of i) chlorination of residual lignin, ii) alkali extraction (with NaOH) of the chlorinated lignin and iii) oxidation, with whitening, with hypochlorite.

Chlorination and alkali extraction may be repeated for pulps containing a large amount of lignin. Alkali extraction produces a dark brine with a high biological oxygen demand, which because of its high lignin content, is unsuitable for biological sewage treatment. Simultaneously, a large proportion of the lignin still present is separated from the product. The final steps of the process are water removal and the formation of cellulose felt.

In paper production, fibres and paper additives are pulped with water and then passed through several refining stages before reaching the paper machine. The fibre content of this slurry is approximately 0.5-1%, the majority water is removed mechanically in the first section of the paper machine which is followed by (thermal) drying, resulting in a product water content of 4-6%. Although most of the removed water is recirculated, the average fresh water consumption still amounts to about 40 m³ of paper per tonne. Membrane process thus have potential applications with wash waters and effluents from pulp and paper industry in the following areas: i) concentration of highly dilute sulphite liquors (wash water), ii) separation of lignins, including desalination of bleach effluents from the alkali extraction stage and iii) treatment of paper machine effluent to assist the internal water recirculation systems.

Ultrafiltration and reverse osmosis have also been tested on a laboratory and pilot plant scale on several of the liquors and waters in the pulp and paper industry including:
 – UF and RO of black liquor
 – UF and RO of bleaching effluents
 – UF of paper machine wash waters

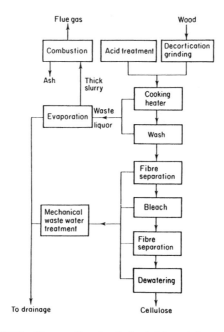

FIGURE 9 – Process flowsheet of sulphite pulping process.

The overall applications of UF (and RO) in the pulp and paper are many as can be seen in the schematic flow diagram of Fig 10.

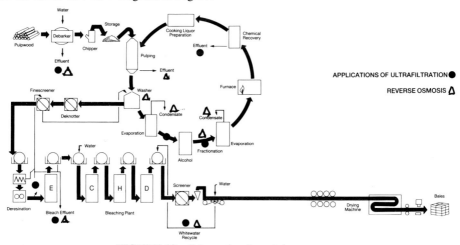

FIGURE 10 – Schematic of a pulping process.

Treatment of Spent Liquors in the Pulp Industry By Ultrafiltration

The production of cellulose is typically based on traditional sulphite or more recent sulphate methods. The characteristics of the respective effluents are given in Table 19.

These liquors are referred to as "spent sulphite liquor" or "black liquor" (from the sulphate processes).

Ultrafiltration processes have proved successful for the treatment of effluents in the pulp and paper industries. The chemical resistance of ultrafiltration membranes has been an important factor as many of the effluents are at high temperature and extremes of pH. The ability of ultrafiltration, to fractionate an effluent into two valuable streams containing high and low molecular weight solutes has improved the economic of several applications.

TABLE 19 – Typical effluents from cellulose production.

Effluent	Parameter	Content (%)
Black liquor	Solid content	17-22
	Alkali lignins	41
	Extracts	3
	Hydroxy acids lactones	28
	Acetic acid, formic acid etc.	9
	Sulphur	3
	Sodium	16
Spent sulphite	Solid content	12-16
	Lignin sulphonate	52
	Extracts	3
	Poly- and oligosaccharides	6
	Monosaccharides	23
	Various organic compounds	11
	Calcium	5

Ultrafiltration has found use in the treatment of Kraft process effluents. The effluent from the first stage of caustic extraction during pulp bleaching contains over half the entire Kraft mill's colour species in the form of large molecular weight lignin molecules. Treatment of these effluents is mandatory. These species can be recovered efficiently by ultrafiltration. Large ultrafiltration plants with membrane areas of 500-1500 m^2 have been

installed in Japan to clean up such bleach effluents. These are used as replacement for lime precipitation. In one such plant at Shikoku Island the bleach effluent is at 45-44°C and pH 10-11.5. It contains 1250-1900 gm^{-3} COD and 20-110 gm^{-3} of suspended solids. The effluent can be concentrated almost twenty-fold by ultrafiltration. In one application the installed membrane area is 672 m^2. The membrane material is polysulphone. A feed of 104 $m^3 h^{-1}$ of bleach effluent is separated into 91 $m^3 h^{-1}$ of permeate and 13 $m^3 h^{-1}$ of concentrate. The concentrate is further treated by conventional evaporation and/or combustion. Although some of the permeate is recycled as wash water, most is discharged. Overall the following performance is achieved for the UF separation:
- COD reduction 82%
- colour reduction 94%
- suspended solids retention 100%.

The membrane service lifetime is reported as 18 months; with membrane cleaning every day.

A second UF unit for treating bleach effluents 145 $m^3 h^{-1}$ also started up in Japan in 1981. This unit was equipped with tubular modules with a total membrane surface area of 1480 m^2. The membrane service lifetime was 18 months. The concentrate is further processed in the mills' recovery system and the permeate, which is completely free of suspended solids (and reduced five-fold in COD), is treated in an activated sludge process prior to discharge. The plant is a six-stage feed and bleed design, and uses tubular, non-cellulosic polysulfone membranes. An average membrane flux of around 3 md^{-1} is reported.

TABLE 20 – Membrane Separations in the Pulp and Paper Industry.

Separation	Application
UF of Kraft	Effluent from the first stage of caustic extraction during pulp bleaching
UF of process effluent spent sulphite liquors	Digested liquors from spent sulphite chemical pulping. Recovery of lignosulphonates and sugars
UF of Kraft black liquor	Recovery of alkali lignins
RO of sulphite liquors	Concentration of spent sulphite liquor
RO of paper machine effluents	Recycling of water
RO of wash waters	Pre-concentration of sulphite contaminated wash water prior to evaporation

There appears to be an increasing use of UF to recover lignosulphonate and alkali lignin from the spent liquors to produce other products. The solids content of the spent sulphite liquors is approximately 60% lignosulphonates and 30% reducing sugars. UF can fractionate (concentrate) lignosulphonates and product a permeate that contains the sugars, low molecular weight lignosulphonates and salts.

The total solids concentration of the liquor can be increased from 8-10% to 22% by ultrafiltration. The lignosulphonate content of the concentrate can be further increased from 80% to almost 95% by adding wash waters (diafiltration). The lignosulphonate stream is used to make vanillin, detergents, binders and adhesives. The sugar-containing permeate can be fermented aerobically, to produce animal feed protein, or anaerobically to produce alcohol. Several large spent sulphite liquor plants, with membrane greater areas than 1000 m^2 reported to be in operation in Scandinavia.

The Kraft black liquor from sulphate processes contains alkali lignins (high molecular weight material) which can be recovered from the liquor by ultrafiltration and subsequent diafiltration. The Kraft liquor is typically concentrated from 16 to 24% total solids, yielding a product which contains more than 90% alkali lignin. About 55% of the lignin content of the effluent can be recovered in this way. the permeate is generally evaporated and combusted. Large plants with membrane areas of over 100 m^2 are processing Kraft black liquors in Scandinavia. The recovered liquor is frequently used in phenol formaldehyde resins.

Example of The Recovery of Lignosulfonate Fractions by Ultrafiltration

The largest ultrafiltration plant in Norway (Borreguard Industries) has been in successful operation for almost five years, recovering valuable lignosulfonate fractions from calcium bisulfite spent liquors. This process uses tubular membrane, which can operate with a minimum of pretreatment and is contained in a robust stainless steel module. The Process is shown in Fig 12. After digestion, the waste pulping liquors pass to chemical recovery, which begins with partial neutralisation by calcium carbonate and fermentation to produce ethanol. After distillation, the fermented spent sulfite liquor contains about 12% total solids, of which 5% are high molecular weight lignin compounds and the remainder are low molecular weight lignins plus pentose sugars and inorganic salts. In the original process this liquor was then evaporated to reduce lignosulfonate products sold for glue or dispersant manufacture. Part of the stream was also passed to vanillin extraction.

The potential existed for ultrafiltration to reduce the volume loading on the vanillin process and, by filtering out some of the low molecular weight materials, not only to improve vanillin extraction efficiency but also to produce higher purity lignosulfonates for new and existing markets. A schematic of the plant is shown in Fig 13.

Fermented spent sulfite liquor passes through rough 0.5 mm filters before entering the UF plant which processes an average of 50 m^3/h of feed liquor with 12% solids, and produces a concentrate stream of 16 m^3/h and 22% solids. The concentrate stream contains all the higher lignosulfonate fractions and a lower percentage of sugars and salts. The exact purity is controlled by the addition of water (diafiltration) into various stages of the plant.

The plant consists of two lines which can operate independently. Each line has six UF stages in series. Within each stage a recirculation pump maintains a high cross-flow velocity in the tubes of 36 modules. Total membrane area is 1120 m^2. The membranes are manufactured from polysulfone. Operating pressure is 10-15 bars and the 12 recirculation pumps consume a total of 260 kW electricity. The temperature range is 60-65°C and the pH normally 4.2 to 4.5.

The plant is fully automatic both in operation and cleaning. It is controlled very simply by flowmeters on the feed and concentrate lines producing a flow ratio signal which compares to an operator set point and acts upon a control valve in the concentrate line. Diafiltration water addition is controlled by the feed flow, with a set-point which corresponds to the feed solids and concentrate purity required. A separate control valve on the delivery feed pump regulates the operating pressure.

The membranes are cleaned once a day by recycling an alkaline detergent solution through the plant. The total process for cleaning takes about two hours. Since the plant was commissioned in June 1981 it has operated continuously without any major problems. The average membrane lifetime is 15 months.

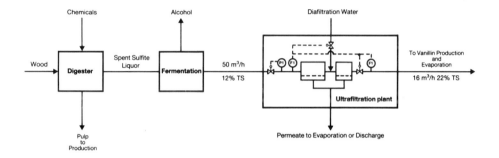

FIGURE 12 – Schematic diagram of lignosulfonate fractions recovery process.

FIGURE 13 – Tubular membrane system.

Reverse Osmosis

Reverse Osmosis is also used in the pulp and paper industry. Reported applications are in Scandinavia, Canada, Italy and Argentina for concentration of sulphite liquor and also possible applications are in the concentration of Kraft black liquor. The typical operating data from two commercial RO plants for the concentration of sulphite liquor are listed in Table 21. Both plants use plate-and-frame modules and cellulose acetate membranes.

Although the pH of the feed is low the service lifetime of cellulose acetate membrane was in excess of 1 year with a cleaning cycle every 1 – 3 days. A satisfactory permeate quality was obtained TDS in permeate, < 0.1 wt %, COD reduction, 97%; BOD_5 reduction, 97%. The retentate is evaporated to concentrate the stream further prior to disposal.

TABLE 21.

Parameter	Toten	Reed
Installation year	1976	1978
Membrane area (m²)	192	448
Congifuration	4-stage	4-stage
Product	NH_4-SSL	CA-SSL
pH	2-2.5	3-3.5
TOS feed (%)	6-10	10-12
TOS concentrate (%)	12	18
Power consumption (kWh m^{-3} permeate)	7.5	8.8

It is reported that replacement of the conventional chemical pulping (sulphite and sulphate process) by a combined chemical and thermo-mechanical attack (Chemo-Thermo-Mechanical Pulping, CTMP) is taking place because of the increased yields of this process. However CTMP effluents contain higher amounts of organic and inorganic substances, in particular resinic acids and represent a high pollution risk. The RO of CTMP effluents is currently being carried out in tubular modules and composite membranes. The dilute liquors with a solids content of 1-5 % can be concentrated, the average permeate flux being 45 dm³ m-²h^{-1} at 60°C.

Reverse Osmosis of Paper Machine Effluents

Paper production requires large amounts of water of the order of 40 t/t paper. It is essential for internal water re-use in paper production to have efficient internal purification systems and buffer tanks to cope with fluctuations in water demand. Reverse osmosis is used in connection with internal water purification of paper-machine effluents. There is a reported

TABLE 22 – Performance of RO of paper machine effluents.

Concentration	Feed (ppm)	Concentrate (ppm)	Permeate (ppm)	Rejection (%)
Total solids	54 400	66 900	372	99 86
Sodium	6 500	8 000	62	99 86
BOD_5	13 900	17 600	1005	98.56
Colour*	112 500	146 800	65	99.94
pH	5.8	5.8	4.1	99.94
Mean permeate flux (41.1 bar, 37°C)				8.7 lm^{-2}h^{-1}
Power consumption per m³ of permeate				21.7 kWh
Feed osmatic pressure 12.5 bar				

* Analysis based on *NCASI Stream Improvement Technical Bulletin*, No. 253, 1971. Reproduced with permission.

use of RO in a NSSC (Neutral Sulphite Semi Chemical) cellulose and paper mill (Green Bay Packaging, Wisconsin, USA). It was originally equipped with 288 tubular modules (membrane surface area 450 m^2) and cellulose acetate membranes and designed for a feed flow rate of 109 m^3 d^{-1}.

The feed to the RO unit is mechanically clarified (filtered) water from the paper machine cooled from 60°C to 38°C because of the cellulose acetate membranes limited temperature stability. The retentate of the reverse osmosis is recycled to the paper machine. The permeate is discharged into the drainage system, and is not recycled, because it has a high acetatic acid content (pH 4.5 in the permeate) which would cause corrosion.

The RO membranes proved to be highly selective (see Table 22). In this installation, internal water recycling was improved and the mean BOD_5 loading of the plant effluents was reduced and peaks in BOD_5 loading were eliminated.

Reverse Osmosis of Paper and Pulp Bleach Effluents

There is potential for the use of reverse osmosis in the treatment of bleach effluent from pulp and paper mills. This effluent must undergo several pretreatment stages prior to RO (see Fig 15), lime addition to remove magnesium and organic material, followed by cross flow microfiltration to remove the precipitate. Sodium metabisulphite is used to remove chlorine from the effluent, sodium carbonate is added to remove excess calcium as calcium carbonate which is removed by crossflow microfiltration. This last step is necessary to prevent fouling of the membrane (by calcium oxalate). It is reported that at 75% water recovery permeate of suitable quality for recycle within the mill is produced.

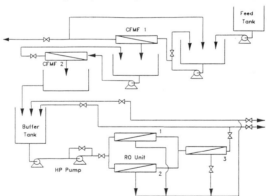

FIGURE 15 – Reverse osmosis and cross flow microfiltration in bleach effluent treatment.

Textile Industry

There are several potential applications of UF and RO in the textile industry including the treatment of size and latex contaminated effluents, wool wash waters and effluents from dyeing operations. The membrane processes are alternatives to classical mechanical – biological and physical-chemical processes such as precipitation, flocculation, flotation and adsorption which offer limited potential for recycling of the components in the waste water. A major use of membranes is in the ultrafiltration of sizing agents, used to coat yarns during weaving.

Size Contaminated Effluents

Size-coated yarns exhibit a greater wear and tear resistance during weaving and therefore the number of fibres broken is significantly reduced. The size materials are natural starch derivatives, semi-synthetic products such as carboxymethyl-cellulose (CMC) and synthetic polymers such as poly (vinyl alcohol) (PVA) and polyacrylate (PAC). After weaving the size must be washed out before any further processing, eg bleaching and dyeing, otherwise the fabric may have a striped appearance. Large volumes of wash water are generated in this operation. Water-soluble size products such as PVA remain chemically unchanged in the wash water. Starch sizes, which are insoluble in water, must be specially decomposed and are therefore not re-usable. Processes capable of size recovery from the various effluents include squeezing, washing and ultrafiltration.

Squeezing was developed for PAC size, which is highly soluble in water. The fabric coating is partially dissolved before washing and removed by squeezing; and 70-75% of the size can be recovered in this way. Washing is used for the recovery of PVA size and about 45 – 55% of the PVA-size can be recovered. Ultrafiltration is used for water-soluble and low-viscosity size and higher recoveries can be achieved although with higher investment costs.

The effluents to be treated typically contain about 1% PVA, and are at a temperature of about 80°C. At this temperature the effluent viscosity is low enough to permit its concentration, by ultrafiltration, to approximately 8% PVA. The concentrate is reused directly within the sizing process, while the hot permeate is recycled to the washing process. Even with the high temperature of the sizing effluent, membrane flux is very low, starting 0.5 m d^{-1} and declining to less than 0.1 m d^{-1} during concentration. Rejection coefficients are greater than 97%. Despite the low flux, the processing of sizing effluents with spiral wound modules is economically attractive. Ultrafiltration operating costs are reported to be no greater than for other effluent applications.

A schematic of the size recycling process is shown in Fig 16. A schematic of the operational characteristics of the UF size recovery process is shown in Fig 17.

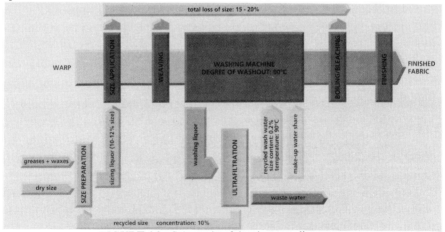

FIGURE 16 – Schematic of the size recycling process.

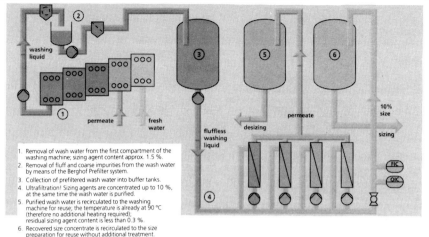

1. Removal of wash water from the first compartment of the washing machine; sizing agent content approx. 1.5 %.
2. Removal of fluff and coarse impurities from the wash water by means of the Berghof Prefilter system.
3. Collection of prefiltered wash water into buffer tanks.
4. Ultrafiltration! Sizing agents are concentrated up to 10 %, at the same time the wash water is purified.
5. Purified wash water is recirculated to the washing machine for reuse; the temperature is already at 90 °C (therefore no additional heating required); residual sizing agent content is less than 0.3 %.
6. Recovered size concentrate is recirculated to the size preparation for reuse without additional treatment.

FIGURE 17 – Structure and operational characteristics of UF operation.

Latex-Contaminated Effluents

Ultrafiltration can be used to treat latex (plastic or rubber) contaminated effluents.
Latex is used in the textile industry as a final coating on fabrics eg carpet backings. Natural latex consists of rubber and proteins while synthetic latex is produced from polystyrene butadiene, poly (vinyl chloride) etc. Latex solutions are essentially emulsions with a latex droplet diameter of 0.5-10 mm.

The traditional method of effluent treatment consists of latex precipitation by metal salts followed by sludge disposal. Ultrafiltration requires suitable membranes with pH resistance (the stability criterion for lactices is pH 10-11), solvent resistances (ketones) and temperature stability (45-50°C). Further a good emulsion stability is essential for the ultrafiltration of latex solutions otherwise latex deposits on the membrane, reducing the flux. Stability can be improved by the addition of stabilisers (eg Tergitol 7) provided they do not interfere with the recovered latex emulsions' re-use.

Typical ultrafiltration size recovery system.

Any destabilised latex which coagulates on the membrane surface can only be removed by a suitable solvent (eg isobutyl-methyl-ketone for PVC and styrene-butadiene, low molecular-weight alcohols for PVA) and not by water flushing or alkali cleaning solutions. UF is practised as a batch operation for concentration latex-contamined waste water. The four module sets, arranged in parallel, can be operated independently, which readily accommodates maintenance and fluctuations in feed flow. Internal circulation of feed ensures high flow-rates in the modules and accordingly high permeate flux.

Concentration of latex (styrene-butadiene) of a factor of 20 are reported using tubular membranes (Abcor) with average membrane fluxrates of 0.65 md^{-1}. The removal of COD was 95% from the permeate thus enabling discharge. Ultrafiltration of PVC latex effluents is reported to give concentrations of over 30% with fluxes in the range of 1-2 md^{-1}.

Water Recovery from Dyehouse Effluent

The textile industry is faced with ever-increasing requirements in the treatment of various waste waters including dyehouse effluent, wool and yarn scouring effluents, as well as effluents containing mothproofers and other pesticides and also rinsing water. There are great potential benefits and much experience in the treatment of bulk dyehouse effluents, and other textile waste waters, using membrane processes. Applications include:

- Bulk effluent treatment and decolourisation
- Individual dye bath liquor
- Scouring liquors (wool, yarn, cotton)
- Treated water recovery for re-use
- Mothproofer/pesticide removal

Up to 80% of warm dyehouse waste water can be recovered for re-use by reverse osmosis. A typical system is shown in Fig 18 and comprises pre-screening and the reverse osmosis. Reverse osmosis splits the waste water into a clean water stream (80% of the total volume) and a concentrated reject stream (20% of the total volume), which is discharged to sewer. The membrane is impervious to the large dye molecules, but permeable water. This reject may be decolourised before discharge as required using adsorption or reduction.

The membrane system has a low space requirement and is usually skid-mounted for ease of installation, and has a low labour requirement. Cleaning is a simple in-place process, effected through the recirculation of a dilute caustic solution. A membrane lifetime of 2 years is typical

Nanofiltration and ultrafiltration are similar processes, but these use more open membranes and a lower operating pressure than RO. They are used where the required filtrate (permeate) quality is moderate, or where desalting of the solution is required. The system operating costs consist of membranes and replacement filters, cleaning chemicals, and electricity. The cost savings comprise water (80%), waste water volume (80%) and steam (the recovered water is typically at 35°C and potential tax benefits. The system economics are illustrated in Table 23.

Overall the advantages of the membrane systems in this application include:-
- Water recovery generates return on investment

INDUSTRIAL WASTE WATER AND EFFLUENT TREATMENT

- Total colour removal – produces high quality water, suitable for re-use in the dyeing process
- Effective removal of most dissolved and suspended materials including COD, BOD, pesticides, all types of dyestuff/colourant, salt
- Low space requirement
- Modular design allows easy expansion
- Energy efficient
- Membrane process tailored to suit the waste water problem

TABLE 23 – Economics of water recovery from dyehouse effluent.

Cost per m³ dye effluent (based on leasing)	
Lease charge (installed and commissioned plant)	13.5 p/m³
Electricity charge @ 5 p'k Wh	13 p/m³
Cleaning chemicals	1.5 p/m³
Membranes and other consumables	14 p/m³
Total cost	**42 p/m³**
Savings per m³ dye effluent (based on water charge 60 p/m³)	
Water (80% of effluent volume)	48 p/m³
Waste water	19 p/m³
Steam @ £7.70/tonne	23 p/m³
Total savings	**90 p/m³**
Net Savings	**48 p/m³**

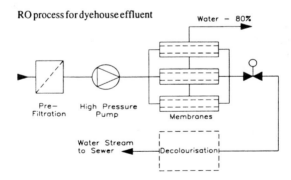

FIGURE 18 – Reverse osmosis process for dtehouse effluents.

Wool Scouring Process Waste Water

Wool scouring (washing process for raw wool) uses large quantities of water, detergents and bleaching chemicals and consequently produces a highly polluting effluent. Ultrafiltration using a polysulphone membrane with a 25,000 nominal molecular weight cut-off is a very effective for concentrating the more significant pollutants into a small volume (typically a 7 fold reduction). It also produces large quantities of permeate suitable for re-use in the scouring process with an equivalent reduction in raw water intake. A typical flow scheme is shown in figure 19 below.

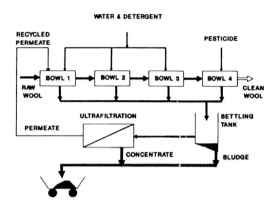

FIGURE 19 – Wool scour process.

Ultrafiltration

Applications in The Nuclear Industry

Ultrafiltration is reported to have been used to a limited extent for the treatment of two types radioactive wastes. One is for the removal of particulate radioactivity from effluents from nuclear power station laundries. The second is for the recovery of actinides, including plutonium, from effluents from reprocessing plants. Plutonium, and other actinides, form insoluble oxides and hydroxides under alkaline conditions, and the insoluble hydroxide polymers and large aggregates can be recovered by ultrafiltration. Fluxes of several meters per day have been reported during effluent concentration by factors of up to 1000 with efficiencies of up to 99.999%.

Reverse osmosis plants have been installed at a number of nuclear power stations to concentrate effluents from the stations' laundries. Such laundry effluents are generally very slightly contaminated with radio-activity, and because of their high surfactant concentration are difficult to treat by evaporative processes. Reverse osmosis concentrates the activity several hundred fold, into a small volume stream that can be disposed of, often after immobilisation. Removal efficiencies of over 98% in activity permit the decontaminated permeate stream to be discharged to the environment. Treatment plants are generally small (< 50 m^2), and utilise tubular reverse osmosis modules because the effluent is dirty.

Electro-dip-coat painting (UF)

Electro-dip-coating is mainly used for the priming of car bodies and for painting household appliances. The principle of dip coating is that in an electric field ionised paint and pigment particles are deposited onto the metal surface. The early preferred mode of operation was anodic precipitation of the paint. In actual operation of a painting line, the part to be painted, the anode, is immersed in the batch following preliminary degreasing phosphating, washing and drying. The excess paint particles are then rinsed off and the part is finally dried prior to further treatment.

Prior to the introduction of ultrafiltration almost universally in this operation, the spent wash waters were treated by precipitation and sludge removal.

A typical composition of an anodic electro-paint bath consists of a suspension of small charged particles of resins, pigments, surfactants and other ionic species.

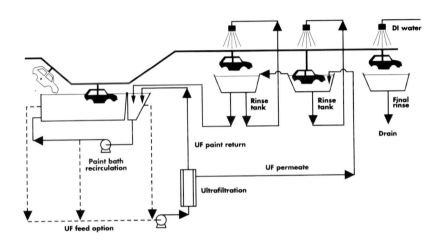

FIGURE 21 A – A typical four zone rinsing electro-painting process.

A typical electro-painting process with four-zone rinsing incorporating ultrafiltration is shown in Fig 21 A. The part which is painted is first rinsed over the painting bath. The water used in this rinse goes directly into the paint bath and to prevent build-up, is continuously separated by ultrafiltration. The water permeate produced by UF is then used in the third rinsing step or returned to the bath, depending upon the UF flux. The concentrate, which has typically been enhanced by 1% in solids concentration, is returned to the paint bath.

The application of ultrafiltration reduces the water consumption and also paint losses which would normally have occurred. In addition the service life of the paint bath is improved as lower molecular weight impurities which are entrained into the baths with the part and which eventually would reduce the quality of the coating, are removed in the permeate.

The economics of ultrafiltration have been well proven since application was introduced in the early 1970's. Several hundred plants of a capacity up to 4.5 m³h⁻¹ with membrane areas up to 200 – 300 m² are in operation. The majority of plants utilise tubular membranes, but hollow fibre, spiral wound and flat sheet plants are also in operation. The early paint processes were anodic, ie the metal object formed the anode and the electrophoretic paint particles were negatively charged. In certain applications dissolution of the metal article can occur which can adversely effect the corrosion protection offered by the paint film. The more recent plants are cathodic the article to be painted is passive, and superior corrosion protection by the paint film is obtained.

For anodic baths ultrafiltration membranes with a small negative surface charge, and therefore capable of repelling anionic paint particles, were developed. These membranes give fluxes of over 1.2 md⁻¹ which can be maintained for several months without chemical cleaning. In cathodic paint processes the negatively charged membrane foul rapidly and thus special positively charged ultrafilitration membranes are used. Such membranes can maintain fluxes of 1 md⁻¹ for many months and membrane lifetimes of over two years have been reported. In some applications regular backwashing is required to remove foulants produced by reaction between phosphate ions in the bath and paint. Membrane cleaning is usually with solutions of organic solvents e.g. butylcellosolve.

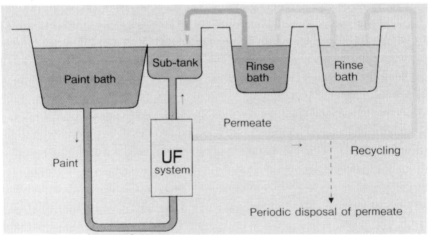

FIGURE 22 – Simplified diagram of a typical electrocoat paint process.

In a typical cationic electrocoat process shown in Fig 22, following degreasing and zinc phosphate treatment, components (eg car bodies) are immersed in a bath filled with a suspension of small charged particles of paint resin in water. The bath also contains pigments and organic solvents. The object forms one electrode of an electrical circuit, so that the charged paint particles are attracted to it and deposit uniformly over its surface. After formation of the paint film the object is removed and rinsed of any non-deposited paint that is dragged out with it.

The UF system separates a small volume of the water-based solution form the paint resin to provide clear permeate for rinsing the electrocoated metal parts. The retentate stream, with the paint solids slightly increased, is returned to the bath. The rinse water together

INDUSTRIAL WASTE WATER AND EFFLUENT TREATMENT

with the recovered drag-out paint is also returned to the bath thereby minimising paint loss and waste disposal.

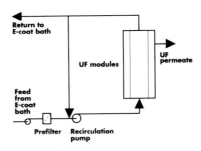

FIGURE 23 – Feed and bleed UF configuration.

In a feed and bleed (or closed loop) system configuration, the recirculation pump provides the required tangential velocity across the membrane filtering surface. The feed pump feeds and pressurises the membrane system with feed flow typically 20 times greater than the system permeate rate. (In the system the feed flow is typically one third as large as the recirculation flow rate). The feed flow that is in excess of the permeate (or filtrate) is returned to the paint bath. The high feed to permeate ratio prevents excessive concentration of paint solids within the closed membrane loop which could adversely effect the filtration rate.

Advantages of this system configuration (which is usually deployed on larger systems) are the reduction in size of the required prefilter and the interconnecting feed and return pipework. There is also a reduction in overall system electrical pumping power consumption.

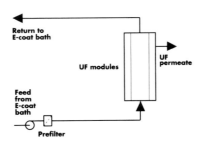

FIGURE 24 – Once through UF configuration.

In a once-through (or open loop) system configuration, the full required recirulation flow is fed to the system via the feed/recirculation pump. Here, the prefilter and the interconnecting feed and return pipework must be sized to handle the full UF system circulation flow. This is a somewhat simpler arrangement which is typically used for smaller UF systems. This arrangement may also be conveniently operated directly from the main paint bath recirculation loop.

UF system for cationic paint and operating parameters.

Oil Contaminated Waters

There are numerous processes which generate oil contaminated wastewaters. The separation of the oil from the water largely depends upon whether or not the oil is in an emulsified state or not. For "free oil" contaminated waters there are a range of separators and treatment methods which can be used to match the kind of oil, the size and proportion of oil contaminant and the presence of other contaminants. A short term solution is essentially tankering and removal by a waste treatment manufacture. Standard plants for waste oil treatment consists of rotary screening (3 mm mesh) to remove course solids, then heating to around 75°C prior to de-sludging in a peeler sedimentation centrifuge and then de-watering in a self cleaning centrifugal concentration (see Fig 25).

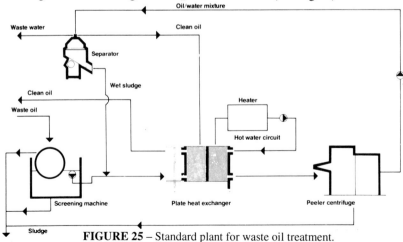

FIGURE 25 – Standard plant for waste oil treatment.

Incineration is also a possible option for disposal of oil contaminated waters but is likely to require special incinerating equipment. Other methods of treatment of oily waste water include (see Filters and Filtration Handbook for further details)
- Gravity settling. Performance depends on the specific gravity of the oil and its size.
- Coalescer/separators. This device utilises a coalescing medium to enable oil globules to combine and agglomerate. The increase in size increases the rate of gravity settling. This rate increases approximately with the square of the average diameter of the oil droplets and thus coalescence of droplets is extremely advantageous.
- Chemical treatment to provide coagulation and/or flocculation followed by sedimentation of flotation.
- Air flotation, acid cracking and thermal splitting
- Filtration.

Emulsified oils such as used in machine tool coolant present a much greater problem in disposal.

Oil Water Emulsions

The production of metal components by cutting, drilling, grinding etc generally requires the use of lubricants/coolants to both lubricate the surface between the work piece and tool and to remove heat generated in the process. The most common choice of fluid is an oil in water emulsion. Such fluids also serve to wash away fine swarf from the work area and to act as corrosion inhibitors.

The water based coolants are generally soluble oils (content 3% to 6%) in water with additives such as emulsifiers, coupling agents (to assist dispersion) corrosion inhibitors, anti-foaming and wetting agents and bactericides. The emulsion droplets in good emulsions are around 1 mm in diameter and positively charged to keep them in a state of suspension. The water used should ideally be softened, to prevent foreign ions neutralising the surface charge and thus causing coalescence, and to reduce the extent of scum formation during metal working. In most tooling operations the "coolants" become contaminated by dirt, metal fines etc, and become degraded by bacteria or by an increase in dissolved minerals due to evaporation. Coolants are generally treated in a recirculation system to remove these contaminants which would otherwise lead to a loss of precision in the work and reduce tool and machine life. The treatment generally involves one or more types of filtration – screens, conventional filters, magnetic filters, hydrocyclones, centrifugal separators. Even with the most efficienct cleaning system the emulsions will reach the end of the useful life and then require disposal. The usual limits in most industrial countries for oil content of liquid wastes discharged to sewage systems is between 10 to 50 ppm which means emulsion have to be split to separate the oil prior to disposal of the water. The recovered oil may then be used as a fuel or as a hydraulic fluid.

The splitting of stable emulsions can only be achieved by chemical or thermal method (see Fig 26). The chemical method involves flocculation which requires the use of sludge handling equipment. The thermal method involves phase formation and is based on the principle of reducing or eliminating the electric charge separating the droplets and thus encouraging agglemenration into larger units. This process requires the addition of a strong electrolyte, eg 2-3% of a saturated magnesium chloride solution, and mixing and heating to 98°C. At this temperature surface tension becomes very low and coalescence occurs eventually splitting the emulsion into an oil and a water phase. The process produces a small amount of sludge, (approximately 0.5%) of the total, formed emulsifying agents, cleaners and other chemicals present in the emulsion.

The treatment with a solution of a salt will not split purely synthetic emulsions, eg emulsions containing aqueous solution of esters, alcohols etc.

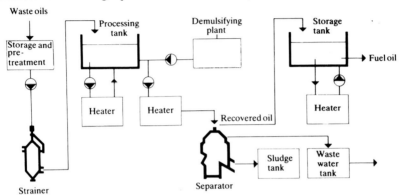

FIGURE 26 – Example of plant for treating miscellaneous waste oils.

Ultrafiltration of Emulsions

Ultrafiltration cannot be used to split emulsions but can be used to concentrate the emulsions up to oil contents of 30% to 60%. After use the emulsions, which are frequently diluted with rinse water used to clean the metal surface after machining contain about 0.2-1.0% of oil, in the form of small droplets, 0.1-1 μm in diameter.

The individual molecules of most cutting oils are small enough to permeate through the pores of an Ultrafiltration membrane. If the oil is present as a separate phase it can be ultrafiltrered as the surface tension at the oil/water interface is sufficiently large to prevent the droplets penetrating a membrane previously wetted by water. Only oil present in solution can pass into the permeate, and permeate oil levels of less than 10 ppm are frequently achieved.

Oil/water emulsions UF membrane system.

INDUSTRIAL WASTE WATER AND EFFLUENT TREATMENT

The typical capacity of UF units for concentration of oil emulsions is between 0.1 to 6 m^3d^{-1} and module types are either tubular, or capillary modules, and plate and frame systems. Membrane manufactures offer a series of fully equipped units for treating oil/water emulsion with capacities of 0.020 – 2.5 m^3/h (check).

Overall there are a range of oil containing wastewaters which are processed by UF and include wastes from steel rolling mills, wastes from metal degreasing and wool and yarn scouring effluents from the textile industry.

Oil Polluted Industrial Effluents

Since the demise of CFCs many areas of industry have been forced to seek alternatives in surface treatment. The degreasing stage of metal finishing is now handled by aqueous cleaning baths, a role previously fulfilled by CFC-based processes. However the grease-laden bath waste requires frequent disposal, as it can become contaminated in relatively short timescales, and the quality of the finished surfaces can also deteriorate quite rapidly.

In the tooling of any metal component, a washing step generally follows, resulting in a spent water wash solution containing oil, suspended matter and detergents. Further treatment of this wash water is required before it can be discharged. UF is used to fractionate this water into a more concentrated solution containing almost all the oil and an almost oil free water permeate. The permeate will have a high detergent concentration and thus a high biological oxygen demand. Thus the question of recycling the permeate arises because of two reasons:
1. to reduce or eliminate the costs associated with effluent treatment prior to discharge
2. recycling would give reduced usage and thus costs of process water and detergent.

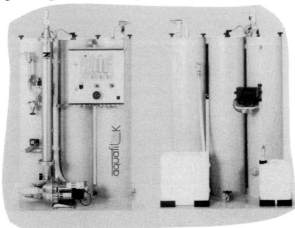

UF operation.

The attraction of recycling generally means that is desirable to separate the oil with minimum separation of the detergent ie to maximise detergent in the permeate. This is best achieved, not by one single membrane separation, either UF or MF, but by a combination of both of these cross flow processes.

The process shown is for a plant effluent containing 2% of detergents, 1% of oil and

0.8% of suspended solids (metal, dirt etc) using tubular modules equipped with a UF membrane and a microfiltration (MF) Membrane (see Table 25). The MF membrane gives the highest flux for detergents but because of the relatively low retention capacity for oil the use of MF membranes is limited to a maximum oil concentration of 10% in the concentrate. The oil concentration in the permeate from MF is about 50 mg dm^{-3} for the UF membranes, where the oil concentration in the permeate is independent of that in the concentrate up to the phase inversion point, a concentration of about 41% in the retentate is achieved.

In both Ultrafiltration and microfiltration the permeate flux is controlled by a gel layer

TABLE 25 – Micro- and ultrafiltration membrane data.

Parameter	Ultrafiltration Membrane	Microfiltration Membrane
Cut-off size	100,000 dalton	300 nm
Membrane material	Hydrophilic polyolefin	Polypropylene
Membrane structure	Asymmetric	Symmetric
Mean pure diameter	20.5 nm	200 nm
Module size (approx)	3000 x 40 mm	1500 x 80 mm
Membrane surface	0.39 m^2	1.0 m^2
Tubular membrane module	4	43
Tubular membrane diameter	11.5 mm	5.5 mm
Dead volume	1.21	1.41
Maximum pressure (feed)	4.2 bar	2.5 bar
Maximum pressure (permeate)	0.2 bar	2.5 bar
Maximum temperature	40°C	40°C
pH	1 - 13	1 - 14
Water flux at bar, 20°C	45 dm3m^{-2} h^{-1}	250 dm3m^{-2} h^{-1}
Velocity ms^{-1}	3.0	3.5

INDUSTRIAL WASTE WATER AND EFFLUENT TREATMENT

and not by the membrane itself. The gel layer can be removed by cleaning, giving a higher average flux. Microfiltration has an advantage in this respect, compared with UF, because microfiltration membranes are available as tubes that withstand internal and external pressures of several bars and which can be easily cleaned by reversing the flux for a short time (5 s every 3 min).

The process gives a recovery of detergent of 85% and a recovery of water of 97.5%. The process includes an evaporation stage to raise the oil concentration up to a value suitable for refining or incineration.

Bath cleaning by carbon microfiltration membranes

An application for crossflow microfiltration is in the purification of degreasing baths, which use hot alkaline solutions to clean oily hydrocarbons from the metal surfaces. Under these conditions a suitable membrane material which can withstand the harsh environment is carbon.

The carbon membranes allow the filtration of extremely corrosive mediums at temperatures up to 165°C. The membranes are formed by a thin, porous layer of carbon, approximately 10 microns thick, applied to the internal surface of narrow-diameter carbon and carbon-fibre support tubes. The inherent strength of the carbon fibre matrix enables the tubes to withstand pressures higher than 50 bars. Membrane cut-off thresholds of 0.1, 0.2 and 0.8 microns are currently available and the system can cater for a range of pH values form 0-14. The modules' corrosion performance, combined with their temperature and pressure handling abilities, also permits regular chemical washing or steam cleaning. This should allow their performance to be maintained, without replacement, over several years. The cross flow design, assisted by the low service requirements, also indicated the method's suitability for continuous on-line filtration.

In one example, a bath filled with a 5% solution of a pH 9 proprietary cleaner, called Paroclean 3812, remained effective for over 5 weeks. Previously, it had been changed daily. The bath's efficiency remained constant and the permeate exhibited less than 1 ppm of oil. Unlike some other membrane materials, the carbon surfaces do not need to be kept constantly wet and the process can be stopped and started as required. The flux across the membrane was found to stabilise at around 300 L/h/m^2, holding this value until the membrane was cleaned at the end of the 5 week period.

A typical installation for a degreasing bath is illustrated by Figure 1. Part of the bath fluid is drawn off via a centrifugal pump and taken to a pre-treatment tank, where any free oil and large solids can be removed. The fluid is then passed to a concentration tank, from whence it is continuously re-circulated though the microfiltration module. Periodically, the concentration tank is emptied to remove the waste oil, with the volume of disposable waste being reduced by factors of up to several hundred, depending on the bath's constituents. This type of installation can also be used on a batch basis, in order to regenerate several baths sequentially.

The carbon/carbon fibre composite membranes achieve high flux levels, which, having stabilised, remain steady for long periods. Additionally, the membranes appear to have excellent selectivity, whilst at the same time suffering virtually no fouling. The membrane's characteristics encourage the bath's additives – wetting agents, detergents, antifoaming agents – to pass through the membrane and be returned to the degreasing bath. This aspect of the process helps the bath to maintain its long-term performance, in addition to showing down the rate of oil contamination quite drastically.

Examples of the use of carbon membranes ainclude:

A Belgian gearbox manufacture, based in Bruges, installed a crossflow microfiltration unit to purify a 1500 litre immersions degreasing bath. The useful life of the bath was extended by ten times and the weekly waste output was also diminished by a large factor.

A marine engine company, in the Netherlands, installed a unit for purifying a 500 litre spray bath. The company previously had problem's with particular dirty bath waste, which required frequent disposal. A total of 18 degreasing plants in the Netherlands have now installed a carbon crossflow microfiltration system.

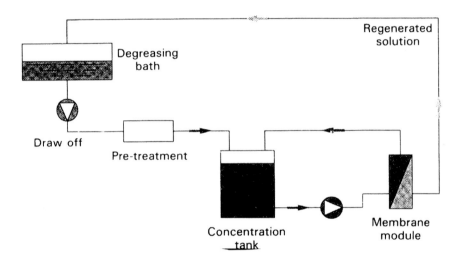

FIGURE 28 – Flow diagram for degreasing bath treatment.

Can Washing

Ultrafiltration is reported in use in a can washing operation to remove oil and grease from the wash waters (see Fig 28). The particular wastewater contains between 2000 – 300 ppm of oil and grease and other impurities and with tubular PVDF, UF membranes is converted into a concentrate of 5 to 8% oil and grease and a permeate with less than 50 ppm grease. The concentrate grease is recovered.

Refinery Fuels

The contamination of refinery fuels by water is a major problem as the water in the fuel can corrode engine parts and cause blockages and also contributes to tank bottom corrosion and bacterial growth. Water concentrations as low as 100 ppm are sufficient to make the fuel off-specification. The water may be dissolved or suspended as droplets in the size range 0.1 to 10 mm. The free water is an emulsion and the use of surfactant materials, eg detergents and additives which lower the interfacial tension make separation of the water droplets more difficult. Other factors which affect separation are viscosity, relative density and temperature. Higher temperatures decrease interfacial tension and water droplet size decreases. Lower temperatures encourage more dissolved water to come out of solution in a suspended state. Separation of water should ideally be done at the coldest point of operation for the fluid. In many cases water droplet removal can be achieved using conventional glass fibre coalescers providing the interfacial tension is greater than 20 dyne cm^{-1}. The function of the glass fibre is to bond the water molecules with the silenol functional group (Si-O-Si) of the fibre. Incoming water molecules then coalesce with the water molecule on the fibre eventually becoming heavy enough to drain and thus exposing fresh fibre surface for the process to repeat. The presence of surfactants disarms the fibre coalescer due to preferential bonding of surfactant with the silenol group. Water removal efficiency is greatly reduced and coalescrer cartridge life is shortened. Other water removal technologies such as tank settling, sand filters and salt dryers are either unreliable, inefficient or can cause operational problems such as corrosion.

The use of a stacked coalescer/separator configuration with a polymeric medium has been shown to improve suspended water removal from fuels in comparison to salt dryers. The system consists of a 3 μm particulate filter and a coalescer/separator stack.

Power Station Effluents

In coal-fired power stations water requirements are of several types
- ultrapure water for the boiler-turbine circuit
- low quality water used for slurry transport of ash and for ash conditioning
- intermediate quality water in the cooling circuit

Cooling circuit water quality is determined by requirements to avoid corrosion and fouling in the condensers and cooling towers. The continuous loss of water by evaporation in this circuit means that in operation the water is kept close to the quality limit. This is achieved by blowdown and treatment and is a major source of effluent in the power station. The overall operation of coal-fired power stations aims for a zero discharge of effluent whereby the intake water is balanced by evaporation and water sent to ash dumps to stabilise them. Reverse osmosis is used in a number of locations (Australia, South Africa) to treat the cooling water blow down. The water permeate is recycled to the cooling circuit and the concentrate is evaporated to further recover water and reduce its volume. The use of spiral wound cellulose acetate modules is reported with a pretreatment plant consisting of a lime-soda softener-classifier, dual-media filters, cartridge filters, acid-SHMP closing and chlorination.

A second source of effluent generated from power stations is the spent regeneration solutions from the ion-exchange plant producing the ultrapure boiler water. Reverse osmosis can substantially reduce the amount of regeneration solution effluent when used as a pretreatment for the IX demineralisers.

Mine Drainage

A process using seeded precipitation and recycling by reverse osmosis (SPARRO) is reported for the desalination of mine water. The RO operates in a suspersaturation mode using tubular cellulose acetate membranes. Seed crystals are introduced to act as preferential sites for the crystal growth and thus to prevent scaling of the membrane. Most of the crystallisation occurs in an external crystallise. The use of seeded-slurry RO to remove calcium sulphate from power station cooling tower blow down is also reported.

Hybrid membrane wastewater treatment process

The petrochemical, food and machining industries produce large amounts of wastewater containing oily substances, halogenated organics, organic solvents and other pollutants. In oil and gas production, large quantities of water emerge with the hydrocarbon streams. This "produced" water must be treated before discharge into the ocean or other receiving water bodies. Another increasing concern is the presence of polychlorinated biphenyls (PCB) in wastewater, the build up of groundwater concentrations and the threat of PCB intrusion into the food chain. In all parts of the world the tighter regulation of water treatment systems requires more effective cleaning of these waste streams. For example, in Europe the current North Sea standard for produced water is 40 mg/l, but it is likely that this will be reduced to 30 mg/l in many countries in the near future. In Norway standards as low as 15 mg/l are being considered, and in Saudi Arabia the planned standard is 7 mg/l. Oil and grease limits for US offshore produced water are currently 48 mg/l monthly average and 72 mg/l daily maximum, but the US EPA has proposed much stricter limits of 7 mg/l and 13 mg/l, respectively, for structures within four miles of the shoreline. For both onshore and offshore plant, the requirement is now to remove hydrocarbons and other pollutants from waste streams at source.

One promising technology, now in the early stages of commercialisation, is the "hybrid" dual membrane process being commercialised by a US company which is claimed to have significant advantages for the treatment of wastewaters streams containing oily substances, halogenated organics or organic solvents. The first stage uses cellulosic hollow fibre membrane and the permeate then passes to a separate nanofiltration or RO system. In some applications, the second stage permeate is further treated by activated carbon to remove trace organic elements. For the most difficult waste streams containing appreciable quantities of soluble hydrocarbons a four-stage process is advise comprising pretreatment prior to hollow fibre membrae separation, a further nanofiltration or reverse osmosis membrane stage, and final treatment with activated carbon to remove any remaining low molecular weight dissolved organics. Only the first two steps are required for waste streams which consist of only insoluble components, such as produced waters and PCB-contaminated wastewaters. In pilot and commercial projects, the modules have reduced PCB concentrations in water from as high as 132 ppm to less than 0.5 ppb.

The hollow fibre module uses regenerated pure cellulose fibres with an inside diameter of 400 m and wall thickness of about 30 µm and form a highly hydrophilic membrane. The selection of pure cellulose for the membrane is because it is impervious to practically all solvents and has the additional characteristic that oils and hydrocarbons do not stick to its surface. The use of pure cellulose also means that the feed stream does not require extensive chemical pretreatment, and this is generally limited to pH management (keeping pH between 5 and d9). Biocides can be added, but experience to date is that bacteria do not seem to harm the membrane when under flow conditions. The membrane module consists of thousands of separate hollow fibre gilaments contained in a suitable housing. In operation the feed flow is passed through the lumen (inside) of the hollow fibres and the purified water passes through the fibre wall. The rate of permeate flow is largely a function of transmembrane pressure – commonly about 2.8 bar (40 psi) – similar to ultrafiltration.

One of the major characteristics of the rejection mechanism has been found to be the solubility of the solutes in the solvent – in most cases hydrocarbons in water. In all cases the concentration of the soluble hydrocarbons in the permeate is one or two orders of magnitude below the solubility levels, especially when the feedwater consists of a mixture of soluble and insoluble components. It is known that a solvent extraction type of process can take place – the soluble hydrocarbons are preferentially absorbed by the insoluble hydrocarbons and are rejected by the membrane. Materials that are highly soluble, such as alcohol, pass readily through the membrane. Materials such as PCBs and unrefined hydrocarbons, which are relatively insoluble in water are rejected over 99%. The rejection mechanism does not appear to be a function of concentration gradient effects in the boundary layer or osmotic pressure. Thus, osmotic substances such as sodium chloride or ferrous iron can pass through the membrane along with the water. This is important when treating produced waters, for instance, where the salinities can exceed 100,000 ppm. Since osmotic pressure does not have to be overcome the permeate continues to be forced through the membrane at low pressures. In addition, the solubility of the hydrocarbon is very low at high salinities, which increases the rejection efficiency.

An application of the system is in refined petroleum storage and pipeline facilities, where a certain amount of water accumulates in the storage tanks – rain water by-passes the seal between sidewalls and the floating roof, and water can also enter via the pipeline. This water becomes polluted with hydrocarbon components and requires treatment before discharge into the municipal sewage system or into receiving water bodies. A schematic diagram of the mobile hybrid treatment system is shown in the figure, and average results obtained at three of the Phillips sites are given in Table 2. The permeate was not only below existing standards, but also below the more rigid stands that have been proposed.

The combination of liquid separation technologies in this way results in a highly efficient separation process that is well suited to the treatment of a variety of difficult wastewaters. The on-site treatment of contaminated water reduces the need for transportation and disposal of hazardous waste and provides appreciable cost savings.

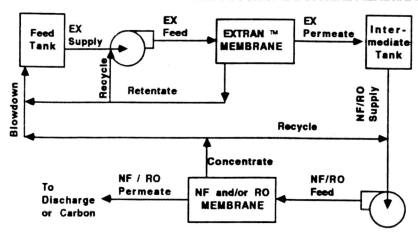

FIGURE 29 – Schematic diagram of hybrid membrane process.

TABLE 26 – Rejection of hydrocarbons in produced water at various field sites using the Extrain membrane only.

Site location	Feed (ppm)	Permeate (ppm)
Orkney Test Centre, Scotland	360 (average)	7.4 (average)
ARAMCO, Saudi Arabia	12 000	3
BANOCO, Bahrain	12 000	3
Hybrid system performance (all values in ppm)						
Parameter	Discharge limits	Site 1 Feed	Permeate	Site 2 Feed	Permeate	Site 3 Feed	Permeate
BTEX	<1.0	22.0	0.03	79	<1.0	~80	<0.01
Benzene	<0.05	8.6	0.005	24	<0.001	112	<0.001
Oil & grease	<15.0	17	<1	148	<1.0	200	1
TOC	<55	1120	19	953	<37.0	986	5
MTBE	—	44	0.02	170	<1.0	—	—

Waste Water Treatment by Electrodialysis

Electrodialysis has many potential applications in the treatment of industrial wastewaters and effluents which contain ionic constituents. The prime function of ED is the separation and thus concentration of a salt or ionic solution. Typically this may mean the removal of a contaminating ionic species from a product solution in, for example, the removal of salt from a dyestuff solution. The method is being widely applied for purification, recovery and recycling processes.

One application of ED is to the recovery of nitric acid from the waste water of an acrylic fibre manufacturing facility. The technology can also be applied to the recovery of other mineral acids notably sulphuric acid and is described further in section 16..

Applications of ED with other membrane separations

There several proposed operations where ED is used in conjunction with other membrane process, such as ultrafiltration, and which can lead to improvements in the process operation. One such example is in the separation of organic acids from Kraft Black liquors. In this process the Ultrafiltration precedes the ED and separates off the high molecular weight fraction from the black liquor to produce a permeate for electrodialysis. The ED then separates this permeate into a concentrate, containing the inorganic acid component, and a deionised stream, containing sodium lignin and aliphatic organic acids. Following further processing this stream is later subjected to electrodialytic water splitting to recover the organic acids and also to generate caustic for the pulping process.

A process referred to as barrier, anolyte, liquor catholyte process (BALC) combines anion and cation exchange membranes with a neutral membrane for the processing of pulping spent liquors, for the renovating of cooking liquors and pulping chemicals and to recover lignin, carbohydrate and other wood derived chemicals. The process consists of four streams, barrier, anion, liquor and cation, separated by four membranes cation, neutral, cation and anion. The barrier stream supplies the ions to the process. In this example the neutral membrane removes the anions from the feed (along with low molecular weight components, organic and inorganic acids and sugar complexes) and with a suitable design of neutral membrane high molecular weight base free lignosulfonic acid is retained in the feed.

Treatment of pulping spent liquors to recover pulping chemicals, sugars and low molecular weight organic and lignosulphonic acids. The applications include, the use of electrodialysis to desalinate waste water in low waste viscose manufacturing technology. One application is at a plant in Russia with an output of 2500 m^3/day. The purified waste can be used for industrial secondary use and the concentrated brine is also returned to the plant for re-use in the precipitation bath.

ED was used to recover silver from spent photographic bleach-fixer solutions. In this process iron is used to oxidise the silver and remains in the bleach-fixer solution, because it forms a complex with EDTA. The silver permeates through the cation-exchange membranes into the iron-free concentrate stream from whence it is reclaimed by electrodeposition in a separate cell.

SECTION 12

Absorption, Desorption and Extraction of Membranes

The inexpensive solution for membrane and cross-flow microfiltration

Ceramic 1, 7-, 19- duct membranes

- sturdy design
- high flux rates
- effective cleaning

For detailed information, training and advice just contact

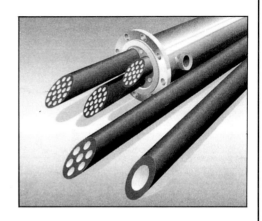

atech innovations gmbh

Teilungsweg 28, 45329 Essen, Germany.
Tel: +49 201/34 10 24 25
Fax: +49 201/34 10 26.

ABSORPTION, DESORPTION AND EXTRACTION WITH MEMBRANES

Methods of phase contact usually involve dispersion of one fluid phase, as droplets, bubbles etc., into another phase. After mass transport the dispersed phases are then separated by a method utilising the difference in their phase density. Equipment designs can have limited interfacial areas and limited versatility for process changes.

Membrane technology which uses modules with, for example, microporous hollow fibre membranes, can establish a consistent, stable interface between the phases. The interface is immobilised at the pores by the hydrophobicity of the membrane and by controlling the differential pressure between the phases. The separation principle differs from other membrane separations such as filtration and gas separation as there is no convective flow through the pores, instead, the membrane acts as an inert support to facilitate diffusive transfer. The two immiscible phases are in direct contact without dispersion, and the mass transfer between the two phases is governed entirely by the equilibrium chemistry of each phase. Because of the hollow fibres pore structure and porosity, very high surface area per unit volume can be achieved.

Membrane high efficiency phase contact technology offers additional advantages over conventional, dispersion technology based on its ability to provide a constant, fixed interfacial area and allow for predictable scale up from single module to multi-module systems.

There are many diverse applications of hollow fibre membrane technology in phase contacting (see Table 1) from waste recovery, food and pharmaceutical industries to analytical and medical applications.

Liquid/Liquid Extraction

Liquid/Liquid Extraction (LLE) is the transfer of a solute from one immiscible fluid to another. Conventional methods involve dispersion of one phase, as droplets, into the second phase to increase the interfacial area of contact. After extraction phases are separated by their differing densities using gravitational or centrifugal forces. Disadvantages of conventional methods include:

Liquid systems having low density differences and/or low interfacial tensions often form stable emulsions.

Liquid systems with high interfacial tensions are difficult to disperse by conventional methods, leading to large extractor volume requirements.

Extractor designs (columns, mixer/settlers, centrifugals) have limited versatility for changes in flow rates or liquid compositions because of loading and flooding problems.

Membrane high efficiency contactors, using microporous hollow fibre membranes, offer a means of contacting aqueous and organic phases without mixing. The contactors immobilise the interface in the pores of the hollow fibre. The aqueous phase will not

TABLE 1 – Phase Contact Application Guide.

Industry	Application	Gas Transfer	Liquid/Liquid Extraction
Pharmaceutical/ Biotechnology	Culture Media Oxygenation	*	
	Antibiotic Extraction		*
	Protein Extraction		*
	Degassing/Deforming	*	
	Ultrapure Water	*	
	Chromatography Buffer Degassing	*	
	pH Control by CO_2 Absorption	*	
	Integrated Membrane Bioreactors	*	*
	Haemoglobin Deoxygenating	*	
Food and Beverage	Process Water Deareation	*	
	Flavour Extraction		*
	Metals Removal from Edible Oils		*
	Carbonation	*	
	Aromas Extraction		*
Analytical/Medical	Process Water Degassing	*	
	Humidification/Dehumidification	*	
	Chromatography Buffer Degassing	*	
Industrial/Chemistry	Metals Extraction		*
	Extractive Reaction Processes		*
	Gas Scrubbing	*	
	Stripping of Volatiles	*	
	Boiler Feedwater Degassing	*	
	Process Water Deoxygenating	*	
	Chemical Liquid/Liquid Extraction		*
Environmental/ Waste Recovery	Removal of Priority Pollutants	*	*
	Stripping of VOCs	*	
	Vent Gas Scrubbing	*	
	Radon Stripping	*	
	Metals Recovery		*
	Waste Stream Aeration	*	

ABSORPTION, DESORPTION AND EXTRACTION WITH MEMBRANES 635

peretrate or "wet" the pore because the membrane is hydrophobic and the organic phase will readily wet the pore and directly contact the acqeous phase. The interface is immbobilised at the pore entrance on the aqueous side of the membrane by applying a higher pressure on the aqueous phase.

The advantage of this technology over existing liquid/liquid extraction processes include:

the ability to vary flow rates independently without flooding or channelling;
the absence of emulsion formation;
no phase separation is required;
higher surface/volume ratios;
direct scale-up due to a modular design.

Contained Liquid Membranes (CLM) and Immobilised Liquid Membranes

In the pharmaceutical and chemical industries, a sequential two-step extraction process is used in a variety of applications to concentrate a solute from one aqueous phase to another

via an organic enrichment step. In the first step, a solute in an aqueous fluid is extracted with a large volume of organic extractant in a conventional contactor. In the second step the solute-loaded extractant is then pumped to a second contactor where a stripping aqueous stream back-extracts the solute from the organic liquid.

The Contained Liquid Membrane (CLM) module combines this two step process into one module that contains two bundles of hollow fibres. The feed stream passes through one bundle of fibres while the stripping steam passes through a second bundle of fibres. Both bundles are contained in a single case which is filled with the organic extractant. The solute passes from the aqueous feed into the organic extractant, which is continuously back-extracted into the aqueous stripping stream. The interface between the organic and aqueous phases is stabilised by maintaining a higher pressure on the aqueous phases.

The total solvent inventory is contained in the module, thus reducing the quantity of solvent used and the scale of equipment.

ABSORPTION, DESORPTION AND EXTRACTION WITH MEMBRANES

Immobilised Liquid Membrane Technology (ILM) can alternatively be used in sequential two step extractions if the organic extractant is extremely insoluble in the aqueous phases. The ILM is a two step process of extraction and recovery using a single bundle of hollow fibre membranes. Typically organic extractant priming phase is pumped through the lumen side of the membranes to impregnate the pores of the membranes. The squeous feed stream passes over one side of the membranes and the aqueous stripping stream passes over the other side without displacing the organic phase in the pores. The solute is extracted from the aqueous feed stream into the organic phase immobilised in the pores and then back-extracted by the aqueous stripping phase.

Immobilised Gas Membranes

Membrane contactors also can be used in coupled gas transfer applications involving the concentration and recovery of volatile contaminants, such as hydrogen sulphide, sulphur dioxide and ammonia, from an aqueous stream. Conventional processes us a two column system which involves volatilising a solute into an air stream and then re-absorbing the solute into an acidic or basic stripping stream. This requires a large air stream, large contact area, large column volume and is expensive to operate.

In contrast, an immobilised gas membrane module combines this two step process into one module using a single bundle of hydrophobic hollow fibre membranes. The aqueous feed stream passes over one side of the membrane but does not wet the pores and the aqueous stripping stream flows on the other side of the membrane trapping small pockets of gas in the membrane pores. Thus these pores act as a stable interface between the fluids. In the first step, the solute in the aqueous feed stream is volatilised into the membrane pores. The solute diffuses across the gas membrane and is then extracted (acid/base) from the pores into the aqueous stripping stream.

This method offers the ability to vary flow rates independently without flooding or channelling; a higher surface area/volume ratio; direct scale-up due to a modular design; one simple system to operate; and lower energy cost.

Membrane Gas Stripping

An example of gas stripping is in de-aeration. Conventional de-aeration technology uses two-stage spray tanks under vacuum to remove oxygen to less than 0.1 ppm. The water must be pumped after each stage to compensate for the pressure drop through the spray nozzles. CO_2 must be added to facilitate removal of air in the final stage and is subsequently removed with air in the second stage, adding significant operating cost.

In gas stripping applications, a vacuum or stripping gas is applied against the aqueous stream to remove dissolved or entrained gas from the liquid. The membrane interface provides larger surface per volume ratios than can be achieved using conventional technology. Membrane high efficiency contactors provide a single stage operation with minimal pressure loss through the system. Low oxygen levels can be achieved without the addition of other gases (such as CO_2) to facilitate air removal.

Equipment

Commercial phase contact units come in a range of module sizes offering effective membrane surface areas from 0.23 m^2 to 19.3 m^2, which is equivalent to greater than 3000 m^2/m^3 contactor volume. Process scale modules are 33 cm or 71 cm in length and utilise polypropylene hollow fibres, 240 μm diameter, 30 μm wall thickness, 30% porosity and 0.05 μm effective pore size. The units can operate at up to 60°C, with maximum transmembrane differential pressure of 4.1 bar and operating pressures of 6.9 bar. The housing material is either polypropylene, polycarbonate or stainless steel.

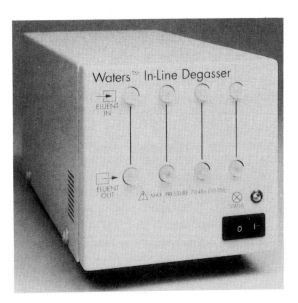

Example Applications

Commercial Water Degassing System

One application of Liqui-Cel Membrane Contactors is degassing of water in building applications. The application covers corrosion and red water inhibition in the fields of feed water treatment, feed hot water treatment, and closed heating/cooling systems in Japan. Because the contactors are extremely efficient, the oxygen level exiting the system is less than 0.5 ppm, and the opportunity for pipe corrosion and red water generation is significantly reduced. Due to the high efficiency the system is smaller in size, approximately one third, than the size of conventional systems.

The Membrane Contactors use microporous hollow fibre polypropylene membranes. Both five and ten ton systems, rated at 5000 and 10,000 l/h respectively, have been installed; each system consists of Liqui-Cel Extra-Flow Membrane Contactors, a vacuum pump, a circulation pump and several meters.

Continuous removal of dissolved gasses for HPLC

The use of membranes in in-line degassing offers an automatic, continuous method of removing dissolved gases from HPLC eluents. The membrane system is designed as an alternative to traditional offline degassing methods such as sonication, vacuum filtration or online degassing which can use relatively large quantities of expensive helium, the membrane degasser provides two independent fluid channels that can be further extended to three or four allowing up to four eluents to be degassed simultaneously.

The basis for eluent degassing is the diffusion of dissolved gases through ultrathin membranes exposed to vacuum pressure. Using the degasser improves solvent delivery reliability for gradient and fluent blending applications, and may also improve detector performance by reducing baseline drift and noise. Superior performance is achieved when up to four eluents are degassed simultaneously at optimal flow rates of 0.2 – 5.0 ml/min.

The unit makes use of a chemically-resistant internal vacuum pump, a vacuum sensor, and tubular membranes housed in vacuum chambers. It serves as a freestanding accessory that can be used with any HPLC pump, including the analytical pumps.

SECTION 13

Waste Water Treatment and Liquid Membranes

WASTE WATER TREATMENT AND LIQUID MEMBRANES

Despite all the research and development activity in recent years on liquid membranes there are very few reported applications of the technology at a commercial level or even at an industrial pilot plant level. There are only two reported applications in wastewater treatment, in China and Austria, the latter is believed to no longer operate. The first of these was for the recovery of phenol from a wastewater, the second was for the recovery of zinc from wastewater. In addition to these there have been two other reported applications of liquid membranes at the pilot scale (0.2 m^3/h) for heavy metal recovery.

Viscose Fibre Industry

The viscose rayon industry uses zinc ions in the spinbaths to improve both the spinning process and the properties of the fibres. Further processing of the fibre requires the removal of the zinc by rinsing with water which generates a waste. This wastewater should ideally be treated to recover the zinc and re-use the water. Much of this wastewater produced is quite dilute and the high volumes generated do not make ELM generally applicable. However certain of the wastewaters contain significant quantities of zinc to economically justify the use of emulsion liquid membranes.

The first reported large-scale application of liquid membranes (approximately 10 years ago) used emulsion liquid membrane for rinse recovery from the waste waters of a rayon production facility. The wastewater contained 400-600 ppm of Zn^{2+} ions, with calcium, sulphuric acid (5-8 gdm^{-3}), sodium sulphate (25 gdm^{-3}) and various modifiers and solids. The alternatives to liquid membranes for treatment of this stream included precipitation with sulphide (H$_2$S, NaHS), or Ca(OH)$_2$, ion-exchange or liquid-liquid extraction. The economics of ELM for selective recovery of Zn were shown to be superior to the other techniques and the operation was proven on a 1 m^3 h^{-1} capacity pilot plant before implementation on full scale, capacity 75 m^3 h^{-1}.

The liquid membrane pertraction was achieved using di-2-ethyl-hexyl dithiophosphoric acid (D2EHDTAPA) – an efficient metal extractant. This extractant is rarely used in liquid extraction due to slow kinetics of stripping, but in ELM the extraction rate is satisfactory due to the high interfacial area. D2EHDTPA is able to extract Zn^{2+} at a pH of approximately 0.5, whereas a similar extractant at D2EHPA cannot be used below a pH of 2. For the stripping phase a minimum concentration of sulphuric acid of 2 mol dm^{-3} is required, 250 gdm^{-3} used in practice.

A typical process plant for recovering zinc is shown schematically in Fig 1. The emulsion in the process is produced by pumping the organic phase and stripping phase at a pressure of 1 MPa through an homogeniser equipped with a special nozzle. After emulsification the zinc solute is extracted from the wastewater in a multistage countercurrent extraction column. The stripped wastewater from this extraction is sent to an oil separator to remove any entrained organic droplets, before discharge as effluent. The emulsion membrane phase (with solute) from this column is then split in an electrostatic coalescer to give a final product phase and the organic phase for recycle. The electrostatic coalescer is based on 2000 V, 10 kHz. AC voltage applied between two insulated (with enamel) electrodes.

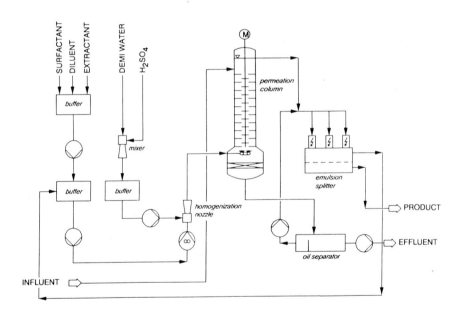

FIGURE 1.

In the plant the flowrate of the membrane phase was 5 $m^3 h^{-1}$ and that of the stripping phase 0.5 $m^3 h^{-1}$. The average process efficiency for treating the 0.5 gdm^{-3} zinc solution was 99.5% and the zinc concentration in the final stripping liquor was 60 gdm^{-3}.

The above application points to the potential of liquid membrane technology in waste treatment and various economic studies of applications indicate both SLM and ELM can be competitive. The cost effectiveness is however reported to be less attractive with more dilute solutions at solute concentrations of below 0.1 gdm^{-3} (100 ppm).

Phenol Recovery

The recovery of phenol from wastewater has been reported in an industrial plant in China. The emulsion used in this process is a 5 wt% NaOH solution in an organic phase consisting of 3.5 wt% surfactant and 6.7 wt% liquid paraffin in kerosene. The emulsion which has an aqueous phase to organic phase ratio of 1:1, is prepared in a mechanical pre-agitator and a colloid mill.

Extraction of the phenol (100 gm^{-3} in concentration) is achieved in a two stage rotary disc column extractor with the final concentration 0.5 gm^{-3} in the waste water and 55 gdm^{-3} in the aqueous stripping phase.

The extraction of phenol is based on a simple diffusion across the liquid membrane. The phenol in the wastewater is in an undissociated form, at a pH below 9, and is soluble in the aqueous phase. The stripping phase contains the sodium hydroxide which reacts with the phenol to produce phenolate (C_6H_5ONa) which is dissociated and insoluble in the organic membrane enabling high concentration factors. The recovered phenolate has little value and is incinerated. Overall the economics of the process are extremely favourable and superior to other processes for phenol recovery.

Oil Field Applications

Well control fluids are used to prevent well blowout and to seal loss zones in oil and gas wells. The concept of emulsion liquid membranes has been applied as a well control fluid. The fluid is a water in oil emulsion containing particles of clay in the oil phase. When the emulsion is subjected to high shear at drill bit nozzles, the oil film separating the water droplets and the clay particles ruptures. This allows direct contact between the clay particles and water, which results in a great swelling of the clay and consequently the formation of a high strength paste.

Other suggested oil field applications of the three phase emulsion are as encapsulation processes:
  i) injection of viscous emulsions into oil and gas wells for hydraulic fracturing
  ii) multiple liquid membrane emulsions for simultaneous injection of reactive components for fracturing
  iii) injection of acid-containing emulsions into wells to dissolve rock, for example carbonates, for fracturing

These application use the emulsion membrane to keep separate the active agent until the point of use.

Applications

The potential applications of liquid membranes cover the separation of hydrocarbons, metals, inorganic substances, organic substances and biotechnology, and medical applications. The appropriate pertraction of these species depends on a suitable mechanism of transport, which broadly is one of two types
  – simple transport based on the use of the membrane as a physical solvent
  – facilitated transfer in which the membrane is a substrate for a selective carrier of the permeate

TABLE 1 – Liquid membrane separations of hydrocarbons.

Solute	Donor (F)	Acceptor (R)	Ref.
Toluene	heptane/toluene	hexane; S-100N	[2,3]
		kerosene, trichlorobenzene	[4,5,79,149, 150]
		kerosene	[159]
		decahydronaphthaline	[1,64]
		o-xylene	[96]
		isooctane+carbon tetrachloride+oil	[86]
		butane;dodecane	[160,415]
	chlorobenzene/toluene	kerosene	[4]
	cyclohexane/toluene		[4,5]
	hexane/toluene		[3,157,161]
	dimethylpentane/toluene		[2,161]
	p-xylene/toluene	hexane	[414]
benzene	benzene	trimethylpentane, carbon tetrachloride	[162]
	cyclohexane/benzene	kerosene, trichlorobenzene	[150]
		dodecane	[54]
	hexane/benzene	chlorobenzene	[122,163]
		hexane	[2,3]
		isooctane, carbon tetrachloride	[94,97,164]
		mineral oil	[155,157]
		isooctane+add.	[151,165]
		mineral oil	[166]
	heptane/benzene	kerosene, pentanol	[167]
	toluene/benzene	o-xylene	[96]
ethylbenzene	actane/ethylbenzene	hexane, nitrogen	[87,154]
	ethylcyclohexane/ethylbenzene	pentane	[154]
	styrene/ethylbenzene		[62,154]
methylnaphthol	dodecane/methylnaphthol	n-heptane	[167]
o-xylene	m-xylene/o-xylene	organic solvent	[154]
o-xylene + p-xylene	o-xylene		[154]
benzene/cyclohexane	hexane/benzene/cyclohexane	hexane	[3]
hexane/hexadiene	hexane/hexene/hexadiene		[3]
aromatic hydrocarbons	paraffins/cycloparaffins/ aromatic hydrocarbons		[3]
benxene/toluene	hexane/benzene/toluene		[3]
benzene/toluene	dimethylbutane/ dimethylpentane/ benzene/toluene	hexane	[3]
hexane	hexane/hexene	S100N	[2]
hexadiene	hexane/hexadiene		[2]
hoptene	hexane/heptene	hexane	[161]
cyclohexane	hexane/cyclohexane		[2,3,157,161]
styrene	ethylbenzene/styrene	pentane	[62]
ethylcyclohexane	actane/ethylcyclohexane		[62]
carbon tetrachloride	hoptane/carbon tetrachloride	kerosene	[4,5]
hexane	hexane/hoptane	n-actane	[102]
methylpentane	actane/methylpentane	S100N	[2]
1-methylnaphthalene	dodecane/1-methyl-naphthalene	—	[413]

TABLE 2 – A selection of carriers and solvents for liquid membranes.

Carriers and solvents	
D2EHPA	di(2-ethyl-hexyl)phosphoric acid
DBBT	1,1-di-*n*-butyl-3-benzoyl-thiourea
DB18C6	dibenzo-18-crown-6(2,3,11,12-dibenzo-1,4,7,10,13,16-hexa-oxacyclooctadeca-2,11-diene)
DC18C6	dicyclo-18-crown-6
DIDPA	diisodecylphosphoric acid
DTPA or D2EHDTPA	bis(2-ethylhexyl)dithiophosphoric acid
EDTA	ethylendiaminetetraacetic acid
TAA	trialkylamine
TBAH	tributyl aceto hydroxamic acid
TBP	tri-*n*-butylphosphate
TIOA	tri-iso-octylamine
TLAHCI	trilaurylammonium chloride
TOA	tri-*n*-octylamine
TOPO	tri-*n*-octylphosphine oxide
Carriers and solvents (trade names)	
ACORGA®	5-nonylsalicylaldoxime + nonylphenol (ICI)
Adogen®464	tri-alkylmethylammonium chloride (Shering)
Alamine®336	tri-octyl amine (Henkel)
Aliquat®336	methyl-tri-octylammonium chloride (Henkel)
Amberlite®LA2	*N*-lauryl-1,1,3,3,5-hexamethyl-hexyl amine (Rohm & Haas)
CYANEX®272	bis(2,4,4-trimethylpentyl) phosphinic acid (Cyanamid)
KELEX®100	7-dodecenyl-8-hydroxyquinoline (Shering)
LIX®34	*n*-8-quinolyl-*p*-dodecylbenzenesulfonamide
LIX®54	phenyl-alkyl-b-diketone (Henkel)
LIX®64N	mixture of two oxime compounds LIX®63 and LIX®65N (Henkel)
LIX®65N	2-hydroxy-5-nonylbenzophen oxime (Henkel)
LIX®70	5,8-diethyl-7-hydroxy-6-dodecanone oxime + 3-chloro-5-nonane-2-hydroxyphenone oxime
S100N®	isoparaffinic neutral oil (Exxon)
SME®529	2-hydroxy-5-nonylacetophenone oxime (Shell)

Hydrocarbons

The separation of a range of hydrocarbons (see Table 1) has been studied in applications for the petrochemical and chemical industries, much of which is for the separation of aromatic hydrocarbons from hydrocarbon mixtures. The system utilises an oil-in water-in oil triple emulsion, in which the aqueous phase is the membrane. Transport is by simple selective extraction and diffusion of one component into the aqueous membrane phase driven by a concentration gradient.

Metals

The broad spectrum of metal extractions which have been studied with liquid membranes includes noble metals, alkaline and earth alkaline metals, rare earth and radioactive metals and heavy and toxic metals (notably copper).

TABLE 3 – Metal ion extraction using liquid membranes.

Na^+, K^+, Pb^{2+}	Cyclic polyethers (crown ethers), carboxylic acids and amino derivatives of crown ethers
Li^+	Non-cyclic polyether derivatives
Mg^{2+}, Ca^{2+}, Ba^{2+}	Crown ether and alkanoid acid
Ag^{2+} (as $AgBr_2^-$)	Macrocyclic ether
Ag^{2+}	D2EHPA CYANEX 471
Au, Ag	Kryptofix 22 DD (marcocyclic complex)
Pd^{2+}	thio-18-crown-6 (sulphur macrocyclic) CYANEX 471
U^{6+}	TOPO, TBP, KELEX 100, D2EHPA
U^{4+}	OPAP (di-octylphenyl phosphoric acid)
Sc, Sm, Er, Gd, La, Y	D2EHPA
La, Nd, Sm, Eu, Gd, Dy and Yb	2-ethyl hexylphosphoric acid mono-2-ethylhexyl ester
Am	D2EHPA
Eu^{3+}	DIDPA
Pu (IV), 2r (IV)	TBAH
Cu	ACORGA, LIX (64N, 65N, 54, 70, 34) L, KE2EX, 5ME 529, D2EHPA
Cr(IV)	Tertiary amines, quaternary ammonium salts, TOA, TAA, Aliquat 336
Hg (as $Hg\ I_4^{2-}$)	Amines, quaternary ammonium salts
Hg^{2+}	Tributyl phosphine sulfide
Cd (as $(Cd\ CN)_4^{2-}$)	DBBT Aliquat 336, D2EHPA
Zn^{2+}	DEPHA, D2EHDTPA
Co^{2+}/Ni^{2+}	D2EHPA, Cyanex 272
Fe^{2+}	TBP
In^{3+}	D2EHPA
Ga^{3+}	ammonium salt of N-(alkly phenyl-N nitrosohydroxylamine)
Tl	DC18C6
W^{6+}	Aliquat 336
Ce^{3+}	TOPO

Separation of metals is usually by facilitied transfer sometimes referred to as carrier mediated. This mechanism does not require the permeate to be soluble in the membrane liquid, as the latter contains an additive (X) which can react selectively and reversibly with the permeating species

$$A + X \Leftrightarrow AX$$

The carrier essentially shuttles the permeate species (A) from the donor feed to the membrane/receptor interface where conditions are such that the equilibrium of reaction shifts to the left and the permeate is released into the reception phase and the carrier returns to the membrane to recombine with permeate on the feed side of the membrane. In this way only relatively small amounts of carrier are required. In some applications a "countercurrent" transport can be utilised in which an ion-selective carrier is used. The carrier complexes with the ionic permeate (A) being transported which, when released into the reception phase, is replaced by an equivalent amount of ions of a similar type stored in the reception phase. A typical example is the counter-transport of protons, in the extraction of metal ions from neutral (or slightly acidic) media with oleophilic chelating agents. The complexation and disassociation is essentially controlled by pH differentials between donor and receptor.

Inorganic Compounds

The pertraction of ammonia has been extensively investigated as a means of treating wastewaters to reduce concentrations of nitrogen. In alkaline solution (pH > 9) ammonia is primarily in the form of undissociated ammonia molecules (NH_3) and is soluble in oil membranes, of iso paraffins (paraffin oil) and thus can be extracted. When the reception phase is acidified (with H_2SO_4, HCl) the ammonia transferred into the oil membrane is transformed to NH_4^+ ions

$$NH_3 + H_2O \Leftrightarrow NH_4OH \Leftrightarrow NH_4^+ + OH^-$$

which are thus removed from the wastewater feed. The process can extract 90% to 99% of ammonia in less than 20 minutes and competes with aerobic biological denitrification. There are no reports of the process being used beyond pilot scale.

Other inorganic species which have been studies include inorganic acids (HNO_3, HCl, H_2SO_4), PO^{3+} ions, Cl^- ions and I_2.

Organic Substances

Many industries (oil refineries, coal cooking, synthetic resin production) produce waste waters which contain toxic phenol which must be removed. Biochemical dephenolisation is effective at low concentrations (< 100 to 200 ppm) whereas liquid extraction can be applied to concentrations ranging from 100 to 10,000 ppm. The polar organic solvents usually used in liquid extraction have a solubility in the dephenolised feed which necessitate further treatment to remove them. By using non-toxic organic solvents (eg normal paraffins with 9 or more carbon atoms) with low water solubilities this problem is overcome. However the distribution coefficients of phenol between the paraffins and water are low (<1) and thus liquid extraction is not generally suitable. Liquid membranes can in these cases offer specific advantages. A range of hydrocarbons (and mixtures) have been used as liquid membranes eg silicon oil, kerosene, naphtha, hydraulic and transformer oils. The transport by diffusion through the membrane is based on the concentration gradient, which is usually maintained by an irreversible chemical reaction of the solute with a strong base, eg sodium hydroxide, resident in the reception phase. The effect of organic membrane phase on the diffusion coefficient of the solute (phenol) is an important factor in the selection of the membrane material.

Overall liquid membranes have a wide potential for removal of toxic organic species from industrial effluents and wastewaters. The species are typically weak organic acids and bases which are relatively soluble in the membrane phase when present as undissociated molecules. The following are typical of species investigated
- o-chlorophenol, 2-, 3- and 4-methylphenol, nitrophenol
- acetic acid, lactic acid, propionic acid, acrylic acid, citric, fumaric acid, l-malic acid
- xylene isomers
- aromatic acids (benzoic, phenylalanine, w-aminosalphonic)
- aromatic and aliphatic amines

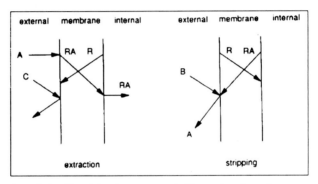

A, solute; B, stripping reagent; C, contaminant; R, extractant.

FIGURE 2 – Principle of separation of membrane microcapsules.

- saccharides (fructose, galactose, glucose); recovery is effected by means of phenylboric acid in a solution of TOMAC in dichloroethane.

Chromium extraction using microcapsules

Mircocapsules consist of particles or droplets immobilised within ultrathin membranes composed of natural or synthetic polymers with varying thickness and degree of permeability. Microcapsules can be used as compact systems for use in separations, due to their large surface area, ultrathin membranes and extremely small volume.

Polyurea microcapsules prepared by interfacial polycondensation are used to separate hexavalent chromium ions from aqueous solution. There are three steps in the process –

preparing the microcapsules in which an extractant for the solute is encapsulated; dispersion of the microcapsules in aqueous solution containing the solute to be separated; and following formation of a complex between the extractant and the solute, dispersion of the microcapsules into another aqueous solution containing a stripping reagent to release the solute.

Various extractants can be used e.g. N235 (R_3N, $R - C_{7-9}$) as extractant gives a greater rate of extraction of chromium (VI) than cobalt (II) and copper (III, trioctylamine (TOA) and tributyl phosphate (TBP). With the latter extractant, the mean diameter of the microcapsules was smaller due to lower interfacial tension at the oil/water interface. The extraction rate could be enhanced by decreasing the pH in the aqueous phase, increasing the extractant concentration in the microcapsules, or reducing the microcapsule diameter to increase the interfacial area and to decrease the mass transfer resistance in the microcapsules.

The membrane formulations were also found to have an impact on the extraction rate. Stripping rate increased with an increase in pH, which was attributed to the increased rate of reaction between the complexes and the hydroxide in the stripping solution.

SECTION 14

Biotechnology and Medical Applications

BIOTECHNOLOGY AND MEDICAL APPLICATIONS

Biotechnology is defined as the technological exploitation or manipulation of biological systems for the production of a useful economic or therapeutic result. Activities cover the biomedical applications of membrane separations including hemodialysis, production of foods and beverages, extraction of components from plants and animal derived materials and the controlled manufacture of biologically derived compounds in a biochemical process. The biochemical process (see Table .1) utilises a bioreactor of some sort followed by appropriate separation units for product(s) recovery and separation (bioseparation).

Biotechnology and Membrane Processes

The scope for membrane processes in the biotechnology industry is large, with many

TABLE 1 – Membrane and non-membrane operations in biotechnology.

Application	Non-membrane operation	Membrane operation
Bioreactor	CSTR suspension	Entrapped – microcapsules – hollow fibres – flat sheets
Particulate Removal	Centrifuge Sedimentation/Flocculation Filtration Decanting	Microfiltration
Macromolecule Separation	Centrifuge, electrophoresis	Ultrafiltration
Solvent Extraction	Rotary (Podbielniak)	Membrane pretraction – coupled transport
Sorption Affinity Desalting	Columns Size exclusion chromatography	Porous membranes Electrodialysis

applications well established. The clarification, concentration of and purification of macromolecular products with membranes is widely used. Newly developed membrane schemes are seen in the separation and purification of macro and micro solutes, the integration into bioreactors and fermentors, and as separating components in analytical sensors.

The general area of membranes in Biotechnology is outlined in this section.

There is a diverse range of products (see Table 2) and processes and correspondingly a wide range of separations which are used for product recovery. The range of bioproducts have been put into three categories:

1. Low molecular-weight products. These are termed "microsolutes", derived principally from conventional fermentations, of which there are more than 200 of commercial significance.
2. Macromolecular products. These species are produced by microbial fermentation and/or mammalian cell culture.
3. Cell biomass. Products range from yeasts to single cell protein.

As a generalised process Fig 1 gives a block diagram of a fermentation process consisting of the following steps

- steam sterilisation of the fermenter vessel and materials charged to it
- feedstock pretreatment (e.g. solubilisation, screening, etc)
- addition of essential nutrients
- adjustment of temperature and pH
- filtration and/or heating of process air to combat contaminating micro-organisms and phase
- inoculum growth from a seed culture
- the fermentation
- product recovery
- product purification

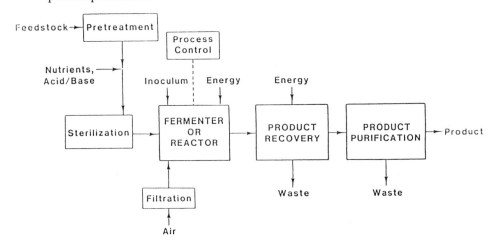

FIGURE 1 – Flowsheeet of bioseparations.

TABLE 2 – The range of products from bioprocessers.

Product	Disease/condition
Lymphokines	Viral infections
Erythropoietin	Anemia, hemodialysis
Recombinant insulin	Diabetics, insulin dependent
Beta cells	Diabetics
Urokinase	Blood clots
Granulocyte stimulating factor	Wounds, severe
Tissue plasminogen activator	Heart attacks, survive to hospital
Transfer factor	Multiple sclerosis
Protein C	Hip surgery, protein C deficiency
Epidermal growth factor	Burns
Factor VIII	Hemophilia
Human growth hormone	Pituitary deficiency
Orthoclone	Kidney transplant rejection
Alpha interferon	Hairy-cell leukemia
Low-molecular-weight fermentation products	
Amino acids	glutamic acid
	lysine
Organic acids	citric acid
	acetic acid
	lactic acid
	gluconic acid
Ketones and alcohols	acetone
	butanol
	ethanol
Vitamins	
Antibiotics	penicillins
	cephalosporins
	cyclosporin
	thienamycin
Other secondary metabolites	
Marcromolecular products from fermentation and cell culture	
Therapeutic proteins	enzymes
	monoclonal antibodies
	interleukins
	interferons
	recombinant plasma proteins
	vaccines
Peptide hormones	recombinant human insulin
	bovine and human growth hormones
Diagnostic enzymes and antibodies	
Polysaccharides and biopolymers	xanthan gums
	dextrans
	polyhydroxybutyrate
Industrial enzymes	proteases
	amylases
	glucose isomerase

Fermenters are designed to vigorously agitate and aerate their contents, to facilitate maintenance of aseptic operation, and to provide close temperature control and are typically simply baffled and sparged stirred-tank reactors.

The fermentation itself utilises a complex mixtures of ingredients (carbon sources, nitrogen source, nutrients, microorganisms etc) and consequently the product mixture is even more complex (see Table 3) consisting of particulates, colloidal, marcomolecular and low molecular weight components not surprisingly in a very wide range of sizes. Additionally the fermentation broths are quite dilute aqueous solutions, with product concentrations at the highest only around 7 to 12% (ethanol) and at the lowest 0.002% (Vitamin B12). The factors of diluteness and broth complexity in addition to the high solid content and potentially high viscosity presents several problems in the design and selection of down stream processing and product recovery.

TABLE 3 – Typical components in whole fermentation broths.

Class	Examples	Molecular weight	Size range (nm)
Particles	Suspended soilds	—	10,000-1,000,000
	Colloidal materials	—	100-1,000
Microorganisms	Yeasts	—	1,000-10,000
	Bacterial cells	—	500-3,000
	Viruses	—	30-300
Marcromolecular solutes	Proteins and polysaccharides	10^4-10^6	2-10
	Enzymes	10^4-10^5	2-5
Microsolutes	Common antibiotics	300-1000	0.6-1.2
	Mono- and di-saccharides	200-400	0.-1.0
	Organic acids	100-500	0.4-0.8
	Inorganic ions	10-100	0.2-0.4
	Water	18	0.2

The use of membranes in separations can be seriously affected by poor resistance to fouling and poor recovery rates, although there are several areas where they are effectively utilised. These include product recovery and product polishing, sterile filtration and more recently product purification.

Fermentation/Digestion Processes

The use of MF, UF or NF enables fermentation and digestion processes to be accelerated. The single-step membrane processes, which are inherently simple allows cells to be concentrated and recycled, and products simultaneously removed. This therefore reduces product inhibition and often enhances the effectiveness of the process through biomass enrichment. This is useful in both the isolation of fermentation products, and in the biological treatment of waste waters.

Clarification of whole broths by MF and UF replaces rotary vacuum filtration or centrifugation. In some cases ultrafiltration may replace solvent extraction and precipitation.

BIOTECHNOLOGY AND MEDICAL APPLICATIONS

If the purified product is concentrated, reverse osmosis can concentrate the dilute product streams at a low temperature (down to 5°C), achieving greater than 99% product yield with low energy requirement. Where RO is used to replace evaporation in the concentration of purified antibiotics, payback times of only a few months have been realised.

Bioseparations

A generalised process flowsheet for bioseparations of the three products of fermentations cellular, extracellular and intracellular is given as Fig 2.

In this we see that most bioseparation processes include one or more of the following
a) removal of insolubles (e.g. cell harvesting or clarification)
b) product recovery and isolation (e.g. concentration and partial enrichment of products)
c) purification, frequently multi-step
d) product polishing (eg final contaminant removal, product formulation, and sterile filtration).

For cellular products e.g. single cell protein, yeasts, separation is relatively straightforward, requiring filtration, washing and drying.

For extracellular product, the fermentation product must be filtered to remove insoluble species before product recovery from the resulting solution.

For intercellular product, the cells are harvested (filtered from the broth) washed and dewatered. The entrapped product in the cells must then be released by disruption of the cell walls, either mechanically or chemically. The complex mixtures must then be filtered to remove the cell debris to produce a clear liquid ready for product isolation.

We see in the flowsheet of Fig 2, that solid/liquid separations are prominent and this is the area where membrane separations, notably UF and MF, have been successfully applied in the concentration of enzymes, the harvesting of cells and clarification of whole fermentation broths and cell culture supernatants.

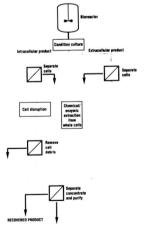

FIGURE 2 – Membrane separations in fermentation/bioreactors ▱ = membrane unit.

For the separation of cell from fermentation there are essentially four methods which can be used, dead end filtration, flocculation/sedimentation, centrifugation and cross flow filtration. Cross flow filtration offer several advantages over the other technologies (see Table 4) which, because of the nature of the cells, highly hydrated, low specific gravity, rather glutinous, tend to be inefficient and present operation problem. Cross flow filtration reduces accumulation of cells and macromolecules at the membrane surface and relatively high fluxes may be achieved over substantial periods of operation. Additionally filter aids and flocculating agents are not used and there is little aerosol formation.

TABLE 4 – Comparison of methods for cell recycle.

Technology	Remarks
Dead-end Filtration	– Sharp reduction of flux due to filter clogging – Need for filter aids
Flocculation-Sedimentation	– Gentle handling of the microorganisms: low shear stress conditions – Use restricted to flocculating microorganisms or need for flocculating agents – Inefficient cell retention. Loss of free cells – Limited to facultative biologies
Centrifugation	– Efficient cell recycle – High shear stress conditions. Potential effect on microorganisms activity and viability – Containment of aerosols is problematic – Costly equipment
Cross-flow filtration	– Efficient cell recycle – Relatively high fluxes can be maintained over long periods – Medium to high shear stress conditions. Potential effect on microorganisms activity and viability

Separation of Microbial Cells

One of the earlier applications of ultrafiltration was for separation of cells from an extracellular product. In this process the secondary metabolite was an antibiotic cephamycin C, produced by the organism Nacardia sp.

The facility produced twenty-eight 55 m^3 batches/week of a difficult-to-clarify broth. The organism is filamentous and is several micrometers long while the antibiotic has a molecular weight of only a few hundred. The broth required extensive pretreatment prior to recovery and purification of cephamycin C by ion exchange. Rotary drum filtration and centrifugal separation were both problematic. Rotary drum vacuum filtration suffered from low filtration rates even with large amount of filter aid. The recovery of Cephamycin C was poor even with reslurrying the cake and filtering for a second time. Centrifugal separation exhibited slow sedimentation rates and the formation of dilute sludge that retained a considerable amount of the product. Overall several resuspension/centrifugation cycles were required to achieve adequate yields.

Because of these limitations ultrafiltration was evaluated as the antibiotic easily passed through a 30,000 molecular weight cut-off membrane. Three membrane configuration were tested; spiral wound, hollow fibre and flat sheet.

Spiral-wound and hollow-fibre module designs were rejected due to their tendency to become clogged with broth solids, either at the flow-channel inlet or within the module as dewatering proceeded. Thin-channel flat-leaf modules were eventually specified, containing about 1070 m^2 of a 24 kD MWCO Dynel (PVC/PAN copolymer) membrane. Pretreatment upstream of the bank of UF modules consisted of chilling and screening the fermentation (30-mesh screen).

The UF modules operated at an inlet pressure of 4 bar with a feed-to-permeate flow rate ratio of 10-to-1. Each batch was initially concentrated to about one-third of its original volume by ultrafiltration. As the filtrate flux began to fall, due to the increasing solids content, the batch was then diafiltered with water. Overall 98% cephamycin C recovery was achieved, albeit at the expense of dilution of the antibiotic with the diafiltrate. Membrane flux averaged around 0.5 mol^{-1}. Membrane lifetime ranged from 600 to 1000 batches (6 to 12 months).

Overall ultrafiltration was found superior to rotary drum vacuum filtration in this application for several reasons:
- UF afforded 98% antibiotic recovery compared to the 96% recovery
- UF system material costs (including membrane replacement) were one-fourth those of precoat filtration
- UF required one-third the operating labour
- the UF system capital cost was 20% lower drum vacuum filtration capacity required, even disregarding the addition cell mass disposal costs, were lower due to absence of filter aids
- cell mass disposal costs were lower due to the absence of filter aids.

Microfiltration has been widely studied in the separation of microbial cells. Both MF, and UF, are particularly attractive due to their ability to concentrate the cells with low density i.e. densities approaching that of the fermentation broth e.g. oleaginous yeast cells. Centrifugal separators suffer from poor volumetric productivity in this application.

One reported study by Merck evaluated three microporous membranes for the harvesting and diafiltration of low- and high-cell density fermentations of a recombinant yeasts (s. cerevisiae) and a recombinant bacteria (E. coli). They used hollow-fibre and flat-sheet modules containing, 0.5-2.5 m^2 of various 0.05 to 0.45 µm pore-size membranes, operated at feed pressures of 70-230 kPa. The test system was designed to process whole, unconditioned fermentation broth. Yeast and bacteria cells were first concentrated four-fold, diafiltered with cold (4C) buffer and then concentrated to a final cell density of about 50%. The intermediate diafiltration step removed soluble components that might otherwise have interfered with other downstream processing operations.

Fermentation batches of 0.2 m^3 (corresponding to five) could be successfully processed in this manner in two hours or less. Short processing times are desirable to minimise potential product degradation.

Filtrate fluxes varied, depending upon the membranes and fermentation systems, from about 25 to 100 dm^{3}m^{-2} h^{-1}. E. coli harvests were generally slower and somewhat more sensitive to cell density than harvests of the yeast S. cerevisiae.

Overall, although the application of microfiltration is competitive with centrifugation, the greater reliability of the latter and the problems associated with membrane fouling have tended to favour centrifugal separation.

Mammalian Cell Separation

The separation of mammalian cells from cell culture supernatants is conducted to produce a clarified steam suitable for recovery of an extracellular product, usually a therapeutic protein. The macromolecular nature of these products, generally means that microfiltration rather than ultrafiltration is required. The fragility of mamalian cells distinguishes them from bacteria and yeast. In processing as particular care must be taken not to stress (shear) them beyond the point at which sublethal cell damage or cell lysis occurs. The lysis of cells will release their contents into the supernatant and introduce troublesome contaminants.

These contaminants will either be unacceptable in the final product or be severe foulants of synthetic polymer membranes. The design and selection of the membrane system for these separation must therefore ensure that cell rupture does not occur. Factors which have the potential to cause cell damage are

a) fluid shear
b) transmembrane pressure induced deformation and lysis in the pores of the membrane
c) damage by pumps
d) fluid turbulence.

Of these factors only a) and b) are critical in membrane systems, damage by positive displacement pumps is minimal when they are not run at too high a speed. There are critical regions of wall shear and transmembrane pressure in which cell damage may occur (see Fig 3). The data shown here is representative of that for cultured animal cells and is not generally applicable to other types of cells, although may certainly be used as a guide. It is interesting to note that with these biosepartions the use of controlled permeate flux

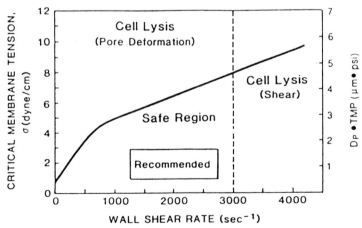

FIGURE 3 – Operating regimes for cross flow microfiltration of mammalian cells.

control is advised and preferable to constant pressure filtration. By limiting the initial flux, which would tend to be high, rapid accumulation of solids on the surface of the membrane is avoided. This reduces the tendency for excessive membrane fouling and pore blocking, in that the throughout is maximised during each run. By operating at moderate transmembrane pressure values high cell concentration can be achieved at reasonable flux values (Fig 4).

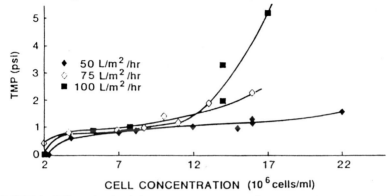

FIGURE 4 – Effect of maintaining a constant permeate flux on the microfiltration of monoclonal antibody from cell culture supernatant using a hydrophilic 0.65 μm PVFD membrane.

A commercial application of MF has been applied to the clarification of a CHO cell culture supernatant containing recombinant t-PA (a high value therapeutic protein) by Genetech. MF membrane system operates at high wall shear rates (2000-8000 s^{-1}) – but at modest pressure drops ($\leq$ 35 kPa or 5 psi), in the laminar flow regime (Re < 2100), and at a reasonably high permeate-to-feed flow rate ratio (ie $Q_f/Q > 0.1$) using a relatively open-channel (0.6 – mm ID), 0.2-m polypropylene hollow-fibre membrane.

Pilot-scale separations with 12 m^3 batches of cell culture suspension under conditions of constant filtrate flux showed the effect of operating parameters on process capacity, achievable cell concentration factors, and yield of t-PA recovered in the filtrate (Fig.5).

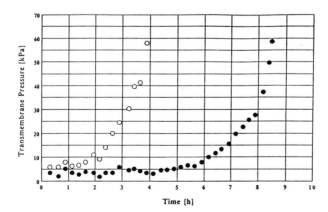

FIGURE 5 – Effect of permeate flux and shear rate on transmembrane pressure in t-PA recovery. Open circles refer to operating conditions of high inlet wall shear rate (8000 s^{-1}) and high flux (50 dm3/m2h-1). Closed circles represent operation at low-shear (4000 s-1), low-flux (27dm3/m2h-1) conditions.

A cell concentration factor of 9.5 was attained at low-shear, low-flux conditions before the transmembrane presure increased to its limit value of 70 kPa whereas a concentration factor of only 2.8 was reached at the higher shear and pressure. Ultimately, a concentration factor of fifteen was realised in routine, full-scale t-PA production.

High yields is the most importance factor in the recovery of protein t-PA. Several factors may tend to reduce protein yields in microfiltration: i) adsorption of protein onto the membrane, ii) rejection of protein due to membrane sieving, iii) denaturation of protein, iv) loss of protein in the cell concentrate

Of these factors only the fourth proved significant in this particular application; 7% of the t-PA protein would be discarded with the cell concentrate if simple concentration by MF is used.

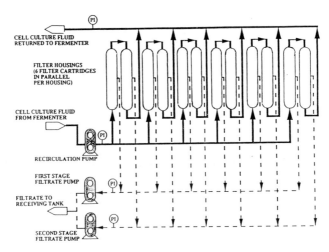

FIGURE 6 – Flow diagram for cross-flow microfiltration system for t-PA recovery.

To avoid this a stage process is implemented wherein the initial cell concentration step was followed by constant-volume diafiltration to improve t-PA recover. Product yields in excess of 99% were achieved with less than 7% product dilution by initially concentrating fifteen-fold and then diafiltering with two volumes of phosphate-buffered saline.

A process flow diagram of the 180 m^2 cross-flow MF system operated at Genetech for this application is shown in Fig 8.13. The membrane unit has an installed area of 180 m^2 and gives a permeate rate of 27 dm3m^{-2} h^{-1}. Permeate rate is controlled by positioning peristaltic pumps on the permeate side. This system processes t-PA containing cell suspensions at a feed flowrate of about 33 m^{3h-1}. t-PA in kilogram quantities is routinely recovered at yields in excess of 99%.

Protein Recovery and Concentration

Ultrafiltration is widely used for the recovery and concentration of enzymes and proteins produced by fermentations. The need to concentrate proteins can arise as a result of several factors.
  i) post clarification, when protein concentration is low
  ii) post purification by any process which significantly dilutes the protein eg column chromatography.

iii) as a final concentration step before product isolation and formulation.

The attraction of UF in protein concentration lies in the energy efficiency of the concentration of dilute fermentation broths and the gentle nature of separation which minimises protein denaturation, i.e. the loss of protein activity.

Consequently ultrafiltration has been used in industrial bioseparations for some years, principally for the concentration of enzymes such as glucose oxidise, amyglycosidase, trypsin, rennin and pectinase. Enzymes are protein molecules that can catalyse specific biochemical reactions. They are often produced by fermentation, the enzyme is excreted from the cell into the fermenter broth. The enzyme is recovered from the broth as a dilute solution, containing 0.5-2.0 wt% protein, by clarification. Ultrafiltration has now been widely adopted for the preconcentration of the enzyme dilute solutions prior to final drying of the enzyme.

Ultrafiltration plant are usually designed to process batches of enzyme solution following the batch fermentations. Concentration cycles typically last 10 – 20 hours, and are carried out at low temperatures (10-20°C), to minimise bacterial growth. Membrane fluxes are typically an average only 0.5 m d^{-1} during the concentration of the enzyme up to 20 wt% using 10,000 molecular weight cut-off membranes. Enzyme retention efficiencies of over 99% have been achieved. Overall enzyme recoveries in excess of 95% have been quoted for ultrafiltration, compared to recoveries of 60-90 % for the alternative vacuum evaporation technique.

Enzyme concentration by ultrafiltration is often followed by a diafiltration (de-ashing step) in the same equipment, to further purify the concentration enzyme solution.

At the end of each batch cycle, membranes are usually cleaned with an alkaline detergent at pH 10-13 and 40-60°C. This is often followed by acid cleaning and sanitisation with 150-300 gm^{-3} sodium hypochlorite solution at a temperature of 40-50°C.

As the biotechnology industry has progressed especially towards the production of high value therapeutic proteins derived from fermentations from genetically engineered microorganisms or from mamalian cell cultures, the emphasis of ultrafiltration has shifted towards the concentration of these important new agents.

These agents include interferon, human insulin and human growth hormone. In the case of interferon 98% interferon activity recovery has been reported using hollow fibre ultrafiltration modules.

The separation of serum albumin from human blood plasma (the Cohn process) involves a series of precipitation stages, performed using cold ethanol at different values of pHs. Fraction V from the separation process is a solution containing 3% albumin and about 20% ethanol and other low molecular weight species. The fraction is concentrated and the protein purified of alcohol by a three-stage ultrafiltration process using membranes with a 30,000 molecular weight cut-off (see Fig 7).

In the first stage the albumin is concentrated to about 5 wt% protein. The process has to be carried out at a low temperature (4-10°C) to prevent denaturation of the protein by the alcohol. The concentrate is then diafiltered to remove alcohol. In the final stages the albumin concentration is increased to 28% by UF.

Fluxes of about 0.5-0.7 m d⁻¹ have been achieved during the first (and second) stage of the process, but are less than 0.1 m d⁻¹ during the final concentration stage. The albumin concentration is suitable for freeze-drying.

Ultrafiltration has been used on the laboratory scale for the concentration of viruses (size range from 30-300 nm) using very open membranes with 100,000 molecular weight cut-off. Although there is some loss of viral activity during ultrafiltration concentration, but the loss of activity is less than in alternative techniques, eg ultracentrifugration and ammonium sulphate precipitation.

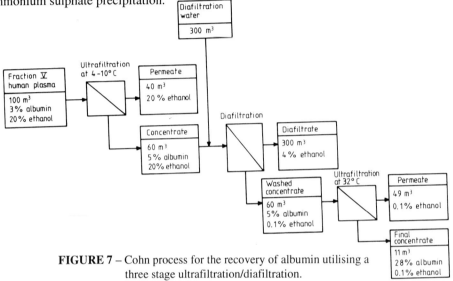

FIGURE 7 – Cohn process for the recovery of albumin utilising a three stage ultrafiltration/diafiltration.

An emerging important application of ultrafiltration is as a concentration step before and after affinity chromatography. The benefits of ultrafiltration are two fold
- throughputs of column chromatography are frequently limited by flowrate and thus pre-concentration can increase productivity and reduce column size
- the elution of protein from the purification column frequently results in significant dilution, and thus concentration of the eluate is necessary.

An example of this application is in the purification of a monoclonal antibody, shown in Fig 8.

The process uses 100,000 Dalton MWCO polysulphone ultrafiltration membranes in a plate and frame cassette to concentrate 0.28 m³ batches. The UF membrane concentrates the IgG_{2a} antibody (146,000 D molecular weight) by approximately a factor of 10 to 20 and achieves a 95%, and greater, antibody recovery.

Following purification by affinity chromatography the eluent fractions, containing the pooled antibodies, are concentrated to a total protein concentration of 20 g dm⁻³ by ultrafiltration.

In the above application areas the use of elution buffers that differ significantly in terms of pH, composition, etc, from the solution loaded to the chromatography may cause problems associated with denaturing product or with the next bioprocessing step. In this context diafiltration may be employed to transfer the product, protein etc, into a more suitable solution environment.

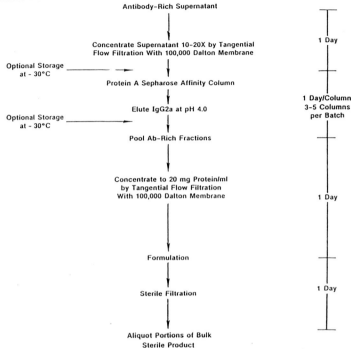

FIGURE 8 – Antibody purification processes utilising UF pre and post column chromatography.

Diafiltration of biological solutions

Diafiltration is a method which can increase the purity or improve the separation, typically of proteins. In ultrafiltration as the concentration of a protein containing solution proceeds, the viscosity of the feed solution increases and will eventually be such that concentration polarisation becomes flux limiting. Further concentration by UF will be no longer viable and therefore any further enrichment of protein must be achieved by removing the microsolutes rather than concentrating the macromolecular solutes. Diafiltration is a feed and bleed processes in which the retentate is dilutated with water (or fresh buffer) while it flows passed a protein impermeable, contaminant permeable UF membrane. Simultaneously permeate is withdrawn at the same rate that the water is added and so the microsolutes are washed out of the product solution.

A compact, laboratory-scale cross flow hollow fibre system is available, capable of rapid processing of volumes up to 5 litres and is designed for fast, efficient concentration and/or diafiltration of a wide range of biological solutions.

The system includes a cartridge support stand, inlet and outlet pressure gauges, and 400 ml and 1 litre reservoirs that can be pressurised for gentle reciculation of labile solutions. It incorporates a precision control backpressure valve and a process solution sampling/

drain valve, and an optional peristaltic pump with variable speed drive and nominal maximum reciruclation rate of 2 l/min is also available.

The system accommodates hollow fibre ultrafiltration (UF) cartridges ranging from 3000 to 500,000 nominal molecular weight cut off and microfiltration (MF) cartridges in pore sizes from 0.1 to 0.65 m. Membrane areas range from 24 cm^2 to 1070 cm^2. The low hold-up volume design allows concentration to as low as 25 ml to be achieved and its use offers advantages over stirred cells or conventional dialysis methods.

Applications include concentration of proteins, enzymes and other dissolved macromolecules, continuous buffer exchange, bacteria/pyrogen removal, blood plasma purification, virus concentration/removal, as well as cell clarification and washing.

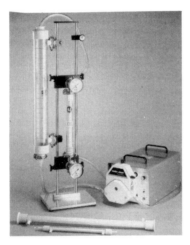

Laboratory scale H.F. membrane module.

Sterile filtration

The exclusion of viruses from cell culture medium is in principle easily achievable by UF as the size of viruses is large, of the order of 10 manometers or greater. However the required virus removal, routinely specified as a 12 log reduction in count, means that almost perfect membranes, free from defects or an occasional large pore size, are required. This is of course a great difficulty in membrane production and product control and consequently has seen limited application of UF filters in this area. Recently Millipore have developed a composite membrane which may find applications in a one or two stage ultrafiltration for virus.

Enzyme Membrane Reactors

The interest in the use of purified enzymes as biocatalysts in laboratory and industrial applications is expanding quite rapidly. The enzymes are globular proteins able to catalyse the complex of biological reactions of living organism i.e. metabolism. They do this under very mild operating conditions (low temperature, low pressure, no extreme pH), very selectively and are thus very attractive for biochemical synthesis. The enzymes are usually confined within living cells and are generally obtained by extraction and purification. The

TABLE 5 – Applications of Enzyme Membrane Reactors in the Agro-Food, Biomedical and Pharmaceutical Industries.

Enzyme or Biocatalyst	Reaction	Purpose
Arginase, Asparginase	Hydrolysis of arginine and aspargine	Care and prevention of leukaemia and cancer
Tripsin, Pronase, (PROTEASE)	Hydrolysis of blood proteic texins	Removal of blood toxins in dialytic patients
Langerhans islets	Insulin secretion	Bioartificial pancreas
Tripsin, Chymotripsin (PROTEASE)	Hydrolysis of whey proteins	Production of peptides for medical and pharmaceutical use
b-Galactosidase	Hydrolysis of lactose in glucose and galactose	Milk or whey delactosization Sweetening
Tripsin, Chymotrypsin, Papain	Hydrolysis of milk proteins	Production of peptides
Cellulose	Hydrolysis of cellulose	Alcohol production
a-Amylase	Hydrolysis of starch	Starch thinning/Maltose production
Polygalacturonase, Pectinesterase	Hydrolysis of pectins	Fruit juice clarification
Glucose Isomerase	Conversion of glucose to fructose	High content Fructose Syrups
–Cyclodextrin	Removal of limonin and naringin	Reduction of joices bitterness
Lipase	Hydrolysis/Snythesis of glycerides	Dietetic nutrients peoduction

enzyme is typically then introduced into the appropriate bioreactor and its activity utilised in the reaction. The removal of the enzyme form the product solution then becomes a problem associated with product contamination. The immobilisation of the biocatalyst in a fixed location/material is consequently attractive as it can in principle allow product and enzyme to remain separate. This therefore allows the enzyme to be used repeatedly and continuously and in addition offers greater stability for the membrane under immobilisation.

There have been several techniques devised for the immobilisation of enzymes
- confinement in gels
- encapsulation in cellulose or alginate beads
- adsorption on solid surfaces

- chemical binding to insoluble supports
- copolymerisation with proteic carrier
- entrapment within polymeric matrix
- immobilisation (bound) on membranes

The latter of these offers an environment to the enzyme which is similar to that "in vivo" i.e. bound to the cellular membrane of the living organism. More importantly for processing it opens up the possibility of coupling the enzyme catalyst reaction with a membrane separation process. This in principle allows product separation to occur simultaneously during its production and would be advantage to reactions which are either thermodynamically unfavourable (eg reversible) or suffer from product inhibition. In this way processes could in principle be operated continuously with high yields and selectivities. The technology is in its early days but several industrial applications have been identified in the food and pharmaceutical industries (see Table 5).

There are a number of methods used to immobilise or contain enzymes on membranes.

1. *Enzyme Membrane Reactors (EMR).*

   In this application ultrafiltration membranes with the appropriate MWCO are used to give complete rejection of the enzyme while permitting permeation of the reaction products. Thus the membrane is either built into the bioreactor or placed in a recycle loop (see Fig 9) through which the fermentation medium can be pumped continuously in cross-flow. The latter enables a much higher membrane surface to volume ratio, when hollow fibre membranes modules are used, and improves ultrafiltration characteristics, both of which are generally advantageous for industrial applications. Both of the methods suffer from the problem of denaturation of the enzyme in its native state in homogeneous solution.

2. *Segregation in Hollow Fibres.*

   As with the EMR the function of the membrane is as a barrier to the transport of enzyme. The membrane however segregates the enzyme from the bulk of the reaction medium by, for example, locating the enzyme in the tube side or shell side of a capillary module. The reaction substrate therefore diffuses towards the enzyme across the porous membrane wall.

3. *Immobilisation in the Membrane.*

   The immobilisation of the enzyme by adsorption in the "porous" walls of asymmetric capillary membranes represents an extension of segregation in hollow fibres. The attraction with this is that the barrier between enzyme and reaction substrate is restricted to the thin membrane "skin" layer.

4. *Immobilisation by Gelation on UF Membranes.*

   The ultrafiltration of enzyme from solution will, under suitable conditions, lead to the accumulation of the enzyme on the membrane and then to a possible gelation of the enzyme. Thus by this way enzyme can be immobilised onto the membrane in a very simple and convenient manner. As the enzyme layer, formed on the membrane, is subject to cross flow or hydrodynamic disturbances from the reaction medium techniques have been proposed to enhance the mechanically stability of attachment – co-deposition of gel with an inert macromolecule (co-gelation) and linking via a bridge molecule, to an inert protein and then deposition as a gel.

5. *Immobilisation by covalent bonding.*
   An alternative method of immobilisation is by chemical bonding (covalent bonding) to the membrane surface using active bridge molecules CNBr or bi/multi functional agents such as glutaraldehyde. The bound enzyme is readily available to react with the chemical substrate, although there is some loss of activity due to the bridging.
   Examples of the use of covalent binding of enzymes to membranes are in the field of membrane electrodes for analytical methods and in the food industry (Table 6).
6. *Immobilisation of whole cells.*
   When the enzymes are intracellular and unstable after extraction and purification it may be suitable to immobilise the complete bacterial cells on the membrane. The catalytic activity of the enzymes is still retained even though the cells are in the dead state. One method of preparation is by entrapment of the cells in the polymeric membrane during the production of the polymer membrane itself. This has successfully applied in the formation of polysulphone capillary membranes, by wet spinning and phase inversion, in which by lophilised whole cells of "sulfolobus saolfactaricus" (a thermophilic and acidophilic micro-organism) were mixed with the polymer solution.

Membrane Fermentors

The majority of fermentations are batch processes using microbial cells in their free form. These processes are relatively inefficient especially when the micro-organisms are slow growing or are significantly affected by product inhibition. There is increasing interest in continuous fermentations, in particular high rate processes, for the production of cells and primary metabolites, in view of the reduced capital and operating costs related to fermentation. Such continuous processes can be performed using microporous membranes to separate the product stream form the fermentation broth. A typical process circulates the fermentation broth to the membrane, either microfiltration or ultrafiltration, in cross flow at a desired flowrate and the retentate recycles back to the fermentor. The permeate stream may be partially or completely removed from the system and flows are adjusted to maintain a constant volume. The most important feature of this process is the possibility of keeping viable cells at a higher concentration (compared to batch), the removal of the end product which is frequently an inhibitor of the fermentation and the dosage of nutrients on an ad hoc basis. In addition if the cells produce growth inhibiting co-metabolites they also can be removed continuously in the permeate. It may also be possible to inactivate or remove these co-metabolites by a physical treatment (thermal denaturing, adsorption, extraction) integrated into the permeate recycle loop.

Membrane modules typically used in the recycle fermentors are hollow fibre or capillary membranes, to give high surface areas and therefore high permeation rates, with the cells/broth retained in the shell of the module. Alternatively the cells are retained by dialysis membranes and separated from a dialysate solution (water or appropriate substrate solution) flowing in a dialysate circuit.

The performance of membrane fermentors is affected in particular by circulation rate of cell medium, dilution rate and fermentor bleed rate. Circulate rate must be adjusted to achieve the correct balance between flux and fouling, micro-organism viability (deactivation) which may be shear induced and operating cost.

During fermentation the cell concentration increases continuously due to growth. This growth increases the broth viscosity and density, eventually leading to operating difficulties associated with membrane performance, higher pressures and foaming. To maintain steady operating conditions it is beneficial to continuously bleed the fermenter of solution which maintains the cell concentration at a suitable high level.

The problems which are associated with the operation of recycle "membrane fermentors" are summarised in Table 7.

Industrial applications of continuous membrane fermentation include the following:
- ethanol production form lactose or glucose
- acetic acid production from ethanol
- lactic acid from glucose
- citric acid from glucose

TABLE 7 – Problems of Cell Recycle in Membrane Fermentors.

1	Recycle and build-up of inhibitory products
2	Recycle of undesirable feedstock constituents
3	Recycle of unused feedstock causing physical as well as chemical imbalances
4	Recycle of selected microorganisms when mixed cultures are used which may cause imbalances of compositions and metabolic activity
5	Mechanical impact in pumping and shear stress affecting culture viability and activity.
6	High cell concentration and increase in viscosity affecting mixing and mass transport
7	Membrane performance and fouling

There is, in particular, interest in the two step fermentations of raw substrate, e.g. starch or cellulose. The two steps, initial hydrolysis to low molecular weight "intermediates" which are then fermented in the second step, can be coupled in a single step.

Alternative Membrane Fermentors

It is perhaps not surprising that other membrane separations have been coupled together in continuous fermentations. These are applied with aims of reducing the effect of product inhibition and/or reducing the cost of downstream processing. Examples of these include
  a. Preparation to remove volatile products (eg ethanol) or inhibitory co-metabolites (e.g. butanol) during fermentation
  b. Membrane extraction couple to acidogenic fermentation to remove non volatile products.

Small Scale Cell Culturing

Many processes in biotechnology are operated at a very small scale, for example culturing of mammalian cells and also plant and insect cells. These processes are complex involving many intricate steps as shown in a typical manufacturing process for recombinant DNA-derived biological or monoclonal antibody products (Fig 11).

The medium for culturing mammalian cells is very complex (essential amino acids, vitamins, lipids, trace elements, inorganic ions, carbon sources, antibiotics, whole or dialysed serum) and recovering desired product from extracellar or intracellular (cell disruption) processes is difficult and expensive. Furthermore maintaining the constituents at optimal concentrations to maximise cell viability, productivity and concentration is difficult. In the production of high value biologicals (e.g. Factor VIII) the most important factor is achieving a high purity product and a high product titre. In this respect membrane bioreactors are attractive and have been applied in three forms as either microcapsules, hollow fibres or flat sheet reactors. The cells are retained in the bioreactor by the membrane barrier, perfused with a steady flow of medium (supplied with O_2 and nutrients) while desired products and wastes are removed.

Microcapsule technology for mammalian cell culture for both biotechnology and biomedical uses is a commercialised method (Abbott Biotech, Needham MA).

Hollow fibre immobilised enzyme reactors have been made with glass, ceramic and synthetic polymer membranes. The design of hollow fibre bioreactors needs to ensure a well defined constant cell region so that mass transport limitations, e.g. of oxygen, are not significant. Consequently alternative designs based on concentric fibres and hollow fibre/flat sheet combination have been used. In the latter design the cells are entrapped between the space formed between the outside of the hollow fibre and the flat sheet membrane. The medium flows through the lumen of the fibres while oxygen and carbon dioxide are supplied through the hydrophobic flat sheet membrane. The final bioreactor is made of multilayers of individual elements.

Affinity Membranes

In the separation of high value products membrane separations frequently suffer from a lack of specificity and resolution and chromatographic methods are commonly used as an alternative. Utilising the principles of biological interactions the method referred to as affinity-chromatography has been developed, commonly based on packed bed sorbents. There are limitations associated with packed beds, notably low flowrates and diffusional (internal) mass transfer which can be overcome by the use of membranes. The affinity membranes used in chromatography are either single or stapled flat sheets, spiral wound, cassettes and hollow fibres which are activated and compled with appropriate ligands using similar method to prepare packed bed systems. The membranes are generally composites – a basic polymer (e.g. cellulose) and a grafted co-polymer (e.g. gycidyl-methacrylate) – activated for coupling of ligands e.g. reactive dyes protein A/G, lectins, ion-exchange groups etc.

There are two modes of operation of affinity membranes; termed cross flow affinity filtration (CFAF) and membrane affinity chromatography (MAC) which is essentially analogous to direct filtration. The applications of affinity membranes (see Table 8) include protein purification from very dilute solutions, eg animal cell cultures, contaminant removal (e.g. proteases, endotoxins) and combined separation and purification including the treatment of particulate suspensions (e.g. "whole-broth" adsorption).

There are many more applications of MAC than of CFAF, which is a developing method. The aim of CFAF is a solid-liquid separation coupled with simultaneous purification using a highly selective ligand. The performance of CFAF has been evaluated at 0.1m³ scale for the recovery of enzymes from crude suspensions (homogenates), e.g. malate dehydrogenase from E.coli homogenate. The membrane used was a cibracon blue modified polyamide membrane (0.45 m pore size) in either a cassette or a spiral flow module.

TABLE 8 – Applications of affinity membranes.

Membrane System/Ligand	Type of Operation	Application
HF/ Protein-A	MAF/CFAF	MAb's
HF/ Gelatin	MAF/MAC	Fibronectin
Radial Flow/ ProteaseInhib.	MAF/MAC	Proteases from Plasma
Flat Sheet/ Dyes	MAC	Enzymes, PDC, G6PDH, FDH
Flat Sheet/ Cassettes/ Dyes	MAF CFAF	Malate-dehydrogenase
HF/ Metal-Chelates	MAF/MAC	BSA
Radial Flow/ Metalchelate	MAF/MAC	Urokinase
Flat Sheet/Protein-A	MAF/MAC	MAb's
Pleated Membrane Cartridge/Ion-Exchange	MAF/MAC	MAb's

The overall operation of affinity membrane and sorption is similar to standard chromatography procedures; adsorption, washing elution but with an additional cleaning step. Washing is generally carried out by back flushing through the membrane, whereas elution depends on the mode of operation – back flushing for particle free suspensions, in the forward filtration direction for suspensions.

Liquid Membranes

Biotechnology and medicine are a potential field for liquid membrane applications. The following illustrate this potential although it is not known whether the application have progressed beyond the laboratory stage.
  i) ELM membranes have been used to extract toxins from biological liquors eg removal of cholesterol and amino acid (lysine) from blood.
  ii) Extraction of poisons salicyclates (Aspirin) and barbiturates from solutions simulating the stomach content using ELM.
  iii) SLM for the removal of local anaesthetics (methyl-, ethyl-, butyl-p-aminobenzoate) from aqueous donor liquids.
  iv) SLM to remove products from acidogenic fermentation. Metabolic species such as acetic acid and butyric acid are selectively removed continuously during fermentation, with a TOPO in kerosene membrane phase.

v) ELM for the separation of citric acid from fermentation broth.
vi) SLM for isolation of benzyl penicillin (antibiotic) from cultural solution utilising tetrabutylammonium cation as a carrier.
vii) SLM for separation of optical isomers (D- and L-glycine, L-leucine and L-tryptophan) using crown ethers as carriers.
viii) ELM as a method of controlled release of drugs.
ix) SLM and ELM for artificial oxidation of blood as so-called "artificial lungs". Liquid fluorocarbons simultaneously enable oxygen to transport into the blood, by first dissolving in the membrane phases, while carbon dioxide is transferred in the opposite direction and removed from the blood.
x) Liquid-membrane enzyme reactors. The method utilises enzymes in the acceptor liquid to assist the separations, typically for removing toxins from physiological liquids. The enzyme biochemically act on the transferred permeate, for example oxidation of phenol, and the transferred species cannot transport back. The method is also proposed as a method for simultaneous extraction, enzyme reaction and accumulation of fermentation products e.g. transformation of penicillin G to 6-amino penicillin acid.

Electrodialysis

Electrodialysis has important applications in the pharmaceutical and biochemical industries where gentle processing conditions are required for materials such as human blood plasma and interferon. The production of essential amino acids will require various demineralisation steps. Certain waste streams in biochemical and pharmaceutical operations contain ammonium sulphate, urea and guanidine hydrochloride which can be recovered by ED and eliminate certain large BOD problems.

There are several other applications of ED for salt removal which are important and include:

1 *Production of Protein Fractions*
   – Desalting plasma to produce an anti-haemophilic factor concentrate.
   – Whey protein purification from other food processes eg. cereals, rice.
   – Separating salts from mixtures containing proteins
   Mother liquor of amino acids after salting out is usually discharged without recovering useful components. The removal of salt constituents enables their re-use in the process and also reduces effluent treatment costs due to reduced BOD and COD.
2 *Separation of amino acids into acidic, basic and neutral group*
   Electrodialysis has been used in the extraction of cytoplasmic leaf proteins from alfalfa leveas. These proteins are denatured by conventional treatment where heat is generated during pH adjustment. Electrodialysis is used to desalt the protein solutions (of K^+ ions) and acidify with H^+ ions generated at the anode. The lowering in pH is used to induce precipitation while the electrodialysis process minimises temperature rise.

More recently the use of ED has been applied in the isolation and purification of iminodiacetic acid [IDA] from its sodium salt. Iminodiacetic acid is a key intermediate in the synthesis of glyphosphate, a broad spectrum herbicide. ED is an effective replacement to neutralisation and crystallisation of the synthesised product. The ED unit is capable of producing an IDA purity of 99.2%, at a current efficiency of 90.5%.

Electrodialytic Salt Splitting

The main use of electrodialytic salt splitting in biotechnology is in the recovery of organic acids from fermentations. One potential application is in the recovery of itanonic acid. Itaconic acid is used as an intermediate in the production of polyacrylonitrile fibres and as a plasticiser in certain polymer products.

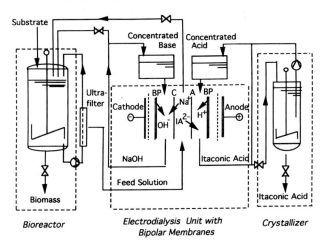

FIGURE 9 – Electrodialytic salt splitting for itaconoic acid production.

In the conventional batch fermentation process the pH in the fermenter which tends towards lower values during the acid production, is kept constant by adding sodium or ammonium hydroxide. The free acid is then recovered form the spent medium by adding sulphuric acid which generates a significant amount of salts with the desired product and thus the process requires further purification steps. By applying bipolar membranes this further purification is avoided, the base is recycled and the itaconic acid recovered continuously. The fermenter is fed with substrate and the bioreactor constituents are passed through an ultrafiltration unit. The product containing ultra filtrate is fed to the central cell of the bipolar electrodialyser. From this compartment sodium ions, migrate towards the cathode and form NaOH with the OH⁻-ions generated in the bipolar membrane and the itaconate ions migrate towards the anode and form itaconic acid with protons generated at the bipolar membrane. The concentration of itaconic acid is then built up by recirculating through the acid compartment until its saturation level is reached (approximately 0.63 mol dm^{-3}, 10°C). It is then recovered by precipitation in a crysatlliser. Thus the itaconic acid is produced and continuously removed from the fermenter without the addition of acids or bases, ie without the production of additional salts. The generated sodium hydroxide is recycled to the fermenter for pH control.

Reconditioning after Separation of the Enzyme from Recamat Splitting of a N-Acetylaminocarbonacid

A process for reconditioning the solution left after separation of the enzyme from recamat splitting of a N-Acetylaminocarbonacid in the presence of L-Aminosaureacylase can be based on the application of bipolar membranes. In this process the separation of amino acids and the reconditioning of the salt in one operation is possible in a 3-compartment arrangement and results in the following dilute solutions of L-Aminoacid, the existing cations as hydroxides and of N-Acetyl-D(L)-Aminoacid and acetic acid in the form of a free acid. These solutions can be recycled, after a further treatment, and thus eliminate wastewater.

The Splitting of Amino Acids

Amino acids have amphoteric properties. At a fixed pH-value (isoelectric point) amino acids exist as neutral components. If the pH-value of the solution is lower than the isoelectric point of the appropriate acid it is positively charged and will be transported as a cation. If the pH-value of the solution is higher than the isoelectrical point, the amino acids are present as anions.

This property can be used for the separation of amino acids by ionselective membranes if there is a pH-value in the raw solution which is between the respective isoelectrical points of the several amino acids. The important requirement for the separation of amino acids by ion-selective membranes is maintaining the appropriate pH-value. For a relatively simple regulation of the pH-value, bipolar membranes can also could be used, and thus avoid the addition of chemicals.

Electrodialytic Dissociation of Alcohols

Bipolar membranes can also be applied for the dissociation of alcohols and thus for the production of alcoholates. Methanol, like water, is both a weak base and a weak acid with a dissociation constant lower than that of water. Therefore, methanol can be converted to its conjugated base typically by a conjugated base of a weaker acid eg ammonia although the reaction is difficult to perform.

$$CH_3OH + NaNH_2 \; CH_3ONa + NH_3$$

An alternative method to produce sodium methylate from methanol and sodium acetate in non-aqueous media is by the use of bipolar membranes as shown in the cell scheme of Figure 14 which shows a electrodialysis stack consisting of a two compartment cell system of repeating cation and bipolar membranes. Water free methanol and sodium acetate are fed into the cells formed by the bipolar and the cation-exchange membranes, while water-free methanol is passed through the other cell.

Methanol is split in the bipolar membrane into protons and CH_3O^- -ions which form CH_3ONa with sodium ion migrating from the sodium acetate- containing cell. The acetate ions recombine on the other side of the bipolar membrane with the protons from acetic acid. Thus in the process sodium acetate and methanol is converted into sodium methylate.

Continuous Deionisation Systems

Continuous Deionisation (CDI) is a membrane separation technology developed to remove ions from process streams by a combination of ion-exchange resins, membranes, and electrical fields. CDI has the removal efficiency of mixed-bed deionisation (MBDI) without the need for chemical regeneration cycles associated with single ion-exchange resin beds. This results in niminal product loss and liquid hold-up is also much less lower using CDI technology. CDI unlike standard ion-exchange is a continuous operation, it requires minimal power consumption and with reduced product loss, results in lower

TABLE 9 – CDI Separations.

Ions Removed		Molecular Weight
Cations	ammonium	18
	sodium	23
	potassium	29
Anions	chloride	35
	formate	45
	nitrite	46
	nitrate	62
	phosphate	95
	sulfate	96
	iodide	127
	gluconate	195
	cholate	408
	antihypertensive	442
Non-ionised Solutes Retained		Molecular Weight
Product	urea	60
	ethylene glycol	62
	amide	95
	vitamin C	176
	glucose	180
	di-peptide	200
	lactose	342
	sucrose	342
	disaccharide	403
	contrats agent	800
	contrast agent	1550

BIOTECHNOLOGY AND MEDICAL APPLICATIONS 679

operating costs. In operation a feed solution is fed to an ion-exchange resin bed sandwiched between two ion-exchange membranes, one cation one anion. The CDI module draws ions from the feed solution into the resin bed (see section 10), and ions then migrate out of the resin bed into separate salt concentrate streams through the action of an electrical field.

The economics of CDI are particularly attractive at feed salt concentrations of > 100 ppm (see Fig 12).

Typical applications (see Table 9) are in deionization/desalting of aqueous solutions of

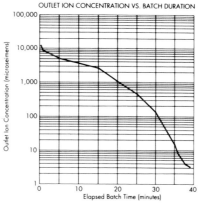

FEED: 10% Contrast Agent (~800 MW) 1% NACL
COND: ~ 14,450 microseimens

FIGURE 11 – The performance of continuous de-ionisation.

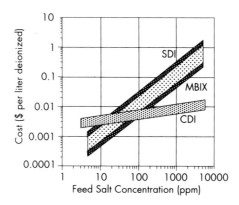

FIGURE 12 – The economics of continuous de-ionisation.

non-ionic solutes, urea recovery for recycle, urea desalting and contrast agent desalting. For example CDI can deionise 7 μm urea solutions to les than 1.5 ppm in a single pass and can effectively reduce the salt concentration in non-ionic injectable contrast agents over 99.9% with 95% yield of the contrast agent. CDI also removes 200 to 300 dalton ionic impurities that are often present.

SECTION 15

Medical Applications

MEDICAL APPLICATIONS

This section describes the membrane processes used in the direct treatment of patients. Medical applications of membranes processes relating to the filtration of drugs and to direct patient care are described in section 8. The use of RO to supply pure water for hemodialysis is discussed in section 10.

TABLE 1 – Medical Devices With Synthetic Polymer Membranes.

Environment	Device	Permeable solute
Ex vivo	Oxygenator Hemodialyzer Hemofilter Plasma separator Plasma fractionator	O_2, CO_2 Small molecules, ions Middle molecules, H_2O Plasma Plasma proteins
In vivo	Contact lens Biosensor Novel bioreactor Drug delivery systems Wound dressing Artificial grafts	O_2 } Biologically active molecules Medicines O_2, H_2O Physiological solution, cell
In vitro	Blood filter Plasma separator Substrate for cell culture	Blood except WBC Plasma Physiological solution

Dialysis

Hemodialysis can be used to completely replace the function of the kidney in removing toxic lower molecular weight components from the blood. These waste products include urea (end product of protein catabolism) creatinue, (end product of muscle metabolism), phosphates, uric acid and so called "middle molecules" which are larger solutes of 13 KD in size which accumulate very slowly in patients with end stage renal disease (ESRD).

Hemodialysis is the dialysis of the blood against a physiological saline solution using ultra filtration membranes. The physiological saline solution is required to prevent the transfer of vital non-toxic low molecular solutes e.g. potassium, sodium which would occur if water alone was used. A second function of hemodialysis is the restoration of electrolyte and acid-base balances.

The major requirement for the membrane material (and module) is compatibility with blood. Blood anticoagulant (heparin) is frequently added to the blood before it enters the dialysis unit (see Section 1).

The process of hemodialysis is a slow process, patient treatment times are 3 to 5 hours per session, three times per week. The ultrafiltration capacities of the membrane vary from 5 to 70 cm^3/hm^3 mmHg. Minimum transmembrane pressure differences are usually in the range of 80-170 mmHg, due to pressure losses in the blood channel and venous resistance. For typical low permeability dialysers with 1 m^2 membrane area this will mean a fluid loss amounting to 400 – 850 cm^3h^{-1}. The water balance control during dialysis is therefore an important factor during operation.

Reverse osmosis system for home dialysis.

Hemofiltration

The limitations of diffusion in dialysis, prompted the use of ultrafiltration i.e. hemofiltration as a faster method for the removal of waste metabolites. The process requires the blood to be pre-diluted with saline solution before ultrafiltraion in the hemofilter. The dilute solution/plasma that is ultrafiltered contains the microsolutes that in dialysis normally permeate by diffusion. Hemofiltration membranes are generally more permeable (porous) to high molecular weight species than dialysis membranes and overall transport rates are higher, giving corresponding shorter treatment times. The major disadvantage of the procedure is the relatively large requirement for endotoxin-free saline solution.

To reduce the volume of diafiltration required, the hemofiltration can be combined with a purely dialytic step, in a process, termed *hemodiafiltration*. The blood first passes through a dialyser – with its countercurrent dialysate – where small solutes are removed efficiently. It is then diluted with a smaller volume of sterile saline and then passed through a hemofilter to remove the dilution fluid. In this step, larger solutes not removed efficiently by dialysis are filtered by convection.

Liver Support System

It is reported that between 85 and 95% of patients with fulminant hepatic failure who do not receive a liver transplant die, even though the liver has an extremely high capacity for regeneration. The major problem is that this regeneration is too slow and requires the use of liver support systems during recovery, to remove toxins formed in the metabolic processes which accumulate in the body. The metabolites and toxins are of two generally types water soluble and fat soluble, which during heptic failure are dominating. The water soluble toxins can be removed by hemodialysis, but the fat soluble lipophilic toxins must be converted to water soluble species by chemical reaction. This is in fact the main function of the liver to produce water soluble species, which can be treated by the kidneys. The purpose of a liver support system is to keep a patient alive during liver recovery or during the period until transplantation can be performed. A range of detoxification methods have undergone clinical trials (resin sorbents, plasmapheresis, perfusion over baboon or human (liver) but success has been limited.

A liver support system based on the use of a dialyer (hydrophobic large pore hollow fibre membrane) and a selective lypophilic liquid supported membrane (hollow fibre membrane support) has been developed (see Fig 1). The patients blood in this system is

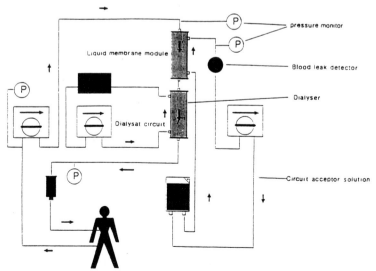

FIGURE 1 – A SLM based liver support system.

first pumped to the liquid membrane module ands then to the dialyser before returning to the patient. As shown the hydrophilic and lipophilic parts have separate flow circuits. The hydrophobic operation is identical to normal hemodialysis operation for removal of water soluble toxins. The membrane is typically a 200 μm diameter hydrophobic polysuphone fibre with a 30 kD MWCO.

The supported liquid membrane is an asymmetric hydrophobic polysulphone hollow fibre membrane (internal diameter 220 μm, 50 μm wall thickness) filled with liquid oil (paraffin mixture). The inner skin has a 80 KD MWCO whereas the outside pore size is 1 – 2 μm. The receptor phase used is a sodium hydroxide solution which formed salts by reaction with many of the lipophilic toxins which are highly soluble in the oil and diffuse across the membrane due to the concentration gradient. The use of enzyme mixture with detoxification properties for the toxins is also reported. Stable operation of the system without paraffin transfer to the blood and with minimal blood leakage to the receptor phase is achieved by controlling the pump pressure in relation to the blood pressure. Successful human trials of the system are reported in 1993.

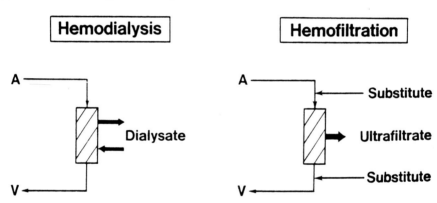

FIGURE 2 – Scheme of hemodialysis and hemofiltration through extracorporeal circulation.

Humidification of Respiratory Gases

The requirements for the humidification of air during respiratory therapy include sterility, temperature control and humidity control. This is especially true in artificial ventilation of new born infants where, commonly, humidification of in spiritroy gas is by water vaporisatation. The operation of these devices is not always satisfactory and consequently a membrane based unit has been developed and is in clinical use. The unit is shown schematically in Fig 2 and is based on a hollow fibre membrane module which is 7 cm long utilising wide bore (1.5 mm) capillary fibre membrane of hydrophilic polysulphone. The system operates by flowing dry air through the fibre lumens, with heated water recirculated through the shell of the module. The relatively large surface area of the membrane gives sufficient capacity for the exchange of heat and humidity while the membranes limit bulk flow of water and also give security against contamination by bacteria

The membrane module operates with a low pressure drop of air (2.5 m bar) with flow

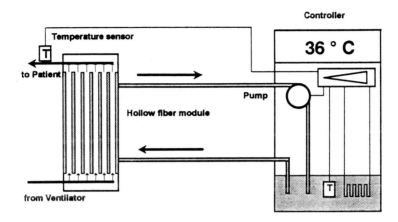

FIGURE 3 – Hollow based humidification system.

in the range of 2-100 dm^3 min^{-1} at 37C and achieves 90% relative humidity (from 5% relative humidity) in 25 ms contact time.

It is essential that bulk flow of water in operation is avoided, otherwise plugging of the fibre will occur with an increase pressure drop and a loss of flux, typically 10 cm^3/min m^2 bar, and also a risk of droplet intrusion into the respiratory system pipework. Because of the risk of bacterial contamination of the air, from the water, over long periods of operation the membrane is a sterilising grade membrane, free of defects, which satisfies bacterial challenge tests.

Hemoconcentration by Ultrafiltration

In recent years there have been a reduction in the use of donor blood in coronary bypass operations as a result of the risk of infections. The use of techniques such as preparative hemodilution and plasmapherisis, priming the heart lung machines with saline solution, retransfusion of drainage blood and salvage of blood left in the extracorporeal circuit, decrease blood requirements and reduces the risk of pulmonary and renal failure. Before retransfusion, the diluted blood must be concentrated to avoid side effects such as fluid overload and reduction in oxygen transport capacity.

The most popular technique for hemoconcentration is based on centrifugation followed by saline washing of blood cells. However this method means that noncellular blood components are discarded and platelets are removed. In addition the equipment is expensive and consequently ultrafiltration is applied as an alternative process. In practice UF is applied to a bypass of the blood flow from the oxygenator. The membranes used are the same type as used in hemodialysis ultrafilters. Ultrafiltration causes a smaller change in the normal blood composition than centrifugation and washing and is much cheaper and easier to operate.

Blood Oxygenation

During open heart surgery, blood oxygenators serve as artificial lungs to maintain oxygen content. Hollow fibre membrane contactors in which blood is pumped over the outside (shell) of the fibres and air, with increased oxygen content, in the fibre lumens are used for this purpose. The blood absorbs oxygen and is stripped of carbon dioxide by mass transfer processes across the membrane. The compactness of the hollow fibres gives a small volume oxygenator which minimises the need for blood transfusions during surgery. Module performance is determined by mass transport in the blood side of the membrane, with better performance from shell side flow of blood in rectangular bundles (see Fig 4 acid gas removal)

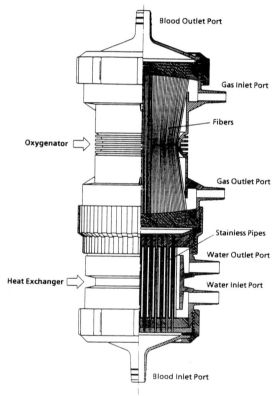

FIGURE 4 – Structure of Capiox capillary membrane oxygenator.

Medical Oxygen

The production of medical oxygen for asthmatic and hypoxic patients by small scale in-situ generation using a hollow fibre module is useful. This offers a failsafe method of oxygen delivery where, at worst only a 50% O_2 product, or air, is supplied rather than a potentially dangerous 100% O_2. The product is also sterile, humid and carbon dioxide enriched. Maintenance requirements of in-situ generators are also minimal.

MEDICAL APPLICATIONS

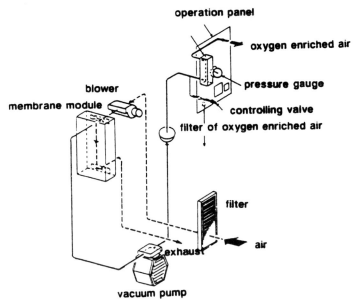

FIGURE 5 – Flow diagram of oxygen enrichment apparatus for medical use.

SECTION 16

Recovery of Salts, Acids and Bases

Aqualytics

A DIVISION OF THE GRAVER COMPANY

POSITIVE SOLUTIONS WITH ION EXCHANGE MEMBRANES

BIPOLAR MEMBRANE ELECTRODIALYSIS
ELECTRODIALYSIS • DIFFUSION DIALYSIS

AQUALYTICS, formerly AQUATECH Systems, pioneered the development of bipolar membrane electrodialysis (or water splitting) for the conversion of salts into their acids and bases. This system can also be used to acidify or basify process streams without the addition of chemicals. AQUALYTICS has unique experience in designing, building, and operating full-scale installations with bipolar membranes. Our expertise and capabilities include a large scale bipolar and monopolar membrane manufacturing plant and proprietary stack hardware.

CONTACT US AT:
7 Powder Horn Drive
Warren, New Jersey 07059-5191 USA
(908) 563-2800
FAX (908) 563-2816

MARMON
A member of The Marmon Group of companies

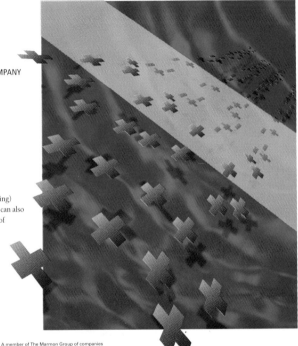

RECOVERY OF SALTS, ACIDS AND BASES

Dialysis and Electrodialysis Applications

The applications of dialysis in industry (see Table 1) are mainly to recover acid or alkali from aqueous solution. Material transfer is by diffusion across an appropriate thin membrane which is generally a slow process and has thus limited the scope for use of dialysis in industry. The early commercial processes were used to recover sodium hydroxide from colloidal hemicellulose during viscose manufacture and for the separation of nickel sulphate from sulphuric acid in electrolytic copper refining. The use of ion-exchange materials expanded the application of dialysis, offering selective transport of acid or alkali. More recently the use of 'dialysis' to enable extraction between aqueous/organic solvent is attracting some interest and is described in section .

The application of an electric field in electrodialysis is one method of increasing the rate of transport of species across the membrane.

There are many potential applications of electrodialysis in desalting and these include
- concentration of acetic acid
- concentration of Glauber's Salt
- concentration of pulp processing waste liquor
- desalting of radioactive waste water
- purification of electrophoresis coating bath
- recovery of electrolytes from plating lines

Treatment of Plating Bath Rinse Waters

Conventional methods of treatment of galvanising effluents generally decompose the effective chemicals remaining in the effluent or separate them out as a precipitated sludge. In recent years the use of electrodialysis (ED) in the recovery of metals from dilute waste liquors has increased rapidly, particularly in the USA. The main area of application is in the processing of rinse waters from the electroplating industry. Using ED, complete recycling of the water and metal ions can in principle be achieved. The recovery of HF and

TABLE 1 – An Overview of Dialysis Application.

Use	Application
Alkali (NaOH) collection	Pulp-pressing solution used in the rayon industry Mercerising waste water Silking waste water Grease-free waste water from the metal industry
Acid collection	Waste acid used for pickling Waste acid from alumite processing Waste acid from aluminium etching Rare metal refining processes Metal refining processes Reproduction of EL sulphuric acid and collection of acid Treatment of waste acid for etching against radioactive contamination from atomic power plants Refinement of by-produced NaOH from organic synthesis processes Removal of acid from organic matter
Salts	Removal of NaCl in producing salt-reduced soy sauce Desalination of enzymes containing cellulose Removal of oil from greasy water Desalination of vegetable proteins Desalination of soluble high molecular material Desalination of waste sugar solution from cane sugar processing Desalination of extract of fishes and shells Removal of ash from milk
Medical care	Artificial kidneys
Alcohol	De-alcoholisation of beer
Electrodialysis membrane	
Concentration	Concentration and desalination of sea water Collection of valuable components in waste water Concentration of diluted solution
Desalination	Desalination of salty water Desalination of industrial waste water Application to fermentation chemistry Sterilisation of water
Separation and refinement	Separation of ion material from non-ion material Desalination of whey Desalination of amino acids Desalination of cane sugar Refining process in food and medical industries Displacement reaction using membrane Production of acid and alkali using bi-polar membrane
Electrodialysis	Electrodialysis of alkaline metal as in electrodialysis with salt Electrodialysis with NaOH Application to organic electrodialysis synthesis

RECOVERY OF SALTS, ACIDS AND BASES

H_2SO_4 from pickle rinse solutions and the removal of heavy metals from electroplating rinse waters by electrodialysis is especially attractive since a substantial recycling of rinse water and rinse water constituents can often be achieved. In a typical plating line, shown in the flow diagram of Fig 17 the plated metal parts are transferred from the plating bath to a still-rinse and then to the regular rinse bath.

Due to the relatively high salt concentration in the plating bath there is a considerable drag-out of salts from the plating bath to the still-rinse. The concentration in the still-rinse therefore increases rapidly and the drag-out of metal from the still-rinse becomes significant. Further treatment of the rinse solution is needed and by using electrodialysis the salts are removed and concentrated and recycled directly to the plating bath. The dilute is fed back to the still-rinse keeping the salt concentration in the still-rinse to a low level.

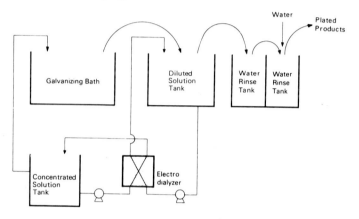

FIGURE 1 – A typical electroplating line.

The ED process is being used successfully on plating baths for metals such as Au, Pt, Ni, Cu, Ag, Pd, Cd, Zn and Sn/Pb. In general ED can concentrate the solution up to bath strength, for example Ni from 1 g/dm³ to 60 g/dm³. This is approximately an order of magnitude greater than that possible by reverse osmosis, which is a competitive membrane technology for this type of recycling operation. A disadvantage of ED is that it will not remove nonionic solutes (eg organics) and thus the purified permeate rinsewater is not as pure as that obtained by RO.

Recovery of Ni

Unlike many others nickel plating requires much more metal because, nickel is plated to several tens microns in thickness as primer for chromium plating. The cost of chemicals for the nickel plating is higher than that for copper, chromium or zinc plating. The loss of nickel taken out from the plate cell is large and thus the recovery of nickel, as well as the treatment of the waste water, are important.

Traditionally the nickel plated materials are rinsed with water in the recovery cell for nickel recovering purposes. In fact, only a small quantity of nickel can be recovered by this method, because the recovered nickel solution is diluted with water to one-tenth in concentration of the normal plating solution. This makes it impossible to feed back the whole quantity of solution of the plate cell. To solve this problem, a closed system for almost complete Ni recovery has been developed by applying electrodialysis. This system consists of the automatic plating process and close waste water treating process. The electrodialysis permits the following features to be obtained.

TABLE 2 – Performance of Electrodialysis in Galvanising Bath Effluents.

Effluent	Feed Solution (gdm^{-3})	Diluted Solution (gdm^{-3})	Concentrated Solution (gdm^{-3})
i) Silver Effluent AgCN	36	0.5	25 - 35
KCN	60	0.8	50
Free-KCN	40	0.6	30
K_2CO_3 (g/l)	45	0.6	40
pH	11 - 12	9 - 10	10 - 11
ii) Copper Solution Cu^{2+}	0.16	1.56	0.01
H^+	0.088	0.66	0.05
$NiSO_4\ 6H_2O$	300	14.0	288
$NiCl_2\ 6H_2O$	50	5.8	61
H_3BO_3	50	11.1	13

- The Ni concentration in the recovery cell can be kept at as low as 1.5 to 3 dm^{-3}, which results in a high recovery efficiency of the recovery column.
- Concentrated solution after electrodialysis, 60 dm^{-3} in Ni concentration and between 5 and 6 in pH values can be sent back to the plate cell without any other treatment

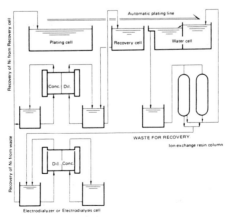

FIGURE 2 – Flowsheet of a closed system for nickel plating system.

- The complete closed system includes ion exchange and virtually eliminates the need of expenses for waste water treatment.

Ni recovery by electrodialysis.

The typical performance of ED on nickel galvanising effluents is shown in Table 2. Electrodialysis is also applied to the treatment of copper galvanising and zinc galvanising liquors. The performance of a zinc galvanising operation is shown in Table 3.

TABLE 3 – Specifications of Facilities for Treatment of Zinc-Galvanization Effluent.

Model of Electrodialyzer	TS-10
Galvanizing Solution (g/l)	100 as Zn
Diluted Solution (g/l)	2.5~3.0 as Zn
Concentrated Solution (g/l)	80~100 as Zn
Capacity of Concentrated Solution	50 Liter/Hour
Recovery of Zinc (kg/Hour)	4~6 (120 kg/day)
Number of Membrane-pairs	330
Operation	Full automatic, partial-circulation system
Space Requirement (M X M)	2.5^L X 2.2^W X 2.0^H

ED has also been used to maintain the plating bath quality and thus eliminate the need for periodic replenishment. . The process uses an uncharged EDTA-Cu complex solution and the reduction of copper is with formaldehyde. The formate ions from the oxidation of formaldehyde and the sulfate ions introduced with the make-up copper are undesirable by-products that are removed by ED. NaOH is added to adjust pH sufficiently to ionise the formic acid. The uncharged EDTA-Cu complex remains in the diluate, which is returned to the plating bath.

Recently the use of bipolar membranes has been incorporated in an ED cell as a means of reducing the metal ions content in the rinse waters to meet discharge levels, enabling the metal ions to be returned to the plating line. The use of bipolar membranes is discussed in more detail later in this book.

Concentrating recovery of zinc sulphate.

The use of ED can be used in the treatment of process solutions from plating baths containing chromium. Dilute rinsewater solutions can be concentrated by the removal of the CrO_7^{2-} ions across an anion exchange membrane into the chromic acid concentrate. ED is also used as a method of purification of plating solutions by the removal of contaminant ions which are picked up during the plating operation. The contaminant ions, for example copper may be removed from the catholyte by electrodeposition. The operation can also be combined with the regeneration of the Cr(VI) by the anodic oxidation of the Cr(III).

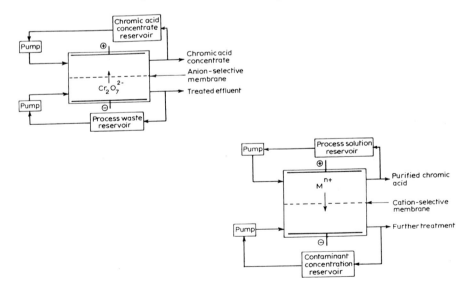

FIGURE 3 – Electrodialysis used to concentrate chromic acid liquors.

RECOVERY OF SALTS, ACIDS AND BASES

Other Applications

Certain industries require industrial waste liquor to be recycled for used of industrial water. In certain cases a simple desalting by ED is sufficient to provide the necessary water. Dump leach waters containing heavy metal ions have been treated by electrodialysis. The removal of nitrates (and nitrites) from drinking water by electrodialysis has also been demonstrated.

A number of studies have used electrodialysis for the recovery and upgrading of phosphoric acid. One of these processes employed an electrodialysis cell fitted with a supported liquid membrane of a trialkylamine in amyl alcohol. This liquid membrane allowed selective transfer of the phosphate ion and prevented transfer of cationic impurities, thus giving a purified phosphoric acid solution. A second example considered the application of ED to the concentration and recycling of a dilute waste phophoric acid solution produced in the generation of phosphoric acid from phosphate rock. This traditional application of ED was able to concentrate reagent grade acid by a factor of 2 -3, up to concentration of 1.0 mol/dm^3 with energy consumptions of 1.73-2.5 kWh/kg of P_2O_5. The cost of ED at that time were approximately 17% higher than evaporation. The process can achieve higher concentrations but at the expense of significantly higher energy costs.

A process referred to as electro-electrodialysis or EED offers a means of recovering metals and reconcentrating acids from wastewaters simultaneously. The principle of operation involves the simultaneous deposition of the metal ions, while the anion transports through the anion-exchange membrane to form acid, with the hydrogen ions liberated by the anodic evolution of oxygen. High levels of acid generation can be achieved with new low-proton-leakage anion-exchange membranes.

A second example of a process using combined membrane separations is described for the recovery of ammonium sulphate (and some sulphite) from a wastewater emanating from a plant to produce p-aminophenol. In this process the ED unit concentrates the retentate from the RO unit prior to evaporation. The membranes in the ED unit were able to tolerate the presence of the organic constituents.

A process for the recovery and reuse of sodium hydroxide from industrial effluents typically from ion-exchange resin generation, pulp and paper, textile and various washing industries combines electrochemical recovery with neutralisation, microfiltration and nanofiltration. The process concentrates the effluent to produce sodium hydroxide, of a suitable quality and concentration for reuse (100 to 200 g/dm^3), reusable water, hydrogen and oxygen gases and two low volume organic concentrates.

The process involves 4 overall stages:-
(i) effluent neutralisation using carbon dioxide gas
(ii) cross-flow microfiltration to remove suspended and colloidal, particulate and complexed and waxy contaminants
(iii) nanofiltration to remove soluble impurities including organics, colour and polyvalent inorganic ions
(iv) electrochemical membrane depletion of the sodium salt in a cell stack with the recovery of sodium hydroxide and carbon dioxide.

The cell uses cation exchange membranes (Nafion) for Na^+ ion transport into the catholyte for NaOH recovery. The anode generates O_2 and H^+ ions which cause a shift in the equilibrium of the carbonate species liberating CO_2 gas, which is recycled to the neutralisation stage. At optimum operating conditions the sodium hydroxide can be recovered at a current efficiency of between 70% and 80%, with a power consumption of 3500 kWh/tonne of NaOH.

Metathesis Reaction

The metathesis reaction is a double decomposition between two electrolytes. The reaction is carried out by means of ion exchange membranes

$$AX + BY \rightarrow AY + BX$$

Examples of this process are generation of sodium hydroxide and ammonium salts from ammonium hydroxide and sodium salts. Another example is the generation of photographic emulsion.

$$NaBr + AgNO_3 = AgBr + NaNO_3$$

The process can be continuously achieved in an electrodialysis cell comprising alternating anion and cation exchange membranes. One compartment is fed with the NaBr solution and the Br^- ions and Na^+ ions are transferred in opposite directions to the adjacent compartments. The compartments next to these are fed with $AgNO_3$ solution from which Ag^+ ions and NO_3^- ions are electrically separated through the cation and anion exchange membranes respectively, into the adjacent compartments. Thus a photographic emulsion of AgBr is formed.

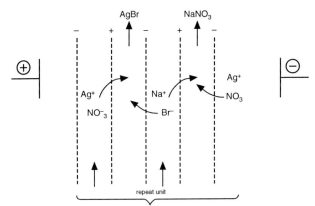

FIGURE 4 – The recovery of photographic reagents by membrane ion exchange.

Reverse Osmosis Treatment of Electroplating Solutions

Effluents from electroplating processes contain contaminants which are both toxic and valuable, such as a range of metals (Ni, Cr, Zn, Cu, Cd etc) and cyanide. These effluents are generally quite acidic or alkaline and in need of significant treatment prior to discharge. The effluents usually arise as a result of the rinsing of the plated articles with water. Such rinsing takes place in a countercurrent series or finishing baths and results in an effluent which is too dilute to be returned directly to the plating process. The use of RO offers in

RECOVERY OF SALTS, ACIDS AND BASES

principle a means of recovering the plating chemicals as a concentrated solution to be re-used in the plating bath and also recovering pure water.

The largest electroplating operation in terms of tonnage is tin, followed by nickel, copper and chrome in approximate equal capacities.

Reverse osmosis is successful in the treatment of Watts nickel-plating baths see Fig 5 This operation is carried out at pH of 3.0 to 4.0 using divalent nickel in a solution also containing buffers (boric acid) and brighteners. The pH of this operation means the cellulose acetate membranes are not rapidly degraded. The rejection of boric acid depends upon its degree of ionisation and is typically 40%, which means monitoring and control of the concentration during plating is necessary.

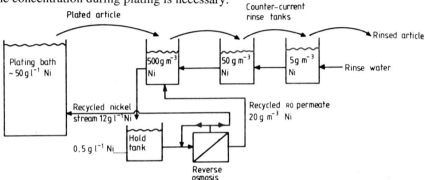

FIGURE 5 – Nickel recovery by RO from electroplating rinse waters.

The major factor hindering greater adoption of RO in electroplating processes is the extremes of pH typically encountered. Copper, zinc and cadmium baths use cyanide, as a complexing agent, in alkaline solution. The pH is between 11 to 13 and is thus generally unsuitable for polyamide membranes. Plating baths with lower pH of 10 (copper cyanide) and 8.6 to 9.2 (copper pyrophosphate) are amenable to RO using polyamide membranes. In acidic baths, for example chrome plating, the solution requires neutralisation prior to treatment by RO. This is less attractive as the chemicals, typically sodium hydroxide, used in neutralisation mean additional processing is required. A large chromic acid recovery plant is known to have been installed at the Rock Island Arsenal in the USA

One process is known to operate where the reverse osmosis concentrate is treated by a cation exchange process to remove the sodium ions. Reverse osmosis is also used in situations where recovery of effluent is not economic such as from mixed plating effluent, where it's major use is as a preconcentrator to reduce the loading on final evaporators.

Diffusion Dialysis

Diffusion dialysis is a process which can recover either acids or bases from waste solutions. The method is widely used to recover acids used in the pickling of metals such as iron, copper and aluminum.

It is a relatively simple process which uses ion-exchange membranes which are similar to those used in electrodialysis but which have specific permselectivity characteristics. The principle of operation utilises the difference in concentrations of species in solutions partitioned by ion-exchange membranes. Diffusion dialysis requires no external forces to promote separation. The only energy required is that to supply the pumping force for the flow of liquids between the membranes. This compares favourably to process such as reverse osmosis or electrodialysis where high energy inputs are required.

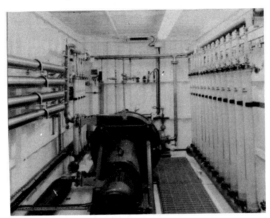

Reverse osmosis plant for wastewater treatment.

The diffusion dialysis process is carried out in a cell stack consisting of a series of anion-exchange membranes (see Fig 6) separating alternating compartments containing either the waste acid solution to be treated or water. The waste acid solutions to be treated or water. The waste acid solutions flow up through the cell while the water solutions flow simultaneously down through the cell compartments. The high concentrations of ions in the waste acid gives rise to high osmotic pressures which cause transport of the ions into the adjacent water streams through the membranes. The anion-exchange membranes function to prevent the transport of free cations and also anions associated with salts, eg $Fe_2(SO_4)$, but permits permeation of anions associated with hydrogen, eg H_2SO_4. Thus overall acids are transported into a water stream which is continuously replaced by flowing through the cell thereby maximising the osmotic driving force. The attractive characteristics of diffusion dialysis are:

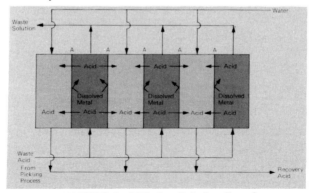

FIGURE 6 – Diffusion dialysis cell stack.

RECOVERY OF SALTS, ACIDS AND BASES

- low running costs with minimal energy requirements for heat or electricity
- almost any acid can be recovered efficiently
- reduction in waste treatment neutralisation costs
- simplified and simple process operation
- stable product quality assumed by continuous recovery of the acids.

Diffusion dialyser.

The features of diffusion dialysis acid recovery mean certain composition and concentrations ranges are more attractive for commercial operations. Consequently the treatment has largely been used in the following applications

- H_2SO_4 from steel plate and wire pickling
- H_2SO_4 from aluminium anodising
- HCl from aluminium etching
- HNO_3 (and HNO_3/HF) from stainless steel pickling.
- H_2SO_4 from TiO_2 (titanium) plant

A typical installation is in acid pickling of metals see Fig 7. In this process the effective acid concentration deteriorates while the dissolved metal concentration increased continuously. This results in a decrease in efficiency of the pickling and eventually the pickling solutions life expires and would have to be replaced. The spent solution would then be neutralised and disposed. This practice is unattractive from the point of economics, waste disposal and quality control during the pickling operation. By introducing diffusion dialysis within this operation these disadvantages are largely rectified, recovered acid from the diffusion dialyser is continuously returned to pickling and the amount of waste neutralisation required is drastically reduced. A typical performance in sulphuric acid recovery will see the pickling acid solution concentration reduced from 191 dm^{-3} to 3.9 dm^{-3} with a recovered acid concentration of 165 dm^{-3}. The amount of metal leaked into the recovered acid solution is not large.

The capacities of diffusion dialysis plant of which over 50 are reported in use, vary from 0.5 to 50 m^3/day. Dialysis is performed in simple filter press types of stacks. Membranes are separated by plastic spacers, 1 to 2 mm in thickness, with, in large installations, several hundred cell pairs used, giving membrane areas of several hundred square metres. The membranes used are between 0.1 mm to 0.3 mm containing ion exchange groups in the matrix and are usually reinforced by inert material. The typical performance of diffusion dialysis plant are reported in Table 4.

Diffusion dialysis as a process completes with a reciprocating flow process using ion-exchange resins (ion retardation). Other applications include purification of battery waste acid and treatment of almite process waste acid.

TABLE 4 – Example of Acid Recovery by Diffusiondialysis No. 2.

Waste Acid (Free Effective Acid and Concomitant Metal)			HNO_3 $-Al$	HNO_3 HF $-Fe$
Temperature (°C)			25	30
Capacity (L.Hr – m²)		Water	1.0	0.9
		Waste Acid	1.0	0.9
Concentration of Waste Acid (g/100ml)		Acid	102.0	HNO_3 : 150.0 HF : 24.0
		Metal	23.9	38.0
Concentration of Recovery Acid (g/100ml)		Acid	126.0	HNO_3 : 131.0 HF : 14.0
		Metal	0.9	1.5
Concentration of Waste Solution (g/100ml)		Acid	4.1	HNO_3 : 15.5 HF : 9.9
		Metal	18.2	38.5
Recovery Ratio of Free Acid (%)			97	HNO_3 : 90 HF : 60
Leak Ration of Metal (%)			3	4
Specifications of Facilities for Recovery of Sulfuric Acid				
Model of Diffusiondialyzer			TSD–25	
Capacity (M³/Day)			1.3	
Amount of Removal Fe			1400 kg/Month as Fe	
Contents of Waste Acid	H_2SO_4 (g/l)		368	
	Fe (g/l)		45	
	Temp. (°C)		40	
Recovery Ration of Free Acid			80	
Leak Ratio of Metal (%)			5	
Number of Membrane			450 sheets as NEOSEPTA AFN	
Space Requirement (M X M)			4^L X 2^W X 4^H	
Operation			Full-automatic system	

RECOVERY OF SALTS, ACIDS AND BASES

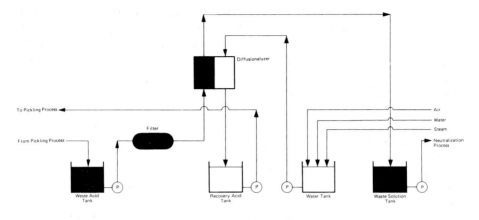

FIGURE 7 – Flow diagram of diffusion dialysis.

Caustic Recovery

A relatively recent development has been the commercialisation of diffusion dialysis for the recovery of caustic. The membranes used in this case are cation-exchange membranes which permit the passage of free base while blocking anions and salts, as shown in Fig 8.

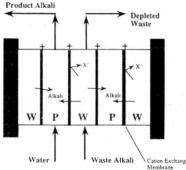

FIGURE 8 – Principle of operation of alkali diffusion dialysis stack.

One known specific application is in the recovery of sodium hydroxide from spent aluminium chemical baths. Spent alkali etchant consists mainly of NaOH and alunminate metal ions in the form of $NaAlO_2$, formed by the solution of aluminium metal in NaOH. $NaAlO_2$ is an unstable, soluble salt that decomposes when the pH is lowered by the removal of NaOH, according to

$$NaAlO_2 + 2H_2O \Leftrightarrow Al(OH)_3 + NaOH$$

The precipitation of $Al(OH)_3$ is slow; which allows diffusion dialyssi to first recover NaOH enabling $Al(OH)_3$ recovery downstream of the DD stack by precipitation.

The DD process for alkali recovery operates in a similar manner described for acid recovery. The alkaline etchant flows upwards, separated from the water flow downwards; by a cation exchange membrane permeable only to free bases and water. NaOH diffuses across the membrane from the high concentration (etchant) to the low concentration in the water, while water, migrates into the waste stream compartment by osmotic pressure.

The practice the outlet product concentration can equal or slightly exceed that of the inlet feed; as a result of the common ion effect. Here a higher driving force is created by the total sodium ion concentration, including the sodium ions associated with the aluminium ions, thus making a base product more concentrated than the feed alkali. The final base product contains typically less than 1% aluminium and is suitable for recycling to the etching tank.

The overall process is shown in Fig 9. The etchant is first filtered to remove smut and is then sent to the DD for caustic recovery an dsimultaneous formation of a supersaturated etchant solution. This solution is sent to a simple stirred tank crystalliser where $Al(OH)_3$ is crystallised. The underflow from the stirred tank, a 35% slurry, is dewatered in a plate and frame filter press. The recovered alkali and the supernatant solution from the crystalliser are returned to the etching tank for recycling. More than 99% of the caustic is recovered.

The precipitate from the crystalliser is pure white crystalline $Al(OH)_3$, which is sold after washing and reduction of pH to 10 or below. The smut is essentially CuS, a solid used as an additive to the milling bath to control the etching rate, and is returned to the etching tank. The remaining product from the process is water, which can be eliminated by evaporation or which can be recycled as make-up for the DD stack.

Practical application of the process with a DD stack using 90 m² of membrane is at Caspian Chemicals in San Diego, California. The installation is reported to be processing approximately 130 dm³/h of spent etchant contain 16.5% NaOH and 10.6% aluminium. The following streams are produced

Caustic product stream: 35% of free caustic removed from the solution, 17% NaOH produced at a rate of 15.5 dm³.

Treated salt product stream: 9% NaOH and 8% alumni at a rate of 59.5 dm³.

The process realises major savings as a results of reduced fresh caustic purchases and virtual elimination of the need to dispose of spent caustic.

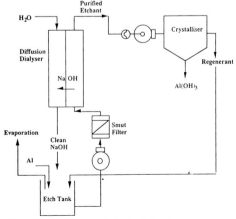

FIGURE 9 – Diffusion dialysis etchant recovery system.

Donnan Dialysis

Donnan dialysis is a process used to exchange ions between two solutions separated by an ion-exchange membrane. The principle of operation is depicted in Fig. 10 for the case of a copper sulphate solution separated from a sulphuric acid solution. In this example the concentration of H^+ ions in the acid is much higher than in the copper sulphate solution (pH 7) and thus there is a driving force for the transport of protons from acid to the copper sulphate solution. The transport of H^+ ions through the cation exchange membrane induces an electrical potential difference which will counter balance the concentration driving force of protons. The potential difference will instigate a flux of Cu^{2+} ions from the copper sulphate solution to the sulphuric acid solution. The transport of Cu^{2+} ions will be maintained until a point when the copper ion concentration difference is of the same order of magnitude as the H^+ ion concentration difference.

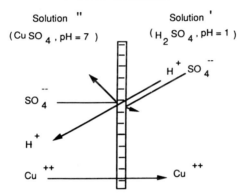

FIGURE 10 – Principle of donnan dialysis between Cu^{2+} and H^+ ions across a cation exchange membrane.

The process can be applied equally well to anions using anion-exchange membrane. Examples of Donnan dialysis include
1) Softening of hard water. Divalent ions, eg Ca^{2+}, Mg^{2+} or SO_4^{2-} are exchanged for monovalent ions such as Na^+ or Cl^-. The process competes directly with conventional ion-exchange water softening technology where it offers a reduction in waste salts and a easier operation.
2) Sweetening of citrus fruits. In this process citric ions in the juice are replaced by OH^- ions in a caustic solution.

Ion Exchange Membrane Processes for Salts and Acids Recovery

In many process industries the use of acids and alkalies results in the generation of salts, both inorganic and organic. The regeneration of the salts to their original constituents by the addition of hydrogen ions and hydroxide ions is highly desirable to minimise chemical consumption, storage costs and effluent treatment costs. The use of electromembrane process based on ion-exchange membranes is of increasing interest and several applications are being realised. The methods used are referred to as electrohydrolysis, electrolytic water dissociation (with bipolar membranes), diffusion dialysis and donnan dialysis.

Electrohydrolysis

The electrolysis of water is used to generate the hydrogen and hydroxide ions. The electrohydrolysis of aqueous streams of sodium sulphate to regenerate sulphuric acid and caustic soda has been known for many years.

- $2H_2O \Leftrightarrow 2H^+ + 2OH^-$
- $2Na^+ + 2OH^- \rightarrow 2NaOH$
- $SO_4^{2-} + 2H^+ \rightarrow H_2SO_4$

The process can be operated in a three compartment membrane cell, shown in Fig 11, in which the sodium sulphate stream is fed to the central compartment and the ionic components transferred across ion exchange membranes. Thus the sodium ions transferred across the cation exchange membrane combine with the hydroxide ion generated through the formation of hydrogen at the cathode. The sulphate ions transferred across the anion exchange membrane combine with the hydrogen ions, formed anodically, to produce the sulphuric acid. This process is limited in terms of the maximum concentrations of acid and base that can be obtained due to salt leakage through membranes. Overall the current efficiency falls off significantly with the formation of higher concentrations of sulphuric acid. In addition, although in principle the process produces an acid stream without sodium sulphate present, the selectivity of the anion exchange membrane is not sufficiently good to eliminate sodium ions from the anode compartment.

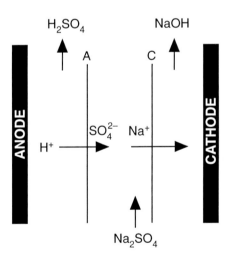

FIGURE 11 – Three compartment electrohydrolysis cell.

As a result of the above limitations an alternative process based on a two compartment cell has been used to carry out this function. In comparison to the two compartment cell this process gives a lower cell voltage than the three compartment unit, avoids the weakness of the anion exchange membranes and gives high current efficiencies at sulphuric acid product concentrations of 15% and greater. Although the sulphuric acid contains sodium sulphate, this can be removed by crystallisation to meet customer requirements.. A purified sodium sulphate solution is saturated by the addition of recycled sodium sulphate from the crystallisation. The saturated electrolyte passes to the anolyte chambers of the electrolyser, where it is enriched with respect to sulphuric acid, the sodium ions being driven through the membrane. A proportion of the exhaust anolyte is passed to anolyte work up whilst the remainder is recycled to the anolyte saturator.

Anolyte work up raises the concentration of sulphuric acid and adjusts the sodium sulphate concentration to meet the requirements of the "host" process. This is achieved by evaporation and crystallisation. Any crystallised solids are returned to the anolyte saturator for dissolution and recycle.

Work up of the catholyte is relatively simple. For concentrations significantly in excess of the exhaust catholyte (20% kg/kg sodium hydroxide) it is usually carried out in a double effect evaporator. Importantly the catholyte product, typically 20% w/w NaOH is pure.

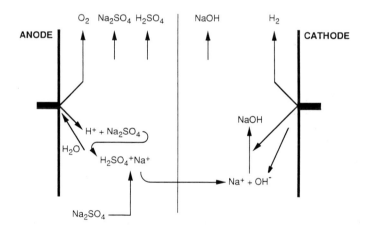

FIGURE 12 – Two compartment electrohydrolysis cell.

An alternative salt splitting process has been developed for the recycling of Na_2SO_4 solution in which ammonium sulphate and sodium hydroxide are generated rather than generate sulphuric acid and sodium hydroxide. This is achieved in a three compartmental cell, in which Na_2SO_4 is fed to the central compartment and SO_4^{2-} ions migrate through an anion exchange membrane to the anode. Ammonia is added continuously to this compartment to maintain a constant pH of 1.5 and form the ammonium sulphate, which is to be used as fertiliser. Maintaining a low pH is important to ensure stability of the DSA anode, which otherwise, along with other materials (stainless steel, Ni, Pt graphite), would corrode at high pH in ammonia solutions. The cell uses Neosepta AMH anion exchange and Nafion 902 cation exchange membranes and a Ni cathode to produce caustic and ammonium sulphate solutions of concentrations greater than 30% and 40% respectively. Current efficiencies of the order of 90% can be achieved.

The use of the salt splitting technique can be applied widely to many inorganic and organic species. For example, an aqueous sodium citrate stream, which may have been pre concentrated by reverse osmosis, can be processed in an electrolytic cell fitted with two cation-selective membranes. Here the protons generated at the anode are transferred across one of the cation exchange membranes into the central compartment. The second membrane transfers the sodium ions present in the citrate feed into the catholye chamber, where they combine with the OH⁻ ions, generated by the cathodic evolution of hydrogen, to form a sodium hydroxide solution. Thus in the central compartment the sodium ions are effectively replaced by the H^+ ions thus forming the citric acid.

Electrodialytic Water Dissociation/Bipolar Membranes

A process for salt splitting is based on the use of bipolar membranes electrodialytic water dissociation. Bipolar membranes as discussed in Section 2 are polymeric materials composed of two homopolar ion-exchange membranes; one cationic and one anionic. When placed in an electrochemical cell, with the cationic layer in contact with catholyte, current is carried by protons moving through this layer and by hydroxyl ions moving through the opposite anionic layer. Because of this property bipolar membranes are often referred to as water-splitting membranes. The process is electrodialytic in nature, involving the change in the concentration of ions already present in the solution. For efficient operation the membrane should have a good water permeability from the external solution to the interface and a thin interface between the cation and anion layers to give efficient, low resistant, transport of the hydrogen and the hydroxide ions.

The free energy for generating one normal ideal product solution is therefore given by

$$\Delta G = -RT \ln a^i_{H+} \cdot a^i_{OH-} = -RT\ln(K_w)$$

where K_w is the dissociation constant of water.

The theoretical potential to therefore achieve the water splitting capability is 0.83 V at 25°C. The actual potential drop across a bipolar membrane is quite close to this being in the range of 0.9 -1.1 V for current densities between 500-1500 A/m², which is the general region of practical interest. The value of the membrane potential drops equate to theoretical energy consumptions of the order of 600-700 kWh/tonne of NaOH. Of course the actual energy consumptions are significantly higher because of the ohmic resistances, in the other cell stack components, in practical operating units.

Bipolar membranes are a unique application of ion-selective membranes for the regeneration of acid and base constituents from aqueous salt solutions. The application of bipolar membranes avoids the production of excess quantities of by-product gases, and reduces the energy costs associated with electrode polarisation in the more conventional electrolytic approach. Bipolar membranes are used with either two- or three-compartment cells. The basic three-compartment cell, shown in Fig 20, consists of one cation exchange,

RECOVERY OF SALTS, ACIDS AND BASES

one anion exchange, and one bipolar membrane. In operation, positive and negative ions migrate through the respective monopolar membranes and concentrate in compartments on opposite sides of the bipolar membranes. Water diffuses through these layers to an interfacial region where it dissociates into the constituent hydrogen and hydroxide ions, these ions diffuse back into the adjacent compartments, in opposite directions, to produce alkali and acid solutions.

Typical use of the bipolar membranes is in the treatment of concentrated salt solutions, such as Na_2SO_4 from the chemical industry to produce H_2SO_4 and NaOH. A cell system consisting of an anion, a bipolar and a cation exchange membrane as a repeating unit is placed between two electrodes. The Na_2SO_4 solution is placed between the cation – and anion-exchange membranes. When direct current is applied, water will dissociate in the bipolar membrane to form the equivalent amounts of H^+ and OH^- ions. The H^+ ions permeate through the cation-exchange side of the bipolar membrane and form H_2SO_4 with the sulphate ions provided by the Na_2SO_4 solution from the adjacent cell. The OH^- ions permeate the anion-exchange side of the bipolar membrane and form NaOH with the sodium ions permeating into the cell from the Na_2SO_4 solution through the adjacent cation-exchange membrane. The net result is the production of NaOH and H_2SO_4 from Na_2SO_4 at a significantly lower cost than by other methods.

Bipolar membranes can also be used in an alternative two cell configuration, regenerating only one base or acid. For example in an anion/bipolar membrane configuration, the anions move through the anion exchange membrane and combine with the H+ ions arising from the bipolar membrane to form the acid product. This type of cell is useful for converting salts of weak bases (e.g. ammonium nitrate) to a salt/base mixture and a relatively pure acid. Conversely the cation/bipolar membrane two compartment cell is useful for the processing of the salts of weak acids (of organic acids) to give a relatively pure base stream and a mixed acid/salt stream. The performance of the two compartment cells can be enhanced by the introduction of a third chamber. For example the multichamber, cation cell now uses two cation exchange membranes. In operation the salt solution is first fed to the chamber between the two cation exchange membranes and then passes to the acid compartment. This gives a salt/acid stream with a higher concentration of acid than the standard two compartment cell. The multichamber anion cell is used in an analogous way to that of the cation cell. Clearly this multicell arrangement incurs a greater electrical energy cost due to the higher cell voltage than that needed in the two compartment cell.

The potential applications of bipolar membranes are numerous and Table 5 summarises the applications under development or implemented.

The consumption of sodium hydroxide (caustic soda) industries is large and this material is usually produced in conjunction with chlorine from the electrolysis of sodium chloride. Concerns over the environmental impact of chlorine in resulting in a steady decline in chlorine used whilst caustic use remains steady, which will see the demand for caustic eventually exceed that of chlorine. Thus alternative process which can produce caustic at competitive levels are of increasing interest.

The following are applications of Bipolar Membranes.

Membrane Technology

an international newsletter

Every month, in just 12 succinct pages, Membrane Technology newsletter brings you an up-to-date international digest to follow all the news and developments affecting Industrial membranes and membrane technology.

Your monthly snapshot of world wide news

Each issue is packed with essential information... from the latest news & views to case studies, and covering the entire range of membrane technologies – from micro filtration to reverse osmosis.

In every issue

- Latest news and views on the development and application of Industrial membranes
- Case studies
- New product launches
- The latest patents – designs and inventions
- Research
- Events

For more information contact:

Elsevier Advanced Technology
The Boulevard
Langford Lane
Kidlington
Oxford OX5 1GB

Tel: (+44) (0) 1865 843842
Fax: (+44) (0) 1865 843971

RECOVERY OF SALTS, ACIDS AND BASES

Regeneration of Spent Pickling Liquors

The stainless steel industry uses hydrofluoric acid and nitric acid as a pickling liquor to remove surface oxides formed during the annealiny process. As the acid solution picks up metals, it becomes inactive and eventually spent. Once spent, the waste pickle liquor must be neutralised and taken to a hazardous waste disposal site. The regeneration of the spent liquors by bipolar membranes is operated at Washington steel, Pennsylvania with a capacity of 6000 m^3/yr of liquor. The solution used in pickling is a mixture of 2-5. % HF (hydrofluoric acid) and 8-15 %. HNO$_3$ (nitric acid). The chemical action of the acids produces metalfluorides and nitrates and disposal is an increasing problem.

In the process (Fig 17) potassium hydroxide neutralises the spent pickle liquor to form a potassium fluoride/potassium nitrate solution. Diatomaceous earth is added as a settling aid to precipitate out the metal hydroxides.

The KF/KNO$_3$ solution is sent to a plate and frame filter press which yields solids – iron, nickel and chromium hydroxides. These solids can possibly be returned to the steel smelting operation. From the back end of the filter press comes a clear KF/KNO$_3$ salt solution which enters the bipolar cell stack where hydrofluoric and nitric acids (typically, 3 Normal) and potassium hydroxide (typically, 2 Molar) are regenerated.

The bipolar stack returns the fresh acid mixture to the pickling plant and sends the potassium hydroxide back to the neutralisation unit. In addition, the system recycles a weak KF/KNO$_3$ salt solution back through the filter as an additional wash, assuring high fluoride, nitrate and potassium recovery.

A water recovery unit uses either reverse osmosis or conventional electrodialysis to eliminate water from the filter cake wash. The newly concentrated salt solution is recycled to the cell stack. Clean water, also from the water recovery unit, is used as make-up water in the cell to maintain proper acid and base concentrations.

Economic estimates of the process put a return on invested capital at 2 years.

Hydrofluoric acid recovery from fluosilicic acid

Fluosilicic acid is a byproduct from wet phosphoric acid manufacturing plants. It is considered a harmful pollutant, and its disposal presents an environmental problem. The water splitting membrane can be used to convert fluosilicic acid into hydrofluoric acid – a valuable, saleable product.

Precipitated silica, recovered during the neutralisation, is another potentially saleable product.

In the process aqueous fluosilicic acid (H_2SiF_6) is combined in Reactor 1 with potassium hydroxide and converted to a soluble potassium fluoride salt (K_2SiF_6). Silica (SiO_2) is added so that the KF in the recycle stream also forms K_2SiF_6. The K_2SiF_6 solids are then filtered out of the solution. The filter also removes other soluble impurities, such as phosphoric acid, from the fluosilicic acid feed.

The solid K_2SiF_6 is sent to Reactor 2 where it combines with additional recycled KOH to produce a soluble KF stream and insoluble SiO_2 stream. The SiO_2 is separated from the KF by sedimentation and filtration. The clear, soluble KF stream enters the bipolar cell stack where potassium and fluoride are converted to HF and KOH. A depleted KF stream exiting from the AQUATECH unit picks up the base generated in the cell stack, and the two are recycled back to the reactors. The dilute HF (8% – 13% by weight) from the cell stack is further concentrated. This step breaks the HF-H_2O azeotrope and allows HF to be recovered by distillation. This is achieved by either electrodialysis or by the use of sulphuric acid.

The overall reaction which occur in the process can be summarised as

Reactor 1	$2KOH + H_2SiF_6 = K_2SiF_6 + 2H_2O$
	$6KF + 2H_2SiF_6 + SiO_2 = 3K_2SiF_6 + 2H_2O$
Reactor 2	$K_2SiF_6 + 4KOH = 6KF + SiO_2 + 2H_2O$
Bipolar Cell unit	$6KF + 6H_2O = 6KOH + 6HF$

Typical performance of the bipolar cell stack system are given in the table for a 40,000 ton/year facility. Economic estimates of the process put a return of capital investment at 2 years.

Flue Gas Desulphurisation

The process is a regenerative wet scrubbing system which produces a concentrated sulfur dioxide stream and is applicable to combustion products from low to high sulfur coals. The sulfur dioxide can be liquefied, converted to sulfuric acid or elemental sulfur. The process is based on the use of a high pH sodium solution as a scrubbing medium to remove the

RECOVERY OF SALTS, ACIDS AND BASES

Cell Stack Battery Limits*.

Average current density	100 amp/ft²
Average HF concentration	10%
KOH concentration	11%
Average current efficiency	80%
Potential drop	2.0 volts/cell
Effective membrane area	57,000 ft²
Power	11.5 megawatts

*Estimate for 5.06 T/hr AHF based upon preliminary research data

sulfur oxides from the flue gas. The product solution from the absorber is regenerated using water splitting membranes in which the bisulfite salt is converted to the original sulfite and hydroxide, that are recycled, and an aqueous solution of SO_2 which is readily stripped to recover concentrated sulfur dioxide.

A diagram of the process is shown in Figure 19. The flue gas first has almost all (99%) of the fly ash particulates removed by use of electrostatic precipitators or bag houses. The SO_2-rich flue gas from the boiler is contacted with sodium sulfite solution in an absorber to substantially recover the SO_2 and convert the sulfite to bisulfite. A pre-saturator is not necessary because incoming impurities, such as chlorides, can be readily purged from the process loop. The reactions occurring in the absorber are:

1. $2NaOH + SO_2 = Na_2SO_3 + H_2O$
2. $Na_2CO_3 + SO_2 = Na_2SO_3 + CO_2$
3. $Na_2SO_3 + H_2O + SO_2 = 2NaHSO_3$
4. $Na_2SO_3 + \frac{1}{2}O_2 = Na_2SO_4$

The reactions take place at the adiabatic saturation temperature of approximately 130F. The first three reactions result in SO_2 absorption and the production of sodium bisulfite ($NaHSO_3$). Reaction 4 represents the oxidation of sulfite to sulfate by the oxygen present in the flue gas. A typical absorber recycle solution would have 0.5% NaOH (or soda-ash equivalent), 17% Na_2SO_3 and 7% Na_2SO_4. Typically 65-68% of sulfite is converted to the bisulfite. Sulfate production is expected to amount to 5% of the SO_2 absorbed.

The product solution from the absorber (mostly bisulfite) is filtered to remove suspended flyash impurities and fed to the bipolar regeneration system. A typical absorber product composition is 2.2% Na_2SO_3, 22.5% $NaHSO_3$, 7.4% Na_2SO_4.

The regeneration system is composed of three parts:
(a) the two-compartment cell system (primary recovery unit)
(b) SO_2 stripping/recovery
(c) the three-compartment cell section (secondary recovery unit)

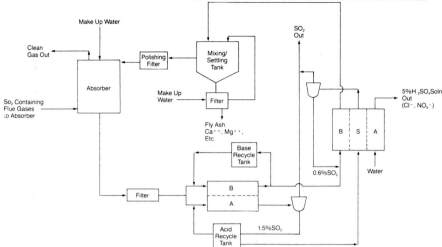

FIGURE 19 – Soxal process for flue gas desulphurisation.

In the regeneration section almost all of the sodium ions are regenerated as sulfite/hydroxide for recycle to the absorber. Concurrently, the SO_2 is recovered as a concentrated (95+%) stream. The small amounts of sulfate values from the oxidation step (reaction 4) along with any chlorides from the absorption step are rejected as a dilute acid stream (HCl + H_2SO_4). This stream is neutralised with limestone and discarded.

The two compartment cell system regenerates the spent bisulfite solution to sulfite and SO_2. In this cell the bipolar membranes are used in conjunction with cation selective membranes as shown in Figure 20, with 200-300 cell units assembled to form a compact water splitting "stack". Spent product solution from the absorber is divided into two streams and fed to the acid and base compartments. The reactions occurring in the cell are:

Acid Compartments

$Na_2SO_3 + 2H^+ - 2Na^+ = H_2SO_3$
$NaHSO_3 + H^+ - Na^+ = H_2SO_3$
$H_2SO_3 = H_2O + SO_2$

Base Compartments

$NaHSO_3 + Na^+ + OH^- = Na_2SO_3 + H_2O$
$Na^+ + OH^- = NaOH$

Thus the sodium bisulfite is converted to SO_2 and sulfite. Any sulfite in the acid compartment feed is also converted to SO_2. However it is desirable to minimise the sulfite in the water splitter feed since its conversion to SO_2 requires twice as much energy as the bisulfite.

The overall current efficiency of the cell depends on bisulfite conversion. Thus the points corresponding to 0 moles/l of $NaHSO_3$ represents complete (100%) conversion of the bisulfite. If the feed solution contained no sulfate (the lower curve) the current efficiency would falls off rapidly as the bisulfite is converted. This is because of the competition between Na^+ and H^+ transport across the cation membranes. Only the transport of Na^+ leads to conversion of $NaHSO_3$ to H_2SO_3. As the conversion of bisulfite proceeds, the ratio Na^+/H^+ will fall to zero at complete conversion. In the process the presence of a second source of sodium formed by oxidation in the absorber (Na_2SO_4) serves to increase the overall efficiency over the entire conversion range.

In the two compartment cell there are two primary sources of inefficiency, H^+ transport across the cation membrane and SO_2 diffusion. The diffusion of SO_2 from the acid to base compartments results in the conversion of caustic soda/sufite to sulfite/bisulfite.

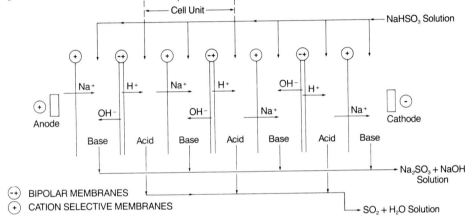

FIGURE 20 – Two compartment cell system for SO_2 recovery.

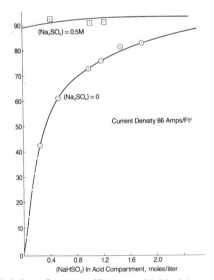

FIGURE 21 – Variation of current efficiency with bisulphate concentration.

The construction of the water splitter units is essentially modular, each primary recovery unit would consist of up to 300 cell units of 1 m^2 membranes. The number of stack modules used is dependent on the amount of bisulfite to be regenerated. It is estimated a

500 MW plant burning $3\frac{1}{2}\%$ S coal (TVA basis) would require 18 operating stacks. The water splitting stacks typically operate at 110F. All of the stacks would be operated in parallel from a common set of acid and base surge tanks.

The second major part of the process is SO_2 stripping and recovery, sulfur dioxide gas is liberated from the dissociation of sulphurous acid in the acid compartment loop of the primary recovery unit. Continuous removal of the gas serves to increase overall process efficiency by decreasing its concentration in solution which in turn minimises diffusion losses through the membranes. As much as 80 – 90% of the regenerated SO_2 can be recovered simply by a sub atmospheric stripping operating (5 – 10 psia). Dissolved SO_2 in the acid recirculation loop is thereby reduced to 2.5 wt% or less which enables the bipolar cells to operate efficiently.

The bleed stream from the primary recovery unit is typically composed of 8-11%, Na_2SO_4, 2.5% SO_2. This stream is further stripped at a higher vacuum (~ $NaNO_3$, $NaNO_2$).

The three compartment cell system (Secondary recovery unit) is used to convert purge sulfate to a dilute acid stream while regenerating the alkali for recycle to the absorber. The unit shown schematically in Figure 6 consists of bipolar membrane, acid compartments, anion selective membrane, salt compartment, cation selective membrane and base compartment with the two compartment cell. The reactions are:

Water Splitting	$H_2O = H^+ + OH^-$
Acid Compartments	$2H^+ + SO_4 = H_2SO_4$
	$H^+ + Cl^- = HCl$
Base Compartments	$Na^+ + OH^- + NaHSO_3 = Na_2SO_3 + H_2O$
	$Na^+ + OH^- = NaOH$

The salt compartments serve as the source of Na^+, SO_4 and Cl^- ions. Only a portion of the salt (equivalent to overall formation rate of sodium sulfate and chloride in the absorber) is converted in secondary cells while the balance is combined with the base product stream and sent to the absorber.

The concentrated SO_2 (95 wt%) is the major "byproduct" of the process. It can be liquefied or converted to sulfuric acid (or elemental sulfur). A relatively small amount of acid is formed as a second byproduct available as a dilute stream (5 wt% is typical). This is suitable for in-house uses such as ion exchange regeneration or cleaning of process units. Alternatively, this acid stream can be neutralised with limestone and disposed of in an environmentally acceptable manner.

The combustion of coal is well known to produce environmental problems worldwide through the generation of SO_2 and also other species. Electrodialytic salt splitting technology has applications in this area and can potentially compete with other processes such as Wellman-Lord and Mag-Ox.

The advantages of the bipolar cell technology are listed in Table 1. The process is modular, making it readily amenable to scale up. The process described regenerates essentially all the sodium base values while recovering the bulk of the sulfur values in the form of concentrated SO_2 stream. It is also compact which makes it attractive for use in retrofit applications and where space is at a premium.

Economic analyses of a 500 MW plant burning 3_% S coal indicate the process is significantly superior to Wellman Lord and Mag-Ox processes.

RECOVERY OF SALTS, ACIDS AND BASES

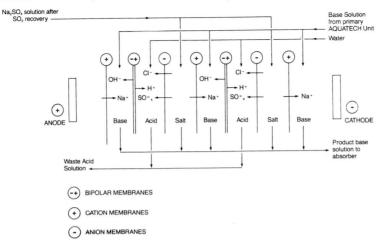

FIGURE 22 – The three compartment bipolar cell unit.

TABLE 6 – Features of SOXAL™ Process.

• High pH Scrubbing Solution Resulting in Very High SO_2 Removal Capabilities
• Sodium-Based Clear Solution for Scrubbing
• Electrically Driven Process
• Operating Flexibility
• Built-in Load Levelling Capability
• Closed-Loop Regenerative Process
• Essentially Eliminates Hazardous Waste Streams
• Saleable Sulfur Products
• Competitive With Throwaway Processes

Bipolar membrane in sodium chloride salt splitting

The bipolar membrane on-site generation of hydrochloric acid and sodium hydroxide from sodium chloride can be an economic alternative to the purchasing, storage and handling of ion exchange regenerates. This application is most suitable to chemical processing applications requiring both acid and base, especially in equimolar quantities.

On-site generation offers many benefits compared to traditional methods of ion exchange regenerate handling:
- shipment and storage of bulk quantities of salt rather than concentrated acid and caustic;
- production of ready-to-use dilute ion exchange regenerates on as-needed basis;
- elimination of hazards associated with diluting and handling concentrated hydrochloric acid and caustic; and
- control over variable caustic pricing.

Figure 24 is a schematic of the pilot bipolar stack used in this testing, which is the typical plate and frame assembly used in electrodialysis. The stack consists of 5 unit cells, each cell containing a bipolar membrane (positioned with the anion layer facing the anode), with a cation exchange membrane on the anode side of the bipolar membrane, and an anion exchange membrane on the cathode side. Sodium from the salt stream transports through the cation exchange membrane and combines with hydroxide generated by the bipolar membrane to produce sodium hydroxide. Similarly, hydrochloric acid is produced on the opposite side of the bipolar membrane when hydrogen ions generated by the bipolar membrane combine with chloride ions transporting through the anion exchange membrane. An electrode rinse is required to provide a conductive medium between the electrodes and the rest of the stack, as well as a means to continuously sweep out gases generated at the electrode surfaces. In this application, sodium hydroxide is used as the electrode rinse.

The pilot plant, shown in Figure 3, is operated on a continuous feed/bleed basis. Acid and base reservoirs are initially primed with their target concentrates, and the salt feed reservoir is primed with the required brine concentration (see below). During operation all channels are continuously recirculated at a high rate to prevent localsied depletion of ions available for transport at the membrane surfaces. A deionised water make-up is provided to the acid and base channels at a rate based on the ion transport rate and target acid and base concentration. Acid and base products are removed as overflow streams from their respective reservoirs, so that the reservoir volume remains constant.

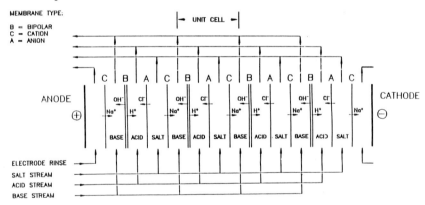

FIGURE 23 – Bipolar cell stock and pilot plant.

Saturated sodium chloride (26 wt%) is produced in the brine saturator which is replenished with solid salt as needed. The saturated salt is continuously supplied to the brine reservoir in order to replace that consumed in the production of acid and base. Overflow of depleted slat is recirculated back to the brine saturator for reuse. A brine softener is located upstream of the brine reciruclation reservoir to remove metals which might otherwise cause precipitation in the stack.

The electrode rinse stream, 6 wt% (1.5 N) sodium hydroxide, is common to both electrodes. After exiting the stack, the two streams are recombined by cascading the cathode rinse stream into the anode rinse tank. This allows degassing of the hydrogen formed at the cathode, before contacting the anode rinse stream which contains oxygen formed at the anode. From the anode reservoir, the solution is recirculated to the two electrode chambers.

Table 7 lists the operating conditions conducted with membranes which have been in operation in excess of one year. Acid and base concentrations were kept approximately equal and ranged from 1N to 2.5N. This covers the typical range of regenerate concentrations for ion exchange resins: 4 wt% or 1 N NaOH for anion regeneration and 4-8 wt% or 1.1-2.2 N HCl for cation regeneration. Note that it is not necessary to produce the same concentration acid and base as was done throughout this study. However, increasing the acid concentration relative to the base can result in a somewhat higher chloride contamination level in the base product.

TABLE 7.

Conditions	
Mode:	continuous
Current density:	60-100 mA/cm2
Product concentration:	1-2.5 N
Effective membrane area:	5 x 200 cm2
Electrode rinse:	1N NaOH (min)
Salt:	1.5N NaCl
Temperature:	30-40°C
Membranes (NEOSEPTA)	
Bipolar	BP-1
Anion Exchange	ACS
	ACM
Cation Exchange	CMS

Current efficiency

Figure 24 shows current efficiency versus acid and base product concentration of a bipolar stack containing anion and cation membranes. As can be seen, current efficiency is highly dependent on the product concentration (about 85% for 1 N product, and decreasing with increasing product concentration to 70% for 2 N product). The loss in efficiency with concentration is due mainly to increasing leakage of hydroxide and hydrogen ions through the monopolar membranes into the salt stream. Hydrogen leakage through the anion membrane is typically higher than hydroxide through the cation. Depending on the membrane, this can result in a slightly lower acid current efficiency than base.

Purity of the acid and base produced is an important consideration for ion exchange regenerates, depending on the application. Leakage of sodium and chloride ions into the product streams occurs mainly through the bipolar membrane, although selection of the monopolar membranes may also affect leakage somewhat. Transport of the contamination ions can occur by diffusion (driven by concentration differences) as well as by electrical transport.

Leakage increases with increasing acid or base concentration due to a phenomenon referred to as "co-ion intrusion" in the bipolar membrane layers, which is directly related to the solution concentrations in the acid and base compartments. Figure 6 gives the actual contaminant values in parts per million of the acid and base products. The 1 N sodium hydroxide product contains 160 ppm chloride; the 1 N hydrochloric acid contains 20 ppm sodium. For many ion exchange applications, these levels are acceptable. However, higher purity base (Rayon grade) is often required in such applications as condensate polishing. A four-cell stack which utilises a blocking stream to reduce chloride leakage into the basehas been developed

Other factors influence purity, including temperature and current density. As temperature is increased, leakage by diffusion also increases as shown in Table 8 opposite. Reducing the current density can also result in increased leakage, since electrical transport of acid and base-forming ions is reduced relative to the diffusional component of leakage, which remains the same. This, together with membrane utilisation considerations, indicates that there is no advantage in operating at current densities less than 100 mA/cm^2.

Figure 25 shows cell voltage for operation at varying product concentration. Voltage

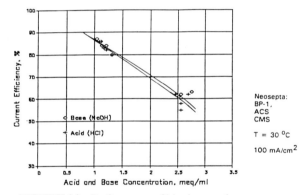

FIGURE 24 – Performance of bipolar membrane.

decreases slightly with increasing solution conductivities. Power consumption for the production per ton of acid and base are shown in Table 3. (Assuming 85% power supply efficiency and includes rectifier, pumping and electrode rinse stream losses.) As seen in the table, power consumption is affected by temperature, which affects cell voltage, and by current efficiency, which is controlled by a number of factors, including membrane type. The difference between acid and base production per kilowatt hour is due to the unequal equivalent weights of sodium hydroxide and hydrochloric acid: the process produces equimolar quantities. In the case of anion membranes, lower current efficiency for hydrochloric acid reduces the acid production rate even further.

Pretreatment of the salt is important to maintain low levels of calcium and iron in the salt feed to the bipolar stack. This can be readily accomplished using a chelating resin. An alternative to the pretreatment is a four channel stack which employs a buffer stream utilising a monovalent selective cation membrane between the salt and base streams. This prevents problems with precipitation of iron hydroxide and calcium carbonate without the necessity of an ion exchange pretreatment.

RECOVERY OF SALTS, ACIDS AND BASES

TABLE 8 – Product contamination, effect of temperature (1N product, 100 mA/cm²).

Monopolar Membrane Type	ACS/	CMS	ACM/	CMS
Temperature, °C	30°	40°	30°	40°
Cl in Base, ppm	160	*	140	250
Na in Acid, ppm	20	*	40	70
KWH/ton NaOH produced	1540	1420	1780	1640
ton HCl produced/ton NaOH	0.92	0.93	0.86	0.86

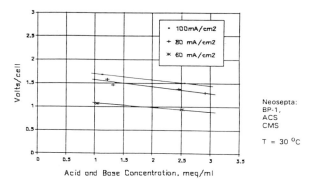

FIGURE 25 – Cell voltage for operation at varying product concentration.

Plating baths

The electrodialytic salt splitting can be used in many application where electrodialysis alone is applied with the potential attraction that valuable acids or bases are generated. One potential area is in plating baths, for example electroless copper plating, where ethyleneediamine tetracetic acid can be recovered.

The wastewater streams from the plating baths consist of low quantities of unused formaldehyde and Cu^{2+} ions and relatively large quantities of Ethylenediaminetetraacetic acid (EDTA) and sodium sulphate. In the alkaline solution, the EDTA exists as a readily soluble sodium salt. By reducing the pH down to 1.7 the sodium salt of the EDTA is transformed in a less soluble acid. With electrodialysis the EDTA can be transported into the ED-cell with the bipolar membrane, where H^+ ions are produced, and then precipitated in an external circuit as the less soluble acid. The precipitated free EDTA can then be recycled without any further treatment. The solution which is left, consists essentially of sulphuric acid and formic acid and can be neutralised by the sodium hydroxide, produced by the bipolar membrane.

Acid Recovery From Stripping Solutions

A second example is shown in Fig 26, for the recovery of amine from a gas scrubbing acid stream. The particular amine, dimethyl propyl amine, is used with air, as catalyst to cure epoxy resin/sand molds for aluminium castings. The amine catalyst is not used and has to be removed from the air by scrubbing with sulphuric acid which converts the free amine

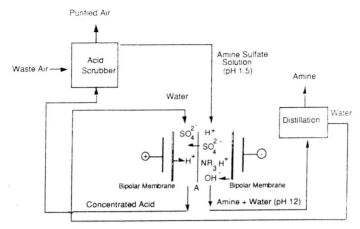

FIGURE 26 – Amine recovery from acid scrubbing solutions.

into amine sulphate. The acid scrubbing solution containing around 10% amine sulphate, is then fed into the electrodialytic water dissociation stack, containing alternate bipolar and anion-exchange membranes. In this system the salt solution simply loses sulphate ions and gains hydroxyl ions which raises the pH to a value which converts the sulphate back in to the free amine. The sulphuric acid produced is recycled to the acid scrubber and the amine water mixture is distilled to recover the amine, the water is recycled to the electrodialysis unit.

SECTION 17

Food Industry

FOOD INDUSTRY

The applications of membrane technology in the food industry are numerous and include:
  microfiltration – for clarification in place of centrifugation, sterilisation in place of heat
  ultrafiltration – for fractionation, concentration and purification
  nanofiltration – for desalting, de-acidification
  electrodialysis – for demineralisation
  reverse osmosis – for concentration, waste treatment

The processes recover valuable components dairy products, fruits, vegetables grains, sugars, animal products and enable purification of many constituents (see Table 1.). The dairy industry is believed to hold the largest share of installed membrane capacity in the food industries.

Reverse osmosis is used in food processing as a means of concentration, purification and recovery of valuable components either alone or in combination with other membrane separations (MF, UF, PV). The food industries include fats and oils, meat by-products, milk, beverages, sugar and fruit and vegetable juices (see Table 2).

The main advantage of RO is food processing is the reduction in the costs associated with evaporation or even the elimination of this step. The energy requirements of RO are approximately 110 kJ/kg of water which compares with 700 kJ/kg for mechanical vapour decompression, the most efficient evaporator used in food processing. The other advantages of RO (see Table 3) must be balanced with several disadvantages
for the use of new membrane materials in food processing.

Milk typically contains 12.5 total solids, 3.3% protein 3.5% fat, 4.9% lactose and 0.7% ash. The osmotic pressure of milk is approximately 600-700 kPa, mainly due to the dissolved salts and lactose present.

There is a considerable reduction in flux due to the osmotic pressure effect and an asymptotic relationship between transmembrane pressure and flux. This latter feature is due to concentration polarisation. Fouling of membranes during RO occurs usually as a

TABLE 1 – Membrane technology in the food industry.

Diary	RO:	(Pre)Concentration of milk and whey prior to evaporation Bulk transport Specialty fluid milk products (2-3X/UHT)
	NF:	Partial demineralization and concentration of whey
	UF:	Fractionation of milk for cheese manufacture Fractionation of whey for whey protein concentrates Specialty fluid milk products
	MF:	Clarification of cheese whey Defatting and reducing microbial load of milk
	ED:	Demineralization of milk and whey
Fruits and Vegetables	Juices:	Apple (UF,RO), apricot, citrus (MF/UF,RO,ED), cranberry, grape (UF,RO), kiwi, peach (UF,RO)
	Pigments:	anthocyanins
	Wastewater:	apple, potato (UF,RO)
Animal Products	Gelatin:	concentration and de-ashing (UF)
	Eggs:	concentration and reduction of glucose (UF,RO)
	AnimaL by-products:	blood, wastewater treatment (UF)
Beverages	MF/UF:	Wine, beer, vinegar — clarification
	RO:	Low-alcohol beer
Suger refining		Beet/cane solutions, maple syrup, candy wastewaters – clarification (MF/UF) desalting (ED), preconcentration (RO)
Grain Products	Soybean processing:	Protein concentrates and isolates (UF) Protein hydrolyzates (CMR) Oil degumming and refining (UF,NF) Recovery of soy whey proteins (UF,RO) Wastewater treatment
	Corn refining:	Steepwater concentration (RO) Light-middlings treatment: water recycle (RO) Saccharification of liquefied starch (CMR) Purification of dextrose (MF/UF) Fermentation of glucose to ethanol (CMR) Downstream processing (MR,UF,NR,RO,ED,PV) Wastewater treatment
Biotechnology		Production of high quality water (MF,UF,RO,ED)
		Downstream processing (MR,UF,RO,ED): cell harvesting, protein fractionation, desalting, concentration
	Bioreactors:	enzyme hydrolysis tissue culture plant cells

ED: electrodialysis, MF: microfiltration, CMR: membrane reactor, NF: nanofiltration, PV: pervaporation, RO: reverse osmosis, UF: Ultrafiltration, UHT: ultra-high temperature.

TABLE 2 – Applications of reverse osmosis in food industries.

Industry	Applications
Dairy	Cheese whey concentration Milk concentration Desalting of salt whey Waste treatment
Grain milling	BOD reduction Waste water recovery and reuse Recovery of by-products from waste water
Beverage	Cold stabilization of beer Removal of colour from wine Removal of alcohol from beer and wine
Fats and oils	Waste water treatment
Suger	Preconcentration of dilute sugar solutions Maple syrup Recovery of sugar from rinse water
Fruit and vegetables	Concentration of tomato juice Waste water treatment Concentrated juices Juice flavor and aroma concentration

TABLE3 – Advantages and Disadvantages of RO in Food Processing

Advantages
  Lower energy requirement
  Product quality separation without heat damage
  Reduced fresh water requirements
  Reduced waste treatment volume and cost
  Potential increased profit margins from new products
  Relatively low floor space and capital requirements

Disadvantages
  Expense and time to document product safety and obtain approval of new membranes
  Uncertainty of membrane durability and operating life and thus replacement cost
  Questionable chemical inertness and pH sensitivity
  Limited operating pressure range in some applications
  Fouling with certain feed stocks

Dairy Industry

There are several applications of RO in the dairy industry; water treatment, fractionation, product and chemical recovery and concentration and denaturing (Table 2). The concentration of process streams from around 10% total solids (TS) to 25% TS can be achieved at a lower cost than by evaporation. Also there is a considerable reduction in volatile flavour component losses and in adverse changes to heat sensitive components (protein denaturation). RO also reduces discharge volumes to water treatment facilities and produces reusable water.

Milk Concentration

The performance of RO in concentrating milk is limited by the osmotic pressure and most commercial modules have operating pressure limits of 3 – 4 MPa which limits the concentration of milk to a factor of 3 to 4. In the production of skimmed or whole milk powder the milk is usually concentrated to 45 – 50% total solids before spray drying. Thus RO cannot substitute conventional evaporation, rather it is used as a pre-concentration step before evaporation to reduce operating costs or to increase capacity of existing plant. The relative energy consumption of RO and thermal concentration methods in the concentration of milk by a factor of 2.5 have been compared (see Table 4) for a capacity of 1t/h feed. Energy consumption are an order of magnitude lower and with thin film composite membranes lower still.

TABLE 4 – Comparison of energy consumption of reverse osmosis and evaporation.

Process	Area(m^2)	Energy (kcal/kg milk)
Thermal concentration	10.4	455
Open-pan boiling	10.4	455
Evaporator:		
Double-effect evaporator	25	209
Mechanical Vapor Recompression	32	136
Membrane Process*		
Batch, single-pump	65	80
Batch, dual-pump	65	7
Continuous, one-stage	206	16
Continuous, three-stage	93	7

*Spiral-wound cellulose acetate membrane. A spiral-wound thin-film composite membrane would use about half this energy (2).

Reverse osmosis plant for the dairy industry.

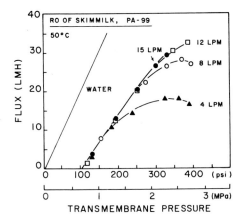

FIGURE 1 – Comparison of energy consumption of reverse osmosis and evaporation.

result of the deposit of protein, but inorganic salts such as calcium phosphate are also a factor.

Milk is essentially 85% water and thus there are economic gains in the concentration of milk prior to transport to manufacturing facilities; reduced haulage costs, storage, refrigeration and processing. Installation of RO can increase the capacities of existing evaporators. There are several unique features to be considered in the concentration of milk by RO.

A – The Effect of Temperature
  i) High temperatures favour increased permeation rate but are limited by whey protein denaturation and thermal resistance of membranes.
  ii) Low temperatures result in low permeation rates and consequently large residence times.
  iii) Intermediate temperatures result in bacterial proliferation.

Heat pretreatment of milk can be used to reduce bacterial count and subsequent- bacterial proliferation, and increases flux. Typical temperatures of operation are 45C for polyacrylamide thin film composites and 30C for cellulose acetate membranes.

B – The Effect of Pressure

The sudden reduction of pressure which can occur on passage of the permeate through pressure relief valves can affect certain physical properties of the milk. The relief valve can act as a homogenizer which damages fat globules in raw milk which can be degraded by the natural enzymes, resulting in, flavour change, foaming and losses of fat in later butter and cheese manufacture. The problem of rancidity in milk products as a result of this sudden pressure reduction can be eliminated by prior heat treatment of the milk before RO

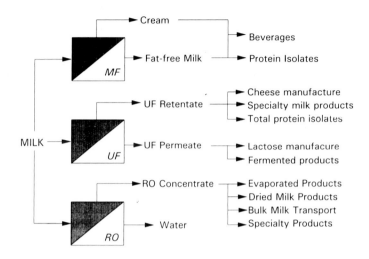

FIGURE 2 – Membrane applications in milk processing.

or by reducing the temperature of the concentrate (to below 10C) before pressure relief of by good design of the outlet pressure relief system. The concentrates(retentates) produced by RO are used in the production of several products

- reconstituted milk, either whole or skimmed, with water
- skimmed milk powder
- cheddar or cottage cheese
- yoghurt
- ice creams

The three cross flow membrane separations, MF, UF and RO are the dominant methods in the dairy industry. The methods are used in the concentration (RO) of, fractionation (UF) of and filtration (MF) of both milk and cheese whey (see Fig 2).

Milk Fractionation (Ultrafiltration)

The ultrafiltration of milk retains the protein and fat and certain insoluble and bound salts while the permeate contains lactose and soluble salts. This fractionation of milk, typically using polysulfone (or polyether sulfone) type membranes, has several applications. These membrane tend to foul more than cellulose based membranes but have higher operating temperature capability (typically temperatures are between 50-55C) and can tolerate chlorine, used for sanitation, and extremes of pH, encountered in preferred cleaning procedures. Concentration polarisation is significant in milk ultrafiltration and generally high cross flow velocities are used. The maximum total solids which can be obtained by UF is the range 38-42%.

The ultrafiltration of milk on farms has been suggested as a means of reducing refrigeration and transport costs but presently there are regulatory and marketing barriers to the adoption. The production of speciality milk based beverages is a potentially large market for the application of UF. In particular beverages which have a high calcium content but relatively low fat and cholesterol are attractive. The ultrafiltration of skimmed milk will give a product of higher calcium content, as calcium is mostly present as an insoluble form, bound or associated with casein micelles, and a higher protein content. The concentration of lactose and sodium and potassium remains unchanged (see Table 5). To improve the taste of this product, fat (cream) separated (skimmed) from the milk is added.

TABLE 5 – Composition of whole milk and PRO-CAL, a low-cholesterol, low-fat, high-calcium milk produced by ultrafiltration [15]. Values are per 100g serving.

Component	Regular whole milk	PRO-CAL One	PRO-CAL Zero
Fat (g)	3.5	0.87	0.1
Protein (g)	3.5	6.1	6.1
Carbohydrate (g)	4.8	4.8	4.9
Sodium (mg)	52	52	52
Potassium (mg)	161	161	161
Calcium (mg)	130	217	217
Cholesterol (mg)*	14	3.5	0.5
Calories	65	52	49
Calories from fat	48	15	4

*Calculated assuming 0.40% of the fat is cholesterol.

Cheese Manufacture (Ultrafiltration)

The manufacture of cheese is a fractional process in which protein (casiun) and fat are concentrated in a curd and lactose, minerals, soluble proteins (and other minor constituents) are formed as whey. Ultrafiltration therefore enables the production of a pre-cheese by the concentration of the protein, fat and solids. The pre-cheese is then used to produce cheese by usual methods. There are several cheese products which can be produced utilising UF (see Fig 3) and include feta cheese, quarg (uripened, soft, smooth curd cheese), camembert, ricotta, mozzarella and cheddar. The technical benefits and disadvantage of UF in cheese making are described in Table 6.

With the manufacture of certain cheese such as quarg the ultrafiltration is performed at the end of the fermentation (see Fig 4).

The production of cheddar cheese was a particular challenge for UF owing to the requirement of a high solids content. The process (see Fig 5) takes standardised and pasteurised whole milk and concentrates this in a multistage UF system utilising polyethersulfone spiral wound modules. Concentration of the milk is by a factor of 5 with the membrane operation including diafiltration to reduce the lactose content in the retentate, which otherwise would be too high for correct texture and flavour development of the finished cheese. The retentate, containing between 38-40% solids, is divided into

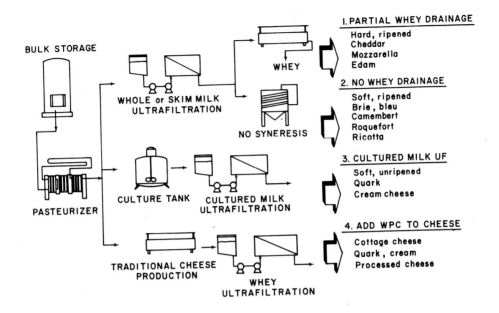

FIGURE 3 – Cheese making options with ultrafiltrations.

TABLE 6 – Technical Characteristics of Ultrafiltration Cheese Making

Benefits	An increase in yield of 10-30% with soft and semi-soft cheese
	A reduced enzyme (rennet) requirement up to 70-85%
	Reduced volume of milk to handle
	Very little whey is produced
	A more uniform product can be obtained
	Ability to use continuous processes
	Improved sanitation and environmental condition
Disadvantages	Increases viscosity if the protein content is greater than 12-14% causes mixing problems with starter culture and rennet
	Increased buffer capacity means desired pH may not be reached
	Recirculation of concentrate can cause partial homogenisation of fat
	Semi-hard and hard cheeses have a low moisture content

FOOD INDUSTRY

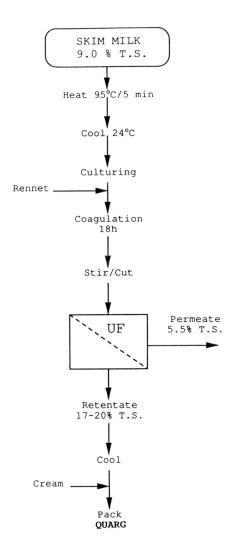

FIGURE 4 – Ultrafiltration in quarg production.

two stream; 10% is repasteurised and fermented and mixed with the main retentate stream and renet (enzyme) in a coagulator. The coagulum produced is cut and cooked in syneresis drums where curd and whey are produced. The curd-whey mixture is sent off for cheddar cheese manufacture.

The permeate from UF is used to produce sterile water for cleaning and diafiltration, by reverse osmosis.

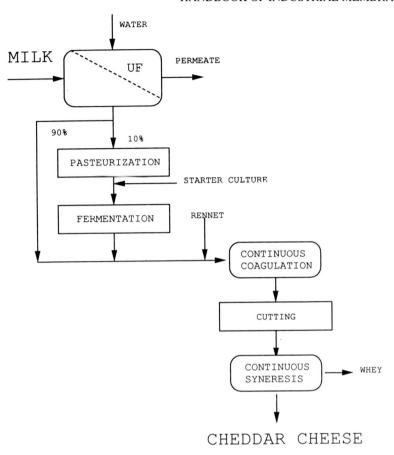

FIGURE 5 – Manufacture of cheddar cheese.

Microfiltration of Milk

Milk consists of many components which have a wide variation of size distribution (see Fig 6). Casein micelles (suspended milk protein) have sizes of 0.5 m and less compared to fat and bacteria which are of the order of 1 m and greater. Thus microfiltration can in principle separate fat and bacteria from milk. Polymeric membranes however quickly form dynamic or secondary membranes comprising of polarised solutes which reject caseins and whey proteins, ie they acted as UF membranes. Secondary membrane formation is minimised by using very high cross flow velocities (> 6 m/s) and low transmembrane pressures (< 40 kPa). However a major difficulty with normal MF operation at high velocities is the large drop in retentate side pressure in the module. This, in conjunction with the usual constant pressure on the permeate side, causes a large variation in transmembrane pressure (see Fig 7) from inlet to outlet, and operational problems.

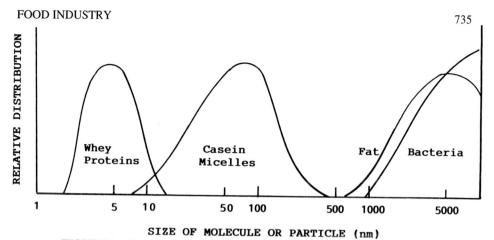

FIGURE 6 – Size distribution of major milk components. Not shown are the carbohydrate (lactose) and soluble salts, which are much smaller than the components shown.

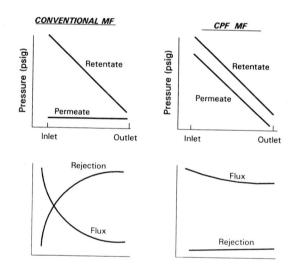

FIGURE 7 – Pressure profiles (top) and flux (bottom) during microfiltration of milk. Left: Conventional microfiltration. Right: Constant-pressure or HFLF microfiltration.

- The high inlet pressure differential causes severe fouling in this region.
- The low exit pressure differential results in under utilisation of the membrane area in this region, ie low flux.

The solution to this problem is to utilise a constant pressure filtration (Bactocatech) in which permeate is pumped co-currently with the feed on the opposite side of the membrane

Co-Current Permeate Flow (CPF) Microfiltration

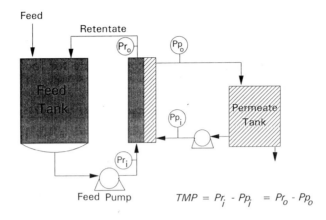

$$TMP = Pr_i - Pp_i = Pr_o - Pp_o$$

FIGURE 8 – Constant-pressure or high-flux, low fouling (HFLF) microfiltration. The feed is pumped through retentate channels at high velocity. The permeate is pumped through the shell side of the module with the same pressure drop as the retentate.

(see Fig 8). By adjusting the parallel flows to achieve similar pressure drops on retentate and feed side, an approximate constant pressure driving force is realised and a much improved flux and rejection performance.

The constant pressure filtration is operated with tubular ceramic membranes with fluxes of $0.5 - 0.7$ m^3 m^{-2} h^{-1} achieved over periods of 6 hours. Operation is particularly sensitive to operating pressure, which can be increased slightly by 0.05 to 0.1 bar to compensate for fouling, if necessary (see Fig 8). Reported bacteria rejection is 99% and so the product is effectively bacteria free skimmed milk.

Concentration of Whey and Milk

Whey is produced in the manufacture of cheese when milk is curdled by addition of lactic acid producing bacteria and rennet (an enzyme) the curd contains the bulk of the milk's protein but some protein is also present in the supernate or whey stream. This whey has considerable nutritional value and consequently a high BOD (approximately 30,000 to 50,000 ppm) which would prevent direct discharge to municipal waterways. A typical composition of a sweet whey is between 5% to 6% total solids (TS) consisting of approximately protein (0.7% TS), lactose (4.3% TS), ash (0.5% TS) and fat (0.05%). Typically whey has an osmotic pressure of 7 bar, which on concentration to 25 wt% TS can be around 35 bar. This limits the applications of RO to preconcentration, prior to evaporation and drying to give a final whey powder used as a food supplement in bakery goods, ice cream and other products.

There are several potential application of membrane processes in the processing of whey including RO, UF, MF, ED and NF (see Fig 9).

FOOD INDUSTRY

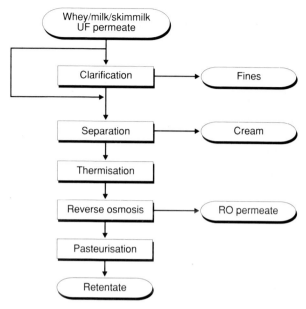

FIGURE 9A – Reverse osmosis of whey an other milk products.

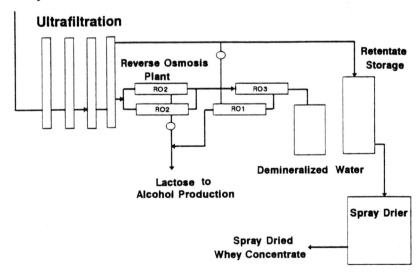

FIGURE 9B – Reverse osmosis of whey and other milk products.

There are two types of cheese whey
- sweet whey formed from cheeses produced by enzyme action (pH >5.6)
- acid whey, formed from cheeses made by acidification (pH 4.6)

Acid whey generally has more lactic acid and twice the calcium phosphate content of sweet whey. The membrane operating conditions are chosen based on the behaviour of calcium salts (calcium phosphate is more soluble at low pH) as well as on considerations of microbial growth, denaturation and viscosity.

Reverse osmosis of whey and ultrafiltration permeates produces whey protein concentrates, lactose and waste streams with reduced BOD. One example of a cheese plant operated in California uses RO (plate and frame modules) to concentrate lactose which is present in an ultrafiltration permeate from whey protein concentrate production (see Fig 9)

TABLE 7 – Applications of RO for concentrating whey and milk.

Feed	Membrane	Pressure MPa	Temperature °C	Concentration	Flux l/m^2h
Whole	PCI T2/15, CA	3.4	30	2x	4
Skim	PCI T2/15, CA	2.8	25	2x	10
Whole	PCI T1/12, CA	4.8	30	2.5x	13
Skim	PCI T1/12, CA	4.8	30	3x	14
SW	Wafilin BV, CA	4.0	30	2x	17
SW	PCI T2/15W, CA	4.5	30	2x	34
Skim	PCI T2/15W, CA	4.5	30	2x	16
Whole	PCI T1/12, CA	4	20	2x	—
Whole	PCI T2/15W, CA	—	30	1.25x	—
Whole	Ultrapore CA	4.0	30	2x	16
Whole	Ultrapore NCA	4.0	50	2x	3
Whole	PCI ZF99, NCA	2.7	50	2x	15
Whole	Koch NCA	4.2	50	2x	7
Skim	Koch NCA	4.2	50	2x	14
SW	Koch NCA	4.2	50	2x	29
Skim	Ultrapore NCA	3.0	50	2.6x	—

PCI = Paterson Candy International
ZF99 = noncellulosic polyamide membrane
CA = cellulose acetate
NCA = noncellulosic acetate
Whole = whole milk; Skim = skim milk; SW = sweet whey

FOOD INDUSTRY

The RO set up consists of two parallel lines rated at approximately 9,200 gph per line. Lactose is concentrated, at a rate of 4,500 gph per line, for alcohol production by fermentation. A second RO treats the mineral permeate to produce demineralised water. A third RO system (ROI) recovers the diafiltration water (permeate) which can be returned to ultrafiltration.

The preconcentration of whey, using RO to approximately 12% TS prior to vacuum evaporation is standard practice in industry. This practice is adopted because of the considerable savings in energy for preconcentration. Preconcentration in the range of 6% to 24% TS is also practiced. The whey concentrates are more efficient feeds for UF and demineralisation by electrodialysis or gel filtration. Reported conditions of operation for typical industrial tubular RO systems for whey concentrate production are 40-45C, 29 – 49 bar, 2-2.75 ms^{-1} linear velocity. The RO system on concentrating whey from 6.5% to 25% TS, produces a permeate with only 0.26% TS.

Specially designed hygienic modules based on either flat plate or tubular systems are used in the RO concentration of whey. These modules are generally of a cost comparable with sea water desalination modules and generally have similar associated operating costs. A large proportion of the operating cost is in membrane replacement, typically yearly. Membranes also have to be frequently cleaned, sometimes daily, which is also a significant operating cost.

Ultrafiltration of Whey

The fractionation of whey by UF to produce protein concentrates has a wide range of application (see Table 8).

TABLE 8 – Uses of whey protein concentrates.

Reported uses of whey protein concentrates [101]		
Baked custard	Coffee whitener	Meat analogs
Beverages	Cream fillings	Meat extenders
Acid-clear	Cream icings	Meat loaf
Acid-turbid	Cream sauces	Meringues
Neutral	Cream desserts	Noodles
Biscuits	Cultured beverages	Pasta
Breads	Doughnuts	Potato flakes
Cakes	Egg white replacer	Puddings
Cake fillings	Egg yolk replacer	Hot dogs
Candy	Gravies	Sausage
Caramel	Hot dogs	Sherbet
Milk chocolate	Ice cream	Snack foods
Nougats	Imitation cream cheese	Tortillas
Canned refried beans	Imitation milk	Whipped toppings
Chocolate drinks	Macaroni	Yogurt

Ultrafiltration can increase the initial protein content form 10-12% (dry basis) to 35%, 50% or 80% protein products with little if any loss in whey protein functionality.

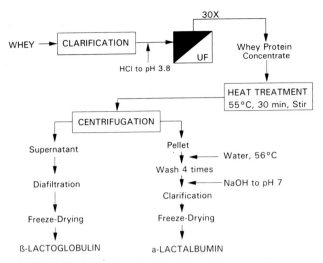

FIGURE 10 – Membrane fractionation of whey proteins.

Following UF the components can be fractionated. Following UF the components can be fractionated into -lactoglobulin and -lactalbumin fractions (see Fig 10).

A problem with whey ultrafiltration is fouling of the membrane by proteins and salts, with -lactalbumin a principal foulant. Polysulphone membranes are more susceptible to fouling than cellulosic membranes. Although various pretreatment methods, e.g. pH adjustment are possible to increase flux values there is a danger of a loss in protein functionally. Pretreatment by centrifugal classification and/or microfiltration can be effective, having reduced detrimental affects on the products. In general the pH of the whey should be as far from the isoelectric point of the proteins as possible. The temperature should not be around 30C, as this gives minimum flux values brought about by a combination of reduced calcium solubility at higher temperature and improved viscosity and diffusivity (improved mass transfer).

Pretreatment by microfiltation, of whey can remove the small quantities of fat, present as globules (0.2 – 1.0 m) and the casein present as fine particulate (5 – 100 m). The components have a detrimental affect on the functional properties of the whey protein (ultrafiltration) concentrates. In addition MF also removes some precipitated salts and a considerable amount of bacteria. MF is performed in the Bactocatch process using 0.2 m tight membranes.

Whey Demineralisation

Demineralisation of whey can be achieved by ion-exchange, electrodialysis and nanofiltration.

Ion-exchange requires short cycle times due to the high salt content of the whey. Thus large amounts of chemical regenerates and water are required and potential protein destabilisation may occur.

Electrodialysis of whey is usually carried out after pre-concentration of the whey which is done to increase the ion concentration and thus reduce cost. However ED does not

FOOD INDUSTRY

achieve completed demineralisation, due to a rapid reduction in conductivity, and thus can be usually followed by ion-exchange to achieve the required ion concentration.

Nanofiltration is a relatively new process for the demineralisation of whey (and of milk). It is competitive process comparable with electrodialysis or ion exchange for desalination. By nanofiltration it is possible to obtain a degree of desalination of up to 40%, which makes it possible to utilise whey, which for example was made worthless due to salt addition during the cheese making process.

With nanofiltration the same pretreatment is necessary as for ordinary RO concentration. This means clarification and separation of the whey and furthermore a pasteurisation to obtain maximum yield and capacity. A pH-adjustment of the whey would also be an advantage.

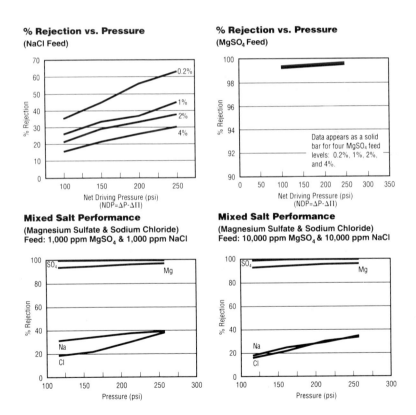

Nanofiltration of whey.

The concentration itself and the desalination take place at a temperature of 10-30C and at a working pressure of 25-32 bar. Pressure and temperature depend on the required degree of concentration and desalination. Nanofiltration allows the univalent ions to pass, such as Na^+, Cl^-, K^+, whereas the larger ions, such as Ca^{++} or phosphates are retained. In order to obtain maximum desalination, diafiltration may be used. This will, however, cause a larger loss of lactose and other components.

The advantages of desalination by nanofiltration include
- cheaper plants than conventional desalination
- gives worthless salty whey the same value as normal whey (1st whey)
  possibility of production of special products with reduced content of ash components
- can be used as predesalination before electrodialysis, or ion exchange, in for example the production of whey concentrate for baby-food.

An "ultra-osmosis" process has been developed to convert salty whey to a product which effectively resemble sweet whey. The process is based on centrifugal clarification and filtration and gives a product with a net 95% removal of the salt.

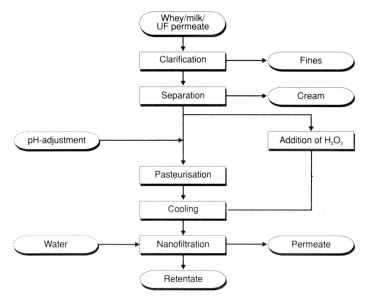

FIGURE 11 – Process for desalting whey by nanofiltration.

Other Applications

Other protein containing wheys can be processed by reverse osmosis. For example an important application is in plants producing starch form potatoes where wheys contain up to 40% of the solids as highly nutritious protein. Reverse osmosis, typically with tubular modules concentrates the whey from 4% to 8% wt dissolved solid with reported average fluxes of 1 m/day. The RO permeate is re-used to wash potatoes and the protein enriched retentate is coagulated by heat treatment and the proteins are then removed.

Ultrafiltration is used to fractionate and concentrate proteins form potato processing waste waters.

Vegetable Protein Processing

An important source of protein is soybean. The processing of soybean to recover the protein (and fat) requires the removal or reduction in content of some undesirable components e.g. oligosaccharides, phytic acid, trypsin inhibitors. Traditional processing methods utilise extraction, heat treatment and centrifugation to separate the protein and fat from other compounds. These processes are not always satisfactory giving products with less than the desired properties and generating a "whey" stream containing a significant amount of protein, which constitutes a waste stream.

The processes developed for full-fat soy protein concentrates and soy isolates both utilise hollow fibre ultrafiltration modules. The undesirable compounds have smaller molar masses than proteins and fats and thus UF can remove these compounds in the production of either purified protein isolate or lipid-protein concentrates. The products from these processes have a higher protein yields than obtained with conventional processes.

Reworked, updated and extended to include the latest developments in sealing technology

Seals and Sealing Handbook

4th Edition

By Mel Brown

For further details post or fax a copy of this advert complete with your address details to:

Elsevier Advanced Technology
PO Box 150, Kidlington,
Oxford OX5 1AS, UK
Tel: +44 (0) 1865 843848
Fax: +44 (0) 1865 843971

ELSEVIER
ADVANCED
TECHNOLOGY

Packed full of useful reference information and data this book is the essential reference guide for all those involved in sealing technology.

Find out how to...

- Cut down on time-consuming selection procedures
- Maximise seal life and reduce maintenance costs
- Ensure compatibility of seal materials
- Comply with environmental guidelines and legislation
- and much more!

With full coverage in all these areas:

- Mechanics of sealing
- Materials
- Static seals
- Dynamic seals
- Mechanical seals
- Fluid power seals
- Special seal types
- Seal selection
- Engineering data

**600+ pages
2000 diagrams, charts, tables and illustrations
ISBN: 1 85617 232 5**

Publication date: July 1995

Animal Products

Abattoirs produce large amounts of by-product rich in protein e.g. blood, offals and gut tissues. The waste product of greatest volume produced in abattoirs is blood, which is typically 70% plasma/serum and 30% red blood cells. The red blood cells contain approximately 28% protein (70% water) and the plasma 7% protein (92% water). Ultrafiltration is used to concentrate the protein content of the two fractions.

Animal product processing, as with any food industry, generates waste waters which contain large amounts of biologically degradable matter (proteins and fats from blood and carcasses). Ultrafiltration can be used to remove this material from the waste water to give a permeate suitable for discharge to the sewage system. The concentrate can be used as animal feed or fertiliser or sold as "brown grease".

Gelatine is a widely used product from the animal processing industry obtained by high temperature extraction with acid or alkali of skins, hides and bones. It has uses as glue, pharmaceutical preparations, photographer products and in foods. The extract typically contains 2-5% protein and a high quantity of ash. A final gelatine product containing 90% protein, 0.3% ash is generally required which typically could be achieved by ion exchange and evaporation. The use of ultrafiltration can be applied as a pre-concentrate step giving 18-20% protein content. Above this value UF membrane flux becomes too low. UF simultaneously achieves a reduction in ash content and removes lower molecular weight components, which detract from the gelling properties of the gelatine.

Fruit Juice Processing

The use of membranes in fruit juices processing is in three primary areas; clarification using microfiltration or ultrafiltration, concentration using reverse osmosis and deacidification using electrodialysis.

Clarification by Ultrafiltration

The production of clear single-strength fruit juices is traditionally carried out batch-wise by a series of filtrations and separations to remove suspended solids, colloids, proteins and condensed polyphenols (see Fig 14).

The traditional method for clarifying fruit juice is to add a fining agent and then either decant the clear juice, or to filter the juice on a preacoat filter e.g. as a rotary vacuum filter. However there are a number of limitations to these methods:
- Decantation is a slow process that requires large numbers of tanks.
- Filter aid leaves a slight after tastes in the juice.
- The filtration process is costly in terms of labour, energy consumption, enzymes and fining agents.

The application of UF as an alternative greatly simplifies the process operation, overall several advantages result
- a single step process reduces processing time to 2 – 4 hours
- increase in juice yield due to elimination of filter acid and fining losses
- eliminate of filter presses and filter aids

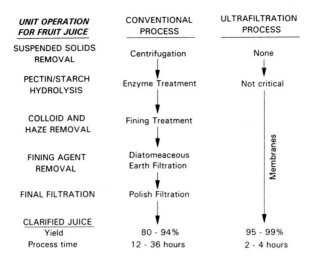

FIGURE 14 – Filtration and separation processes in the processing of single strength fruit juice.

- elimination of fining agents and gelatine etc
- improved product quality through reliable and consistent removal of haze forming components
- reduction in enzyme use, around 50-70% depending upon the juice.

Ultrafiltration, using open-channel (tubular) membranes, produces a high quality clear juice with a much reduced use of consumables with a much higher juice recovery, in a single operation.

The principle commercial application is the clarification of apple juice, although there are commercial plants operating on citrus fruits, such as lime or lemon, pineapple juice and pear juice. In general, the juice yield is controlled by the concentration factor that can be achieved in the plant and the characteristics of the juice, such as the pulp level. The upstream processing can effect the performance of the membrane notably, the presence of pectins will reduce the flux and yield significantly. Therefore juices high in pectins are usually depectinised before they are clarified. For example, greater than 95% recovery is possible using UF to clarify depectinised apple juice.

The most usual plant configuration used for clarification is the topped batch as the process is simple to operate. The juice is first depectinised and then fed to a batch tank and the level in the tank kept constant. The permeate is continually drawn off whilst the retentate is returned to the tank.

The viscosity of the juice continuously increases which leads to a drop in flow across the membrane. When the maximum pulp level in the retentate has been reached, the remaining juice in the plant is displaced with water and the plant is chemically cleaned. A typical operating cycle would be between 20-22 hours on process, then cleaning for between 2-3 hours. It is normal to uses a chlorinated caustic clean for apple juice and an acid and caustic clean for citrus fruit juices.

Applications

The biggest application of membranes in fruit juice production has been the clarification of apple juice by ultrafiltration. Unless the juice is depectinised, at least partially, the pectins foul the membrane and the rate at which juice is processed falls off very rapidly and hence UF is not generally economic. The pectinases can be added to the feed tank of the process and remain within the system during processing. This reduces enzyme usage by up to 70 per cent. The ultrafiltration performance on apple juice depends upon the apple variety and processing method, 95% to 98% of the juice can be recovered as clarified juice. With a 25,000 molecular weight cut-off membrane, the tannins remain in the clarified juice importing a brownish colour and a sharp flavour. With a smaller MWCO (tighter) membrane, the tannins are removed and the clarified juice has light golden colour. The juice clarity and colour is much better than that produced by a rotary vacuum filter. The purchase of expensive filter aids is also avoided.

Although UF of the apple juice may proceed without depectinization (pretreatment with enzymes) it is recommended as it helps to reduce viscosity, and fouling, resulting in higher flux and lower energy use in pumping. With depectinisation a doubling of the UF flux with polysulfone hollow fibre membranes is possible. The molecular weight cut-off has a significant effect on initial flux, turbidity and phenolic compounds. Noticeably after 6 months of standing the product obtained by UF with high MWCO membranes showed a large increase in turbidity. Microfiltration with 0.1 m ceramic filters is reported to give similar values of flux as the 50 K MWCO UF membranes although the product was much darker but with a preferred taste.

A process based on 30-40 K MWCO zirconia (formed in place) membranes on sintered stainless steel tubular substrates (25 mm diameter and greater) have been used in a process which combines juice pressing and filtration in one step. The so called "ultrapress" process takes the puree of whole apples, which have been treated with cellulose and pectinase at 50C for 2 hours to reduce viscosity and pumps the slurry through a 70 m long single pass tubular membrane system. Operating pressure are typically 13 to 24 bar, for 1.25 inch diameter tubes. The yields obtained were 80-85% with a single pass, although these can be increased with multiple pass operation.

TABLE 9 – Performance of UF of apple juice.

	Molecular weight cut-off of membrane			
	10K	50K	100K	500K
Initial flux (LMH)	28	100	210	280
Permeate solids (%)	12.39	12.53	12.57	12.57
Turbidity (NTU):				
Initial	0.4	0.6	2.2	2.3
After 6 months	–	–	4.0	6.0
Total phenolics (mg l^{-1})	350	366	377	369
Hunter L values	51.6	51.0	50.4	50.4

Romicon module for juice processing.

There are several reported plants clarifying lime juice at the point of production. Lime juice is very aggressive and has a tendency to break down rubber and therefore, care has to be taken in the choice of rubber seals and membrane polymers. A bonus from the use of UF is the recovery of lime oil, a valuable by-product. Lime juice is usually processed in batches, typically 20 m^3 and the recovery is typically 80-90%.

The limit to the recovery of other fruit juices by UF is set by the viscosity of the concentrate – the rate of processing depends upon the viscosity of the clarified juice; a high viscosity reduces the rate of which juice permeates the membrane.

Overall in juice ultrafiltration process generates savings in cost and manpower due to:-
- reduction in the quantity of enzymes required to depectinise the juice;
- reduction in filter aids, bentonite, Kieselguhr usage and final filtration steps;
- improved overall extraction efficiency, typically 95 – 97% against 90 – 93% for the traditional processes.

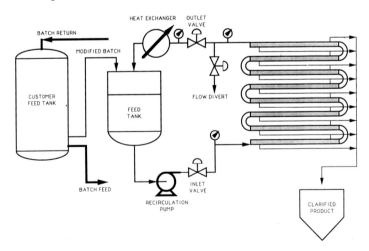

FIGURE 15 – Manufacture of apple juice.

FOOD INDUSTRY

The plants are compact and simple to operate. The technology is no more complicated than ordinary filtration and maintenance needs are usually lower.

The capital costs of equipment and the hold-up volume of, for example, apple juice in UF is reduced compared to the conventional sedimentation, centrifuge, filter press (with filter aids) and final filter press sequence.

The reported economic gains in utilising UF in apple juice clarification are £350,000 per year for a 500 m^3/day plant, due to elimination of diatomaceous earth (350,000 kg/yr), a major reduction in labour and an increase in yield (4%).

The method used in the production of the juice has an effect on the solids content and thus the yield of clarified juice. Suspended solids content may vary from 0.5% to 10% (see Fig 16).

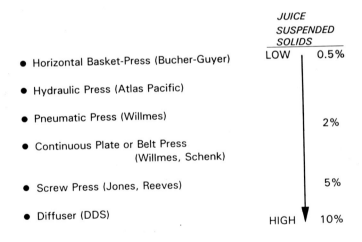

FIGURE 16 – Methods of juice extraction.

The configurations most often used in juice clarification are tubular or hollow fibre although plate and frame systems using flat sheet membranes are also available. Tubular membranes are made in 7 mm to 25 mm diameter sizes and cope well with the high pulp content of many fruit juices. Hollow fibre membranes are between 1 and 1.5 mm diameter and are potted into bundles containing hundreds of fibres in parallel.

TABLE 10 – Fruit juice clarification applications.

Fruit	Status
Apple	c
Blackcurrant	c
Lemon	c
Lime	c
Orange	t
Peach	c
Pear	c
Pineapple	c
Strawberry	c
Raspberry	c
Black Cherry	c
Melon	c
Chinese plum	c
Passion fruit	t
Apricots	c

TABLE 11 – Fruit juice concentration applications.

Fruit/	Clarified	Status	Feed Solids (°Brix)	Maximum Conc.	Limiting Factor
Apple	y	c	6-12	25-30	o
Apple	n	t	10	20	v
Apricot	n	t	11.5	19	v
Grape	y	c	14-17	25	o
Mango	n	t	9	18	v
Guava	n	t	9	19	v
Orange	n	c	11	25	o
Lemon	y	t	7	20	o
Peach	n	t	11.5	17	v
Pear	n	t	11.5	20	v
Pineapple	y	t	11	20	o
Tomato	n	c	4.5	12	v
Cranberry	n	c	6	12	o
Strawberry	n	c	8	16	v
Raspberry	n	c	9	16	v

c = commercial   t = trial   0 = osmotic pressure   v = viscosity

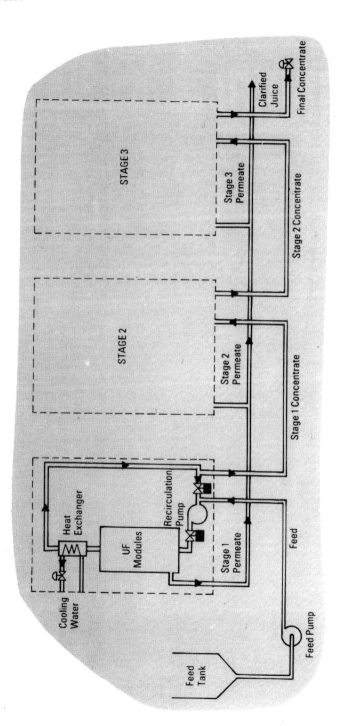

FIGURE 17 – Flow diagram of a continuous multistage ultrafiltration operation.

Apple juice before and after filtration.

TABLE 12 – Performance data for RO of fruit juices.

	Feed (ppm)	Permeate (ppm)	Retention (%)
Dry matter	122,700	327	99.7
Sucrose	12,100	20.7	99.8
Glucose	28,700	57.5	99.6
Fructose	65,400	130	99.8
Malic acid	3,910	21.2	99.5
Citric acid	341	0.9	99.8
Free amino acids	276.4	<0.1	
Methyl-ethyl butyrate	0.97	<0.04	
Butyl butyrate	0.1	<0.04	
Hexyl acetate	0.04	<0.04	
Total acidity (meq/l)	46.1	0.38	99.2
pH	3.95	5.13	

FOOD INDUSTRY

Reverse osmosis plant for fruit juice clarification.

Ultrafiltration of fruit concentrates for cosmetics

A manufacture of natural cosmetic products, has established a facility for the production of selected product concentrates. As part of the new facility, the company has installed a ultrafiltration system, for the clarification of the concentrates. The company is using the facility to produce concentrates for its passion fruit, peach and salad based products, which include cleansers, moisturisers, bath product and face creams.

The installation of the ultrafiltration system separates and removes over 99% of unwanted species, including particles, colloids, enzymes and bacteria which are present in most fruit. The concentrate is produced from a cold infusion of the prepared fruit. This is passed through a coarse sleeve to remove the large chunks of waste fruit, with the resulting liquid being centrifuged to separate any solids down to approximately 10 m in size. The concentrates then exits the centrifuge to a collecting tank, for ultrafiltration. The system is fitted with three ultrafiltration membrane modules, which provide a combined throughput of 500 1 of concentrate per hour. In addition, the system is designed in a modular format, which provides the option to introduce further membrane modules, with minimal disruption, should additional capacity be required at a later date.

Each of the ultrafiltration membrane modules features thousands of hollow fibres, 1 mm in diameter and have a molecular weight cut-off of 500,000. The fibres are manufactured from a polysulphone material and are produced using a spinning process that creates single piece fibres, which are asymmetric in design and are free from the macrovoids.

The concentrate is pumped longitudinally through the core of the hollow fibres. The filtrate passes out through the pores of the fibres and is ducted directly to a collecting barrel, where it is sealed and stored until required by production. The bacteria, enzymes and any other particles suspended in the concentrate are repelled by the membrane surface of the hollow fibres and are retained within its core.

Concentration by Reverse Osmosis

The traditional method used to concentrate juices and purees has been evaporation and the competing membrane technology is reverse osmosis. Although evaporator designs have improved giving lower energy consumption and reduced operating costs, there are still several advantages associated with reverse osmosis;
- improved products quality in terms of flavour and colour
- lower capital and operating costs
- expansion of capacity to preconcentrate upstream of an existing evaporator.

Reverse osmosis is an attractive method for the recovery of valuable apple solids from apple pomace, waste streams in peeling and coring operations and apple handling. The principle can be applied to other fruit juices to give several benefits;
- recovery of valuable fruit solids
- reduction of BOD in waste streams with reduction in sewerage charges
- provides high quality water which can be re-used as CIP water, leach water or flume water and boiler feedwater after treatment
- reduced running costs in comparison to evaporation (approximately £2/m^3)

RO plant performance depends primarily on three factors, juice viscosity, the osmotic pressure and the constraints imposed by the need for product quality. As the temperature increases the viscosity will decrease resulting in higher permeate fluxes. However, operating at high temperature can impair the quality of the juice. The operating temperature of a RO plant depends on the juice being processed. For example, tomato juice can be processed at 70 centigrade with little damage, whilst orange juice can be damaged if the temperature rises much above 25 centigrade.

TABLE 13 – Plant specification.

Tomato juice RO plant	
Feed:	126 t/hr, 4.5 °Brix
Water removal:	59.3 t/hr
Concentrate:	8.5 °Brix
Power consumption (approx):	450 kW
Operating Costs (1985):	
i) 3-effect evaporators:-	£7.6/tonne water removed
(based on steam & electricity costs)	
ii) PCI RO Plant:	£0.9/tonne water removed
(based on electricity & membranes)	
Saving using RO £8,500/day (based on 21.5hr/day operation)	
Membrane Lifetime:	4-5 seasons
Capital cost (1985):	(1993 cost: approx £2.0 million)
Orange Juice RO Plant	
Feed:	5 t/hr, 10-12°Brix
Water removal:	1.8 t/hr
Concentrate:	18 °Brix
Cleaning:	Every 6 hours
Power consumption:	41 kW
Membrane Area:	145 m^2
Operating temperature:	20 Centigrade
Operating pressure:	up to 60 bar

Osmotic pressure is in most cases the limiting factor when concentrating fruit juices. It varies from fruit to fruit but typically is around 40 bar maximum, which corresponds to about 20 to 30% sugars. This means that for most clear juices which need to be concentrated to 60 or 70 Brix, reverse osmosis cannot be used alone. However, some products can be made using RO alone, such as Passata, an 8.5 Brix tomato product used in Italian cooking.

When non clarified juices are processed the viscosity, not osmotic pressure, often limits the concentration achieved. This is particularly true for pulpy tropical fruits.

Unlike most clarification plants, the normal configuration for RO plants in juice concentration is a continuous "feed and bleed" arrangement. Although the average flux of such a system is lower than a batch operated plant, the residence time of the concentrate, which is the product, in the plant is lower which is generally beneficial to product quality.

Where fruit processing is seasonal, both clarification and concentration plants can be designed to process more than one fruit, thus maximising the use of the plant.

Orange Juice

The production of orange juice concentrates is mostly by multi-stage evaporation. This results in stripping of most of the citrus essences in the first stage and destruction of compounds and transformation into undesirable compounds eg furfurals in later stages.

Overall a problem in the concentration of orange juice by evaporation, is the reduction in quality aroma and flavour by the loss of various volatile alcohols, aldehydes and esters and chemical alterations of flavour and aroma compounds by lipid oxidation and Browning reaction. The low temperature operation of RO is a means of minimising this effect. In the case of citrus fruits many of the flavour compounds are sparingly soluble in water and reside in an oil emulsion phase in the juice. Thus RO should achieve high flavour retention and consequently RO has been used to preconcentrate feed juice from around 11% Brix to values of 14% to 20% without loss of taste and flavour.

The osmotic pressure of orange juice rises rapidly with sugar content (Brix) and at 42Brix and 60Brix in 100 bar and 200 bar respectively. Such high osmotic pressures, coupled with high viscosities puts RO beyond current state of the art plant. A process has been devised (see Fig 18), based on ultrafiltration and reverse osmosis, which can produce highly concentrated orange juice (42 – 60Brix).

The process takes a 12Brix orange juice feed and separates out the pulp (bottom solids) by UF. The permeate serum contains the sugar and flavour compounds. The serum, which is approximately 90 – 95% of the feed volume is concentrated using fine, aromatic polyamide hollow fibre RO membranes operating at between 65 – 130 bar pressure. The reverse osmosis operation is a multistage process using first, high retention (HR) membranes and later low retention membranes. The low retention membranes are used when the serum concentration becomes to high to enable an effective driving force to be used in the HR membranes. The "leaky" LR membranes enable concentration to be achieved at lower pressures although sugars are lost to the permeate. However, this permeate is returned to HR RO membrane modules and thus the sugar is not lost. The

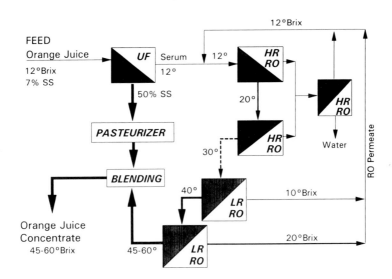

FIGURE 18 – Freshnot process for the production of highly concentrated fruit juices using a combination of UF and high retention (HR) and low-retention (LR) reverse osmosis membranes.

concentrated serum containing 45-60Brix can now be blended with the pulp produced by ultrafiltration. This pulp is however, first subjected to pasteurisation, to destroy spoilage microorganisms and improve stability of the finished product, before blending. Overall there is a good market potential for this product when cost and flavour comparisons are made. Similar processes have been considered for apple juice.

An alternative process to produced highly concentrated juices based on osmosis utilising 100 MWCO tubular RO membranes has been explored. In this a concentrated sugar solution flows on the opposite side of the membrane to the juice feed and osmosis causes transfer of the water from juice to feed. The diluted sugar (osmotic agent) can be reconcentrated or used within the plant.

Reverse osmosis has been investigated as a means of concentrating a variety of fruit juices (eg pear, pineapple) although with limited commercial success. A successful application has been in concentrating tomato juice.

Tomato Juice

Tomato juice has a high pulp content (25% fibre) and a high viscosity. It behaves as a non-Newtonian fluid whose rheological properties depend upon the method of juice manufacture. The temperature to which tomatoes are exposed after chopping, can give different extents of enzyme inactivation, which affects the final viscosity. The degree of concentration of tomato juice by RO is thus limited by the viscosity as well as the osmotic pressure. The concentration of natural tomato juice is around 4.5 – 5Brix.

Generally in the concentration of tomato juice there is less concern over heat damage which occurs during final evaporation and in some cases a slightly cooked flavour is preferred. Thus RO operation must compete purely on an economic level with evaporation (and centrifuge) concentrations. The development of thin film composite membranes (aromatic polyamide) in hygienically designed modules has resulted in commercial operation of RO in tomato juice concentration. At temperatures of 65 – 75C flux rates of 1 md^{-1} at pressures of 60 bar for concentrating juices with osmotic pressures of 30 bar are achievable. The degree of concentration which can be achieved is restricted by viscosity, osmotic pressure and membrane fouling. However retentates are produced, which are suitable for tomato sauces (8 – 12Brix) at a cost significantly less than evaporation. This operation is of particular interest in Italy.

The production of tomato paste, with a 28-29Brix requires concentration by evaporation. The reduced time the juice spends in the evaporator means that the colour quality is very good, showing little browning usually associated with evaporation alone.

There is some interest in the production of concentrates of vegetables such as carrots, celery as well as tomato and cucumber (fruits). A simplified flow diagram is given in Fig 19. In this process the screened juice extract is ultrafiltered and the clarified juice permeate is concentrated by RO using thin film composite membranes. The product is of excellent quality because of the absence of heat treatment in the process.

Concentration of Grape Must by Reverse Osmosis

Reverse osmosis is a non-thermal process consisting of dewatering by the separation of pure water from liquid solutions, such as grape must, by the application of an elevated

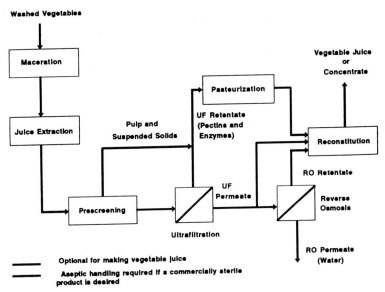

FIGURE 19 – Simplified flow diagram of process for fruit juice concentration.

pressure. RO with tubular membranes is suited to grape must concentration prior to vinification as pretreatment is not required. The RO system uses tubular membranes of 12.5 mm diameter which prevents blockage from without the need for prefiltration of the must. These tubular membranes can be easily and effectively cleaned-in-place.

Overall a high product quality is realised at economical cost. The grape must concentration process operates at ambient temperature thus avoiding losses of volatile aromas, and ensuring that the organoleptic qualities of the must are not modified. Red wine must from Cabernet, Sauvignon and Merlot grape varieties can be concentrated by RO up to a sugar content equivalent to 12 – 13 alcohol.

The potential advantages of concentration by RO included

– Concentrated musts are rich in tannin and in organoleptic components
– Addition of sugar and rectified grape must prior to vinification may not be necessary, and in any case the quantity added is substantially reduced
– The process does not affect the delicate balance of aroma compounds in the must.

RO membrane systems have been used on the pilot scale in the concentration of grape must in the great wine districts of France grape varieties. The wines produced from these musts show excellent body and aroma; comparable with wines produced by traditional methods. The typical performance of RO is illustrated in Table 15.

Maple Syrup

It is reported that there are over 100 RO plants used to produced maple syrup from maple sap. The RO removes approximately 60% of the water from the sap to produce a concentrate. The concentrate is then boiled in open pan evaporators to impart the characteristic flavour and colour.

TABLE 15 – Case study and typical chemical composition of RO concentration of grape must

Case study:	
Initial must volume	170 hl at 10° (potential alcohol)
Initial batch volume	100 hl at 10°
Water removed	28 hl
Concentrated must	72 hl at 14.4°
Final must volume	142 hl at 12°
Processing temp	18-20°C
Batch time	6 hours approx.
Membrane type	AFC99
Membrane area	70 m²
Overall dimensions	L=3.7m, W=0.9m, H=2m
Absorbed Power	26 kW
Cleaning Procedure	Daily 0.25% Ultrasil 11

Note: Investment and operating costs are a function of the membrane surface area employed, which is in turn dependent upon:
* potential alcohol of must
* volume of must processed
* processing temperature
* potential alcohol required

Typical Chemical Composition, White Sauvignon Must

	Raw Must	Chaptalised Must	RO Must
Reducing sugar, g/l	185.6	211.9	213.7
Acidity, g/l H_2SO_4	4.53	4.46	4.92
pH	3.31	3.31	3.30
Tartaric acid	7.42	7.38	7.73
Malic acid	3.22	3.22	3.80
Potassium, mg/l	1,402	1,374	1,428
Free SO2 mg/l	2	2	2
Total SO2, mg/l		33	35

	Chaptalised Wine	RO Wine
Degrees alcohol	12.93	13.02
Density 20/20	0.9916	0.9920
Residual sugar	0.9	0.9
Dry matter	21.9	23.4
pH	3.35	3.34
Total acidity	4.43	4.70
Tartaric acid	4.13	4.30
Malic acid	2.30	2.67
Potassium mg/l	725	703
Volatile acid g/l H_2SO_4	0.53	0.52
Free SO2	20	18
Total SO2	66	66
Total iron	1.0	1.1
Optical density	8.0	10.0
Folin-Ciocalteu index	9.0	10.8

Beverages

There is considerable demand for the production of low alcohol beer, the objective being to supply two full bodied products a low-alcohol beer (around 2% wt ethanol) and a "non-alcoholic" beer, < 0.5% wt ethanol. Low-alcohol beers can be produced in two ways

i) alteration of the brewing process to ferment less alcohol
ii) removal of the ethanol by, for example, dialysis, distillation and RO.

Distillation changes the taste of the beer and is very expensive. Dialysis results in a high percentage of the small molecular components being lost in the permeate, giving a light bodied beer.

The extent of de-alcoholisation depends to a large extent on the type of membrane used. Most RO membranes have relatively low ethanol rejection compared to other constituents (see Table 16). Cellulose acetate rejection of ethanol are as low as 5% which results in a very low alcohol content beer. The cellulose acetate membrane however also has a relatively low rejection of volatiles. Thin film composite membranes can reject some 70 to 75% of the ethanol and almost all the volatiles in the stock beer, thus retaining the quality of the low-alcohol beer.

TABLE 16 – Composition of Feed and Permeate of Beer Treated by Reverse Osmosis.

Composition	Cellulose Acetate		Thin Film Composite	
	Feed	Permeate	Feed	Permeate
Acetaldehyde	1.1	0.9	1.2	0.3
n-Propanol	16.2	14.5	11.7	0.0
Ethyl acetate	41.0	40.4	26.0	1.0
Isobutanol	25.3	19.4	14.4	0.0
Amy alcohol	23.0	17.8	20.3	0.0
Isoamyl alcohol	82.9	66.2	42.8	0.4
Isoamyl acetate	4.8	4.3	2.7	0.0
Diacetyl	0.02	0.01	0.02	0.0
2.3-Pentanedione	0.02	0.01	0.02	0.0
Ethanol	5.3	5.4	3.7	1.0

From Coldstein et al. 1986

There are also potential applications for RO in the production of low and high alcohol content wines and also non-alcoholic wines (0.01%).

Cereal Processing

Potential applications of membrane processes in cereal production are in the treatment of stillage, in corn wet milling operations, evaporation of steep water, concentration of dilute sweet waters and in polishing of RO permeate and evaporator overheads.

Corn proteins can be recovered from stillage solubles of dry milled fractions of corn,

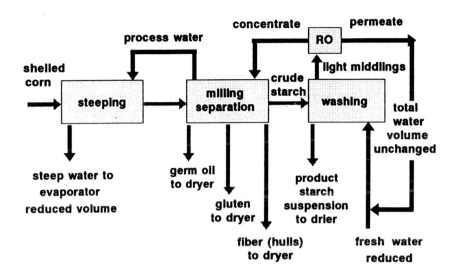

FIGURE 21 – Use of RO in wet corn milling processing.

grits, flour, degerminated meal etc by reverse osmosis. After production of the alcohol by fermentation, and separation of the insoluble fractions by filtration and centrifugation, the soluble fractions are ultrafiltered and the UF permeate processed by reverse osmosis. The RO permeate contains 70 to 80% of the original stillage volume.

RO has been used to reduce the major load on evaporators in wet corn milling operations (see Fig 21).

The wet milling operation utilises fresh water in a starch washing step. The water leaves this stage containing all proteins and some starch and flows countercurrent to the flow of corn fractions to the front of the process. The used "waste water" or sleep water is then processed in an evaporator. By introducing RO to separate the light middling, into a permeate water for recycling as wash water and a concentrate for wet milling, significant benefits are achieved

    reduced fresh water consumption
    reduced waste water load
    reduced energy and cooling water demands at the evaporator
    improved starch quality
    capacity expansion without evaporator and boiler expansion
    installation in existing plant without operational shut down

It is reported that a nominal 400 m^2 cellulose acetate tubular membrane installation is operating in a 300 ton/day grind wet corn milling plant in the US.

Vegetable Proteins

Protein containing whey streams formed in the processing of several vegetables can be treated by reverse osmosis. One application is in the production of starch from potatoes where the whey contains 4% of dissolved solid with up to 40% as highly nutritious protein. Tubular RO modules have been used to concentrate this whey to 8% dissolved solids at fluxrates (average) of 1 md^{-1}. The protein containing retentate is coagulated by heat treatment before recovery of the proteins. The RO permeate is suitable for washing potatoes.

RO in combination with UF can also be used to produce low-intermediate- and full-fat protein production from full-fat soy bean and peanut flours. In the process, defatted flours are contacted with warm, slightly alkaline water to extract proteins. This extract liquid is then ultrafiltered and the permeate treated by RO. Ultrafiltration separates smaller protein molecules that are tightly bound to flavour and colour components. RO recovers the smaller proteins as soluble carbohydrates and the permeate water is very low in TDS and can be recycled.

Oil and Fats Processing – Waste Water Treatment

Waste waters from vegetable oil extraction plants and refineries contain high amounts of inorganic and organic purities, COD and BOD content (see Table 17). These waste waters can be sent directly to municipal treatment systems or treated on site. On site treatment involves acidification, gravity separation, pH adjustment to a neutral state, chemical complexation or coagulation, biological treatment and possibly filtration through granular media. On site treatment is not, as yet, widely practised although with stricter legislation this will rise and thus so will the need for alternative, lower cost water treatment.

TABLE 17 – Characteristics of a typical waste water from an edible oil refinery

BOD	500 – 2,000 ppm
COD	1,500 – 10,000 ppm
TSS	400 – 1,600 ppm
TDS	1,100 – 2,900 ppm
oil and grease	200 – 700 ppm
total phosphorous	1,000 – 2,500 ppm
iron	1 – 6 ppm
silicon	1 – 6 ppm
manganese	0.01 – 0.25 ppm
pH	4 – 6

The most common applications of RO and UF membrane systems are after conventional waste treatment plants which include, screens, settling tank and fat skimmer.

The membrane system separates the oils from the waste streams and concentrates the organic and inorganic substances. Gelling substances in the wastewater can foul the membranes and thus pretreatment, with calcium chloride, may be required.

Production of Salt

Sodium chlorine is the major ionic constituent of seawater and is thus a potential feedstock for the manufacture of food grade salt. The use of electrodialysis as a pre concentration step in the production of salt, prior to evaporation, is practised almost exclusively in Japan.

FOOD INDUSTRY

Salt production by electrodialysis.

It is reported that over 350,000 tons/annum of salt are produced in this way requiring around 0.5 million m² of installed ion-exchange membrane. The success of ED in Japan in this application was due to the development of low cost highly conductive membranes with a high selectivity to monovalent ions. Prior to this, salt was produced in large salt fields utilising solar evaporation. The production of salt in Japan is highly subsidised by the government, otherwise it would not be an economic operation on the world market.

A typical flowsheet of salt manufacture by electrodialysis is shown above. The system consists of two major sections: electrodialysis and evaporative crystallisation. The former process concentrates filtered seawater (3% salt content) taken in to produce brine at approximately 20% for the latter process of which crystallisation of salt from the brine. A multi-effect system is utilised for the evaporative crystallisation and the waste heat of the final evaporating crystalliser is used to heat seawater for electrodialysis. A typical pretreatment of seawater may be a conventional sand filtration or two stage sand filtration depending on the quality of raw seawater. Either a single stage or two stage electrodialysis can be utilised for seawater concentration. Steam turbine generation is generally used for the system and the electricity generated by the turbine is used for electrodialysis while low pressure steam from the turbine is used for evaporative crystallisation.

The pre-concentration of sea water, prior to evaporation, by membrane separation requires a product with a concentration of 200 kg/m^3 and thus a high osmotic pressure. This factor and also the scaling problems which can arise make RO quite unsuitable and thus ED is the preferred method.

Electrodialysis in The Food Industries

Membrane processes in general play a significant part in the food industry and in particular the dairy industry. In particular electrodialysis is used to de-ionise or de-acidify fruit juices, wines, milk and whey. Desalting of molasses in the production of sugar is an economically attractive process. In these applications electrodialysis is often competing directly with ion-exchange, but in general, because of the continuous process operation, ED is more economic.

HANDBOOK OF INDUSTRIAL MEMBRANES

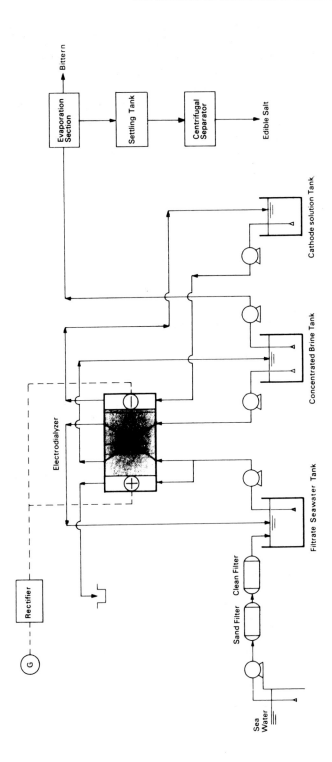

FIGURE 22 – Process flowsheet for production of salt from seawater.

The scale of operation of ED in the dairy industry is small compared to desalination although the value of the product in these applications is usually much greater. The desalting of whey is the largest area of use for ED in the food industry. The total capacity of whey desalting by ED is about 100,000 tones per year of 90% demineralised whey powder and requires over 25,000 m² of installed membrane area.

Crude cheese whey contains calcium, phosphorous and other inorganic salts (approximately 6% TDS), which must be removed to enable production of foods such as ice-cream, cake, bread and baby foods.

The whey deashing process consists of 4 general steps:
1. Concentration from 6% solids to 20-30%.
2. pH adjustment and clarification to remove insoluble proteins.
3. Electrodialysis to remove 25-90% of the salts.
4. Concentration and spray drying to produce a free-flowing powder.

In whey de-ashing, a 90% demineralisation is considered to be the practical limit due to decreasing electrical conductivity with decreasing demineralisation. In an electric field, milk and whey behave as weak electrolytes. The conductivity of whole milk is about 5 sm⁻¹, while fully demineralised milk and whey have negligible electrical conductivity. This is why electrodialysis cannot be completed to near 100%. Fig 23, show typically data of desalting by electrolydialsys. ED can also be utilised for similar processing of skimmed milk.

Electrodialysis is performed in equipment similar to that used for water desalination. However, stacks contain fewer cell pairs to limit the voltage across a stack and sanitary piping is used, arranged for ease of cleaning. In large plants, with 50% or less deashing, the whey is de-ashed continuously, otherwise batch operation is used.

The commercial value of whey can be increased by its fermentation to lactic acid. This

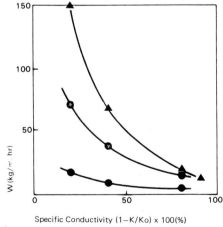
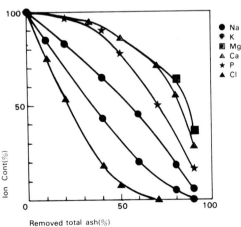
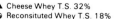

FIGURE 23 – Performance of desalination of milk products.

fermentation is inhibited and a method is required which can continuously extract the lactic acid as soon as it is produced. ED has been proposed for this process. There are whey electrodialysis plants in the USA, Mexico, the Netherlands, Ireland, Japan, and France with the estimated production of desalted whey solids of 10^5 tonne/year.

Grape Musts and Wine

The precipitation of tartrates from grape musts and wines, is a normal and common phenomenon in the evolution of a wine. It is caused by the insolubilisation of the salts of the tartaric acid naturally present in the must. Electrodialysis can be used to extract the excess of these salts from musts and wines, the amount of salt removed appears to be dependent on the type of wine. For the stabilisation of wine, it is considered sufficient to reduce the quantity of potassium by about 100 -250 mg/dm^3.

The salt removal produces a slight increase in the total acidity and a slight decrease in alcohol content. Additionally malic acid can be removed and some sugar can be lost in the process. During de-ashing, sulphur dioxide is also extracted as HSO_2^- ions at a very high rate. The current efficiency in grape juice stabilisation is around 68-75%. The energy consumption is typically around 5.0 kW h per kg of K^+ removed, at a current density of 100 A m^{-2}.

Electrodialysis also has an application in the production of concentrated grape must although the precipitation of the tartrates is more problematic since the salts are present in higher concentrations.

Electrodialysis can also be used in wine production to harmonise the wine by adjusting the ratio between sugar and acid. For this, a special process is used, called ion substitution, in which the ED unit works as an ion exchanger.

Fruit Juices

As well as the applications in the stabilisation of grape musts, ED can also be used in de acidification of fruit juices such as grape, orange, pineapple, apple, lemon. A commercially developed process unit for deacidifiction of orange juice, consists of only anion-exchange membranes. The sections are alternating depleting compartments, for the orange juice, and enriching compartments for KOH. As only anions are able to pass through the membranes, the net effect is the extraction of citrate ion from the juice and their replacement by hydroxide ions. To limit the deposition of colloidal material on the surface of the membranes, the voltage is periodically reversed. The energy requirement for juice deacidifiction varies between 0.020 and 0.100 kWh/equiv., and the current efficiency is between 52% and 90%.

Demineralisation of sugar-beet and sugar-cane juices.

Refined sugar is made from crude sugar through resolution, decolourising and crystallisation. After 4 or 5 cycles or recrystallisation, etc. the waste molasses mother liquor is frequently used as animal feed. However ED can lower the ash content of the waste molasses liquor to obtain a refined sugar of some value (see Table 18).

FOOD INDUSTRY

TABLE 18 – Performance of ED in food processing and other applications.

Amino Acid Liquor Treatment		
	Before Treatment	After Treatment
Desalted Sauce		
• NaCl(g/?)	60	2.4
• pH	5.0	5.2
• Conductivity (ms/cm)	84.5	9.5
• Viscosity (c.p.)	2.0	
• Concentrated solution	0.1% as NaCl	
• Electrode solution	1% as NaCl	
Desalted NaCl (%)		96%

Separation of Organics Example 1		
	Before Treatment	After Treatment
Desalted Sauce		
• Organic matter	1.0N	1.2N
• NH_4Cl	1.2N	126ppm as Cl-
• pH	6.3	6.4
• Conductivity (ms/cm)	94.0	
• Temp. (°C)	25.0	
Concentrated Solution	1N NH_4Cl	
Electrode Solution	NaCl	
Desalted NaCl (%)		99
Lost Organic matter (%)		4.6

TABLE 19 – Desalting of soy sauce.

Sugar Desalting (Polysaccharide Group) 4% Aqueous Solution	Before Treatment	After Treatment
Desalted Sauce		
• Cl" (ppm)	482.1	34.1
• SO_4^{2-} (ppm)	346.6	9.7
• pH	7.37	6.85
• Conductivity (ms/cm)	2.64	0.17
• Visosity (c.p.)	15	—
• Temp. (°C)	55	—
Concentrated Solution	0.05N-NaCl	
Electrode Solution	0.25N-NaCl	
Desalted Ratio of NaCl (%)		94%
Leakage of Polysaccharide group		0
Sugar Desalting (Polysaccharide Group) 4% Aqueous Solution		
	Before Treatment	After Treatment
Desalted Sauce		
• Cl" (ppm)	482.1	34.1
• SO42" (ppm)	346.6	9.7
• pH	7.37	6.85
• Conductivity (ms/cm)	2.64	0.17
• Viscosity (c.p.)	15	—
• Temp. (°C)	55	—
Concentrated Solution	0.05N-NaCl	
Electrode Solution	0.25M-NaCl	
Desalted Ratio of NaCl (%)		94%
Leakage of Polysaccharide group		0

Desalting of soy sauce

Naturally brewed soy sauce has a high salt content, 16-18%, soy sauce of lower salt content is required in certain diets or for salad dressings. Table 19 shows an example of the treatment.

De-alcoholisation of Beer by dialysis

Dialysis has been reported to be in use for the removal of alcohol from beer on the scale of 1500 dm³/h. The process uses hollow fibre modules with a total area of 90 m². A 40% reduction in alcohol content is achieved although the process is limited owing to the removal of other low molecular weight constituents along with the beer which influence both taste and aroma. It is possible to reduce this affect by using a dialysate which has a

similar composition to the beer to be de-alcoholised, but which is alcohol free. Operation of the dialysate in conjunction with vacuum distillation allows the ethanol to be removed continuously.

SECTION 18

Membranes for Electrochemical Cells

MEMBRANES FOR ELECTROCHEMICAL CELLS

Electrolysis is a widely used method in industry for the manufacture of a range of chemicals, both inorganic and organic. Many of these processes rely on the use of membranes in the cells to achieve high selectivity and efficiency. The notable use of membranes in electrolysis is in chlor-alkali for the production of chlorine and caustic soda, perfluoropolymer cation exchange membranes were specially researched and developed for this application. Since then these membranes have found many used in electrosynthesis and in other applications of effluent treatment, recycling and energy generation ie fuel cells.

Cell separators

In electrochemical processes it is frequently necessary to separate the processes occurring at the anode and cathode with membranes for reasons of high yield and selectivity, product separation and safety in operation.

In a cell both oxidation and reduction occurs simultaneously and if starting materials, or products, are susceptible to undesirable reaction at the counter electrode then separation of anode and cathode is required to achieve high yield of desired product. Furthermore if starting reagents or products of one electrode process can chemically react undesirably with species generated at the counter electrode then again separators will be required. For example in the case of the chlor alkali electrolysis in the absence of a cell separator, the contact of brine solution, saturated with chlorine, with the hydroxide ion generated by the cathodic decomposition of water would form hypochlorous acid and other species, and thus result in a loss of efficiency and selectivity.

Cell separators are often required to separate gaseous products obtained at the two electrodes in a cell, eg in the electrolytic production of hydrogen and oxygen by water electrolysis where the requirement of safety in operation is important.

The separators used in cells are generally classified as permeable or semi permeable
1) Permeable membranes permit the bulk flow of liquors through their structure and are thus nonselective regarding transport of ions or neutral molecules.

In electrochemical processes these are frequently referred to as Diaphragms.

2) Semi-permeable membranes permit the selective passage of certain species by virtue of molecular size or charge. In electrochemical processes these are frequently termed Membranes and separation is based on the charge carried by the molecule.

Although there has been a steady move away from the use of porous diaphragms towards membranes (ion exchange materials) the former are still used in several industries. An effective separator material ideally should exhibit a range of desirable properties as listed in Table 1.

TABLE 1 – Desirable Properties of Separators.

Correct permeability and selectivity	Good current efficiency
Inert to cell environment: good temperature and chemical stability	Long life
Uniform properties across its surface and homogeneous flow	Good CE, good quality products
Low voltage drop	Lower energy performance
Finite thickness	Overcomes diffusion
Some physical strength: mechanical stability	Long life – easy to install n cells
Resistance to gas blinding	Low voltage – good energy performance
Low cost	Economic operation

The porous diaphragm represents a compromise between the demands of separation of anolyte and catholyte and effective electrical conductivity between anode and cathode via the ions in solution. A good degree of separation is achieved by using a uniformly fine porous structure which permits diffusion of material, but not mass flow. The conduction of electrical current is by the solution of ions in the porous structure, which gives rise to a higher electrical resistance than that of the bulk electrolyte solution. The higher the porosity (size and/or number of pores) the greater the electrical conductivity of the diaphragm, but the poorer the separation of the anolyte from the catholyte. The transport of species across a porous diaphragm will be greater the thinner the material and the higher the concentration gradient across it. Since a diaphragm is positioned in a voltage gradient the material it is made of should be an electrical nonconductor.

Membrane Materials

There is potentially a wide range of materials which can be used as cell separators [1]. Table 2 lists some of the materials used as separators; which are divided into three types, organic, inorganic and composites.

Overall the materials used as separators may either simply be microporous separators or may posses specific ion transport characteristics.

One of the cell designs for the production of chlorine and caustic soda uses an asbestos diaphragm constructed onto the electrode itself. Asbestos fibres are widely used in the chlor-alkali industry (and in water electrolysis) because they possess the required chemical and physical stability in alkaline media and can be engineered to give the required permeability. Chrysotile asbestos (approximate formula Mg_3 (Si_2O_5) $(OH)_4$) fibres are slurried with caustic soda solution and the slurry sucked on to a mild steel mesh cathode which collects the asbestos fibre. The quantity (wt) and quality of fibre is controlled to give a required thickness and permeability of diaphragm.

TABLE 2 – Materials Used as Separators and Diaphragms.

Organic	*Inorganic*	*Composites*
Porous plastics (polypropylene, PTFE, Ryton, polysulphones)	Asbestos Paper, felt, fibre	Asbestos fibres and glass fibres Asbestos fibres and fluorocarbon binder Asbestos sheet and composite fibre sheet Asbestos and polyacid Asbestos on metal screen (cathode) Asbestos and silicate coat (anolyte side) PMX (PTFE/ZrO_2 or TiO_2) Hyrdophobic polymer with ZrO_2/MgO/K_2TiO_3 coating
	Ceramics	Sb_2O_5/Anolyte: ZrO_2/Catholyte polysulphone and Sb_2O_5
Ion exchange:	Al_2O_3, SiO_2	Resin and $CaSO_4$
Flemion, Nafion (perfluoro) or porous PTFE	ZrO_2	
Cationic: – SO_3^-	Glass fibres	
Anionic: – $N^+(CH_3)_3$	b alumina Fe phosphate	

As a general diaphragm asbestos is not an ideal material as it is not resistant to very acid conditions, not robust physically and environmentally unacceptable. Thus for general applications alternatives materials with the required chemical resistance are required. This generally restricts the choice to perfluorinated plastics or ceramics. For a chlor-alkali cell the choice is restricted to perfluorocarbon plastics, e.g. PTFE, and certain ceramics, e.g. TiO_2, ZrO_2.

In the case of ceramics processing the material into a porous structure suitable for a diaphragm can be difficult with the resultant material quite brittle. Installation in parallel plate electroysers presents mechanical difficulties, which outweigh advantages of mechanical strength, and temperature and chemical stability. In the case of certain electrolyses, e.g. alkaline water electrolysis, some flexibility in the structure is introduced by sintering the porous ceramic onto a supporting metal net.

The fabrication of polymer membranes for separators does not create unusual problems. Operating conditions in electrolysis usually involve extremes of pH and/or organic solvents and the materials are limited to polymers of ethylene, propylene, vinyl chloride and tetrafluoro ethylene etc. In manufacture an open porous structure can either be created at the time of fabrication of the sheet or by the incorporation of a leachable filler. Several polymer materials have been tested in alkaline water electrolysis and in chlor alkali cells. In these applications the materials used must be hydrophilic so that they are completely wetted and thus not blocked by gas bubbles. In the case of fluoropolymers this can be achieved by adding suitable wetting agents e.g. ZrO_2, into the diaphragm structure. A major limitation is the operating temperatures of organic materials, generally to 120°C, although materials such as Ryton and PTFE are stable at temperatures up to 160°C.

The limitations of permeable polymer separators in some electrochemical cells has generated interest in ion exchange membranes. Ion-exchange membranes have the characteristic property of being able to distinguish between cations and anions and can be used to keep, selectively, either anions or cations, from transferring from one cell compartment to the other (see Fig 1) and they allow electrolysis to be carried out under close control of pH. For example the use of an anion exchange membrane will prevent the transfer of H^+ ions, generated at an anode during oxygen evolution, into the catholye chamber and thus allow a pH differential to be set up in the cell.

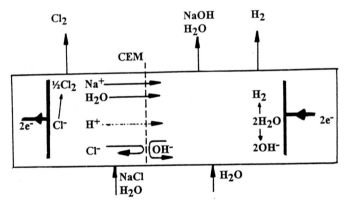

FIGURE 1 – Cation exchange membranes in electrochemical cells.

The main properties required of ion-exchange membrane for them to be successful in technical processes are;

1. Low electrical resistance. The permeability for the counter-ions under the driving force of an electrical potential gradient should be high to minimise the membrane IR losses.
2. High permselectivity. It should be highly permeable for counter-ions, but should be highly impermeable to co-ions, and to non-ionised molecules and solvents.
3. Good mechanical stability. It should be mechanically strong, to prevent high

degrees of swelling or shrinking due to osmotic effects, when transferred from concentrated to diluted salt solutions and vice versa, and be dimensionally stable.
4. Good chemical stability. It should be stable over a wide pH-range and in the presence of oxidising agents.
5. Good operating characteristics. It should be capable of operation over a wide range of current densities and under varying conditions of temperature, current density, pH etc.

The stability of a membrane is of paramount importance, membrane materials, because of their high cost, are required to operate for periods of several years. They should exhibit good chemical stability over a wide range of pH in the presence of oxidising agents and over a wide range of temperatures. The ion-exchange material should be mechanically strong, to withstand compression between gaskets, abrasion, high degrees of swelling by the electrolyte solvent, or shrinking due to osmotic effects, when transferred from diluted salt solutions to concentrated salt solutions and vice-versa. It is also important to prevent the membrane from drying out, to avoid wrinkling or stress in the material.

A factor in the operation of cells with membranes is the transport of solvent eg water which accompanies the transferring ions. In aqueous systems the transport of water can

TABLE 3 – Applications of Membranes in Electrosynthesis and Effluent Treatment.

Inorganic	
Chlorine, Sodium Hydroxide (and Potassium hydroxide)	
Peroxidisulphate	Electrorefining
H_2/O_2 (water electrolysis)	Potassium Stannate
Hydrogen Peroxide	Chlorine Dioxide
Ozone	Dinitrogen Pentoxide
Chromic Acid	Arsine
Electrowinning	Sodium Dithionite
Electroplating	
Organic	
Hydrodimerisation of Acrylonitrile	
Hydrogenation of Heterocycles and Nitriles	
Reduction of Carboxylic Acids	
Cathodic Cleavages	
Oxidation of Polynuclear Aromatic Hydrocarbons	
Indirect Oxidations with Cr(IV) and Ce(IV)	

be significant e.g. 3-5 water molecules accompanies one sodium ion in chlorine cells. If hydrogen ions are transferred then typically two molecules of water are transferred per ion.

Electrochemical Synthesis

The production of organic and inorganic chemicals by electrosynthesis is widely practised and many cell designs require the use of membranes to separate the anode and cathode processes. Historically inorganic electrosynthesis is well established with the largest being the chlor-alkali industry where major developments in membrane materials has taken place. Other inorganic synthesis which utilise membrane cells operate at much smaller production capacities. Table 3 lists some applications of membranes in electrosynthesis and also in effluent treatment and recycling.

Effluent Treatment and Recycling

Metal recovery from eg	electroplating rinse waters
	etching solutions and rinse waters
	metal cleaning solutions
	scrap reprocessing and refining
	catalyst liquors
	chemical-processing reagents
	photographic processing solutions
	reprocessing spent batteries
	primary ore leaching
	mine dump leaching or runoff water
	liberator cells in electrowinning and electrorefining
Chromium liquors from eg	electroplating baths, spent solutions, metal finishing, pickling, etching and stripping solutions.

Chlor-Alkali Industry

The chlor-alkali industry is the largest tonnage electrochemical process with estimates of the energy consumption in the USA alone at 2% of all electrical energy consumed. There are three electrochemical cells used to produce chlorine and caustic, one uses mercury cathodes without diaphragms or membranes, one uses diaphragm cell as previously discussed and the third most recent design uses ion-exchange membranes. Membrane cells have replaced mercury cells and some diaphragm cells in many countries, and the growing trend is in this direction, due to improving performance and environmental pressures against the use of the mercury cells.

In membrane cells chlorine gas is produced by the oxidation of chloride ion at the anode, typically a dimensionally stable anode (DSA) comprising a titanium substrate with a thin coating of ruthenium oxide and other valve, precious or transition metals. At the cathode hydrogen gas and hydroxide are generated from water. In the membrane cell this hydroxide combines with the sodium ion transferred through the membrane to give the product sodium hydroxide.

Membrane cells are similar to diaphragm cells in some respects, although no bulk flow of electrolyte occurs through the separator, which in this case is a cation exchange membrane. The cell design has gone as far as is physically possible to minimise the internal resistance by operating in the "zero-gap" mode. This entails the anode and the cathode both being in physical contact with the membrane (see Fig 2). This in turn requires special design of both the electrodes and the membrane.

A cation exchange membrane in principle represents the ultimate separator for the chlor alkali industry, that is one which will allow the transport of Na^+ ions into the catholyte with the exclusion of Cl^- ions. To be effective this membrane must be stable to both the desired 50% caustic soda concentration produced in the catholyte and to the wet chlorine produced at the anode and have a low electrical resistance.

In the chlor-alkali industry a crucial factors is electrolysis power consumption which is inversely and directly proportional to current efficiency and voltage respectively. This has led to development of membranes capable of operating to 95 – 96% current efficiency with low electrical resistance to minimise the membrane contribution to the electrolyser voltage. The membranes display a high level of cation permselectivity, impede the back migration of OH^- ions, the source of inefficiency and are facile ionic conductor.

The introduction of the ionic groups into the polymer has significant influence on the structural, physical and electrochemical properties. In aqueous solutions the polymer adopts an unique morphology in which ion-exchange sites, counterions and absorbed water are phase separated from the fluorocarbon backbone material The concentration of the ion-exchange groups is a key variable in optimising membrane properties and is quantified in terms of the equivalent weight of the polymer film i.e. the mass, in grams, of the dry polymer in the acid form that would neutralise one equivalent of base. In the development of membrane cells for chlor alkali, improvements with membranes containing sulphonic acid ion-exchange groups included low electrical resistance due to high water uptake and resistance to protonation as a result of their low pKa value. Carboxylic acid ion-exchange groups were known to have improved OH^- ion rejection properties and companies developed synthetic routes to perfluorinated carboxylic copolymers and have since demonstrated the attainment of 95-96% current efficiency for the production of caustic soda in the range 30-35% w/w.

No single membrane offers all the required characteristics for operation in chlor alkali cells and thus cell designs are based on modern bilayer membranes (see Fig 3), which consist of four components.

1. A thin weak acid cation exchange polymer, capable of operating at caustic concentrations of 30-0% without a significant loss in current efficiency. The thinness counteracts the relatively poor electrical conductivity.
2. A thicker, strong acid polymer which in principle offers superior properties regarding resistance and efficiency in comparison to the weak acid type. However, the performance is limited to relatively low caustic concentrations (12%), and thus it is placed away from the cathode side.
3. A reinforcing fluorocarbon net in the strong polymer membrane side to give mechanical stability.

4   A surface coating on the anolyte side of the membrane to encourage gas release from an otherwise hydrophobic surface and the contacting anode surface.

The electrode design in the zero gap configuration must ensure easy gas release from the back, otherwise the bubbles would constitute a large resistance to the flow of current. Typically this is achieved by using expanded metal or by louvered, or otherwise contoured anodes. Both the anode and the cathode will have appropriate electrocatalyst coatings.

The search for the most appropriate ion-exchange group has influenced the structure and design of the commercial chlor-alkali membranes. The structure of some membranes exploits the merits of both ion-exchange groups, whereas other membranes have tended to rely on the carboxylate function.

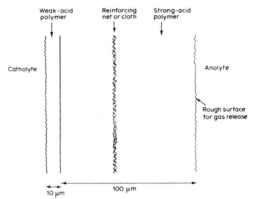

FIGURE 3 – A bilayer chlor-alkali membrane.

Water Electrolysis

The electrolysis of water produces H_2 and O_2 gases. Water electrolysis is used in applications where typically a high purity of hydrogen gas is required in for example production of foodstuffs where catalyst poisoning is problematic in the hydrogenation of edible oils and fats (for margarine and shortening products). Other uses include the manufacture of semiconductors, refining of high purtiy and ammonia synthesis. There are also applications of water electrolysis on a small scale for the supply of hydrogen and oxygen due to the convenience of continuous production, at a controlled rate without the reliance of bottle gas supply (and replacement), especially at remote industrial sites. The applications include hydrogen to cool electric generators in power stations, hydrogen in meteorological bolloons, reducing atmospheres for metallurgical heat-treatment processes, gases to laboratories and oxygen to life-support systems in spacecraft and submarines.

The scale of water electrolysis varies from units to produce 2 dm^3/h H_2 for gas-liquid chromatographs to 10^4 m^3/h H_2 for ammonia plant.

The majority of commercial water electroysers are based on filterpress or tank designs. The separators are predominantly asbestos based serving merely as a means of segregating the hydrogen. and oxygen gases. For smaller for example point of use applications there are two alternatives, solid polymer electrolyte cells and palladium membrane electrode cells.

The first solid polymer electrolyte cells were developed for space and military applications. The solid polymer electrolyte is a thin perfluorinated cation exchange membrane (sulphonic acid type). The structure of the SPE electrolyser is shown in Fig4 , water (demineralised) is supplied to the anode which produces oxygen and protons. The protons pass through the membrane, with associated water of hydration, and are reduced to hydrogen gas. The electrodes are based on fuel cell type electrode structures being porous PTFE/graphite/precious metal (Pt) catalyst composition. In practice several cell elements (ie current collectors – electrodes – supports – membrane) are sandwiched together in parallel to give power ratings of 7 to 50 kW.

The performance of the electrolyser is particularly good (see Table 4) with a typical cell voltage of -2V at a current density of approximately 1000 mAcm^{-2} (50C).

TABLE 4 – Performance characteristics of solid polymer electrolyte cell.

Parameter	Value
Cell module:	
Cell active area	0.093 m^2
Number of cells	>7-51
Current density	1.075 A cm^{-2}
Maximum current	1000 A
Initial cell voltage	—2 V/cell
Gas production:	
Hydrogen gas flow rate	0.42 m^{3}h^{-1}/cell
Oxygen gas flow rate	0.21 m^{3}h^{-2}/cell
Hydrogen gas purity	H$_2$>99.995% by volume
	O$_2$<0.005% by volume
	Halogens and halides <4 v.p.m.*
	Total impurities <50 v.p.m.
Oxygen gas purity	O$_2$>99% by volume
	H$_2$<1% by volume
	Total impurities <500 v.p.m.
Maximum differential pressure (H$_2$ over O$_2$)	7 bar
Process water:	
Demineralized water conductivity	<0.25 μScm^{-1}
Inlet temperature	50°C
Outlet temperature	65°C

*Volumes per million (volumes)

Ozone Production

The concept of solid polymer electrolysis has more recently been applied to the production of ozone from the electrolysis of water

$$3H_2O - 6e^- \; O_3 + 6H^+$$

The cell uses a perfluorinated cation exchange membrane as electrolyte sandwiched between a porous lead dioxide anode and a porous platinum based cathode. The cell is fed with pure water and generates approximately 30% ozone gas with 70% oxygen at cell voltages between -3 V to -5V at current densities between 0.5 to 2 Acm^{-2}. The operation of the elctrolyser under pressure can produce concentrated solution (> 100 mgdm^{-3}) of ozone in ultrapure water. The process water quality for the cell must be pure with a conductivity less than 20 µscm^{-1}.

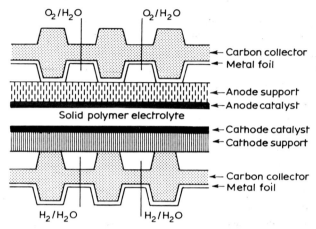

FIGURE 4 – Structure of a solid polymer electolyte cell.

Applications of ozonisers are in pharmaceutical, electronic and other industries which require ultrapure water. Many ultrapure water plants (see Section 10) are designed on a closed loop basis in which periodic ozone generation can main system sterility and minimise microbial growth. The SPE ozonisers are used in conjunction with or compete with other methods of purification eg ultrafiltration, UV radiation in pure water production, air-phase carona discharge ozonisers and other electrochemical cells.

Other applications of electrochemical ozonisers include, sterilisation in cooling tower water and swimming pools, oxidation of industrial process liquors containing cyanides or organics, PCB's.

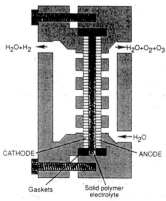

FIGURE 5 – Solid polymer electrolyte ozone generators.

Other Synthesis

Membranes used in the chlor-alkali industry rely on the transport of Na⁺ ions for efficient operations whereas in cells the ion transported is H⁺. Clearly these two ionic species are not the only ones which will be transported across ion-exchange membranes. For example in the production of potassium stannate (see Fig6) the species transported is K⁺ ions.

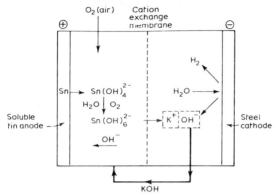

FIGURE 6 – The production of potassium stannate ($K_2Sn(OH)_6$) from tin.

$$Sn + 2\ OH^- + 3\ H_2O + 1/2\ O_2\ Sn(OH)_6^{2-} + H_2$$

The cell consists of a basket, which holds the tin anode bars, a cation exchange membrane and a steel cathode for the evolution of oxygen in potassium hydroxide electrolyte. Under ambient conditions the tin dissolves in the potassium hydroxide electrolyte, to stannite, which reacts rapidly with the oxygen in air, sparged to the cell, to form the stannate.

$$Sn(OH)_4^{2-} + H_2O + 1/2\ O_2\ Sn(OH)_6^{2-}$$

The potassium stannate solution is continuously withdrawn form the anolyte while the potassium hydroxide catholyte is passed into the anolyte. The migration of K⁺ ions from the anolyte, into the catholyte, through the membrane provides part of the KOH required for the overall reaction.

In the synthesis of dinitrogen pentoxide the membrane transports both N_2O_4 and H_2O under severe oxidising conditions. In the process both electrode reactions are utilised, in the following reactions.

Anode $N_2O_4 + 2\ HNO_3\ 2\ N_2O_5 + 2\ H^+ + 2e^-$
Cathode $2\ HNO_3 + 2\ H^+ + 2\ e^-\ N_2O_4 + 2\ H_2O$

The N_2O_4 generated at the cathode assists in the splitting of nitric acid into N_2O_5 and water. The water formed is separated form the anolyte by the membrane. Any water which is present in the anolyte will be converted to nitric acid and thus in the process the cathode product N_2O_4 must be purified of water before it is fed to the anode reaction.

Other Applications of Membranes In Electrochemical Cells

The following are provided to illustrate the diverse use of membranes in electrochemical cells. Further details can be found in the text Electrochemical Processing for Clean Technology, K. Scott, RSC Publishers, 1995.

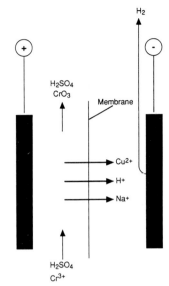

FIGURE 7 – Regeneration of chromium (VI) solutions by the anodic oxidation of spent Cr(III) liquors.

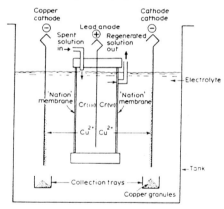

FIGURE 8 – Regeneration of chromium (VI) solutions and the recovery of metal cations.

Anode (PbO_2) reaction
$$2\ Cr^{3+} + 7H_2O - 6e^-\ Cr_2O_7^{2-} + 14\ H^+$$
The relevant cell process are
    anode $2\ Cu\ Cl_3^{2-} - 2e^-\ 2\ Cu^{2+} + 6\ Cl^-$
    membrane transport Cu^{2+} anode to cathode
    cathode $Cu^{2+} + 2e^-\ Cu$ (flake)

Organic Electrosynthesis

The use of ion-exchange membranes is widely practised in electro-organic synthesis. The choice of a membrane for electrosynthesis application is frequently affected by the fact

MEMBRANES FOR ELECTROCHEMICAL CELLS

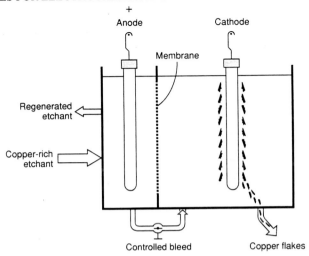

FIGURE 9 – The Capenhurst electrolyte etchant regeneration (CEER) cell for the continuous recovery of copper as a flake deposit from printing board manufacturers etchant.

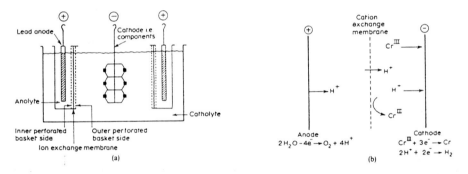

FIGURE 10 – The electroplating of decorative chrome.

that although a high degree of ion-exclusion is achieved, the membranes are particularly susceptible to transport of water and organic molecules. This can be by co-transport with ions, by diffusion or by electro-osmosis and the molar flux can be greater than that of ion-transport alone. It is depended both on the type and material of the membrane and on the molecular species. Transport of organic molecules into counter-electrode compartments can lead to several problems;
- destruction of organic at the anode
- polymerisation and subsequent electrode and membrane fouling
- need for purification of electrolyte
- corrosion of the counter-electrode.

The application of ion-exchange membranes in electrosynthesis in general relies on appropriate screening of materials at laboratory and pilot scale tests. The screening is required to ascertain membrane stability, permselectivity and potential transport of other species.

The largest scale electro-organic synthesis is the production of adiponitrile, from acrylonitrile, which is an intermediate step in the production of Nylon ((-NH(CH$_2$)$_6$ NH-CO-(CH$_2$)$_4$ CO)$_n$). The process is based on the cathodic reduction of acrylonitrile.

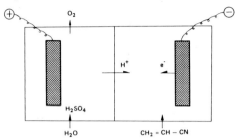

This method electroreduces acrylonitrile and at the same time dimerizes it by cathodic reaction.

2CH$_2$ = CH–CN + 2H$^+$ + 2e$^-$ → NC–(CH$_2$)$_4$–CN
acrylonitrile   adiponitrile

FIGURE 10 A – Electrolytic hydrodimerization of acrylonitrile.

The anode reaction is the evolution of oxygen. In the original process development proton conducting cation exchange membranes were used in the cells to prevent oxidation of organic species.

Subsequent development of the process lead to the abondment of the membrane cell in favour of an undivided (membrane free) cell by one company. However the use of membrane electrolysers is still believed to be operational in Japan.

Fuel Cells

A fuel cell is an electrochemical system which converts the free energy change of an electrochemical reaction into electrical energy. The usual fuel which supplies the energy is hydrogen. There are five main types of fuel cells which operate under quite different conditions. Alkaline cell based on a 6 M concentration of potassium hydroxide, phosphoric acid cell, Molten carbonate cell, Solid oxide cells and solid polymer electrolyte cells.

The relative performance of the fuel cells together with their current and projected applications is illustrated in Table 5. The factors which must be considered in the assessment of fuel cells are fuel efficiency, power density which is important to minimise volume, weight and capital cost, related power levels, life time and cost.

The first four of these utilise a form of separator; porous diaphragms, in the case of the alkaline cells, and an electrolyte matrix to act as a reservoir for electrolyte. The last of, these the solid polymer electrolyte fuel cell (SPEFC) utilises a solid polymeric membrane as the electrolyte and not a liquid or molten salt. In operation the cell requires humid gas, typically hydrogen and oxygen, to ensure the membrane is hydrated and electro-osmotic transport can take place. The membrane used is a proton conducting cation exchange membrane (see Fig 11) for the transport of hydrogen ions formed when hydrogen is oxidised

TABLE 5 – Types, characteristics and applications of fuel cell power plant.

Type of fuel cell and of fuel	Fuel efficiency (%)		Power density ($mW\ cm^{-2}$)		Rates power level projected (kW)	Lifetime projected (h)	Capital costs projected ($/kW)	Applications, time frame
	Present	Projected	Present	Projected				
Alkaline H_2	40	50	100-200	>300	10-100	>10,000	>200	Space 1960- Transportation 1996- Standby power 1966-
Phosphoric Acid; CH_4, CH_3OH	40	45	200	250	100-5000	>40,000	$1000–	Onsite integrated energy systems peak sharing 1992-
Molten carbonate; CH_4; coal	45	50-60	100	200	1000-100,000	>40,000	$1000	Base load and intermediate load power generation, cogeneration 1996-
Solid oxide fuel cell CH_4; coal	45	50-60	240	300	100-100,000	>40,000	$1500	Base load and intermediate load power generation, cogeneration 2000- Regenerative 2010- Space and terrestrial Space
Solid polymer H_2; CH_3OH	45	50	350	>600	1-1000	>40,000	>200	Space 1960- Transportation 1996- Standby power 1992- Underwater 1996-
Direct methanol $CH_3\ O + 1$	30	40	40	>100	1-100	>10,000	>200	Transportation 2010- Remote power 2000-

$$2H_2 \rightarrow 4H^+ + 4e^-$$

At the cathodic oxygen is reduced

$$O_2 + 4H^+ + 4e^- \rightarrow 2H_2O$$

The theoretical potential of this system is 1.23 V at 25C and 1 bar pressure of H_2 and O_2.

Early versions of the SPEFC used phenol sulfonic, polystyrene sulfonic and polytrifluorostyrene sulfonic acid membranes. Power densities of these systems were relatively low. The development of the perfluorosulfonic acid membrane brought radical improvement in power density to levels of 0.8 W/cm². The reactions in the fuel cell use Pt supported on carbon as electrocatalayst, with typical Pt loading of 0.4 mg cm². At a temperature of 80°C and with 3 bar H_2/5 bar air, the cell gives 0.78 V at 200 mAcm². The most significant recent development has been the use of a perfluorosulfonic acid membrane which has a shorter side chain (see Fig). This family of functional ionomers can be synthesised to lower equivalent weights than previously achieved whilst still maintaining physical strength. Power densities of SPEFC using Dow membranes are reported as a high as 2.5 W/cm² and 1.4 W/cm² on H_2-O_2 and H_2-Air respectively.

The attraction of solid polymer membranes in fuel cells lies in the following characteristics:
- Solid non corrosive electrolyte
- High efficiency
- Low temperature and rapid start-up
- High power density
- Ability to withstand differential pressures
- Long life
- Potable liquid product water
- Versatility of application
- Wide power range covered
- Simple design and ease of manufacture

Disadvantages of SPEFC are the generation of low quality waste heat, sensitivity to carbon monoxide, high catalyst loading and high membrane cost with a limited number of suppliers.

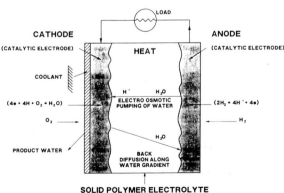

FIGURE 11 – Principle of operation of solid polymer electrolyte fuel cell.

NAFION

$(CF_2CF_2)_x(CF\!-\!CF_2)_y$
|
O
|
CF_2
|
CF_3CF
|
O
|
CF_2
|
CF_2
|
SO_3H

DOW

$(CF_2CF_2)_x(CF\!-\!CF_2)_y$
|
O
|
CF_2
|
CF_2
|
SO_3H

FIGURE 12 – Structure of perfluorinated sulfonic acid ionomer membranes.

There are several different types of SPFC systems under development and use for space and military applications (H_2 as fuels), power generation and transportation (methanol fuel, steam reformed to H_2) and fixed power plants (natural gas, steam reformed). A recent development is the use of methanol as a vapour or liquid fuel (without reformation) in the direct methanol solid polymer electrolyte fuel cell. The cell itself represents only a small part of the overall cell module. Supporting controllers, gas supply, heat exchangers are essential for the complete unit, as shown in Fig 13.

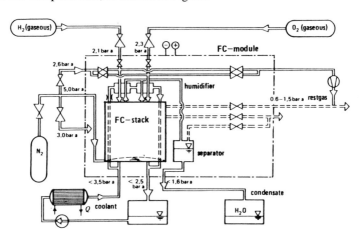

FIGURE 13 – Schematic of fuel cell module.

Batteries

A battery is an electrochemical device containing chemical energy which can be readily converted to electrical energy to drive an external circuit. The electrical energy is derived from a spontaneous chemical (redox) reaction which has a negative free energy. The diversity of types of batteries and their size is large, varying from small button cells to large units, many cubic diameters in size. In general however the componts of batteries are the same: an anode and a cathode current collector with an attached active anode and cathode masses (the electroactive species), electrolyte(s), a separator. These three essential components are then sealed in a container, which contains appropriate vents and external terminals for connection into the external circuit.

The separator in batteries can be manufactured from a wide range of materials. The materials must clearly be chemically stable to the active components and electrolyte under the conditions used. The separator must offer appropriate properties of low resistivity, selectivity, wetability and flexibility and also low cost. In some batteries, eg Ni-Cd high discharge rate pocket plate cells, the separator is a simple open physical barrier ie a series of thin plastic pins. In other batteries, eg sintered plate cells, the separator is a composite structure of three layers of nylon (or cellulose based) felt reinforced by a membrane to act as an oxygen barrier.

Generally the most common battery separators are microporous (0.01 – 10 μm pore size) sheet or macroporous (30 – 70 μm pore size) sheet. The materials vary from cheap polymers eg polyethylene to resin impregnated paper, stiffened paper etc. Sophisticated

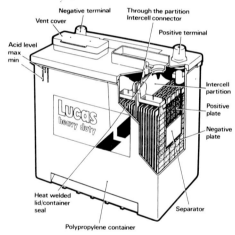

FIGURE 14 – Cutaway of a lead acid starter battery.

Table 6 – A Selection of Batteries Utilising Membranes.

Lead-acid starter battery	Miroporous polyethylener
Ni-Cd sintered plate cells	Nylon/membrane composite cellophone/polyamide
Na-b-alumina	Ion exchange
Redox batteries (Zn/Halogen)	Perfluorosulfonic acid cation exchange
Lithuim-MnO$_2$ (and other Li types)	Microporous polymer

membrane materials are generally not used. Other batteries under development use ion exchange membranes, such as perfluorosulphonic acid cation exchange membranes.

Overall the current use of "membrane" in batteries is mainly as the cheap microporous polymers.

SECTION 19

Electrokinetic Separations

ELECTROKINETIC SEPARATIONS

Electrokinetics is used to describe the movement of charged particles and water molecules in an applied D.C. electric field. There are two main application areas of electrikinetics in separation; electro-osmosis and electrophoresis, one being the converse of the other. That is to say that the movement of charged particles is always in the opposite sense to the electro-osmotic flow of water (see Fig 1). In general, any two phase interface (solid-liquid, liquid-liquid) is charged and the electrokinetic phenomena are related to the distribution of charge at surfaces. Due to this charge the particle is surrounded by a cloud of liquid containing ions with the opposite charge. This cloud is the diffuse double layer which serves to maintain electroneutrality in the system. In low ionic strength media the thickness of the diffuse double layer can be extensive (100 nm for a 10^{-5} mol/dm^3 1:1 electrolyte). In electro-osmosis the flow velocity of water in a pore is given by the Helmoltz-Smoluchowski equation

$$v_f = j\varepsilon\zeta / \mu\kappa$$

where, ζ, is the zeta potential which is potential at the outer Helmholtz plane, the locus of the closest approach of hydrated ions to the solid/liquid interface. ε is the permittivity of the electrolyte, μ is the viscosity of the electrolyte, j is the current density and κ is the specific conductivity.

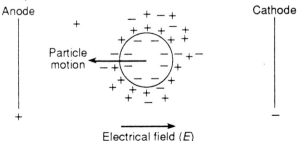

FIGURE 1 – Schematic representation of electrokinetic processes.

The zeta potential can be defined as the work necessary to take a unit positive charge from the bulk liquid phase up to the absorbed rigid layer, and is generally < 0.1 V. The volume of liquid displaced per second through the pore is related to the current I flowing through the capillary.

The equations used to describe the electrokinetic effects of electro-osmosis and electrophoresis are essentially the same. In electrophoresis the velocity of the particle, V_p, is determined by the electrophoretic mobility, U_p, and the electric field strength E,

$$V_p = U_p E$$

The electrophoretic mobility depend upon the thickness of the diffuse double layer, the viscosity of the surrounding fluid and the zeta potential.

Electro-osmotic and electrophoretic membrane separations

Electro-osmosis (EO) is used in a membrane based separation processes. The electrical potential facilitates the transport of mobile ions and liquid through the pores of the membrane. The mechanism involves the formation of an electrical double layer at the surface, due to the ability of the membrane to acquire a charge when immersed in an aqueous solution. This double layer is a region of varying electrical potential, caused by the adsorption of ionic solutes and the resultant fixed-charge layer at the surface of the membrane. Further into the solution, although only a relatively small distance from the surface, a mobile diffuse layer of opposite charge is also formed. Upon experiencing an electrical potential gradient, this mobile layer of ions and water moves through the pores. Applications of this process are seen mainly in dewatering, and for the treatment of colloidal suspensions and sludges in effluent and waste streams.

Electrophoresis is also used in membrane separation process driven by an electrical potential for dewatering and thickening. The physical process is more commonly associated with electrophoretic painting, where small particles of paint resin form a double (charged) layer. Upon application of a potential, the particles are attracted to an appropriate electrode (the article to be painted), forming an electrical circuit. Electrophoresis has also been used for the formation of coatings of metal oxides, carbides, borides and alloys onto metal substrates.

Overall electrokinetics can be extremely effective in many dewatering applications. Electrokinetic separations have been applied over a wide range of materials, eg PVC thickening, "low tech" consolidation of soils etc, blood serum separation and the electrofiltration of clays. Generally if a product slurry (or emulsion) is difficult to dewater by conventional methods, and has a low conductivity ($10^{-2} - 10^{-1}$ Sm^{-1}) and a zeta potential around ±100 mV, then electrokinetics should be considered as an alternative to or a compliment to (by pre-watering) conventional dewatering.

Electro-kinetic Dewatering

The selection of equipment in industry for the removal, or separation, of water from a particulate species depends upon several factors including the particle size and the solid content. For particle sizes of less than 10 micron electrokinetic dewatering can offer an attractive alternative to thermal drying or mechanical dewatering processes. The process can be used to concentrate slurries, or sludges, from 1% up to 40% solids content under the influence of a potential field. In this process, the pores of a non-conducting microporous membrane act as a support for a fine layer of particulate species. The surface then gains a small immobile charge matched by an excess of ions of equal and opposite charge in the adjacent solution. The application of a potential field across the membrane causes the solution, with an excess of mobile counter ions, to transport along the pores. In addition to this effect, the dispersed particles are transported away from the membrane surface by electrophoresis.

Figure 2 shows an electrofilter for dewatering fine particulate suspensions of mineral materials. The principle of operation of the unit, shown in Figure 3, combines vacuum cake filtration with electro-osmosis, electrophoresis, and electrophoretic deposition. There are three important elements to this device;

i) *Anode element.* This is composed of three parts, the anode (typically DSA), the surrounding electrolyte and a solid polymer electrolyte, which separates the anolyte from the media to be electrofiltered, whilst enabling the passage of the charge.

ii) *Cathode element.* This is also composed of three parts, the cathode, the electrolyte and a filter medium which is in direct contact with the filter cloth. The filtrate which is drawn through the filter, under an applied vacuum, is the catholyte for the cathode reaction (hydrogen evolution).

iii) *The slurry.* The application of the potential field across the slurry causes the solid and liquid to separate. The solid normally carries a negative charge and thus migrates towards the anode, where it is collected as filter cake and subsequently removed. In operation the formation of a cake at the cathode will reach an equilibrium thickness, when the velocity on the particles, due to drag, equals the

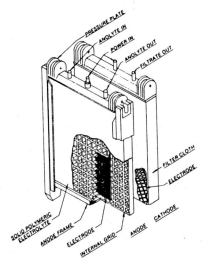

FIGURE 2 – Electrofilter for dewatering.

particle electrophoretic mobility. The filtrate rate is determined by the applied field and the differential, pressure applied across the filter cloth.

The energy consumption of electrokinetic dewatering is typically an order of magnitude lower than for thermal dewatering. Values are quoted in the range of 0.06 kWh dm^{-3} to 0.18 kWh dm^{-3} dewatered feed for applied voltages to 10 to 30 V. This compares with approximately 2 kWh dm^{-3} for dewatering by evaporation. The application of potential fields can considerably increase dewatering fluxrates. For example the electrofiltration of ion hydroxide colloids using 0.45 micron size membrane gives fluxrates of between 200 to 600 dm^3 h^{-1} m^{-2} at voltages of 10 V to 30 V. The commercial unit has an area of approximately 60 m^2 and has seen commercial service at a kaolin production plant.

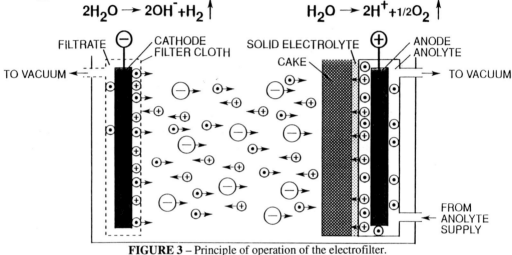

FIGURE 3 – Principle of operation of the electrofilter.

TABLE 1 – Electrokinetic dewatering cells.

Vertical electrode with moving fine polymer filter belt.	Applications are in dewatering free flowing materials which form a thin filter cake on a fine mesh under gravity prior to onset of blinding, eg sewage.
Horizontal polymer filter belt cell.	Suitable for dewatering materials which can be described as semi solid.
Concentrator parallel plate cell.	Suitable for dewatering low solid content using vertical parallel electrodes contained within dialysis membrane bags.
Parallel plate cell with a chain driven scraper removal system.	Applications are in clay dewatering.
Parallel plate cell with polymer collector mesh removal system. Rotating drum electrode.	Applications in the drying of emulsion PVC. Applications in the electro-osmotic dewatering of sewage and in electrophoretic deposition of titanium dioxide.

There are several types of cells which carry out electrokinetic dewatering as shown in Table 1.

Not all the devices listed in Table 1 utilise membranes. Typically the concentrator cells and the parallel plate cells do.

Drying of PVC

Conventional emulsion phase PVC, with a particle size of 0.1 to 0.2 µm has to be spray dried. The feed to the spray dryer contains a large volume of water typically 55%. By reducing the water content of the feed to the drier large saving in thermal energy can arise. In dewatering PVC it is necessary to control the pH of the electrophoretic deposit as below a pH of 2.3, the heat stability of the PVC product is impaired. Therefore in the parallel plate cell, shown schematically in Fig 4, the pH around the anode (platinised titanium) region is controlled by utilising either an ion-exchange or semi-permeable membrane. In this way the PVC slurry cannot contact the anode, where acid is generated.

In operation the deposited PVC formed at the anode membrane is allowed to grow through a polypropylene collector mesh to form a thick (2 cm) hard solid deposit. The cathode in this device serves to "electro-osmotically" pump the water from the slurry through the filter cloth. In the case of PVC slurries which are cationic, it may be necessary to isolate the cathode from the "electro-osmotic" pumped water using a membrane. The cell is capable of dewatering 150 kg/h/m^2 electrode area of slurry from 42% solids to approximately 84% solid.

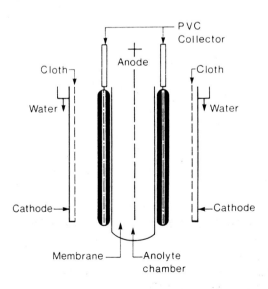

FIGURE 4 – Schematic diagram of parallel plate for PVC dewatering.

Clay Dewatering – Low Level Solid Content Water

There are many industrial processes in which large amounts of water are found which contains relatively small proportions of solid (< 1%). This solid content must be removed, or significantly reduced, to enable the water to be re-used or discharged. One method of water removal termed "electrodecantation" using an arrangement of one vertical anode and one cathode (platinised titanium) contained in separate dialysis membrane bags. The membrane bags enable the electrode reactions to be isolated, preventing back mixing of evolved gases and allowed pH to be controlled.

The dewatering cell contains many dialysis membranes 5 mm apart, between the isolated anode and cathode, to give a total membrane area of 3 m^2/dm^3 cell volume. In operation particles migrate under a potential field in the direction of the anode and collect on the anode sides of the membranes. On the cathode side of the dialysis membranes a water rich phase is formed. As the solids accumulate on the membrane they fall under gravity and are thus separated form the slurry. Power requirements of the cell are estimated at 17.6 kW/m^2 of liquid, for solid contents between 0.037 to 0.33%.

Electrophoretic Separations

Electrophoresis is also used for several applications of dewatering and thickening which includes a proposed method for the removal or the concentration of colloidal particles (which are usually negatively charged in biological systems and in polluted waters) referred to as forced flow electrophoresis. By using microfiltration membranes, which are permeable to water and ionic species, but not the colloidal particles, a solution free of colloidal material is produced on simultaneous application of an applied pressure and an electrical potential. An alternate version of this device, replacing the microporous membranes by cation exchange membranes, has been proposed in water desalination although it is believed industrial applications of this process have still to be developed.

One example of electrophoresis is to concentrate rubber latex particles to concentrations up to 50%. The latex particles are negatively charged and move towards the anode of a cell under an applied potential field. Placing a semi-pemerable membrane in front of the anode enabled the latex to concentrate at the anode near the upper part of the cell owing to the lower density.

A small scale application of electrophoresis is in the separation of biochemical species from natural sources. Applications include fraction of blood plasma, isolation of enzymes and concentration of platelet cells from red blood cells. The electrophoretic separator shown in Fig 5 carries out the separation in an annulus formed between an inner and outer cylinder of the separator. The device uses a vertical rotating outer cylinder to give stable laminar flow of the buffer electrolyte and avoid convective turbulence which would arise from resistive heating. The mixture to be fractionated is injected into the carrier solution and is separated in the electric field as it passes up the annulus. Because the components of the mixture move at different rates under the influence of the electric field, they form concentric bands within the annulus. A splitter at the outlet of the device enables up to 30 of the annular fractions to be collected. The membranes used in this application are of synthetic cellulose supported in inert porous tubes.

ELECTROKINETIC SEPARATIONS

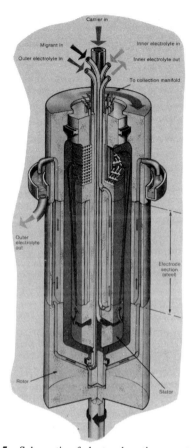

FIGURE 5 – Schematic of electrophoretic separator.

TABLE 2 – Biostream general information

Nominal throughput: 100 g/h protein (dry weight basis) or 10^{12}/h cells	
Typical operating parameters:	
Carrier electrolyte flow	2 litre/min (pulse free)
Anode electrolyte flow	0.1 litre/min
Cathode electrolyte flow	0.1 litre/min
Migrant flow	Up to 3.0 litre/h
Migrant concentration	Up to 50 gm/litre protein
Rotor speed	Fixed in the range 130 to 160 rpm
D.C. voltage	Up to 40 volt
D.C. current	Up to 100 A
Temperature increase (carrier)	0 to 25°C
Materials of construction	Stainless steel and biologically compatible plastics
Diaphragm material	Regenerated cellulose tubing
Diaphragm support	Resin bonded cellulose
Typical carrier electrolytes	Tris citrate (pH 7-9)
	Ammonium acetate (pH 9.6 to 11.5)
	Tris acetate (pH 4-6)
	Ammonium formate (>pH 3)

Typical conditions of operation are given in table 2. The electrodes are of stainless steel and the unit typically operates with an electric field of 30 V and with a current of 60 A.

SECTION 20
Appendix

ASTM Standard Methods Relevant to Microfiltration Membranes
Chemical Resistance of Membrane Filters & Filter Holders
Chemical Compatibility of Filter Components
Characteristics of Membranes for Ultrafiltration and Microfiltration
Selected Trade/Registered Name Index
Dead End Microfiltration/Membrane Applications Guide
Manufacturers Classified Index by Product Category
Representative Range of Contaminant Removal
General Guide to Contaminant Sizes
Ion-Exchange Membranes & Manufacturers
A Summary of Contaminant Removal and Alternative Technologies
Cleaning Formulations for Processing Biologicals
Polysulfone Membrane Cartridge Chemical Compatibility
List of Commercial Crossflow Microfilter Media
A Selection of Commercially Available Membranes for Cross Flow Filtration
Gas Permeation Module Information.
Characteristics of Reverse Osmosis Membranes
Characteristics of Selected Nanofiltration Membranes
Membranes for Ultrafiltration
Selected Water-Permeable Ethanol/Water Pervaporation Membranes
Selected Organic-Permeable Organic/Water Pervaporation Membranes
Unit Conversion Chart

APPENDIX

TABLE 1a – ASTM Standard Methods Relevant to Microfiltration Membranes.

Method Number	Date	Title
Gases		
D 3267-88	1988	Standard Test Method for Separation and Collection of Particulate and Water Soluable Gaseous Fluorides in the Atmosphere (Filter and Impinger Method)
D 3268-89	1989	Standard Test Method for Separation and Collection of Particulate and Water Soluable Gaseous Fluorides in the Atmosphere (Sodium Bicarbonate-Coated Glass Tube and Particulate Filter Method)
D 4240-83	1989	Standard Test Method for Airborne Asbestos Concentration in Workplace Atmosphere
D 4765-88	1988	Standard Test Method for Fluorides in Workplace Atmosphere
F 25-68	1988	Standard Test Method for Sizing and Counting Airborne Particulate Contamination in Clean Rooms and Other Dust Controlled Areas Designated for Electronic and Similar Applications
F 318-78	1989	Standard Practice for Sampling Airborne Particulate Contamination in Clean Rooms for Handling Aerospace Fluids
Liquids		
D 2276-88	1988	Standard Test Method for Particulate Contaminant in Aviation Fuel
D 3830-79	1984	Standard Practice for Filter Membrane Color Ratings of Aviation Turbine Fuels
F 311-78	1983	Standard Practice for Processing Aerospace Liquid Samples forParticulate Contamination Analysis Using Membrane Filters
F 312-69	1980	Standard Methods for Microscopial Sizing and Counting Particles from Aerospace Fluids on Membrane Filters
F 313-78	1983	Standard Test Method for Insoluble Contamination for Hydraulic Fluids by Gravimetric Analysis
Other		
F 24-65	1983	Standard Method for Measuring and Counting Particulate Contamination on Surfaces
F 51-68	1984	Standard Test Method for Sizing and Counting Particulate Contaminants in and on Clean Room Garments

TABLE 1b – ASTM Standard Methods Relevant to Microfiltration Membranes.

D 3861-84	Standard Test Method for Quantity of Water-Extractable Matter in Membrane Filters
D 3862-80	Standard Test Method for Retention Characteristics of 0.2 μm Membrane Filters Used in Routine Filtration Procedures for the Evaluation of Microbiological Water Quality
D 3863-97	Standard Test Method for Retntion Characteristics of 0.40 to 0.45 μm Membrane Filters Used in Routine Filtration Procedures for the Evaluation of Microbiological Water Quality
D4196-82	Standard Test Method for Confirming the Sterility of Membrane Filters
D 4197-82	Standard Test Method for Percent Porosity of Membrane Filters
D 4198-82	Standard Methods for Evaluating Absorbent Pads Used with Membrane Filters for Bacteriological Analysis and Growth
D 4200-82	Standard Test Method for Evaluating Inhibitory Effects of Ink Grids on Membrane Filters
E 1294-89	Standard Test Method for Pore Size Characteristics of Membrane Filters Using Automated Liquid Porosimeter
F 316-86	Standard Test Method for Pore Size Characteristics of Membrane Filters by Bubble Point and Mean Flow Pore Test
F 317-72	Standard Test Method for Liquid Flow Rate of Membrane Filters

APPENDIX

TABLE 2a.

	Acrylate	Buna	Ethylene-Propylene	Nylon	Polycarbonate	Polyethylene	Polypropylene	Polystyrene	Polysulphone	Polyvinyl Chloride	Polytetrafluoroethylene	Silicone	Stainless Steel	Tygon	Viton
ACIDS															
Acetic acid, glacial	●	●	●	●	■	■	■	▲	●	■	■	●	■	●	■
Acetic acid, 5%	▲	●	■	▲	■	■	■	■	■	■	■	■	■	●	■
Boric acid	■	■	■	■	■	■	■	■	■	■	■	■	■	■	■
Hydrochloric acid (conc.)	▲	●	■	●	●	■	■	▲	●	▲	■	●	●	●	■
Hydrofluoric acid	●	●	●	●	●	■	■	▲	●	●	■	●	●	●	■
Nitric acid (conc.)	●	●	●	●	●	▲	●	●	●	●	■	●	■	●	■
Sulphuric acid (conc.)	●	●	●	●	●	■	▲	●	●	●	■	●	●	●	●
BASES															
Ammonium hydroxide (6N)	■	●	■	▲	●	■	■	▲	●	■	■	■	■	▲	●
Sodium hydroxide (conc.)	■	●	■	▲	●	■	■	▲	●	■	■	■	■	●	●
SOLVENTS															
Acetone	●	●	■	■	●	■	■	●	●	●	■	●	■	●	●
Acetonitrile	●	●	■	■	●	■	▲	—	—	■	■	●	■	—	●
Amyl acetate	●	●	■	■	●	■	▲	●	●	●	■	●	■	●	●
Amyl aocohol	●	●	■	■	—	■	■	▲	■	▲	■	●	■	●	●
Benzene	●	●	●	■	●	■	▲	●	●	●	■	●	■	▲	■
Benzyl alcohol (1%)	▲	■	■	■	▲	■	■	▲	■	■	■	■	■	▲	■
Brine (sea water)	■	■	■	■	■	■	■	■	■	■	■	■	●	■	■
Butyl alcohol	●	■	●	●	▲	■	■	■	■	■	■	●	■	▲	■
Carbon tetrachloride	●	●	●	■	●	●	▲	●	●	▲	■	●	■	▲	■
Cellosolve (ethyl)	—	●	●	■	●	■	■	●	●	▲	■	●	■	—	●
Chloroform	●	●	●	▲	●	■	▲	●	●	●	■	●	■	▲	●
Cyclohexanone	●	●	●	■	●	■	■	●	■	●	■	●	■	●	●
Dimethylacetamide	—	●	●	●	●	■	■	●	●	■	■	■	■	—	●
Dimethylformamide	—	●	●	▲	●	■	■	●	●	■	■	■	■	●	●
Dioxane	●	●	●	■	●	■	■	●	●	●	■	●	■	—	●

TABLE 2a (continued).

	Acrylate	Buna	Ethylene-Propylene	Nylon	Polycarbonate	Polyethylene	Polypropylene	Polystyrene	Polysulphone	Polyvinyl Chloride	Polytetrafluoroethylene	Silicone	Stainless Steel	Tygon	Viton
DMSO	—	●	●	■	●	■	■	■	●	●	●	●	●	—	●
Ethyl alcohol	▲	■	■	▲	■	■	■	■	■	▲	■	■	■	▲	■
Ethers	●	●	●	■	●	▲	▲	●	■	●	■	●	■	▲	●
Ethyl acetate	●	●	●	■	●	■	■	●	■	●	■	●	■	●	●
Ethylene glycol	■	■	■	■	■	■	■	■	■	■	■	■	■	■	■
Formaldehyde	■	●	●	▲	▲	■	■	■	■	■	■	●	■	▲	●
Freon TF or PCA	—	■	●	■	▲	■	●	●	■	■	■	●	■	●	■
Gasoline	—	■	●	■	●	■	▲	■	●	■	■	●	■	▲	■
Glycerine (glycerol)	■	■	■	■	▲	■	■	■	■	■	■	■	■	■	■
Hexane	—	■	●	■	▲	■	■	●	■	■	■	●	■	▲	■
Hydrogen Peroxide (3%)	—	■	■	●	■	■	■	■	■	■	■	■	■	▲	■
Hypo (photo)	■	■	■	■	—	■	■	■	■	■	■	■	■	▲	■
Isobutyl alcohol	▲	●	■	●	■	■	■	■	■	■	■	■	■	—	■
Isopropyl acetate	—	●	●	■	●	■	▲	●	●	●	■	●	■	—	●
Isopropyl alcohol	▲	●	■	●	■	■	■	■	■	■	■	■	■	—	■
Kerosene	■	■	●	■	■	■	▲	●	■	■	■	●	■	●	■
Methyl alcohol	●	■	■	●	■	■	■	▲	■	■	■	■	■	■	●
Methylene chloride	●	●	●	▲	●	▲	●	●	●	●	■	●	■	●	●
MEK	●	●	■	■	●	■	■	●	●	●	■	●	■	●	●
MIBK	—	●	●	■	●	■	▲	●	●	●	■	●	■	—	●
Mineral spirits	—	■	●	■	●	▲	▲	▲	■	■	■	●	■	▲	■
Nitrobenzene	●	●	●	▲	●	▲	●	●	●	●	■	●	■	●	●
Paraldehyde	—	■	■	—	—	—	—	—	■	—	■	■	■	—	■
Ozone (10 ppm in water)	■	●	●	■	■	■	■	▲	●	■	●	●	●	—	●
Pet base oils	—	■	●	▲	—	■	—	—	■	—	■	●	■	—	■
Pentane	—	■	●	■	▲	■	■	●	■	■	■	●	■	▲	■
Perchloroethylene	●	■	●	▲	●	●	●	●	●	●	■	●	■	●	■
Petroleum ether	▲	■	■	■	▲	■	▲	●	■	■	■	■	■	—	■

TABLE 2a (continued).

	Acrylate	Buna	Ethylene-Propylene	Nylon	Polycarbonate	Polyethylene	Polypropylene	Polystyrene	Polysulphone	Polyvinyl Chloride	Polytetrafluoroethylene	Silicone	Stainless Steel	Tygon	Viton
Phenol (5%)	●	■	■	▲	●	■	■	●	■	▲	■	■	■	▲	■
Phenol (10%)	●	■	■	●	●	■	▲	●	■	●	■	■	■	●	■
Pyridine	●	●	●	■	●	■	▲	●	●	●	■	●	■	●	●
Silicone oils	■	■	●	■	■	■	■	▲	■	■	■	●	■	—	■
Toluene	●	●	●	■	●	▲	■	●	●	●	■	●	■	●	■
Trichloroethane	●	●	●	▲	●	●	●	●	●	●	■	●	■	—	■
Trichloroethylene	●	●	●	▲	●	●	●	●	●	●	■	●	■	—	■
TFA	—	▲	▲	●	▲	▲	▲	●	▲	●	▲	▲	▲	—	▲
THF	●	●	●	■	●	▲	▲	●	●	●	●	●	●	—	●
Xylene	●	●	●	■	●	▲	▲	●	●	●	■	●	■	●	■

Notes

This chemical compatibility chart is intended as a guide only. Because of variations in temperature (recommendations in the table are based on 20°C), concentration, duration of exposure and other factors outside of our control, which may affect the compatibility of the materials, no warranty is given, or is to be implied, with respect to such information.

The compatibility information takes into consideration that the exposure of filter construction materials to fluids occurs normally over relatively short periods only. The information may, therefore, deviate from compatibility data which are provided by other sources for applications in which extended exposure to fluids occurs.

Codes

- ■ = Recommended
- ▲ = Limited applications; testing prior to use is recommended
- ● = Not recommended
- — = No data available

TABLE 2b – Chemical Compatibilities of Plastics/Membrane Filters.

Chemical Compatibility Guide
F = Fair   G = Good   P = Poor

	Styrene	Polyprop	PVC
Acids:			
Hydrochloric acid, (25%)	G	G	G
Hydrochloric acid (concentrated)	F	G	F
Nitric acid (concentrated)	P	P	P
Nitric acid (25%)	P	G	F
Alcohols:			
Butanol	G	P	G
Ethanol	G	G	G
Methanol	G	G	G
Amines:			
Aniline	G	G	P
Dimethylformamide	P	G	F
Bases:			
Ammonium hydroxide (25%)	F	G	G
Ammonium hydroxide (1N)	F	G	G
Sodium hydroxide	G	G	G
Hydrocarbons:			
Hexane	P	G	F
Toluene	P	G	P
Xylene	P	F	P
Dioxane	P	G	P
Dimethylsulfoxide (DMSO)	P	G	P
Halogenated Hydrocarbons:			
Carbon Tetrachloride	P	G	G
Chloroform	P	G	P
Methylene chloride	P	F	P
Ketones:			
Acetone	P	G	P
Methyl ethyl diketone	P	G	P

TABLE 2b – Chemical Compatibilities of Plastics/Membrane Filters.

R = Recommended   N = Not Recommended
L = Limited Resistance   O = Testing Advised

	CA	PC	CN	NY	MCE	PTFE
Acids:						
Hydrochloric acid, (25%)	N	R	R	N	O	R
Hydrochloric acid (concentrated	N	R	N	N	N	R
Nitric acid (concentrated)	N	R	N	N	N	O
Nitric acid (25%)	N	R	L	N	O	R
Alcohols:						
Butanol	R	R	R	R	R	R
Ethanol	R	R	N	R	O	R
Methanol	R	R	N	R	O	R
Amines:						
Aniline	N	N	R	R	N	R
Dimethylformamide	N	N	N	R	N	R
Bases:						
Ammonium hydroxide (25%)	R	N	R	R	O	N
Ammonium hydroxide (1N)	N	N	R	R	O	N
Sodium hydroxide	N	N	N	R	N	R
Hydrocarbons:						
Hyxane	R	R	R	R	R	R
Toluene	R	O	R	R	R	R
Xylene	R	R	R	R	R	R
Dioxane	N	N	N	R	N	R
Dimethylsulfoxide (DMSO)	N	N	N	R	N	R
Halogenated Hydrocarbons:						
Carbon Tetrachloride	L	O	R	R	O	R
Chloroform	N	N	R	R	N	R
Methylene chloride	N	N	R	R	N	R
Ketones:						
Acetone	N	O	N	R	N	R
Methyl ethyl diketone	N	O	N	R	O	R

- CA — Cellulose Acetate
- PC — Polycarbonate
- PTFE — Polytetraflyorethylene (Teflon)®
- CN — Cellulose Nitrate
- NY — Nylon
- MCE — Mixed esters of Cellulose

R = Recommended
N = Not recommended
L = Limited resistance
D = Testing advised

TABLE 3 – Chemical Compatibility of Filter Components.

	Membranes		
Reagent Families	Nuclepore® Polycarbonate	Membra-Fil® Cellulosic	Filinert™ PTFE
Acids			
Acetic (10%)	+	0	+
Glacial Acetic	0	-	+
Boric (5%)	+	+	+
Formic (5%)	+	+	+
6N Hydrochloric	+	0	+
Concentrated Hydrochloric	+	-	+
Hydrofluoric (35%)	+	+	+
6N Nitric	+	0	+
Concentrated Nitric	+	-	0
Perchloric (60%)	-	0	+
Phosphoric (85%)	-	+	+
6N Sulfuric	+	+	+
Concentrated Sulfuric	-	-	+
Alcohols			
N-Amyl Alcohol	+	+	+
Butanol	+	+	+
Ethanol	+	0	+
Ethylene Glycol	+	+	+
Glycerol	+	+	+
N-Hexanol	+	0	+
Isobutanol	+	0	+
Isopropanol	+	0	+
Methanol	+	0	+
Propanol	+	+	+
Propylene Glycol	+	0	+
Butyl Cellosolve	0	-	+
Methyl Cellosolve	-	0	+
2.2 Ethoxyethoxy Ethanol (carbitol)	-	-	+
Polyethylene Glycol 1000	+	+	+
Benzyl Alcohol	0	0	+
Aldehydes			
Butyraldehyde	0		+

APPENDIX

TABLE 3 – Chemical Compatibility of Filter Components.

	Holders				O-Rings				
	Glass	Poly-carbonate	Poly acetal	Stainless Steel	Viton® A	Buna N	Silicone	EP	Teflon®
Acids									
Acetic (10%)	+	+	+	+	+	+	+	+	+
Glacial Acetic	+	-	-	+	-	+	+	+	+
Boric (5%)	+	+	+	+	+	+	+	+	+
Formic (50%)	+	+	0	+	+	-	-	0	+
6N Hydrochloric	+	0	-	+	+	0	-	+	+
Concentrated Hydrochloric	+	0	-	-	+	-	-	0	+
Hydrofluoric (35%)	-	+	-	-	+	0	-	+	+
6N Nitric	+	+	0	+	+	-	-	0	+
Concentrated Nitric	+	-	-	+	+	-	-	-	+
Perchloric (60%)	+	-	-	+	+	-	-	0	+
Phosphoric (85%)	+	-	-	+	+	-	0	+	+
6N Sulfuric	+	+	-	+	+	-	-	0	+
Concentrated Sulfuric	+	-	-	0	+	-	-	-	+
Alcohols									
N-Amyl Alcohol	+	+	+	+	+	+	-	+	+
Butanol	+	+	+	+	+	+	+	+	+
Ethanol	+	+	+	+	0	0	+	+	+
Ethlene Glycol	+	+	+	+	+	+	+	+	+
Glycerol	+	+	+	+	+	+	+	+	+
N-Hexanol	+	+	+	+	+	+	+	0	+
Isobutanol	+	+	+	+	+	+	+	+	+
Isopropanol	+	+	+	+	+	+	+	+	+
Methanol	+	-	+	+	-	+	+	+	+
Propanol	+	+	+	+	+	+	+	+	+
Propylene Glycol	+	0	+	+	+	+	+	+	+
Butyl Cellosolve	+	0	0	+	-	0	-	+	+
Methyl Cellosolve	+	-	0	+	-	0	-	+	+
2.2 Ethoxyethoxy Ethanol (carbitol)	+	-	0	+	+	+	+	+	+
Polyethylene Glycol 1000	+	+	+	+	+	+	+	+	+
Benxyl Alcohol	+	-	0	+	+	-	+	+	+
Aldehydes									
Butyraldehyde	+	0	0	+	-	-	-	+	+

Chemical Compatibility Guide Legend: + = Recommended   0 = Testing Advised   - = Not recommended

TABLE 3 – Chemical Compatibility of Filter Components.

	Membranes		
Reagent Families	Nuclepore® Polycarbonate	Membra-Fil® Cellulosic	Filinert™ PTFE
Formaldehyde (37%)	+	0	+
Amines			
Aniline	-	-	+
Diethyl Acetamide	0	-	+
Dimethylformamide	-	-	+
Triethylamine	0	-	+
Bases			
6N Ammonium Hydroxide	-	0	-
6N Potassium Hydroxide	-	-	+
6N Sodium Hydroxide	-	-	+
Esters			
Amyl Acetate	+	0	+
Butyl Acetate	+	0	+
Ethyl Acetate	0	-	+
Methyl Acetate	0	-	+
Methyl Formate	0	-	+
Ethers			
1.4 Dioxane	-	-	+
Ethyl Ether	+	0	+
Isopropyl Ether	+	+	+
Petroleum Ether	+	+	+
Fuels			
Gasoline	+	+	+
Jet Fuel 640A	+	+	+
Kerosene	+	+	+
Halogenated Hydrocarbons			
Bromoform	-	+	+
Carbon Tetrachloride	0	0	+
Chloroform	-	-	+
Ethylene Dichloride	-	0	+
Methylene Chloride	-	-	+
Tetrachloroethylene (perchloroethylene)	+	+	+
1,1,1-Trichloroethane	0	0	+

TABLE 3 – Chemical Compatibility of Filter Components.

	Holders				O-Rings				
	Glass	Poly-carbonate	Poly acetal	Stainless Steel	Viton® A	Buna N	Silicone	EP	Teflon®
Formaldehyde (37%)	+	+	+	+	0	+	+	+	+
Amines									
Aniline	+	-	0	+	0	-	-	+	+
Diethyl Acetamide	+	-	-	+	-	0	+	0	+
Dimethylformamide	+	-	-	+	-	+	+	0	+
Triethylamine	+	-	-	+	-	0	0	+	+
Bases									
6N Ammonium Hydroxide	+	-	0	+	+	+	+	+	+
6N Potassium Hydroxide	0	-	0	+	-	+	0	+	+
6N Sodium Hydroxide	0	-	0	+	0	+	+	+	+
Esters									
Amyl Acetate	+	-	-	+	-	-	-	0	+
Butyl Acetate	+	-	-	+	-	-	-	0	+
Ethyl Acetate	+	-	-	+	-	-	0	0	+
Methyl Acetate	+	-	-	+	-	-	-	0	+
Methyl Formate	+	-	-	+	-	-	-	0	+
Ethers									
1,4 Dioxane	+	-	0	+	-	-	-	+	+
Ethyl Ether	+	+	+	+	-	0	-	0	+
Isopropyl Ether	+	+	+	+	-	+	-	-	+
Petroleum Ether	+	+	+	+	+	+	+	+	+
Fuels									
Gasoline	+	+	0	+	+	+	-	-	+
Jet Fuel 640A	+	+	0	+	+	+	-	-	+
Kerosene	+	+	0	+	+	+	-	-	+
Halogenated Hydrocarbons									
Bromoform	+	-	-	+	+	-	-	-	+
Carbon Tetrachloride	+	-	-	+	+	+	-	-	+
Chloroform	+	0	-	+	+	-	-	-	+
Ethylene Dichloride	+	-	-	+	+	-	-	0	+
Methylene Chloride	+	-	-	+	+	-	-	-	+
Tetrachloroethylene (perchloroethylene)	+	-	-	+	+	+	-	-	+
1,1,1-Trichloroethane	+	-	-	+	+	-	-	-	+

Chemical Compatibility Guide Legend: + = Recommended   0 = Testing Advised   – = Not recommended

TABLE 3 – Chemical Compatibility of Filter Components.

	Membranes		
Reagent Families	Nuclepore® Polycarbonate	Membra-Fil® Cellulosic	Filinert™ PTFE
1,1,2-Trichloroethane	-	0	+
Monochlorobenzene	-	+	+
Trichlorobenzene	-	+	+
Trichlorethylene	-	+	+
Hydrocarbons			
Benzene	0	+	+
Cyclohexane	+	+	+
Hexane	+	+	+
Pentane	+	+	+
Toluene	0	+	+
Xylene	+	+	+
Ketones			
Acetone	0	-	+
Cyclohexanone	0	-	+
Methyl Ethyl Ketone (MEK)	0	0	+
Oils			
Silicones	+	+	+
Petroleum Oils	+	+	+
Skydroll 500	+	0	+
Photo Resists			
Kodak: KMER, KTFR	+	-	+
Shipley (AZ-111,340,1350)	+	-	+
Waycoat 59	+	0	+
Miscellaneous			
Hydrogen Proxide (30%)	+	+	+
Diacetone Alcohol	+	-	+
Nitrobenzene	-	-	+
1-Nitropropane	0	-	+
Pyridine	-	-	+
Tetrahydrofuran	-	-	+
Dimethylsulfoxide (DMSO)	-	-	+
Freon TF™	+	+	+
Mineral Spirits	+	+	+
Turpentine	+	+	+
Acetonitrile	0	0	+

APPENDIX

TABLE 3 – Chemical Compatibility of Filter Components.

	Holders				O-Rings				
	Glass	Poly-carbonate	Poly acetal	Stainless Steel	Viton® A	Buna N	Silicone	EP	Teflon®
1,1,2-Trichloroethane	+	-	+	+	0	-	-	-	+
Monochlorobenzene	+	-	-	+	+	-	-	-	+
Trichlorobenzene	+	-	-	+	+	-	-	-	+
Trichloroethylene	+	-	-	+	+	0	-	-	+
Hydrocarbons									
Benzene	+	-	-	+	+	-	-	-	+
Cyclohexane	+	0	0	+	+	+	-	-	+
Hexane	+	+	0	+	+	+	-	-	+
Pentane	+	+	0	+	+	+	-	-	+
Toluene	+	-	-	+	+	-	-	-	+
Xylene	+	-	-	+	+	-	-	-	+
Ketones									
Acetone	+	-	-	+	-	-	-	+	+
Cyclohexanone	+	-	-	+	-	-	-	+	+
Methyl Ethyl Ketone (MEK)	+	-	-	+	-	-	-	+	+
Oils									
Petroleum Oils	+	+	+	+	+	+	-	-	+
Silicones	+	+	+	+	+	+	0	+	+
Skydroll 500	+			+	-	-	0	+	+
Photo Resists									
Kodak: KMER, KTFR	+	0	0	+	0	-	-	-	+
Shipley (AZ-111, 340, 1350)	+	0	0	+	-	-	-	+	+
Waycoat 59	+	0	0	+	-	-	-	+	+
Miscellaneous									
Hydrogen Peroxide (30%)	+	+	+	+	+	+	+	+	+
Diacetone Alcohol	+	0	0	+	-	-	-	+	+
Nitrobenzene	+	-	-	+	+	-	-	-	+
1-Nitropropane	+	-	-	+	-	-	-	+	+
Pyridine	+	-	-	+	-	-	-	+	+
Tetrahydrofuran	+	-	-	+	-	-	-	+	+
Dimethylsulfoxide (DMSO)	+	-	-	+	-	-	0	+	+
Freon TF™	+	+	+	+	+	+	-	-	+
Mineral Spirits	+	+	0	+	+	+	+	0	+
Turpentine	+	-	-	+	+	+	-	-	+
Acetonitrile	+	0	0	+	0	-	-	-	+

Chemical Compatibility Guide Legend: + = Recommended   0 = Testing Advised   – = Not recommended

TABLE 4 – Characteristics of Membranes for Ultrafiltration and Microfiltration.
a) Membrane microfilters.

PTFE MEMBRANE
Chemical Composition

Fluoropore (FG, FH, FA type) : polytetrafluoroethylene membrane with polyethylene backing (except FHUP: pure PTFE).

Mitex (LC, LS) : polytetrafluoroethylene (PTFE).

SEM Pictures

Fluoropore   Mitex
Membrane (1500 x)   Membrane (300 x)

Main Characteristics
Pore Sizes : See Centre Chart
Thickness : 125 μm - 175 μm
Porosity : 85% (FA) - 70% (FG) - 68% (LC/LS)
Hydrophobic
T. max: Fluoropore (FG, FH, FA) : 121°C (AUT)
            130°C (FIL)
            Mitex (LC, LS): 260°C
Protein: N/A
Compatibility: (1 2 3 4 5)
Availability:
Ø: 13 – 25 – 47 – 90 –142 – 293 mm
‡: 30 x 300 cm

Typical Applications
- Sterilizing filtration of gases and organic solutions (FGLP).
- Clarifying filtration of HPLC samples (FHLP) and solvents (FHUP).
- Amplification of bacterial DNA by PCR (FGLP, FHLP).
- Particulate removal (FHLP, FALP, LSWP) or analysis (FHWG) from aggressive chemical solutions.
- Filtration of cryogenic fluids (LSWP, LCWP: stable down to – 100°C).
- Air contaminants analysis: dibutylphosphate (FHLP), NaOH (FALP).
- Collection of organic precipitates (LSWP, LCWP).

PVDF MEMBRANE
Chemical Composition

Polyvinylidene difluoride.

SEM Picture (5000 x)

Main Characteristics
Pore Sizes : See Centre Chart
Thickness : 125 μm
Porosity : 70%
Hydrophobic
T. max : 126°C (AUT)
Protein : 150 μg/cm^2 (CH$_3$OH wetted)
Compatibility : 1 2 3 4 **5**
Availability :
Ø: 13 – 25 – 47 – 90 – 142 – 293 mm
‡: 30 x 300 cm

Typical Applications
- Sterilizing filtration of air (GVHP).
- Particulate removal from solvents (GVHP, HVHP).

APPENDIX

TABLE 4 – Characteristics of Membranes for Ultrafiltration and Microfiltration.
a) Membrane microfilters.

PET MEMBRANES

Chemical Composition

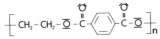

Polyethylene terephtalate.

SEM Picture (10,000x)

Main Characteristics
Pore Sizes : See Centre Chart
Thickness : 12 µm - 18 µm
Porosity : 8% - 14%
Hydrophobic
T. max : 121°C (AUT) - 140°C (FIL)
Protein : N.D.
Compatibility : 1 2 3 **4**
Availability :
Ø : 13 – 25 – 47 mm

Special Benefits
- Transparency.
- Flat surface : ideal for SEM.
- Calibrated pores allow sieving and liposomes extrusion.
- Low thickness: fast diffusion.

Typical Applications
- Isopore PET membranes are chemically more resistant than Isopore PC. They will be used in the same applications when chemical compatibility is an issue.
- Specific application: Gunpowder Residue Test (HPET).

CELLULOSE ESTER MEMBRANES

Chemical Composition

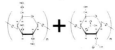

Mixed cellulose esters (nitrate + acetate).

SEM Picture (5,000 x)

Main Characteristics
Pore Sizes : See Centre Chart
Thickness : 150 µm
Porosity : 84% (8 µm) – 70% (0.025 µm)
Hydrophilic
T. max : 121°C (AUT) – 75°C (FIL)
Protein : 150 µg/cm^2
Compatibility : 1 2
Availability :
Ø: 13 – 25 – 37 – 47 – 50 – 82 – 85 – 90 – 100 – 137 – 142 – 293 mm
‡: 19 x 42 mm
§: 30 x 300 cm

Typical Applications
- Sterilizing filtration of aqueous solutions (GSTF).
- Colony hybridization (HATF).
- Nucleic acids/protein blotting (HAHY).
- Air asbestos analysis (AAWG).
- Exfoliative cytology (SMWP).
- Particulate analysis in fuels, hydraulic fluid, parenterals (HAWG).
- Bacteriological Analysis (HAWG).
- Virus removal (VSWP, VMWP).
- Fast drop (50 µl) dialysis of DNA and proteins (VSWP, VMWP).
- Vitamin A deficiency detection (HAWP).

TABLE 4 – Characteristics of Membranes for Ultrafiltration and Microfiltration.
a) Membrane microfilters.

PVDF HYDROPHILIC MEMBRANES
Chemical Composition

Modified polyvinlidene difluoride.

SEM Picture (5,000 x)

Main Characteristics
Pore Sizes : See Centre Chart
Thickness : 125 µm
Porosity : 70%
Hydrophilic
T. max : 126°C (AUT) - 85°C (FIL)
Protein : 4 µg/cm^2
Compatibility : 1 2 3 4
Availability :
Ø : 13 – 25 – 47 – 50 – 90 – 100 – 142 – 293 mm
‡ : 30 x 300 cm.

Special Benefits
- High mechanical strength can be folded or pleated without breaking).
- Wide solvent compatibility, low protein binding, high temperature autoclaving.

Typical Application
- Sterilizing filtration of protein solutions (GVWP).
- Filtration of aqueous, alcoholic and THF mobile phases in LC/HPLC (HVLP).
- Particulate removal from protein solutions (HVLP/SVLP).
- Ultrapure (Milli-Q®) water filtration at point of use.

POLYCARBONATE MEMBRANES
Chemical Composition

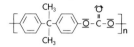

Bisphenol polycarbonate.

SEM Picture (10,000 x)

Main Characteristics
Pore Sizes : See Centre Chart
Thickness : 6 - 10 µm
Porosity : 5% - 20%
Hydrophilic
T. max : 121°C (AUT) - 140°C (FIL)
Protein : 3 µg/cm^2
Compatibility : 1 2 3
Availability :
Ø : 13 – 25 – 37 –47 – 90 – 142 mm

Special Benefits
- Transparency.
- Flat surface: ideal for SEM.
- Calibrated cylindrical pores allow sieving.
- Low thickness = fast diffusion.

Typical Applications
- Support for SEM analysis of cells/particulates.
- Size fractionation of cells/particulates (phyto-plankton analysis).
- Epifluorescence bacteriology (HTBP).
- Exfoliative cytology (TSTP).
- Schistosoma, canine heartworm microfilariae and other parasites detection by microscopy (TMTP).
- Erythrocyte deformability test (TMTP).
- Chemotaxis analysis (TMTP).
- DNA Alkaline Elution (TSTP).
- Airbourne particulates analysis by X-ray fluorescence (ATTP).
- Adsorbable Organic Halides (AOX) analysis (HTTP 025 OX).
- Production of calibrated liposomes by extrusion (GTTP, HTTP).

APPENDIX

TABLE 4 – Characteristics of Membranes for Ultrafiltration and Microfiltration.
b) Membrane ultrafilters.

CELLULOSE
Chemical Composition

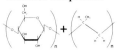

Regenerated cellulose membrane; polypropylene support.

SEM Picture (300 x)

Main Characteristics
Retention Rating : See Centre Chart
Thickness : 130 - 230 μm
Porosity : N/A
Hydrophilic
T. max : 35 - 50°C   P. max : 7 bar
Protein : 1 μg/cm^2
Compatibility : 1 pH : 2-13

Availability
Ø : 13 – 25 – 44.5 – 47 – 63.5 – 76 – 90 – 142 – 150 mm

Typical Applications
- Concentration of protein solutions.
- Separation of low from high molecular weight molecules (with a factor 10 difference, minimum).
- Protein binding studies.
- Buffer exchange.

POLYSULPHONE
Chemical Composition

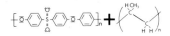

Polysulphone membrane; polypropylene support.

SEM Picture (300 x)

Main Characteristics
Retention Rating : See Centre Chart
Thickness : 130 - 230 μm
Porosity : N/A
Hydrophilic
T. max : 50°C   P. max : 7 bar
Protein : <100 μg/cm^2
Compatibility : 1 pH : 1-14
Availability :
Ø : 13 – 25 – 44.5 – 47 – 63.5 – 76 – 90 – 142 – 150 mm

Typical Applications
- Concentration of protein solutions.
- Separation of low from high molecular weight molecules (with a factor 10 difference, minimum).
- Protein binding studies.
- Buffer exchange.

ULTRAFILTERS

NMWL (Dalton)	Cellulosic UF Membranes	Polysulphone UF Membranes
300,000	PLMK	PTMK
100,000	PLHK	PTHK
30,000	PLTK	PTTK
10,000	PLGC	PTGC
5,000	PLCC	
3,000	PLBC	
1,000	PLAC	

KEY TO TABLES 4a & 4b:
Thickness: (μm).
Porosity: Filter porosity (% of void space).
T. max: Maximum temperature during autoclaving (AUT) and filtration (FIL).
Protein: Protein binding capacity, expressed in g/cm^3, for BSA, in standard test conditions.
Compatibility: 1 Water and aqueous solutions (pH 2-11), 2 Alcanes, aromatic hydrocarbons and halogenoalcanes, 3 Alcohols, 4 Ketones, esters and organic acids, 5 Concentrated, strong acids. (1,2,3, 4,5,6). Bold figures indicate limited compatibility.
Availibility: Ø = diameter of standard cut filter discs (mm), § = Size of standard filter sheets (cm), ‡ = Size of standard filter rolls (cm).

TABLE 5 – Selected Trade/Registered Name Index.

AbsoLife	absolute cartridge and filters – Gelman science
ADUF	membrane assisted anaerobic digestion process – Membratek Ltd
ACROFLOW	membrane cartridge filter – Gelman sciences
BIOFLOW	microfilter with nylon membrane – PTI Technologies Inc.
CERAFIL	ceramic gas filtration element – Cerel Ltd
CARBOSEP	mineral ultra filtration membrane system – Ultra-Tech Services Ltd
CELLOTON	porous ceramic media – Fairey Industrial
CELGARD	microporous polypropylene fibre membrane – Hoechst Ceramics Ltd
CONRTICON	unit for concentration and desalting of biological samples – Amicon
CORALITH	porous ceramic media – Fairey Industrial Ceramics Ltd
CHEMOMAT	metal recovery system based on electrodialysis reversal system – Ionics Incorporated
CYCLOPORE	non-nuclear method of producing track etch membranes – Cyclopore S.A.
CENTRISTART	ultrafiltration units for small volume used in a standard laboratory centrifuge – Sartorius A.G.
DURAPORE	polyvinylidene difluoride membrane filter discs – Millipore
ENVIROMAT	integrated membrane filtration system – Ionics Incorporated
FLOSEP	ultra filtration hollow fibre modules – Ultra-Tech
GENERON	air separation systems based on gas permeation – Dow Chemicals
FILTROX	range of filtration equipment, including cross flow (ceramic and polymer) technologies – APV Baker Ltd
FLOUROFLOW	microfilter with PTFE membrane, 100% Fluorocarbon support structure – PTI Technologies Inc.
KERASEP	ceramic ultra filtration membrane system – Ultra-Tech Services Ltd
ISOPORE	track etched polycarbonate and polyethylene teraphtalate UP membrane – Millipore
MICROCON	membrane system for purifying microsamples of proteins and nucleic acids – Amicon
MICROPURE	particle separator for microsamples – Amicon
LIQUI-CEL	hollow fibre contactors for gas/liquid and liquid/liquid systems – Hoechst-Celanese
MICROFLOW	cross flow microfilters – Seitz Filter Werke
IRIS	organic flatsheet membranes – Tech Sep.
FILLINET	PTFE membrane filters – Costar
NUCLEPore	polycarbonate and PVC membrane filters – Costar Corp.
MEGAPAK	teflon filter cartridge for tonic, corrosive and reactive bases – MEMTEC American Corporation
Maxcell	Microfiltration and ultrafiltration cartridge – A/G Tech
Milli	Prefix a range of filtration and membrane products -Millipore Corp.
Minitan	Crossflow flat sheet microfiltration and ultrafiltration systems – Millipore Corp.
MEMBRALOX	ceramic membrane – S.C.T.
MEMBRA-FIL	cellulose membrane – Costar
MEMTUF	tubular ultra filter – Membratek Ltd
MIDISART	air filter for use as sterile vent, bioisolation or sterile air and gases – Sartorius A.C.
MINISART	small volume ultra filter for sterile syringe – Sartorius A.G.

APPENDIX

TABLE 5 – continued.

MICROSTAR	crossflow systems – Carlson Filtration Ltd
Nuclepore	Polycarbonate and PVC membrane filters – Costar Corp
PLEIADE	organic membrane ultrafiltration system – Ultra-Tech
POLYFLOW	calandered meltblown polypropylene cartridge filter – PTI Technologies Inc.
POLYGARD	cartridge water and gas filters – Millipore
PROFLOW	microfilter with PTFE membrane and polypropylene support structure – PTI Technologies Inc.
PERVAP	fluid separation system based on pervaporation – Le Carbone-Lorraine
PROFLUX	Tangential flow filtration system-Amicon
PRISM	gas permeation hollow fibre module units – Monsanto
PUREMET	refactory ceramics – Fairey Industrial Ceramics Ltd
PYROLITH	porous ceramic media – Fairey Industrial Ceramics Ltd
PHOTOLINK	membrane surface treatment – B.S.I. Corporation
RAVLEX	microporous TFE Terpolymer Coatings – The Ravesworth Lab
SEMPAS	device for filtration of liquids, gases and mixtures – Sempas Membran technit gmbh
PRISM ALPHA	gas permeation module for air separation – Monsanto
SEMPHI	ceramic microfiltration membranes – Holland Industrial Ceramics B.V.
SEPHI-MATIC	small ceramic membrane crossflow filtration unit in stainless steel housing, designed primarily for filtration tests or small process streams – Hoogovens Industrial Ceramics B.V. and Holland Industrial Ceramics B.V.
SPIRA-CEL	spirally wound ultra filtration cartridges – Hoechst
TECRAMICS	ceramic "star" section channel crossflow microfilter – Fairey Industrial Ceramics Ltd
TFC	reverse osmosis spiral wound element – Fluid systems
TFCL	reverse osmosis spiral wound element – Fluid systems
ULTRAC	oil removal filters – Ultra filter gmbh
ULTRACHECK VARIO	test devices – Ultra filter gmbh
ULTRACHEM	sterile filters for liquids and gases – Ultra filter gmbh
SARTO	prefixes a range of products from the Sartorius A.G.
SARTOLAB	filters for sterile filtration of tissue culture media
SARTOBRAN	low absorption heterogeneous membrane filter (double layer) for pharmaceutical industry – Sartorius A. G.
SWUF	low cost tubular ultra filter – Membratek Ltd
SARTOCON	gas flow membrane filters for laboratory and small scale ultra filtration – Sartorius A.G.
ROGA	reverse osmosis spiral wound element – Fluid systems
SUPORE	hydrophilic polysulfane membrane – Gelman Science
UPILEX	hollow fibre polyimide reverse osmosis membrane – U.B.E. – Japan
PLAIDE	flat sheet polymer membrane module – Tech-Sep
TUFFRYN	low protein binding polysulfone membrane – Gelman Science
R02	aqua clear reverse osmosis plant – Culligan
VARAFINE	membrane cartridge filters – MEMTEC America Corporation
VERSAPOR	acrylic polymer membrane – Gelman Science
AQUAPORE	tubular microfiltration membranes – Ionics Inc.

TABLE 5 continued.

V.SEP	vapour permeation – MTR
FILMTEC	composite R0 membranes – Dow
BACTOCATCH	constant pressure or high flux, low fouling microfiltration – Alfa Laval
SELEX7	polypropylene micro sequencing membrane – Schleicher and Schuell
NYTRAN-N	nylon-6 membrane filter (Schleicher and Schuell
AVIRR	point of use air separation unit for biopharmaceutical application (A/G Technology Corporation)
XPRESS	Hollow fibre ultrafiltration cartridge – A/G Tech.
MAXIFIBER	Hollow fibre cartridge – A/G Tech.
CERATREX	Controlled porosity ceramic microfiter – Osmonics
Duratrex	Stainless steel filter elements – Osmonics
Selro	Chemically stable Uf and Nf membranes – Membrane Products.
Microza	Series of ultrafiltration modules – Asahi Chem. Ind.
CONSEP	ultrafiltration oily waste water pollution control device- Koch Membrane systems
Advantage	Dual asymmetric PTFE membrane filter cartridge – Parker microfiltration.
MiniBUS	Benchtop ultrafiltration system- CUNO
Varafine	Disposable double layer membrane capsule – Memtec America Corp.
Winefilter	Hollow fibre membrane ultrafiltration unit for wine filtration – Koch Membrane Systems
Quick Rinse	Membrane filtration cartridge – Costar.
Maxifibre	Hollow fibre cartridges – A/ G Tech.
XAMPLER	Hollow fibre membrane unit for continuous fermentation monitoring. – A/G Tech.
SPIRA-CELL	Spirally wound modules for UF and NF – HOECHST
ULTIMEM	laboratory filtration using ceramic metal mesh membraenes – NWW Acumem Ltd.
SUPER-CORE	Ultrafiltration membrane module – Koch membrane Systems,

APPENDIX

TABLE 6 – Dead End Microfiltration/Membrane Applications Guide.

Application	Recommended Filter Media			Reference
	Description	Pore Size (μm)	Diameter (mm)	
AIR POLLUTION ANALYSIS				
Asbestos, Airborne	Mixed-Esters of Cellulose	0.45, 0.8, 1.2	25	NIOSH Methods 7400 & 7402: EPA-CFR763-Fed Reg. 1987, pp. 41826-41905, 322375 (TEM): 322575 (PCM).
Cadmium	Mixed-Esters of Cellulose	0.8	37	NIOSH Method 7048 (322544)
Carbon Black	PVC	5.0	37	NIOSH Method 5000 (3000044 240810/361850)
Cyanides	Mixed-Esters of Cellulose	0.8	37	NIOSH Method 7904 (322544)
Lead	Mixed-Esters of Cellulose	0.8	37	NIOSH Method 7082 (322544)
Lead Sulfide	PVC	5.0	37	NIOSH Method 7505 (361844)
Nuisance Dust	PVC	5.0	37	NIOSH Methods 0500 & 0600 (361844)
Quartz in Coal Dust	Mixed-Esters of Cellulose	0.8	37	NIOSH Method 7602 (322544)
Silica, Crystalline	PVC	5.0	37	NIOSH Method 7601 (361844)
Welding & Brazing Fume	Mixed-Esters of Cellulose	0.8	37	NIOSH Method 7200 (322544)
Zinc Oxide	PVC	0.8	25	NIOSH Method 7502 (300015 240610/361520)
Trace Elements	Polycarbonate Aerosol-Type	0.2-8.0	25-47, 8x10 in	Lows trace metal contamination (Br, Pb, Zn, etc.) with Aerosol Holders (430200, 430400)
BACTERIAL ANALYSIS				
Total Coliform Count	Mixed-Esters of Cellulose	0.45	47	*Standard Methods/Water & Wastewater*, 17th Ed., 9222B, (148818)
Fecal Coliform	Mixed-Esters of Cellulose	0.7	47	*Standard Methods/Water & Wastewater*, 17th Ed., 9222D (148822)
Legionella	Polycarbonate	0.2	37, 47	
Heterotrophic Plate Count (HPC) – formerly known as standard plate count	Mixed-Esters of Cellulose	0.45	47	*Standard Methods/Water & Wastewater*, 17th Ed., 9215D (148818/241110/ 414410/920100)
Direct Total Microbial Count	Polycarbonate (black) Mixed-Esters of Cellulose	0.2 5.0	25 25	*Standard Methods/Water & Wastewater*, 17th Ed., 9216B (110656/140613/414220)

TABLE 6 continued.

Application	Recommended Filter Media			Reference
	Description	Pore Size (μm)	Diameter (mm)	
BEVERAGE & FOOD QUALITY CONTROL - See Bacterial Analysis				
Escherichia coli	Mixed-Esters of Cellulose	0.45	85	For E. coli, use REC-85™ Food Micro Membrane, Direct Plating Technique in Petri Dish (request Applications Bulletin)
Yeast and Molds	Mixed-Esters of Cellulose (black or green), Polycarbonate (black)	0.65-1.2 0.6, 0.8	13-47 13-47	
Sterility Testing	Mixed-Esters of Cellulose	0.22, 0.45	47	For Sterility Testing, use gridded, Sterile Hydrophobic Edge Membranes (see *Code of Federal Regulations*, Tital 21 #436) 149718, 149728
BLOOD FILTRATION				
RBC Deformability	Polycarbonate Hema-Fil	4.7-5.0	13, 25	110124, 110624
Plasmapheresis	Polycarbonate	0.6-1.0	90	Use Fluid Cross-Flow, Thin Channel Technique
CELL CULTURE	Polycarbonate	0.4-3.0	13-47	Fit Membrane to Petri Dish or Culture Plate Request Transwell Bibliography
CHEMOTAXIS				
Epithelial, Fibroblasts, Neutrophils, Polymorphonuclear leukocytes (PMNL)	Polycarbonate (Chemotaxis Membrane, PVP-free)	2.0-8.0	13	For Chemotaxis, use BlindWell or modified Boyden Chambers
Macrophage	Polycarbonate (Chemotaxis Membrane)	2.0-8.0	13	
CYTOLOGY				
Cytopreparative & Cyto-diagnostic Methods	Polycarbonate. Mixed-Esters of Cellulose	2.0-8.0 3.0-5.0	25, 47 19x42	Use with Swin-Lok Holder or Vacuum Filtration
EPA TESTING				
EPA Toxicity Characteristic Leaching Procedure (TCLP)	Glass Fiber	0.7	90, 142	EPA Ref. 40CFR Part 268 Fed. Reg. 53:18795 May 24, 1988
FUEL TESTING	Mixed-Esters of Cellulose Fuel Monitor	0.45, 0.8 0.8	47 37	410502 ASTM D2276 8245

TABLE 6 continued.

Application	Recommended Filter Media			Reference
	Description	Pore Size (µm)	Diameter (mm)	
GENERAL FILTRATION				
General Clarification or Prefiltration	Mixed-Esters of Cellulose. Glass Fiber	0.8-5.0 D49, D59 D79	13-293 13-293	
Beverage Stabilization	Mixed-Esters of Cellulose	0.45-1.2	47-293	
Particulate Removal	Polycarbonate Mixed-Esters of Cellulose Glass fiber (D49, D59, D79, 0.7 nominal)	0.1-5.0 0.1-5.0	13-293 13-293 10-293	
Fine Clarification Aqueous	Polycarbonate Mixed-Esters of Cellulose	0.6-1.0 0.65-1.2	13-293 13-293	
Bacterial Removal	Mixed-Esters of Cellulose	0.22-0.45	13-293	
Adsorbable Organic Halogens (AOX)	Polycarbonate	AOX®	25, 47	Ask for Applications Bulletin (quantitative assay via microcoulometry)
Alkaline Elution, DNA	Polycarbonate	0.8, 2.0	25, 47	Poretraits® (NUCLEPORE®) Winter 1988.
Forensic Analysis	Polycarbonate	0.4	13	Sample collection for S.E.M.
Liposome Extrusion	Polycarbonate	0.1-0.4	25-76	Use with high-pressure holder (i.e.534819)
HPLC SOLVENT PURIFICATION				
Samples, Aqueous Samples, Organic	Mixed-Esters of Cellulose PTFE	0.45 0.45	13, 25 4-25	140418/420100/140667 134567/180498/130667
Solvents	PTFE	0.45	47	410502/131120
PARASITOLOGY				
Microfilariae (Dirofilaria immitis)	Polycarbonate	5.0	25	Use Swin-Lok Holder (420200) or Stainless Steel Syringe Holder (421500)
Schistosoma haematobium	Polycarbonate	12.0	13	110416 & Pop Top Holder (420100)
PHARMACEUTICALS (Human or Veterinary)				
Small Volume Parenterals Prefiltration	Syrfil®-MF Glass Fiber	0.22 D49-D79	25 10-293	Parenteral Processing must conform with FDA GMPs;
Sterilization	Mixed-Esters of Cellulose	0.22	13-293	21 CFR 210 and 211.

TABLE 6 continued.

Application	Recommended Filter Media			Reference
	Description	Pore Size (μm)	Diameter (mm)	
PROTEIN OR VIRUS ASSAY & PURIFICATION				
Fractionation or Collection	Polycarbonate	0.015-0.1	25-293	For Protein or Virus Filtration, use Stirred Cell Series - S25, S43, S76 Swin-Lok™ Holders, or Stainless Steel Holders
Purification	Polycarbonate	0.015-0.4	25-293	
Colony Hybridization	Mixed-Esters of Cellulose	0.45	25, 85	
Low Binding	Polycarbonate	0.4	25	
SERUM FILTRATION				
Prefiltration	Glass Fiber Mixed-Esters of Cellulose	D49-D79 0.3-1.2	10-293 13-293	
Bacterial Removal	Mixed-Esters of Cellulose	0.1-0.22, 0.45	13-293	Request Product Bulletin MF0985-1
Mycoplasma Removal	Mixed-Esters of Cellulose	0.1	13-293	Request Product Bulletin MF0985-1
STERILIZING FILTRATION				
Air Venting	Syrfil®-FN (PTFE)	0.2	25, 50	130666, 131266
Fluids, Aqueous	Mixed-Esters of Cellulose	0.22, 0.45	90-293	
Air or Gas	PTFE	0.2	25-293	
TISSUE CULTURE MEDIA FILTRATION				
Prefiltration	Glass Fiber Mixed-Esters of Cellulose	D49 0.3-1.2	47-293 47-293	
Bacterial Removal	Mixed-Esters of Cellulose	0.22	90-293	
Mycoplasma Removal	Mixed-Esters of Cellulose	0.1	90-293	
WATER MICROBIOLOGY				
Escherichia coli	Mixed-Esters of Cellulose	0.45	47	*Standards Methods/Water & Wastewater,* 17th Ed., 9260F
Fecal Coliform	Mixed-Esters of Cellulose	0.7	47	*Standards Methods/Water & Wastewater,* 17th Ed., 9222D
Fecal Streptococcus	Mixed-Esters of Cellulose	0.45	47	*Standards Methods/Water & Wastewater,* 17th Ed., 9230C

TABLE 6 continued.

Application	Recommended Filter Media			Reference
	Description	Pore Size (μm)	Diameter (mm)	
Fine Particulates	Mixed-Esters of Cellulose	0.45	47	Request Product Bulletin MF0985-1
Giardia lamblia	Polycarbonate	5.0	293	Cyst Concentration and Analysis, EPA 600/S2-85/027 Sem. 1985
Leptospires	Mixed-Esters of Cellulose	0.45	13, 25	*Standards Methods/Water & Wastewater,* 17th Ed., 92601
Plytoplankton	Mixed-Esters of Cellulose	1.2-5.0	47	*Standards Methods/Water & Wastewater,* 17th Ed., 10200C
Heterotrophic Plate Count (HPC) - formerly Standard Plate Count	Mixed-Esters of Cellulose	0.45	47	*Standard Methods/ Water & Wastewater,* 17th Ed., 9215D (148818/241110/ 414410/920100)
Salmonella	Mixed-Esters of Cellulose	0.45 0.45	142 47	*Standard Methods/Water & Wastewater,* 17th Ed., 9260B
Suspended Particulates	Mixed-Esters of Cellulose Polycarbonate Glass Fiber	1.2-5.0 1.0-5.0 0.7 nom	47 47 47	211125 (pure glass fiber, no binder)
Direct Total Microbial Count	Polycarbonate (black) Mixed-Esters of Cellulose	0.2 5.0	25 25	*Stdnard Methods/Water & Wastewater,* 17th Ed., 9216B (110656/140613/414220)
Total Coliform Count	Mixed-Esters of Cellulose	0.45	47	*Standard Methods/Water & Wastewater,* 17th Ed., 9222B (148818)
Vibrio cholerae	Mixed-Esters of Cellulose	0.45	142	*Standard Methods/Water & Wastewater,* 17th Ed., 9260H
Virus Concentration	Mixed-Esters of Cellulose Glass Fiber, D49	0.45	47, 90, 142	*Standard Methods/Water &Wastewater,* 17th Ed., 9510B 141118, 141718, 142118, 210914, 211414, 211714

TABLE 7 – MANUFACTURERS CLASSIFIED INDEX BY PRODUCT CATEGORY

Absolute Filters
  Carlson Filtration Ltd
  GAF Filter Systems
  Osmonics, Inc
  Penair Filtration Ltd
  Polymer Papers Ltd
  Sartorius Ltd
  Schenk Filterbau GmbH
  Vokes Ltd

Air Filters
  3M United Kingdom plc
  British Filters Ltd
  Fairey Industrial Ceramics Ltd
  Faudi Feinbau GmbH
  Filtrair BV
  Haver & Boecker
  Osmonics, Inc
  Millipore
  Penair Filtration Ltd
  Polymer Papers Ltd
  Sartorius Ltd
  Sempas Membrantechnik GmbH
  ultrafilter graph
  Vokes Ltd

Air Drying
  Delair BV
  MTR

Air Filter Media
  3M United Kingdom plc
  Fairey Industrial Ceramics Ltd
  Filtrair BV
  Fratelli Testori
  JC Binzer Papierfabrik GmbH
  Kureta GmbH
  Lydall, Inc
  Millipore
  Penair Filtration Ltd
  P and S Filtration
  Polymer Papers Ltd
  SAATI SpA
  Sartorius Ltd
  Verseidag Industrietextilien GmbH
  Vokes Ltd

Bipolar Membranes
  Stantech GmbH
  Graver Water
  Tokuyama
  WSI

Carbon Membrane/Filter
  Le Carbone-Lorraine,
  Ametek
  GFT
  Toray

Cartridge Filters
  3M United Kingdom plc
  Airpel Filtration Ltd
  British Filters Ltd
  Carlson Filtration Ltd
  C C Jensen A/S
  Eurofiltec Ltd
  Facet Industrial UK Ltd
  Fairey Industrial Ceramics Ltd
  Faudi Feinbau GmbH
  Filters srl
  GAF Filter Systems
  G Bopp & Co AG
  Haver & Boecker
  Osmonics, Inc
  Paul GmbH & Co
  Sartorius Ltd
  Sparkler Filters

Cellulose Membranes
  Carlson Filtration Ltd
  Osmonics Inc
  Sartorius Ltd
  Schenk Filterbau GmbH

Cermanic Membranes
  ALPMA Haina nd Co KG
  Atech Innovations GmbH
  Berghof GmbH Laboretechnik
  Berkefeld-Filter Annlagenbau GmbH
  BHS Werk Sonthofe, Filtration Technology
  Blotech Systems Ltd
  Cuno Europe

APPENDIX

TABLE 7 continued.

Carlson Filtration Ltd	Carbone Lorraine
CeramMem Separations	Carlson Filtration Ltd
Chemitreat Pte Ltd	Chemitreat Pte Ltd
Facet Najade BV	Filtra GmbH
Fairey Industrial Cermanics Ltd	FIS Corporation Inc
Filtro Pte Ltd	Fluid Systems Corporation
Filtron GmbH	General Waters (UK) Ltd
Hatenboer-Demi BV	Hatenboer-Demi BV
Hoogovens Industrial Cermaics BV	Hydro Air Research
Hydro Air Research	Mannesmann Demag Sack GmbH
MannesmannDemag Sack GmbH	Matt-Son Inc
Membraflow Filtersysteme GmbH and Co KG	Membrane Products Kiryat Weizmann Ltd
Memtech (UK) Ltd	Memtech (UK) Ltd
Memtech America Corp	Millipore (UK) Ltd
Merrem and La Porte BV	Norwet A/S
Millipore (UK) Ltd	Pall Corporation
Millipore Corporation	Pall Europe Ltd
Millipore SA	Pall Filtratationstechnik GmbH
NWW Acumen Ltd	PCI Membrane Systems Ltd
Osmonics Inc	Schumacher Umwelt-u Trenntechnik GmbH
Pall Corporation	Sempas Membran technik GmbH
Pall Europe Ltd	Sempas Membrantechnik GmbH
Pall Filtrationstechnik GmbH	Sepam BV
Refractron Technologies Corporation	Stork Friesland BV
Schenk Filterbau GmbH	Tech-Sep
Schumacher Umwelt-u. Trenntechnik GmbH	Tech-Sep GmbH
Sempas Membrantechnik GmbH	Ultra-Tech Services Ltd
Sepam BV	Wolftechnik Filtersysteme GmbH
Stark Friesland BV	**Crossflow Filters**
Tech-Sep	Berghof GmbH Laboretchnik
Tech-Sep GmbH	Carlson Filtration Ltd
Ultra-Tech Services Ltd	Fairey Industrial Ceramics Ltd
US Filter Corporation	Faudi Feinbau GmbH
US Filter/IW T	Hoogovens Industrial Cermanics BV
US Filter/Membralox	Osmonics, Inc
Vampipe Process Systems Ltd	Sartourius Ltd
Velterop	Schenk Filterbau GmbH
Wolfechnik Fitersystem GmbH	Sempas Membrantechnik GmbH
Zander UK Ltd	**Coatings**
Composite Membranes	BSI Corporation
ALPMA Hain and Co KG	The Ravensworth Laboratory
Berghof Filtrations - u Anlagentechnick GmBH	

TABLE 7 continued.

Dialysis
 Berkefeld-Filter Anlagnebau GmbH
 Chemitreat Pte Ltd
 Fileder Filter GmbH
 Filtron GmbH
 General Waters (UK) Ltd
 Graver Water
 Hoechst AG, Separations Products
  Division
 Mannesmann Demag Sack GmbH
 Microgon Inc
 Mupor Ltd
 Osmota Membrantechnik GmbH
 Pall Europe Ltd
 Pall Corporation
 Pall Filtrationstechnik GmbH

Drinking Water Filters
 Airpel Filtration Ltd
 British Filters Ltd
 Carlson Filtration Ltd
 Eurofiltec Ltd
 Fiarey Industrial Cermanics Ltd
 Norddeutsche Seekabelwerke AG
 Osmonics, Inc
 Sartorius Ltd
 Sempas Membrantechnik GmbH
 Wilhelm Kßpp GmbH

Effluent Filters
 Hoogovens Industrial Ceramics BV
 Osmonics, Inc.
 Sartorius Ltd
 Schenk Filterbau GmbH
 Wemco GB

Electrochemical Cells
 ICI Chemical and Polymers
 Electrocatalytic Inc.
 Electrocell AB

Electrodeionization
 Ionics Incorporated
 Millipore

Electrodialysis - Lab Units
 Berghof GmbH Labortechnik

Electrodialysis
 Asahi Chemical Industry Co Ltd
 Berghof Filtrations -u Anlagentechnick
  GmbH
 Berghof GmbH Labortechnik
 Chemitreat Pte Ltd
 EIVS SA
 Freudenberg Nowovens Ltd
 General de Filtras
 Graver Water
 Hydro Air Research
 Ionics (UK) Ltd
 Ionics Inc
 Mannesmann Demay Sack GmbH
 Material Perrier
 Millipore (UK) Ltd
 Osmota Membrantechnik GmbH
 Prosep Technologies Inc
 Solvay S A
 Stan Tech GmbH
 Stantech GmbH
 Tokuyama
 Tri-Sep Corporation

Filter Media
 3M United Kingdom plc
 Carlson Filtration Ltd
 Cyclopore SA
 Fairey Industrial Cermaics Ltd
 Filtrair BV
 Fratelli Testori
 Fugafil - saran GmbH & Co
 G Bopp & Co AG
 JC Binzer Paierfabrik GmbH
 Haver & Boecker
 Lydall, Inc.
 Polymer Papers Ltd
 Precision Textiles Ltd
 Sartorius Ltd
 SST Thal
 Tripette et Renaud SA
 United Wire Ltd
 Verseidag Industrietextilien GmbH

APPENDIX

TABLE 7 continued.

Vokes Ltd
ZBF AG

Filter Cartridges
  3M United Kingdom plc
  Airpel Filtration Ltd
  British Filters Ltd
  Carlson Filtration Ltd
  Eurofiltect Ltd
  Facet Industrial UK Ltd
  Fairey Industrial Cermanics Ltd
  Faudi Feinbau GmbH
  Filters srl
  G Bopp & Co AG
  GAF Filter Systems
  Haver & Boecker
  Millipore
  Norddeutsche Seekabelwerke AG
  Osmonics, Inc.
  Paul GmbH & Co
  Polymer Papers Ltd
  Satorius Ltd
  Sempas Membrantechnik GmbH
  Sparkler Filters (GB) Ltd
  Vokes Ltd
  Wilhelm Köpp GmbH

Filter Tubes
  Berghof GmbH Laboretchnik
  Carlson Filtration Ltd
  Fairey Industrial Cermaics Ltd
  Hoogovens Industrial Cermanics BV
  Norddeutsche Seekabelwerek AG
  Paul GmbH & Co
  Precision Textiles Ltd
  Sempas Membrantechnik GmbH
  Sparkler Filters (GB) Ltd

Filtration Equipment
  Amsel Engineering Ltd
  Ionics Incorporated
  Netzsch Filrrationstechnik
  Renovexx Technology Limited

Gas Filters
  British Filters Ltd

Cerwel Ltd
FAcet Industrial UK Ltd
Fairey Industrial Cermanics Ltd
Haver & Boecker
Penair Filtration ltd
Polymer Papers Ltd
Sartorius
Sartorius Ltd
Sempas Membrantechnick GmbH
ultrafilter gmbh
Vokes Ltd

Gas Permeation
  A/G Technology
  Aluminiom Rheinfelden
  Asahi Glass
  Cynara (DOW)
  Dow
  Dainippon Ink. Chem. Co
  Di Pont/ L'Air Liquide
  Grace Mmebranes
  GKSS
  Hoechst Celanese
  International Permeation
    Membrane Technolgy and Research
  Nippon Kokan K.K.
  Nitto electric Co.
  Osaka Gas
  Oxygen Enrichment Co.
  Perma Pure
  Permea (Air Products)
  Techmashexport
  Teijin Ltd
  Toyobo
  Ube Industries
  Union Carbide
  UOP Union Carbide

Hollow Fibre Filters
  Berghof GmbH Laboretchnik
  Sempas Membrantechnik GmbH
  ultrafilter gmbh

Ion Exchange Equipment
  Berghof GmbH Labortechnik
  Osmonics, Inc.

TABLE 7 continued.

Ionics
Stantech
Tokuyama
Eurodia
 	
Gas Separation
  ACS Filtertechniek BV
  A/G Technology Corporation
  Delair BV
  Hoechst AG, Separations Products
    Division
  Merren and La Porte BV
  Millipore Corporation
  Nitto Denko
  Norwet A/S
  Pall Europe Ltd
  Pall Corporation
  Pall Filtrationstechnik GmbH
  Rellumit Filtration Ltd
  Schumacher Umwelt-u Trenntechnik
    GmbH
  Sempas Membrantechnik
  Technische Textilien Lorrach GmbH
  Ube Industries Ltd
  US Filter/IWT
  US Filter Corporation
  Wolftechnik Filtersysteme GmbH
 	
Ion Exchange Membranes
  Berghof GmbH laboretchnik
  Asahi Chemical Industry Co Ltd
  Asahi glass Co LtdBerghof Filtrations -u
    Anlagentechnick GmbH
  Du Pont
  EIVS SA
  Eurodia
  General de Filtras
  Graver Water
  Hydro Air Research
  Ionac Chem Co.
  Ionics (UK) Ltd
  Ionics Inc
  Mannesmann Demay Sack GmbH
  Millipore (UK) Ltd
  Osmota Membrantechnik GmbH
  Prosep Technologies Inc
	RAI Research
	Solvay S A
	Stan Tech GmbH
	Stantech GmbH
	Tokuyama Soda Co
	Tri-Sep Corporation
	
	Laboratory Filters
	  Carlson Filtration ltd
	  GAF Filter Systems
	  Haver & Coecker
	  JC Binzer Papierfabrik GmbH
	  Krupp Buckau Machinenbau
	  Millipore
	  Osmonics, Inc
	  Penair filtration Ltd.
	  Sartorius AG
	  Sartouris Ltd
	  Schenk Filterbau GmbH
	  Sparkler Filkters (GB) Ltd
	
	Media, Porous
	  3M United Kingdom plc
	  Carlson Filtration Ltd
	  Fairey Industrial Ceramics Ltd
	  G Bopp & Co AG
	  JC Binzer Papierfabrik GmbH
	  haver & Boecker
	
	Membranes (all types)
	  Berghof GmbH Labortechnik
	  Bioteck Systems Ltd
	  Carlson Filtration Ltd
	  Chemitreat Pte Ltd
	  Costar Corporation
	  CS & S Filtration
	  Eberhard Hoesch (UK) Ltd
	  Fairey Industrial Cermaics Ltd
	  Fileder Filter Systems Inc
	  Fluid Technologies Inc
	  Hoogovens Industrial Cermacis
	  Ichikawa Co Ltd
	  IFTS
	  Osmonics, Inc
	  Sartorius Ltd
	  Schenk Filterbau GmbH

TABLE 7 continued.

Sempas Membrantechnik GmbH
SF Air Filtration AG
The Durrion Company Inc

Metal Membranes
Altenburger Electronic GmbH
Bedford Steer End and Co Ltd
Facet Najade BV
Hydro Air Research
INA Filtration Corporation
LCI Corporation Inc
Memcor
merrem and La Porte BV
Millipore (UK) Ltd
NV Bekaert SA
Osmoncis Inc
Pall Corporation
Pall Europe Ltd
Pall Filtrationstechnik GmbH
Porvair Technology Ltd
Potter and Soar Ltd
Schenk Filterbau GmbH
Wolftechnik Filtersysteme GmbH

Macrofiltration
NU Bekaert
Cuno Europe
Dieme SpA
Filterteck Inc
Filtron GmbH
Chemitreat Pte Ltd
Dieme SpA
Freudenberg Nonwovens Ltd
graver Water
Hatenboer-Demi BV
Hoogovens Industrial Ceramics BV
Mannesmann Demag Sack GmbH
Microgon Inc
Millipore SA
Facet Najade BV
Osmonics, Inc
PTI Technologies Inc
Schenk Filterbau GmbH
Schumacher IDMF
Tetko Inc

Veratec, Divison of International Paper
Whatman Scientific Ltd

Microfiltration Equipment
A/G Technology Corp
Amazon Filters Ltd
Amicon Inc
APV Baker Ltd
Asahi Chemical Industry Co Ltd
Berghof GmbH Laboretchnik
Berkefeld Filter Anlagenbau GmbH
BHS Werk Sonthofen Filtration Technology
Biotek Systems Ltd
Carbone Lorraine
Carlson Filtration
Carlson Filtration Ltd
CeraHem Separations
Chemitrent Pte Ltd
Consept Solutions Ltd
Costar Corporation
Costar UK Ltd
CUNO Inc
Cyclopore SA
Deutch Carbone AG Geschaftseinheit GFT
Diemme SpA
Domnick Hunter Ltd
EIVS SA
Elga Ltd
Epoc Water Inc
Facet Industrial UK Ltd
Facet Najade BV
Filtertek Inc
Filtres Philippe
Filtro Pte Ltd
Freudenberg Nonwovens Ltd
Gelman Sciences
Gelman Sciences Ltd
Graver Water
GST De Geest Separation Technology BV
Haten boer - Demi BV
Heinrich Frings GmbH and Co KG
Helapet Ltd
Hermans Flexibles BV
HET Filtertechnick GmbH

834   HANDBOOK OF INDUSTRIAL MEMBRANES

TABLE 7 continued.

Hoogovens Industrial Ceramics BV	Sartorius AG
Hoogovens Industrial Ceramics BV	Sartorius Separation Technology
Hydro Air Research	Satorius Ltd
Indufilt GmbH	Scheicher and Schuell GmbH
Ionics (UK) Ltd	Schenk Filterbau GmbH
Ionics Inc	Schenk Filterbau GmbH
Jet Airtechnologoies	Schumacher Umwelt -u Trenntechnik
Kalspe Ltd	GmbH
Koch International (UK) Ltd	Schumacher/DMF
Koch membrane Systems	Scienco/Fast Systems
Kuss Filtration	Seitz Filtration (GB) Ltd
Le Carbone GB Ltd	Seitz-Filter-Weke GmbH and Co
Mahle Inustriefilter Knecht Filterwerke	Sempas Membrantechnik GmbH
Mannesmann Demag Scak GmbH	Sempas Membrantechnik GmbH, Sepan
Matt-Son Inc	BV
Membraflow Filtersysteme GmbH and	Stan Tech GmbH, Stork Friesland BV
Co KG	Tech-Sep Tech-Sep GmbH, Tecnornara
Memcor	AG
Memtec America Corporation	Ultra-Tech Services Ltd, US Filter/IWT
Memtec Ltd	US Filter/Membralox, US Filter Corpora-
Memtech (UK) Ltd	tion
Merrem and La Porte BV	Vanpipe Process Systems Ltd, Vokes Ltd
Microgon Inc	W.L. Gore and Associates (UK) Ltd
Micropure Filtraation Inc	W.L. Gore and Associates GmbH
Millipore (UK) Ltd	Whatman Scientific Ltd, Wolftechnik
Millipore Corporation	Filtersysteme GmbH
Millipore SA	X-Flow BV
Netzsch Filtrationstechnik	
Nitto Denko	**Microfiltration Membranes**
Norddeutsche Seekabelwerke AG	ALPMA Hain and Co KG
Norwet A/S	Atech innovations GmbH
NSW Corporation	Berghof GmbH Laboretchnik
NWW Acumem Ltd	Carlson Filtration Ltd
Osmonics, Inc.	CUNO Inc
Osmonics, Inc	Cyclopore SA
Osmota Membrantechnik GmbH	Hoogovens Industrial Cermaics BV
Pall Corporation	Osmonics, Inc.
Pall Europe Ltd	Sartorius Ltd
Pall Filtrationstechnik GmbH	Schenk Filterbau GmbH
PCI Membrane Systems Ltd	Sempas Membrantechnik GmbH
Polyfiltronics Ltd	
Prosep Technologies Inc	**Micro Separations**
PTI Technologies Inc	Amicon
Refactron Technologies Corporation	Millipore
Rellumit Filtration Ltd	Sartorius
Renovexx Technology Ltd	

TABLE 7 continued.

Membrane Reactors
 F.M. Velterop BV

Nanofiltration Membranes
 ALPMA Hain and Co KG
 Berkefeld-Filter Anlagenbau GmbH
 Biokek Systems Ltd
 Film Tec Corp (Dow)
 PCI
 NWW Acumen Ltd

Nitrogen System
 Belair BV
 Chemitreat Pte Ltd
 Consept Solutions Ltd
 Costar Corporation
 Culligan International (UK) Ltd
 Culligan International
 Cuno Europe
 Cuno Inc
 Fluid Systems Corporation
 Freudenberg Non Wovens Ltd
 FSI Corporation Ltd
 General Waters (UK) Ltd
 Graver Water
 Haten boer-Demi BV
 Hoogovens Industrial Ceramics BV
 Intereco snc di Boaglio Ing F+C
 Ionics (UK) Ltd
 Ionics Inc
 Matt-Son Inc
 Membrane Products Kiryat Weizmann Ltd
 Memtech (UK) Ltd
 Millipore Corporation
 Millipore SA
 MWT AG
 Facet Najade BV
 NORWET A/S
 NWW Acuem Ltd
 Osmonics Inc
 PCI Membrane Systems Ltd
 Science/Fast Systems
 Sempas Membrantechnik GmbH, Sepam BV
 Separem SpA, Stork Friesland BV

 Tech-Sep, Tri-Sep Corporation
 US Filter/Membralox, Vanpipe Process Systems Ltd

Odour Removal Filters
 3M United Kingdom plc
 Carlson Filtration Ltd
 Osmonics, Inc.
 Penair Filtration Ltd
 ultrafilter gmbh

Oil Filters
 Airpel Filtration Ltd
 Berghof GmbH Laboretchnik
 British Filters Ltd
 Carlson Filtration Ltd
 C C Jensen A/S
 Faudi Feinbau GmbH
 GAF Filter Systems
 Haver & Boecker
 Hoogovens Industrial Ceramics BV
 Howden Wade Ltd
 Paul GmbH & Co
 Polymer Papers ltd
 SAATI SpA
 Sartorius Ltd
 Vokes Ltd

Oil/Water Separators
 Facet Industrial UK Ltd
 Hoogovens Industrial Ceramics BV
 Koch
 Osmonics, Inc.
 Sempas Membrantechnik GmbH
 Stefield Ltd
 ultrafilter gmbh
 Vokes Ltd
 Wemco GB
 Westfalia Separator AG

Paint Filters
 Airpel Filtration Ltd
 Eurofiltec Ltd
 Fairey Industrial Ceramics ltd
 Filters srl
 GAF Filter Systems

TABLE 7 continued.

Penair Filtration Ltd
SAATI SpA
Sartorius Ltd
Schenk Filterbau GmbH

Particulate Filtration
Berghof GmbH Labortechnik
Carlson Filtration Ltd
Fairey Industrial Ceramics ltd
Howden Wade Ltd
Lydall, Inc.
Osmonics, Inc.
SAATI SpA
Sartorius Ltd
Schenk Filterbau GmbH
Stefield Ltd
ultrafilter gmbh
Vokes Ltd

Pervaporation Equipment
Bekaert Corproation
Berkefelf-Filter Analagenbau GmbH
Carbone Lorraine
Deutsche carbone AG, Geschaftseinheit GFT
Hoechst AG, Separtions Products Division
Kalsep
Le Carbone GB Ltd
Lurgi
Mitsui
MTR
Pall Corporation
Pall Europe Ltd
Pall Filtrationstechnik GmbH
Sempas Membrantechnik GmbH
Tokuyama Soda
Wolftechnik Filtersysteme GmbH

Polyaramide Membranes
Du Pont de Nemours International SA

Pleated Filters
Carlson Filtration Ltd
Facet Industrial UK Ltd
Fairey Industrial Ceramics Ltd

Osmonics, Inc.
Paul GmbH & Co
Vokes Ltd

Polyaramide Membranes
Ametek-Plymouth Productions Division
Chemitreat Pte Ltd
General Waters (UK) Ltd
Koch International (UK) Ltd
Memtech (UK) Ltd
Monofil Trading Co

Polyester Membranes
ALPMA Hain and Co KG
APV Baker Ltd
Chemitreat Pte Ltd
Costar Corporation
Costar UK Ltd
Cuno Inc
Cyclopore SA
D Barry Couper
Epoc Water Inc
ExxFlow Ltd
Fileder Filter Systems Inc
Hermans Flexibles BV
Hyck Austria GmbH
Jongerius Aerob BV
Memtech (UK) Ltd
Menardi-Criswell, Veratec, Divison of International Paper
Monofil Trading Co

Porous Ceramics
Berghof GmbH Labortechnik
Fairey Industrial Ceramics Ltd
Hoogovens Industrial Ceramics BV
Osmonics, Inc.
Sartorius Ltd
Sempas Membrantechnik GmbH

Polycarbonate Membrane
Ametek - Plymouth Products Division
Chemitreat Pte Ltd
Costar UK Ltd
Cycolpore SA
Merrem and La Porte BV

APPENDIX

TABLE 7 continued.

Nuclepore GmbH
Pall Corporation
Pall Europe Ltd
Pall Filtrationstechnik GmbH
Poly Filtronics
Schleicher and Schuell GmbH
Schumacher Umwelt-u Trenntechnik
   GmbH, Schumacher/DMF

PTFE Membranes
Amazan Filters Ltd
Bellerini Filtrazione
Chemitreat Pte Ltd
Du Pont Nemours Int S.A.
Cuno Europe
Fielder Filter Systems Inc
Filtra GmbH
Gelman Science
General Waters (UK) Ltd
Helapet Ltd
Hermans Flexibles BV
Memcro Filtertechnik GmbH
Memtec (UK) Ltd
Memtec America Corporation
Memtec Ltd
Micro Pore Filtration Inc
Millipore
Mupor Ltd
Osmonics Inc, Schumacher GmbH
Pall Corporation
Pall Europe Ltd
Pall Filtrationstechnik GmbH
PTI Technologies Inc
Sartorius Separation Technology
Schenk Filterbau GmbH
Schleicher and Schuell GmbH
Umwelt-u Trenntechnik
Vokes Ltd
W.L. Gore and Associates
Whatman Scientific Ltd

Reverse Osmosis
Berghof GmbH Labortechnik
Carlson Filtration ltd
Osmonics Inc
ADY

Amafilter BV
Ametek - Plymouth Products Divison
APV Baker Ltd
Aquate SA
Berkefeld - Filter Anlagenbau GmbH
Biotek Systems Ltd
Bird Machine Company
Chemitreat Pte Ltd
Culligan International (UK) Ltd
Culligan International
Cuno Europe
Diemme SpA
Elga Ltd
Domnick Hunter Ltd
Fielder Filter Systems Ltd
Flowgen Instruments Ltd
Fluid Systems Corporation
Freudenberg Non wovens Ltd
FSI Corporation Ltd
Gaco Systems
General Water (UK) Ltd
Graver Water
GST De Geest Separation Technology BV
Hankook Jungsoo Industries Co Ltd
Haten boer- Demi BV
Hydronautics
Hydro Air Research
ICI Plc, Membrane Products Business
Intereco snc di Boaglion Ing F + C
Ionics (UK) Ltd
Ionics Inc
Koch Membrane Systems
Mannesmann Demag Sack GmbH
Matt-Son Inc
Membratek (Pty) Ltd
Memco
Membrane Products Kiryat Weizmann Ltd
Memtech (UK) Ltd
Millipore Corporation
Millipore SA
Millipore (UK) Ltd
MWT AG
Facet Najade BV
Nitto Denko Corporation
Nordlys France
Norwet A/S

TABLE 7 continued.

NWW Acummen Ltd
Osmonics Inc
Osmota Membrantechnik GmbH
Pall Europe Ltd
Pall Corporation
Pall Filtrationstechnick GmbH
PCI Membrane Systems Ltd
PenaFla Ltd
Material Perrier
Prosep Technologies IncScienco/Fast Systems
Sempas Membrantechnick GmbH
Sepam BV
Separem SpA
Serck Baker Ltd
Stan Tech GmBH
Stella-Meta Filters Ltd
Stork Friesland BV
Tech-Sep
Toyobo Co Ltd
Tri-Sep Corporation
US Filter/IWT
US Filter Corporation
Vanpipe Process Systems Ltd
Veratex Division of International Paper
Water Equipment Technologies Inc.
Wolftechnik Filtersysteme GmbH
 	
Reverse Osmosis Membranes
  Ametek - Plymouth Products Division
  UOP
  Osmonics
  DDS
  Toray
  Envirogenics
  Universal Water
  Du Pont
  Dow
  Nitto
  Membrane Products
  Kiryat Weizmann
  Desalination
  Celfa
  Kalle
	Single Membranes
	  Biotech Systems Ltd
	  Chemitreat Pte Ltd
	  Costar Corporation
	  Costar UK Ltd
	  Fileder Filter Systems Ltd
	  Memtech America Corporation
	  Memtech (UK)
	  PTI Technologies Inc
	  Pall Europe Inc
	  Pall Corporation
	  Pall Filtratrionstechnik GmbH
	
	Ultra Filtration
	  A/G Technology Corp
	  Altenburger Electronic GmbH
	  Amicon
	  APV Baker Ltd
	  Aquasource
	  Auatec SA
	  Asahi Chem. Industry Co Ltd
	  Atech Innovations GmbH
	  Berghof Filtrations -u Anlagentechnik GmbH
	  Berkefeld - Filter Anlagenbau GmbH
	  Biotek Systems Ltd
	  Carbone Corraine
	  Carlson Filtration Ltd
	  Cera Mem Separations
	  CUNO Inc
	  Cuno Europe
	  Chemitreat Pte Ltd
	  Consept Solutions Ltd
	  Costar Corporation
	  Culligan International (UK) Ltd
	  Culligan Europe NV
	  Culligan International
	  Deomed Ltd
	  Deutsche Carbone AG Geschaftseinheit GFT
	  Diemme SpA
	  Dominck Hunter Ltd
	  Elga Ltd
	  Faudi Feinbau GmbH
	  Fielder Filter Systems Ltd
	  Filtermist International Plc

APPENDIX

TABLE 7 continued.

Filtro Pte Ltd	Le Carbone GB Ltd
Filtron GmbH	Mannesmann Demag Sack GmbH
Flowgen Instruments Ltd	Matt-Son Inc
Freudenbergh Non wovens Ltd	Membraflow Filtersysteme GmbH and Co KG
Heinrich Frings GmbH and Co KG	
FSI Corporation Ltd	Membrane Products Kiryat Weizmann Ltd
Gaco Systems	Membrateck (Pty) Ltd
General Waters (UK) Ltd	Memcor Filtertechnik GmbH
Graver Water	Memtec America Corporation
GST de Geest Separation Technology BV	Memtec Ltd
	Memtech (UK) Ltd
Haten boer - Demi BV	Merrem and La Porte BV
HET Filtertechnick GmbH	Microgon Inc.
Hoechst AG, Separations Products Division	Millipore Corporation
	Millipore SA
Hoogovens Industrial Ceramics BV	Millipore (UK) Ltd
Hydac Filtertechnik GmbH	Mupor Ltd
Hyde (UK) Ltd	MWT AG
Hydranautics	Facet Najade BV
Hydro Air Research	Netzsch Mastermix Ltd
ICI Plc, Membrane Products Business	New Logic International
Inchcape Testing Services ICS	Nitto Denko Corporation
Indu Filt GmbH	Nardlys France
Intereco snc di Boaglio Ing FTC	Norwet A/S
Sempas Membrantechnik GmbH	NWW Acument Ltd
Sepam BV, Separem SpA, Serck Baker Ltd	Osmonics, Inc.
	Osmota Membrantechnik GmbH
Serfilco Europe Ltd, Stan Tech GmbH, Stetfield Ltd	Pall Europe Ltd
	Pall Corporation
Stork Friesland BV, Tech-Sep, Tech-Sep GmbH	Pall Filtrationstechik Ltd
	Sartorius Separation Technology
Tri-Sep Corporation, Ultrafilter Ltd, Ultrafilter gmbH	Sartorius AG
	Schenk Filterbau GmbH
Ultra-Tech Services Ltd	Schleicher and Schuell GmbH
US Filter/IWT	Scienco/Fast Systems
US Filter/Membralox	SCT
US Filter Corporation	
Vanpipe Process Systems Ltd	**Ultra Filtration Membranes**
Veratec, Division of International Paper	Alcoa
Wolftechnick Filtersysteme GmbH	Alcan/Anapore
X-Flow BV	Amicon(W R Grace)
Ionics (UK) Ltd	Asahi
Ionics Inc	Gerghof
Kalsep Ltd	DDS
Koch International (UK) Ltd	Desalination Systems
Koch Membrane Systems	Dorr-Oliver (W R Grace)

TABLE 7 continued.

	Water Purification
Filtron	3M United Kingdom plc
Fluid Systems	Airpel Filtration Ltd
W. Gore	British Filters Ltd
Hoechst	Carlson Filtration Ltd
Kalle	Culligan
MilliporeKoch Membrane Systems	Fairey Industrial Ceramics Ltd
Nitto	GAF Filter Systems
Hydronautics	Kureta GmbH
Osmonics	Norddeutsche Seekabelweke AG
Rhone Poulenc	Osmonics, Inc
Carbosep	SAATI SpA
PCI Membrane	Sartouris Ltd
Romicon	Schenk Filterbau GmbH
Sartorius	Sempas Membrantechnik GmbH
Schott	ultrafilter gmbh
TeijinTeijin	Vokes Ltd
Waflin	Millipore
Zenon	
Alpma Hain and Co KG	
CUNO	
Sartorius	
Millipore	

APPENDIX

TABLE 8 – Representative Range of Contaminant Removal.

Type of filter or separator	1000	100	50	20	10	1	0.1	0.01
Membrane				12 ———————————————————→ 0.005				
Electrostatic					10 ———————————→ 00.1 · → 0.001			
Microglass					10 ———————→ 0.01			
Asbestos fibre		100 ————————————————————→ 0.1						
Paper		100 ————————————————→ 1						
Sintered metal		100 ——————————→ 10 ·········→ 1						
Woven wire:								
Square weave	300 ————————————————→ 20							
Plain Dutch		100 ——————→ 20						
Reverse Dutch	115 ———————————→ 15							
Twilled Dutch		100 ——————————————→ 6						
Wire cloth:								
Coarse	950 ←—————————							
Medium	950 ——→ 180							
Fine	200 ——→ 160							
Perforated plate (strainers)	1000 ——→ 100							
Wire mesh (strainers)	150 ———————→ 40							
Screens	250 ———————————→ 25							
Settling tank	1000 ——————→ 100							
Cyclone	1000 ————————————————→ 10							
Multi-cyclone	1000 ———————————————————→ 5 ············→ 1							
Wet collector		100 ——————————————→ 5 ·············→ 1						
Venturi scrubber		100 ————————————————————————→ 0.1 ········→ 0.01						
Fabric dush collectors		100 ————————————————————————→ 0.1 ········→ 0.01						
Panel filters		100 ———————————————→ 1						
Viscous panels	1000 ———————————————→ 1							
Filter cells		100 ——————————————————————→ 0.1						

Particle size range μm

TABLE 9 – General Guide to Contaminant Sizes.

Contaminant	Particle size μm					
	under 0.01	0.01-0.1	0.01-1	1-10	10-100	100-1000
Haemoglobin	X					
Viruses	X	X				
Bacteria			X-X	X		
Yeasts and fungi			X-X	X		
Pollen				X-X	X	
Plant spores					X	
Inside dust	X	X-X	X-X	X-X	X	
Atmospheric dust				X-X	X-X	
Industrial dusts				X	X-X	X-X
Continuously suspended dusts	X	X-X	X-X			
Oil mist		X	X-X	X-X		
Tabacco smoke		X-X	X-X	X		
Industrial gases			X-X			
Aerosols	X	X-X	X-X	X		
Powdered insecticides				X-X		
Permanent atmospheric pollution	X	X-X	X-X			
Temporary atmospheric pollution				X-X	X-X	X-X
Contaminants harmful to machines				X	X-X	X-X
Machine protection normal				X	X-X	X-X
Machine protection maximum				X-X	X-X	X-X
Silt control				X (3-5)		
Partial silt control					X (10-15)	
Chip control					X (25-40)	
Air filtration, primary					X-X	X-X
Air filtration, secondary				X-X		
Air filtration, ultra-fine		X	X-X			
Staining particle range			X-X	X		

TABLE 10 – Ion-Exchange Membranes & Manufacturers.

Type (Trade Name)	Producer
cation-P (XF-OCF—$_2$CO$_2$H/SO$_3$H) NP (Aciplex)	Asahi Chemical Ltd
cation-P (XF-O(CF$_2$)$_n$CO$_2$H) NP (Flemion, Selemion)	Asahi Chemical Ltd
cation-P (XF-OCF$_2$CF$_2$SO$_3$H) (Nafion (1000 series))	Du Pont
cation-P (XF-OCF$_2$CF$_2$SO$_2$NH$_2$) (Nafion Sulfonamide)	Du Pont
cation-P (XF-OCF$_2$CF$_2$CO$_2$H/SO$_3$H) (Nafion 901)	Du Pont
cation-P (XF-O(CF$_2$)nCO$_2$H/SO$_3$H) NP (Neosepta-F, Neosepta)	Tokuyama Soda
anion-P (R$_1$NR$_3^+$X$^-$) (Tosflex)	Tosoh
P$_1$ B	WSI Tech
P A cation, anion	Solvay-Morgane
P$_1$ B, cation, anion	Aquatech
P$_1$ B, cation, anion	Stantech
NP, cation, anion	Ionics Inc
NP, cation, anion	Membranes International
NP, cation, anion RAI	RAI Research Corp
NP, cation, anion Ionac	Sybron Chemicals Inc

P: Perfluorinated, NP: non-perfluorinated, B: Bipolar

TABLE 11 – A Summary of Contaminant Removal and Alternative Technologies.

Technology/description	Stage of development	Economics	Types of waste streams	Separation efficiency[a]	Industrial applications
Physical separation:					
Gravity settling: Tanks, ponds provide hold-up time allowing solids to settle; grease skimmed to overflow to another vessel	Commonly used in wastewater treatment	Relatively inexpensive: dependent on particle size and settling rate	Slurries with separate phase solids, such as metal hydroxide	Limited to solids (large particles) that settle quickly (less than 2 hours)	Industrial wastewater treatment first step
Filtration: Collection devices such as screens, cloth, or other; liquid passes and solids are retained on porous media	Commonly used	Labor intensive: relatively inexpensive; energy required for pumping	Aqueous solutions with finely divided solids; gelatinous sludge	Good for relatively large particles	Tannery water
Flotation: Air bubbled through liquid to collect finely divided solids that rise to the fsurface with the bubbles	Commercial application	Relatively inexpensive	Aqueous solutions with finely divided solids	Good for finely divided solids	Refinery (oil/water mixtures); paper waste; mineral industry
Flocculation: Agent added to aggregate solidsd together which are easily settled	Commercial practice	Relatively inexpensive	Aqueous solutions with finely divided solids	Good for finely divided solids	Refinery; paper waste; mine industry
Centrifugation: Spinning of liquids and centrifugal force causes separation by different densities	Practiced commercially for small-scale systems	Competitive with filtration	Liquid/liquid or liquid/solid separation, i.e. oil/water; resins; pigments from lacquers	Fairly high (90%)	Paints

APPENDIX

TABLE 11 (continued).

Technology/description	Stage of development	Economics	Types of waste streams	Separation efficiency[a]	Industrial applications
Component separation:					
Distillation: successfully boiling off of materials at different temperatures (based on different boiling points)	Commercial practice	Energy intensive	Organic liquids	Very high separations achievable (99+% concentrations) of several components	Solvent separations: chemical and petroleum industry
Evaporation: Solvent recovery by boiling off the solvent	Commercial practice in many industries	Energy intensive	Organic/inorganic aqueous streams; slurries, sludges, i.e. caustic soda	Very high separations of single, evaporated component achievable	Rinse waters from metal-plating waste
Ion exchange: Waste streat passed through resin bed, ionic materials selectively removed by resins similar to resin adsorption. Ionic exchange materials must be regenerated	Not common for HW	Relatively high costs	Heavy metals aqueous solutions; cyanide removed	Fairly high	Metal-plating solutions
Ultrafiltration: Separation of solecules by size using membrane	Some commercial application	Relatively high	Heavy metal aqueous solutions	Fairly high	Metal-coating applications
Reverse osmosis: Separation of dissolved materials from liquid through a membrane	Not common: growing number of applications as secondary treatment process such as metal-plating pharmaceuticals	Relatively high	Heavy metals; organics; inorganic aqueous solutions	Good for concentrations less than 300 ppm	Not used industrially

TABLE 11 (continued).

Technology/description	Stage of development	Economics	Types of waste streams	Separation efficiency[a]	Industrial applications
Electrolysis: Separation of positively/negatively charged materials by application of electric current	Commercial technology; not applied to recovery of hazardous materials	Dependent on concentrations	Heavy metals; ions from aqueous solutions; copper recovery	Good	Metal plating
Carbon/resin absorption: Dissolved materials selectively absorbed in carbon or resins. Absorbents must be regenerated	Proven for thermal regeneration of carbon; less practical for recovery of adsorbate	Relatively costly thermal regeneration; energy intensive	Organics/inorganics from aqueous solutions with low concentrations, i.e. phenols	Good, overall effectiveness dependent on regeneration method	Phenolics
Solvent extraction: Solvent used to selectively dissolve solid or extract liquid from waste	Commonly used in industrial processing	Relatively high costs for solvent	Organic liquids, phenols, acids	Fairly high loss of solvent may contribute to hazardous waste problem	Recovery of dyes
Chemical transformation:					
Precipitation: Chemical reaction causes formation of solids which settle	Common	Relatively high costs	Lime slurries	Good	Metal-plating waste-water treatment
Electrodialysis: Separation based on differential rates of diffusion through membranes. Electrical current applied to enhance ionic movement	Commercial technology, not commercial for hazardous material recovery	Moderately expensive	Separation/concentration of ions from aqueous streams; application to chromium recovery	Fairly high	Separation of acids and metallic solutions

APPENDIX

TABLE 11 (continued).

Technology/description	Stage of development	Economics	Types of waste streams	Separation efficiency[a]	Industrial applications
Chlorinolysis: Pyrolysis in atmosphere of excess chlorine	Commercially used in West Germany	Insufficient U.S. market for carbon tetrachloride	Chlorocarbon waste	Good	Carbon tetrachloride manufacturing
Reduction: Oxidative state of chemical changed through chemical reaction	Commercially applied to chromium; may need additional treatment	Inexpensive	Metals, mercury in dilute streams	Good	Chrome-plating solutions and tanning operations
Chemical dechlorination: Reagents selectively attack carbon-chlorine bonds	Common	Moderately expensive	PCB-contaminated oils	High	Transformer oils
Thermal oxidation: Thermal conversion of components	Extensively practiced	Relatively high	Chlorinated organic liquids; silver	Fairly high	Recovery of sulfur, HCl

[a] Good implied 50 to 80 percent efficiency, fairly high implied 80 percent, and very high implies 90 percent.
SOURCE: Office of Technology Assessment.

TABLE 12 – Cleaning Formulations for Processing Biologicals.

Process	Foulants	Alternate Cleaning Procedures in Order of Preference
Mammalian Cell solution at 50°C	Cell Debris	A. 1. Flush with clean water, buffer or saline 2. Circulate 500 ppm NcOCI at 50°C, pH 10-11 for 1 hour. 3. Flush with clean water. B. 1. Flusfh with clean water, buffer or saline solution at 50°C. 2. Circulate 0.5N NaOH at 50°C for 1 hour. 3. Flush with clean water. C. 1. Flush with clean water, buffer or saline solution at 50°C. 2. Circulate 0.2% Terg-A-Zyme® at 50°C, pH 9-10 for 1 hour. 3. Flush with clean water.
Bacterial Cell Whole Broths	Proteins, Cell Debris, Poly-saccharides, Lipids, Antiforms	A. 1. Flush with clean water, buffer or saline solution at 50°C. 2. Circulate 0.5N NaOH at 50°C for 1 hour. 3. Flush with clean water. *Optional:* 4. Circulate 500 ppm NaOCI at 50°C, pH 10-11 for 1 hour. 5. Flush with clean water. B. 1. Flush with clean water, buffer or saline solution at 50°C. 2. Circulate 0.2% Terg-A-Zyme® at 50°C, pH 9-10 for 1 hour. 3. Flush with clean water. C. 1. Flush with clean water, buffer or saline solution at 50°C. 2. Circulate 0.5% Henkel P3-11 at 50°C, pH 7-8 for 1 hour. 3. Flush with clean water.
Bacterial Cell Lysates	Proteins, Cell Debris	A. 1. Flush with clean water, buffer or saline solution at 50°C. 2. Circulate 0.5N NaOH at 50°C for 1 hour. 3. Flush with clean water. *Optional:* 4. Flush with H_3PO_4 at 50°C, pH 4 for 1 hour. 5. Flush with clean water. B. 1. Flush with clean water, buffer or saline solution at 50°C. 2. Circulate 500 ppm NaOCI at 50°C, pH 10-11 for 1 hour. 3. Flush with clean water. *Optional:* 4. Circulate H_3PO_4 at 50°C, pH 4 for 1 hour. 5. Flush with clean water. C. 1. Flush with clean water, buffer or saline solution at 50°C. 2. Circulate 0.1% Tween 80® at 50°C, pH 5-8 for 1 hour. 3. Flush with clean water.
Blood/Serum Products, Enzymes, Vaccines, Products	Proteins, Lipoproteins, Lipids	A. 1. Flush with clean water, buffer or saline solution at 50°C. 2. Circulate 0.5N NaOH at 50°C for 1 hour. 3. Flush with clean water. 4. Circulate 500 ppm NaOCI at 50°C, pH 10-11 for 1 hour. 5. Flush with clean water.

TABLE 12 – Cleaning Formulations for Processing Biologicals.

Process	Foulants	Alternate Cleaning Procedures in Order of Preference
		B. 1. Flush with clean water, buffer or saline solution at 50°C. 2. Circulate 0.1% Tween 80® at 50°C, pH 5-8 for 1 hour. 3. Flush with clean water. C. 1. Flush with clean water, buffer or saline solution at 50°C. 2. Circulate 0.2% Terg-A-Zyme® at 50°C, pH 9-10 for 1 hour. 3. Flush with clean water.
Juice and Beverage Clarification	Proteins, Pectin, Colloids, Tannins, Polyphenolics	A. 1. Flush with clean water. 2. Circulate 0.5N NaOH at 50°C for 1 hour. 3. Flush with clean water. *Optional:* 4. Circulate 500 ppm NaOCl at 50°C, pH 10-11 for 1 hour. 5. Flush with clean water. B. 1. Flush with clean water. 2. Circulate 0.5% Henkel P3-11 at 50°C, pH 7-8 for 1 hour. 3. Flush with clean water. *Optional:* 4. Circulate 500 ppm NaOCl at 50°C, pH 10-11 for 1 hour. 5. Flush with clean water. C. 1. Flush with Clean water. 2. Circulate 0.2% Terg-A-Zyme® at 50°C, pH 9-10 for 1 hour. 3. Flush with clean water. *Optional:* 4. Circulate 500 ppm NaOCl at 50°C, pH 10-11 for 1 hour. 5. Flush with clean water.
Dairy	Protein, Insoluble Calcium Complexes	A. 1. Flush with clean water. 2. Circulate H_3PO_4 at 50°C, pH 3.5-4 for 20 minutes. 3. Flush with clean water. 4. Circulate 0.5N NaOH at 50°C for 20 minutes. 5. Flush with clean water. 6. Circulate 500 ppm NaOCl at 50°C, pH 10-11 for 1 hour. *Monitor and maintain chlorine level.* 7. Flush with clean water.
Water Treatment	Iron Complexes	A. 1. Flush with clean water. 2. Circulate Citric Acid at 50°C, pH 2-2.5 for 1 hour. 3. Flush with clean water. *Optional: if low water flux,* 4. Circulate a low foaming alkaline cleaner for 20 minutes. 5. Flush with clean water.
	Mineral Scale	A. 1. Flush with clean water. 2. Circulate HNO_3 at 50°C, pH for 1 hour. 3. Flush with clean water.

TABLE 12 (continued).

Process	Foulants	Alternate Cleaning Procedures in Order of Preference
		Optional: 4. Repeat Step 2 and leave soaking overnight. 5. Flush with clean water. B. 1. Flush with clean water. 2. Circulate H_3PO_4 at 50°C, pH 4 for 1 hour. 3. Flush with clean water. *Optional:* 4. Repeat Step 2 and leave soaking overnight. 5. Flush with clean water.
Edible Oils	Oil, Grease, Colloids	A. 1. Flush with clean water. 2. Circulate 0.2% Micro® at 50°C, pH 9-10 for hour 1 hour. 3. Flush with clean water. *Optional:* 4. If ironing fouling is suspected, wash with Citric Acid, pH 2-2.5, as noted above. B. *Substitute alternate detergent cleaners for Micro®.* *Increase detergent concentration.*

Flushing time is typically 30 minutes. Tighter membrane pore sizes and systems with large hold-up volumes may require longer flushing time.

TABLE 13 – A/G Technology Polysulfone Membrane Cartridge Chemical Compatibility.

Reagent	Usage	Reagent	Usage
Acetic acid (<5%)	Acceptable	Isopropyl Alcohol (≤10%)	Acceptable
Acetic Acid (>5%)	Short Term Only	Kerosene	Not Recommended
Acetic Anhydride	Not Recommended	Lactic Acid (≤5%)	Acceptable
Acetone	Not Recommended	Mercaptoethanol (≤0.1M)	Acceptable
Acetonitrile (≤10%)	Short Term Only	Methyl Alcohol (≤10%)	Acceptable
Aliphatic Esters	Not Recommended		
Amines	Not Recommended	Methylene Chloride	Not Recommended
Ammonium Chloride (<1%)	Acceptable	Methyl Ethyl Ketone	Not Recommended
Ammonium Hydroxide (<5%)	Acceptable	N-Methyl Pyrolidone	Not Recommended
Benzene	Not Recommended	Nitric Acid (≤1%)	Short Term Only
Butanol (<1%)	Acceptable	Nitrobenzene	Not Recommended
Bytyl Acetate	Not Recommended	Oleic Acid (≤5%)	Short Term Only
Butyl Cellosolve	Not Recommended	Oxalic Acid (≤1%)	Acceptable
Calcium Chloride	Acceptable	Phenols (<0.25%)	Acceptable
Chloroform	Not Recommended	Phosphoric Acid (≤0.1N)	Short Term Only
Citric Acid (≤1%)	Acceptable	Sodium Azide (≤1%)	Acceptable
Cyclohexanone	Not Recommended	Sodium Chloride	Acceptable
Dichlorobenzene	Not Recommended	Sodium Dodecyl Sulfate (≤0.1%)	Acceptable
Diethanolamine (≤5%)	Acceptable		
Dimethyl Acetamide	Not Recommended	Sodium Hydroxide (≤1N)	Acceptable
Dimethylformamide	Not Recommended		
Dimethyl Sulfoxide	Not Recommended	Sodium Hypochlorite (≤300 ppm)	Acceptable
Disodium Salt of EDTA (≤10%)	Acceptable	Sodium Hypochlorite (>300 ppm)	Short Term Only
Ethanol (≤10%)*	Acceptable		
Ethyl Acetate	Not Recommended	Sodium Nitrate (≤1%)	Acceptable
Formaldehyde (≤1%)	Acceptable	Sulfuric Acid (≤1%)	Acceptable
Formic Acid (≤1%)	Acceptable	Terg-A-Zyme® (≤1%)	Acceptable
Furfural	Not Recommended	Tolyene	Not Recommended
Glutaldehyde (≤0.5%)	Acceptable	Tris Buffer (pH 8.2, 1M)	Acceptable
Glycerine (≤2%)	Acceptable		
Guanidine HCL (6M)	Acceptable	Triton X-100 (<200 ppm)	Acceptable
Hydrochloric Acid (≤0.01N)	Acceptable		
Hydrogen Peroxide (≤1%)	Short Term Only	Urea (≤25%)	Acceptable
		Xylene	Not Recommended
Isopropyl Acetate	Not Recommended		

Chemicals noted as "Short Term Only" are typically acceptable for membrane cleaning.

*Higher alcohol concentrations acceptable depending on operating conditions.
100% alcohol acceptable for non-pressurized exposure.

TABLE 14 – List of Commercial Crossflow Microfilter Media.

Material	Geometries					
	Pleated sheet	Tubular	Spiral wound	Tubular MC*	Hollow fiber	Flat sheet
1. *Polymers*						
Celluylosics	●		●		●	●
Polysulfone		●	●		●	●
Polyvinylidenefluoride		●	●			
Acrylic					●	
Polytetrafluoroethylene		●				●
Polybenzimadazole			●			●
Polypropylene					●	●
Nylon	●				●	●
2. *Ceramics*						
Alumina				●		●
Zirconia/alumina				●		
Zirconia/sintered metal		●				
Zirconia/carbon		●				
Silica		●				
Silicon carbide				●		
3. *Sintered metal*						
Type 316 stainless steel		●				
Other alloys		●				

*MC = multichannel monolithic elements.

TABLE 15 – A selection of commercially available membranes for cross flow filtration.

Material	Application		
	MF	UF	RO
Cellulose acetate (CA)	X	X	X
Cellulose triacetate	X	X	X
CA/Triacetate blend			X
Cellulose esters (mixed)	X		
Cellulose nitrate	X		
Cellulose (regenerated)	X	X	
Gelatin	X		
Polyacrylonitrile (PAN)		X	
Polyvinylchloride (PVC)	X		
Polyvinylchloride copolymer	X	X	
Polyamide (aromatic)	X	X	X
Polysulfone	X	X	
Polybenzimidazole (PBI)			X
Polybenzimidiazolone (PBIL)			X
Polycarbonate (track-etch)	X		
Polyester (track-etch)	X		
Polyimide		X	X
Polypropylene	X		
Polyelectrolyte complex		X	
Polytetrafluoroethylene (PTFE)	X		
Polyvinylidenefluoride (VF)	X	X	
Polyacrylic acid + zirconium oxide (skin layer of dynamic membrane)		X	X
Polyethleneimine + toluene diisocyanate (skin of thin-film composite)			X

TABLE 16 – Gas Permeation Membrane Module Manufacturers.

Common gas separations	Application	Suppliers
O_2/N_2	Nitrogen generation, oxygen enrichment	Permea (Air Products), Linde (Union Carbide), A/G Technology, Generon (Dow Chemical/BOC), Asahi Glass, Osaka Gas, Oxygen Enrichment Co.
H_2O/Air	Air dehumidification	Permea, Ube Industries, Perma Pure
H_2/Hydrocarbons	Refinery hydrogen recovery	Permea, Grace Membrane Systems (W.R. Grace)
H_2/CO	Syngas ratio adjustment	As above
H_2/N_2	Ammonia purge gas	As above
CO_2/Hydrocarbons	Acid gas treating, landfill gas upgrading	Grace Membrane Systems, Cynara (Dow Chemical), Separex, (Hoechst Celanese), Permea
H_2O/Hydrocarbons	Natural gas dehydration	As above
H_2S/Hydrocarbons	Sour gas treating	As above
He/Hydrocarbons	Helium separations	As above
He/N_2	Helium recovery	As above
Hydrocarbons/Air	Pollution control, hydrocarbon recovery	Membrane Technology and Research, Aluminium Rheinfelden/GKSS, NKK

TABLE 17a – Characteristics of Reverse Osmosis Membranes.

Test Module[a]	Solute	Test Conditions	Flux x 10^4 cm^3/cm^{2}s (gfd)	% Rejection
Cellulose acetate				
F	NaCl	50,000 ppm, 8 MPa	9.17 (19.4)	98
	Methanol	1.7 MPa		7
	Ethanol	23-138 ppm, 1.7 MPa		10
	Phenol	1.7 MPa		0
T	NaCl	5000 ppm, 25°C, 4.1 MPa, $r = 0\%$	4.8 (10.2)	98
	Methanol	1000 ppm, 25°C, 4.1 MPa, $r = 0\%$		<0
	Ethanol	1000 ppm, 25°C, 4.1 MPa, $r = 0\%$		2
	Urea	1000 ppm, 25°C, 4.1 MPa, $r = 0\%$		26
	Phenol	1000 ppm, 25°C, 4.1 MPa, $r = 0\%$		17
S-40 x 7.9 in.	NaCl	2000 ppm, 25°C, 2.9 MPa, $r = 10\%$	[0.456 L/s (10,4000 gpd)][b]	90
	NaCl		[0.355 L/s (8100 gpd)][b]	95
	NaCl	pH 5.0-6.0	[0.280 L/s (6400 gpd)][b]	97.5
	NaCl	2500 ppm, 25°C, pH 7, 4 MPa	19.5 (41.3)[b]	90-92
	NaCl	2500 ppm, 25°C, pH 7, 4 MPa	13.9 (29.5)[b]	95-97
	NaCl	2500 ppm, 25°C, pH 7, 4 MPa	5.57 (11.8)[b]	98-99.5
	NaCl	1500 ppm, 25°C, 1.5 MPa	3.47 (7.37)	96
	Methanol	1000 ppm, 25°C, 1.5 MPa		5
	Ethanol	1000 ppm, 25°C, 1.5 MPa		9
	Urea	1000 ppm, 25°C, 1.5 MPa		26
	Phenol	1000 ppm, 25°C, 1.5 MPa		0
Cellulose diacetate and triacetate				
	NaCl	2500 ppm, 20°C 3 MPa	23.7 (50.2)	26-34
	NaCl	2500 ppm, 20°C, 1 MPa	16.7 (35.4)	26-34

TABLE 17a (continued).

Test Module[a]	Solute	Test Conditions	Flux x 10^4 cm^3/cm^{2}s (gfd)	% Rejection
	NaCl	2500 ppm, 20°C, 3 MPa	19.5 (41.3)	55-65
	NaCl	2500 ppm, 20°C, 4 MPa	16.7 (35.4)	>85
	NaCl	2500 ppm, 20°C, 4 MPa	16.7 (35.4)	>90
	NaCl	2500 ppm, 20°C, 4 MPa	11.1 (23.6)	>94
Cellulose triacetate				
F	NaCl	5000 ppm, 25°C, 4.1 MPa, $r = 0\%$	2.31 (4.9)	98
	Methanol	1000 ppm, 25°C, 4.1 MPa, $r = 0\%$		<0
	Ethanol	1000 ppm, 25°C, 4.1 MPa, $r = 0\%$		23
	Urea	1000 ppm, 25°C, 4.1 MPa, $r = 0\%$		38
	Phenol	1000 ppm, 25°C, 4.1 MPa, $r = 0\%$		<0
Cellulose acetate butyrate				
F	NaCl	5000 ppm, 25°C, 4.1 MPa, $r = 0\%$	0.65 (1.4)	>99
	Methanol	1000 ppm, 25°C, 4.1 MPa, $r = 0\%$		<0
	Ethanol	1000 ppm, 25°C, 4.1 MPa, $r = 0\%$		1
	Urea	1000 ppm, 25°C, 4.1 MPa, $r = 0\%$		8
	Phenol	1000 ppm, 25°C, 4.1 MPa, $r = 0\%$		10
Aromatic polyamide				
HF - 4 ft x .5 in	NaCl	5000 ppm, 25°C, 5.2 MPa, $r = 75\%$	[0.197 L/s (4500 gpd)]	99
	Methanol	1000 ppm, 25°C, 5.2 MPa, $r = 75\%$		10
	Ethanol	1000 ppm, 25°C, 5.2 MPa, $r = 75\%$		15
	Urea	1000 ppm, 25°C, 5.2 MPa, $r = 75\%$		41

TABLE 17a (continued).

Test Module[a]	Solute	Test Conditions	Flux $\times 10^4$ cm³/cm²s (gfd)	% Rejection
	Phenol	1000 ppm, 25°C, 5.2 MPa, $r = 75\%$		64
Cross-linked aromatic polyamide				
S-40 in. x 7.9 in.	NaCl	35,000 ppm, 25°C, pH 8. 5.5 MPa, $r = 10\%$	5.71 (12.1) [0.175 L/s (4000 gpd)]	99.5
S-40 in. x 7.9 in.	NaCl	2000 ppm, 25°C, pH 8, 1.6 MPa, $r = 15\%$	10.7 (22.7) [0.329 L/s (7500 gpd)]	98
	Methanol	2000 ppm, 25°C, pH 7, 1.6 MPa		25
	Ethanol	2000 ppm, 25°C, pH 7, 1.6 MPa		70
	Urea	2000 ppm, 25°C, pH 7, 1.6 MPa		70
F	Phenol	51 ppm, pH 7.4, $r = 83\%$, 2.1 MPa		90
S-40 in. x 3.9 in.	NaCl	2000 ppm, 25°C, pH 8, 1.6 MPa, $r = 15\%$	10.6 (22.5) [0.079 L/s (1800 gpd)]	98
	NaCl	2500 ppm, 25°C, 4 MPa	20.9 (44.3)	95-97
	Methanol	1000 ppm, 25°C		42
	Ethanol	9000-26,000 ppm, 25°C, 4 MPa		70-75
	Urea	1000-20,000 ppm, 25°C, 4 MPa		70
	Phenol	30-1000 ppm		80-90
S-40 in. x 3.9 in.	NaCl	2000 ppm, 25°C, pH 8, 1.6 MPa, $r = 10\%$	[0.079 L/s (1800 gpd)]	98
Polyvinylalcohol (TFC)				
	NaCl	1500 ppm, 25°C, pH 6-7, 0.99 MPa	15.1 (32)[b]	92
	Ethanol	1500 ppm, 25°C, pH 6-7, 0.99 MPa		25
	NaCl	1500 ppm, 25°C, pH 6-7, 0.99 MPa		95
	Ethanol	1500 ppm, 25°C, pH 6-7, 0.99 MPa		30

TABLE 17a (continued).

Test Module[a]	Solute	Test Conditions	Flux x 10^4 cm^3/cm^2s (gfd)	% Rejection
Aryl-alkyl polyamide/polyurea				
S	NaCl	5000 ppm, 25°C, 4.1 MPa	26.7 (56.5)[b]	98.5
S	NaCl	1500 ppm, 25°C, 1.5 MPa	29.0 (61.5)	90
	NaCl	1500 ppm, 25°C, 1.5 MPa	11.6 (24.5)	98-99
	Methanol	1000 ppm, 25°C, 1.5 MPa		9
	Ethanol	1000 ppm, 25°C, 1.5 MPa		34
	Urea	1000 ppm, 25°C, 1.5 MPa		28
	Phenol	1000 ppm, 25°C, 1.5 MPa		26
PEC 100 (polyfuran)				
	NaCl	25,000 ppm, 25°C, pH 7, 5.5 MPa	4.05 (8.59)	99.92
	Methanol	55,000 ppm, 25°C, pH 6.9, 5.5 MPa	4.40 (9.33)	41
	Ethanol	60,000 ppm, 25°C, pH 6.9, 5.5 MPa	2.67 (5.65)	97
	Urea	10,000 ppm, 25°C, pH 6.9, 5.5 MPa	6.46 (13.7)	85
	Phenol	10,000 ppm, 25°C, pH 5.2, 5.5 MPa	2.78 (5.89)	99.0
Cross-linked polyethylenimine				
F	NaCl	5000 ppm, 25°C, 4.1 MPa, $r = 0\%$	4.58 (9.71)	99
	Methanol	1000 ppm, 25°C, 4.1 MPa, $r = 0\%$		17
	Ethanol	1000 ppm, 25°C, 4.1 MPa, $r = 0\%$		61
	Urea	1000 ppm, 25°C, 4.1 MPa, $r = 0\%$		79
	Phenol	1000 ppm, 25°C, 4.1 MPa, $r = 0\%$		64
F	NaCl	5000 ppm, 25°C, 4.1 MPa, $r = 0\%$	3.16 (6.69)	97

TABLE 17a (continued).

Test Module[a]	Solute	Test Conditions	Flux $\times 10^4$ cm^3/cm^{2}s (gfd)	% Rejection
	Methanol	1000 ppm, 25°C, 4.1 MPa, $r = 0\%$		36
	Ethanol	1000 ppm, 25°C, 4.1 MPa, $r = 0\%$		84
	Urea	1000 ppm, 25°C, 4.1 MPa, $r = 0\%$		78
	Phenol	1000 ppm, 25°C, 4.1 MPa, $r = 0\%$		85
OTHER MEMBRANES				
TFC-LP				
S	NaCl	BW, 1.4 MPa net	10.3 (21.8)	97
	NaCl	5530 ppm, 2.8 MPa	9.43 (20)	98
	NaCl	35,000 ppm, 6.9 MPa	9.43-11.8 (20-25)	99.4
	Ethanol	700 ppm, 25°C, pH 4.7, 6.9 MPa		90
	Urea	1250 ppm, 25°C, pH 4.9, 6.9 MPa		80-85
	Phenol	100 ppm, 25°C, pH 4.9, 6.9 MPa		93
	Phenol	100 ppm, 25°C, pH 12, 6.9 MPa		>99

Notes: [ ] = Quantities in brackets represent module production of permeate water. SW = Seawater. BW = Brackish water. gfd = gallons/ft^{2}day. gpd = gallons/day. $r(\%)$ = % recovery. ppm = parts per million by weight.
[a]F—flat-sheet membrane; S—spiral-wound module; T—tubular membrane; HF—hollow-fiber module.
[b]Pure water flux.
TFC—Thin Film Composites
NF—Nano Filtration Membrane

TABLE 17b – Characteristics of Selected Nanofiltration Membranes.

Test Module[a]	Solute	Test Conditions	Flux x 10^4 cm^3/cm^2s (gfd)	% Rejection
NF40HF				
	NaCl	2000 ppm, 25°C, 0.9 MPa	12.0 (25.4)	40
	MgSO$_4$	2000 ppm, 25°C, 0.9 MPa		95
	Glucose	2000 ppm, 25°C, 0.9 MPa		90
NTR-7450				
	NaCl	5000 ppm, 25°C, 0.99 MPa	25.9 (55)[b]	51
	Na$_2$SO$_4$	5000 ppm, 25°C, 0.99 MPa		92
	MgSO$_4$	5000 ppm, 25°C, 0.99 MPa		32
	Glucose	25°C, 1 MPa		93
S	NaCl	1500 ppm, 25°C, 1.5 MPa	28.3 (60)	70-80
	MgSO$_4$	2000 ppm, 25°C, 1.5 MPa	11.1 (23.6)	>94
	Glucose	1000 ppm, 25°C, 1.5 MPa		85
MPT-20				
	NaCl+ low MW organics	35,000 ppm (5% organics), 45°C, 2.5 MPa	11.8 (25)	
	NaCl	50,000 ppm, 25°C, 2.5 MPa		0
	Glucose	10,000 ppm, 25°C, 2.5 MPa		75
Desal-5				
	NaCl	1000 ppm, 1 MPa	12.8 (27.1)	47
	Glucose	25°C, 1 MPa		83
DRC-1000				
	NaCl	3500 ppm, 1.0 MPa	13.9 (29.5)	10
HC50				
	NaCl	2500 ppm, 4.0 MPa	22.3 (47.2)	60

TABLE 17b (continued).

Test Module[a]	Solute	Test Conditions	Flux $\times 10^4$ cm^3/cm^{2}s (gfd)	% Rejection
NF-PES-10/PP 60				
	NaCl	5000 ppm, 4.0 MPa	111.3 (236)	15
NF-CA-50/PET 100				
	NaCl	5000 ppm, 4.0 MPa	33.4 (70.8)	55
SU200HF				
	NaCl	1500 ppm, 1.5 MPa	41.7 (88.5)	50
SU600				
	NaCl Glucose	500 ppm, 0.35 MPa 25°C, 1 MPa	7.79 (16.5)	55 93
SU700				
	Glucose	25°C, 1 MPa		99

Notes: [ ] = Quantities in brackets represent module production of permeate water. gfd = gallons/ft^{2}day. gpd = gallons/day. r% = % recovery. ppm = parts per million by weight.
[a] F—flat-sheet membrane; S—spiral-wound module; T—tubular membrane; HF—hollow-fiber module.
[b] Pure water flux.

Pump Users Handbook

4th Edition

By Ray Raynor

500 pages

numerous tables and diagrams

ISBN: 185617 216 3

Publication date: May 1995

This handbook places emphasis on the importance of correct interpretation of pumping requirements, both by the user and the supplier. Completely reworked to incorporate the very latest in pumping technology, this practical handbook will enable you to...

Avoid costly mistakes when ordering pump equipment

Understand the principles of pumping, hydraulics and fluids

Define the various criteria necessary for pump and ancillary selection

Recognise fundamental operational problems and avert failures

Contents include...

- Hydraulic fundamentals
- Process fluid properties
- Criteria for pump selection
- Head determination: static and friction
- Pump types/classifications
- Performance characteristics vs system head
- Specific speed - optimum BEP criteria
- Recirculation/cavitation
- Radial and Axial thrust loads
- Loss analysis
- Capacity control
- Types of impellers/casings
- Packing, seals and chambers
- Pipeline inertia and surge
- Vibration and balance

For further details post or fax a copy of this advert complete with your address details to:
Elsevier Advanced Technology, PO Box 150, Kidlington, Oxford OX5 1AS, UK
Tel: +44 (0) 1865 843848 Fax: +44 (0) 1865 843971

ELSEVIER
ADVANCED
TECHNOLOGY

APPENDIX

TABLE 19 – Membranes for Ultrafiltration.

Supplier/Trade Name	Module Type	Membrane Type	Separation Characteristics MWCO or Pore Size	Flux Characteristics Water Flux	Maximum Operating Conditions		
					Pressure (bar)	Temperature (°C)	pH Range
ALCOA/ Membralox	4- or 6-mm-i.d. tubes in multi-channel elements with 1-19 tubes/element; 0.2-3.6 m² area/element. Single tubes 7 mm and 15 mm i.d. also available	g-alumina on a-alumina support	4-100 nm	10 l/m²·h·bar at 20°C for 4-nm membrane	8-25 (may be limited by module construction)	300	
ALCON/Anopore	Disk	Al$_2$O$_3$	25 nm				
AMICON (WR Grace)	Flat sheet disks. Hollow-fiber modules. Fiber i.d. = 0.2, 0.5, 1.1 mm; length = 20, 64 cm. Module area = 0.3-0.5 ft²	Polysulfone		F = flux mL/min·cm² of sheet at 55 psi H = 0.3 ft² module, 1.1 mm i.d. typical flux mL/min		50	
			10K	2.5-4.0 (F) 40-100 (H)	25 psi for most modules		
			30K	6-10 (F) 40-100 (H)	16 psi for 0.2-mm-i.d. modules		
	Spiral-wound modules 0.9-10 ft² with 0.8-mm feed channel spacers. Available only with YM membranes		100K	170-220 (H)			

TABLE 19 (continued).

Supplier/Trade Name	Module Type	Membrane Type	Separation Characteristics MWCO or Pore Size	Maximum Operating Conditions			
				Flux Characteristics Water Flux	Pressure (bar)	Temperature (°C)	pH Range
ASAHI		Hydrophilic polymer	1K	0.02-0.035 (F)		75	
			5K	0.06-0.12 (F)			
		Acrylic vinyl polymer	50K	1.0-1.8 (F)		75	
			100K	0.4-2.0 (F)			
	0.8– or 1.4–mm–i.d. hollow-fiber modules; 35 or 112 cm long. Areas: 0.2-4.7 m²	Polyacrylonitrile		1/m²·h at 1 atm, 25°C for 4.7 m², 0.8-mm–i.d. module			
			6K	36	3	50	2-10
			7K	170	3	50	2-10
BERGHOF	BTU series: 11.5-mm tubular modules with 0.2-3.5 m² area/module	Noncellulosic polymers	20K	PEG soln. flux L/m²·h at pressure (bar)			
		BTU-1020		80/4	8	80	2-12
	BMK series: 0.6–, 1.1– or 1.5–mm–i.d. hollow-fiber modules with 0.2-4.0 m² area/module	Polyamide or polysulfone	2K, 10K, 30K, 50K, 100K	—	2	50-60	—
DAICEL	14.5-mm-i.d. flat sheet and tubular modules			F: flat sheet T: tubular L/m²·h	atm	10	45 —

APPENDIX

DDS	Plate-and-frame modules with 0.5- to 1.0-mm feed channel spacers	Polyacrylonitrile HH Polyethersulfone -40	5K 40K	13(F), 21 (T) 4(T)	10	90	—	
		Cellulose Acetate FS 60P	30K	$1/m^2 \cdot h$ at 5 atm, 20°C 200	15	80	0-14	
DESALINATION SYSTEMS INC/DSI	2-, 4-, and 8-in.-diameter spiral-wound modules, 12-40 in. long. Test modules: 6-20 ft², production modules: 60-350 ft², feed channel spacers: 30, 50, 90 mils. Available in tape or FRP wrap	Polysulfone type (E series)	Flat sheet % Rej. of 1% dextran at 7 psi 95% of 35K dextran 96% of 500K dextran	Test results water perm. $L/m^2 \cdot h \cdot atm$ 36-108 180-360	2-6 (limited by module)	100 (membrane) 50-60 (module)	1-13	
		TFC type (G series)	NaCl % rej. 85	gal/day 1400 (310 psi)	40 (module)	50	2-10	
DORR-OLIVER (WR Grace)/ Ioplate.	Plate-and-frame modules with 1.0- or 2.5-mm feed spacers for industrial processes	Cellulose C series	1K, 5K, 10K 30K, 100K	at 2 atm, 50°C $L/m^2 \cdot h$ 210 (10K) 500 (100K)	75	3.5-10		
		Dynel D series	50K, 100K	500 (50K) 850 (100K)	60	2-12		
		Polysulfone S-30	30K	700				
		Polyamide MFA series	20K, 200K	425	60	2-12		

TABLE 19 (continued).

Supplier/Trade Name	Module Type	Membrane Type	Separation Characteristics MWCO or Pore Size	Flux Characteristics Water Flux	Maximum Operating Conditions Pressure (bar)	Temperature (°C)	pH Range
FILTRON	Flat sheet disks & cassettes (individual or pre-assembled) (0.14-25.7 m² area) using either screen channel or open channel feed separators	NOVA series (polyethersulfone)	Nominal MWCO 1K, 3K, 5K, 8K, 10K, 30K, 50K, 100K.	Expected DI water flux in stirred cells at 25°C & 55 psi (mL/min/cm²)	Cassette operation		
					5	60	1-14
		PES, modified to resist fouling by antifoam agents)	Same as NOVA series plus 300K, 1000K.		5	60	1-12
FLUID SYSTEMS	3.8- to 4.3-in.-diameter spiral-wound modules 33-39 in. length. designed to 3A standards for dairy applications. Feedchannel spacers: 30, 41, 80 mils.	Polyethersulfone	6-10K	RO water flux 65 gal/ft²·day 50 psi, 20°C in modules	5.5 (75) 8 (50) 10 (25)	75	2-9 (75) 1-12(50) 1-13 (25)
GORE-TEX®	Disks, 3-mil thick	PTFE	0.02 μm (350 psi min. H₂O entry pressure)	1.2 mL/min MeOH at 27.5 in. Hg pressure drop & 21°C	cm²	to 315°C	-230
NADIR (HOECHST)	Flat-sheet membrane in either sheet stock, plate-and-frame, or spiral-wound module form						

APPENDIX

	MWCO/Avg. % Rejection of solute	Aq. solution flux L/m²·h			
PES (hydrophilic) UF-PES-4/PP 60	4K/75% of 5K inulin.	30-50 (1% inulin)	40	90	1-14
PES (hydrophobic)	25K/92% of 49K PVP	150-250 (2% PVP)	10	90	1-14
Polysulfone (Hydrophilic)	100K/85% of 2000K detran	300-500 (1% 2000K dextran)	10	90	1-14
Cellulose Acetate	1K/98% of 5K inulin	20-25 (1% inulin)	20	40	2-8
	100K/88% of 110K dextran	250-400 (1% dextran)	10	40	2-8
Regen. cellulose	30K/72% of 49K PVP	400-900 (2% PVP)	10	60	1-12
Aromatic Polyamide	20K/95% of 49K PVP	150-250 (2% PVP)	10	80	2-12
Polyvinylidene-fluoride	30K/75% of 49K PVP	250-400 (2% PVP)	15	80	1-10
Polyacrylonitrile	30K/80% of 49K PVP	500-1000 (2% PVP)	10	60	1-10

TABLE 19 (continued).

Supplier/Trade Name	Module Type	Membrane Type	Separation Characteristics MWCO or Pore Size	Flux Characteristics Water Flux	Maximum Operating Conditions		
					Pressure (bar)	Temperature (°C)	pH Range
	1-in. tubular membrane on PET support	Polysulfone types	30K/87% of 49K PVP	200-300 (2% PVP)	15	90	2-10
		Cellulose Acetate	10K/70% of 10K dextran	40-70 (1% dextran)	20	40	2-8
KALLE	Tubular and flat sheet	Cellulose acetate	2K-100K		20	10	2-8
		Polyamide	20K-100K		10	60	2-12
		Polysulfone	8K-25K		10	90	1-14
MILLIPORE	13- to 150-mm disks. Cassettes (0.5-5, 15, 25 ft² area) with screen (20, 50 mesh) or linear path feed spacers.		Nominal MWCO	Water flux gal/min at 50 psi, 25°C for 50 ft² spiral(2) (1)	80-100 psi For cassette operation	50	2-14
		Polyethersulfone	10K	2.0			2-14
		Polysulfone	300K				2-14
		Mixed cellulosic esters					2-14
		Cellulose	1K	0.45			4-8
		Polyvinylidene-fluoride	10K	2.0			2-13

Manufacturer	Module	Membrane	MWCO	Flux		Membrane limits at pH6 at 25°C**	2-14 (pH range is restricted in spirals).
KOCH MEMBRANE SYSTEMS	2-, 4-, and 8-in.–diameter spiral-wound modules with feed channel spacers from 20-80 mils. 0.5- and 1-in.-i.d. tubular modules available in various designs.	Polyethersulfone	300K	gfd at 50 psi, 40°C S: 4-in. spiral flux F: Flat sheet flux 30-50(S), 80-100(F)	10	90	1-13
			1-10K		10	90	1-11
		PVDF HFM-100	10-30K	50-90(S), 300-400(F) 1/m² h at 3.3 atm and 25°C			
		PVDF (cationic)	20-80K	~170	10	60	2-11
		PVDF (anionic)	110-600K	~1000	10	90	1-13
NITTO/ Hydranautics	4-in.–diameter spiral-wound modules	Polysulfone 3150	50K (Based on PEG)	386 gal/h in 4- x 40-in. spiral with RO water at 29 psi, 25°C	10	40 (limited by module)	2-11
	11.5-mm-i.d. tubular modules with 4 or 18 tubes/shell of length 1.3-2.9 m long	Hydrophilic Polyolefin Hydrophilic polyolefin	20K 20K, 100K	—	4-10	40	1-13 (25°C) 3-10 (40°C)
	11.5-mm-i.d., 18 tubes, 1.3 m long shell	Polyimide	8K, 20K	—	10	40	2-8

TABLE 19 (continued).

Supplier/Trade Name	Module Type	Membrane Type	Separation Characteristics MWCO or Pore Size	Flux Characteristics Water Flux	Maximum Operating Conditions Pressure (bar)	Temperature (°C)	pH Range
	Hollow-fiber module i.d./o.d. = 0.55/1.0 mm	Polysulfone Polysulfone type	8K, 20K, 100K. 20K	Water flux at 14.2 psi & 25°C >0.44 for 4.3 ft² module			
	i.d./o.d. = 1.1/1.9 mm. shell 3.5–11.4 cm diameter by 0.5–1 m. long	3250	20K	—do—			
OSMONICS/ SEPA	Spiral-wound modules 5, 10, 21 cm diameter and 66, 102 cm long. Feed spacers: 24, 34, 45 mils	Cellulose acetate		L/m² · hr, at 35 atm, 25°C			
		Polysulfone	1K	85	14	40	2-8.5
		Fluoro-polymer	2K	130	14	100	1-13
			2K	130	14	90	1-12
RHONE-POULENC/Iris	Flat sheet used in plate frame modules 0.5– and 1.5–mm feed spacers. Areas from 0.11-10 m²	PAN	copolymer		L/m² · h, at 2 atm, RT		
		3038	25K	830	—	50	3-10
		PVDF 3065	20K	1670	—	80	1-10
Carbosep (SFEC)	6-mm–i.d. tubular membranes up to 1.2 m long. Available in areas from 0.1-5.7 m²	Sulfonated Poly-fulfone 3026	15K	500	—	80	3-14
PCI Membrane	12.5-mm–i.d. tubular modules, 2 x 9 tubes in series, 3.66 m long	ZrO₂/carbon composite M1	60-80K	L/m² · h in tubular modules at		300°C	0-14

APPENDIX

Company	Description	Membrane	MWCO/pore	Flux			pH
		Polyethersulfone ES 404	4K	4 atm, 25°C	30	70	2-12
		Polysulfone PU 608	8K	800	1	70	2-12
		Polyacrylonitrile AN 620	25K	180	10	60	2-10
		PVDF FP 100	100K	4000	10	70	2-12
		Cellulose Acetate CA 407	7K	100	20	30	3-6
ROMICON	0.5– and 1.1-mm-i.d. hollow-fiber modules 31, 63, 109 cm long. Membrane areas from 0.17-4.9 M^2	Polysulfone	1K, 2K, 5K, 10K, 30K, 50K, 100K	Water flux at 1 bar ml/min. cm^2			
SARTORIUS	Flat-sheet membranes for stirred cell, plate-and-frame or centrifugal devices	Cellulose triacetate	5K 10K 20K	0.02 0.05 0.4			4-8
		Polysulfone	100K	1.6			
		Cell. Nitrate	10K, 50K				
		Cell. Acetate	20K, 70K, 160K				1-14
		Regen. Cellulose	20K, 70K, 160K				
SCHOTT/Bioran	Hollow-fiber 0.3-mm-i.d. 0.05–m^2 test module	Glass (>96% SiO_2) can be modified to be philic	10, 13.4, 19, 27, 44, 90 nm pore diameter	10 $L/m^2 \cdot h \cdot bar$ for 44-nm pore diameter			
TEIJIN	Tubular	Polysulfone Tu series			10	90	
WAFILIN	14–mm tubular modules	Non-cellulosic polymers 8010	20K	1/$m^2 \cdot h$ at 1 atm, 25°C 30-50	10		
ZENON	Tubular Modules: i.d./area 24 mm/2.2 ft^2	ZM-1	Low-medium	Waste water 100-200 gfd	4	40	2-12

TABLE 20 – Selected Water-Permeable Ethanol/Water Prevaporation Membranes.

Membrane Material	Ethanol Feed Concentration (wt.%)	Temperature (°C)	Selectivity (a)	Flux (kg/m$_2$-h) for H$_2$O
Polyvinylalcohol (GFT)	92 to 100	90 to 100	—	0 to 0.9
Polyvinylalcohol (GFT)	0 to 100	60	High	0 to 4
Polyvinylalcohol (GFT)	60 to 100	75 to 100	50 to 2000	0 to 2
Polyvinylalcohol	0 to 100	60	High	0 to 2.4
Cellulose acetate	0 to 100	25	5 to 12	0.1 to 0.5
Cellulose triacetate	5 to 95	20	1 to 3.6	0.3 to 1.2
Carboxymethylcellulose	81 to 95	25	2400 to 5900	0.005 to 0.1
Polysulfone (asymmetric)	15 to 95	20 to 50	3 to 6	—
Acrylic acid-acrylamide GPC	0 to 90	40	1 to 20	0 to 10
Polyacrylic acid-polycation	20 to 100	70	<1 to 2000	0.5 to 20
Polyvinylfluoride/acrylic acid	80	70	—	1.8
Polyvinylidenefluoride-N-vinylimidazole	0 to 95	70	—	0 to 6
Nafion™	30 to 98	40	Low	<0.5

TABLE 21 – Selected Organic-Permeable Organic/Water Pervaporation Membranes.

Membrane Material	Organic Feed Concentration (wt.%)	Temperature (°C)	Selectivity (a)	Organic Flux (kg/m^2-h)
Polypropylene	Acetone (45)	30	3	0.1 to 1.2
Silicone rubber	Butanol (0 to 8)	30	45 to 65	<0.035
Silicone rubber	IPA (27 to 100)	25	0.5 to 12	—
Silicone rubber	IPA (9 to 100)	25	9 to 22	0.03 to 0.11
Polyetheramides	Acetic acid (1.5 to 9)	50	—	0.18 to 0.28
Polyacrylic acid	Acetic acid (48)	15	2 to 8	0.4 to 0.55
Silicone rubber	Ethyl acetate (0.5 to 4)	30	High	—
GFT ethanol membrane (PDMS)	Ethanol (87 to 100)	60	150 to 10,000	0 to 1.6

APPENDIX

TABLE 22 – Unit conversion charts.
METRIC CONVERSION TABLE

Multiply	By	To obtain
Atmospheres	76.0	Centimetres of mercury
Atmospheres	29.92	Inches of mercury
Atmospheres	33.9	Feet of water
Atmospheres	10,333	Kilograms per square metre
Atmospheres	14.7	Pounds per square inch
Bar	0.9869	Atmospheres
Bar	1.020×10^{-4}	Kilograms per square metre
Bar	2,089	Pounds per square foot
Bar	14.5	Pounds per square inch (psi)
Centimetres	0.3937	Inches
Cubic centimetres	3.531×10^{-5}	Cubic feet
Cubic centimetres	2.642×10^{-4}	Gallons
Cubic centimetres	2.199×10^{-4}	Imperial gallons
Cubic feet	1728	Cubic inches
Cubic feet	0.02832	Cubic metres
Cubic feet	7.48052	Gallons
Cubic feet	6.229	Imperial gallons
Cubic feet	28.32	Litres
Cubic metres	35.31	Cubic feet
Cubic metres	1.308	Cubic yards
Cubic metres	264.2	Gallons
Cubic metres	219.969	Imperial gallons
Feet	12	Inches
Feet	0.3048	Metres
Feet of water	0.0295	Atmospheres
Feet of water	0.8826	Inches of mercury
Feet of water	304.8	Kilograms per square metre
Feet of water	62.43	Pounds per square foot
Gallons	0.1337	Cubic feet
Gallons	3.785×10^{-3}	Cubic metres
Imperial gallons	4564	Cubic centimetres
Imperial gallons	0.1605	Cubic feet
Imperial gallons	4.546×10^{-3}	Cubic metres
Imperial gallons	5.946×10^{-3}	Cubic yards
Imperial gallons	4.546	Cubic liters
Grams	0.03527	Ounces
Grams	0.03215	Troy ounces

TABLE 22 (continued).

METRIC CONVERSION TABLE – contd.

Multiply	By	To obtain
Inches	2.54	Centimetres
Inches of mercury	0.03342	Atmospheres
Inches of mercury	1.133	Feet of water
Inches of mercury	345.3	Kilograms per square metre
Inches of mercury	70.73	Pounds per square foot
Inches of water	0.002458	Atmospheres
Inches of water	0.07355	Inches of mercury
Inches of water	25.4	Kilograms per square metre
Inches of water	5.202	Pounds per square foot
Kilograms	2.205	Pounds
Kilograms per square metre	9.678×10^{-5}	Atmospheres
Kilograms per square metre	98.07×10^{-6}	Bar
Kilograms per square metre	3.281×10^{-3}	Feet of water
Kilograms per square metre	2.896×10^{-3}	Inches of mercury
Kilograms per square metre	0.2048	Pounds per square foot
Litres	0.03531	Cubic feet
Litres	10^{-3}	Cubic metres
Litres	0.2642	Gallons
Litres	0.21998	Imperial gallons
Millimetres	0.03937	Inches
Millimetres	39.37	Millilitres
Millilitres	0.002540	Centimetres
Millilitres	10^{-3}	Inches
Pounds per square inches	0.06804	Atmospheres
Pounds per square inches	2.307	Feet of water
Pounds per square inches	2.036	
Pounds per square inches	703.1	Kilograms per square metre
Square centimetres	1.076×10^{-3}	Square feet
Square feet	144	Square inches
Square feet	0.0929	Square metres
Square feet	3.587×10^{-8}	Square miles
Square inches	645.2	Square millimetres

TABLE 22 (continued).

BASIC EQUIVALENTS

Imperial gallons	Pints	Quarts	US gallons	Cubic inches	Cubic feet	Cubic yards	Litres
1	8	4	1.20095	277.42	0.16054	0.005946	4.54596
0.125	1	0.5	0.15012	34.68	0.02007	0.000743	0.56825
0.25	2	1	0.30024	69.355	0.04014	0.001486	1.13649
0.83267	6.66136	3.33068	1	231.0	0.13458	0.004951	3.785
0.003605	0.02884	0.01442	0.004329	1	0.000579	0.00002143	0.016387
6.2288	49.831	24.915	7.48	1728	1	0.03704	28.3161
168.178	1355.424	672.712	201.974	46,656	27	1	0.7645
0.219975	1.7598	0.8799	0.2642	61.026	0.035316	0.001308	1

VOLUME CONVERSIONS – CUBIC FEET TO CUBIC METRES

Cubic feet	0	10	20	30	40	50	60	70	80	90
	—	—	0.28317	0.56634	0.84951	1.13267	1.41584	1.69901	1.98218	2.26535	2.54852
100	2.83169	3.11485	3.39802	3.68119	3.96436	4.24753	4.53070	4.81387	5.0970	5.3802
200	5.6634	5.9465	6.2297	6.5129	6.7960	7.0792	7.3624	7.6456	7.9287	8.2119
300	8.4951	8.7782	9.0614	9.3446	9.6277	9.9109	10.1941	10.4772	10.7604	11.0436
400	11.3267	11.6099	11.8931	12.1762	12.4594	12.7426	13.0258	13.3089	13.5921	13.8753
500	14.1584	14.4416	14.7248	15.0079	15.2911	15.5743	15.8574	16.1406	16.4238	16.7069
600	16.9901	17.2733	17.5565	17.8396	18.1228	18.4060	18.6891	18.9723	19.2555	19.5386
700	19.8218	20.1050	20.3881	20.6713	20.9545	21.2376	21.5208	21.8040	22.0872	22.3703
800	22.6535	22.9367	23.2198	23.5030	23.7862	24.0693	24.3525	24.6357	24.9188	25.2020
900	25.4852	25.7683	26.0515	26.3347	26.6178	26.9010	27.1842	27.4674	27.7505	28.0337
1000	28.3169	—	—	—	—	—	—	—	—	—

VOLUME CONVERSION – CUBIC METRES TO CUBIC FEET

Cubic metres	0	1	2	3	4	5	6	7	8	9
0	—	35.3147	70.629	105.944	141.259	176.573	211.888	247.203	282.517	317.832
10	353.147	388.461	423.776	459.091	494.405	529.72	565.03	600.35	635.66	670.98
20	706.29	741.61	776.92	812.24	847.55	882.87	918.18	953.50	988.81	1024.13
30	1059.44	1094.75	1130.07	1165.38	1200.70	1236.01	1271.33	1306.64	1341.96	1377.27
40	1412.59	1447.90	1483.22	1518.53	1553.84	1589.16	1624.47	1659.79	1695.10	1730.42
50	1765.73	1801.05	1836.36	1871.68	1906.99	1942.31	1977.62	2012.94	2048.25	2083.56
60	2118.88	2154.19	2189.51	2224.82	2260.14	2295.45	2330.77	2366.08	2401.40	2436.71
70	2472.03	2507.34	2546.66	2577.97	2613.28	2648.60	2683.91	2719.23	2754.54	2789.86
80	2825.17	2860.49	2895.80	2931.12	2966.43	3001.75	3037.06	3072.38	3107.69	3143.00
90	3178.32	3213.63	3248.95	3284.26	3319.58	3354.89	3390.21	3425.52	3460.84	3496.15
100	3531.47	—	—	—	—	—	—	—	—	—

SECTION 21

Advertisers Index with Addresses, Telephone and Facsimile Numbers

Classified Index of Advertisers by Product Category

Trade Names Index

Editorial Index

ADVERTISERS INDEX WITH ADDRESSES, TELEPHONE AND FACSIMILE NUMBERS

A/G Technology Corporation, 101 Hampton Avenue, Needham, MA 02194-2628, USA
Tel: +1 617 449 5774   Fax: +1 617 449 5786 .... Facing p. 25

Aqualytics, Division of the Graver Company, 7 Powder Horn Drive, Warren, NJ 07059-5191, USA
Tel: +1 908 563 2800   Fax: +1 908 563 2816 .... Facing p. 691

Aquilo Gas Separation B.V., Oudekerhstraat 4, 4878 AA, Ettenheur, The Netherlands
Tel: +31 7650 85300   Fax: +31 7650 85333 .... Facing p. 271

atech innovations gmbh, Teilungsweg 28, 45329 Essen, Germany
Tel: +49(0)201/34 10 24   Fax: +49(0) 201/34 10 26 .... p. 632

domnick hunter, Durham Road, Birtley, Co. Durham, DH3 3SF, UK
Tel: +44 (0)191 410 5121   Fax: +44 (0)191 410 5312 .... Facing p. 379

Fairey Industrial Ceramics Ltd, Filleybrooks, Stone, Staffordshire, ST15 OPU, UK
Tel: +44 (0)1785 813241   Fax: +44 (0)1785 818733 .... Facing p. 373

Gelman Sciences Ltd, Brackmills Business Park, Caswell Road, Northampton, NN4 7EZ, UK
Tel: +44 (0)1604 765141   Fax: +44 (0)1604 761383 .... p. 7

Graver Separations, Inc., 200 Lake Drive, Glasgow, Delaware 19702, USA
Tel: +1 302 731 1700   Fax: +1 302 731 1707 .... Facing p. 5

Hoechst Aktiengesellschaft GFP. Membranen, Building D512,
Rheingastr. 190, 65174 Wiesbaden, Germany
Tel: +49 611 962 6418   Fax: +49 611 962 9237 .... Facing p. 188

Inceltech, 15 allees de Bellefontaine, Toulouse 31100, France
Tel: +33 61 40 8585   Fax: +33 61 41 5178 .... Facing p. 95

Koch Membrane Systems, Inc., 850 Main Street, Wilmington,
MA 01887-3388, USA
Tel: +1 508 657 4250   Fax: +1 508 657 5208 .... Facing p. 195

Membrane Products Kiryat Weizmann Ltd, P.O. Box 138,
Rehovot 76101, Israel
Tel: +972 8 407557   Fax: +972 8 407556 .... Facing p. 189

Microdyn Modulbau GmbH & Co. KG, Ohder Str. 28,
42289 Wuppertal, Germany
Tel: +49 202 602092   Fax: +49 202 603087 .... Facing p. 373

Millipore S.A., 39 route de la Hardt, BP 116, Molsheim 67120,
France
Tel: +33 88 38 90 00   Fax: +33 88 38 91 93 .... Facing p. 378

New Logic International, 1155 Park Avenue, Emeryville,
CA 94608, USA
Tel: +1 510 655 7305   Fax: +1 510 655 7307 .... Facing p. 94

Osmonics, 5951 Clearwater Drive, Minnetonka, Minnesota,
55343-8995, USA
Tel: +1 612 933 2277   Fax: +1 612 933 0141 .... Facing p. 24

Osmota Membratechnik GmbH, Jahnstrasse 4/1, Postbox 1368,
D-70809, Germany
Tel: +49 0711 831091   Fax: +49 0711 834755 .... p. 8

Refractron Technologies, 5750 Stuart Avenue, Newark, New York,
14513, USA
Tel: +1 315 331 6222   Fax: +1 315 331 7254 .... Facing p. 196

SCT, BP1, 65460 Bazet, France
Tel: +33 62 389595   Fax: +33 62 389550 .... Facing p. 187

Serck Baker, 380 Bristol Road, Gloucester, GL2 6X7, UK
Tel: +44 (0)1452 421561   Fax: +44 (0)1452 423414 .... Facing p. 25

Tech Sep, 5, Chemin du Pilon, Saint-Maurice de Beynost - 01703
Miribel F-01703, France
Tel: +33 7201 2727   Fax: +33 7225 8899 .... Facing p. 4

Union Filtration a/s, Sandvikenvej 7, DK-4900 Nakskov, Denmark
Tel: +45 54 95 13 00   Fax: +45 54 95 13 01 .... p. 8

X-Flow BV, Bedrijvenpark Twente 289, NL-7602, KK Almelo,
The Netherlands
Tel: +31 5496 75202   Fax: +31 5496 75102 .... Facing p. 195

CLASSIFIED INDEX OF ADVERTISERS BY PRODUCT CATEGORY

MEMBRANE TYPES

BIPOLAR MEMBRANES
Aqualytics

DEHYDRATION
Aquilo Gas Separation B.V.
Graver Separations Inc.

DIALYSIS
Gelman Sciences
Union Filtration a/s

ELECTRODIALYSIS
Aqualytics

GAS PERMEATION
Aquilo Gas Separation B.V.
Serck Baker

LIQUID MEMBRANES
Graver Separations Inc.
Refractron Technologies Corp.

MICROFILTRATION
A/G Technology Corporation
atech innovations gmbh
domnick hunter
Fairey Industrial Ceramics Ltd
Gelman Sciences Ltd
Graver Separations Inc.
Inceltech
Koch Membrane Systems Inc.
Microdyn
Millipore S.A.
New Logic
Osmonics
Refractron Technologies Corp.
SCT
Serck Baker
Union Filtration a/s

NANO FILTRATION
Graver Separations Inc.
Hoechst
Membrane Products Kiryat Weizmann Ltd
Millipore S.A.
New Logic
Osmonics
SCT
Serck Baker
Union Filtration a/s

PERVAPORATION
Aquilo Gas Separation B.V.
Hoechst

REVERSE OSMOSIS
Millipore S.A.
New Logic

Osmonics
Serck Baker
Union Filtration a/s

ULTRA FILTRATION
A/G Technology Coroporation
atech innovations gmbh
Gelman Sciences Ltd
Graver Separations Inc.
Hoechst
Inceltech
Koch Membrane Systems Inc.
Membrane Products Kiryat Weizmann Ltd
Millipore S.A.
New Logic
Osmonics
SCT
Serck Baker
Union Filtration a/s

VAPOUR PERMEATION
Aquilo Gas Separation B.V.
Refractron Technologies Corp.

GAS SEPARATION
A/G Technology Corporation

MODULES FOR MEMBRANE TYPES

BIPOLAR MEMBRANES
Aqualytics

DEHYDRATION
Aquilo Gas Separation B.V.

DIALYSIS
Aqualytics
Union Filtration a/s

ELECTRODIALYSIS
Aqualytics

GAS PERMEATION
Aquilo Gas Separation B.V.
Serck Baker

LIQUID MEMBRANES
Graver Separations Inc.

MICROFILTRATION
A/G Technology Corporation
atech innovations gmbh
Fairey Industrial Ceramics Ltd
Gelman Sciences Ltd
Graver Separations Inc.
Inceltech
Koch Membrane Systems Inc.
Microdyn
Millipore S.A.
Osmonics
SCT
Serck Baker
Union Filtration a/s

NANO FILTRATION
Graver Separations Inc.
Hoechst
Membrane Products Kiryat Weizmann Ltd
Millipore S.A.
Osmonics
SCT
Serck Baker
Union Filtration a/s

PERVAPORATION
Aquilo Gas Separation B.V.
Hoechst

REVERSE OSMOSIS
Millipore S.A.
Osmonics
Serck Baker
Union Filtration a/s

ULTRA FILTRATION
A/G Technology Corporation
atech innovations gmbh
Graver Separations Inc.
Hoechst
Inceltech
Koch Membrane Systems Inc.
Membrane Products Kiryat Weizmann Ltd

Millipore S.A.
Osmonics
SCT
Serck Baker
Union Filtration a/s

VAPOUR PERMEATION
Aquilo Gas Separation B.V.
Hoechst

MODULE CONFIGURATIONS

CARTRIDGE
Gelman Sciences Ltd
Graver Separations Inc.
Inceltech
Microdyn
Millipore S.A.

CASSETTE
Gelman Sciences Ltd
Hoechst
Millipore S.A.
Serck Baker

ELECTRODIALYSIS
Aqualytics

FILTERS
Gelman Sciences Ltd
Hoechst
Inceltech
Microdyn
Millipore S.A.
Osmonics
Serck Baker

FLAT SHEET
Gelman Sciences Ltd
Hoechst
Koch Membrane Systems Inc.
New Logic
Osmonics
Union Filtration a/s

HOLLOW FIBRE
A/G Technology Corporation
Aquilo Gas Separation B.V.

Hoechst
Inceltech
Koch Membrane Systems Inc.
Serck Baker
Union Filtrtion a/s

PLATE AND FRAME
Aqualytics
New Logic
Serck Baker
Union Filtration a/s

SPIRAL WOUND
Hoechst
Koch Membrane Systems Inc.
Membrane Products Kiryat Weizmann Ltd
Millipore S.A.
Osmonics
Serck Baker
Union Filtration a/s

TUBULAR
atech innovations gmbh
Fairey Industrial Ceramics Ltd
Graver Separations Inc.
Hoechst
Inceltech
Koch Membrane Systems Inc.
Membrane Products Kiryat Weizmann Ltd
Microdyn
SCT
Union Filtration a/s

VIBRATORY SHEAR
New Logic

TUBULE
A/G Technology Corporation

MEMBRANE MATERIALS

CELLULOSE
domnick hunter
Hoechst
Inceltech
Koch Membrane Systems Inc.
Millipore S.A.
New Logic

Osmonics
Serck Baker
Union Filtration a/s

CERAMIC (ALUMNIA ZIRCONIA)
atech innovations gmbh
Fairey Industrial Ceramics Ltd
Graver Separations Inc.
Inceltech
Refractron Technologies Corp.
SCT
Union Filtration a/s

METALS
domnick hunter
Graver Separations Inc.

POLYCARBONATE
New Logic
Osmonics

POLYDIMETHYSILOZANE
Serck Baker

POLYETHERIMIDE
Serck Baker

POLYIMIDE
Inceltech
Koch Membrane Systems Inc.
New Logic
Osmonics
Serck Baker

POLYPROPYLENE
Gelman Sciences Ltd
Hoechst
Microdyn
Millipore S.A.
New Logic
Osmonics
Union Filtration a/s

POLYSULPHONE
A/G Technology Corporation
Aqualytics
Gelman Sciences Ltd
Hoechst
Inceltech

Koch Membrane Systems Inc.
Membrane Products Kiryat Weizmann Ltd
New Logic
Osmonics
Union Filtration a/s

P.T.F.E
domnick hunter
Gelman Sciences Ltd
Millipore S.A.
New Logic
Osmonics

P.V.D.F
Gelman Sciences Ltd
Koch Membrane Systems Inc.
Millipore S.A.
New Logic
Osmonics
Union Filtration a/s

ACRYLIC
Koch Membrane Systems Inc.

SIC
atech innovations gmbh

POLYETHERSULPHON
Aquilo Gas Separation B.V.

POLYPHENYLEENOXIDE
Aquilo Gas Separation B.V.

POLYSTYRENE
Aqualytics

APPLICATIONS/INDUSTRIES

BEVERAGES
A/G Technology Corporation
atech innovations gmbh
Aquilo Gas Separation B.V.
domnick hunter
Fairey Industrial Ceramics Ltd
Gelman Sciences Ltd
Graver Separations Inc.
Hoechst
Inceltech

CLASSIFIED INDEX OF ADVERTISERS BY PRODUCT CATEGORY

Koch Membrane Systems Inc.
Membrane Products Kiryat Weizmann Ltd
Microdyn
Millipore S.A.
Osmonics
Refractron Technologies Corp.
SCT
Union Filtration a/s

BIOCHEMICAL/PHARMACEUTICAL
A/G Technology Corporation
Aqualytics
atech innovations gmbh
domnick hunter
Fairey Industrial Ceramics Ltd
Gelman Sciences Ltd
Graver Separations Inc.
Hoechst
Inceltech
Koch Membrane Systems Inc.
Membrane Products Kiryat Weizmann Ltd
Millipore S.A.
Osmonics
Refractron Technologies Corp.
SCT
Union Filtration a/s

BREWERY
atech innovations gmbh
Aquilo Gas Separation B.V.
domnick hunter
Fairey Industrial Ceramics Ltd
Gelman Sciences Ltd
Graver Separations Inc.
Hoechst
Inceltech
Koch Membrane Systems Inc.
Membrane Products Kiryat Weizmann Ltd
Microdyn
Millipore S.A.
Osmonics
Refractron Technologies Corp.
SCT
Union Filtration a/s

CHEMICAL PROCESS
A/G Technology Corporation
Aqualytics
Aquilo Gas Separation B.V.

atech innovations gmbh
domnick hunter
Gelman Science Ltd
Graver Separations Inc.
Hoechst
Koch Membrane Systems Inc.
Membrane Products Kiryat Weizmann Ltd
Microdyn
New Logic
Osmonics
Refractron Technologies Corp.
SCT
Serck Baker
Union Filtration a/s

CHLOR-ALKALI
Aqualytics
Refractron Technologies Corp.

DAIRY
Aqualytics
atech innovations gmbh
domnick hunter
Fairey Industrial Ceramics Ltd
Gelman Sciences Ltd
Graver Separations Inc.
Hoechst
Inceltech
Koch Membrane Systems Inc.
Membrane Products Kiryat Weizmann Ltd
Millipore S.A.
Osmonics
Refractron Technologies Corp.
SCT
Union Filtration a/s

DISTILLERS
atech innovations gmbh
domnick hunter
Gelman Sciences Ltd
Graver Separations Inc.
Millipore S.A.
Refractron Technologies Corp.

ELECTRONIC
atech innovations gmbh
domnick hunter
Gelman Sciences Ltd
Graver Separations Inc.

Hoechst
Microdyn
Millipore S.A.
New Logic
Osmonics
SCT
Serck Baker

FOOD
Aqualytics
Aquilo Gas Separation B.V.
atech innovations gmbh
domnick hunter
Fairey Industrial Ceramics Ltd
Gelman Sciences Ltd
Graver Separations Inc.
Hoechst
Inceltech
Koch Membrane Systems Inc.
Membrane Products Kiryat Weizmann Ltd
Microdyn
Millipore S.A.
Osmonics
Refractron Technologies Corp.
SCT
Serck Baker
Union Filtration a/s

FUELS
Aquilo Gas Separation B.V.
Fairey Industrial Ceramics Ltd
Graver Separations Inc.
Refractron Technologies Corp.
SCT
Serck Baker

MEDICAL
A/G Technology Corporation
atech innovations gmbh
domnick hunter
Gelman Sciences Ltd
Graver Separations Inc.
Hoechst
Inceltech
Microdyn
Millipore S.A.
Osmonics
Union Filtration a/s

METALLURGICAL
Aqualytics
atech innovations gmbh
Graver Separations Inc.
Hoechst
New Logic
Osmonics
Refractron Technologies Corp.

NUCLEAR
atech innovations gmbh
Graver Separations Inc.
Microdyn
Millipore S.A.
New Logic
Refractron Technologies Corp.
SCT

PAPER
atech innovations gmbh
Fairey Industrial Ceramics Ltd
Graver Separations Inc.
Hoechst
Koch Membrane Systems Inc.
Membrane Products Kiryat Weizmann Ltd
Microdyn
New Logic
Osmonics
Refractron Technologies Corp.
SCT
Serck Baker
Union Filtration a/s

PETROCHEMICAL
Aqualytics
Aquilo Gas Separation B.V.
atech innovations gmbh
Graver Separations Inc.
Hoechst
Membrane Products Kiryat Weizmann Ltd
New Logic
Refractron Technologies Corp.
SCT
Serck Baker

PULP
Aqualytics

CLASSIFIED INDEX OF ADVERTISERS BY PRODUCT CATEGORY

atech innovations gmbh
Graver Separations Inc.
Hoechst
Koch Membrane Systems Inc.
Membrane Products Kiryat Weizmann Ltd
New Logic
SCT
Serck Baker
Union Filtration a/s

TEXTILES
Fairey Industrial Ceramics Ltd
Graver Separations Inc.
Hoechst
Inceltech
Koch Membrane Systems Inc.
Microdyn
New Logic
Osmonics
Refractron Technologies Corp.
SCT
Serck Baker
Union Filtration a/s

UTILITIES
Aqualytics
Graver Separations Inc.
Hoechst
Koch Membrane Systems Inc.
Millipore S.A.
New Logic
Osmonics
Refractron Technologies Corp.
SCT
Serck Baker

WINE
A/G Technology Corporation
Aqualytics
Aquilo Gas Separation B.V.
atech innovations gmbh
domnick hunter
Fairey Industrial Ceramics Ltd
Gelman Sciences Ltd
Graver Separations Inc.
Hoechst
Koch Membrane Systems Inc.
Microdyn

Millipore S.A.
Osmonics
Refractron Technologies Corp.
SCT
Union Filtration a/s

BIOTECHNOLOGY
Inceltech

KAOLIN CLAY DEWATERING
New Logic

TITANIUM DIOXIDE DEWATERING
New Logic

CALCIUM CARBONATE DEWATERING
New Logic

COAL FINES DEWATERING
New Logic

CAR INDUSTRY
Hoechst

GALVANIC INDUSTRY
Union Filtration a/s

MACHINE COOLANT RECYCLE
A/G Technology Corporation

ALKALINE CLEANER BATH RECYCLE
A/G Technology Corporation

OIL/WATER SEPARATION
A/G Technology Corporation

FILTRATION

ANALYTICAL APPLICATIONS
atech innovations gmbh
Fairey Industrial Ceramics Ltd
Gelman Sciences Ltd
Graver Separations Inc.
Koch Membrane Systems Inc.
Millipore S.A.
Osmonics
Refractron Technologies Corp.

BOILER FEED (POWER STATION)
Hoechst
Koch Membrane Systems Inc.
Millipore S.A.
New Logic
Osmonics
Refractron Technologies Corp.
SCT
Serck Baker
Union Filtration a/s

CELL HARVESTING
A/G Technology Corporation
Fairey Industrial Ceramics Ltd
Gelman Sciences Ltd
Graver Separations Inc.
Hoechst
Inceltech
Koch Membrane Systems Inc.
Microdyn
Millipore S.A.
SCT

CEMENT SLURRY
atech innovations gmbh
Graver Separations Inc.
Hoechst

CLARIFICATION
A/G Technology Corporation
atech innovations gmbh
domnick hunter
Fairey Industrial Ceramics Ltd
Graver Separations Inc.
Hoechst
Inceltech
Koch Membrane Systems Inc.
Microdyn
Millipore S.A.
New Logic
Osmonics
SCT
Serck Baker
Union Filtration a/s

COMPRESSED AIR
domnick hunter
Refractron Technologies Corp.

DAIRY
atech innovations gmbh
domnick hunter
Fairey Industrial Ceramics Ltd
Graver Separations Inc.
Hoechst
Inceltech
Koch Membrane Systems Inc.
Membrane Products Kiryat Weizmann Ltd
Millipore S.A.
Osmonics
Refractron Technologies Corp.
SCT
Union Filtration a/s

DOMESTIC WATER
atech innovations gmbh
Fairey Industrial Ceramics Ltd
Gelman Sciences Ltd
Hoechst
New Logic
Osmonics
Refractron Technologies Corp.
Serck Baker
Union Filtration a/s

ELECTROPAINT
Koch Membrane Systems Inc.
Graver Separations Inc.
Hoechst
New Logic
Osmonics

EMULSIONS
atech innovations gmbh
Fairey Industrial Ceramics Ltd
Graver Separations Inc.
Hoechst
Koch Membrane Systems Inc.
Microdyn
New Logic
Osmonics
SCT
Serck Baker
Union Filtration a/s

FOOD INDUSTRY
atech innovations gmbh

domnick hunter
Fairey Industrial Ceramics Ltd
Gelman Sciences Ltd
Graver Separations Inc.
Hoechst
Inceltech
Koch Membrane Systems Inc.
Membrane Products Kiryat Weizmann Ltd
Microdyn
Millipore S.A.
Osmonics
Refractron Technologies Corp.
SCT
Serck Baker
Union Filtrtaion a/s

METALLURGY
atech innovations gmbh
Graver Separations Inc.
Hoechst
Membrane Products Kiryat Weizmann Ltd
Microdyn
Osmonics
SCT

RADIOACTIVE LIQUID
atech innovations gmbh
Graver Separations Inc.
Microdyn
New Logic
Refractron Technologies Corp.
SCT

STERILISATION
atech innovations gmbh
domnick hunter
Gelman Sciences Ltd
Graver Separations Inc.
Hoechst
Microdyn
Millipore S.A.
Osmonics
Refractron Technologies Corp.
SCT

ULTRAPURE WATER
Gelman Sciences Ltd
Hoechst
Koch Membrane Systems Inc.

Millipore S.A.
New Logic
Osmonics
SCT
Serck Baker

DYES & PIGMENTS
Koch Membrane Systems Inc.

OILY WATER
Fairey Industrial Ceramics Ltd

MINERALS
Fairey Industrial Ceramics Ltd

MEDICAL

DIAFILTRATION
Graver Separations Inc.
SCT

DIALYSIS
Gelman Sciences Ltd
Microdyn

HEMODIALYSIS
Gelman Sciences Ltd
Union Filtration a/s

STERILE FILTRATION
atech innovations gmbh
domnick hunter
Gelman Sciences Ltd
Graver Separations Inc.
Hoechst
Inceltech
Millipore S.A.
SCT

SPECIALTY DEVICES
A/G Technology Corporation

BIOMEDICAL/ FERMENTATIONS

ANTIBIOTICS
Fairey Industrial Ceramics Ltd

Gelman Sciences Ltd
Graver Separations Inc.
Hoechst
Inceltech
Koch Membrane Systems Inc.
Membrane Products Kiryat Weizmann Ltd
Millipore S.A.
SCT
Union Filtration a/s

BIOREACTORS
atech innovations gmbh
Fairey Industrial Ceramics Ltd
Gelman Sciences Ltd
Graver Separations Inc.
Hoechst
Inceltech
Koch Membrane Systems Inc.
Millipore S.A.
Refractron Technologies Corp.
SCT
Union Filtration a/s

ENZYME REACTORS
Fairey Industrial Ceramics Ltd
Gelman Sciences Ltd
Graver Separations Inc.
Hoechst
Koch Membrane Systems Inc.
Microdyn
Millipore S.A.
Union Filtration a/s

FACILITIES FILTRATION
Fairey Industrial Ceramics Ltd
Gelman Sciences Ltd
Graver Separations Inc.
Hoechst
Koch Membrane Systems Inc.
Millipore S.A.
SCT

SEPARATION OF MICROBIAL CELLS
A/G Technology Corporation
atech innovations gmbh
Fairey Industrial Ceramics Ltd
Gelman Sciences Ltd
Graver Separations Inc.

Hoechst
Inceltech
Koch Membrane Systems Inc.
Millipore S.A.
Refractron Technologies Corp.
SCT
Union Filtration a/s

AMINO ACIDS
Union Filtration a/s

MAMMALIAN CELLS
A/G Technology Corporation
Fairey Industrial Ceramics Ltd
Gelman Sciences Ltd
Graver Separations Inc.
Hoechst
Inceltech
Koch Membrane Systems Inc.
Millipore S.A.

ORGANIC /AMINO ACIDS
Aqualytics

LABORATORY

ANALYSIS
atech innovations gmbh
Gelman Sciences Ltd
Graver Separations Inc.
Millipore S.A.

DE-IONISED WATER
Gelman Sciences Ltd
Osmonics

OPTHALMICS
domnick hunter
Gelman Sciences Ltd
Millipore S.A.
SCT

PURE WATER
domnick hunter
Gelman Sciences Ltd
Hoechst
Koch Membrane Systems Inc.

Millipore S.A.
Osmonics
SCT
Serck Baker
Union Filtration a/s

REAGENT GRADE WATER
Gelman Sciences Ltd
Millipore S.A.

S.V.P. FILTRATION
Gelman Sciences Ltd
Millipore S.A.
SCT

SMALL VOLUME CONCENTRATION
A/G Technology Corporation

CONTINUOUS SAMPLING
A/G Technology Corporation

DESALINATION

BOILER FEED WATER
Gelman Sciences Ltd
Koch Membrane Systems Inc.
New Logic
Osmonics
Refractron Technologies Corp.
Serck Baker
Union Filtration a/s

BRACKISH WATER
Aqualytics
Fairey Industrial Ceramics Ltd
Gelman Sciences Ltd
New Logic
Osmonics
Refractron Technologies Corp.
Serck Baker
Union Filtration a/s

DRINKING WATER
Aqualytics
atech innovations gmbh
Gelman Sciences Ltd
Koch Membrane Systems Inc.
New Logic

Osmonics
Serck Baker
Union Filtration a/s

SEA WATER
Fairey Industrial Ceramics Ltd
Gelman Sciences Ltd
New Logic
Osmonics
Serck Baker
Union Filtration a/s

ULTRAPURE WATER
Gelman Sciences Ltd
Koch Membrane Systems Inc.
New Logic
Osmonics
Serck Baker

WASTE WATER
Aqualytics
Fairey Industrial Ceramics Ltd
Gelman Sciences Ltd
Graver Separations Inc.
Koch Membrane Systems Inc.
New Logic
Osmonics
Refractron Technologies Corp.
Serck Baker
Union Filtration a/s

SALT WASTES
Aqualytics

WASTE WATER AND EFFLUENT TREATMENT

COOLING WATER
atech innovations gmbh
Graver Separations Inc.
Hoechst
Koch Membrane Systems Inc.
New Logic
Osmonics
Refractron Technologies Corp.
SCT
Serck Baker

HEAVY METALS
Fairey Industrial Ceramics Ltd
Graver Separations Inc.
Hoechst
Koch Membrane Systems Inc.
Membrane Products Kiryat Weizmann Ltd
Microdyn
New Logic
Osmonics
SCT
Union Filtration a/s

LAND FILL
Hoechst
Microdyn
New Logic

LEACHATE
atech innovations gmbh
Fairey Industrial Ceramics Ltd
Graver Separations Inc.
Hoechst
New Logic
Osmonics
SCT

SEWAGE (EFFLUENT)
atech innovations gmbh
Fairey Industrial Ceramics Ltd
Hoechst
Membrane Products Kiryat Weizmann Ltd
Microdyn
New Logic
SCT
Serck Baker
Union Filtration a/s

SEWAGE (RAW)
Hoechst
Microdyn
New Logic
SCT

SEWAGE (SLUDGE)
Hoechst
New Logic
Union Filtration a/s

SEWAGE (TREATED)
Fairey Industrial Ceramics Ltd
Hoechst
New Logic
Refractron Technologies Corp.
SCT
Serck Baker

SLURRY
Graver Separations Inc.
Hoechst
New Logic
SCT

SALT, ACID & BASE WASTES
Aqualytics

GASES

AIR
Aquilo Gas Separation B.V.
Gelman Sciences Ltd
Millipore S.A.
Refractron Technologies Corp.

CO_2 SEPARATION
Aquilo Gas Separation B.V.
Serck Baker

GAS FILTRATION
domnick hunter
Gelman Sciences Ltd
Millipore S.A.
Refractron Technologies Corp.
SCT

GAS STERILISATION
domnick hunter
Gelman Sciences Ltd
Millipore S.A.

HELIUM RECOVERY
Aquilo Gas Separation B.V.

NITROGEN PRODUCTION
A/G Technology Corporation
Aquilo Gas Separation B.V.

CLASSIFIED INDEX OF ADVERTISERS BY PRODUCT CATEGORY

REAGENT GASES
SCT

OXYGEN ENRICHED AIR PRODUCTION
A/G Technology Corporation

GAS DEHYDRATION
Aquilo Gas Separation B.V.

STEAM
Fairey Industrial Ceramics Ltd

TRADE NAMES INDEX

Abcor – *Spiral wound/tubular products*
Koch Membrane Systems, Inc.

Acrodisc syringe filter – *syringe filters for sterilization and analysis*
Gelman Sciences Ltd

AEGIS – *String wound and pleated backwashable filter cartridges*
Graver Separations, Inc.

Alkasave – *Recovery of spent caustic soda & acids*
Membrane Products Kiryat Weizmann Ltd

atech-SiC-membranes – *Ceramic membranes for micro/ultra-filtration*
atech innovations GmbH

atech-A1$_2$O$_3$-membranes – *Ceramic membranes for micro/ultra-filtration*
atech innovations GmbH

atech housings – *Stainless steel housing for membranes*
atech innovations GmbH

AVIR – *Air separation system*
A/G Technology Corporation

Biotrace blotting membranes – *blotting membranes for separation/purification*
Gelman Sciences Ltd

CARBOSEP – *Tubular mineral membrane on carbon support*
Tech-Sep

CARRE – *Tubular stainless steel inorganic membranes and systems*
Graver Separations Inc.

CELGARD – *Microporous polypropylene film*
Hoechst

CELGARD – *Microporous polypropylene hollow fibres*
Hoechst

CEL-TAN – *Cassette modules for ultrafiltration*
Hoechst

CERAFLO – *Ceramic crossflow microfiltration membranes*
SCT

EXTRA FLOW – *Phase contactors for Gassing/Degassing*
Hoechst

FASTEK – *Spiral wound membrane elements*
OSMONICS

FICL STAR – *Ceramic microfiltration membrane and modules*
Fairey Industrial Ceramics Ltd

FlexStand – *Benchtop pilot hollow fiber system*
A/G Technology Corporation

FLOSEP – *Hollow fiber membrane*
Tech Sep

GN Metricel membrane – *ideal membrane for water and waste water testing*
Gelman Sciences Ltd

HiFLO Sol-Vent cartridge – *hydrophobic cartridge for sterile venting*
Gelman Sciences Ltd

KERASEP – *Monolithic ceramic membrane*
Tech Sep

LIQUI-CEL – *Phase contactors for Gassing/ Degassing*
Hoechst

MaxCell – *Process scale hollow fiber cartridge*
A/G Technology Corporation

MaxiFiber – *Open channel hollow fiber cartridge*
A/G Technology Corporation

MEMBRALOX – *Ceramic crossflow micro-, ultra and nanofiltration membranes*
SCT

MEMTREX – *Pleated filter cartridge*
OSMONICS

MOLSEP – *Hollow fibre modules for UF*
Hoechst

MOLSEP – *Multitubular modules for UF*
Hoechst

MOLSEP – *Hollow fibre pervaporation modules*
Hoechst

NADIR – *Flat sheet/tubular/membranes*
Hoechst

Nanospin Plus centrifugal device – *centrifugal device for MWCO separation*
Gelman Sciences Ltd

Nitro/Drypoint/Produce – O_2/N_2 *separation*, CO_2 *removal, dehydration*
Aquilo Gas Separation B.V.

OSMO – *Spiral wound membrane elements*
OSMONICS

PLEIADE – *Organic plate membrane*
Tech Sep

QuixStand – *Benchtop hollow fiber system*
A/G Technology Corporation

Romicon – *Hollow fibre cartridges*
Koch Membrane Systems, Inc.

RO PLANTS, UF PLANTS, NF PLANTS, MF PLANTS
Union Filtration a/s

SCEPTER – *Tubular stainless steel inorganic membranes and systems*
Graver Separations, Inc.

Selro – *Nanofiltration membranes for process & waste stream treatment*
Membrane Products Kiryat Weizmann Ltd

SEPA – *Flat sheet membrane*
OSMONICS

SPIRA-CEL – *Spirally wound UF/NF Modules*
Hoechst

Spiral/Cap capsule – *capsule for sterilization of air and gas venting*
Gelman Sciences Ltd

SuporCap capsule – *capsule for sterilizing-grade filtration*
Gelman Sciences Ltd

SuporFlow cartridge – *hydrophilic cartridge for sterilizing-grade filtration*
Gelman Sciences Ltd

Supor membrane – *superior membrane with high flow rates*
Gelman Sciences Ltd

TurboTube – *Open channel membrane cartridge*
A/G Technology Corporation

VacuCap device – *bottle-top vacuum filter for lab bench*
Gelman Sciences Ltd

VENTREX – *Pleated vent filter cartridge*
OSMONICS

VSEP – *Vibratory Shear Enhanced Process*
New Logic

TRADE NAMES INDEX

Xampler – *Laboratory scale hollow fiber cartridge*
A/G Technology Corporation

XL-1000 – *Long life spiral wound modules*
Koch Membrane Systems, Inc.

Xpress – *High pressure ultrafiltration membranes*
A/G Technology Corporation

Z-Spin Plus centrifugal device – *centrifugal device for microfiltration separation*
Gelman Sciences Ltd

EDITORIAL INDEX

A

Acetone, 52, 59, 61, 202, 209, 226, 242, 265, 319, 332-3, 335-6, 342-3, 345-8, 365, 576, 657
Acid compartment, 676, 709, 714, 716
Acid gas, 16, 271, 273, 276, 302, 324, 326, 575, 688
Acid recovery, 699, 701-3, 712, 721
Acid scrubbing solutions, 722
 amine recovery from, 722
Activated carbon filtration, 85, 530, 582, 584, 589
Aerobic treatment, 585, 588, 592-3
Aerosol monitoring kit, 434
Affinity membranes, 13, 673-4
 applications of, 674
Agar culture media, 450
Air and gas sterilisation, 313
Air compressor condensate, 418
Air monitoring, 435, 468
Air sterilisation cartridge filters, 315
Alcohol dehydration, 334, 360
Alternative membrane fermentors, 672
Alternative microfiltration configurations, 160
Alumina, 75, 120, 143, 146, 179, 271
Aluminium asymmetric membranes, 220
Aluminium etched metallic membranes, 222
Amine absorption, 277, 281
Amino acids, 479, 674
 liquor treatment, 767
 separation of, 675
 splitting of, 677
Ammonia, 55, 82, 246, 292, 677, 707-8
 stripping, 588
 synthesis, 293
Anaerobic
 digestion, 300, 592-3, 596-7
 treatment, 585, 592, 597
Anhydrous alcohol, 336
Animal products, 725, 745
Anion exchange membranes, 68-9, 173-4, 257-9, 261-2, 265, 508, 512, 696, 698, 706-7, 709, 718
Anode elements, 795
Anodised aluminium, 221
Antibiotics
 production, 157
 processing, 378
Antibody purification processes, 667
Apple juice, 13, 747
Aroma compounds, 346, 348-50, 756, 758
Asymmetric membranes, 15, 24, 26, 47-8, 54, 112, 117, 189, 191, 201, 210, 220
Azeotropic composition, 331-3

B

Back flushing, 76, 80-81, 85, 96, 142, 525, 674
Bacterial Challenge Test, 235
Bacterial contaminants
 relative sizes of, 20
Bactocatech, 735

Balance system, 358
Base compartments, 714-6, 720
Batteries, 16, 789
  utilising membranes, a selection of, 790
Beer, 20, 132, 287-8, 313, 335, 406, 449, 596, 692, 726-7, 760, 768-9
  de-alcoholisation of, 692, 768
Beta ratio, 238
Beverages, 21, 26, 114, 119, 131-2, 151, 312, 373, 433, 435, 446, 481, 527, 531, 533, 597, 634, 655, 725, 726-7, 731, 739, 760
Binding, 117
  and transfer media, 459
Biochemistry, 127, 128, 462
Biofouling, 44, 91-3, 528, 539
Biological
  pretreatment, 584, 587, 590-91
  processes, 576-7, 580, 585
  solutions, diafiltration of, 667
Bioreactor effluent, 585
Bioreactor effluent (UF), 427
Biosafety tests, 116
Bioseparations, 341, 655-6, 659, 665
  flowsheet of, 656
Biostream, 800
Bipolar membranes, 70, 174, 261-2, 676, 678, 696, 705, 708-9, 711, 714, 716-9, 720-21
Block copolymer, 192, 350
Blood oxygenation, 688
Blotting, 433, 458, 460, 461, 463
Boil-point, 228
Boundary layer, 31, 71, 156, 158, 172, 174, 219, 220, 255, 627
  formation of, 72
British Gas Stretford Process, 285
Bubble point, 106, 111, 120-21, 227-32, 236-8, 245, 247-9, 251, 254, 444, 467
Bubble Point Test, 227
Buffer exchange, 475, 668
Bulk chemicals, 336, 373

C

Cake retention, 114, 117
Can washing, 624
Capenhurst electrolyte etchant regeneration *(see CEER)*
Capillary modules, 96, 97, 621, 670
Carbon membranes, 141, 143, 408, 623-4

Carbon microfiltration membranes, 623
  bath cleaning by, 623
Carbon monoxide production, 296
Cartridge filters, 80, 128, 315, 320, 377, 537, 539, 626
Cartridge properties, 248
Cartridge units, 137, 180
Cascade operation, 104-5
Cathode element, 795
Cation exchange membranes, 68-70, 174, 257, 260-1, 705
Caustic recovery, 703-4
CDI, 565, 572, 678-9, 680
  separations of, 678
  systems, 678
CEER, 785
Cell
  culturing, small scale, 672
  growth, 433, 457, 472
  recycle, comparison of methods for, 660
  separators, 773
Cell stack battery limits, 713
Cellulose production, 602, 604
Centrifugal filtration units, 463
Centrifugal ultrafiltration concentration, 474
Ceramic filter elements, 179, 324
  and systems, 179
Ceramic microfiltration, 412, 419
Cereal processing, 760
CFAF, 673-4
Chain
  flexibility, 195
  interactions, 195
Cheddar, 730-1, 733-4
Cheese, 13, 46, 313, 726-7, 729-34, 736, 738-9, 741, 765
Cheese whey, 46, 396, 726-7, 730, 738, 765
Chemical
  processes, 288, 336, 344, 542, 576-7, 589, 609, 717
  resistance, 14-15, 25, 45, 114, 116, 131, 159, 189, 195, 217, 263, 316, 320-1, 604
  transformation, 579, 580
Chemically stable polymers, 196
Chlor-alkali industry, 778
Chlorine wash, 83
Chromium extraction
  using microcapsules, 650
Cider production, 407

Clay dewatering, 796, 798
Cleaning (DMC), 81
Cleaning agents, 27, 83, 87, 341
CLM, 636
CO_2 Separation, 277
Coatings, 48, 57, 90-91, 131, 191, 210-11, 213, 217-8, 221-2, 226, 242, 261-2, 318, 322, 342, 348, 371, 373, 456, 578, 610, 611, 614-5, 691, 794
Cohn process, 665-6
Coliform, 449, 450, 454-5, 468
Commercial water degassing system, 639
Component separation, 19, 578, 580
Composite membranes, 14, 43, 51, 54, 58, 191, 201, 209, 210-12, 214, 216-7, 219, 257, 343, 357, 363, 365, 515, 529-31, 534, 537, 539, 581, 588, 601, 606, 608, 623, 668, 727-8, 757, 760
Compressed air, 236, 302, 309, 311, 313, 315, 325, 600
   installation, 311
   cleaning and sterilisation, 309
   dryer, 318
Compression-condensation, 362-3, 365, 367
Concentration
   polarisation, 28, 31, 40, 60, 71-7, 79, 92, 96-7, 100-1, 103, 105, 113, 123, 152-3, 155-7, 177, 181, 254, 258, 361, 499, 502, 667, 725, 730
   waves, 271
Connections, 131
Contained liquid membranes *(see CLM)*
Contamination analysis, 434-5, 468
Contamination control, 435, 469
Continuous deionisation *(see CDI)*
Conversion, 37
Copolymers, 54, 68, 90, 119, 192, 224, 260, 261, 342, 350, 661, 670
Cosmetics, 13, 21, 127-8, 373, 449, 481, 527, 540, 753
Cottage cheese, 730
Covalent bonding, 192
   immobilisation by, 671
Cross flow
   affinity filtration *(see CFAF)*
   filters, 176-8, 399
   filtration, 25, 138, 175
      electrochemically enhanced, 75
   microfiltration, 117, 133-4, 600, 609, 662
   microfiltration in Bio-processing, 106
   operations, 102-3
   rotational membrane filters, 160
   rotational membrane filtration applications of, 161
   rotational membranes, 161
Cryogenics, 271-2, 282, 285, 294
Culture medium determination, 448
Cylindrical pores, 15, 189, 191, 205, 243

D

Dairy industry, 312, 317, 725, 727-8, 730, 763, 765
Dead-end, 24, 102-3, 106, 113, 120, 125, 175, 232, 252, 319, 374
   filtration, 176, 660
   microfiltration cartridges, 374
Dean vortices, 73, 76
Degradation of flow, 124
Degreasing bath, 418, 623, 624
Dehydration, 335
Depth filters, 110-12, 377, 443, 465, 467
Depth filtration, 23, 110, 117, 314, 466
Depyrogenation, 84, 177
Desalting, 9, 70, 166, 170-71, 191, 210-11, 475-7, 479, 499, 506, 523, 536, 543, 566, 612, 655, 675, 679, 691, 697, 725-7, 763, 765, 768
Detergent removal, 476
Dialyser modules, 158
Dialysis, 4, 5, 14, 72, 158-9, 188-9, 201, 227, 263, 269, 462, 467, 475, 478, 546-7, 550-52, 554, 556, 580-82, 668, 671, 683-5, 691-2, 699-705, 760, 768, 796, 798
Dialysis and electrodialysis applications, 691
Diffusion dialysis, 184, 582, 699, 700-5
Diffusion testing, 231, 249
   principle of, 231
Digester gas, 300
Digestion process, 658
Dip coating, 210-11, 614
Direct membrane, 81
Direct patient care, 465
Direct product recovery, 342
Dirt loading, 111
Dirt loading/unloading, 117

Disc plate, 135
Disposal filters, 389
Distillation, 4, 6, 9, 11, 56-7, 61-4, 157, 272, 331, 335-8, 340-1, 344-5, 348, 350, 362, 472, 489, 492, 523, 526, 541, 547, 557, 576-8, 580-81, 582, 606, 712, 760, 769
Distribution, 62, 10-1, 112, 114-15, 122-3, 127, 136, 143, 157, 159, 198, 204, 206, 218, 220, 228, 232, 238, 242-3, 248, 252, 254, 274, 277, 316, 357, 464, 527, 547, 552, 557-9, 570-1, 601, 649, 734-5, 793
DMC, 81
Domestic
 and industrial wastewater, 597
 sewage treatment, 598
Donnan dialysis, 705

E

Economics, 398
ECTFE, 131
ED, 3, 12, 14, 46, 68, 70, 79, 85, 94-5, 99-100, 164, 167, 171, 173, 188, 257, 262, 492, 498, 500, 502-6, 509-10, 512, 523, 543, 628-9, 675-6, 691, 693, 695-7, 721, 726, 736, 740, 763, 765-7
Edible oil, 286, 762
Effluent treatment
 and recycling, 778
Effluents
 latex-contaminated, 611
Electrical resistance, 68, 99-101, 258, 259, 261, 266, 540
Electro-kinetic dewatering, 794
Electro-osmosis *(see EO)*
Electro-osmotic
 and electrophoretic membrane separation, 794
Electrochemical
 membrane cells, 183
 membranes, 302, 697
 processes, 15, 131
 redox reaction, 305
 synthesis, 778
Electrocoat paint process, 616
Electrode membranes, 81
Electrodialyser module, 184
Electrodialysis *(see also ED)*, 4, 9, 12, 14, 68, 70, 72, 77, 99-100, 163, 165, 168-72, 174, 183, 188, 257-8, 263, 266, 489, 49-3, 497-508, 513-4, 516, 523, 542-3, 565, 577, 579-82, 628-9, 655, 675, 678, 691-700, 711, 712, 718, 721-2, 725-6, 739-42, 745, 762-3, 765-6
 in the food industries, 763
Electrodialytic
 dissociation of alcohols, 677
 salt splitting, 676
 water issociation/bipolar membranes, 708
Electrodialytic, 629, 676, 708, 721-2
Electrofiltration, 4, 16, 76, 794, 796
Electrohydrolysis, 582, 705-7
Electrokinetic
 dewatering cells, 796
 processes, 793
Electron microscope, 43, 240, 242
Electronics, 11, 23, 111, 114, 128, 247, 309, 312-3, 340, 373, 515, 527, 541, 543, 545, 563
Electrophoresis, 74-5, 263, 458, 477, 479, 655, 691, 793-5, 798
Electrophoretic separations, 798
EMR, 668-70
Emulsion
 break-up, 67
 liquid membranes, 66-7, 643, 645
 preparation, 67
Enhanced oil recovery *(see EOR)*
Environmental legislation, 355
Enzyme membrane reactors *(see EMR)*
EO, 75, 793-5
EOR, 278, 280
Essential oils, 350
Etching, 189, 199, 201, 205, 206, 222, 238, 257, 316, 546, 692, 701, 704
Ethanol dehydration, 338-9, 361
Ethene-propene terpolymer, 343, 344
Evaporation, 6, 9, 156-7, 162, 199, 207-8, 259, 262, 489-91, 498, 523, 564, 578, 580-2, 584, 587, 589-91, 595, 605, 619, 623, 625, 659, 665, 697, 704, 707, 725-9, 736, 739, 745, 754-5, 757, 760, 762-3, 796
 and granulation, 589
Extractables, 23, 112, 114, 116, 119, 131, 247-8, 457, 469, 473
Extrusion, 188-9, 204, 225, 286, 477

F

Facilities filtration, 382
Farm waste, 601
Fast gases, 301
Feed flowrate, 34, 664
Feed pretreatment, 78-9, 85, 87, 96, 493, 495-6
Fermentation, 12-13, 32-33, 64, 79, 106, 110, 134, 177, 179, 276, 291, 312, 336, 342, 344, 346, 347-8, 350, 383, 454, 457, 479, 656-61, 664-5, 670-2, 674-6, 692, 726, 731, 739, 761, 765-6
   marcromolecular products from, 657
   and digestion processes, 658
Filter cake, 24, 113-14, 319, 795-6
Filter
   cartridge technologies, 375
   cartridge tests, 243
   design and selection, 317
   medium stability, 114
   stability, 116
   fabrics for, 321
Filtration
   factors in, 114
   housings, 130
Fine chemicals, 373, 557
Flat sheet modules, 154
Flat sheets, 66, 99, 149, 655, 673
Flowrate, 3, 19, 24, 31-4, 45, 47, 96, 98, 101, 120, 124, 157, 166, 170, 182, 278, 291, 314, 360-1, 467, 477, 523, 563, 582, 584, 664, 666, 671, 673
Flue gas desulphurisation, 713-4
Fluid contamination
   analysis monitor, 434
Fluorocarbon membranes, 261
Flurocarbon elastomer, 344
Flushing, 76, 80-1, 83, 85-6, 96, 107, 142, 291, 447, 523, 525, 551, 556, 612, 674
Food, 6, 9, 11-13, 20-1, 26, 36, 46, 68, 70, 82, 98, 114, 119, 127-8, 131-2, 151, 161, 163, 243, 274, 276, 286, 291, 309, 312, 313, 331, 335, 373, 433, 446, 449, 457, 481, 485, 527, 531-4, 542, 557, 582-3, 597, 626, 655, 669-71, 675, 723, 725-7, 736, 739, 742, 745, 762-3, 765, 767
   processing, reverse osmosis in, 727

   beverages and pharmaceuticals, testing of, 446
Foulants, 32, 42, 79-80, 82, 85-9, 97-8, 263, 523, 616, 662
Fouling, 1, 25, 28, 31-4, 40--5, 71, 73, 77-83, 85, 96, 99, 101-3, 105, 113-4, 123, 136, 147, 149, 154, 160, 167, 175, 220, 232, 257-8, 262-3, 324, 587, 594, 599-601, 609, 623, 625, 658, 662-3, 672, 725, 727, 735-6, 740, 747, 757
   control and correction, 79
Fractionation, 9, 62, 177, 333, 338, 476, 582, 600, 725-7, 730, 739-40
Free From Protein-Bound Microsolute, 478
Freshnot process, 756
Fruit juice, 224, 286, 406, 669, 745-8, 750, 752-8, 763, 766
Fuel
   cells, 16, 303, 786
   return system, 359

G

Gas
   cleaning, 319, 325
   high temperature, 323
   filtration, 114, 116-7, 233, 309, 316, 324, 435, 467, 469
   line filter holder, 434
   membranes
      immobilised, 637
   permeation, 4, 6, 9-10, 12, 14, 47, 48, 51, 55, 57, 71, 154-55, 157, 184, 188-9, 201, 272-3, 276-8, 282, 284-5, 287, 291-2, 294, 297, 300-1, 576
   permeation and pervaporation, module design, 154
   sensing electrodes, 484
   separation, 47
   transfer contactor, 275
Gaseous effluents, 575
Gasoline
   station, 358
   tank farms, 357
Gel layer, 31-4, 71, 77-8, 92, 216
   resistance, 78
Gel-purified samples
   recovery of, 477

Gelatine, 397
General industrial housings, 130
Generic membranes, 87
Grafted copolymers, 192
Grafting, 59, 84, 210, 259
Granulation, 584, 589-90, 595
Grape must, 757, 759

H

Heavy metals, 16, 128, 246, 393, 468
Helium recovery, 301-2
Helmoltz-Smoluchowski equation, 793
Hemodiafiltration, 685
Hemodialysis, 158, 655, 657, 683-7
Hemofiltration, 684-6
Heterogeneous, 100
Hollow based humidification system, 687
Hollow fibre
  membrane absorber, 274
  fibre modules, 73, 95, 98, 101, 107, 147-8, 151-2, 155-6, 180, 274
Hollow fibres, 63, 66, 106-7, 109, 151-2, 155, 207, 213, 273, 315, 655, 670, 673, 688
  segregation in, 670
Homogeneous membranes, 6, 41, 48, 68, 259
Homopolymers, 192, 224
Hospital pharmacies, 463
Hot gas effluents
  sources of, 324
Housings, 132
HPLC, 435-43, 462, 469, 472-3, 478-9, 569, 639
  analysis, 478
  solvent filtration, 467
Hybrid systems, 295
Hydrocarbon
  gas dewpointing, 285
  liquids, CO_2 from, 285
Hydrocarbons, 647
Hydrocracker, 297-9
Hydrofluoric acid
  recovery from fluosilicic acid, 712
Hydrogen recovery, 292, 300
Hydrogen sulphide
  and water vapour, 284
Hydrogen/Hydrocarbon separations, 297
Hydrogenation, 297
Hydrolytic, 114, 224

stability, 116
Hydrophilic, 19, 32, 57, 66, 73, 83, 116-9, 158, 198, 201-3, 232, 236, 259, 315, 460, 465, 469-70, 472-4, 622, 627, 663, 686
  membranes, 57, 66, 117-9, 460, 470, 473, 627
Hydrophilicity or hydrophobicity, 467
Hydrophobic, 116
  membranes, 25, 63, 114, 119, 201, 233, 465
  porous membranes, 63

I

Ice cream, 730, 736, 739
Immobilon membranes, 461
Impingement, 114, 117
Industrial modules, 158
Industrial waste water, 393, 575
Inertial impact, 117, 314
Injectable drugs (parenterals), 384
Inorganic compounds, 649
Inorganic membrane materials, 268
Inorganic membranes, 216
Inorganic substances, 608, 762
Integrity test, 109-10, 228, 230, 232-4, 236, 249, 254, 256, 317, 600
  equipment, 230
Interfacial polymerisation, 210
Ion exchange, 14
  permselectivity, 265
  IEC, 265
  membranes, 68-70, 99-100, 173-4, 189, 257-8, 260, 262, 265, 507, 698, 705-6
  characterisation, 265
  heterogeneous, 258
  special properties of, 262
Ion selective electrodes, 481
IPA, 341, 362
Isopropanol
  dehydration of, 362
Isopropyl alcohol, 334, 336, 340-1
Isotherm, 252, 271
Isothermal operation, 155, 158, 361

J

Joule-Thomson, 272
Juice extraction, 749
JUMBO cartridge filtration, 404

EDITORIAL INDEX

K

Kaolin clay, 426

L

Laboratory
  filtration and membranes, 466
  membrane filtration units, 176
Landfill, 300
Landfill leachate
  treatment of, 583
  treatment technologies, 592
Landfill site treatment, 590
Latex
  polymer (UF), 426
  sphere retention, 240
  sphere test, 237
  spheres, 238-9, 254
Leaching, 23, 188, 207, 215-6, 558, 563
Lignosulfonate fractions
  recovery of, 606
Limulus ameobocyte lysate, 116
Liquid membranes, 4, 10, 16, 64-7, 187, 189, 481-2, 484, 636-7, 641, 643-50, 674, 686, 697
  immobilised, 636
Liquid mixtures, 332
Liquid/liquid extraction (LLE), 633
Liver support system, 685
Low binding membranes, 475
  MWCO of, 475
Low-molecular-weight
  fermentation products, 657
LVP filtration, 386

M

MAC, 673
Macromolecules
  blotting of, 458
Mammalian cell separation, 662
Maple syrup, 726-7, 758
Materials, 4, 131, 187
Mechanical and chemical properties, 266
Mechanical strength, 41, 58, 68, 116, 141, 153, 258-9, 470, 523
Medical Use, 159
  devices, 433, 683

  oxygen, 688
Membrane
  immobilisation in the, 670
  affinity chromatography *(see MAC)*
  and feed separation, 67
  belt filters, 427
  carbon dioxide, removal from liquid, 281
  cleaning, 80
  coated bag filters, 324
  distillation, 61
  fermentors, 671
  filters, 111, 125
  filtration
    system selection, 123
  fluxrate, 81
  fouling, 78
  gas absorption, 273
    air cleaning by, 318
  gas stripping, 638
  materials, 198, 774
  modification, 85
  module designs, 94
  modules
    characteristics of, 105
  performance, 40, 48
  performance characteristics, 120
  phase contactors, 67
  preparation, 217
  properties, 83, 248
  reactor ceramic units, 184
  selection, 41
  unit
    spiral flow, 77
Membranes
  characterisation of, 227, 254, 256
  characteristics of, 107
  in electrochemical cells, 783
Mercury intrusion method, 232
Metal finishing industry, 410
Metals, 14-16, 25, 30, 113, 120, 187, 189, 216, 221, 246, 286, 647
Metathesis reaction, 698
Methane, 50
  recovery, 55, 279-80, 300
MF, 175, 197
Micro-electroelutor, 478
Microbial cells
  separation of, 660

Microbiological
  assay, 446
  investigations, 467
Microbiology, 119, 128, 456
  reliability, 398
Microcapsule membrane, 226
Microfilters, 143
Microfiltration
  characteristics and applications of, 119
  clarification by, 397
  in the food industry, 395
  of adsorbates, 411
  of milk, 396, 734
  operating factors in, 115
  membranes, morphology of, 118
  plate modules, 138
Microporous:
  ceramic, metal, glass, 14
  sintered polymer, 14
  stretched polymer, 14
  track-etched PC, 14
Milk, 12, 669, 725-41, 744, 763, 765
  fractionation, 730
Mine drainage, 626
Module design, 1, 73, 81, 94-5, 98, 103, 105-6, 139-40, 143, 154-6, 178, 278, 281, 291, 360, 492
Molecular
  biology, 462
  diffusion, 64
  sieve, 271
  weight, 255
Monitors, 93, 287, 443, 446, 481, 556
Monolayer technology, 375
Morphology, 43-4, 115, 227, 232, 455, 556
MTBE, 335, 346-7, 628
Multi layer technology, 376
Multiple in-line steam sterilisation cycles, 245
Multistage membrane processes, 280
Mycoplasma removal, 462, 472, 474
  retention, 247

N

N_2 generation, 287
Nanofiltration, 4-6, 12, 34, 46, 154, 523, 526, 588, 612, 626, 725, 726, 740-3
  membranes, 46
Natural gas, 16, 55, 105, 274, 276-85, 292, 29-7, 368-9, 541, 575
  sweetening, 280
Nernst, 303
NGL feed stream, 283
Ni, recovery of, 693
NIOSH Reference, 435
NIOSH/OSHA Analytical Method, 435
Nitrile butadiene rubber, 344
Nitrogen
  applications of, 286
  from air, 285
NMWCO = nominal molecular weight cutoff, 177
Non porous membranes, 47-8, 61, 191, 227, 256-7
  characterisation tests for, 257
Noninfiltrated ceramic membranes, 219
Nuclear industry
  applications in, 373, 614
Nucleic acids, 458, 460, 467, 475-9
Nutrient pad sets, 447, 451-3

O

Off-gas treatment, 357
Oil
  and fats processing, 762
  contaminated waters, 618
  field applications, 645
  polluted industrial effluents, 621
  and water emulsions, 417, 620
Olive oil, 419
Open vacuum assisted system, 358
Open-channel membranes, 746
Ophthalmic filtration, 387
Orange juice, 755
Organic
  electrosynthesis, 784
  phase, 64, 644-5
  separations, 344
  substances, 473, 552, 554, 556, 608, 645, 649, 762
  vapours, 50, 52, 196, 302, 355, 357, 363, 365, 575
  recovery of, 355
Organics, 342
  separation of, 767
OSHA Permissible Explosure Limit, 435
Oxychemicals, 346, 348

Oxygen enriched air, 288
Ozone production, 781

P

Paints, 373, 421
PAN, 14-15, 57, 60, 156, 661
Paper
  and pulp bleach effluents, 609
  machine effluents, 608
Particle size
  contamination of, 1, 16
Particulate contamination, 433-4, 445, 464, 471-3
  detection and analysis of, 433
Particulates, 3, 19, 23, 30, 73, 78, 98, 105, 115, 122, 133-4, 153, 160, 246, 443, 495, 502, 508, 543, 575-6, 581-2, 658
Patch test kit, 434
PEA, 131
Performance, 27
  criteria, 36
Perishable food storage, 291
Permeability, 3, 14, 24, 27-8, 36-7, 40, 47, 48, 50-1, 57, 242, 256
Permeate flux, 32, 82-3, 109, 143, 155-6, 360-1, 596, 608, 622, 655, 661, 663
Personal monitoring, 434
Pertraction, 64
  systems, 65
Pervaporation, 10, 14, 56, 154-5, 188-9, 331, 333-48, 576, 726
  applications of, 331
  enhanced esterification by, 350
  and vapour permeation, 101
Petrochemicals, 373
Pharmaceutical – Antibiotic Concentrations, 427
Pharmaceutical Filter, 243, 474
Pharmaceuticals, 340, 373
Phase contact application guide, 634
  interface polymerisation, 216
  inversion, 189, 207
  inversion membrane preparation, 208
Phenol recovery, 645
Physical processes, 577
  separation, 578
Pigments, 427
Plasma polymerisation, 210, 212

Plate modules, 99
  units, 135
Plating bath rinse waters
  treatment of, 691
  baths, 721
Pollutants present in waste waters, 576
Pollution control, 344
Poly-acrylate rubber, 344
Polymer characteristics, 191
Polymerase chain reaction (PCR), 477
Polymorbornene, 344
Polyoctenamer, 344
Polyolefin resin degassing, 368
Polysulphone membrane, 28-30, 73, 83, 348, 613, 740
Polyurethane, 344
Pore size, 115, 146, 179, 465, 468, 474
Pore size, 17, 21, 24, 26-7, 32-3, 35, 62, 64, 83, 94, 111-4, 120, 122-4, 143, 160, 177, 196, 199, 204, 206, 209, 217-8, 220, 222, 227-8, 232, 236-40, 242-3, 245, 247-49, 252, 254
Porous membranes, 198, 227
Power station effluents, 625
PPS, 197, 225
Prefilters, 380
Preparation methods, 204
Pressure
  the effect of, 729
  filter holders, 126, 128
  regulator, 283
  swing adsorption (PSA), 357
PRO-CAL, 731
Process continuous de-ionisation (CDI), 382
Product quality assurance, 398
Properties of ion-exchange membranes, 263
Protein fractions
  production of, 675
Protein recovery
  improving, 477
  and concentration, 664
PSA, 357-8
PTFE membranes, 314
Pull-down capacities, 291
Pulp
  and paper industry, 602
  bleaching effluent, 161
PVA, 15, 57-8, 60, 156, 197, 338, 340, 362
  membranes, 340

PVC
  drying of, 797
PVDF, 131
Pyrogen test, 247
Pyrogenicity, 116

Q

Quarg production, 733

R

Radioactive waste sludges, 420
Rectangular plate and frame, 136
Refinery fuels, 625
Respiratory gases
  humidification of, 686
Reverse osmosis, 10, 34, 181, 725
  concentration by, 583, 754
  membrane cleaning, 85
  modules, 151
  system design, 45
  treatment of electroplating solutions, 698
  and evaporation, energy consumption of, 728-9
Romicon module, 748
Rotating biological contactor, 588
  cylindrical filter, 76

S

Salt
  from seawater, 764
  splitting, 174, 676, 707-8, 716-7, 721
  and acids, recovery of, 705
Sample filtration, 473
Sampling device, 455
Sanitary housing, 130
Santisation, 84
Screen/membrane filter, 377
Seeded precipitation and recycling by reverse osmosis, 626
SEM, 241
Semiconductor process chemicals
  filtration of, 388
Sensors, 481
Separation curve, 58
Separators
  desirable properties of, 774

and diaphragms, materials used as, 775
Shape, 19, 115
Shear rate, 33-4, 107, 134, 160, 663
Shedding, 116
Sheet filters, 129
Sieve
  retention, 117
  filtration, 112
Silica gel, 271, 284
Silicone rubber, 342-4, 347, 350, 484
Siloxane, 54
Single flow path design, 100
Single stage membranes, 279
Sintered membranes, 114
Size
  contaminated effluents, 610
  related performance of filters, 320
Sludge dewatering, 389
Small molecules
  removal of, 476
Solid polymer electrolyte cells
  Performance characteristics, 781
Solid polymer electrolyte fuel cell (SPEFC), 786
Solute
  passage, 252
  pertraction, 67
  rejection, 252
Solvent
  filtration, 472
  prefilters, 378
  recovery, end of pipe, 360
  recycling, 340
  sterile filters, 380
  for absorption of CO_2, 281
Soxal process, 714
Soy sauce
  desalting of, 768
Spent pickling liquors
  regeneration of, 711
Sphere retention profiles, 240
Spiral flow
  element, 76
  membrane unit, 77
Spiral wound module, 79, 91, 97-8, 105-6, 149, 151, 180, 222, 339, 610, 731
Stacked disc membrane filters, 126
Stainless steel

filter holder, 434
filters, 411
Steam and air
  filtration of, 402
Steam reforming, 292, 295-6
Stereoisomerism, 192
Sterile
  filtration, 467, 472, 668
  venting, 402
Sterilising grade, 238
Stirred cell, 175-6
Storage, 84
Stretched membranes, 114
Stretching, 189, 204
Surface
  charge, 43
  filter, 377
  filtration, 110
  morphology, 43
SVP filtration, 386
Sweep gas technology, 282
Symmetric
  membranes, 15, 24, 48, 112, 114, 189, 191
  microporous phase inversion, 14
  porous, 191
Synthesis gas, 292, 295-6
Synthetic polymer membranes, 683
Syrups and sweeteners, 397
System Design, 41, 45, 101-2, 105, 124, 170, 227, 557

T

Temperature
  the effect of, 729
Thermal
  stability, 116, 245
  swing adsorption (TSA), 271
Thermally stable polymers, 196
Thermopervaporation, 61
Thermoplastic polymer ultrafiltration membranes, 224
Thin channel cross flow, 176
Thin film
  composite membranes, 43, 212, 515, 529-31, 534, 537, 539, 601, 727, 757, 760
  membranes, 42-5, 530, 534, 536
Titanium dioxide

  recovery of, 324
Tomato juice, 755, 757
Tortuous flow path design, 100
Track-etched, 14, 188, 204, 206-7, 248, 457-8
Transfer media, 459
Trimix, 301
TSA *(see thermal swing adsorption)*
Tubular inorganic and ceramic modules, 141
Tubular membranes, 71, 85, 98, 107, 154, 179-81, 587, 591, 597, 606-7, 612, 616, 622, 639, 746-7, 750, 758, 761
  pressing, 389
  modules, 98, 139
Tubular RO
  laboratory module, 182
  modules, 154
Two stage membrane processes, 279

U

UF hollow fibre modules,
  operating parameters of, 148
  membranes,
  immobilisation by gelation on, 670
Ultrafilters, 143
Ultrafiltration, 4-5, 10-11, 13, 17, 26-7, 32, 72, 84, 96, 99, 133, 175, 197, 254, 462, 474, 476-9, 523, 525, 560, 578, 582, 585, 587, 593, 596, 602-4, 606, 610-4, 620, 622, 624, 629, 655, 664-6, 687, 725-6, 730-3, 739, 743, 745-6, 753, 762
  and bioreactors, 585
  clarification by, 745
  hemoconcentration by, 687
  in cheese making, 732
  membranes, 224, 251
  modules, 133
  of bioproducts, 106
  of emulsions, 620
  of wastewaters, 596
  of whey, 739
  systems, 382
Ultrasonic cleaning, 81
Unfermented must
  cold sterilisation of, 402

V

Vacuum filter holders, 447

Vacuum filtration, 480
Vapour permeation (VP), 60, 157, 355
   product, 346
   separation systems, 368
Vegetable protein, 761
   processing, 743
Vibratory shear enhanced filtration (VSEP), 425
Virus removal filters, 382
Viscose fibre industry, 643
VOC, 355-6, 362-3, 365, 367, 369, 372, 634
   from air, 372
   recovery, 362
Volatile bioproducts
   recovery of, 346
Volatile organic compounds (VOC), 355, 582
VP, 3, 157, 361
VSEP, 425

W

Waste water treatment
   by electrodialysis, 628
   hybrid membrane, 626
Water
   dissociation, 101, 705, 708, 722
   electrolysis, 780
   flux values, 30
   integrity test, 232
   integrity test unit, 233
   microbiology, 454
   penetration value, 235

   purification, 64
   quality, 83
   recovery from dyehouse effluent, 612
   vapour removal, 302
Wet corn milling processing, 761
Whey, 13, 46, 669, 692, 726-7, 729-43, 761-3, 765-6
   desalting, 743
   nanofiltration of, 742
   and milk, concentration of, 736
   demineralisation, 740
Whole cells
   immobilisation of, 671
Wine, 12-13, 21, 132, 286, 335, 405, 449, 468, 758-60, 763, 766
   filtration before storage of, 400
   clarified by CFMF, last cellar filtration of, 403
   making, filtration processes in, 400
   production, 766
      applications in, 398

Y

Yoghurt, 730

Z

Zeolites, 271, 344
Zeta potential, 115
Zinc-galvanization effluent, 695